XVth International Conference on Calorimetry in High Energy Physics

(CALOR2012)

Journal of Physics: Conference Series Volume 404

Santa Fe, New Mexico, USA
4-8 June 2012

ISBN: 978-1-62276-870-7
ISSN: 1742-6588

XVth International Conference on Calorimetry in High Energy Physics

(CALOR2012)

Journal of Physics: Conference Series Volume 404

Santa Fe, New Mexico, USA
4-8 June 2012

ISBN: 978-1-62276-870-7
ISSN: 1742-6588

IOP Institute of Physics

Printed from e-media with permission by:

Curran Associates, Inc.
57 Morehouse Lane
Red Hook, NY 12571

Some format issues inherent in the e-media version may also appear in this print version.

Printed by Curran Associates, Inc. (2013)

Institute of Physics
Dirac House, Temple Back
Bristol BS1 6BE UK

Phone: 44 1 17 929 7481
Fax: 44 1 17 920 0979

techtracking@iop.org

Additional copies of this publication are available from:

Curran Associates, Inc.
57 Morehouse Lane
Red Hook, NY 12571 USA
Phone: 845-758-0400
Fax: 845-758-2634
Email: curran@proceedings.com
Web: www.proceedings.com

Opening Session

Large Hadron Collider

Calorimetry Techniques

Astrophysics and Neutrinos

Operating Calorimeters

Algorithms and Simulations

Front-end and Trigger

Future Calorimetry

PREFACE

The XVth International Conference on Calorimetry in High Energy Physics, CALOR2012, was held in Santa Fe, New Mexico from 4–8 June 2012. The series of conferences on calorimetry started in 1990 at Fermilab, and they have been the premier event for calorimeter aficionados, a trend that CALOR2012 upheld. This year, several presentations focused on the status of the major calorimeter systems, especially at the LHC. Discussions on new and developing techniques in calorimetry took a full day. Excellent updates on uses of calorimeters or about ideas that are deeply rooted in particle physics calorimetry in astrophysics and neutrino physics were followed by talks on algorithms and special triggers that rely on calorimeters. Finally, discussions of promising current developments and ongoing R&D work for future calorimeters capped the conference. The field of calorimetry is alive and well, as evidenced by the more than 100 attendees and the excellent quality of over 80 presentations. You will find the written contributions in this volume. The presentations can be found at calor2012.ttu.edu.

The first day of the conference was dedicated to the LHC. In two invited talks, Guillaume Unal (CERN) and Tommaso Tabarelli de Fatis (Universita' & INFN Milano Bicocca) discussed the critical role electromagnetic calorimeters play in the hunt for the Standard Model Higgs boson in ATLAS and CMS, respectively. The enhanced sensitivity for light Higgs in the two-gamma decay channel renders electromagnetic calorimeters indispensible. Much of the higher mass region was already excluded for the SM Higgs by the time of this conference, and after less than a month, on 4 July, CERN announced the discovery of a new boson at 125 GeV, a particle that seems consistent with the Higgs particle so far. Once again, without the electromagnetic calorimeters, this would not have been possible.

Professor Geoffrey West from the Santa Fe Institute gave the keynote address. His talk, 'Universal Scaling Laws from Cells to Cities – a Physicist's Search for Quantitative, Unified Theories of Biological and Social Structure and Dynamics,' inspired many interesting questions from the audience both after the talk and throughout the week during informal conversations.

Calorimetry is extremely diverse: many different techniques may be employed in building the detector and also in extracting information from it. The topics of the Calorimeter Techniques sessions included high-rate liquid argon calorimeters, SiPM sensors, highly granular digital calorimeters, new crystals, and beam test and simulation results. In these pages, you will find exciting and sometimes contradicting points of view expressed, for example about fully sampling hadronic calorimeters.

A rare astronomical event, the Venus transit, coincided with the second day of the conference. The participants enjoyed viewing Venus' trail across the sun with a solar telescope (H-alpha line at 656 nm). In Santa Fe, the interior ingress was at 16:23:04 and reached center at 19:27:04. The last transit occurred in 2004, and the next one will happen in 2117. In 1627, Johannes Kepler published data about the planetary orbits that predicted that Venus would pass directly between earth and the sun in 1631.

Unfortunately Kepler died in 1630 and apparently nobody recorded the 1631 transit. The first recorded observation of a transit was in 1638, which Kepler had not predicted. Later, Jeremiah Horracks, an English astronomer, realized Kepler had made an error in his calculations. It was not until the Venus transit observations of 1769 that scientists measured the distance from the earth to the sun to be 95 million miles (actually 93 million miles or 149.7 million kilometers) based on the 1716 triangulation suggestion from Edmund Halley (of comet fame). It's interesting to remember that before the 18th century, one of the most vexing scientific puzzles, not unlike today's Higgs boson quest, was 'How far away is the Sun?'

Although natural media such as Mediterranean water (ANTARES), Arctic ice (ARA, ARIANNA, ANITA, and others) or Utah air (TA) would hardly be choice absorbers in accelerator-based experiments, they are nevertheless successfully exploited in searches for new phenomena, as discussed by the members of these collaborations in several talks. Philippe Bruel (LLR, Ecole Polytechnique) gave an overview of the gamma-ray sky above 20 MeV using the Fermi Gamma-ray Space Telescope and the role of the hodoscopic array of CsI(Tl) calorimeter in an invited talk. In a second invited presentation, Sylvie Rosier-Lees (LAPP-Annecy) described the ECAL design (Pb/scintillating fiber sandwich) of the Alpha Magnetic Spectrometer (AMS-02) that has been operating in the International Space Station since May 2011 and has collected over 15 billion events.

The environs of Santa Fe have long been considered sacred by the first nation Americans and have been a source of inspiration for generations of artists, writers, and scientists. Robert Oppenheimer's love of this area played no small role in establishing what is now the nearby Los Alamos National Laboratory during the Second World War. On the third conference day, some participants visited the lab after an awe-inspiring trip to the Bandelier National Monument, where beautiful canyons and mesas show evidence of a human presence for over 11,000 years. Petroglyphs, dwellings carved into the soft rock cliffs, and standing masonry walls in the park mark the early days of a culture that still persists in surrounding communities. We are grateful to the International Advisory Committee for allowing us the opportunity to hold this Conference in these enchanted lands.

In addition to making advances in calorimeter design, hardware, and front-end electronics, we in particle physics increasingly find ourselves inventing new algorithms to reconstruct physics objects that use the detector information to its maximum capacity. Several presentations provided details of the reconstruction and trigger of jets, missing transverse energy, electrons, photons, and taus. Pile-up, anomalous signals, and noise mitigation techniques were also discussed in the conference.

On the last day, several future R&D initiatives were presented: highly granular CALICE with different technology options and plans for the dual-readout DREAM projects were the main topics. Although these approaches are quite different conceptually, future experiments will certainly benefit from their innovations. Concluding remarks by the chair of the organizing committee, Nural Akchurin (TTU), summarized the highlights of the conference and invited proposals to host the CALOR2014 conference in Europe, as the conference venue rotates between the Americas, Europe, and Asia every two years.

We strived hard to keep the cost of this conference as low as possible without sacrificing the scientific mission. I am delighted to report that we were able to provide support for six junior colleagues to participate in this conference. I am also grateful to the institutions, industrial partners, and agencies that provided the support that made a lot possible: Texas Tech University, the University of New Mexico, Los Alamos National Laboratory, the US Department of Energy, CAEN, and the Wiener Plein & Baus, Corp. I also would like to thank the session conveners who organized sessions and reviewed the papers. The members of the local organizing committee were instrumental to the success of this conference: their experience and attention to detail were invaluable. Most of all, I extend my appreciation to the conference participants and to all my other colleagues who continue to enrich the field of calorimetry through their hard work and creativity. The future is bright.

Nural Akchurin
Chair of the Organizing Committee

International Advisory Committee:
Mikhail Danilov, ITEP Moscow
Marcella Diemoz, INFN Roma I
Antonio Ereditato, Univ. of Bern
Franco L. Fabbri, INFN Frascati
Tomio Kobayashi, ICEPP Tokyo
Michele Livan, Pavia Univ. & INFN
Pasquale Lubrano, INFN Perugia
Steve Magill, ANL
Amelia Maio, LIPP Lisbon
Horst Oberlack, MPI Munich
Adam Para, FNAL
Klaus Pretzl, Univ. of Bern
Yifang Wang, IHEP Beijing
Richard Wigmans, TTU
Ren-Yuan Zhu, Caltech

Local Organizing Committee:
Nural Akchurin, TTU
Debra Boyce, TTU (Secretary)
Xiadong Jiang, LANL
Jon Kapustinsky, LANL
Sung-Won Lee, TTU
Sally Seidel, UNM
Igor Volobouev, TTU

Session Conveners:
LHC I-III:
David Barney (CERN)
Ana Henriques (CERN)
Sally Seidel (UNM)

Calorimetry Techniques I-II:
Francesca Tedaldi (ETH-Zurich)
Tao Hu (IHEP-Beijing)

Calorimetry Techniques III-IV:
Craig Woody (BNL)
Tohru Takeshita (Shinshu)

Astrophysics and Neutrinos:
Don Groom (LBNL)
Steve Magill (ANL)

Operating Calorimeters:
Jordan Damgov (TTU)
Gabriella Gaudio (INFN-Pavia)
Frank Chlebana (FNAL)

Algorithms and Simulations:
Artur Apresyan (Caltech)
Igor Volobouev (TTU)

Front-end and Trigger:
Chris Tully (Princeton)
Kejun Zhu (IHEP-Beijing)

Future Calorimetry:
Michele Livan (Pavia Univ.)
Frank Simon (MPI)
Vishnu Zutshi (NICADD)

List of Participants:
ABOUZEID, Hass University of Toronto
AKCHURIN, Nural Texas Tech University
ANDEEN, Timothy Columbia University
ANDERSON, Jake Fermilab
APRESYAN, Artur California Institute of Technology
AUFFRAY, Etiennette CERN
BARILLARI, Teresa Max-Planck-Inst. fuer Physik
BARNEY, David CERN
BESSON, Dave University of Kansas

BOYCE, Debra Texas Tech University
BRUEL, Philippe LLR, Ecole Polytechnique, CNRS/IN2P3
BUCHANAN, Norm Colorado State University
CARLOGANU, Cristina LPC Clermont Ferrand / IN2P3 / CNRS
CHEFDEVILLE, Maximilien CNRS/IN2P3/LAPP
CHLEBANA, Frank Fermilab
CLARK, Jonathan Texas Tech University
CONDE MUINO, Patricia LIP-Lisboa
COWDEN, Christopher Texas Tech University
DA SILVA, Cesar Luiz Los Alamos National Lab
DAMGOV, Jordan Texas Tech University
DAVYGORA, Yuriy University of Heidelberg
DEMERS, Sarah Yale University
EIGEN, Gerald University of Bergen
EUSEBI, Ricardo Texas A&M University
FERRI, Federico CEA/Saclay Irfu/SPP
FRANCAVILLA, Paolo IFAE Barcelona
GATAULLIN, Marat California Institute of Technology
GATTO, Corrado INFN-Napoli
GAUDIO, Gabriella INFN-Pavia
GERMANI, Stefano INFN-Perugia
Goldenzweig, Pablo University of Rochester
GRAF, Norman SLAC National Accelerator Laboratory
GROOM, Don Lawrence Berkeley Lab
GUARDINCERRI, Elena Los Alamos National Laboratory
HAUPTMAN, John Iowa State University
HENRIQUES, Ana CERN
HUANG, Jin Los Alamos National Laboratory
HU, Tao IHEP-Beijing, CAS
JIANG, Xiaodong Los Alamos National Laboratory
JUI, Charles University of Utah
KAPUSTINSKY, Jon Los Alamos National Laboratory
KIBILKO, Mark SE Technical Sales, Inc.
KIRSCHENMANN, Henning University of Hamburg
KISTENEV, Edouard Brookhaven National Laboratory
KLIMEK, Pawel Stockholm Universitet
KROEGER, Robert University of Mississippi
LECOQ, Paul CERN
LEE, Sehwook Texas Tech University
LEE, Sung-Won Texas Tech University
LIVAN, Michele Pavia University
LUTZ, Benjamin DESY
MAGILL, Stephen Argonne National Laboratory
MATHIS, Mark College of William and Mary
MATTHEWS, John University of Utah
MENKE, Sven Max-Planck-Institut fuer Physik

MOULSON, Matthew INFN-Frascati
NAGEL, Martin Max-Planck-Institut fuer Physik
NAKAMURA, Isamu KEK
NEMECEK, Stanislav FZU AVCR Praha
NESSI-TEDALDI, Francesca ETH Zurich
NOVOTNY, Rainer 2nd Physics Institute, University Giessen
OREGLIA, Mark University of Chicago
PERLOFF, Alexx Texas A&M University
PETYT, David Rutherford Appleton Laboratory
RAHMAT, Rahmat University of Mississippi
RAMILLI, Marco Hamburg Universitaet
ROSIER LEES, Sylvie LAPP- IN2P3-CNRS
RUTHERFOORD, John University of Arizona
SAKUMA, Tai Texas A&M University
SANTIAGO CERQUEIRA, Augusto Federal University of Juiz de Fora
SARRA, Ivano INFN-Frascati
SEIDEL, Sally University of New Mexico
SEIFERT, Frank TU Dresden, Germany
SHAMIM, Mansoora University of Oregon
SIMON, Frank Max-Planck-Institute for Physics
STAFFAN, Paul Wiener Plein and Baus, Corp
Dr. STAROVOITOV, Pavel DESY
TABARELLI DE FATIS, Tommaso Universita' & INFN Milano-Bicocca
TADEVOSYAN, Vardan AANL
TAKESHITA, Tohru Shinshu University
TALAGA, Richard Argonne National Laboratory
TAPAN, Ilhan Uludag University
TERWORT, Mark DESY
TSAI, Oleg UCLA
TULLY, Christopher Princeton University
UNAL, Guillaume CERN
VICHOU, Eirini University of Illinois at Urbana-Champaign
VILASIS-CARDONA, Xavier La Salle - Universitat Ramon Llull
VOLOBOUEV, Igor Texas Tech University
VOLPI, Matteo The University of Melbourne

WANG, Zhigang IHEP-Beijing, CAS
WENZEL, Hans Fermilab
WHITE, Andy University of Texas at Arlington
WIGMANS, Richard Texas Tech University
WINN, David Fairfield University
WOODY, Craig Brookhaven National Lab

YANG, Fan California Institute of Technology
ZABI, Alexandre LLR-Ecole Polytechnique
ZHANG, Liyuan California Institute of Technoogy
ZHAO, Zhiwen University of Virginia
ZHU, Kejun IHEP-Beijing, CAS
ZHU, Ren-Yuan California Institute of Technology
ZUTSHI, Vishnu Northern Illinois University

Journal of Physics: Conference Series **404** (2012) 012001 doi:10.1088/1742-6596/404/1/012001

The Role of the LAr Calorimeter in the Search for H$\rightarrow \gamma\gamma$ in ATLAS

G.Unal[1], **on behalf of the ATLAS Collaboration**

[1] CERN

E-mail: `Guillaume.Unal@cern.ch`

Abstract. In the low mass range (100 to 140 GeV) one of the most promising search channel for the Higgs boson at the LHC is the rare decay in a pair of photons. Photon identification and energy as well as direction measurements are key ingredients to reduce backgrounds and maximize the sensitivity to the Higgs boson signal. The analysis of the Higgs boson search in the two photon decay channel performed by the ATLAS experiment using 4.9 fb^{-1} of proton-proton collisions data recorded in 2011 at the LHC is described. The role of the ATLAS Liquid Argon calorimeter in this analysis is emphasized.

1. Introduction

The Higgs mechanism [1] is one of the best-motivated processes to explain electroweak symmetry breaking. In the Standard Model, this mechanism explains the generation of the W and Z boson masses and predicts the existence of an elementary scalar particle, the Higgs boson. The search for the Higgs boson is one of the main goals of the Large Hadron Collider (LHC). The Higgs boson is copiously produced at the LHC via the gluon-fusion process. At large masses (above $\approx$ 140 GeV mass), decays to vector bosons provide a clean signature with large branching ratio. At lower masses, the decay to $b\bar{b}$ pairs dominates and is difficult to observe in a hadron collider experiment. Thus rare decay modes have to be used. The decay in a pair of photons is considered as one of the most promising. This process was taken as a benchmark process for the optimization of the LHC detectors, especially the electromagnetic calorimeters. Figure 1 illustrates the cross-section times branching ratio as a function of the Higgs boson mass for decay modes considered as observables at the LHC [2]. With a branching ratio around 0.2%, the two photon decay mode gives a total rate of $\approx$ 40 fb.

The ATLAS detector is described in Ref. [3]. The electromagnetic calorimeter is a lead-liquid argon sampling calorimeter with accordion geometry. It is divided into a barrel region, covering pseudorapidity (η) less than 1.475, and two end-caps covering 1.35 to 3.2. It is finely segmented in three longitudinal layers for η less than 2.5. The first layer, around 4-5 radiation lengths thick, is finely segmented in the η direction to provide γ-π^0 separation and γ direction measurement. The cell size in η is for instance 0.003125 in the barrel region. The second layer collects most of the shower energy and has a granularity of 0.025$\times$0.025 in $\eta \times \phi$. The last layer is used to correct for leakage beyond the electromagnetic calorimeter. In addition a thin presampler layer is located between the cryostat and the calorimeter, covering the region η less than 1.8. It is used to correct for fluctuations in the energy loss upstream the calorimeter. A sketch of the calorimeter granularity in the barrel region can be seen in Figure 2. The electromagnetic calorimeter comprises 173312 readout channels, 99.8% of them being operationnal during most of the 2011 data taking period. More details on the electromagnetic calorimeter and its operation in 2011 can be found in Ref. [4].

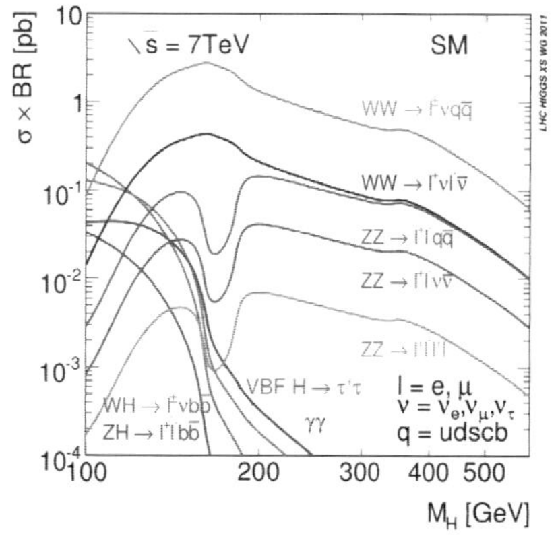

Figure 1. The SM Higgs boson production cross-section times branching ratio at $\sqrt{s}$=7 TeV [2]

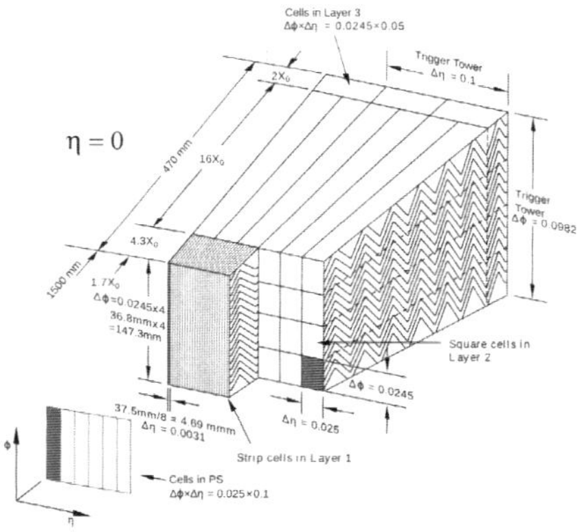

Figure 2. Sketch of the electromagnetic calorimeter granularity in the barrel region

The recipe for the search of the Higgs boson decay into two photons consists in:

- Selecting events with two photon candidates. These events are "easy" to trigger on at the LHC. Jet-jet and γ-jet events should be reduced well below the level of the irreducible γ-γ continuum background and this requires rejection against jets of few thousands. To achieve this a rejection against high energies π^0, from tails of jet fragmentation, decaying into collimated photons is useful.
- Reconstructing the invariant mass of the photon pair as accurately as possible, to separate the signal from the continuum background. For the relevant Higgs boson mass range, the Higgs width is small thus the resolution is completely driven by the experimental resolution. To achieve a good mass resolution both a good energy resolution and angular resolution are needed.
- Optimizing the analysis to improve the sensitivity. This is done by splitting the data in different categories with different signal over background ratios or different mass resolutions. The classification is based on kinematical variables and on the expected photon energy resolution

The analysis described here is based on the full data set collected in 2011 by the ATLAS experiment at $\sqrt{s}$=7 TeV, corresponding to an integrated luminosity of 4.9 fb^{-1} [5].

2. Photon identification

Photon candidates are seeded by energy clusters in the electromagnetic calorimeter. Clusters without matching inner detector tracks are classified as unconverted photon candidates. Clusters matched to either pairs of tracks which are consistent with the hypothesis of a photon conversion or single tracks without hits in the pixel layer nearest to the beam pipe are considered as converted photon candidate. The photon reconstruction efficiency is $\approx$ 98%.

Photon identification is based on the lateral and longitudinal energy profiles of the shower in the calorimeter. The photon candidate is required to deposit only a small fraction of its energy in the hadronic calorimeter. The transverse shower shape in the second layer of the electromagnetic calorimeter needs to be consistent with that expected for a single electromagnetic shower. Using these variables, a set of loose identification criteria is defined. Finally, the high granularity first calorimeter layer is used to discriminate single photons from overlapping photon pairs from neutral meson decays produced in jet fragmentation, which are the main background source. Using cuts on all the shower shape variables, a tight identification selection is defined. This selection is different for converted photons and unconverted photons.

Figure 3 illustrates the distribution of two of the discriminating variables using the first layer of the calorimeter, for candidates fulfilling the loose identification selection. These variables are E_{ratio}, the asymmetry between the first and second maxima in the energy profile of the first layer along η, and F_{side}, the fraction of the energy in seven layer cells centered around the first maximum which is not contained in the three core cells centered around it. The separation between prompt photons and background can clearly be seen.

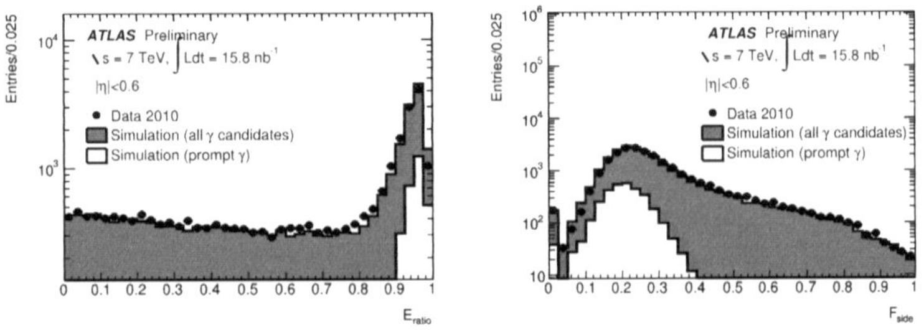

Figure 3. Examples of discriminating shower shape variables based on the energy deposit in the first longitudinal compartment of the electromagnetic calorimeter, for photon candidates passing the loose identification criteria. Left: E_{ratio}. Right: F_{side}. Photon candidates with pseudorapidity $|\eta| < 0.6$ [6].

Clean electron samples from $Z \rightarrow e^+e^-$ are used to compare data and simulation for the distributions of the shower shape variables. An example is shown in Figure 4 for the R_η variable, the ratio of the energy deposits in 3×7 over 7×7 cells in the second layer of the calorimeter. The shift between the data and the simulation corresponds to the fact that shower shape in data are somewhat broader than in the simulation. This difference is reduced in the latest simulation version.

To take into account these small differences, the shower shape variables are shifted in the simulation before the identification cuts are applied, by an amount corresponding to the average difference between data and simulation.

The photon identification efficiency ranges typically from 65% to 95% for transverse energies (E_T) in the range 25 to 80 GeV. The uncertainty on the identification efficiency is assessed using data control samples. Three different samples are used: photons from $Z \rightarrow \ell\ell\gamma$ decays, electrons from $Z \rightarrow e^+e^-$ decays where electron shower shapes are extrapolated to photon shower shapes using simulation and loose photon candidates. The typical uncertainty on the photon identification efficiency is $\pm$ 5%, except at low E_T ($<$ 30 GeV) and high $|\eta|$ ($>$ 1.8) for unconverted photons where it reaches $\pm$ 10%.

To further reduce the fake backgrounds, the photon candidates are required to be isolated. The isolation variable is computed by summing the transverse energy in calorimeter cells in a cone of radius 0.4 in the $\eta \times \phi$ space around the photon candidate. Cells in the electromagnetic calorimeter within 0.125×0.175 from the shower barycentre are excluded from the sum. The

Journal of Physics: Conference Series **404** (2012) 012001 doi:10.1088/1742-6596/404/1/012001

small photon energy leakage outside the excluded cells is evaluated as a function of the transverse energy and is subtracted from the isolation variable. To mitigate the effect from the underlying event and pileup, the isolation is further corrected using a method suggested in Ref. [7] : for each of the two different pseudorapidity regions $|\eta| < 1.5$ and $1.5 < |\eta| < 3.0$, low energy jets are used to compute an ambient energy density which is then multiplied by the area of the isolation cone and subtracted from the isolation energy. Photon candidates are required to have an isolation variable less than 5 GeV. The isolation cut retains $\approx 87\%$ of the Higgs boson signal events, while rejecting $\approx 44\%$ of the data events.

The photon candidates are ordered in E_T and the leading (subleading) candidate is required to have $E_T > 40$ GeV (25 GeV). Both photon candidates should be in the fiducial region $|\eta| < 2.37$, excluding the calorimeter barrel/end-cap transition region $1.37 < |\eta| < 1.52$. After all the identification steps, 22489 candidates are selected in the 2011 data sample. The composition of the sample (prompt diphoton pairs, photon-jet events where one jet fakes a photon, dijet events and residual Drell-Yan background with electrons misidentified as photons) is investigated using data-driven methods. A method based on the use of control regions for two discriminating variables is applied to measure the contributions of fake photon background. This method exploits relaxed isolation and photon identification cuts to estimate the fake component by relying on the fact that the rejections from these two cuts are almost independent. The Drell-Yan background is estimated by measuring the probability for an electron to be reconstructed as a photon candidate with Z events and applying this probability to the observed yield of Drell-Yan events at high mass. The fraction of true diphoton events is estimated to be $(71 \pm 5)\%$. The photon-jet background is $(23 \pm 4)\%$, the dijet background $(5 \pm 3)\%$ and the Drell-Yan background $(0.7 \pm 0.1)\%$. The sample composition as a function of the diphoton invariant mass is shown in Figure 5. At this level, 69 signal events from Higgs boson decays into photon pairs are expected in the sample for a Higgs boson mass hypothesis of 125 GeV, assuming Standard Model production cross-section and branching ratio values. The acceptance of the kinematical cuts is around 60% and the overall acceptance times efficiency of the full selection is 35%. To isolate these signal events in the data sample (dominated by diphoton production) a good mass resolution is needed.

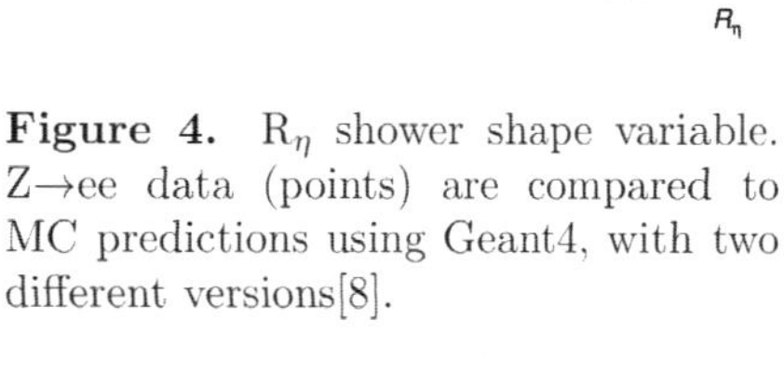

Figure 4. R_η shower shape variable. Z→ee data (points) are compared to MC predictions using Geant4, with two different versions[8].

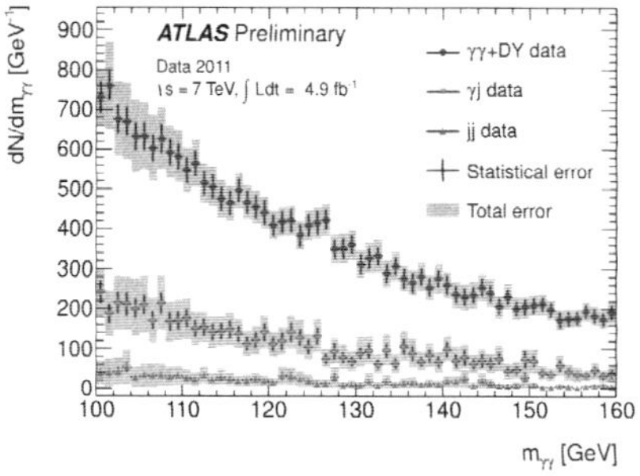

Figure 5. Composition of the inclusive data sample as a function of $m_{\gamma\gamma}$. The various components are stacked on top of each other. The error bars correspond to the statistical uncertainties on each component separately. The gray bands show the overall uncertainty on each component[9].

3. Energy measurement

The LAr calorimeter cell signals are read out through 1524 Front-End Boards with 128 channels each. The FEBs perform analog processing (amplification and shaping), store the signal while waiting for the L1 trigger decision and digitize at 40 MHz the accepted signals [10] [11]. The digitized signals (5 time samples around the trigger time) are transmitted via optical fibers to the Readout Drivers where energy reconstruction is performed. For cells with enough energy, the digitized signals are also stored and energy can be re-computed offline. From the 5 digitized time samples, the cell energy is reconstructed with:

$$E_{cell} = F_{\mu A \rightarrow MeV} \frac{M_{cal}}{M_{phys}} \; R \; \Sigma a_i (ADC_i - P) \tag{1}$$

where ADC_i are the digitized time samples, P is the pedestal, R the conversion factor from ADC to injected current (in μA), $F_{\mu A \rightarrow MeV}$ is the conversion factor from current to energy deposited in the cell, a_i are optimal filter coefficients used to estimate the pulse amplitude while minimizing the noise contribution and $\frac{M_{cal}}{M_{phys}}$ (typically few percent) accounts for the difference between electronics calibration injected signal and ionization signal in the maximum amplitude after shaping for the same initial current. P and R are computed using electronics calibration, injecting calibrated current to simulate energy deposits in the calorimeter. This is done typically once a week and these calibration constants are very stable with time. $F_{\mu A \rightarrow MeV}$ was determined from test-beam calibration and adjusted with the first LHC collision data. This factor also accounts for difference between operating High Voltage and nominal values for few percents of the cells. $\frac{M_{cal}}{M_{phys}}$ and differences between electronics calibration and ionization pulse shapes, important in the computation of the a_i factors, come from the different time shapes of these two signals (exponential for the electronics calibration, triangular for the ionization pulse shape) and from the electrical properties of the cells. The predicted ionization shapes are calculated from calibration pulses by modeling each readout cell as a resonant RLC circuit. The effective LC and RC are estimated from a frequency analysis of the output calibration pulse shape. The prediction of the ionization pulse also requires the knowledge of the electron drift time in liquid argon, which was directly measured from data [12].

To illustrate the good quality of the ionization pulse prediction, Figure 6 shows a comparison between a measured pulse (over 800 ns) from a radiating cosmics muon deposing few GeV in a cell with the pulse shape prediction, scaled to the measured cell energy. The agreement is a the few percent level, within the specifications to achieve the design constant term in the energy resolution of the calorimeter (0.7%).

The overall energy scale of the calorimeter, as measured with collision data using Z and W decays, is very stable with time, as shown in Figure 7.

Photon energy measurement is made using a cluster size which depends on the photon classification. In the barrel, a size of 0.075×0.125 in $\eta \times \phi$ is used for unconverted photons, and 0.075×0.175 for converted photon candidates, to account for the larger spread of the shower in ϕ for converted photons due to the magnetic field. In the end-cap, a cluster size of 0.125×0.125 is used for all candidates. A dedicated energy calibration [15] is applied to account for upstream energy loss, lateral leakage and longitudinal leakage, separately for converted and unconverted photon candidates. The correction for upstream energy loss is parametrized as a function of the energy deposited in the presampler for $|\eta| < 1.8$. These corrections are derived from detailed full simulation samples. For 100 GeV photon energy, the energy resolution in the simulation varies from 1.2% to ≈ 1.7% as a function of η, assuming the design constant term of 0.7%.

The final energy calibration is determined from Z→ ee decays, resulting in η-dependent correction factors of the order of ± 1%. After this calibration procedure, the constant term in the energy resolution is investigated from Z decays. Figure 8 shows the reconstructed Z invariant mass distribution when both electrons are in the barrel calorimeter and Figure 9 when

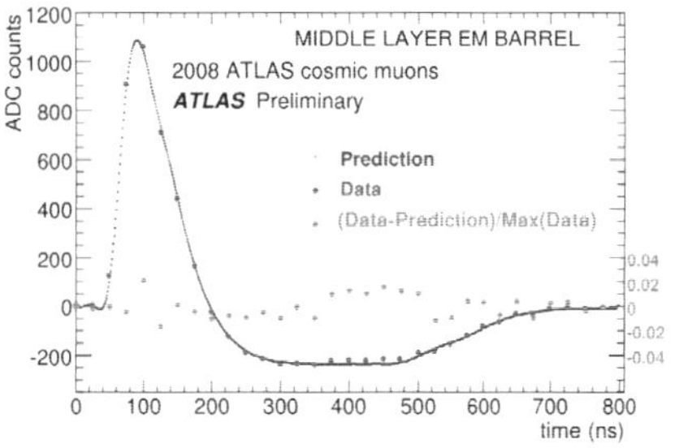

Figure 6. Comparison between the observed pulse shape in the second layer of the barrel electromagnetic calorimeter over 800 ns using cosmics muon data and predictions [13].

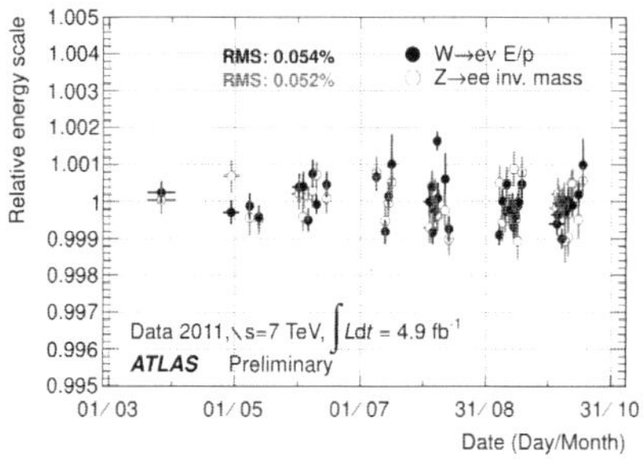

Figure 7. Stability of the calorimeter energy scale with time in 2011, as measured using the Z mass in $Z \rightarrow$ee events and using E/p in $W \rightarrow e\nu$ events [14].

both electrons are in the end-cap calorimeter. The data are compared to simulation, without any additional energy smearing. This comparison leads to estimates of the constant term in the energy resolution which are found to be $1.2 \pm 0.5\%$ in the barrel calorimeter and $1.8 \pm 0.6\%$ in the end-cap calorimeter [17] . These constant terms are then applied to the Higgs boson signal simulated samples. The uncertainty on the photon energy resolution ($\pm 13\%$ relative) arises from the uncertainty on the sampling term of the energy resolution (from test-beam studies), the uncertainty in the constant term and the uncertainty in the electron to photon extrapolation (mostly from uncertainties in material upstream the calorimeter).

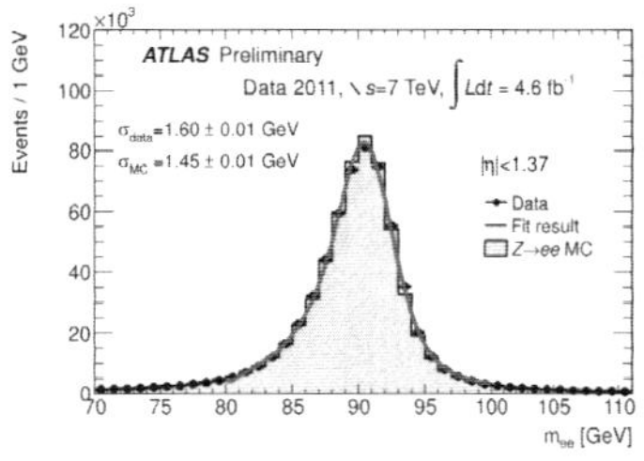

Figure 8. Distribution of the reconstructed Z mass from $Z \rightarrow$ee events, where both electrons are in the barrel region [16].

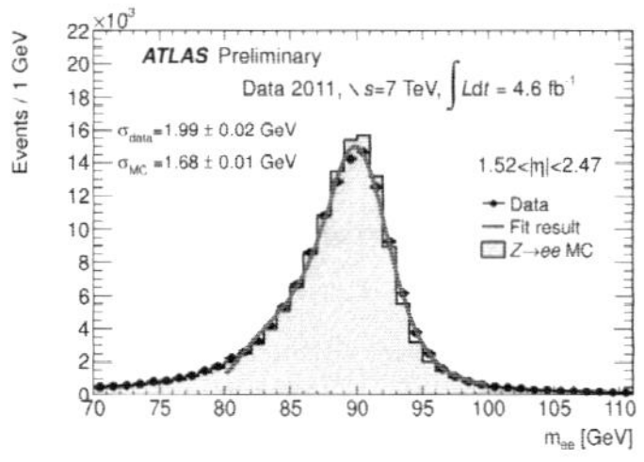

Figure 9. Distribution of the reconstructed Z mass from $Z \rightarrow$ee events, where both electrons are in the end-cap region [16].

4. Direction measurement

In addition to the energies, the angle between the photons is needed for the computation of the diphoton invariant mass. This angle is determined from the interaction vertex position and the photon impact points in the calorimeter. The resolution of the angle measurement is dominated by the reconstruction of the primary vertex z position. The RMS vertex spread in the z direction is ≈ 5.5 cm and a more accurate event-by-event estimate is performed to reduce the impact on the invariant mass resolution. Given the level of pileup, the determination of the vertex position is based only on the photon candidates, without relying on other charged tracks in the event. For converted photons with tracks having a precise measurement in the z direction, the vertex position is estimated from the intercept of the line joining the reconstructed conversion position and the calorimeter impact point with the beam line. For all other photons, the vertex position is estimated from the shower position measurements in the first and second layers of the electromagnetic calorimeter, which can be used to calculate the photon direction. Finally the vertex positions from both photons are combined taking also into account the average beam spot position in z.

Figure 10 shows the difference between the estimates of the vertex positions from the two photons, when they are both unconverted in the barrel calorimeter. When both photons are unconverted, the typical final vertex position resolution is ≈ 1.6 cm in z. The resolution is better in events with converted photons. The resulting impact of the angle measurement on the invariant mass resolution is negligible compared to the contribution from the photon energy resolution, as shown in Figure 11.

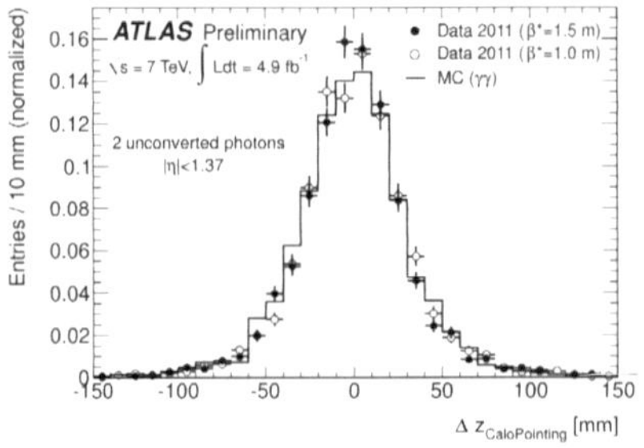

Figure 10. Difference between two estimates of the z vertex position computed separately for each photon using calorimeter pointing, when both photons are unconverted and in the barrel region. The distribution is shown separately for "low" and "high" pile-up periods in the 2011 data [18].

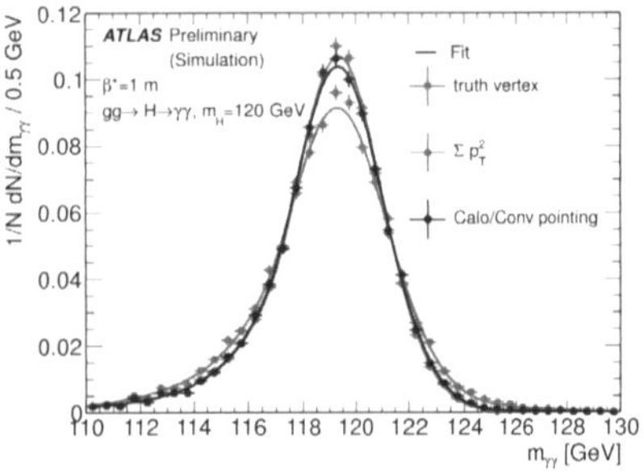

Figure 11. Reconstructed diphoton invariant mass distribution for simulated H→ $\gamma\gamma$ events with different methods to estimate the primary vertex. When using the calorimeter pointing and photon conversion informations, the mass resolution is as good as when using the truth primary vertex position in the simulation [9].

5. Data analysis

The events are separated into nine mutually exclusive categories with different mass resolutions and signal-to-background ratios, to increase the sensitivity to a possible Higgs boson signal. Categories are defined by the conversion status, η of the photons and p_{Tt}, the component of the diphoton transverse momentum that is orthogonal to the thrust axis. Events with two unconverted photons are separated into *unconverted central* ($|\eta| < 0.75$ for both candidates) and *unconverted rest* (all other events). Events with a least one converted photon are separated into *converted central* ($|\eta| < 0.75$ for both candidates), *converted transition* (at least one photon with $1.3 < |\eta| < 1.75$) and *converted rest* (all other events). Except the *converted transition* category, each category is further divided by a cut at p_{Tt}=40 GeV into two categories.

The invariant mass shape of the signal in each category is modeled by the sum of a Crystal Ball function describing the core of the distribution and of a wide Gaussian with a small amplitude describing the tails. In Figure 12, the sum of all signal processes in all categories is shown for a

Higgs mass boson of 120 GeV. The full-width-at-half maximum is 4.1 GeV and the core gaussian of the Crystal Ball function (σ_{CB}) is 1.7 GeV. σ_{CB} varies from 1.4 GeV (in the *unconverted central* categories) to 2.0 GeV (in the *converted rest* categories) and 2.3 GeV (in the *converted transition* category which is the one with the lowest sensitivity).

The background in each category is estimated from the data by fitting the diphoton mass spectrum in the range 100-160 GeV with an exponential function with free shape and normalization parameters. The distribution of the invariant mass of the diphoton events, summed over all categories is shown in Figure 13.

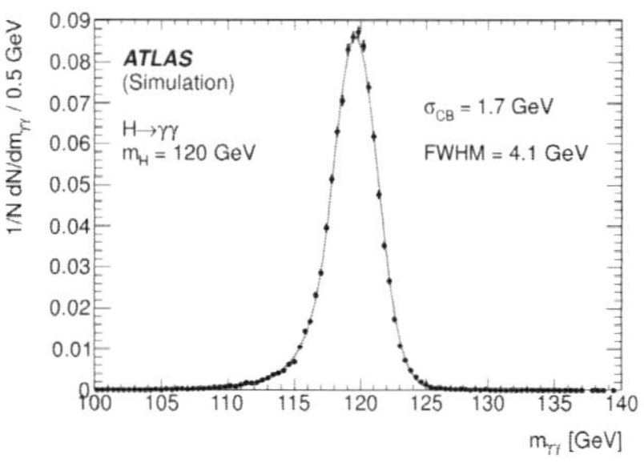

Figure 12. Reconstructed invariant mass distribution for a simulated signal of m_H = 120 GeV, summed over all categories [5].

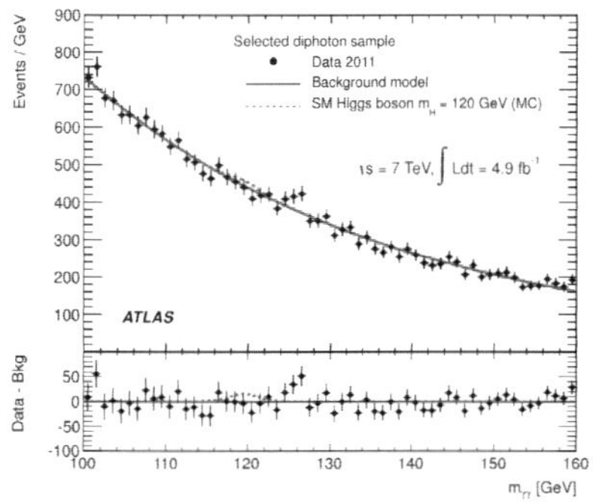

Figure 13. Invariant mass distribution for the selected data samples, overlaid with the total background fit. The bottom inset displays the residual of the data with respect to the total background. The Higgs boson expectations for a mass hypothesis of 120 GeV and the Standard Model cross-section is also shown [5].

The dominant experimental uncertainty on the signal yield is the photon reconstruction and identification efficiency ($\pm$ 11% per event). Pileup also affects the identification efficiency and contributes to the uncertainty ($\pm$4%). Further uncertainties on the signal yield are related to the trigger ($\pm$1%), Higgs boson transverse momentum modeling ($\pm$1%), isolation ($\pm$5%), and luminosity ($\pm$3.9%). Uncertainties on the predicted cross-sections are due to uncertainties on the QCD renormalization and factorization scales ($^{+12}_{-8}$%) and on the parton density functions and α_S ($\pm$8%). The total uncertainty on the signal yield is $^{+20}_{-17}$%. The total uncertainty on the mass resolution is $\pm$14%, dominated by the uncertainty on the energy resolution of the calorimeter ($\pm$12%). Further uncertainties on the mass resolution result from an imperfect knowledge of the material in front of the calorimeter affecting the extrapolation from electron to photon calibration ($\pm$6%), the impact of pileup ($\pm$3%) and the photon angle measurement ($\pm$1%). The uncertainty on the knowldege of the material in front of the calorimeter is used to derive the amount of event migration between the converted and unconverted categories ($\pm$4.5%). Uncertainty in the modeling of the Higgs transverse momentum distribution lead to a $\pm$8% migration between the low and high p_{Tt} categories. Systematic uncertainties in the background modeling arise from possible deviations of the background shape from the assumed exponential shape. This has been estimated by checking how accurately the chosen model fits different predicted diphoton mass distributions, and comparing different function forms for the background model. The uncertainties are $\pm(0.1 - 7.9)$ events in a 4 GeV signal mass window, depending on the category.

6. Results

The data are compared to background and signal-plus-background hypothesis using a profile likelihood test statistic. The exponentially falling invariant mass distribution used for the background model is determined by two nuisance parameters per category which are left free in the unbinned fit. The signal is modelled with the mass resolution functions described above, one per category, fixing the fraction of events in each category to the MC predictions, within the uncertainties discussed above. The fitted parameters for the signal are thus the overall signal strength relative to the Standard Model prediction and the nuisance parameters on the predicted event yield and mass resolution which have Gaussian constraints in the fit. The systematic uncertainty on the background shape is included as another nuisance parameter with a Gaussian constraint in the fit. From this fit, the best estimate of the signal yield is extracted, as well as the likelihood ratio (profile likelihood) between any assumed signal yield (leaving the nuisance parameters free to maximise the likelihood) and the best estimate. A modified frequentist approach (CL_S) for setting limits and a frequentist approach to calculate the p_0 value are used. The p_0 is the probability that the background fluctuates to the observed number of events or higher. The 95% confidence level (CL) limit on the cross-section relative to the Standard Model cross-section is shown in Figure 14. The median expected upper limit for background only experiments varies between 1.6 and 2.7 times the Standard Model cross-section in the mass range 110-150 GeV. The observed limit is between 0.83 and 3.6 times the Standard Model cross-section. A Standard Model Higgs boson is excluded at 95% CL in the mass ranges of 113-115 GeV and 134.5-136 GeV. The local p_0 value is shown in Figure 15. Around 126.5 GeV, an excess is observed in the data with a local significance of 2.8 standard deviations. This becomes 1.5 standard deviation when the look elsewhere effect for the mass range 110-150 GeV is included.

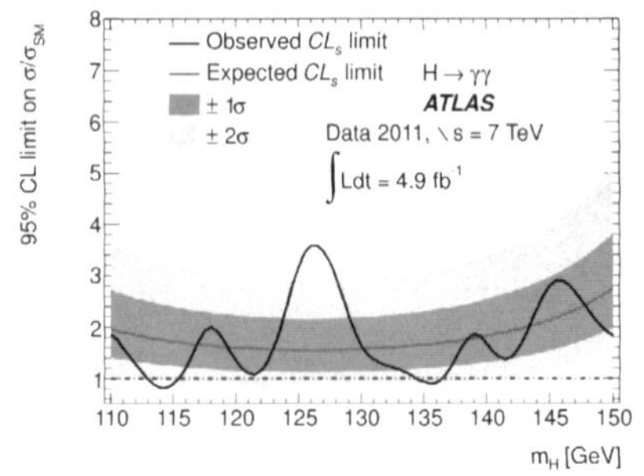

Figure 14. Observed and expected 95% CL limits on the Standard Model Higgs boson production normalized to the predicted cross-section as a function of m_H [5].

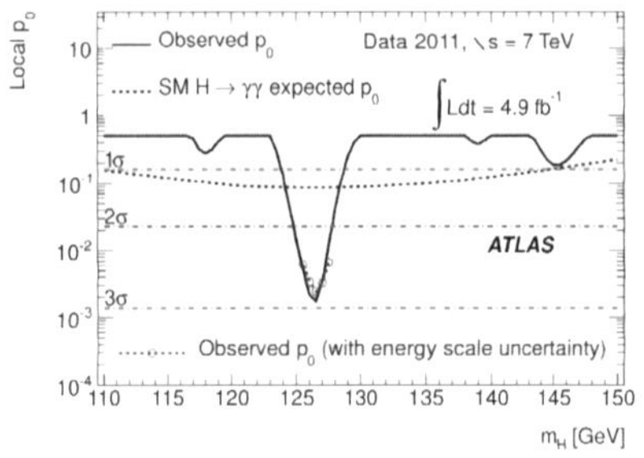

Figure 15. The observed local p_0, the probability that the background fluctuates to the observed number of events or higher (solid line). The open points indicated the observed value when energy scale uncertainties are taken into account. The dotted line shows the expected median local p_0 for the signal hypothesis when tested at m_H [5].

When combining with the other search channels in the 2011 dataset [19], the Standard Model

Higgs boson is excluded in the mass ranges 110-117.6 GeV, 118.5-122.5 GeV and 129-539 GeV. In the non-excluded mass range, an excess with a local significance of 2.5 standard deviations is observed corresponding to a mass hypothesis of 126 GeV. When taking into account the look elsewhere effect, the global p_0 is around 10%. The excess at low mass is compatible with the Standard Model Higgs boson but the statistical significance is not large enough to draw definitive conclusions.

Data taking in 2012 at 8 TeV centre of mass energy and higher luminosity started in April 2012 and, thanks to a very successful LHC operation, allowed ATLAS to observe shortly after this conference a significant signal for a new boson at a mass close to 126 GeV [20].

7. Conclusions

With the 2011 LHC data, the diphoton decay channel is already bringing an important sensitivity to the low mass Higgs boson search in ATLAS. This is possible thanks to the good performance of the liquid argon calorimeter and the ATLAS detector. With the 2011 data, the allowed mass range for the Standard Model Higgs boson has been reduced to a small range, with a small excess observed in the non-excluded mass range. The diphoton decay channel will also play an important role in the search of the Higgs boson with the 2012 data.

References

[1] F. Englert, R. Brout, Phys. Rev. Lett. 13, 321 (1964); P. W. Higgs, Phys. Rev. Lett. 13, 508 (1964); G.S. Guralnik, C.R. Hagen, T.W.B. Kibble, Phys. Rev. Lett.13,585 (1964)
[2] LHC Higgs Cross Section Working Group, CERN- 2011-002 arXiv:1101.0593 (2011) and arXiv:1201.3084 (2012)
[3] ATLAS Collaboration, JINST 3, S08003 (2008)
[4] H.Abouzeid, these proceedings.
[5] ATLAS Collaboration, Phys. Rev. Lett. 108 (2012) 111803, arXiv:1202.1414
[6] ATLAS Collaboration, ATLAS-CONF-2010-077, http://cdsweb.cern.ch/record/1281368
[7] M. Cacciari, G. P. Salam, G. Soyez, JHEP 04 (2008) 005
[8] ATLAS Collaboration, http://atlas.web.cern.ch/Atlas/GROUPS/PHYSICS/EGAMMA/PublicPlots/20111005/ATL-COM-PHYS-2011-1299
[9] ATLAS Collaboration, ATLAS-CONF-2011-161, http://cdsweb.cern.ch/record/1406356
[10] H.Abreu et al, JINST 5 P09003 (2012)
[11] ATLAS Collaboration,Eur. Phys. J. C70 (2010) 723, arXiv:0912.2642
[12] ATLAS Collaboration,Eur. Phys. J. C70 (2010) 755, arXiv:1002.4189
[13] ATLAS Collaboration, http://twiki.cern.ch/twiki/bin/view/AtlasPublic/LArCaloPublicResultsDetStatus
[14] ATLAS Collaboration, http://atlas.web.cern.ch/Atlas/GROUPS/PHYSICS/EGAMMA/PublicPlots/20121403/StabilityPlots/ATL-COM-PHYS-2012-259
[15] ATLAS Collaboration, CERN-OPEN-2008-020 (2008) 44-71.
[16] ATLAS Collaboration, http://atlas.web.cern.ch/Atlas/GROUPS/PHYSICS/EGAMMA/PublicPlots/20110512/CalibratedZee/ATL-COM-PHYS-2011-1637
[17] ATLAS Collaboration, Eur. Phys. J. C72 (2012) 1909, arXiv:1110.3174
[18] ATLAS Collaboration, http://atlas.web.cern.ch/Atlas/GROUPS/PHYSICS/EGAMMA/PublicPlots/20111412/Pointing/ATL-COM-PHYS-2011-1661
[19] ATLAS Collaboration, arXiv:1207.0319, accepted for publication in Phys. Rev. D
[20] ATLAS Collaboration, arXiv:1207.7214, accepted for publication in Phys. Lett. B.

Role of the CMS Electromagnetic Calorimeter in the hunt for the Higgs boson in the two-gamma channel

T. Tabarelli de Fatis *(on behalf of the CMS Collaboration)*

Dipartimento di Fisica, Università di Milano Bicocca and INFN Milano Bicocca, Piazza della Scienza 3, I-20126, Milano, Italy

E-mail: tommaso.tabarelli@cern.ch

Abstract. The distinctive signature of the two-photon decay mode of a low-mass Higgs boson ($H \rightarrow \gamma\gamma$) would be a narrow resonance, smeared by the photon energy and direction resolution, over a non-resonant background of diphotons or spurious events. The sensitivity to this decay mode greatly benefits from the energy and position resolution and photon identification capabilities of the electromagnetic calorimeters at the LHC. In this context, the performance of the electromagnetic calorimeter (ECAL) of the CMS experiment - a hermetic, fine grained and homogeneous calorimeter made of lead-tungstate ($PbWO_4$) crystals, completed by a silicon/lead preshower installed in front of the endcaps - is presented.

1. Introduction

The Compact Muon Solenoid (CMS) experiment [1] at the LHC has a broad physics programme [2], including the investigation of electroweak symmetry breaking through the direct search of the postulated standard model (SM) Higgs boson. The two-photon decay mode ($H \rightarrow \gamma\gamma$) is one of the most sensitive channels in the search for a low-mass SM Higgs boson ($m_H < 150$ GeV) [3]. Its distinctive experimental signature would be a narrow peak – of width determined by the instrumental resolution – in the invariant mass distribution of two isolated photons of high transverse momentum to the beam axis, on a large irreducible background from QCD production of two photons. Events where at least one of the photon candidates originates from misidentification of jet fragments contribute to an additional, reducible background. The electromagnetic calorimeter (ECAL) [4] of CMS has been specifically designed to provide excellent invariant mass – via energy and position – resolution, and fine transverse granularity for photon identification purposes, to enhance the sensitivity to the $H \rightarrow \gamma\gamma$ decay mode.

In this report, we discuss the instrumental and operational aspects of the CMS ECAL relevant in the 'hunt' for the $H \rightarrow \gamma\gamma$ decay channel. Emphasis is given to single channel response stability and uniformity within ECAL, and to the calibration of the energy of electrons and photons in CMS, as these effects directly contribute to the overall energy resolution. The energy resolution is estimated from the analysis of $Z \rightarrow e^+e^-$ events collected during the 2011 LHC run at a centre of mass energy $\sqrt{s} = 7$ TeV and for an integrated luminosity of 4.98 fb^{-1}, and compared to Monte Carlo (MC) simulation. Implications for the $H \rightarrow \gamma\gamma$ analysis are presented. CMS results on the search for the $H \rightarrow \gamma\gamma$ decay mode have been published elsewhere[1] [5, 6].

[1] At the time of writing of these Proceedings, the observation of a new boson, with a two-photon decay mode, compatible with a ~125 GeV SM Higgs, has been reported by the CMS [7] and the ATLAS [8] collaborations.

2. The electromagnetic calorimeter of CMS

The CMS ECAL is a compact, hermetic, fine-grain and homogeneous calorimeter made of 75848 lead-tungstate ($PbWO_4$) scintillating crystals, arranged in a quasi-projective geometry and distributed in a barrel region (EB), with pseudorapidity coverage up to $|\eta| = 1.48$, closed by two endcaps (EE) that extend up to $|\eta| < 3.0$. The scintillation light is readout with avalanche photodiodes (APDs) in EB and with vacuum phototriodes (VPTs) in EE. To facilitate photon identification, the crystals have transverse size comparable to the typical shower size in $PbWO_4$ (Molière radius 21 mm). The front-face cross section of the crystals is 22×22 mm^2 in EB and 28.6×28.6 mm^2 in EE, with crystal depths of about 26 and 25 radiation lengths, respectively. Preshower detectors (ES) comprising two planes of silicon sensors interleaved with lead absorber for a total of three radiation lengths are located in front of each endcap, at $1.65 < |\eta| < 2.6$, to help in π^0/γ separation. Electron/photon separation is limited to the region covered by the silicon tracker ($|\eta| <2.5$), which defines the acceptance region for photons in the $H \rightarrow \gamma\gamma$ search.

The performance of the components of the calorimeter has been extensively tested with electron beams: the stochastic and the electronic noise contributions to the energy resolution of ECAL have been shown to match the design requirements; the overall resolution was proven excellent and well below 1% at high energies, with an irreducible constant term of about 0.3% for particles impinging on the centre of the crystals [9].

In contrast to the test beam setup, in CMS additional contributions to the energy resolution are caused by residual miscalibrations of the channel-to-channel response within ECAL and by channel response changes with time, due to radiation damage of the crystals and environmental instability. These effects are dominant for photons that do not interact in the material upstream of ECAL ('unconverted photons'), and have to be calibrated-out to a fraction of a percent not to spoil the excellent intrinsic resolution of ECAL. The energy resolution is further degraded for electrons and photons that interact in the tracker material in front of ECAL, for which specific reconstruction algorithms and additional calibration factors are needed.

3. ECAL operation and stability

ECAL has been efficiently operating since installation, with a small fraction (about 1% in EB and EE and 4% in ES) of non-operational channels. Triggers for e/γ candidates are provided through the two-level trigger system of CMS. At level-1 (L1), electromagnetic candidates are formed from the sum of the transverse energy (E_T) in two adjacent trigger towers [10] (e.g. arrays of 5x5 crystals in EB). Coarse information on the lateral extent of the energy deposit inside each trigger tower elaborated by the front-end trigger electronics is exploited to suppress spurious triggers, such as those originated by direct ionization in the APD sensitive region [11]. This feature has allowed the single-photon L1-trigger to be operated unprescaled at a low threshold ($E_T = 15$ GeV during 2011 LHC run). From offline data analysis, this trigger has been verified to be efficient (> 99%) for $E_T > 20$ GeV [12], causing no inefficiencies to the $H \rightarrow \gamma\gamma$ analysis, where events are retained for $E_T > 35$ GeV of the leading photon [6].

Fluctuations in temperature directly affect the light yield of the crystals (-2%/°C) and the gain of the APDs (-2.3%/°C). Throughout operations, a cooling system utilising water flow [13] has maintained the temperature stable within about 0.02°C rms in EB and 0.05°C rms in EE, ensuring a contribution below 0.1% to the energy resolution. The stability of the bias voltage supply has also matched specifications [14, 15].

The ECAL response varies under irradiation due to formation of colour centres that reduce the transparency of the lead tungstate. A monitoring system, based on the injection of laser light at 440 nm, close to the emission peak of scintillation light from $PbWO_4$, into each crystal, is used to track and correct for these variations during LHC operation [16]. It provides one monitoring point per crystal every 40 minutes with single point precision better than 0.1% and long-term instabilities of about 0.2%, related to maintenance interventions on the laser. Upon

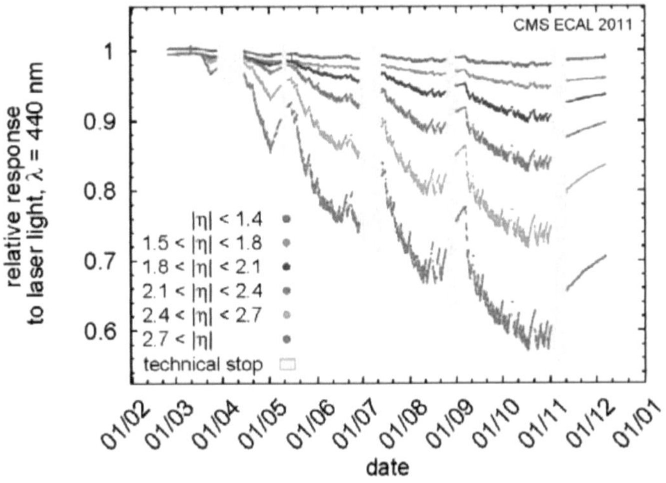

Figure 1. Relative response variation measured by the laser monitoring system in 2011. The response is averaged over the pseudorapidity ranges listed in the legend. Technical stops without monitoring data are shaded. The LHC luminosity varied from 10^{33} cm^{-2}s^{-1} in April to 3.5×10^{33} cm^{-2}s^{-1} in October. Heavy ion collisions took place in November.

quasi-online processing of the monitoring data, response variation corrections are delivered in less than 48 h for prompt reconstruction of CMS data. To strengthen the monitoring system, a new less maintenance intensive laser (447 nm) has been installed for the 2012 run [17].

Response variations to laser light during 2011 are shown in Fig.1 for several $|\eta|$ intervals corresponding to increasing levels of irradiation. Damage and recovery cycles are noteworthy, as well as steady recovery periods during LHC stops due to thermal annealing of the colour centres. The smooth recovery in the last period occurred during heavy ion collisions at low luminosity. The observed losses are consistent with expectations and reach 3% in the barrel and about 15% at the end of the CMS acceptance region for e/γ ($|\eta| < 2.5$). Crystals with radiation tolerance insufficient to stand a dose rate of 0.15 Gy/h with a response loss below 6% were rejected at construction [18]. At the peak luminosity of the LHC in 2011 (3.5×10^{33} cm^{-2}s^{-1}), this dose rate is not exceeded up to $|\eta| \simeq 2.0$. In EE, a contribution to the observed response variations is also expected from photocathode aging of the VPTs [19]. The two effects are not disentangled by the current monitoring system.

4. Electromagnetic shower reconstruction and energy calibration

Electromagnetic particles deposit their energy over several ECAL crystals and the energy estimate implies a sum over the corresponding channels. At test beams, the best energy estimate is obtained by summing the energy deposited in fixed arrays of crystals. In CMS, dynamic 'clustering' algorithms are used: clusters of energy deposits due to secondary emission in the tracker material by bremsstrahlung or photon conversions, and spread azimuthally by the intense magnetic field of CMS, are merged into superclusters [20]. After clusterisation, the energy of the e/γ candidate is estimated from the sum of the signal amplitudes (A_i) [21] of the individual channels in the cluster, weighted with channel dependent coefficients to correct for time response variation ($S_i(t)$), to equalize the channel response C_i (hereafter referred to as

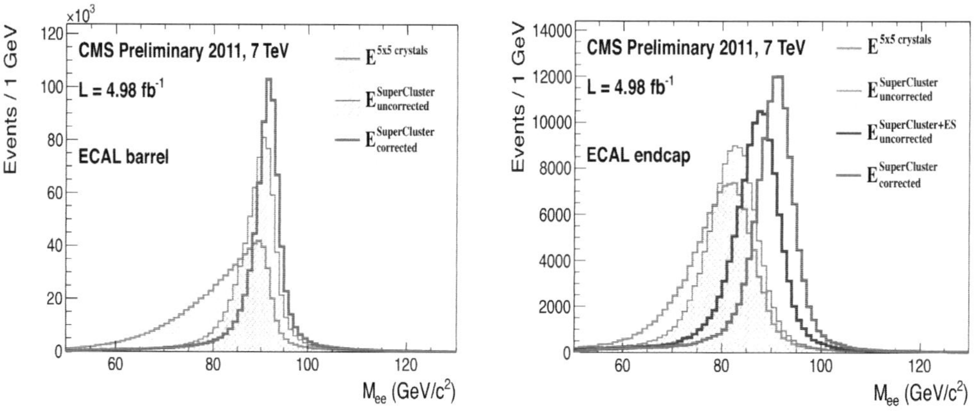

Figure 2. Reconstructed invariant mass of electron pairs from $Z \to ee$ events, using the energy reconstructed in fixed arrays of 5x5 crystals (blue shaded), in the raw supercluster (red shaded) and in the supercluster with algorithmic corrections. For EE the effect of the addition of the preshower energy is also shown.

intercalibration coefficients), to calibrate the ADC-to-energy conversion (G) and to correct for imperfect clustering and geometry effects ($F_{e/\gamma}$). For endcap clusters the preshower energy E_{ES} is also added:

$$E_{e,\gamma} = F_{e,\gamma} \cdot \left[G \cdot \sum_i S_i(t) \cdot C_i \cdot A_i + E_{ES} \right] \tag{1}$$

As an illustration of the clusterisation process, the invariant mass of electron pairs in $Z \to ee$ decays is shown in Fig.2 at different levels of the energy reconstruction, using fixed arrays of 5×5 crystals, the raw supercluster energy, and the supercluster energy including algorithmic corrections (F_e). The figure also suggests that the ratio of the energy in a fixed array of crystals over the energy in the supercluster is a convenient way to identify electrons with little radiation in the tracker or unconverted photons, for which a better energy resolution is expected. To this purpose the variable $R_9 = E_{3\times3}/E_{raw}$, defined as the ratio of the energy in a 3×3 array to the energy in the supercluster before algorithmic corrections, is introduced. According to MC, about 70% of the photons with R_9>0.94 are truly unconverted photons, while all the photons with R_9<0.94 interact upstream of ECAL [22].

In the calibration schema of Eq.(1), single channel calibration and e/γ corrections are factorised. *In situ* calibration, designed in conformity to this factorisation, exploits different event samples to intercalibrate, and to verify and tune monitoring and algorithmic corrections.

4.1. Corrections for channel response variation

According to test beam results [16, 23], transparency variations can be corrected for by a simple power law relating the relative response to laser light $L(t)/L_0$ to relative response to scintillation light generated by electromagnetic showers $S(t)/S_0$:

$$S(t)/S_0 = (L(t)/L_0)^\alpha \,. \tag{2}$$

The spectral index α, related to the optical path of the light in the crystal, has been determined empirically only on a few tens of crystals from different production batches and

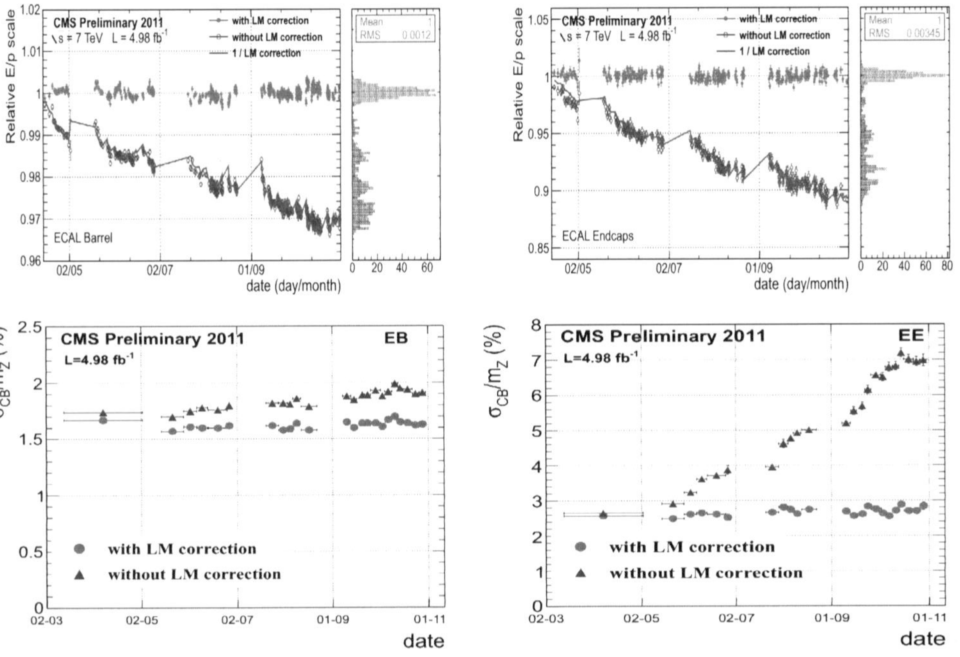

Figure 3. Stability of the energy response (top) and mass resolution (bottom) during the 2011 LHC run for electrons in the ECAL barrel (left) and endcaps (right), upon application of laser monitoring (LM) corrections (green dots). Uncorrected data are also shown (red dots).

manufacturers. Mean values of $< \alpha >= 1.52$ and 1.00 have been measured for BTCP [2], the vast majority in ECAL, and SIC [3] crystals respectively, with a 10% rms spread within the crystals from the same manufacturer [24]. Given this reproducibility, a common value is currently adopted for all the crystals from the same manufacturer. With the observed losses, a 10% spread on α would result in a spread of the single channel corrections of the order of 0.1% in EB and 1% in EE. Verification of the response stability *in situ* using reference physics signatures led to a better determination of the value of α, 1.16, for the BTCP crystals in EE, where deviations from Eq.(2) are expected at large values of transparency loss, or as a consequence of VPT response losses.

In Fig.3-*top*, the evolution of the ECAL response monitored from the energy-to-momentum ratio E/p of isolated electrons is shown. The overall relative response stability throughout the year is about 0.12% and 0.35% in in EB and EE respectively. Additional information is gathered studying the resolution stability, through the analysis of the instrumental contribution to the Z width, reconstructed from the invariant mass of e^+e^- pairs (see Fig.3-*bottom*). The resolution in EB is stable within the measurement accuracy. In EE, a worsening of about 1.5% in quadrature is observed. This is not inconsistent with the observed losses and the uncertainty on α estimated from test beam results, although part of the effect might be associated to the increased number of collisions per beam crossing (*pileup interactions*) during the run. For further optimization of the corrections, studies are being carried out to measure α for each crystal with *in situ* events.

[2] Bogoroditsk Plant of Technochemical Products, 43, per. Vyazovsky, Bogoroditsk, Tula region, 301801, Russia
[3] Shanghai Institute of Ceramics, Chinese Academy of Sciences 1295 Dingxi Road Shanghai, 200050, P.R.China

4.2. Intercalibrations

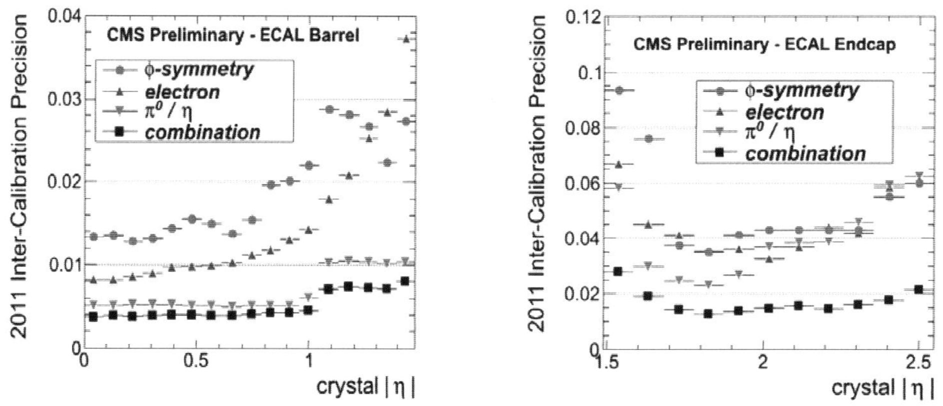

Figure 4. Precision of the various calibration sets used in 2011 in EB (left) and EE (right).

The spread of the channel response affects the constant term of the energy resolution of ECAL with little dilution, as electromagnetic showers involve only a small number of crystals: unconverted photons deposit on average about 70% of their energy in a single crystal. The main sources of channel-to-channel response variation are the crystal light yield variation in EB, about 15% at construction, and the gain spread of the photodetectors in EE, about 25%. To reduce these spreads and provide an acceptable performance at startup, calibration procedures with different levels of accuracy have been adopted during the construction and commissioning phase of ECAL [25]. Refined intercalibration has been derived *in situ* with several techniques exploiting the properties of collision events [26]. These include the invariance around the beam axis of the energy flow in minimum bias events, the π^0/η mass constraint on the energy of the two photons from the $\pi^0/\eta \rightarrow \gamma\gamma$ decays, and the momentum constraint on the energy of isolated electrons from W decays. For fast calibration purposes, special calibration streams have been deployed for the former two methods: designated high-level trigger (HLT) modules unpack, select *online*, and log data only in restricted regions of ECAL upon specific L1 triggers, while all the rest of the event is dropped. With this sagacity, calibration data can be logged at high rate (~ 5 kHz) with little impact on the total CMS bandwidth.

The precision of each method has been estimated through the cross-comparison of the individual results, and cross-checked against precalibration constants. Results obtained in the 2011 run are shown in Fig.4 as a function of pseudorapidity for EB (left) and EE (right). The intercalibration precision upon combination of the different methods and of the precalibration constants is also displayed. Further progress is anticipated for the intercalibration precision with isolated electrons, still statistically limited at variance with the other methods. The residual miscalibration of the channel response already ensures a contribution to the energy resolution below 0.5% in the central part of the barrel ($|\eta| < 1$), and below 2% in the endcaps.

4.3. Algorithmic corrections to electron and photon energies

Dynamic clustering algorithms are used in the energy reconstruction to mitigate the impact of the material upstream of ECAL. However, not all the energy is clustered, due to algorithm inefficiencies, and part of the energy, lost in the tracker and swept by the magnetic field, never reaches ECAL. Thus, algorithmic corrections must be applied. These corrections are particle,

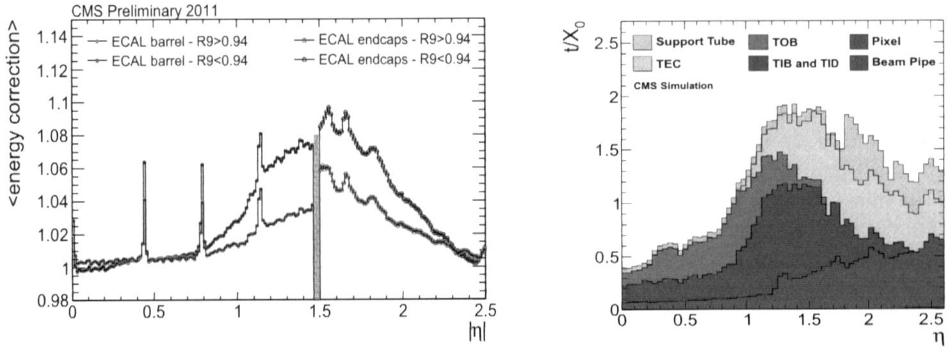

Figure 5. Average energy correction factors for W→ $e\nu$ electrons with R_9>0.94 and R_9<0.94 (left) and amount of radiation lengths upstream of ECAL (right) as a function of η.

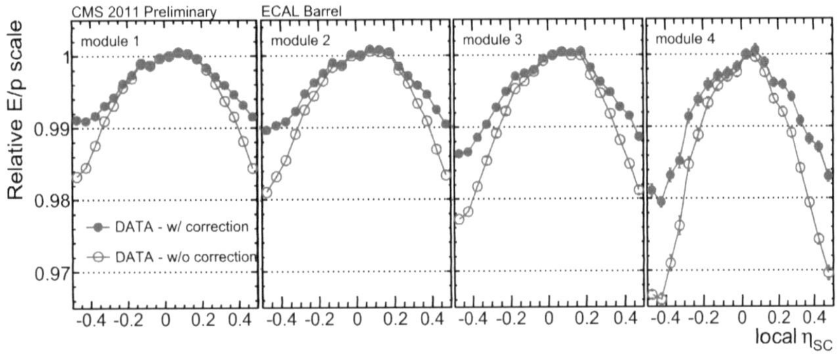

Figure 6. Dependence of the electron E/p on the supercluster centre along η referred to a local coordinate system where crystal edges are at -0.5 and 0.5. Results for EB crystals are shown in the η ranges: $[0, 0.43]$, $[0.43, 0.79]$, $[0.79, 1.10]$ and $[1.10, 1.48]$ from left to right.

energy and position dependent due to the different interaction mechanisms of e/γ upstream of ECAL and to the CMS geometry. Corrections have been optimized separately for electrons and photons on MC events with a multivariate analysis (MVA). Input variables include shower shape information on the azimuthal spread of the energy deposit, shower position in ECAL local and CMS global coordinates, and global event variables sensitive to pileup effects. Corrections closely follow the distribution of the material budget upstream of ECAL (see Fig.5), and are sizable at $1 < |\eta| < 2$, where the tracker material is up to two radiation lengths thick. Local structures are related to the ECAL geometry.

Due to unavoidable imperfections of the MC model, algorithmic corrections must be tested and tuned on collision data. To this end, the stability of the E/p ratio of electrons is studied as a function of the MVA input variables. An identified case of imperfect corrections is shown in Fig.6, displaying the variation of the E/p ratio versus the impact point of the electron on the crystal for four different $|\eta|$ intervals in EB. MC-driven corrections do not fully compensate for the energy leakage in the inter-crystal gaps, yielding a residual response variation of 0.3-0.5% rms. This indicates that the shower width in MC is not exactly matched to data.

5. Energy resolution

In order to study the energy resolution, *in situ* data are compared to the predictions of a full MC simulation of the CMS detector based on Geant4 [27]. The simulation of the ECAL standalone response has been tuned to match test beam results, and includes a detailed description of the single channel noise, a spread of the single channel response corresponding to the estimated residual miscalibration, and a constant term of 0.3% matched to test beam results. Moreover, the few non-operational channels are also simulated. Response instabilities are not simulated. Pileup interactions are simulated for beam crossings up to ±50 ns, which might result in an underestimate of the out-of-time pileup contribution to the noise term.

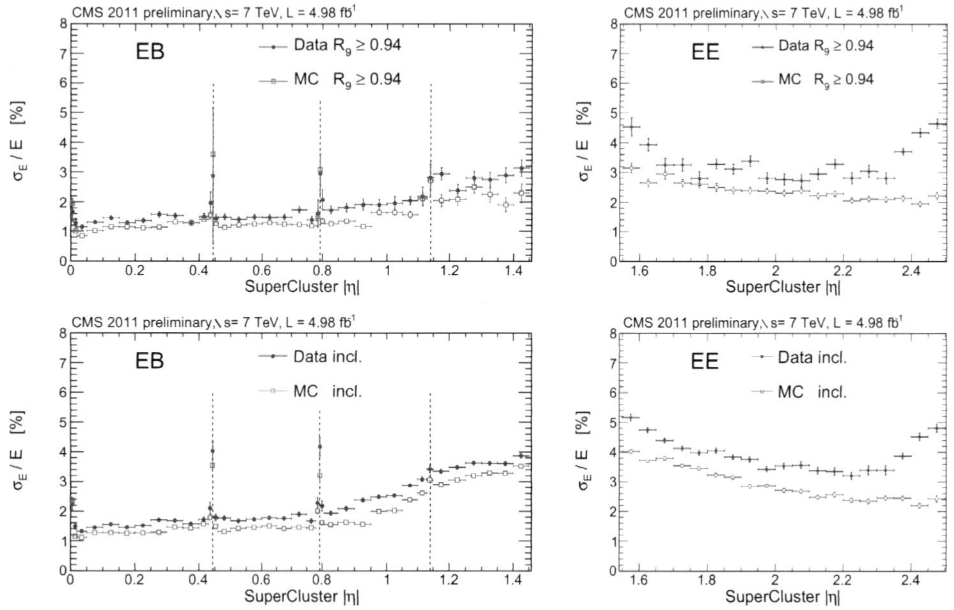

Figure 7. Relative energy resolution for $Z \to ee$ electrons in data and MC unfolded in η bins for EB (left) and EE (right). The resolution is shown separately for pairs with two $R_9 > 0.94$ electrons (top) and for the inclusive sample (bottom).

The energy and mass resolutions are studied with $Z \to ee$ events. The instrumental contribution to the Z width is extracted from a fit to the invariant mass distribution of a Breit-Wigner convoluted with a Crystal-Ball response function [28]. The electron energy resolution is derived from an unbinned maximum likelihood fit to the invariant mass distribution of e^+e^- pairs, where the energy resolution of each electron is floated as a function of $|\eta|$. The inclusive energy resolution in data varies between 1.5% in central EB, to 3-4% in the outer EB and 4% in EE (see Fig.7). For the R_9> 0.94 sample the resolution is better than 1.5% in the central barrel. The corresponding mass resolutions are given in Table 1 for different categories of events.

The impact on the resolution from the material upstream of ECAL, in particular at $|\eta|$ >1, is noteworthy, and it is the main limitation to the resolution in the barrel. Effects not included in the simulation or not perfectly simulated may explain the data/MC difference. In the EE, likely contributions to the discrepancy come from the single channel calibration. In this region

Category:	Both EB & R_9>0.94	Both EB	Both EE
σ_{CB} (Data)	(0.97 ± 0.01) GeV	(1.51 ± 0.01) GeV	(2.36 ± 0.02) GeV
σ_{CB} (MC)	(0.83 ± 0.02) GeV	(1.29 ± 0.02) GeV	(1.78 ± 0.02) GeV

Table 1. Mass resolution of the Crystal-Ball component in $Z \rightarrow ee$ events.

the impact of the material in front of ECAL is also large. As we discuss below, in the $H \rightarrow \gamma\gamma$ analysis the MC is tuned to match the resolution in data.

6. ECAL position reconstruction and photon direction

In diphoton events, an additional contribution to the invariant mass resolution comes from the knowledge of the opening angle between the two photons, determined from the positions of the showers and of the interaction vertex. The precision of the shower position measurement in ECAL [29], including residual ECAL/Tracker misalignment effects, matches the performance of a MC with aligned geometry and meets resolution goals for $H \rightarrow \gamma\gamma$ reconstruction [30]. In 2011 LHC conditions, about 10 interaction vertices are reconstructed for each event with an rms spread of 6 cm along the beam direction. The diphoton vertex must be located to better than about 1 cm to make a negligible contribution to the mass resolution, as compared to the ECAL energy resolution. The vertex assignment relies on the kinematic properties of the tracks associated with that vertex and on their correlation with the diphoton kinematics. The efficiency of correct assignment (within 1 cm of the true vertex) has been estimated with simulation and control data samples to be about 83% for $H \rightarrow \gamma\gamma$ events [5].

7. Photon identification

Photon identification is tuned for the same signal to background ratio in four different categories of 'unconverted' (R_9> 0.94) and 'converted' (R_9< 0.94) photons in EB and EE.

In addition to R_9, used to classify photons of high or low resolution, a key calorimetric variable for π^0/γ separation is the measure of the lateral extent of the ECAL cluster along η ($\sigma_{i\eta i\eta}$): a projection orthogonal to the bending direction of the magnetic field and thus relatively insensitive to conversions upstream of ECAL. The shower shape of photon canidates in $Z \rightarrow \mu\mu\gamma$ decays are well described by simulation (see Fig.8), upon a small reweighting of MC based on the shower shapes observed in $Z \rightarrow ee$ events. In addition, isolation variables based on tracker, ECAL and the hadron calorimeter are used to separate photons from electromagnetic deposits associated with jet fragments [31].

Photon selection efficiencies have been verified with data mainly by means of $Z \rightarrow ee$ events, with electrons reconstructed as if they were photons. A 'tag & probe' technique is adopted, where the topology of the event is used to tag the presence of a supercluster to be probed as a photon candidate. Efficiencies are somewhat smaller in EE, plagued by a larger background, than in EB (Fig.9 left). Simulation studies indicate that the background in the selected sample for $H \rightarrow \gamma\gamma$ analysis is dominated by the irreducible background of diphotons from QCD production and that fewer than 30% of the events contains one or more misidentified photons, mainly from γ + jet production (Fig.9 right). The MC background composition and shape are not used in the search of the $H \rightarrow \gamma\gamma$, but only to tune photon identification selections.

8. Implications for the $H \rightarrow \gamma\gamma$ hunt

The $H \rightarrow \gamma\gamma$ analysis is designed to exploit optimally the dependence of the resolution on the properties of the reconstructed photons and on their location within ECAL. To enhance the sensitivity of the analysis, events with two photon candidates satisfying p_T requirements and

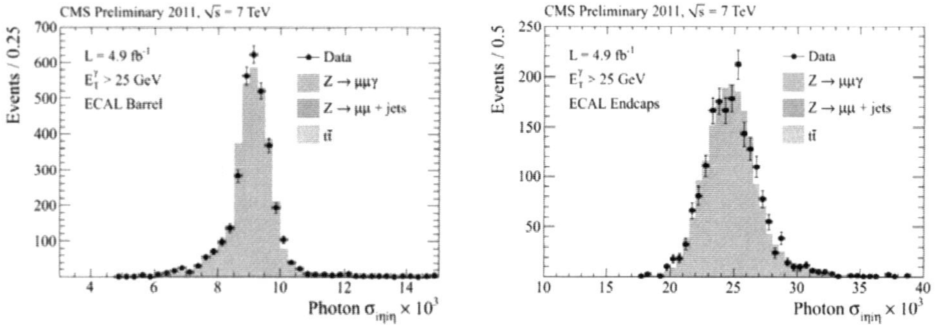

Figure 8. Distribution of $\sigma_{i\eta i\eta}$ in data and MC for photons in $Z \to \mu\mu\gamma$ decays

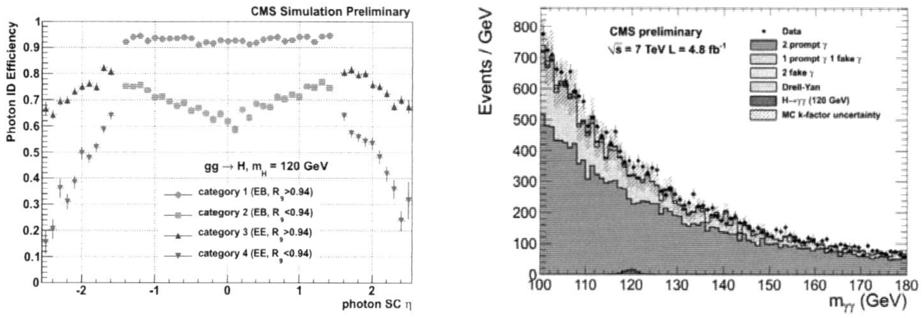

Figure 9. Left: Photon identification efficiency as a function of $|\eta|$ for high and low R_9 candidates. Right: Di-photon spectrum in 2011 data and its composition estimated from MC.

"loose" photon identification criteria are classified into mutually exclusive categories of different mass resolution and signal-to-background ratios. The $H \to \gamma\gamma$ search is performed in each category independently and results are combined.

A simple categorisation separates candidates with both photons in EB from those with at least one photon in EE, and then splits these two sets into candidates with both photons 'unconverted' (R_9>0.94) or with at least one converted photon (R_9<0.94) [5]. In an improved version of the analysis events are categorized by means of a multivariate (MVA) technique using photon, vertex and global event variables to estimate the expected mass resolution and signal likelihood of the diphoton candidates, with category boundaries optimized for sensitivity to an SM Higgs boson [6]. The analysis also includes a category of dijet-tagged events with inclusive diphoton selection. This latter sample has enhanced sensitivity to the Higgs search in the vector boson fusion production (VBF) channel, where two forward jets are associated to the Higgs production. The MVA analysis has a higher efficiency to the signal and a larger background rejection factor than the analysis with simple categorisation, resulting in a sensitivity enhancement of 15%.

In each category, the background is estimated from data with a parametric fit to the invariant mass spectrum of the diphoton candidates on a range wider than the signal search window. The signal model is derived from the MC simulation of the $H \to \gamma\gamma$ decay. The constant term of the energy resolution in the MC is tuned in different regions of ECAL and for the different R_9 categories to match the observed resolution in $Z \to ee$ data, with electrons reconstructed

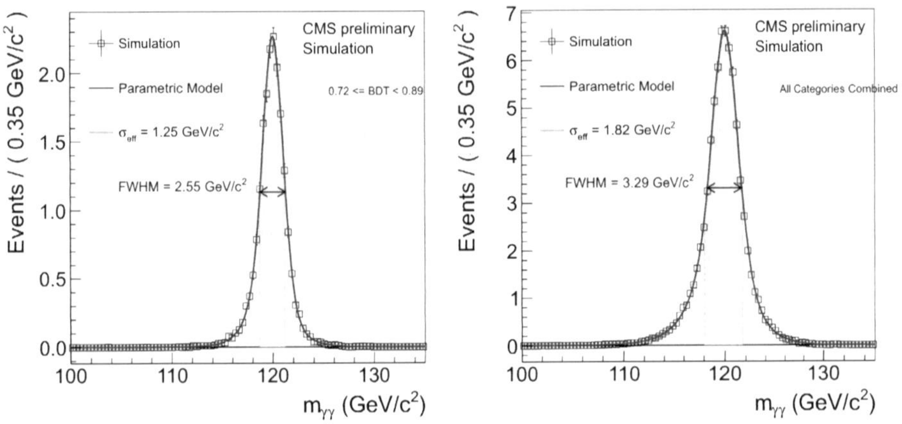

Figure 10. Invariant mass distribution for a Higgs of m_H = 120 GeV decaying into two photons for the category of second-best resolution (25% of the sample) and for the inclusive sample.

	0	1	2	3	Dijet tag
SM signal expected	3.4 (4.4%)	19.3(25.0%)	18.7 (24.2%)	33.0 (42.8%)	2.8 (3.6%)
Data (events/GeV)	4.5 (1.2%)	55.1(14.8%)	81.3 (21.8%)	229.1 (61.6%)	2.1 (0.6%)
FWHM/2.35 (GeV)	1.09	1.09	1.43	2.08	1.32

Table 2. Expected SM $H \to \gamma\gamma$ signal for m_H =120 GeV, selected events and mass resolutions in the different analysis categories. Numbers in parentheses give the percentages of events in each category. Results are given for 4.8 fb^{-1} at $\sqrt{s} = 7$ TeV.

as photons, except for the vertex. The shape of the expected signal at 120 GeV is shown in Fig.10, for the inclusive sample of $H \to \gamma\gamma$ events and for the category of second-best resolution, which includes 25% of the expected signal. A breakdown of the expected signal, the observed candidates and the expected resolutions is given in Table 2 for the five categories of the MVA analysis. Due to the optimization of the category boundaries, the expected signal significances are rather similar in the different categories [4]. Given the variation of the energy resolution and of the photon reconstruction properties within ECAL, the overall significance is, however, driven by events with both photons in the ECAL barrel: diphoton candidates with at least one photon in EE only contribute to less than about 50% of the events of the category of lowest resolution.

The results of the $H \to \gamma\gamma$ search based on 4.8 fb^{-1} of data collected at $\sqrt{s}$=7 TeV in 2011 are summarized in Fig.11, where the exclusion curve normalised to the SM $H \to \gamma\gamma$ production cross section is shown as a function of the mass interval probed by the analysis. On the right panel, the (*p-value*) plot shows the probability for the background to deviate from expectation by at least the observed amount. The exclusion limit is weaker than expected at mass of about 125 GeV, due to a 2.9σ excess of events with contributions from both dijet-tagged and non dijet-tagged events. Additional 10 fb^{-1} of integrated luminosity are deemed sufficient to ascertain the origin of the excess and discover at 5σ significance or exclude the SM Higgs boson.

[4] An approximate estimator of the significance is given by the $S/\sqrt{S+B}$ ratio in a FWHM-wide mass window, with S given by the second row of the table and $(S+B)$ proportional to the product of the third and fourth rows.

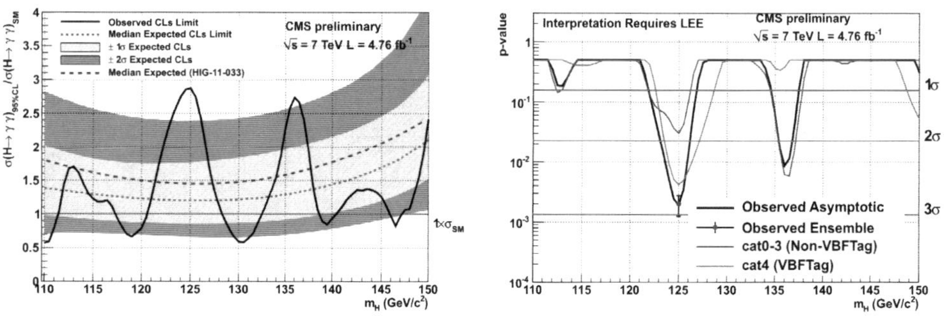

Figure 11. Left: Exclusion limit at 95% of the confidence level for the SM Higgs boson in the $H \to \gamma\gamma$ search mode. The expected exclusion limit for the MVA analysis (red) and the simple categorisation analysis (blue) are also shown. Right: Local probability of the background only hypothesis (*p-value*) as a function of the Higgs mass.

9. Summary and outlook

The hunt for the SM $H \to \gamma\gamma$ decay placed stringent requirements on the energy resolution and photon identification capabilities of electromagnetic calorimeters at the LHC. In CMS these requirements have been translated into the first incarnation of a crystal calorimeter at a hadron collider. Notwithstanding the harsh radiation environment, the calorimeter has been successfully operated and calibrated.

In the ECAL barrel, the achieved resolution and photon identification performance put CMS in an excellent position to be able to observe or exclude the postulated SM Higgs boson through the $H \to \gamma\gamma$ decay with the data sample being collected in 2012 at $\sqrt{s} = 8$ TeV. Limitations to the resolution come predominantly from the impact of the material in front of ECAL, and from imperfect algorithmic corrections, for which further tuning is in progress. The ECAL endcaps, with lower acceptance to $H \to \gamma\gamma$ photons, contribute less to this search because of the lower photon identification efficiency, and non optimal resolution. Further progress in the precision of the single channel calibrations is needed.

At the time of writing of these Proceedings, the observation of a new boson with a mass of 125 GeV, with a local significance of 4.1 σ in the two photon decay mode alone, has been reported by CMS after the analysis of a data sample of 5.1 fb^{-1} at $\sqrt{s} = 7$ TeV (2011) and 5.4 fb^{-1} at 8 TeV (2012) [7]. Similar results have been reported by ATLAS [8].

Acknowledgements

I am thankful to the organizers of this conference for the nice venue and atmosphere, and for the timely, stimulating and challenging topic proposed for this invited talk. I am also indebted to my CMS ECAL colleagues for an extraordinary adventure.

References

[1] Chatrchyan S *et al.*, *The CMS experiment at the CERN LHC'*, JINST 3 (2008) S08004, http://dx.doi.org/10.1088/1748-0221/3/08/S08004

[2] CMS Collaboration, *CMS Physics Technical Design Report, Volume II: Physics Performance*, CERN-LHCC-2006-021, published in J. Phys. G: Nucl. Part. Phys. 34 (2007) 995-1579, http://dx.doi.org/10.1088/0954-3899/34/6/S01

[3] Seez C J *et al.*, *Photon decay modes of the intermediate mass Higgs*, in Proceedings of the Large Hadron

Collider Workshop, G. Jarlskog and D. Rein, eds., p. 474. Aachen, Germany, 1990. CERN 90-10-V-2/ECFA 90-133-V-2, http://cdsweb.cern.ch/record/220524

[4] The CMS Collaboration, *The Electromagnetic Calorimeter Project: Technical Design Report*, CERN/LHCC 1997, CMS TDR 4.

[5] Chatrchyan S *et al.*, *Search for the standard model Higgs boson decaying into two photons in pp collisions at* $\sqrt{s} = 7$ *TeV*, Phys. Lett. B 710 (2012) 403, doi.10.1016/j.physletb.2012.03.003.

[6] CMS Collaboration, *A search using multivariate techniques for a standard model Higgs boson decaying into two photons*, CMS PAS HIG-12-001, http://cdsweb.cern.ch/record/1429931.

[7] Chatrchyan S *et al.*, *Observation of a new boson at a mass of 125 GeV with the CMS experiment at the LHC*, Phys.Lett. B 716 (2012) 30-61.

[8] G. Aad *et al.*, *Observation of a new particle in the search for the Standard Model Higgs boson with the ATLAS detector at the LHC*, Phys.Lett. B716 (2012) 1-29.

[9] Adzic P *et al.*, *Energy resolution of the barrel of the CMS Electromagnetic Calorimeter* JINST 2 (2007) P04004

[10] Paganini P, for the CMS collaboration, *CMS Electromagnetic Trigger commissioning and first operation experiences*, J.Phys.: Conf.Ser. 160 (2010) 012062 , doi.10.1088/1742-6596/160/1/012062

[11] Petyt D *for the CMS collaboration*, *Anomalous APD signals in the CMS electromagnetic calorimeter*, Nucl. Instr. Meth. A (2011), doi.10.1016/j.nima.2011.10.025

[12] Zabi A *for the CMS collaboration*, *Triggering on electrons and photons with CMS*, arXiv:1202.0594 [physics.ins-det], http://arxiv.org/abs/1202.0594

[13] Baillon P *et al.*, *Performance of the cooling system of the CMS ECAL*, Proceedings of the 8th conference on Astroparticle, Particle and space physics, detectors and medical applications, Como, 2003, World Scientific Publishing Co., 2003, 203-207.

[14] Bartoloni A *et et al.*, *High Voltage system for the CMS electromagnetic calorimeter*, Nucl. Instr. Meth. A 582 (2007) 462468.

[15] Chatrchyan S *et al.*, *Performance and Operation of the CMS Electromagnetic Calorimeter*, JINST 5 (2010) T03010.

[16] Anfreville M *et al.*, *Laser monitoring system for the CMS lead tungstate crystal calorimeter*, Nucl. Instr. and Meth. A 594 (2008) 292-320.

[17] Zhang L *for the collaboration*, *A Diode-Pumped DP2-447 Blue Laser for Monitoring CMS Lead Tungstate Crystals Calorimeter at the LHC*, CMS CR-2012/230 (2012) and these Proceedings.

[18] Adzic P *et al.*, *Radiation hardness qualification of* $PbWO_4$ *scintillation crystals for the CMS Electromagnetic Calorimeter*, JINST 5 (2010) P03010.

[19] Leslie D E for the CMS collaboration, *The effect of pulse rate on VPT response and the use of an LED light to improve stability*, CMS-CR-2009-284 (2009).

[20] CMS Collaboration, *CMS Physics Technical Design Report, Volume I*, CERN-LHCC-2006-021

[21] Adzic P *et al.*, *Reconstruction of the signal amplitude of the CMS electromagnetic calorimeter*, Eur. Phys. J. C 46, s1, (2006) 23-35.

[22] CMS Collaboration, *Photon reconstruction and identification at* $\sqrt{s} = 7$ *TeV*, CMS PAS-EGM-2010-005 (2010).

[23] Ghezzi A *et al.*, *Analysis of the response evolution of the CMS electromagnetic calorimeter under electron and pion irradiation*, CMS Note 2006/038 (2006).

[24] Adzic P *et al.*, *Results on the first performance tests of the CMS electromagnetic calorimeter*, CMS Note 2005/020 (2005).

[25] Adzic P *et al.*, *Intercalibration of the barrel electromagnetic calorimeter of the CMS experiment at start-up*, JINST 3 (2008) P10007.

[26] CMS Collaboration, *Electromagnetic calorimeter calibration with 7 TeV data*, CMS PAS-EGM-2010-003 (2010).

[27] Cossutti F, for the CMS Collaboration, *The simulation of the CMS electromagnetic calorimeter*, J. Phys.: Conf. Ser. 119 (2008) 032015.

[28] Skwarnicki T, Ph.D Thesis, DESY F31-86-02(1986), Appendix E, and references therein.

[29] Arcidiacono R, for the CMS Collaboration, *Studies of the CMS Electromagnetic Calorimeter performance in the electron test beam*, J. Phys.: Conf. Ser. 160 (2009) 012048

[30] CMS Collaboration, *ECAL Detector Performance, 2011 Data*, CERN-CMS-DP-2012-007 (2012).

[31] CMS Collaboration, *Isolated Photon Reconstruction and Identification at* $\sqrt{s} = 7$ *TeV*, CMS-PAS-EGM-10-006 (2010).

First years of running for the LHCb calorimeters

Xavier Vilasís Cardona *on behalf of the LHCb calorimeters collaboration*

LIFAELS, La Salle, Universitat Ramon LLull, Barcelona

E-mail: xvilasis@salle.url.edu

Abstract. The design and construction characteristics of the LHCb calorimeter system are described. Strategies for monitoring and calibration during data taking are detailed in all aspects. Scintillating fibres, plastics and photomultipliers suffer from aging due to radiation damage or high currents; different methods which are used to calibrate the detectors and to recover the initial performances are presented. The performances achieved are illustrated in selected channels of interest for B physics.

1. Introduction

The LHCb detector [1] is a single arm spectrometer devoted to the study of CP violation and rare decays in B meson and hadron systems. Its calorimeter is divided in two, a Hadronic Calorimeter (HCAL) and an Electromagnetic Calorimeter which, in turn, splits into a Scintillator Pad Detector (SPD), a PreShower (PS) and the bulk of the Electromagnetic Calorimeter (ECAL) [2]. All four subdetectors work on the same principle: collect scintillation light produced by incoming particles, convert it to an electrical signal using photomultipliers and integrate it to obtain the energy measurement depending on the photo-electron statistics.

The PS-SPD system plays various roles in the detector operation. It provides particle identification for electrons and photons at the hardware trigger level (called L0 trigger). The PS is used for the electron, photon and pion separation while the SPD contributes with the photon-charged particle separation. To cut down too complex events, the SPD also provides a charged multiplicity veto. The ECAL measures the transverse energy of electrons, photons and π^0 for the L0 trigger, which is relevant in selecting B-decay channels such as $B \to K^* e^+ e^-$ or $B \to K^* \gamma$. PS and ECAL are also used for the offline reconstruction of π^0 and prompt photons. HCAL gives the transverse energy of hadrons for the L0 trigger to select decay channels such as $B \to \pi' s$ or $B \to D_s K$. PS, ECAL and HCAL are used, of course, for particle identification.

The LHCb calorimeter system was installed between years 2004 and 2008. The commissioning took place between 2005 and 2009 using the built in monitoring tools, cosmic rays and injection tests splash events. The first trace of a cosmic ray was detected in January 2008 and the system was fully operational for the first collisions in November 2009.

This paper aims at giving an overview of the LHCb calorimeters, their structure, calibration strategies, the effects of aging after three years of operation and finally their performance illustrated through the results on one selected data channel, $B \to X\gamma$. We devote one section to each of these points and we add some conclusions at the end.

Journal of Physics: Conference Series **404** (2012) 012003 doi:10.1088/1742-6596/404/1/012003

2. Description of the LHCb Calorimeters

The four elements in the calorimeter system as found downstream from the interaction point are SPD, PS, ECAL and HCAL.

The SPD and PS are walls of scintillator pads with a wavelength shifting fiber (WLS) coil grooved inside for better light collection as seen on figure 1. They are separated by a 2.5 radiation length thick lead curtain. The SPD delivers a single bit depending on whether the crossing particle is charged or neutral and completes the PS tagging of the electromagnetic shower. The pad size is adjusted to three different values to achieve three zones with higher granularity around the beam-pipe The geometry is projective with respect to ECAL. Overall, SPD and PS have around 6000 cells each. The readout is made by multi-anode photomultipliers Hamamatsu 5900 M64.

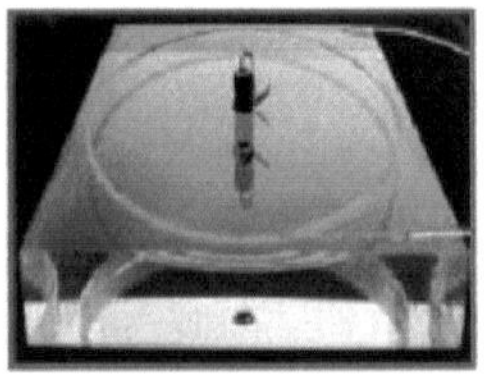

Figure 1. Scintillator pad with the WLS fiber coil grooved in and an LED for testing and calibration purposes.

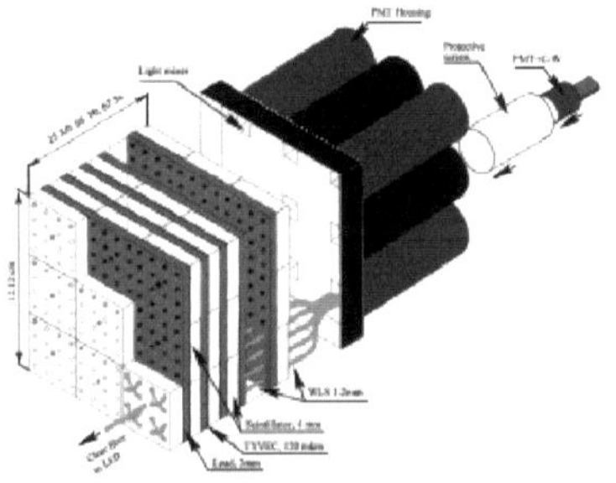

Figure 2. ECAL cells have a shashlik structure read out by photomultipliers.

LHCb ECAL cells have a shashlik structure (figure 2) with 4 mm-scintillator and 2 mm-lead layers equivalent to 25 radiation lengths. The readout is performed by Hamamatsu R7899-20 photomultipliers. The total number of cells is around 6000. As described for the SPD-PS system, three regions, inner, middle and outer have been defined according to the distance to the beam-pipe (figure 3). The energy resolution, as measured in test beam is [1]

$$\sigma(E)/E = 0.090 \pm 0.005/\sqrt{E} \oplus 0.008;$$

where E is the particle energy in GeV. The last term represents the contribution of the electronics noise.

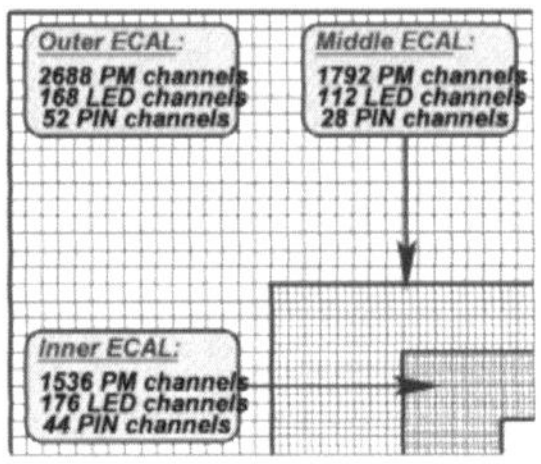

Figure 3. Three regions have been defined to improve the granularity close to the beam-pipe.

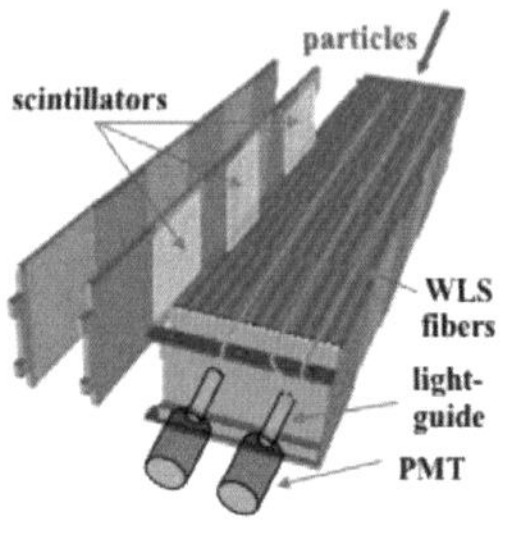

Figure 4. HCAL cells have a tile structure read out by photomultipliers.

The HCAL has a tile structure. The same photomultiplier as in ECAL, namely Hamamatsu R7899-20, is used for the read out. It overall has 1500 cells divided in two regions, inner and outer, corresponding to the distance to the beam-pipe. The energy resolution follows [1]:

$$\sigma(E)/E = (0.69 \pm 0.05)/\sqrt{E} \oplus (0.09 \pm 0.02),$$

where E is the deposited energy in GeV.

The common approach to process the photomultiplier signal is first to make an analog integration of the signal, then sample and finally digitalize it to send it through the trigger and data paths. The physical construction of the calorimeters and the detector frequency cause the signal to spill over more than one clock cycle. Two solutions are adopted to face this problem. ECAL and HCAL electronics performs a clipping of the signal prior to integration so that it fully fits in one clock cycle. The readout is performed by an analog chip hosted in the so called front end boards located in crates on top of the calorimeters. PS and SPD, in turn, subtract a fraction of the signal measured in the previous clock cycle to solve the spill over problem. For this a dual channel integration system is built in an analog chip. Since the readout uses 64 channel multi-anode photomultipliers (MA-PMT), small boards, called very front end boards, located close to the detector, host the MA-PMT and the associated signal processing electronics. The conditioned analog signal of the PS and the SPD bit are sent to front end boards in the crates on top of the detector. Further details can be found in reference [1] for the general structure of the calorimeters and in reference [3] for the strategy of the detector time alignment.

High Voltage (HV) is provided for the four susbsystems by full custom Cockroft-Walton voltage multipliers. PS and SPD share a common design for their 64 channel MAPMTs just as ECAL and HCAL do for their R7899-20. Besides, PS and SPD MAPMT bases are full custom ones, involving active elements to provide the last PMT dynodes with the necessary current to keep the response in the linear regime.

Detectors include built in LED systems for monitoring the detector stability. ECAL and HCAL share a common system [1]. LED's are pulsed simultaneously and their signal is monitored by a pin diode. The driver is designed to control time and temperature stability. It provides a small pulse duration with low dispersion of amplitude. The pulse rate is adjustable as well as the amount of light in order to emulate electromagnetic particles in full physics region. The gain control allows an accuracy better than 1%. The LED system runs in empty bunches to monitor the detector.

3. Calibration tools and methods

Each part of the calorimeter system follows a different strategy for calibration.

The SPD-PS system looks at the tracks pointing to the SPD-PS cells and extrapolates their trajectory. In the case of the SPD, tracks should correspond to positive signals: the efficiency is defined as the fraction of tracks effectively leaving signal on the SPD [4]. Efficiencies show to be around 90%. In the PS case, the MIP signal is fitted and adjusted to a given number of ADC counts.

The calibration approach for ECAL relies on fitting the π^0 mass from two photon signals, using the mass distribution fit method [5]. This is an iterative procedure which consists in selecting 3×3 clusters corresponding to photon signals and fix the seed, corresponding to the central cell. For each cell, we compute the di-photon invariant mass and then we fit the π^0 mass distribution. We correct the calibration of every seed cell and restart the procedure until we obtain stable calibration coefficients. The first fit of the π^0 mass was obtained in November 2009 from the first collisions by setting a uniform ADC count value per transverse energy unit. The result was a mass fit $M_{\pi^0} = 133 \pm 3$ MeV with a width $\sigma = 11 \pm 4$. Figure 5 shows the effect of the calibration algorithm on the π^0 mass fit: the final width is of the order of 6%.

Another available method to monitor and correct the calibration of ECAL cells is through electron E/p. Electrons are identified by estimating the momentum of the extrapolated of tracks

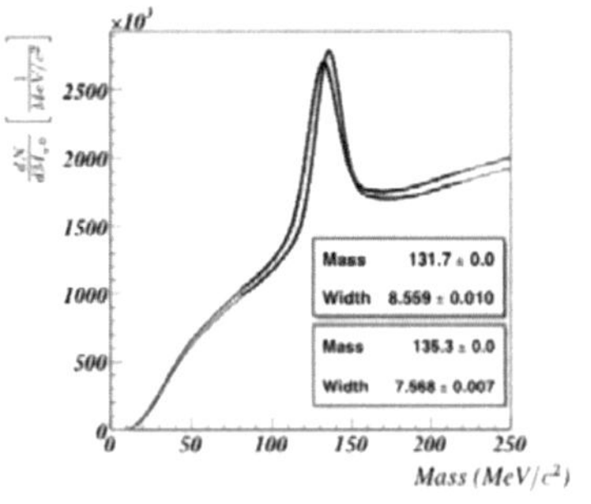

Figure 5. Effect of the calibration algorithm on the π^0 mass fit. The blue curve corresponds to the initial calibration value, while the red curve is the final one.

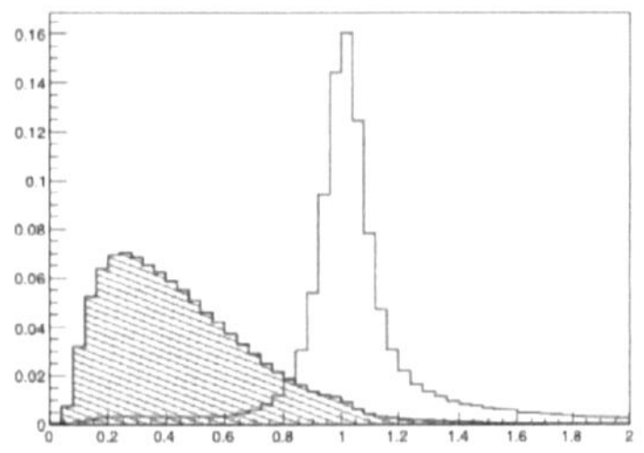

Figure 6. E/p on ECAL for electrons (red) and hadrons (blue) using 2011 data.

and energy of the matching clusters. Figure 6 shows how the electron signal can be clearly separated from the hadron signal on ECAL so that the fit can be used for cell calibration.

The calibration of the hadronic calorimeter relies on a built-in ^{137}Cs source [1]. Scans are performed during technical stops and allow to get an absolute calibration for the HCAL cells. Figure 7 shows the measured signal on a calibration source run normalised by the expected signal from the actual calibration. During normal data-taking, the behavior of HCAL can be monitored efficiently using the LED system. Figure 8 shows the linear relation between the relative variation in gain obtained from two measurements with the ^{137}Cs source with respect to the LED measurement variation.

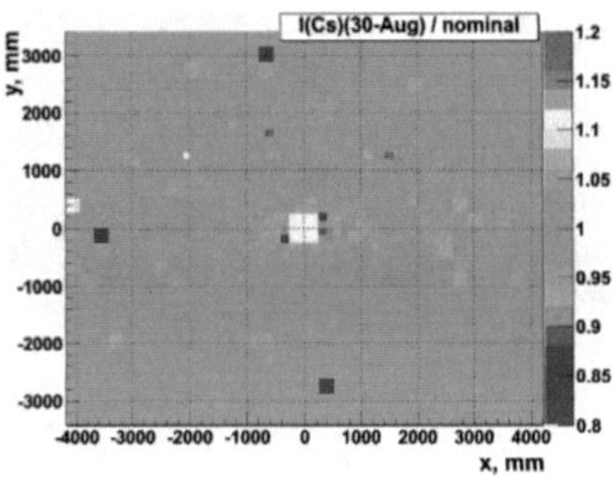

Figure 7. Measured signal on a source run compared to the expected by calibration.

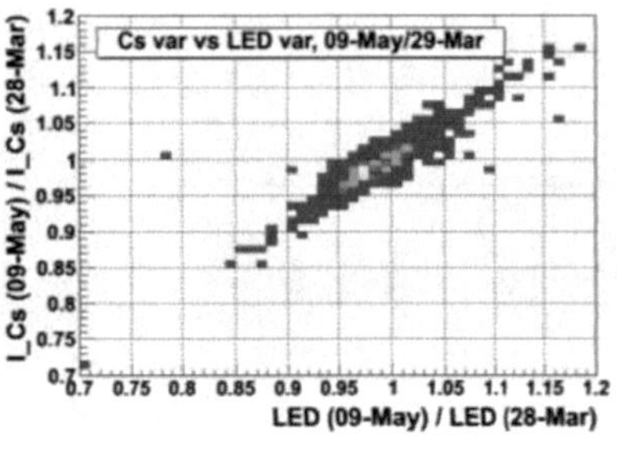

Figure 8. Relative variation in gain obtained from two measurements with the ^{137}Cs source with respect to the LED measurement variation.

4. Aging

The effects of the aging of the detector have been observed from different perspectives. We may look for instance to the impact on the L0 trigger by computing the ratios between the trigger rate on hadron channels and on muon-dimuon channels in which a 5% decrease was observed in 80 pb^{-1} in 2011. The ratio between the electron/photon L0 trigger rate and the muon-dimuon channels decreased a 20% in 1.2 fb^{-1}.

In the case of HCAL, aging can be seen on LED response. We have shown that aging affects both the detector and the photomultiplier. We define the PMT sensitivity variation as the PMT gain variation reduced to the gain at the beginning of the run (March 2011) HV, calculated from the calibration coefficients. Figure 9 shows how the PMTs of the cells closer to the beam-pipe, so with higher occupancies, have a gain loss. This gain loss is compatible with the cumulated charge gain loss measured in the laboratory. Besides, scintillator rows in the tile get affected depending on their depth as shown by figure 10 for the central cells, where the light yield is measured with respect to the delivered luminosity for different rows. First rows, where the hadronic shower is maximum, degradate faster.

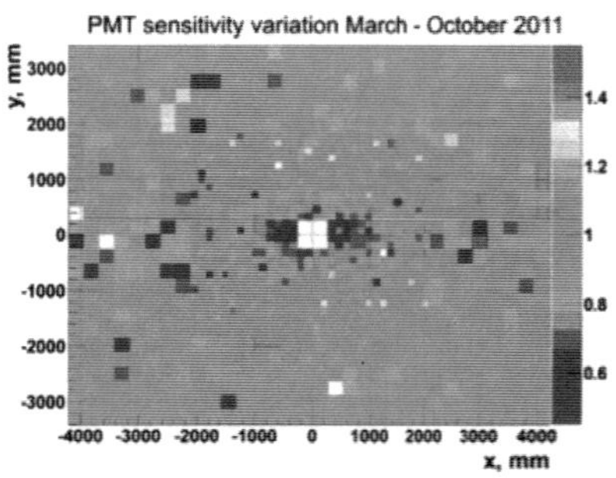

Figure 9. "PMT sensitivity variation" for each HCAL channel.

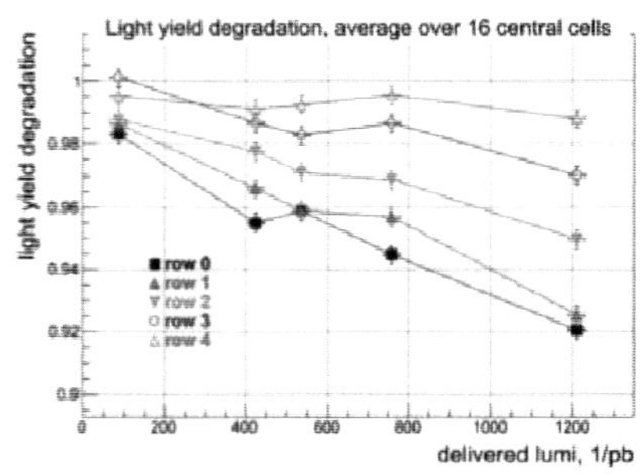

Figure 10. Light yield for 16 HCAL central cells as a function of the delivered luminosity.

To compensate for the detector aging, we follow a two-fold strategy. From the measurements given by the ^{137}Cs source, followed up by the LED monitoring we compute the correct modification of the PMT gain and so, the corresponding new high voltage. To keep the gain variation moderate, we may also modify the calibration coefficients of the HCAL cells.

If the offline accounting for the HCAL ageing would be found necessary, one can use the E/p based calibration on hadron tracks. The E/p calibration gives absolute scale and calibrates the whole signal chain, accounting also for the spread of electronics sensitivities. Figure 11 shows the correlation of ratio of E/p-based calibration coefficients for fill ranges 1883-1901 and 1997-2007 (around 5 weeks in between) and LED amplitude change for the same period. This validates the use of the LED corrections at least for short time scale.

ECAL aging effects can be seen from the variation on the π^0 mass peak fit for calibration over time (figure 12). To solve this problem, we proceed by calibrating ECAL at two levels. First, we apply the fine calibration of each ECAL cell using π^0 mass fit and adjusting its mass on a short period of data taking. Then, on top of fine-calibrated coefficients, data trending coefficients are applied. Since π^0 statistics is not high enough to follow closely the changes, we use the photon conversion and look at the E/p calibration technique.

So far, no relevant aging effect has been observed neither on PS nor on SPD.

5. A sketch of physics results

To illustrate the performance of the LHCb calorimeter systems, we focus onto a result related to decays $B_d^0 \to K^{*0}\gamma$ and $B_s \to \Phi\gamma$, which are particularly sensitive to the ECAL calibration [6]. Reference [7] updates the results of reference [6] to $1\,\mathrm{fb}^{-1}$ of pp collisions at a centre of mass energy of $\sqrt{s} = 7$ TeV. The ratio of branching fractions of $B_d \to K^*\gamma$ and $B_s \to \Phi\gamma$ decays has been measured to be

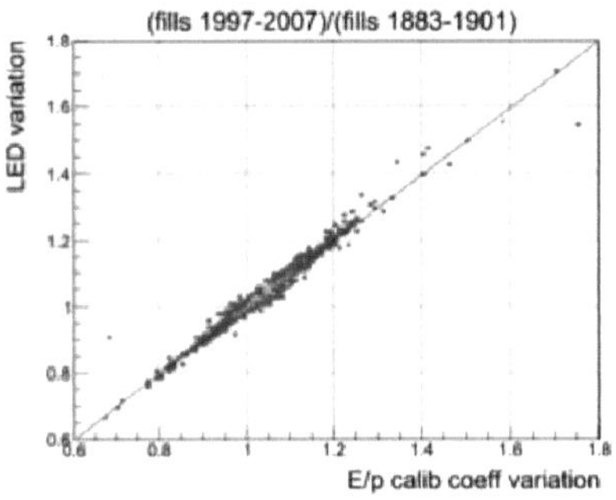

Figure 11. Ratio between E/p-based calibration coefficients in a 5 week time period and the corresponding ratio of the LED corrections for the same period.

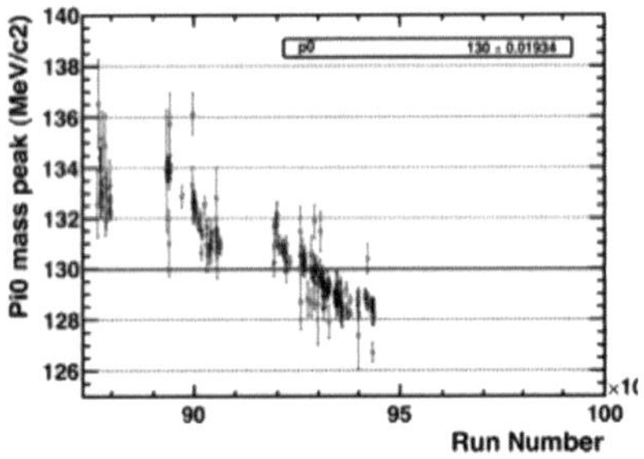

Figure 12. Observed π^0 mass variation on ECAL due to aging.

$$\frac{\mathcal{B}(B_d \to K^*\gamma)}{\mathcal{B}(B_s \to \Phi\gamma)} = 1.23 \pm 0.06(\text{stat}) \pm 0.04(\text{syst}) \pm 0.10(f_s/f_d) \quad (1)$$

in good agreement with the theoretical prediction of 1.0 ± 0.2 [8].
Using $\mathcal{B}(B_d \to K^*\gamma) = (4.33 \pm 0.15) \times 10^{-5}$ [9], one obtains

$$\mathcal{B}(B_s \to \Phi\gamma) = (3.5 \pm 0.4) \times 10^{-5} \quad (2)$$

(statistical and systematic errors combined), which agrees with the previous experimental value. This is the most precise measurement of the $\mathcal{B}(B_s \to \Phi\gamma)$ branching fraction to date.

6. Conclusions

The LHCb calorimeter system are fully functional and deliver precise measurements which contribute to accurate physics results. Aging has been observed in the hadronic and electromagnetic calorimeters. Aging effects are compensated by modifying the photomultiplier gains and the calibration coefficients so that physics results are not affected.

7. Acknowledgments

This work is supported by the Spanish MEC (projects FPA2008-06271, FPA2011-30163 and CPAN CSD2007-00042 of the Programme Consolider-Ingenio 2010) and by the Generalitat de Catalunya (AGAUR 2009SGR01268)

References

[1] LHCb Collaboration. (2008) Journal of Instrumentation Volume 3, S08005 (2008).
[2] The LHCb Collaboration, (2000) *LHCb Calorimeters : Technical Design Report*. CERN-LHCC-2000-0036.
[3] C.Abellan Beteta et al (2012) JINST **7** P08020 doi:10.1088/1748-0221/7/08/P08020
[4] R.Vàzquez-Gómez *et. al.* (2011) *Calibration and alignment of the SPD detector with cosmic rays and first LHC collisions* LHCb-PUB-2011-024
[5] O.Igonkina *et. al.* (2000) *Online π^0 calibration of ECAL* HERA-B-00-103
[6] LHCb Collaboration, (2012) Phys. Rev. D **85** 112013 [arXiv:1202.6267 [hep-ex]].
[7] LHCb Collaboration, (2012) *Measurement of the ratio of branching fractions $\mathcal{B}(B_d \to K^*\gamma)/\mathcal{B}(B_s \to \Phi\gamma)$ and the direct CP asymmetry in $B^0 \to K^{*0}\gamma$*, arXiv:1209.0313 [Unknown].
[8] A. Ali, B. D. Pecjak and C. Greub, (2008) Eur. Phys. J. C **55** 577 [arXiv:0709.4422 [hep-ph]].
[9] D. Asner *et al.* [Heavy Flavor Averaging Group Collaboration], (2010) arXiv:1010.1589 [hep-ex].

Journal of Physics: Conference Series **404** (2012) 012004 doi:10.1088/1742-6596/404/1/012004

Status of the ATLAS Liquid Argon Calorimeter; Performance after 2 years of LHC operation

Hass AbouZeid on behalf of the ATLAS Collaboration

Department of Physics, University of Toronto, 60 St. George St., Toronto, Ontario, M5S 1A7, Canada

E-mail: `ossama.abouzeid@cern.ch`

Abstract. The ATLAS experiment is designed to study the proton-proton collisions produced at the Large Hadron Collider(LHC) at CERN. Liquid argon sampling calorimeters are used for all electromagnetic calorimetry covering the pseudo-rapidity region up to 3.2, as well as for hadronic calorimetry in the range 1.4-4.9. The electromagnetic calorimeters use lead as passive material and are characterized by an accordion geometry that allows a fast and uniform azimuthal response without any gap. Copper and tungsten were chosen as passive material for the hadronic calorimetry; whereas a classic plate geometry was adopted at large polar angles, an innovative one based on cylindrical electrodes with thin argon gaps was designed for the coverage at low angles, where the particle flow is higher. All detectors are housed in three cryostats kept at about 88 K. After installation in 2004-2006, the calorimeters were extensively commissioned over the three years period prior to first collisions in 2009, using cosmic rays and single LHC beams. Since then, around 9 fb^{-1} (as of June, 2012) of data have been collected at a center of mass energy of 7 and 8 TeV. During all these stages, the calorimeter and its electronics have been operating almost optimally, with performances very close to the specifications.

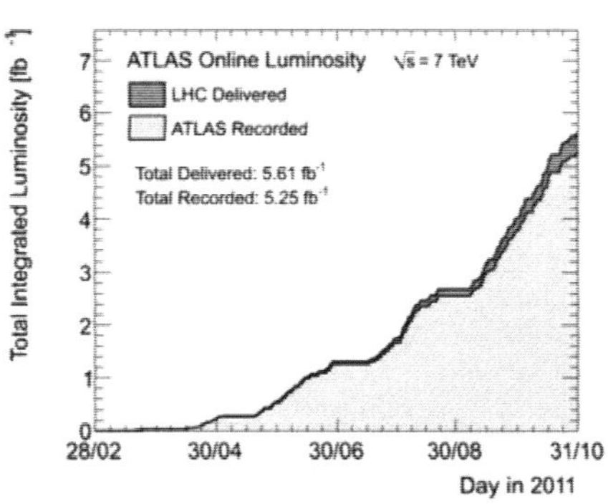

Figure 1. Delivered and recorded luminosity for 2011 proton-proton run [3].

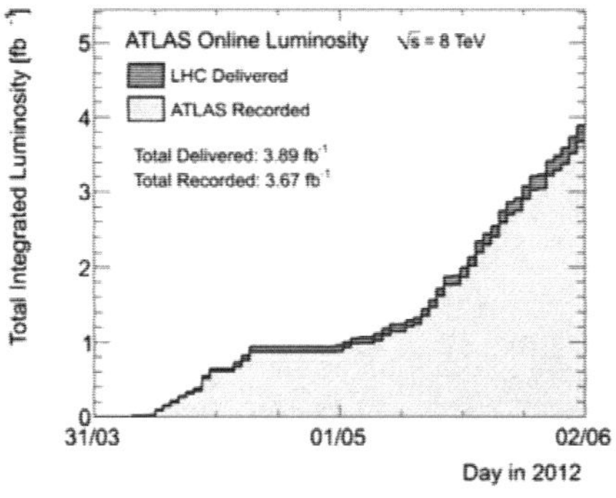

Figure 2. Delivered and recorded luminosity for 2012 proton-proton run (as of June 4th) [3].

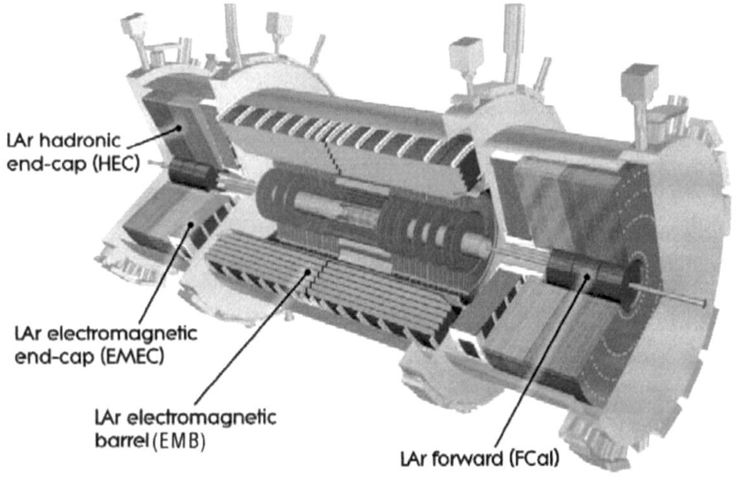

Figure 3. ATLAS Liquid Argon Calorimeter [2]

1. Introduction

The ATLAS [1] Liquid Argon (LAr) calorimeters [2] have been installed and fully operational since the cosmic-ray data-taking in 2006. Since then, with the first LHC collisions coming in 2009, the Liquid Argon calorimeters have collected over 9 fb^{-1} of proton-proton collision data as of June 4th, 2012 (see figures 1,2) [3]. Throughout this time, the LAr calorimeters have also exhibited excellent performance and operational stability.

The LAr calorimeters, shown in figure 3, are sampling calorimeters, with liquid argon as the active medium. They are composed of 4 different sub-systems that cover different regions of pseudo-rapidity and are constructed with different absorber materials. Properties of these various sub-systems are summarized in Table 1. These 4 sub-systems are contained inside three separate cryostats. The barrel cryostat houses the electromagnetic barrel (EMB) calorimeter, while the two end-cap cryostats contain the electromagnetic end-cap (EMEC), hadronic end-cap

Table 1. Summary of LAr sub-system properties.

Sub-system	Coverage	Absorber	Channels	Layers
EMB	$0 < \|\eta\| < 1.475$	Pb	109568	3 (+ presampler)
EMEC	$1.375 < \|\eta\| < 3.2$	Pb	63744	3: $\|\eta\| < 2.5$ (+ presampler $\|\eta\| < 1.8$) 2: $\|\eta\| > 2.5$
HEC	$1.5 < \|\eta\| < 3.2$	Cu	5632	4
FCal	$3.1 < \|\eta\| < 4.9$	Cu/W	3524	3

(HEC) and forward (FCal) calorimeters. In total these cryostats contain 80 m^3 of liquid argon kept at 88 K.

2. Signal Readout

Ionization from charged particles crossing the liquid argon gap in the calorimeters drifts through a high electric field generated by the LAr high-voltage system in between the absorber plates and readout electrodes. This gives rise to a triangle shaped ionization pulse with a steep rise, and slowly falling tail shown in figure 4. This analog triangular signal is then processed by the front-end boards, giving a bi-polar pulse (super imposed on figure 4), and digitized at $40MHz$. A pulse shape from a cosmic muon event, showing the agreement between predicted and actual pulse shape is shown in figure 5. The digitized samples are then sent to the read-out drivers, where the amplitude and timing of the digitized signal are computed. The equations for the amplitude (in ADC counts) and time of the signal are as follows:

$$A = \sum_{i=1}^{N} a_i(s_i - p), \quad \tau = \frac{1}{A}\sum_{i=1}^{N} b_i(s_i - p)$$

where N is the number of samples (5 for standard data-taking), s_i are the digitized samples in ADC counts, p are the pedestals and a_i and b_i are the Optimal Filtering Coeffecients (OFCs) [4]. Both the pedestals and OFCs are obtained by calibration runs taken in between LHC physics fills. The energy (in MeV) of the pulse is then obtained by applying a series of constants – taken from calibration runs, and test beam results – to the amplitude A.

The timing of the signal in the calorimeter is used for both physics studies and to suppress unwanted backgrounds. By studying the timing of the signal one can differentiate between energy deposited in the calorimeter by an event that was triggered on and an event from a neighbouring LHC bunch crossing. The timing can also be similarly used to determine the direction of the energy flow and therefore veto cosmic events, or study long-lived exotic particles which may decay far from the interaction vertex (non-pointing photons, or other particles). The overall timing resolution of the EM barrel calorimeter is shown in figure 6.

The readout channels from the Liquid Argon calorimeter along with the ATLAS Tile calorimeter form trigger tower cells for the Level-1 Calorimeter (L1Calo) trigger. The L1Calo trigger is a hardware based trigger which shares part of its readout with the calorimeter (LAr + Tile) readout. Once the pulse coming from the calorimeter is shaped by the front-end electronics, the readout paths split. The L1 portion of the readout sends the analog pulse to be

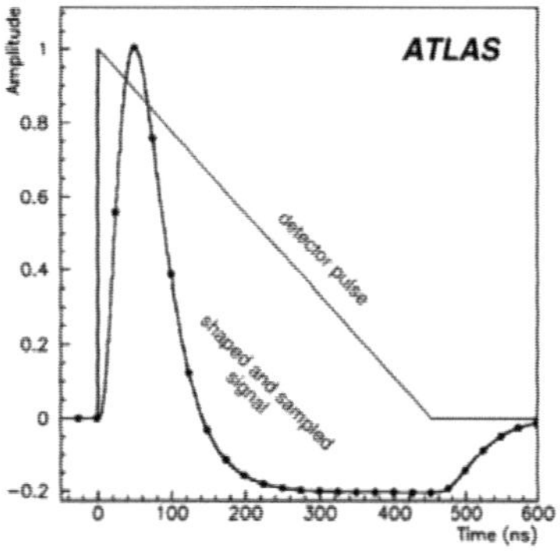

Figure 4. The ionization pulse from the calorimeter before shaping and after shaping [4].

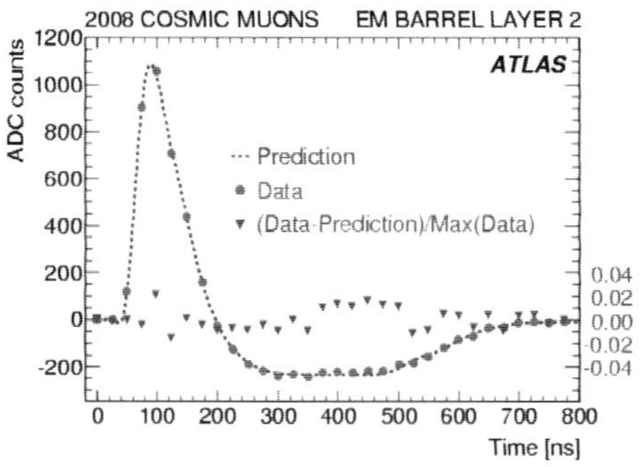

Figure 5. The ionization pulse from the calorimeter after electronics shaping [5].

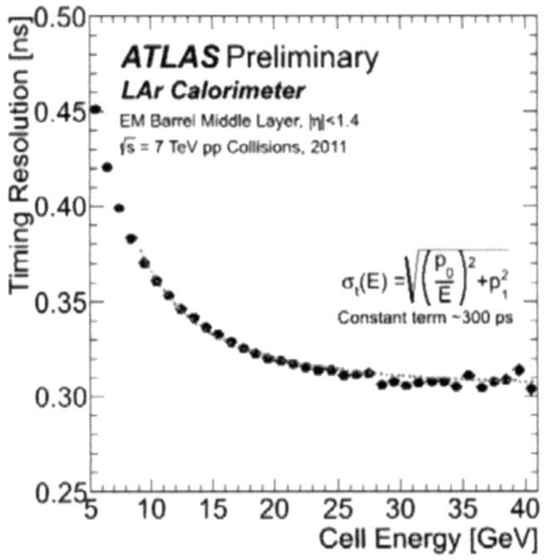

Figure 6. Timing Resolution from Middle Layers Cells from the LAr EM Barrel Calorimeter (2011 data). Includes 200ps contribution from intrinsic beam spread [6].

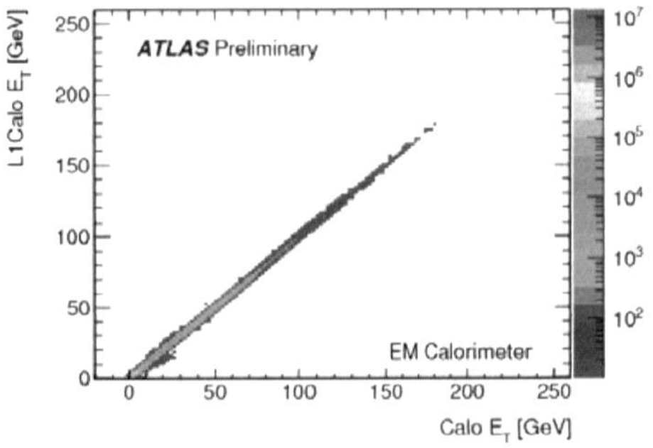

Figure 7. Energy as measured by the L1Calo trigger system vs the full calorimeter readout electronics for the LAr EM Calorimeter [6].

summed with approximately 60 other readout cells, forming a trigger tower with granularity of $\Delta\eta \times \Delta\phi \approx 0.1 \times 0.1$. These summed analog pulses are then sent to the L1 electronics, which must then make a decision whether or not to keep or discard the event within $2\mu s$. Figure 7 shows the agreement in energy calculated between these two different readout paths. The L1Calo energy resolution is designed and operates such that $\sigma_{E_T(L1Calo)} < 5\%$ for $E_T(Calo) > 10GeV$.

3. LAr Monitoring and Stability

3.1. Liquid Argon Temperature and Purity Monitoring

Deviations of the temperature of the liquid argon from $88K$ can have adverse affects on both the energy resolution and the energy response of the calorimeters. The energy response has a linear relationship with the temperature, with a 2% signal decrease for each degree Kelvin.

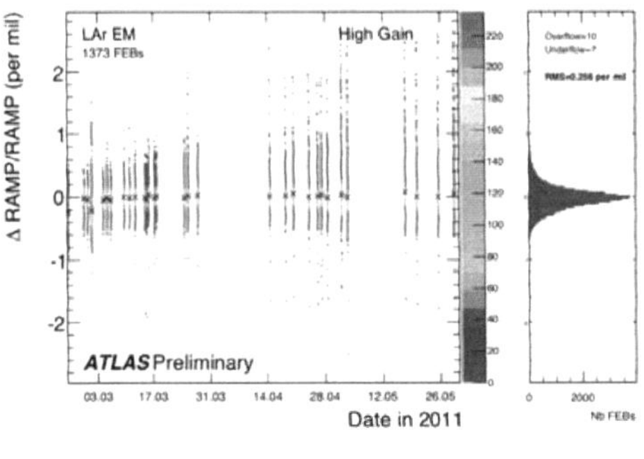

Figure 8. Relative variation in gain per front end board over 3 month period [6].

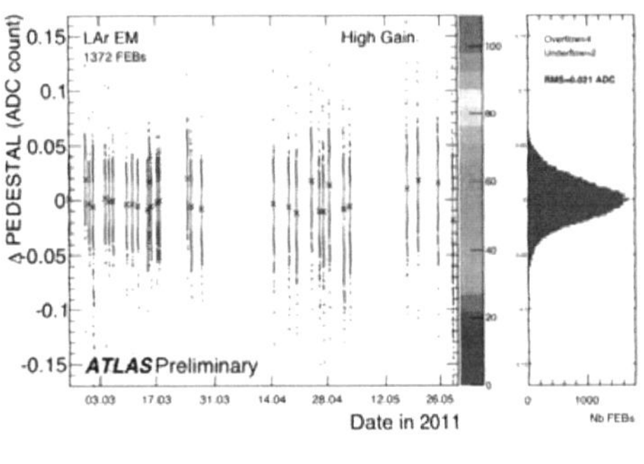

Figure 9. Pedestal variation per front-end board over 3 month period [6].

This is due to both a change in the drift time in the liquid argon and a change in the density that accompanies the temperature variation. In order to monitor the temperature of the liquid argon, > 400 Precision PT100 temperature probes are installed throughout the 3 cryostats. Monitoring of these temperature measurements over long periods of time show that the LAr cryostats display an overall temperature stability of $59mK$, which is well within the design specifications of $< 100mK$.

The liquid argon is also closely monitored for impurities, in particular O_2. The design specifications dictate that the purity be better than 1000 ppb. Each cryostat contains 10 purity monitors that each have two radioactive sources (^{207}Bi and ^{241}Am). Using these purity monitors, it was found that the levels of impurities are far below the operational limit and are approximately 200 ppb in the barrel and 140 ppb in the endcap cryostats.

3.2. Calibration Stability

Figures 8 and 9 show the long term stability of the calibration constants and therefore the electronics (which, as mentioned previously, play an important part in determining the energy deposited in the calorimeters) for the LAr EM calorimeters. The pedestal variations are below 0.03 ADC counts (corresponding to 0.3 MeV for the EM calorimeter) and the relative variation in gain is $< 10^{-3}$ for all LAr sub-systems.

4. Physics with the Liquid Argon Calorimeters

In order to provide meaningful physics results in the challenging LHC environment, the LAr calorimeter must meet the following important design goals: have excellent linear response over a broad range of energies, be able to reconstruct and model an event's missing transverse energy (E_T^{miss}) well, and provide excellent performance with electron and photon objects. The results from the ATLAS physics program over the past 2 years of operations strongly vouches for the excellent performance of the LAr calorimeters in physics analysis.

4.1. Linearity of Response and Electron Performance

Figures 10 and 11 [7] show that the calorimeter response is very good over a wide range of transverse momentum (p_T). While the electrons from the J/ψ decay are relatively low energy compared to the much harder spectrum of the Z-boson di-electrons, the agreement with the Monte Carlo simulation is excellent in both cases.

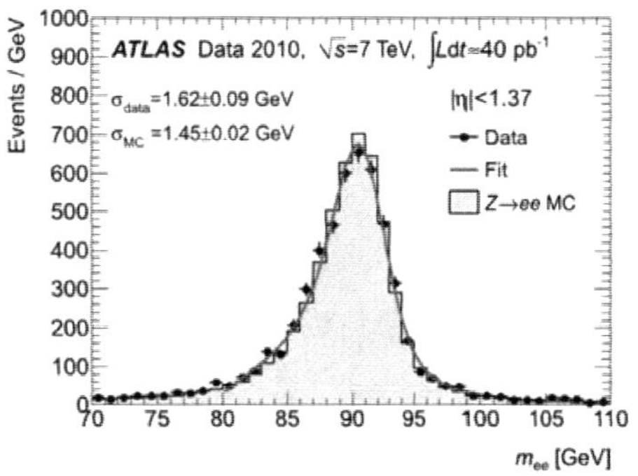

Figure 10. Invariant mass spectrum for electron pairs around 90 GeV, showing a clear Z-boson resonance [7].

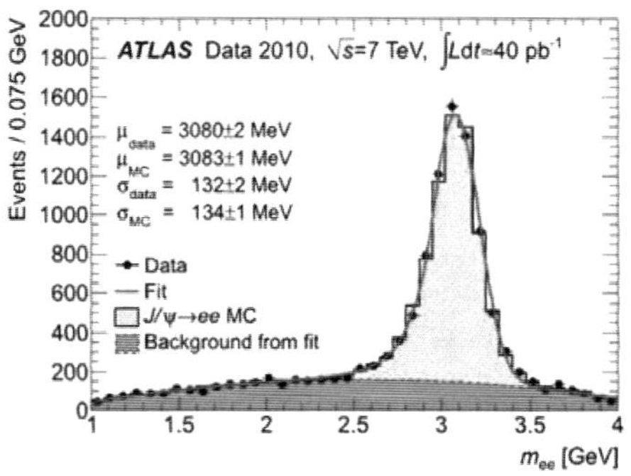

Figure 11. Invariant mass spectrum for electron pairs around the 3 GeV mass window, showing a clear J/ψ peak [7].

4.2. E_T^{miss} Modelling
Similarly, in figure 12, the E_T^{miss} spectrum is shown in the WW diboson analysis [8]. This spectrum is seen to be well modelled, even up into the tails.

4.3. Photon Performance
Finally, figure 13 [9] shows the invariant mass distribution for diphoton events in the Standard Model Higgs search. The figure shows the invariant mass for di-photon events, and also for the main backgrounds that can fake photon-like signals.

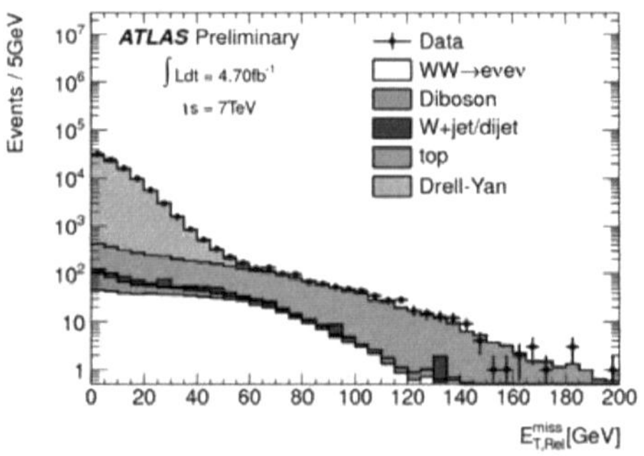

Figure 12. E_T^{miss} spectrum in the W^+W^- production [8].

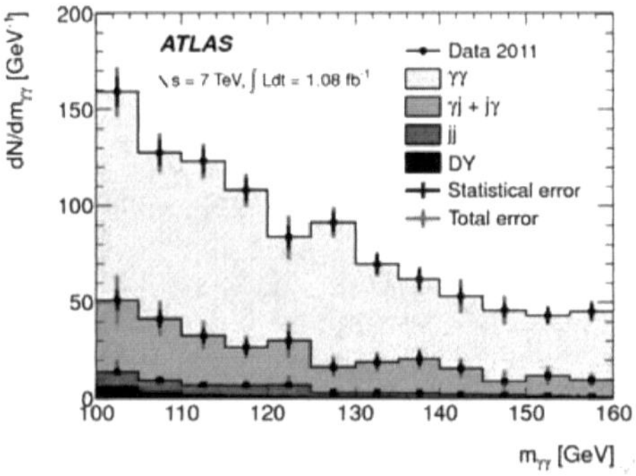

Figure 13. Invariant mass for diphoton events and main backgrounds [9].

5. Summary
Although the LHC has been delivering physics data only since 2009, the ATLAS Liquid Argon calorimeters have been installed and taking cosmic ray data as of 2006. Since that time, the LAr calorimeters have exhibited excellent operational stability and the physics performed with the calorimeters shows excellent understanding and modelling of both the calorimeter and the underlying physics processes at the LHC.

References
[1] ATLAS Collaboration. The ATLAS experiment at the CERN Large Hadron Collider, JINST 3 (2008) S08003.
[2] ATLAS Collaboration. Readiness of the ATLAS Liquid Argon Calorimeter for LHC Collisions, Eur. Phys. J. C 70 (2010) 723.
[3] https://twiki.cern.ch/twiki/bin/view/AtlasPublic/LuminosityPublicResults
[4] H. Abreu et al., Performance of the electronic readout of the ATLAS liquid argon calorimeters, JINST 5 (2010) P09003.
[5] ATLAS Collaboration. Drift Time Measurement in the ATLAS Liquid Argon Electromagnetic Calorimeter using Cosmic Muons, Eur. Phys. J. C 70 (2010) 755-785.
[6] https://twiki.cern.ch/twiki/bin/view/AtlasPublic/ApprovedPlotsLAr
[7] ATLAS Collaboration. Electron performance measurements with the ATLAS detector using the 2010 LHC proton-proton collision data, Eur. Phys. J. C72 (2012) 1909.
[8] ATLAS Collaboration. Measurement of the W^+W^- Production Cross Section in Proton-Proton Collisions at sqrt(s) = 7 TeV with the ATLAS Detector, ATLAS-CONF-2012-025.
[9] ATLAS Collaboration. Search for the Standard Model Higgs boson in the two photon decay channel with the ATLAS detector at the LHC, Phys. Lett. B705 (2011) 452-470.

Operational Experience with the CMS Hadronic Calorimeter in the 2011 LHC run

Pablo Goldenzweig
University of Rochester, Rochester, NY, 14627, USA
E-mail: pablo.goldenzweig@cern.ch

Abstract. The Hadronic Calorimeter (HCAL) of the CMS experiment has successfully recorded data at a center-of-mass energy of 7 TeV during the 2011 LHC run. Performance of the HCAL detector components and operational experience gained will be reviewed, as well as the overall impact of the HCAL on the physics reach of the CMS experiment.

1. Introduction

The CMS detector at the CERN Large Hadron Collider has been constructed to study a range of high-energy processes involving diverse signatures of final states, by using various objects such as electrons, muons, taus, photons, jets, and missing transverse energy [1, 2]. The hadronic calorimeter (HCAL) plays a crucial role in the measurement of hadron jets and exotic particles with missing transverse energy signatures.

The barrel (HB) and endcap (HE) calorimeters are positioned behind the tracker and electromagnetic calorimeters, while inside the 3.8T magnet solenoid. The outer hadron calorimeter (HO) lies outside the solenoid, to identify late starting showers and to measure the shower energy deposited after HB. The barrel, endcap, and outer detectors contain active plastic scintillator tiles interspersed between stainless steel and brass absorber plates. They are read out with custom made hybrid photo-detectors (HPD) [3] designed to operate in magnetic fields up to 4T. The forward hadron calorimeters (HF) are positioned 11.2m away from the interaction point and are made of steel absorbers and embedded quartz fibers, which are read out with conventional photomultiplier tubes (PMTs) [4]. The calorimeters extend to pseudorapity $|\eta| = 5$.

Throughout the first year of operation, timing and calibration measurements were continuously made to optimize HCAL performance. In this note, we will review the initial, and in-situ calibration of HCAL throughout the first full year of the LHC run.

2. Monitoring HCAL Status

The low and high voltage of HCAL is monitored through PCs in the CMS control room. Additionally, HCAL is equipped with laser, LED, and radiation damage monitoring systems. The pedestal values of the individual channels are periodically checked by using dedicated pedestal runs, and are adjusted if significant drifts from the initial pedestal values are observed. The signal synchronization (timing) of the HCAL channels is performed with the laser system, which consists of a single UV laser which can illuminate portions of the sub-detectors at once through a series of optical splitters. The stability of photo-detector gains is monitored using the LED system. Calibration and monitoring data is collected during dedicated calibration session in which there is no beam in the LHC, and during data taking in orbits which do not contain proton bunches. The data collected is used to produce relevant histograms and tables for monitoring the status of HCAL. Experts continuosly monitor this data, and take corrective action when necessary, to ensure quality data taking.

3. HCAL Calibration

Prior to the LHC startup, charge injection calibration (fC/ADC) as part of incoming QC was performed. This was followed by calibration from Co^{60} radioactive sources and cosmic ray muons in the underground hall. The absolute energy scale was determined with pion and electron test beams on selected modules, and carried over using radioactive sources. At the start-up of the LHC, additional calibration was performed with proton bunches hitting collimators upstream of CMS, which resulted in a large horizontal flux of muons traversing CMS detectors (splash events). Collisions data was then used to obtain calorimeter response corrections, which include the following methods discussed below: Φ-symmetry (Sec. 3.2); the method of isolated hadrons for HBHE (Sec. 3.4); and $Z \to ee$ scale for HF (Sec. 3.5).

3.1. Monitoring HPD and PMT Gain Drift

Approximately 4% of barrel and endcap channels exhibited gain drifts of 10% from Oct-2010 to Sep-2011. The remaining channels were stable at the level of 2-3% over a period of one year (Fig. 1). In the forward calorimeters, LED and laser data showed an η-dependent drop in response in the PMTs, with a systematic gain loss vs. integrated luminosity. There was a drop of 3% for $5fb^{-1}$ range and up to 10% for high η in 2011.

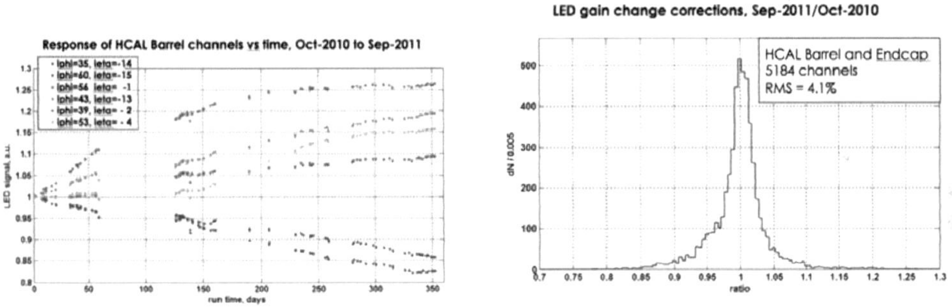

Figure 1. Response of selected barrel channels to LED signals vs time (left), and LED gain change corrections Sept-2011 / Oct-2010 (right).

3.2. Φ-Symmetry Calibration

Two methods were used to determine the relative response of HCAL channels at fixed η, and estimate corrections to produce a uniform response in Φ, without changing the overall energy scale. Corrections to calibration coefficient were found under the requirement that all channels with the same η have their total energy depositions, estimated within chosen thresholds, equal to the average energy of channels at this η (iterative method). The second method (method of moments) uses non-zero-suppressed data to determine corrections for the calibration coefficients, under the requirement that all channels with the same η ring have the same mean or same variance - in the variance case, the contribution from noise is subtracted. The scatter plot on Fig. 2 (left) shows the correlation between the two methods, and the final correction (right), which is obtained from an error weighted average of the two methods, in the initial 2010 data.

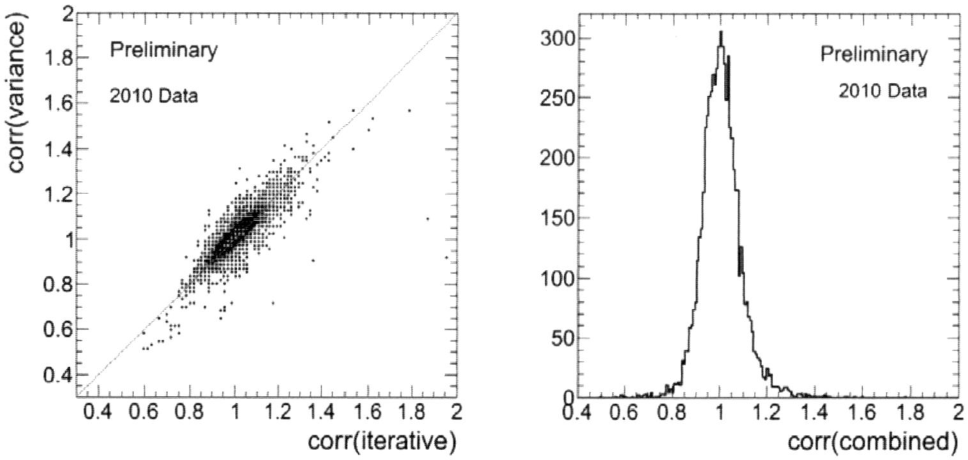

Figure 2. Comparison of corrections obtained using the Φ-symmetry calibration methods (left), and the final corrections determined by an error-weighted average of the two methods (right) with initial 2010 data.

3.3. Φ-Symmetry Calibration with LED Corrections

The gain corrections are based on monitoring the HPD response to LED signals. With the LED, each channel can be individually corrected, while with the Φ-symmetry method, the corrections are obtained in rings of η, and are not as precise. However, solely correcting with LED data is insufficient, as it is not known if the LED pulsers consistently put out the same energy. To verify if there was indeed a gain drift, the Φ-symmetry calibration is performed both with and without the LED corrections. The right plot of Fig. 3, which shows the result of the Φ-symmetry calibration without, and with the LED gain corrections, confirms that the LED corrections reduce the spread of 2011 Φ-symmetry calibration w.r.t. the 2010 results.

3.4. Isolated Hadron Response for HBHE

Isolated tracks are used to apply an η-dependent energy correction to the data in the region with tracker coverage ($|i\eta| \leq 22$), to set the absolute energy scale. A dedicated trigger is used to select events with isolated tracks with $p_T > 40$ GeV. The left plot on Fig. 4 show the response of the tracks used to derive the corrections. On the right, a cross-check is performed with lower p_T tracks (38-40 GeV), showing that the response is corrected. The 2011 response correction was calculated on top of the 2010 absolute scale corrections with isolated tracks.

3.5. Absolute Scale for HF

The absolute scale for HF is derived from $Z \rightarrow ee$ events. The LHC sends many Z bosons far forward, with an electron hitting HF. The electromagnetic calorimeter is used to provide detailed inputs for the calibration and rejection of jet fakes. The position of the electrons are

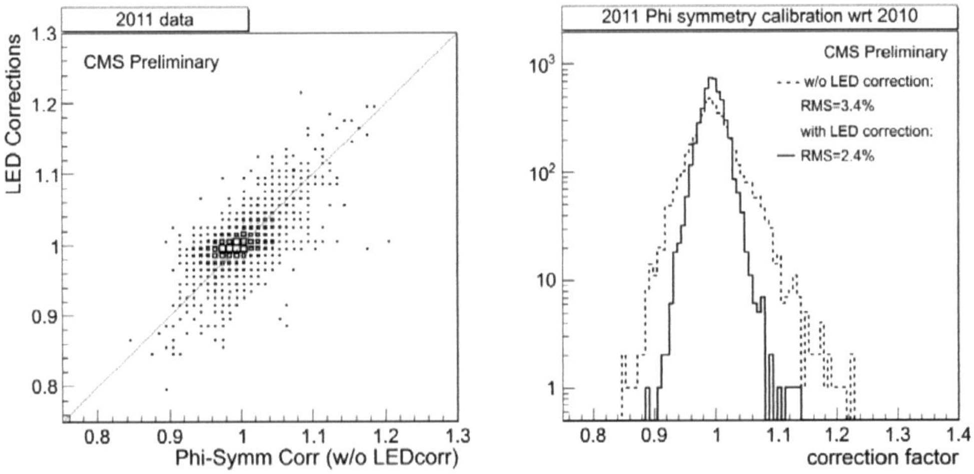

Figure 3. LED correction vs. Φ-symmetry correction before applying the LED correction (left), and the correction factor with and without the LED correction (right), for 2011 data.

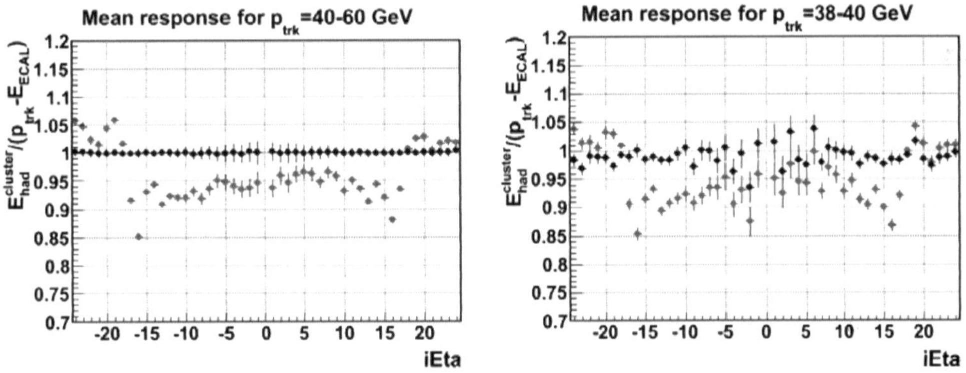

Figure 4. The mean response for $p_T = 40 - 60$ GeV tracks before (red) and after (black) corrections (left histogram). The right histogram shows a cross check performed with lower p_T tracks (38-40 GeV). The sudden change in the response of the uncorrected tracks on the left histogram correspond to the last η ring of HCAL barrel detector ($|\eta| = 1.4$).

reconstruced in HF, and the Z mass constraint is used to derive the expected energy in HF via

$$E_H = \frac{m_Z^2 \, cosh\eta_E \, cosh\eta_H}{2E_E(cosh(\eta_E - \eta_H) - cos(\phi_E - \phi_H))}. \tag{1}$$

A fit to the ratio of the expected and observed energy in HF as a function of the iη of the seed tower is then performed (Fig. 5, left).

The calibration can be affected by pileup energy being added in the HF cluster (particularly neutral-pion energy). This effect is dependent on the number of interactions in the event. A vertex dependent correction is applied by fitting the calibration constant as a function of the vertex count to extrapolate to the one-vertex case (the vertex for the Z production). The correction ratio as a function of the effective number of vertices is shown in Fig. 5, right.

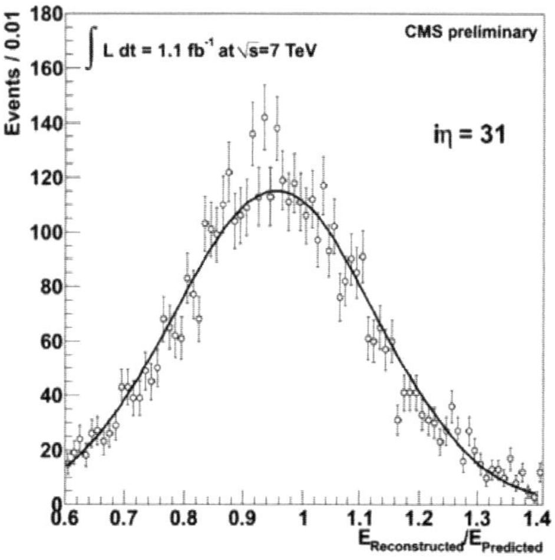

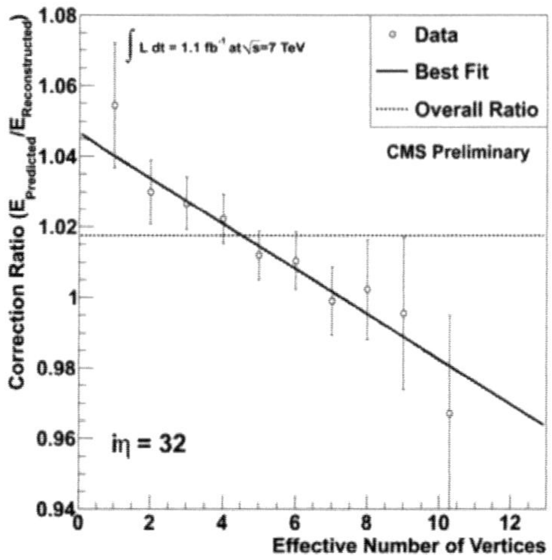

Figure 5. Fit to the ratio of expected and observed energy in HF for $i\eta = 31$ (left), and the dependence of calibration constants on the number of vertices (right).

4. Summary

The CMS HCAL has operated stably during the first years of running, with more than 99% of channels alive. It has, and continues to, deliver excellent quality data with high efficiency. The HCAL detector monitoring and in-site calibration account for the effect of gain shifts and pixel response drift to ensure optimal performance.

References

[1] S. Chatrchyan *et al.*,"The CMS experiment at the CERN LHC", *JINST* **3**, S08004 (2008).

[2] S. Chatrchyan *et al.*, "The CMS hadron calorimeter project: Technical Design Report", CERN-LHCC-97-031.

[3] P. Cushman, A. Heering, A. Ronzhin, "Custom HPD readout for the CMS HCAL", *Nucl. Instrum. Methods* A **442** (2000).

[4] U. Akgun *et al.*, "Complete tests of 2000 Hamamatsu R7525HA phototubes for the CMS-HF Forward Calorimeter", *Nucl. Instrum. Methods* A **550** (2005).

XVth International Conference on Calorimetry in High Energy Physics (CALOR2012) IOP Publishing
Journal of Physics: Conference Series **404** (2012) 012006 doi:10.1088/1742-6596/404/1/012006

HF Status

Rahmat Rahmat
University of Mississippi
E-mail: rrahmat@phy.olemiss.edu

Rob Kroeger
University of Mississippi
E-mail: rrahmat@phy.olemiss.edu

Yasar Onel
University of Iowa
E-mail: yasar-onel@uiowa.edu

Abstract. We report the performance of HF calorimeter and some details about the PMT window hit events. The new 4-anode PMTs will be installed during the 2013 shutdown period. We have developed HF GFlash, a very fast simulation of electromagnetic showers using parameterizations of the profiles in Hadronic Forward Calorimeter. HF GFlash has good agreement to 7 TeV Collision Data and previous Test Beam results. In addition to good agreement with Data and previous Test Beam results, HF GFlash can simulate about 10000 times faster than Geant4.

1. Introduction

CMS has two Hadron Forward (HF) calorimeters on both sides of the interaction point (IP), at about 11 m distance, cover the very forward angles of CMS, in the pseudorapidity range $3 < |\eta| < 5$. Such detectors will optimize the detection of those processes that produce forward jets, especially processes involving heavy Higgs and SUSY particles. An added benefit is to improve the determination of the missing transverse energy. The two cylindrical HF units are placed at each end of the detector at the beam height. They are sampling calorimeters with plastic clad quartz fibers embedded into the iron absorber. It has an active radius of 1.4 m. Each unit is 1.65 m long and composed of 18 slices of 20-degree sections. Long (1.65 m) and short fibers(1.43 m) in the calorimeter sample the energy in hadronic and electromagnetic showers, respectively. About 1000 km of quartz fiber have been used for the two HF modules, which weight about 250 tons each. These devices can measure hadronic jets up to few TeV energies, detecting the Cherenkov light emitted by the shower particles in the quartz fibers, and transported through the same fibers to photomultiplier tubes (PMT), located on the downstream side of the calorimeter modules.

2. PMT Upgrade

During 2010, the anomalous events that had been seen in the HF calorimeter earlier continued to be a problem. These events were mostly due to the particles hitting the PMT window and producing Cherenkov radiation, thereby creating large false energy signals. One of the suggestions to eliminate these false signals is to discard the data that do not show any correlation in the long and short fibers. In principle, long and short fibers should measure similar energies. However, using the correlations in the long and short fibers eliminates some of the anomalous events after the data acquisition; it does not prevent the signals to be produced in the first place.

Even though the correlations between the long and short fibers seem to give us a good way of eliminating most of the anomalous signals, this is not enough because it is not complete and it does not eliminate the erroneous trigger signals. The solution has to reduce these signals at the source such as replacing all the PMTs with thinner window 4-anode PMTs. Thinner window will produce less Cherenkov scintillation in the window and using the correlation between different anodes will eliminate the remaining signals coming from the windows. The real Cherenkov light coming from the calorimeter towers will produce similar signals in all four anodes but not the Cherenkov light produced in the window.

Hamamatsu 4-anode PMTs (R7600) are chosen to be the replacement PMTs for the HF. Initial tests done with these PMTs in the test beam at Fermilab and CERN show that the Cherenkov light produced by particles hitting the PMT windows could be distinguished easily and clearly from the real signals. During the 2013 shutdown, new thin-window high-efficiency 4-anode PMTs will be installed instead of the current PMTs to reduce the anomalous signals to negligible levels. These new PMTs are being tested prior to the installation, as they are delivered by Hamamatsu.

3. HF Simulation

Previous HF simulation (based on Shower Library) in CMS Collaboration has no ability to simulate PMT Window Noise and other noises. The worst part of the previous simulation that it was created using only discrete energy bins (10 GeV, 20 GeV , 30 GeV, 40 GeV, 50 GeV, etc. not continuous) that limit its precision significantly. The HF Shower Library has another problem because it deletes particles that enter HF Detector immediately and replace them with Shower Library that has very limited statistics. Fortunately, we have developed HF GFlash, a fast simulation of electromagnetic showers using parameterizations of the profiles in Hadronic Forward Calorimeter. HF GFlash solves almost all problems that previous HF simulation has.

HF will experience unprecedented particle fluxes because on average, 760 GeV per proton proton interaction is deposited into the two forward calorimeters, compared to only 100 GeV for the rest of the detector. Due to this condition, CMS has changed the HF geometry for simulation because it will need very long computing time if we use full HF geometry for huge number of high energy particles. The new HF geometry does not include detail description of sensitive detectors in HF, but only general description of HF.

HF GFlash has been optimized to the latest HF geometry and condition. In fact, since we have collision data available, our understanding about our detector and physics at 7 TeV improved significantly and we have modified HF GFlash in such a way that it has good agreement to Data and previous Test Beam results. In addition to good agreement to Data and previous Test Beam results [1], we also have improve the speed of HF GFlash significantly. The new HF GFlash can process electron particle guns 2 times faster than previous HF GFlash.

Journal of Physics: Conference Series **404** (2012) 012006 doi:10.1088/1742-6596/404/1/012006

4. Theory

For physics analysis and feasibility studies large number of Monte Carlo events may have to be produced. Using individual particle tracking, the computing time needed for such kind of simulations increases linearly with the energy absorbed in the detector and can easily become prohibitive. Using parameterizations for electromagnetic showers can speed up the simulations considerably, without sacrificing precision. The high particle multiplicity in electromagnetic showers as well as their compactness and the good understanding of the underlying physics makes their parameterization advantageous.

The Gflash package allows the parameterization of electron and positron showers in homogeneous (for the time being) calorimeters and is based on the parameterization described by G. Grindhammer [2] . The spatial energy distribution of electromagnetic showers is given by three probability density functions (pdf),

$$dE(\vec{r}) = E\, f(t)dt\, f(r)dr\, f(\phi)d\phi,$$

describing the longitudinal, radial, and azimuthal energy distributions. Here t denotes the longitudinal shower depth in units of radiation length, r measures the radial distance from the shower axis in Moliere units, and ϕ is the azimuthal angle. A gamma distribution is used for the parameterization of the longitudinal shower profile, $f(t)$. The radial distribution $f(r)$, is described by a two-component ansatz. In ϕ, it is assumed that the energy is distributed uniformly: $f(\phi) = 1/2\pi$.

The center of gravity, $\langle t\rangle$ and the depth of the maximum, T, can be calculated from the shape parameter α and the scaling parameter β according to

$$\langle t\rangle = \frac{\alpha}{\beta}$$

$$T = \frac{\alpha - 1}{\beta}.$$

In the parameterization all lengths are measured in units of radiation length (X_0), and energy in units of the critical energy E_c defined as

$$E_c = 2.66\left(X_0\frac{Z}{A}\right)^{1.1}$$

This allows material independence, since the longitudinal shower moments are equal in different materials. The following equations are used for the energy dependence of T_{hom} and (α_{hom}), with

$$y = \frac{E}{E_c}$$

and

$$t = \frac{x}{X_0}$$

where we define x as the longitudinal shower depth:

$$T_{hom} = \ln y + t_1$$

$$\alpha_{hom} = a_1 + (a_2 + a_3/Z)\ln y.$$

5. Performance of HF GFlash

We simulated the computing time for 10,000 20 GeV electrons and Table 1 shows that HF GFlash can perform simulation faster than Shower Library.

	Shower Library (previous MC)	HF GFlash
Minimum Computing Time	0.0095	0.0059
Maximum Computing Time	3.01	2.81
Average Computing Time	0.62	0.53

Table 1. Comparison of computing time between HF GFlash and Shower Library for 10000 20-GeV electrons.

We also check the longitudinal shower profiles produced using HF GFlash and compare them to the longitudinal shower profiles produced using Shower Library. In Fig.1. we can see that longitudinal profiles produced by HF GFlash(CMSSW_3_11_0) and Shower Library(CMSSW_3_10_0_pre3).

Using the results of energy response ratio from Team Beam data as the reference, we can check the performance of energy response ratio of HF GFlash compared to Shower Library. From Table 2, we found that HF GFlash has better agreement to Test Beam data compared to Shower Library.

	HF GFlash	Test Beam	Shower Library
Se_{50}/Le_{50}	0.24	0.24	0.20
Lp_{50}/Le_{50}	0.67	0.66	0.63
Sp_{50}/Le_{50}	0.51	0.50	0.51
Sp_{50}/Lp_{50}	0.76	0.76	0.80

Le_{50} = Energy deposited in Long Fiber from 10000 50-GeV electrons
Se_{50} = Energy deposited in Short Fiber from 10000 50-GeV electrons
Lp_{50} = Energy deposited in Long Fiber from 10000 50-GeV charged pions
Sp_{50} = Energy deposited in Short Fiber from 10000 50-GeV charged pions

Table 2. Comparison of energy response ratio between HF GFlash, Test Beam (reference) and Shower Library using electrons and pions at 50 GeV.

	HF GFlash	Test Beam	Shower Library
Se_{100}/Le_{100}	0.30	0.30	0.25
Lp_{100}/Le_{100}	0.70	0.69	0.67
Sp_{100}/Le_{100}	0.57	0.55	0.56
Sp_{100}/Lp_{100}	0.82	0.80	0.84

Le_{100} = Energy deposited in Long Fiber from 10000 100-GeV electrons
Se_{100} = Energy deposited in Short Fiber from 10000 100-GeV electrons
Lp_{100} = Energy deposited in Long Fiber from 10000 100-GeV charged pions
Sp_{100} = Energy deposited in Short Fiber from 10000 100-GeV charged pions

Table 3. Comparison of energy response ratio between HF GFlash, Test Beam (reference) and Shower Library using electrons and pions at 100 GeV.

	HF GFlash	Test Beam	Shower Library
Se_{150}/Le_{150}	0.33	0.34	0.28
Lp_{150}/Le_{150}	0.71	0.73	0.70
Sp_{150}/Le_{150}	0.59	0.60	0.56
Sp_{150}/Lp_{150}	0.83	0.82	0.80

Le_{150} = Energy deposited in Long Fiber from 10000 150-GeV electrons
Se_{150} = Energy deposited in Short Fiber from 10000 150-GeV electrons
Lp_{150} = Energy deposited in Long Fiber from 10000 150-GeV charged pions
Sp_{150} = Energy deposited in Short Fiber from 10000 150-GeV charged pions

Table 4. Comparison of energy response ratio between HF GFlash, Test Beam(reference) and Shower Library using electrons and pions at 150 GeV.

For energy resolution, we found that HF GFlash can improve the resolution by 50%. In this case, we define energy resolution as the difference of energy resolution observed in Test Beam data and simulation. For simulation we use HF GFlash or Shower Library, and we found HF GFlash has better resolution compared to Shower Library.

Electromagnetic energy response of electrons is predicted to be linear and Test Beam data has shown that it is linear(within stat. error) up to 150 GeV. We have tuned HF GFlash so that it has linear energy response up to 14 TeV.

Previous MC simulation based on Shower Library can not simulate high energy particle for example, electrons with energy higher than 2 TeV. Fortunately HF GFlash can handle not only low energy particles but also high energy particles. We can prove that HF GFlash can produce nice longitudinal profiles correctly for 1 TeV, 7 TeV and 14 TeV (see Figure 1)

One major breakthrough is the ability of HF GFlash to simulate PMT Window hits for the first time and previous MC simulation can not simulate PMT Window Hits. This achievement, encourage us to simulate other noises such as Fibre Bundle in HF. The early results show that inclusion of PMT Window Hits, Fibre Bundle and Jungle improve the agreement between HF GFlash Data and 7 TeV Collision Data.

We have used HF GFlash to produce some sample datasets, for example: ttbar, Z→ee and MinBias. The internal memory size is very crucial and we have checked and confirmed that HF GFlash used reasonable computer memory size for physics simulation. We use the correct geometry and we have done validation of HF GFlash using CMSSW_3_10_0_pre9 (a standard CMS software combination used for analysis at the end of 2010) when HF GFlash was chosen

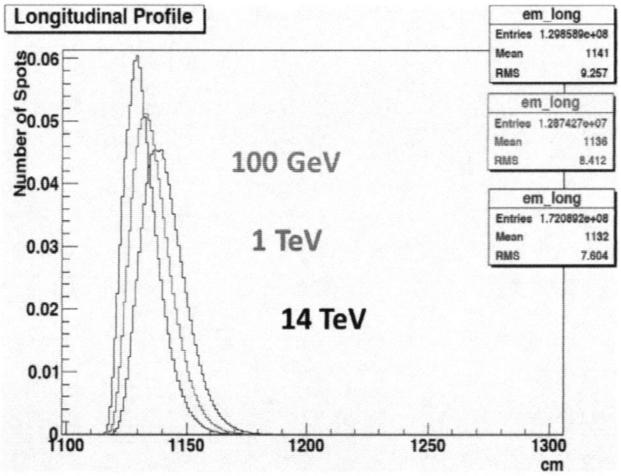

Figure 1. Longitudinal Shower Profile produced by HF GFlash for electron with energy 100 GeV(blue), 1 TeV(red) and 14 TeV(black). This plot shows the capability of HF GFlash to handle very high energy particle simulation.

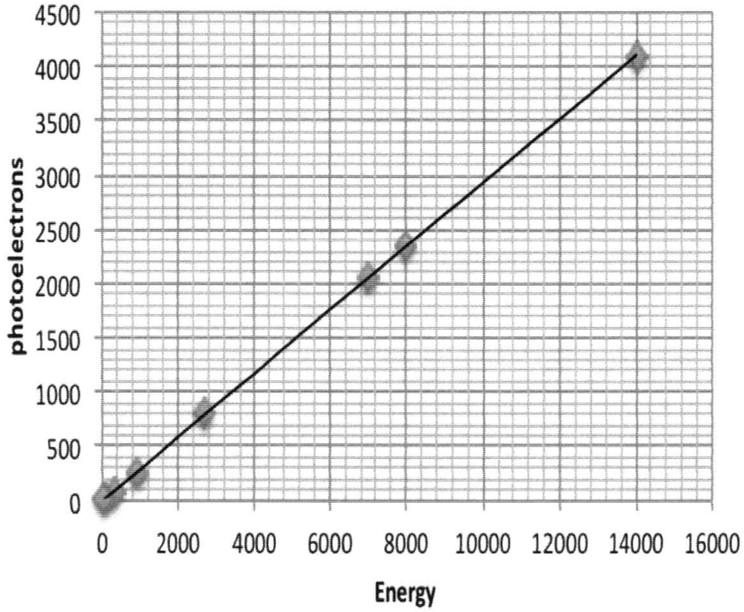

Figure 2. HF GFlash has linear energy response for electron with energy from 50 GeV to 14 TeV

as the default of HF simulation.

We reconstruct SimHit produced by HF GFlash to produce RecHit(reconstructed hits) that will be used for physics analysis in CMS Collaboration. We should be very careful about timing in RecHit and we see that HF GFlash gives reasonable timing information compared to certified Collision Data.

By the end of November 2010 we have collected about 36 pb^{-1} certified Collision Data that can be used to study RecHit(reconstructed hits) energy distribution in HF towers for Long and Short Fibres. Using MinBias generator we can simulate HF RecHit for every tower. We have made 52 comparison plots for every HF tower and we found the HF GFlash has good agreement with 36 pb^{-1} Certified Collision Data and we can see HF GFlash can perform better than previous MC based on Shower Library.

The crucial part of tuning is coming the next few months when we have 100 fb^{-1} Collision Data from CMS Detector. At that time, we will have reasonable number of events to use Z→ee to tune HF GFlash with real physics process. In this analysis we will require one electron in central region as the tag and the other as the probe to check the performance of HF GFlash.

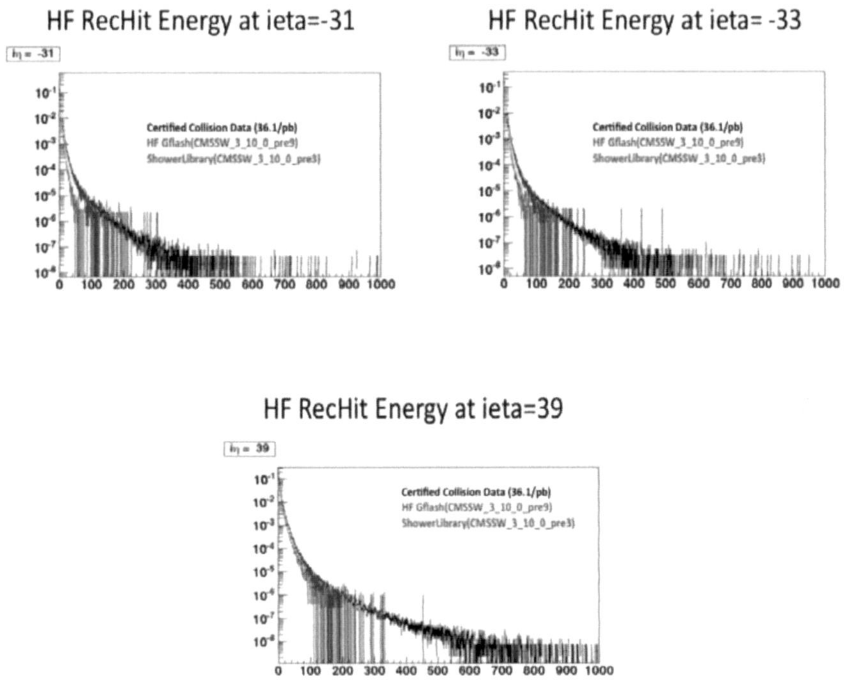

Figure 3. Recontructed hit energy at towers 31, 33 and 39 collected using HF GFlash (Blue), Shower Library (Red) and 2010 Collision Data (Black).

6. Summary

In summary, The HF detectors performed very well during the 2011 run period. Prior to the run, relative timing of the HF channels were improved and light guide sleeves were replaced. Operating parameters of the HF detectors were further optimized. In 2013, all the existing PMTs will be replaced with the 4-anode PMTs.

We have developed the most powerful simulation on earth that can handle very high energetic particles with better performance. HF GFlash has been compared and tested

(i) Test Beam data
(ii) Shower Library
(iii) 36 fb^{-1} Certified Collision Data

Due to its better performance, CMS Collaboration has chosen HF GFlash as the standard HF Detector simulation since 2011. HF GFlash has been tested and the tests showed that it is faster and more accurate so that HF GFlash will be a very useful simulation not only for CMS Detector but also for other physics experiments, such as International Linear Collider, Muon Collider, etc.

[1] The CMS-HCAL Collaboration *Design, performance, and calibration of CMS forward calorimeter wedges.* Eur. Phys. J. C.53, 139-166(2008)
[2] G. Grindhammer, S. Peters, *The Parameterized Simulation of Electromagnetic Showers in Homogeneous and Sampling Calorimeters.* hep-ex/0001020 (1993).
[3] E. Longo and I. Sestili,Nucl. Instrum. Meth. 128, 283 (1975).
[4] Rossi Prentice Hall, New York (1952).
[5] G. Grindhammer, M. Rudowicz, and S. Peters, Nucl. Instrum. Meth. A290, 469 (1990).
[6] The CMS Collaboration. *Technical Proposal*, CERN/LHCC 94-39(1994).
[7] The CMS Collaboration, *The Hadron Calorimeter Project Technical Design Report*, CERN/LHCC 97-31 (1997)

The ATLAS Tile Hadronic Calorimeter performance at the LHC

Paolo Francavilla, On behalf of the ATLAS Collaboration
Institut de Física d'Altes Energies, Edifici Cn, Universitat Autònoma de Barcelona, E-08193 Bellaterra, Spain
E-mail: francav@cern.ch

Abstract. The Tile Calorimeter (TileCal), the central section of the hadronic calorimeter of the ATLAS experiment, is a key detector component to detect hadrons, jets and taus and to measure the missing transverse energy. Due to the very good muon signal to noise ratio it assists the spectrometer in the identification and reconstruction of muons. TileCal is built of steel and scintillating tiles coupled to optical fibers and read out by photomultipliers. The calorimeter is equipped with systems that allow to monitor and to calibrate each stage of the read-out exploiting different signal sources: laser light, charge injection, a radioactive source and the signal produced by minimum bias events. The performance of the calorimeter has been measured and monitored using calibration data, random triggered data, cosmic muons, splash events and most importantly the large sample of pp collision events. Results that are discussed demonstrate how the calorimeter is operated, how it is monitored and what performance has been obtained. These results also show that the Tile Calorimeter is performing well within the design requirements and is giving essential input to the physics results.

ATL-TILECAL-PROC-2012-008
18 September 2012

1. Introduction

The ATLAS experiment at CERN [1] is successfully taking data at the LHC at 0.9 TeV, 2.47 TeV, 7 TeV and 8 TeV center-of-mass energy since 2009. The Tile Calorimeter (TileCal) [2] is the central hadronic section of the ATLAS Calorimeter. TileCal is a sampling calorimeter made of scintillating tiles as active medium and steel plates as absorbers. It is divided into four partitions, two barrels (LB) and two extended barrels (EB), covering in total a pseudorapidity range of $|\eta| < 1.7$ and is segmented into 64 modules along the azimuth ϕ. Wavelength shifting fibers collect the light generated in the scintillating tiles and carry it to photomultipliers (PMT) (see Fig. 1). Two fibers, attached to every tile from different sides in ϕ, go to different PMTs, providing redundant double readout of a signal. Each PMT receives signals from multiple tiles which are grouped into cells of different size depending on their pseudorapidity and depth. Three longitudinal layers A, BC, D are defined inside the modules and the dimensions of the cells are optimized to obtain a structure of projective towers with granularity of $\Delta\eta \times \Delta\phi = 0.1 \times 0.1$ in the first two layers and $\Delta\eta \times \Delta\phi = 0.2 \times 0.1$ in the last layer (see Fig. 2). Cells of an additional special layer E, attached to the extended barrel modules, are read out by a single PMT each. In total, TileCal has 5182 cells and 9852 channels.

Because of the double readout of most of the cells, single channel failure does not impact the energy measurement. More serious failures in on-detector electronics, which cannot be fixed remotely, might require to disable the readout from a whole module - 22 cells in barrel and 18

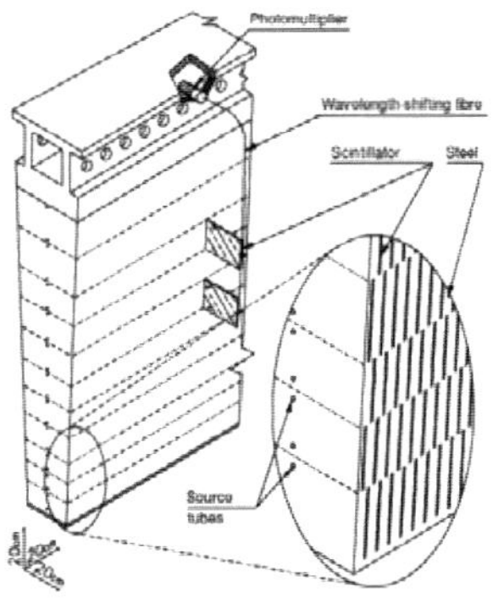

Figure 1. Schematic of one of the 64 azimuthal modules of TileCal showing the system of signal collection.

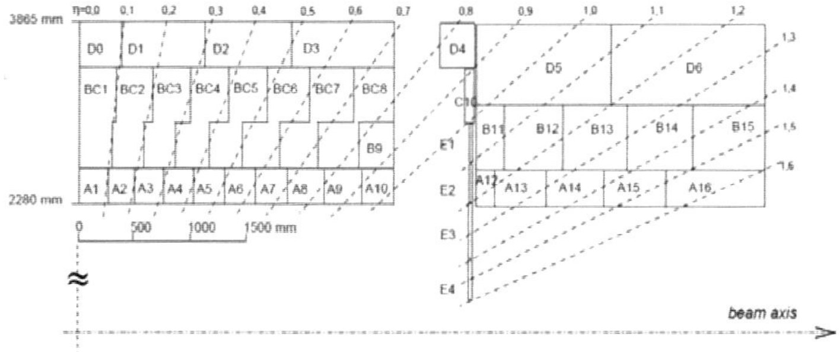

Figure 2. Drawing of a half of the calorimeter divided into a barrel and an extended barrel part with the cell division scheme depicted.

cells in extended barrel are masked in this case. The energy deposited in the masked cells is recovered offline using interpolation between working neighbors, thanks to the good calorimeter spatial granularity. Before the maintenance period of winter 2011-2012 5% of the cells were unusable for physics and all of them were successfully repaired. The amount of unusable cells at the end of August 2012 is 1% (see Fig. 3).

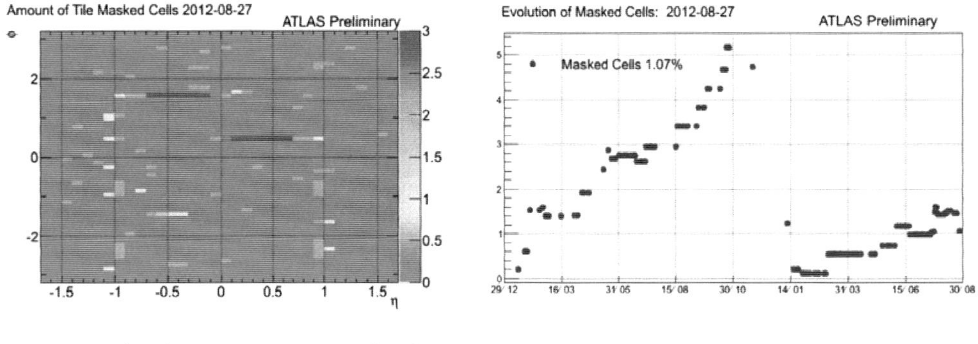

Figure 3. (left) Two dimensional $(\eta;\phi)$ map showing the number of cells masked per tower, each tower being composed by three cells from the layers A, BC, D.
(right) Evolution in time of the percentage of masked cells.

2. The new low voltage power supplies

The low voltage power supplies (LVPS) adopted for the front-end electronics are located in the detector hall, in a high radiation environment. Each module is equipped with an independent LVPS, with a total of 256 LVPS units. In 2011, the collaboration faced a rate of 0.8 LVPS trips per inverse picobarn, and a procedure to restore the data taking after each trip was adopted. After a trip, and during the configuration preceding the normal operation mode, the channels of the module affected by a LVPS trip are dynamically masked, and the energy is recovered offline using interpolation between working neighbors. Despite the high rate of trips, TileCal had in 2011 a very high efficiency in recording high quality data, used for the physics analysis.

In 2012, 40 LVPS units were substituted with a new production of power supplies[1]. Thanks to the new design which gained from the better knowledge of the operation environment and groundings, these new LVPS are more robust against radio-technical noise. Fig. 4(left) shows the number of trips per LVPS unit in 2012. It is evident the improvement in stability of the new LVPS (in green and blue), compared to the old models (in red). Another important improvement of the new LVPS design is the reduction of the non Gaussian noise induced in the front-end electronic, and affecting the measurement of the signals, as shown for a specific channel in Fig. 4(right).

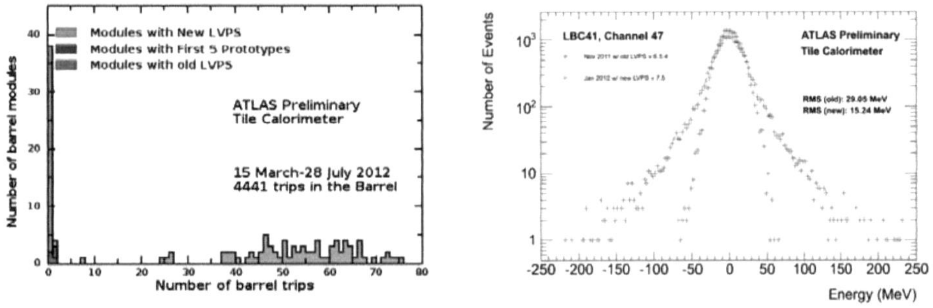

Figure 4. (left) Number of trips in the central barrel per LVPS unit, in the period from March 13th 2012 to July 28th 2012. The old power supplies are shown in red. The new LVPS are shown in green and blue (5 prototypes).
(right) Distribution of the reconstructed energy in two high statistics dedicated pedestal runs, for a given channel. The RMS of the distribution goes down by almost a factor 2 in the channel after changing LVPS.

3. Energy measurement and timing

A fraction of 11% of the TileCal modules were calibrated in the beam tests in 2001-2003. Electron and muon beams were used to establish the electromagnetic scale (EM) and to inter-calibrated the different layers. After installation of the whole TileCal in the ATLAS experimental hall, cell inter-calibration was done with the help of the Cesium calibration system. The Cesium source was moved through every calorimeter cell and high voltage of every PMT was adjusted to have the cell response equal to the response measured during beam tests. The comparison between cosmic ray data and Monte Carlo (MC) prediction and between beam test muons and cosmic muons has confirmed that propagation of the electromagnetic scale from the beam tests to ATLAS was successful. Non-uniformity within one layer as seen by muons turned out to be at the level of 2-3%, the maximal difference between layers is 4%.

3.1. Stability of the detector response

The calorimeter is equipped with systems that allow to monitor and to calibrate each stage of the read-out exploiting different signal sources: laser light, charge injection, a radioactive source and the signal produced by minimum bias events. Fig. 5 shows the drift of the energy response of TileCal cells A14 with respect to D5, both located in one of the extended barrels (EBC).

[1] In addition to the 40 new LVPS, 5 prototypes were installed during the 2010-2011 maintenance period.

Because of its stability, at the level of 0.2%, cell D5 is used as a reference cell in order to be able to study the stability of the detector response combining the various TileCal calibration systems and identify the possible source(s) of the drift. The three monitoring systems shown in the plot (laser, cesium and minimum bias integrator), are sensitive to different possible sources of the drift. Since all of them show a similar behavior, the drifts that are observed can be attributed mostly to a variation of A14 photomultiplier response and not to the scintillator irradiation. The downdrift periods coincide with the periods of data taking with high instantaneous luminosity, while the updrifts coincide with the technical stops (no collisions). The maximum variations in the response of any cell type of the TileCal are below 1% over all 2011 data taking period with an integrated luminosity of $\sim$5.6 fb^{-1}. The cesium calibration system is used every month to recalibrate all the cells and to restore the electromagnetic scale.

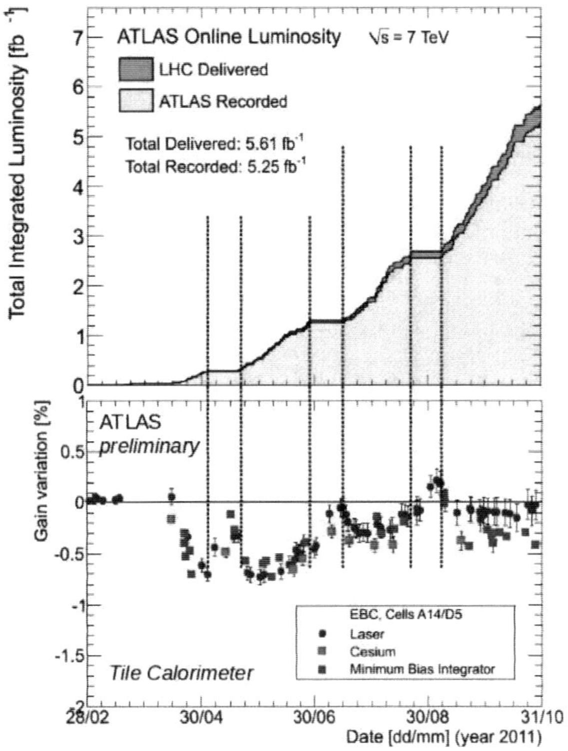

Figure 5. (Top) The evolution of ATLAS total integrated luminosity. (Bottom) The evolution of the ratio between the response of cells A14 (located at η=1.35 and inner radius 2300 mm) and the most outer radius cell D5 (η=1.0 and inner radius 3140 mm) as a function of time in EBC. The response is measured by the Cesium, the Laser and the Minimum Bias calibration systems. Each point on this plot corresponds to an average over 64 modules in ϕ. The measurements are normalized to the first run taken as a reference (taken on 18th February 2011). The error bars correspond to the statistical uncertainty on the average.

3.2. Timing

To reconstruct the signal amplitude correctly [3], the peak pulse time with respect to the electronic sampling clock has to be known with good precision. The cell times were synchronized to a single reference channel in every partition using the laser calibration system. Inter-calibration between partitions was also performed using single-beam splash events. In such events some protons from the beam collide with collimators placed at about 140 m from the nominal interaction point and produce a very large number of minimum ionizing particles reaching the detector parallel to the beam axis and depositing a large amount of energy in the whole TileCal. The signals produced by these particles, and the knowledge of their time of flight, have been used to synchronize all the TileCal cells with the precision of better than 1 ns.

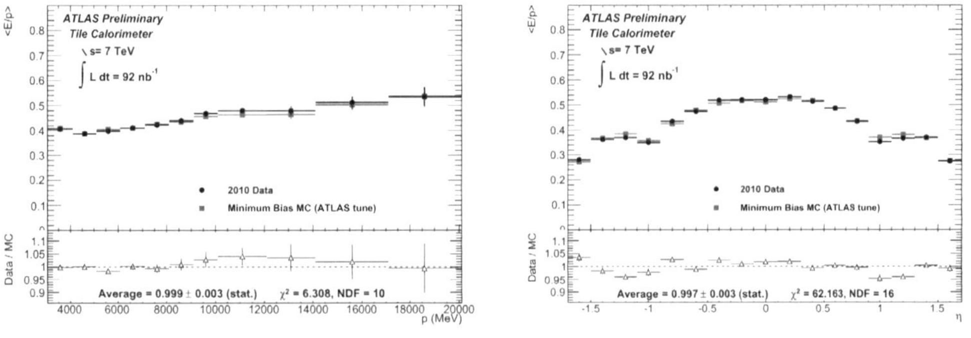

Figure 6. Mean value of the ratio between energy deposited in TileCal and the track momentum p (measured by the Inner Detector) as a function of p (left) and η (right) of the impinging particle.

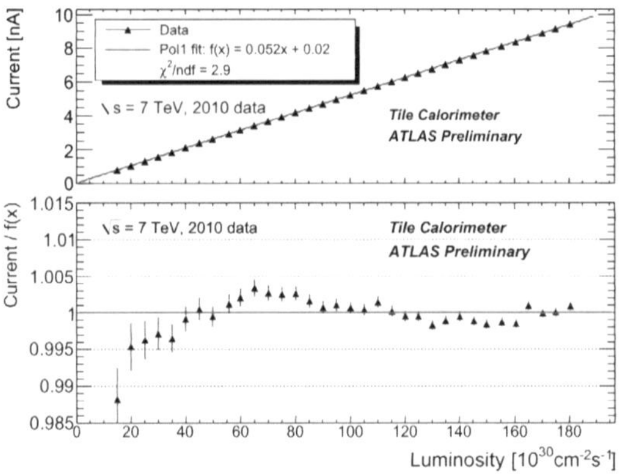

Figure 7. The anode current for a cell of the TileCal is shown as a function of the instantaneous luminosity on full data sample taken in 2010. The errors on the current are the quadratic sum of the statistical and systematic errors. The red line is obtained from the linear fit of the data points.

4. Performance in collisions

The response of TileCal to particles produced in collisions has been extensively studied in proton-proton and heavy ion collisions and for different center-of-mass energies. These studies have the twofold goal of cross-checking the performance of the detector, and of improving its Monte Carlo description. As an example, isolated tracks of momentum p were required to deposit little energy, consistent with minimum ionizing particles in the electromagnetic calorimeter in front of TileCal, to be sure their whole energy E is deposited in the hadronic calorimeter. For these tracks (mostly pions), the ratio E/p has been compared with the Monte Carlo description of the detector. The mean value $< E/p >$ is shown in Fig. 6, as a function of p and η of the impinging particle. In all the cases good agreement between data and MC was found.

4.1. Luminosity measurement

TileCal is also used to monitor the luminosity during physics runs with collisions. This is possible by integrating the signals from particles produced in low-momentum transfer inelastic pp collisions (minimum bias events) whose rate is proportional to the LHC luminosity. A dedicated TileCal read-out provides the anode currents from each photomultiplier and it is used to measure the Minimum Bias current, which is proportional to the interaction rate (and to the luminosity). The top plot in Fig. 7 shows the relation between the anode current for a cell integrated over few ms and the luminosity during collisions using Minimum Bias events. Given the stability and the linearity of the dedicated read-out (highlighted in the ratio in the bottom plot in Fig. 7), TileCal is participating to the on-line and off-line determination of the ATLAS luminosity measurement.

5. Conclusions

The first years of data taking at the LHC demonstrate good performance of the ATLAS Tile Calorimeter, which is well within the design specifications, and its ability to operate in proton-proton collisions at different center of mass energies, and in different pile-up scenario. The TileCal calibration and monitoring systems ensure stability in time of the calorimeter response, and also uniformity within 2-3% in η and ϕ. The energy scale uncertainty, which was successfully extrapolated from the beam tests to ATLAS, is conservatively considered to be 4%. The time synchronization between cells is well below 1 ns and has been verified with single beam, with cosmic muons and in collisions.

References

[1] G. Aad, et al., The ATLAS Experiment at the CERN Large Hadron Collider, JINST 3 (2008) S08003. doi:10.1088/1748-0221/3/08/S08003

[2] G. Aad, et al., Readiness of the ATLAS Tile Calorimeter for LHC collisions, The European Physical Journal C - Particles and Fields 70 (2010) 11931236, doi:10.1140/epjc/s10052-010-1508-y.

[3] G. Usai et al., Signal Reconstruction of the ATLAS Hadronic Tile Calorimeter: implementation and performance, 2011 J. Phys.: Conf. Ser. 293 012056 doi:10.1088/1742-6596/293/1/012056

[4] A. Solodkov, The ATLAS Tile Calorimeter performance at the LHC in pp collisions at 7 TeV, proceedings for the 13th ICATPP conference on Astroparticle, Particle, Space Physics and Detectors for Physics Applications, (2011)

XVth International Conference on Calorimetry in High Energy Physics (CALOR2012) IOP Publishing
Journal of Physics: Conference Series **404** (2012) 012008 doi:10.1088/1742-6596/404/1/012008

Single Hadron Response Measurement in ATLAS

Pavel Starovoitov on behalf of the ATLAS Collaboration
Deutsches Elektronen-Synchrotron (DESY), Notkestraße 85, 22607 Hamburg, Germany
E-mail: Pavel.Starovoitov@desy.de

Abstract. Single hadron response measurement in minimum bias proton-proton collisions at a center of mass energy of $\sqrt{s} = 7$ TeV are presented. Together with test-beam results, these measurement form the basis to evaluate the calorimeter energy response uncertainty of jets at high transverse momenta. The single hadrons response is measured in the momentum range of 0.5 to about 20 GeV in-situ, by comparing the calorimeter response of all energy deposits in a cone around an isolated track with the more precisely measured track momenta. The agreement between data and Monte Carlo simulation is on the level of a few percent. Using kaon and Λ particles, the calorimeter response of identified pions, proton and anti-proton is studied. The MC simulation describes the energy response of pions and protons well, but differences are observed for anti-protons. It is discussed how the jet calorimeter response uncertainty and its correlation between transverse momentum bins is determined from these measurements.

1. Introduction

Hadron jets are the dominant physics objects produced in high-energy proton-proton collisions at the Large Hadron Collider (LHC) at CERN. They are the key ingredient of many physics analyses from the di-jet cross-section to the top quark measurements as well as for searches for new phenomena with jets in the final state. The uncertainty on the jet energy scale (JES) is the largest source of detector-related systematic uncertainty. It is thus the subject of an extensive and detailed study.

The jet energy measured by the calorimeter is corrected for calorimeter non-compensation and energy loss in dead material. The corresponding jet energy scale correction factor is referred to as the JES. The JES is derived from Monte Carlo (MC) simulations by comparing the calorimeter energy of an isolated reconstructed jet to that of the particle jet that points to it.

The uncertainty on the calorimeter energy response is a significant component of the total uncertainty on the JES. It is derived by convolving the measured uncertainty on the single charged hadron energy response and the estimated uncertainty on the neutral particle energy response with the expected particle spectrum within a jet.

The calorimeter response to single isolated charged hadrons, and the accuracy of its Monte Carlo simulation description, can be evaluated from the ratio of the calorimeter energy E to the associated isolated track momentum p. The aim of the measurement is to estimate the systematic uncertainty on jet calorimeter response and therefore the focus is on data-to-MC comparison. The ratio E/p is measured using proton-proton collisions at centre-of-mass energies of $\sqrt{s} = 7$ TeV over a wide range of track momenta in the central region of the calorimeter. Possible additional uncertainties introduced by certain particle species are addressed by measuring the response to hadrons identified through the reconstruction of known short-lived particles.

A sample of about 20 million non-diffractive proton-proton collision events at $\sqrt{s} = 7$ TeV are generated using PYTHIA 6.421 [1] with the ATLAS minimum bias tune 1 (AMBT1) [2]. For isolated, high-momentum ($p_T > 15$ GeV) tracks the MC sample has $\sim 60\%$ of the number of events containing a track with $p_T > 15$ GeV than in the data. All the events are run through a full detector simulation [3] based on GEANT4 [4]. The GEANT4 physics model used is QGSP_BERT [5]. The reconstruction and analysis software used for the MC simulation is the same as for the data.

2. The ATLAS detector

The ATLAS detector covers almost the whole solid angle around the collision point with layers of tracking detectors, calorimeters and muon chambers and is described in detail in Ref. [6]. Here, the features relevant for this analysis are summarised.

The inner detector (ID) is immersed in a 2 T axial magnetic field and provides tracking for charged particles with $|\eta| < 2.5$. The ID consists of a silicon pixel tracker and silicon microstrip tracker (SCT) covering $|\eta| < 2.5$ and a transition radiation tracker (TRT) covering $|\eta| < 2.0$.

The calorimeter system covers $|\eta| < 4.9$, using a variety of technologies. High granularity liquid-argon (LAr) electromagnetic (EM) sampling calorimeters, with excellent performance in terms of energy and position resolution, cover $|\eta| < 3.2$. They use accordion shaped electrodes and lead absorbers and consist of a barrel (EMB, $|\eta| < 1.475$) and an end-cap (EMEC, $1.375 < |\eta| < 3.2$). They are longitudinally segmented in depth into three layers, with a pre-sampler behind the solenoid. For $|\eta| < 1.7$ hadronic calorimetry is provided by a sampling calorimeter made of iron and scintillating tiles (TileCal). TileCal comprises a large barrel ($|\eta| < 0.8$) and two smaller extended barrel cylinders ($0.8 < |\eta| < 1.7$). It is segmented longitudinally into three layers, with a total thickness of about eight interaction lengths at $\eta = 0$. The hadronic end-cap calorimeters (HEC, covering $1.5 < |\eta| < 3.2$) are LAr sampling calorimeters with copper absorbers. The copper/tungsten-LAr forward calorimeters (FCal) provide both electromagnetic and hadronic energy measurements and extend the coverage to $|\eta| < 4.9$.

3. Calorimeter response to charged hadrons

For every selected event, each track candidate is extrapolated to the second longitudinal layer of the EM calorimeter. A track is defined as isolated if its impact point has a distance $\Delta R = \sqrt{(\Delta\eta)^2 + (\Delta\phi)^2} > 0.4$ from all other track candidate impact points. Events are required to have at least one reconstructed vertex with at least four associated tracks.

The isolated tracks must also have a transverse momentum of $p_T > 500$ MeV, a minimum of one hit in the pixel detector and six hits in the SCT, and small transverse and longitudinal impact parameters computed with respect to the primary vertex, $|d_0| < 1.5$ mm and $|z_0| \sin\theta < 1.5$ mm. The above requirements ensure a good quality of the track and reduce contributions from fake tracks to a negligible level.

The sum of the energy deposits in layers of calorimeter cells associated to a selected track is computed using topological clusters [7] at the electromagnetic (EM) scale, i.e., without applying any correction for the calorimeter non-compensation or for energy loss in dead material. The purpose of the topological clustering algorithm is to identify areas of connected energy deposits in the calorimeter, based on the significance of the energy deposits in cells with respect to the expected noise level. Topological clusters are formed around cells with energy $|E_{\text{cell}}| > 4\sigma_{\text{noise}}$ ("seeds"), where σ_{noise} is the RMS of that cell noise. Then, iteratively, the cluster is expanded by adding all neighbouring cells with $|E_{\text{cell}}| > 2\sigma_{\text{noise}}$. Finally, the cells surrounding the resulting cluster are added, regardless of their energy. The $\eta - \phi$ position of a cluster i in a given calorimeter layer j, $(\eta_{cl}^{ij}, \phi_{cl}^{ij})$ is computed as the energy–weighted position of the cells in layer j belonging to the cluster.

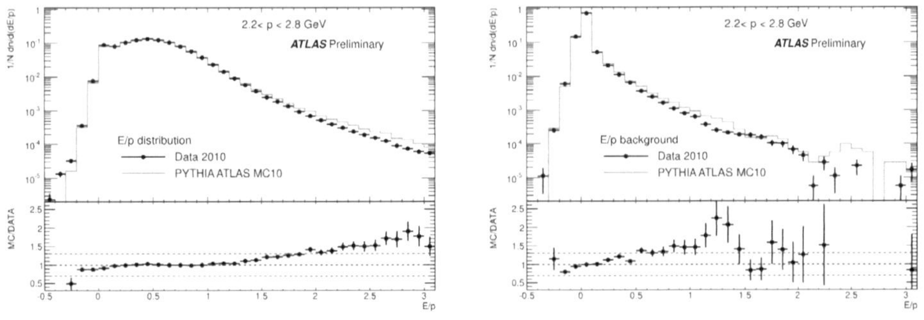

(a) (b)

Figure 1: (a) The E/p distribution for isolated tracks with an impact point in the region $|\eta| < 0.6$ and with a momentum in the range $2.2 \leq p < 2.8$ GeV. (b) The same as (a) for background [8].

The position of the track k extrapolated to the layer j is $(\eta_{tr}^{kj}, \phi_{tr}^{kj})$. The energy of a cluster in the layer j (E_j) is associated to the track if:

$$\sqrt{(\eta_{tr}^{kj} - \eta_{cl}^{ij})^2 + (\phi_{tr}^{kj} - \phi_{cl}^{ij})^2} < R_{\text{coll}}. \tag{1}$$

The parameter R_{coll} is set to 0.2 based on a trade-off between maximising the particle shower containment and minimising the background contribution coming from neutral particles produced close to the track. Roughly 90% of the shower energy is collected in a cone of such size.

The energy E associated to a track is computed as the sum of the associated energy deposits in all layers, $E = \sum_j E_j$, and the ratio E/p is formed with the reconstructed track momentum. Note that because of calorimeter noise fluctuations, E (and therefore E/p) can assume negative values.

3.1. *E/p distributions*

The E/p distribution in the one representative region of η and track momentum is shown in Figure 1a. The large number of entries with $E/p = 0$ corresponds to isolated tracks that have no associated cluster in the calorimeter. Several effects may be responsible for this:

- Particles can interact hadronically before reaching the calorimeter (in the ID, cryostat or solenoid magnet). Such particles can change their direction, or produce a large number of low momentum secondary particles.
- A cluster is created only if a seed is found. Hadrons with low momentum and an extended shower topology sometimes do not have a single cell energy deposit large enough to seed a topological cluster.

The cases where the calorimeter response is compatible with zero have been further studied. The probability that the calorimeter response is suppressed by noise threshold requirements, $P(E = 0)$, is shown in Figure 2a as a function of the amount of material (in nuclear interaction lengths) in front of the active volume in the central ($|\eta| < 1.0$) calorimeter region. When the hadron passes through more material, the probability that no energy is associated to the track increases.

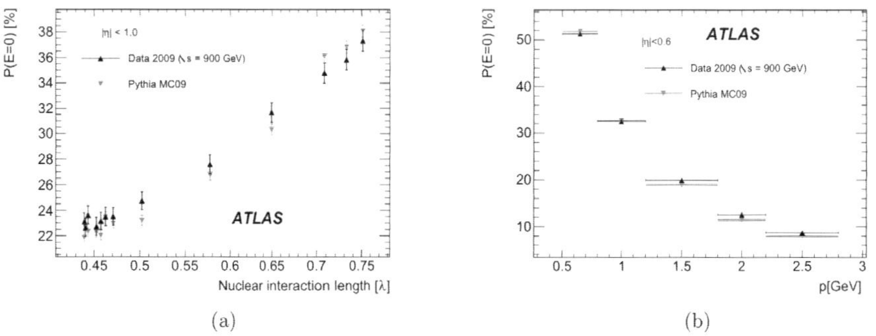

Figure 2: (a) Probability to measure a calorimeter response consistent with zero $P(E = 0)$ as a function of the amount of material in nuclear interaction lengths in front of active volume of the calorimeter for $|\eta| < 1.0$. (b) $P(E = 0)$ as a function of track momentum for $|\eta| < 0.6$ [9].

Figure 2b shows $P(E = 0)$ as a function of the track momentum in the central region of the calorimeter ($|\eta| < 0.6$). In this region, the dead material in front of the calorimeter is approximately constant. The probability decreases with increasing track momentum. In general, $P(E = 0)$ is well predicted by the MC simulation.

3.2. Background

The energy measured inside the cone of $\Delta R < R_{\mathrm{coll}} = 0.2$ centered around the track impact point may be contaminated by energy deposits from the showers of close-by particles produced in the proton-proton collision. The track isolation requirement suppresses possible shower contamination from charged particles. There is no obvious way to suppress shower contamination from photons, mostly produced in $\pi^0 \to \gamma\gamma$ decays, and neutral hadrons. The neutral particle background contribution to the E/p measurement in the MC simulation depends on the event generator settings of the parameters governing non-perturbative QCD processes and on the modelling of the calorimeter response to low momentum neutral particles, and it is therefore difficult to model correctly.

The background subtraction relies on the assumption that the EM energy from photons and neutral hadrons is independent of the energy deposited by the selected track. Charged hadrons are selected that behave like minimum ionising particles in the EM calorimeter and start their shower in the hadronic calorimeter (late-showering hadrons). Excluding a narrow region around the late-showering hadron track, the remaining EM energy is mainly due to showers from neutral particles.

Late-showering hadrons are selected by requiring a small amount of EM energy in a cone

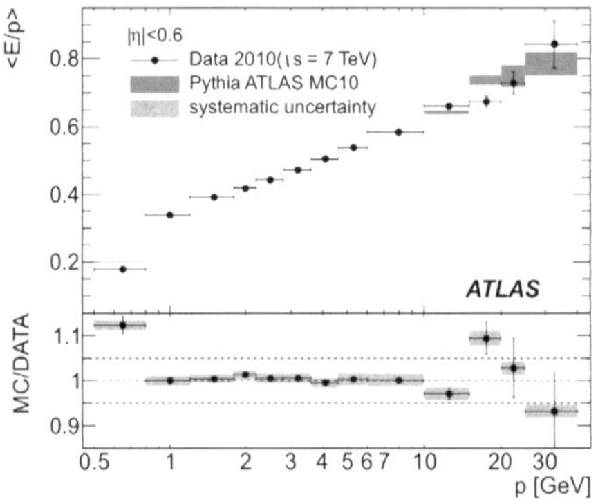

(a)

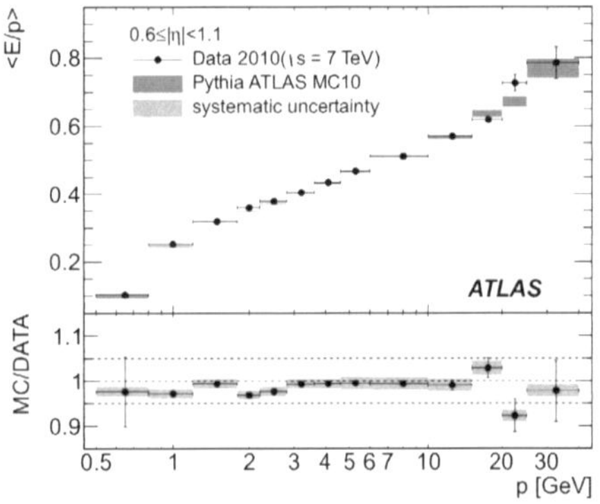

(b)

Figure 3: $\langle E/p\rangle$ at $\sqrt{s} = 7$ TeV as a function of the track momentum for (a) $|\eta| < 0.6$ and (b) $0.6 \leq |\eta| < 1.1$. The black markers represent the collision data, while the green rectangles represent the MC prediction, with the vertical width showing its statistical uncertainty. The lower panes show the ratio of the MC simulation prediction to collision data. The grey band indicates the size of the systematic uncertainty on the measurement. The dotted lines are placed at $\pm 5\%$ of unity and at unity [9].

of $\Delta R < 0.1$, $E_{\rm EM}^{0.1} < 1.1$ GeV, and a large HAD energy fraction, $E_{\rm HAD}^{0.1}/p > 0.4$. The background is measured in the EM calorimeter in an annulus around the late-showering charged hadrons. The mean of the background distribution over many events in a given momentum and pseudorapidity bin estimates the energy deposition of photons and neutral hadrons showering in the EM calorimeter:

$$\langle E/p\rangle_{\rm BG} = \left\langle \frac{E_{\rm EM}^{0.2} - E_{\rm EM}^{0.1}}{p} \right\rangle. \tag{2}$$

where $E_{\rm EM}^{0.2}$ is the EM energy in a cone of $\Delta R < 0.2$. The background E/p distribution in the one representative region of η and track momentum is shown in Figure 1b.

The background contribution from neutral hadrons depositing their energy in the hadronic calorimeter was estimated with a similar technique applied to different hadronic calorimeter layers and found to be negligibly small.

3.3. Results

The single charged hadron response is defined as

$$\langle E/p\rangle = \langle E/p\rangle_{\rm raw} - \frac{4}{3}\langle E/p\rangle_{\rm BG} \tag{3}$$

where $\langle E/p\rangle_{\rm raw}$ is the mean value of the E/p distribution before background subtraction.

The background is rescaled by the factor of 4/3, which is the ratio of the area of the full $\Delta R < 0.2$ cone to that of an annulus with $0.1 \leq \Delta R < 0.2$, to take into account the background contribution in $E_{\rm EM}^{0.1}$, since the background is only measured in this annulus. A uniform energy density of the background in the $\Delta R < 0.2$ cone is assumed.

Several systematic uncertainties on the measurement are estimated. Each is taken to be completely correlated between all pseudorapidity and momentum bins in the $\langle E/p\rangle$

measurement. The dependence of $\langle E/p \rangle$ on the track selection requirements has been estimated by varying the number of silicon tracker hits required and the impact parameter selection with respect to the primary vertex within reasonable ranges. The MC-to-data ratio of $\langle E/p \rangle$ was found to be almost unaffected by variations in the track selection. The maximum variation found in the ratio (0.5%) is taken as a systematic uncertainty. The uncertainty on the momentum scale p as measured by the inner detector is negligibly small for $p < 5$ GeV. For $p > 5$ GeV, a conservative 1% uncertainty has been assumed on the momentum scale. The 1% difference between the background estimate described in and the alternative one is taken as a systematic uncertainty.

The mean E/p value after background subtraction is evaluated in bins of momentum and pseudorapidity. Figure 3 show $\langle E/p \rangle$ as a function of the track momentum, in two different $|\eta|$ bins up to $|\eta| = 1.1$, at $\sqrt{s} = 7$ TeV. The lower parts of the figures present the ratio of MC simulation to data. The maximum momentum that can be probed with the data considered is approximately 30 GeV. The agreement between data and MC simulation is within $\sim$2% for particles with momenta in the 1–10 GeV range, and it is around 5% for momenta in the 10–30 GeV range. Below 1 GeV, where tracks are just at the kinematic threshold of entering the calorimeter volume, large differences of $\sim$10% or more between data and MC simulation are visible. However, due to the low absolute calorimeter response to very low momentum particles, these differences are not critical for the JES determination.

4. Calorimeter response to identified hadrons

The extrapolation of the previous single particle response studies into the environment of a jet requires understanding of two additional effects. A jet includes a variety of hadrons that may differ from the inclusive sample of isolated hadrons. Therefore, measuring the average response to different species of particles is valuable to ensure that the Monte Carlo correctly models all aspects of the jet shower. Additionally, the hadrons in a jet are not isolated. Threshold effects and hadronic shower widths affect the calorimeter response to the multi-hadron system.

Single hadrons are identified using decays of K_S (for positive and negative pions), Λ (for protons), and $\overline{\Lambda}$ (for anti-protons) particles. These single hadrons are required to be isolated from all other charged particles in the event. Single pions will have manifestly lower energy response distributions from those of anti-protons, because of the eventual annihilation of the anti-proton. By identifying and isolating single pions, single protons, and single anti-protons, the effects of hadronic interactions and annihilation can be separated at low to moderate energies, where they are most important.

Figure 4 shows the difference in response between π^- and $\bar{p}$ in the central ($|\eta| < 0.6$) pseudorapidity bins as a function of available energy (E_a). The difference shows a $\sim 30\% \times \langle E/p \rangle$ disagreement between data and MC simulation for $E_a < 3$ GeV, though they are consistent for $E_a > 4$ GeV. At these low available energies, the anti-proton response should be dominated by the annihilation and the subsequent shower. The difference indicates large contributions from processes not well-modelled by GEANT4.

When the mother particle is highly boosted, the decay products are more collimated. For track momenta of $\sim$2–6 GeV from K_S decays, the range of opening angles of the decay products allows a measurement of calorimeter response as more tightly collimated pairs of pions are selected. There are two related effects probed by such a measurement. Energy deposited in the calorimeter may fall below the thresholds for reconstruction and be neglected as consistent with noise. As two showers overlap, the addition of energies that might have individually been below threshold can produce a signal above these noise thresholds.

The ratio E/p as calculated here for a single pion should increase as another particle approaches it and more of the second particle's energy is included in the calorimeter energy, E. As shown in Figure 5, the response does not vary with separation for large extrapolated

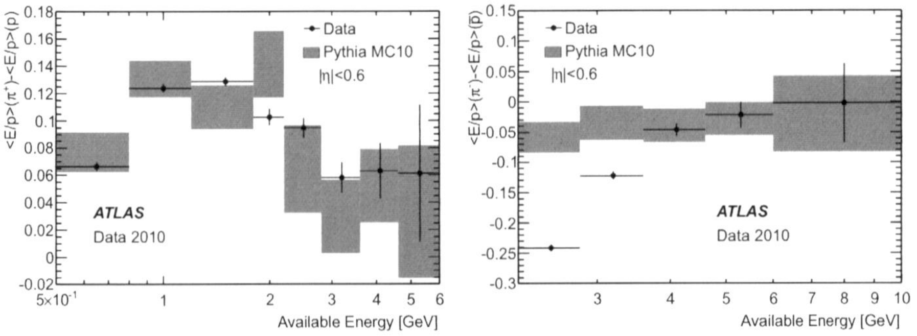

(a) (b)

Figure 4: The difference in $\langle E/p \rangle$ between $\pi^+(\pi^-)$ from K_S candidates and p ($\bar{p}$) from $\Lambda(\bar{\Lambda})$ candidates in tracks with (a) $|\eta| < 0.6$ [9].

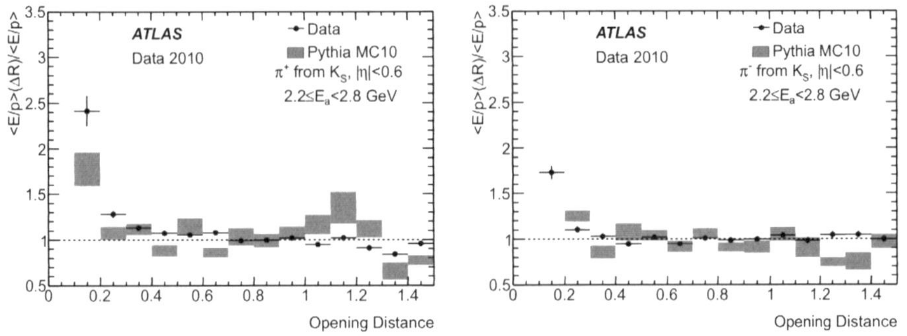

(a) (b)

Figure 5: $\langle E/p \rangle$ for pions as a function of the extrapolated distance between that pion and a pion of the opposite sign. In all cases the pions are daughters of a reconstructed K_S candidate and are required to be isolated from all tracks in the event except the other daughter of the decay. The response is shown for low energy ($2.2 \leq E_a < 2.8$ GeV) (a) π^+ and (b) π^- in the central pseudorapidity bin ($|\eta| < 0.6$). The MC simulation has no tracks passing the selection in the smallest bin (smallest two bins) of opening distance for π^+ (π^-) [9].

distances, where the single pion is isolated.

5. Calorimeter jet energy scale uncertainty

The jet energy scale calibration corrects the measured jet energy for several effects, including calorimeter non-compensation and energy loss in dead material. The calibration itself is derived from MC simulation. The calorimeter uncertainty on the JES is calculated from the uncertainty on the energy response of all particles contributing to a jet. Within the MC simulation, the energy contribution of each individual particle to a given jet can be separated. The convolution of the uncertainty on the single particle energy response with the MC jet particle composition is then used to calculate the calorimeter uncertainty on the jet energy scale.

The calorimeter JES uncertainty is derived for the well-understood central region of the calorimeter. The main reasons for the restriction to the central calorimeter region are the smaller amount of material in front of the calorimeter and the existence of combined test beam measurements with a setup very similar to the final ATLAS configuration. Therefore, the calorimeter JES uncertainty is only evaluated using the single particle response for $|\eta| < 0.8$.

The numerical evaluation of the uncertainty on the jet energy scale is performed with Monte Carlo pseudo-experiments. In each pseudo-experiment, the jet energy scale is calculated after randomly changing the Monte Carlo single particle energy response within the appropriate uncertainty range given by the measured data/MC ratio. The final uncertainty on the jet energy scale is then given by the spread of the distribution of the jet energy scale over all pseudo-experiments. Within each pseudo-experiment, all randomly changed parameters are kept fixed, such that the energy response correlations are properly taken into account.

The uncertainties on the particle energy response functions entering the calculation are taken from the in-situ E/p measurements, ATLAS combined test beam (CTB) measurements [10] and GEANT4 Monte Carlo predictions.

When the single charged hadron response from E/p is used to assess the uncertainty on the jet energy scale, further systematic uncertainties that might affect the propagation of the response to the jet have to be taken into account. The probability to find $E/p = 0$ is strongly correlated to the amount of upstream material and hence a good measure of the E/p acceptance. A fully correlated (in p and η) 28% uncertainty is derived from the maximal observed difference between data and MC simulation in this probability. Given an amount of energy released in the calorimeter, the energy collected in the topological clusters may differ from the energy released in the calorimeter, depending on how isolated the energy deposit is. This can introduce a bias in the particle response in a jet with respect to the response measured for isolated hadrons. A conservative systematic uncertainty has been computed by comparing the results of the $\langle E/p \rangle$ measurement obtained using topoclusters to those obtained by repeating the measurement using all calorimeter cells in a cone of size $\Delta R < 0.2$. The double ratio of data/MC cluster response to data/MC cell response is used to estimate the uncertainty. The relevant result for the central calorimeter region is presented in Figure 6, showing discrepancies of $\sim 5\%$ at low p that disappear within the statistical uncertainties for $p > 10$ GeV.

Statistically consistent results were found for the E/p data to Monte Carlo ratio for calorimeter energy measured in $\Delta R < 0.3$. Hence no additional uncertainty is assumed due to the out-of-cone $\Delta R > 0.2$ energy deposits.

The E/p measurements only cover the response of charged hadronic particles with momenta less than ~ 20 GeV. However, depending on the jet momentum, on average between 35% and 90% of the energy in jets is carried by particles that are not measured in situ using the isolated track analysis (mostly photons from π^0 decays, neutral hadrons and high momentum charged hadrons). Hence, the uncertainty on the energy response to these particles is needed in order to obtain the total calorimeter uncertainty on the jet energy scale.

In 2004, an ATLAS Combined Test Beam (CTB) program was carried out at CERN. A "slice" of the ATLAS detector composed of the final versions of all sub-detectors in the barrel region was exposed to Super Proton Synchrotron (SPS) test beams. The layout of the sub-detectors was designed to be as close to that of ATLAS as possible. The setup was used to measure the combined calorimeter response to single charged pions of energies between 20 and 350 GeV for pseudorapidity values of 0.20, 0.25, 0.35, 0.45, 0.55 and 0.65 [10, 11, 12]. From the measurements in Ref. [10], the ratio of data to MC simulation predictions is used to supplement the in-situ E/p measurements with a larger energy range.

For single particle momenta above 400 GeV no direct measurements in a test beam or in situ exist. Therefore an additional uncertainty of 10% is added in quadrature to that of the 350 GeV

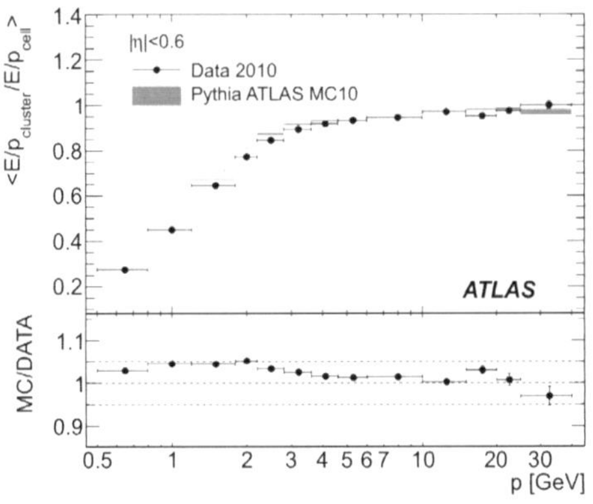

(a)

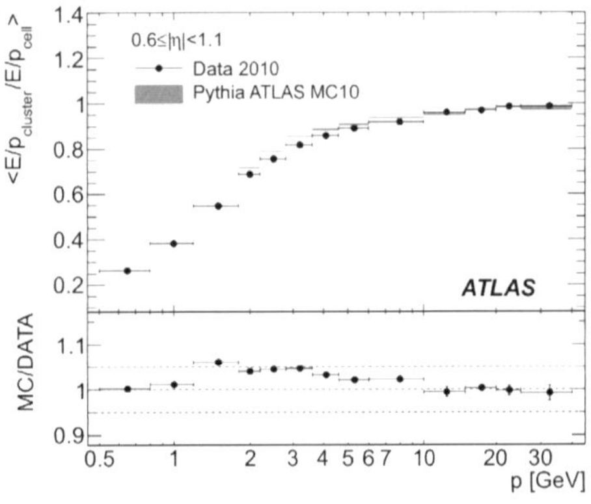

(b)

Figure 6: Ratio of the $\langle E/p\rangle$ measurement obtained with topological clusters to that obtained using all calorimeter cells in (a) the central ($|\eta| < 0.6$) and (b) forward ($0.6 \leq |\eta| < 1.1$) regions. The inset shows the ratio of the MC prediction to data [9].

measurement uncertainty in order to cover possible effects from calorimeter non-linearities at high energy densities and longitudinal leakage [13].

The absolute electromagnetic energy scale in ATLAS has been established using $Z \to ee$ decays for the electromagnetic LAr calorimeters and using the energy loss of minimum ionising muons in the TileCal. For the bulk of the electromagnetic LAr barrel calorimeter, the uncertainty on the cell energy measurement is 1.5%, and for the LAr presampler the uncertainty is 5% [14]. For the TileCal, the scale uncertainty is 3% [15]. This uncertainty does not affect charged particles with E/p measured in situ, but needs to be considered for all other particles contributing to jets.

Test beam measurements of protons [10, 11, 12, 13] and E/p measurements for identified pions and protons have shown that the agreement between data and MC simulation for protons is similar to the data to MC agreement for charged pions. Hence no additional uncertainty for charged baryons is assumed.

No test beam measurements for neutral hadronic particles have been carried out. Moreover, the GEANT4 models have large uncertainties. On average 10–12% of the jet energy is carried by neutral hadrons, mostly K_S, K_L and neutrons. Most of the K_S decay to pions before they reach the calorimeter. Hence the E/p and CTB measurements can be used for K_S.

For neutrons and anti-neutrons GEANT4 studies comparing alternative GEANT4 hadronic physics models to the ATLAS-default hadronic physics model QGSP_BERT [5] show that the (anti-)neutron to (anti-)proton response ratio is determined at the 10% level for particle momenta below 3 GeV and at the 5% level for higher momenta. Hence the (anti-)neutron response can be related to the sufficiently well simulated (anti-)proton response with these additional uncertainties.

The individual components of the calorimeter uncertainty on the jet energy scale together with its expected shift are summarised in Figure 7 for anti-k_t [17] jets with $R = 0.6$ in the pseudorapidity range $|\eta| < 0.8$. No single component is dominant and depending on the jet momentum several components contribute at approximately the same level to the total uncertainty. The envelope of the shift and uncertainty on the calorimeter JES is taken as the

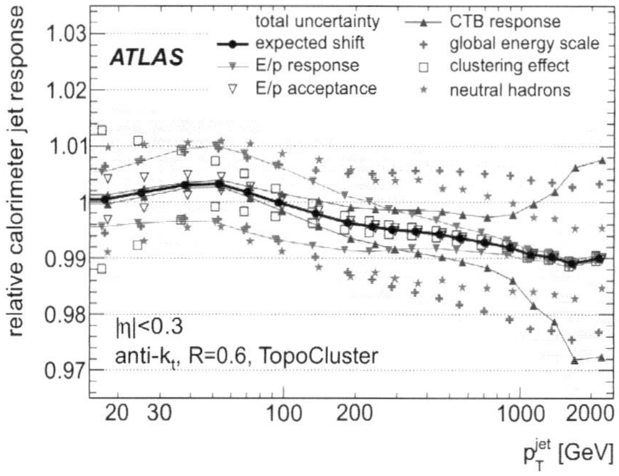

Figure 7: Expected total shift (black dots) and uncertainty (light blue band) on the relative calorimeter jet response with respect to the MC simulation for jets reconstructed with the anti-k_t jet algorithm ($R = 0.6$) in the range $|\eta| < 0.3$ as function of the jet transverse momentum [9]. The x-axis is the jet transverse momentum calibrated from the EM scale to the hadronic scale using an MC-based JES calibration factor [16].

contribution to the total JES uncertainty discussed in Ref. [16].

6. Conclusions

The average calorimeter response to isolated hadrons with respect to the track momentum $\langle E/p \rangle$ has been measured in minimum bias events $\sqrt{s} = 7$ TeV. A background from neutral hadrons of 4–8% is subtracted from data and Monte Carlo with a systematic uncertainty of below 1%. After the background has been removed, the agreement between data and Monte Carlo simulation is within 2% for particles with momenta up to 10 GeV and is around 5% for momenta in the 10–30 GeV range, where the statistical uncertainty dominates the total uncertainty.

The calorimeter response of identified single charged hadrons has been measured using short-lived particle decays. A good agreement between data measured at $\sqrt{s} = 7$ TeV and the Monte Carlo simulation is found for charged pions and protons. However, a disagreement of up to 10% is found between the Monte Carlo simulation and data for the difference of responses to low momentum anti-pions and anti-protons ($\langle E/p \rangle^{\pi^-} - \langle E/p \rangle^{\bar{p}}$). This difference is attributed to the poor modelling of the anti-proton response in the Monte Carlo simulation.

The ATLAS calorimeter jet energy scale uncertainty has been determined for the well understood central detector region by propagating the energy response uncertainty of all particles contributing to a jet. For charged hadron momenta below 20 GeV, the single charged hadron response has been used, while for higher momenta, the response measured in the ATLAS combined test beam has been included. For the response to neutral pions ($\pi^0 \rightarrow \gamma\gamma$), the uncertainty on the electromagnetic calorimeter energy scale is dominant, while for all other neutral hadrons an additional uncertainty due to the limited knowledge of the calorimeter response has been incorporated.

References

[1] Sjostrand T *et al.* 2006 *JHEP05* **026** (*Preprint* hep-ph/0603175)
[2] ATLAS Collaboration 2011 *New J.Phys.* **13** 053033 (*Preprint* 1012.5104)
[3] ATLAS Collaboration 2010 *Eur. Phys. J. C* **70** 823–874 (*Preprint* 1005.4568)
[4] Agostinelli S *et al.* 2003 *Nucl. Instr. and Meth. A* **506** 250 – 303 ISSN 0168-9002
[5] Ribon A *et al.* 2010 *CERN-LCGAPP-2010-01*
[6] ATLAS Collaboration 2008 *JINST* **3 S08003**
[7] ATLAS Collaboration 2008 *CERN-OPEN-2008-020* **1** 268
[8] ATLAS Collaboration 2011 *ATLAS-CONF-2011-028* **1** 17
[9] ATLAS Collaboration 2012 Submitted to Eur. Phys. J. C (*Preprint* 1203.1302)
[10] Abat E *et al.* (ATLAS) 2010 *Nucl. Instrum. Meth.* **A621** 134–150
[11] Cojocaru C *et al.* (ATLAS Liquid Argon EMEC/HEC) 2004 *Nucl. Instrum. Meth.* **A531** 481–514
[12] Dowler B *et al.* (ATLAS Liquid Argon HEC) 2002 *Nucl. Instrum. Meth.* **A482** 94–124
[13] Adragna P *et al.* 2010 *Nucl. Instrum. Meth.* **A615** 158–181
[14] ATLAS Collaboration 2011 Submitted to Eur. Phys. J. C (*Preprint* 1110.3174)
[15] ATLAS Collaboration 2010 *Eur. Phys. J. C* **70** 1193–1236 (*Preprint* 1007.5423)
[16] ATLAS Collaboration 2011 *CERN-PH-EP-2011-191* Submitted to Eur. Phys. J. C (*Preprint* 1112.6426)
[17] Cacciari M, Salam G P and Soyez G 2008 *JHEP* **04** 063 (*Preprint* 0802.1189)

XVth International Conference on Calorimetry in High Energy Physics (CALOR2012) IOP Publishing
Journal of Physics: Conference Series **404** (2012) 012009 doi:10.1088/1742-6596/404/1/012009

Calorimeter Performance for Tau Reconstruction and Identification at ATLAS

Matteo Volpi on behalf of the ATLAS Collaboration
The university of Melbourne, Melbourne, Australia
E-mail: matteo.volpi@unimelb.edu.au

Abstract. The ATLAS physics program involving tau final states ranges from Standard Model measurements involving W, Z, and top pair production, to searches for the Higgs, Supersymmetry and other beyond the Standard Model signatures. The ATLAS calorimeter plays a crucial role in the reconstruction and identification of hadronically decaying tau leptons in ATLAS. This proceeding discusses the role of the calorimeter in reconstructing the tau energy, as well as methods to measure the systematic uncertainties on the tau energy scale. The calorimeter is further a key component in building identification variables used to differentiate tau candidates from hadronic jets. The role of these variables is presented, and the use of variables in high luminosity environments is also discussed. A brief overview of the impact of the tau energy scale and identification efficiency uncertainties in searches for new physics with tau-based signatures is also given.

1. Introduction

Tau leptons play a significant role in the LHC [1] physics program. For example, it is an important signature for the Higgs boson. Tau leptons decay hadronically 64.7% of the time, predominately to one or three charged pions, a neutrino and often additional neutral pions. Hadronically decaying tau, are categorised by the number of charged decay products observed or prongs. Hadronic 1-prong decays are the most common (BR = 49.5%) followed by 3-prong decays (BR = 15.2%) [2].

The challenge in identifying hadronic tau decays at the LHC is that the cross section for multijet QCD production, which can be falsely identified as hadronic tau decays, is many orders of magnitude above the cross sections for weak interaction processes involving tau leptons. The most discriminating features for identifying taus among this background are the characteristic 1 or 3-prong hadronic tau signature, consequently low track multiplicity, and relatively narrow clustering of tracks and depositions in the calorimeters. Calorimeter and tracking information are combined into hadronic tau decay identification variables capturing these distinct features.

2. Reconstruction

Hadronic tau candidates [3] are seeded by jets with a transverse energy at the electromagnetic scale higher than 10 GeV. Tracks in the angular vicinity of the jet barycenter (ΔR [1] =0.2) are associated with the hadronic tau candidate. The number of tracks within $\Delta R < 0.2$ is used

[1] $\Delta R = \sqrt{(\Delta\eta)^2 + (\Delta\phi)^2}$ where the pseudorapidity, η, is defined as $\eta = -\ln(\tan\frac{\theta}{2})$ and the cylindrical coordinates ϕ and θ are defined in the usual way.

to classify the hadronic tau candidate into 1 or 3-prong categories. The collected tracks and calorimeter clusters from the jet are then used to calculate a number of identification variables. Variables based on calorimeter information are calculated from calorimeter cells in $\Delta R < 0.4$ around the hadronic tau candidate axis.

3. Identification

The track and calorimeter variables are combined into multivariate discriminants to reject misidentified jets and electrons. These algorithms take as inputs various quantities built from the calorimeters or the inner detector. While the tracking variables are less sensitive to pile-up[2] due to vertex association, the calorimeter variables have higher sensitivity to the activity from pile-up interactions. To reduce the impact of the pile-up in calorimeter quantities, the hadronic tau cone is choosen to be as small as possible ($\Delta R < 0.4$). An example variable used for jet-tau discrimination is R_{cal}, the distribution of which is shown in Fig. 1. The fraction of transverse energy of the hadronic tau candidate deposited in the electromagnetic calorimeter is an example of a variable used to reject electrons (Fig 2).

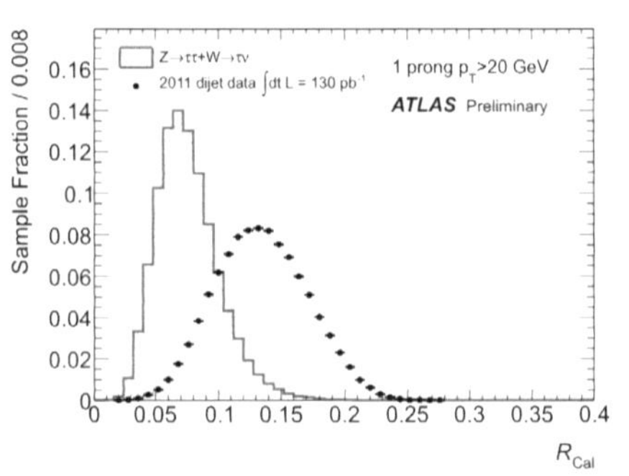

Figure 1. Energy weighted shower width in the calorimeter, R_{cal}, for hadronic tau signal Monte Carlo (red) and compared to QCD di-jet data (black) [3].

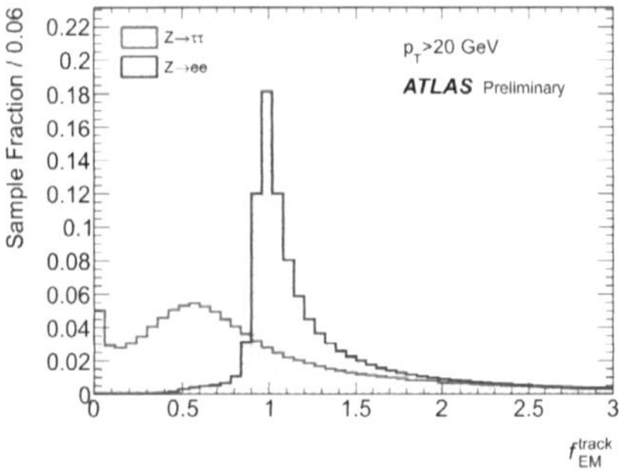

Figure 2. The ratio of the transverse energy deposited in the electromagnetic calorimeter over the transverse momentum of the leading track for signal ($Z \to \tau\tau$ sample) and background ($Z \to ee$ sample) [3].

There are three independent methods for tau identification in ATLAS [4]: a cut based approach, placing cuts on variables, a projective likelihood method, using the log likelihood ratio of signal and background, and boosted decision trees (BDT), to find the optimal separation in a multi dimensional phase space. The methods have a different sets of identification variables and are separately trained for 1 and 3-prong hadronic tau candidates. Three dedicated working points with signal efficiencies with target efficiencies of approximately 60%, 45% and 30% (loose, medium, tight) are provided for all hadronic tau vs QCD jets identification. For the training of the identification algorithms, the QCD background is obtained from data, while the hadronic tau decay signal is simulated in $W \to \tau\nu$ and $Z \to \tau\tau$ Monte Carlo samples. The inverse background efficiency versus signal efficiency for all three methods is shown for 1-prong in Fig 3. Electrons can also be misidentified as a hadronic tau due to the signature in the detector. They will be reconstructed mostly as a 1-prong hadronic tau decays.

[2] Pile-up in the following refers to the contribution of additional pp collisions superimposed on the hard physics process.

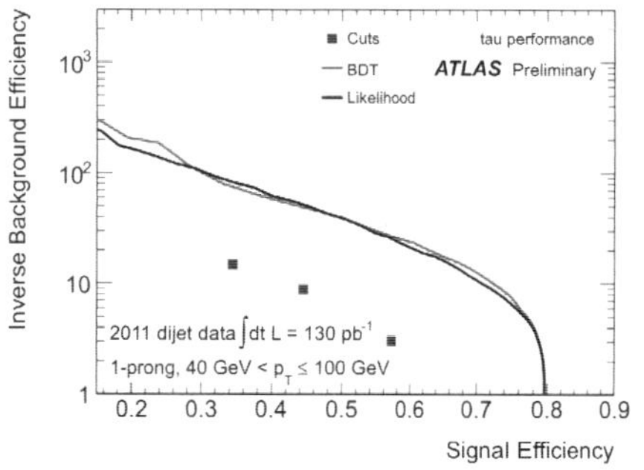

Figure 3. Background rejection power with respect to QCD jets is shown as a function of signal efficiency for each discriminant for 1-prong hadronic tau in the region $20\ GeV < p_T < 40\ GeV$ [3].

Selection on the output of the discriminants is used to select a sample of hadronic tau candidates with the desired level of background rejection or signal efficiency.

4. Identification efficiency measurements in data

The performance and systematic uncertainties of the hadronic tau identification methods are evaluated on data using two different signal channels [3]. The first method uses $Z \rightarrow \tau\tau$ events in 800 pb^{-1} of ATLAS data and relies on a tag-and-probe method [5]. Events are tagged with a muon from a tau decay, and the other tau lepton in the event is required to decay hadronically, forming the probe that is used to measure the identification efficiency. The background is dominated by QCD multijet events estimated using a data driven method. The other backgrounds are estimated through simulation. The signal efficiency is derived by counting events before and after the hadronic tau identification is applied. The visible mass of the muon and the hadronic tau is shown for data before (Fig 4) and after (Fig 5) applying the tight BDT tau identification and agrees well with Monte Carlo predictions. The second method to measure the hadronic tau identification efficiency uses $W \rightarrow \tau\nu$ events collected in 1.37 fb^{-1} of ATLAS data [3]. The measured efficiencies in both methods are in good agreement with Monte Carlo predictions within 5% (8 - 12%) for the $W \rightarrow \tau\nu$ ($Z \rightarrow \tau\tau$) method.

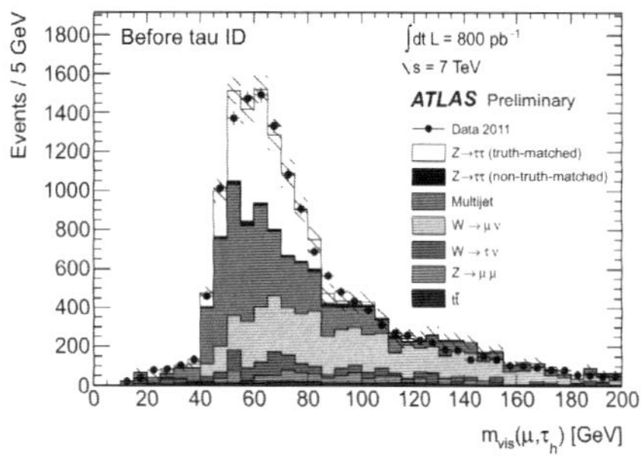

Figure 4. Visible mass distribution after full event selection without any identification requirement on the hadronic tau [3].

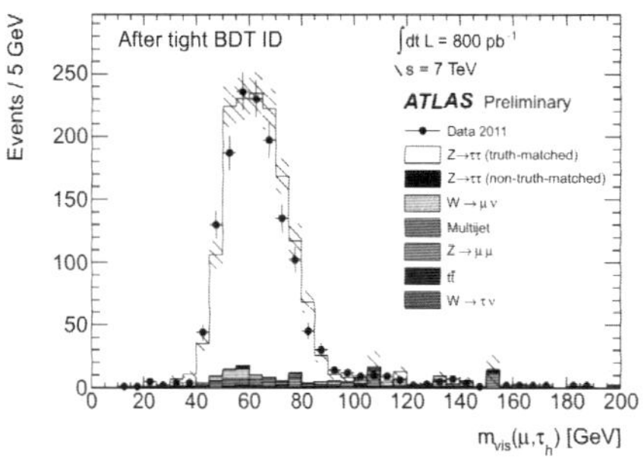

Figure 5. Visible mass distribution after full event selection and requiring the hadronic tau to pass tight BDT identification [3].

5. Energy calibration

The energy of the visible hadronic tau decay products[3] is calculated using all calorimeter clusters calibrated at the Local Hadron Calibration (LC) [6], within a core of $\Delta R < 0.2$ around the 4-vector sum of clusters associated with the jet seed. LC takes into account the difference of the response of the ATLAS calorimeters to hadronic and electromagnetic energy depositions due to non-compensation of the ATLAS calorimeters.

The LC improves the hadronic tau decay energy resolution (Fig. 6) but does not fully restore the hadronic tau energy scale, due to particles that do not reach the calorimeters and imperfections within the LC itself.

An additional correction is derived from simulation of various physics processes with hadronic taus and applied on top of LC calibration. Calibration factors [7] are derived from response functions, which are the ratio of reconstructed tau energy to true visible tau energy. Response functions (Fig. 7) depend on the reconstructed tau energy at LC scale, and are calculated separately for 1 and 3-prong hadronic tau, as well as for different detector regions.

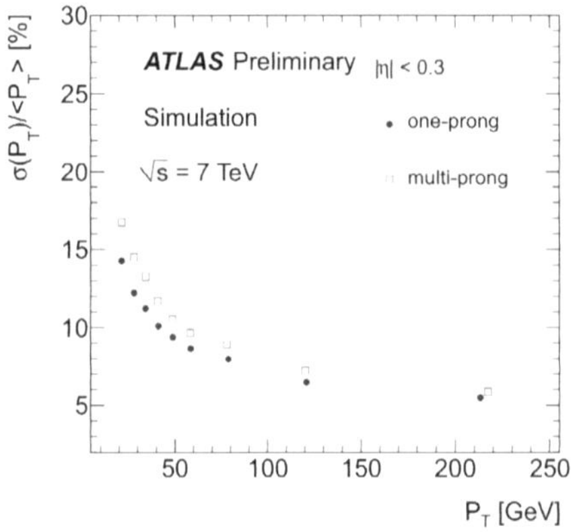

Figure 6. The hadronic tau transverse momentum resolution is calculated as the difference between the reconstructed and generated P_T. The resolution is obtained from a Gaussian fit by dividing the σ of the Gaussian by the mean value of reconstructed P_T [7].

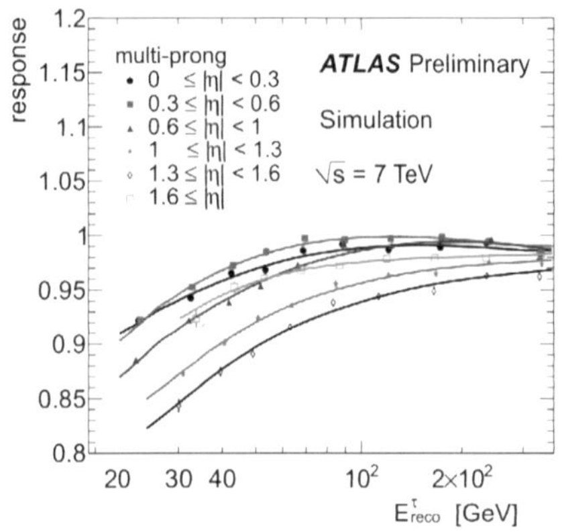

Figure 7. Response curves as a function of the reconstructed tau energy at LC scale for 3-prong tau decays for different pseudorapidity points. Uncertainties are statistical only [7].

The systematic uncertainty on the tau energy scale (TES) [7] is derived from a combination of simulation and data.

Single particle response measurements can be used to determine the calorimeter response uncertainty by decomposing the reconstructed hadronic tau decay into its decay products and convolving the constituents' response with the hadronic tau particle composition.

We use the simulation to determine the composition of the decay products of the reconstructed tau, namely charged and neutral pions, then we fluctuate them according to measured responses in data.

The uncertainties on particle responses in the calorimeter is derived by varying the response of single particles within uncertainties measured in-situ with isolated single hadrons ($< E/P >$[4])

[3] All tau decay daughters, except neutrinos.

[4] The mean energy response for low momentum charged hadrons is measured by comparing their calorimeter energy (E) depositions to the momentum (P) measured by the inner detector ($< E/P >$).

and combined test beam data. Unfortunately the combined test beam data is not available besides $|\eta| > 0.8$. Then the systematic uncertainty on the calorimeter response is given by a mixture of in-situ $< E/P >$ and combined test beam measurements in the central region and a combination of in-situ $< E/P >$ measurements and Monte Carlo simulation in the other detector regions.

The comparison of the mean energy response $< E/P >$ between data and Monte Carlo is used to determine a relative response bias in the simulation. The $< E/P >$ analysis uses data collected in 2011 and follows the analysis method described in [8] [9]. Fig. 8 shows the E/P distribution for particles at low momentum. Fig. 9 shows the background-subtracted $< E/P >$ for isolated tracks with at least one associated cluster. This figure shows also that the Monte Carlo simulation describes the calorimeter energy response well, with deviations up to 5% at low momentum.

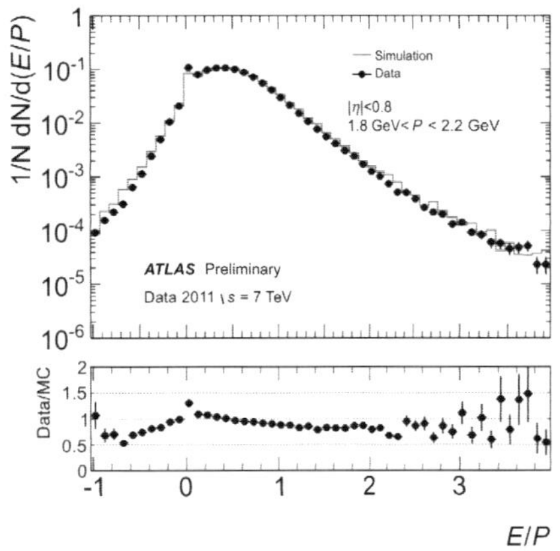

Figure 8. E/P distribution for isolated tracks with momentum $1.8 < P < 2.2\ GeV$.The bin at 0 in the left plot contains tracks which have no associated cluster [7].

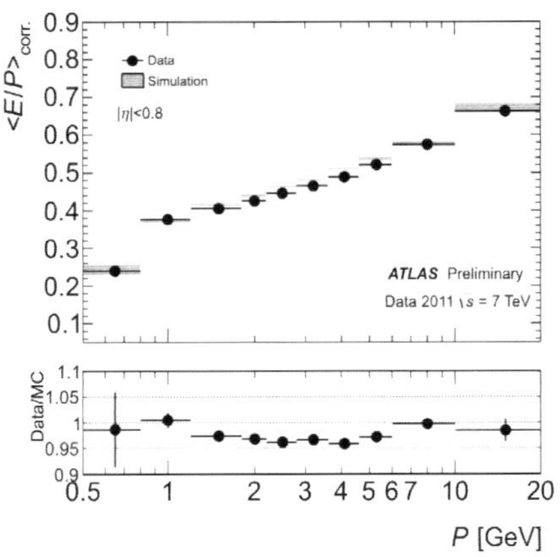

Figure 9. The background subtracted $< E/P >$ distribution as a function of the track momentum [7].

Additional uncertainties on the hadronic tau energy scale are fully derived from Monte Carlo and they are evaluated from five distinct sources: knowledge of the dead material in front of the calorimeters, underlying event model, shower shape modeling, non-closure of the calibration method and pile-up. The uncertainty due to the dead material, underlying event model, shower shape modeling are obtained by comparing the TES of the nominal sample with a dataset generated using a dedicated Monte Carlo with these various systematic effect varied. Uncertainties related to the numerical inversion used in determining the calibration constants are referred to as non-closure uncertainties. These account for deviations in the reconstructed hadronic tau candidates kinematics from the true kinematics. The pile-up uncertainty is estimated by dividing up the nominal samples into seven bins according to the average number of primary vertices. The difference in the highest and lowest closure [5] is taken as the pile-up uncertainty. The final systematic uncertainty (Fig. 10) is found to be less than 3% for 1-prong hadronic tau decay central region and slightly larger than 3% for $1.3 < |\eta| < 1.6$. The final

[5] We define the closure as the difference between the reconstructed transverse momentum and the true visible transverse momentum divided by the true visible transverse momentum.

systematic uncertainty for 3-prong, is on average 3%, and up to 5% at low momentum for $1.3 < |\eta| < 1.6$.

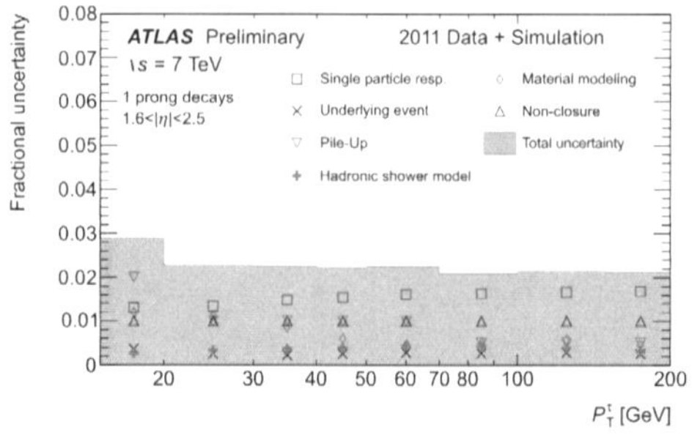

Figure 10. TES uncertainty for 1-prong decays in the forward region. The individual contributions are shown as points and the combined uncertainty is shown as a filled band [7].

5.1. *In-situ TES cross check systematic*

The reconstructed visible mass peak of Z decays can be used to measure both the TES and its uncertainty in-situ. However, with the luminosity available in 2011, there are not enough data for a competitive data driven TES determination. Instead, this method is used to verifying that the TES systematics in the region where no combined test beam data are available covered any possible uncertainty.

6. Summary

ATLAS has an important physics program with tau lepton final states, and a well performing tau identification is a essential part of these analyses. Different techniques are used to separate tau leptons from the quark and gluon initiated jet background. The corresponding efficiencies and systematic uncertainties of the tau identification methods have been studied using Standard Model processes. The energy scale of tau leptons decaying to hadrons is also well understood, with an average uncertainty of 3% calculated on data driven and Monte Carlo driven methods.

References

[1] L. Evans and P. Bryant, JINST 3, (2008) S08001
[2] J. Beringer et al. (Particle Data Group), Phys. Rev. D 86, 010001 (2012).
[3] ATLAS Collaboration, *Performance of the Reconstruction and Identification of Hadronic Tau Decays with ATLAS*, ATLAS-CONF-2011-152, https://cdsweb.cern.ch/record/1398195.
[4] ATLAS Collaboration, 2008 JINST 3 S08003.
[5] ATLAS Collaboration, *Measurement of the $Z \to \tau\tau$ cross section with the ATLAS detector*, Phys. Rev. D 84, (2011) 112006.
[6] ATLAS Collaboration,*Local Hadronic Calibration*, ATL-LARG-PUB-2009-001-2, https://cdsweb.cern.ch/record/1112035.
[7] ATLAS Collaboration,*Determination of the tau energy scale and the associated systematic uncertainty in proton-proton collisions at sqrt(s) = 7 TeV with the ATLAS detector at the LHC in 2011*, ATLAS-CONF-2012-054, https://cdsweb.cern.ch/record/1453781.
[8] ATLAS Collaboration, *Response of the ATLAS calorimeters to single isolated hadrons produced in proton proton collisions at a center of mass energy of sqrts = 900 GeV*, ATLAS-CONF-2010-017.
[9] ATLAS Collaboration, *ATLAS Calorimeter Response to Single Isolated Hadrons and Estimation of the Calorimeter Jet Scale Uncertainty*, ATLAS-CONF-2011-028.

Missing E_T Reconstruction with the CMS Detector

Tai Sakuma for the CMS collaboration
Mitchell Institute for Fundamental Physics and Astronomy
Department of Physics & Astronomy, Texas A&M University, College Station, TX 77843, USA
E-mail: sakuma@fnal.gov

Abstract. The CMS experiment uses missing E_T to both measure processes in the Standard Model and test models of physics beyond the Standard Model. These proceedings show the performance of the missing E_T reconstruction evaluated by using 4.6 fb^{-1} of proton-proton collision data at the center-of-mass energy $\sqrt{s} = 7$ TeV collected in 2011 with the CMS detector at the Large Hadron Collider. Missing E_T was reconstructed based on a particle-flow technique. Jet energy corrections were propagated to missing E_T. After anomalous signals and events were addressed, the missing E_T spectrum was well reproduced by MC simulation. The multiple proton-proton interactions in a single bunch crossing, pile-up events, degraded the performance of the missing E_T reconstruction. Mitigations of this degradation have been developed.

1. Introduction

In the CMS experiment, *missing* E_T (MET) in proton-proton collisions is reconstructed and used in a wide range of physics analyses. MET is the imbalance in the transverse momentum of all *visible particles*, particles which interact with the *electromagnetic* or *strong forces*, in the final state of proton-proton collisions. Because momentum is conserved in each direction, MET is the transverse momentum that must have been carried by something *invisible*. *Neutrinos*, for example, are invisible particles; therefore, MET is an estimate of transverse momentum of neutrinos. We use MET in measurements of *W bosons*, *top quarks*, and *tau leptons* as these particles can decay into neutrinos. Further, many models of *physics beyond the Standard Model* predict the existence of particles or something else which are invisible and can carry momentum; e.g., *Dark Matter* models, *supersymmetric* models, *unparticle* models, and models with *large extra dimensions*. For this reason, we use MET to test such models.

Accurate reconstruction of MET is demanding because it entails reconstruction of all visible particles in an event with precision. This requires a *hermetic* detector which can detect all particles which electromagnetically or strongly interact with matter. The CMS detector, located at one of two *high luminosity interaction points* of the *Large Hadron Collider* (LHC), meets this requirement. The subsystems of the CMS detector include *highly granular electromagnetic calorimeters*, *hermetic hadronic calorimeters*, *redundant muon systems*, and *all silicon trackers* in a *strong magnetic field*. A thorough description of the CMS detector can be found in Ref. [1].

Based on 36 pb^{-1} of data collected at the center-of-mass energy $\sqrt{s} = 7$ TeV in 2010, the CMS collaboration published the results of comprehensive studies of the MET reconstruction performance [2]. In 2011, the CMS detector collected a considerably larger amount of data. These proceedings summarize preliminary results of the MET reconstruction performance study based on the data collected in 2011 [3, 4].

Published under licence by IOP Publishing Ltd

With the increase in the LHC luminosity, the number of multiple proton-proton interactions in the same *bunch crossing* (*in-time pile-up*) and overlapping detector signal from previous or following bunch crossings (*out-of-time pile-up*) significantly increased. The pile-up events worsen the performance of the MET reconstruction. We have developed several techniques to mitigate the effect of pile-up events.

2. The 7 TeV proton-proton collision run in 2011

These proceedings use 4.6 fb^{-1} of proton-proton collision data collected with the CMS detector at the center-of-mass energy $\sqrt{s} = 7$ TeV in 2011. We divided the data collection period into two *eras*: Run 2011A and Run 2011B. For proton-proton collisions, Run 2011A started in March 2011 and ended in August 2011; Run 2011B started in September 2011 and ended in October 2011. Of the 4.6 fb^{-1} of data, 2.0 fb^{-1} was collected in Run 2011A and 2.6 fb^{-1} was collected in Run 2011B. Since the LHC luminosity rapidly increased over the year, Run 2011B was shorter than Run 2011A but more data were collected. In addition, the number of pile-up events also increased. The average number of the reconstructed *vertices* in a single bunch crossing, which indicates the average number of in-time pile-up events, increased from 5.5 in Run 2011A to 9.2 in Run 2011B. The distributions of the number of the reconstructed vertices in a single bunch crossing are shown in Figure 1.

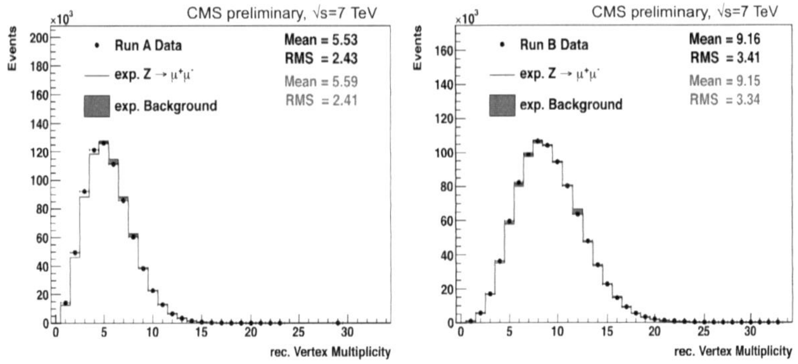

Figure 1. The distributions of the number of reconstructed vertices in a single bunch crossing in Run 2011A (left) and Run 2011B (right). The events contained Z bosons decaying into dimuons. The detail of the reconstruction and selection of events can be found in Ref. [4].

3. MET reconstruction algorithms and corrections

MET used in these proceedings is based on a *particle-flow algorithm* and includes the *Type-I* MET *correction*, a propagation of *jet energy corrections*. This MET, both with and without the correction, is called *particle-flow* MET (pfMET) in Ref. [2]. Ref. [2] contains more detailed description of pfMET and the Type-I MET correction as well as other definitions of MET and corrections to MET.

First, *raw* MET was defined as the imbalance in the *transverse energy*[1] of all particles in the event reconstructed by a particle-flow algorithm. The detail of the particle-flow algorithm at CMS can be found in Ref. [5]. In short, the particle-flow algorithm uses information from all CMS detector subsystems, i.e., trackers, calorimeters, and muon systems; then, it reconstructs

[1] We will use the transverse momentum instead of the transverse energy in the near future.

Journal of Physics: Conference Series **404** (2012) 012010 doi:10.1088/1742-6596/404/1/012010

four momenta of all visible particles, each of which is identified as one of five particle types, i.e., *electrons*, *photons*, *muons*, *charged hadrons*, and *neutral hadrons*.

The raw MET is systematically different from *true* MET, i.e., the transverse momentum carried by invisible particles, for many reasons including the non-compensating nature of the calorimeters. To make MET a better estimate of true MET, the Type-I MET correction was applied. The Type-I MET correction replaces the transverse energy of particles which can be clustered as jets with the transverse momentum of the jets to which jet energy corrections are applied. The detail of jet energy corrections at CMS can be found in Ref. [6].

4. False MET and event cleaning

Large MET is caused not only by interesting physics processes in proton-proton collisions such as production of invisible particles. In fact, large MET has more often uninteresting causes such as detector noise, *cosmic rays*, and *beam-halo* particles. MET with uninteresting causes is called *false* MET or *anomalous* MET. For an accurate reconstruction of MET, it is, therefore, not sufficient to reconstruct all visible particles produced in proton-proton collisions.

We developed several algorithms to identify false MET. These algorithms, for example, use timing, pulse shape, and topology of signal. After the identified false MET was removed, the agreement of the MET spectrum with simulation, in which causes of false MET were not explicitly simulated, significantly improved (Figure 2).

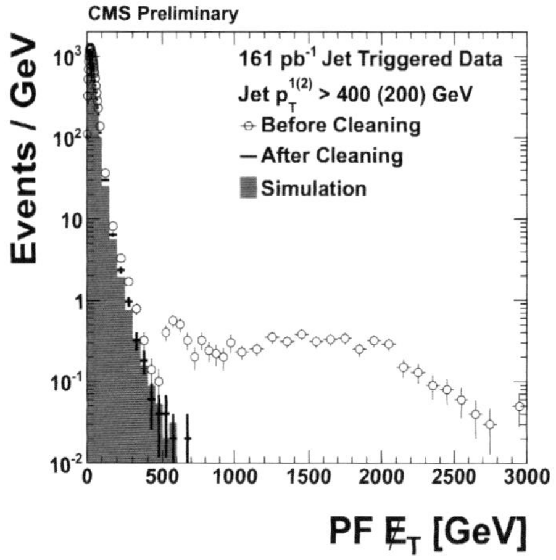

Figure 2. MET spectrum in dijet events collected in early 2011 before and after the cleaning. The highest bin includes the overflow. The simulation includes QCD events, top pair production events, W + jets production events, and Z + jets production events. The simulation does not explicitly include sources of false MET such as cosmic rays or beam-halo particles. The detail of the event selection and cleaning can be found in Ref. [3].

5. Performance of MET reconstruction

MET *response* and *resolution* are measures of MET reconstruction performance. We evaluated them by artificially inducing MET in events very likely to have only little or no true MET. We used events with *vector bosons*, either *Z bosons* decaying into *dimuons* or photons. These vector bosons are predominantly produced in interactions with no true MET, such as $qg \to q\gamma$, $q\bar{q} \to Z$, $qg \to qZ$, $q\bar{q} \to gZ$. Therefore, an event is primarily composed of a vector boson and its *hadronic recoil*. The dimuons and photons were measured with precision by, respectively, the trackers and muon systems, and the electromagnetic calorimeters. To induce MET for the performance evaluation, we excluded the dimuons or photons from the MET reconstruction; the dimuons or photons played the role of invisible particles. Then, the MET performance can be indicated by how close reconstructed MET is to the transverse momentum of the excluded dimuons or photons.

The transverse momentum of the vector boson is $\vec{q}_T$. The vector sum of the transverse energy of all reconstructed particles except the vector boson, i.e. the hadronic recoil, is $\vec{u}_T$. Thus, the induced MET is $\vec{\not{E}}_T = -\vec{u}_T$. The projection of $\vec{u}_T$ on $\vec{q}_T$ is $u_\parallel$, which is typically negative. The component of $\vec{u}_T$ perpendicular to $\vec{q}_T$ is $u_\perp$. These kinematic variables are illustrated in Figure 3.

MET response is defined as $-u_\parallel/|\vec{q}_T|$, the ratio of the MET component parallel to the transverse momentum of the vector boson and the magnitude of the transverse momentum of the vector boson. Figure 4 shows the MET response in events with Z bosons decaying into dimuons. The response is close to unity at above certain $|\vec{q}_T|$ and has little dependence on the number of pile-up events.

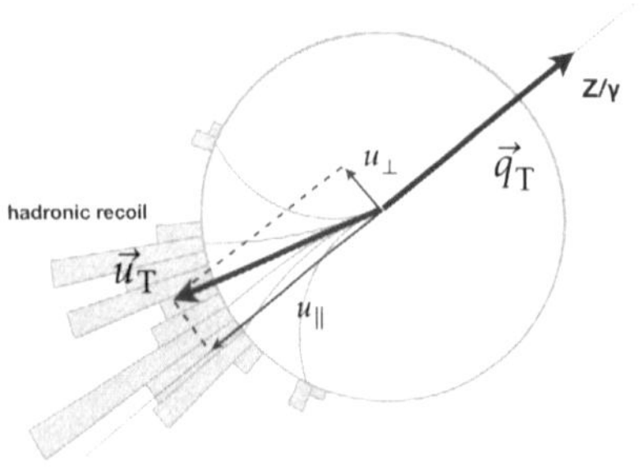

Figure 3. The transverse momentum of the vector boson and the transverse energy of its hadronic recoil on the plane perpendicular to the beam axis.

MET resolutions are defined as the RMS of $u_\parallel$ and $u_\perp$ about their mean values, denoted, respectively, as RMS($u_\parallel$) and RMS($u_\perp$). Figure 5 shows the MET resolutions in Z events as a function of the number of reconstructed vertices for data collected in Run2011A and Run2011B. Figure 6 shows the MET resolutions as functions of $|\vec{q}_T|$ in photon events for four different

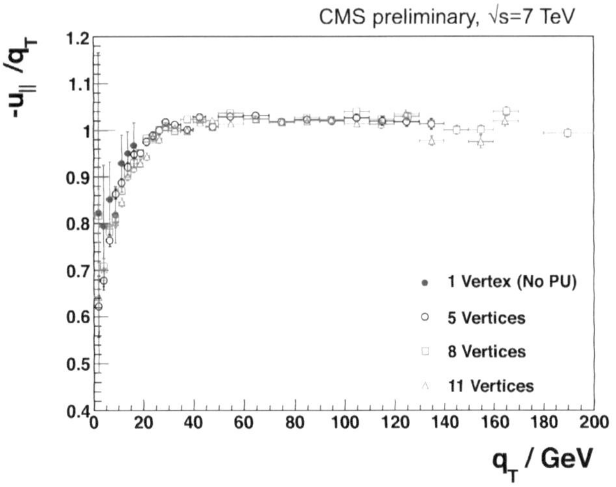

Figure 4. MET response ($-u_\parallel/|\vec{q}_T|$) as a function of $|\vec{q}_T|$ in events with Z bosons decaying into dimuons. Z bosons with transverse momenta $\vec{q}_T$ were excluded in MET reconstruction. The projection of $\vec{\not{E}}_T$ on to $\vec{q}_T$ equals $-u_\parallel$. Events with four different numbers of vertices are shown as different markers. The detail of the reconstruction and selection of events can be found in Ref. [4]

numbers of reconstructed vertices for data collected in early Run2011A. The resolutions degrade as the number of reconstructed vertices (in-time pile-up events) increases. The resolutions are worse in events for Run2011B than events with the same number of in-time pile-up events for Run2011A because the out-of-time pile-up increased.

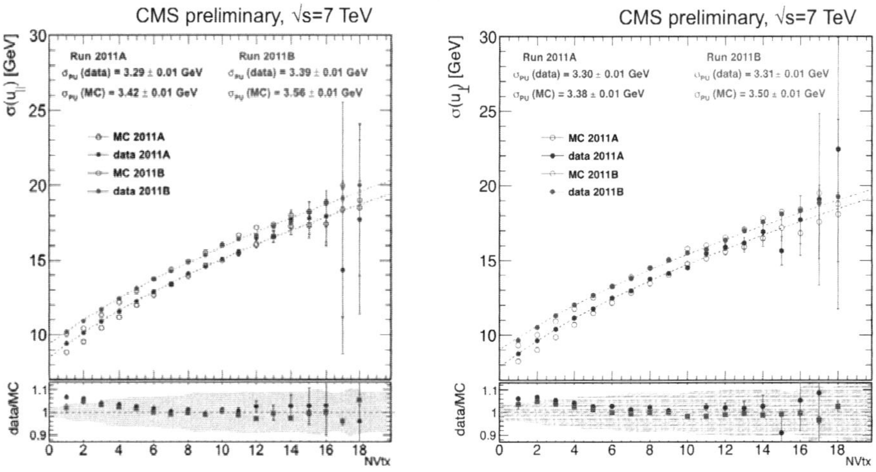

Figure 5. MET resolutions RMS($u_{\parallel}$) (left) and RMS($u_{\perp}$) (right) as functions of the number of reconstructed vertices (N_{vtx}) in events with Z bosons decaying into dimuons. MET resolutions are separately shown for data collected in Run2011A and Run2011B and for MC simulation for each era. The detail of the reconstruction and selection of events can be found in Ref. [4]

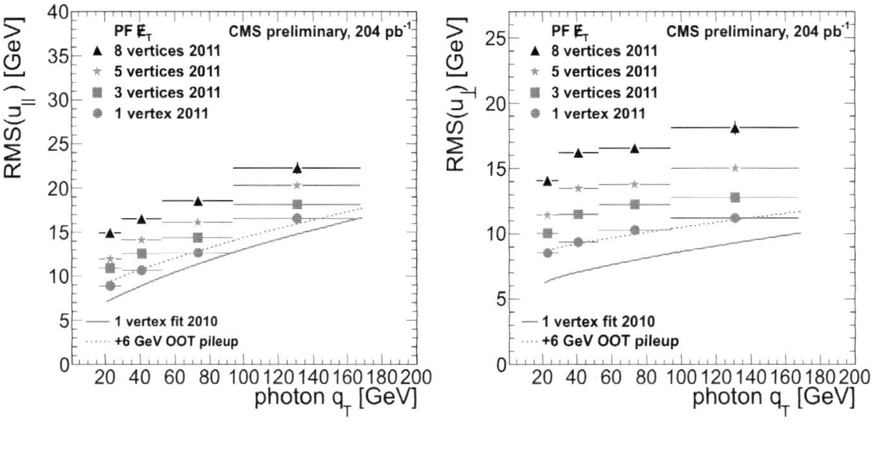

Figure 6. MET resolutions RMS($u_{\parallel}$) (left) and RMS($u_{\perp}$) (right) as functions of the photon $|\vec{q}_{\mathrm{T}}|$ in events with photons for data collected in early Run2011A, shown separately for events with four different numbers of reconstructed vertices. A curve which fit data for events with one vertex collected in 2010 is also shown. The detail of the reconstruction and selection of events can be found in Ref. [3]

6. Mitigation of MET degradation in high pile-up events

This section briefly introduces one of the techniques that CMS used to mitigate the degradation of the MET performance caused by the pile-up events.

As shown in the previous section, the MET performance becomes worse as the number of pile-up events increases. Then, when MET is used in event selections in physics analyses, the efficiency of selecting signal events decreases. In order to prevent the efficiency from decreasing, an analysis of the search for the *Higgs boson* decaying into a pair of W bosons [7] defined another kinematic variable similar to MET, which is called *TrackMET* in Ref. [4]. TrackMET is the imbalance in the transverse momentum of charged particles originating from the primary vertex of the high-p_{T} event. By its construction, it depends little on the number of pile-up events. It was shown that using a combination of TrackMET and MET improved the signal efficiency in this analysis [7].

7. Summary

MET is an estimate of the transverse momentum of neutrinos and other invisible particles or some stuff with momentum. In the CMS experiment, MET plays a central role in both precision measurements of Standard Model physics and searches for physics beyond the Standard Model.

Using 4.6 fb^{-1} of proton-proton collision data at the center-of-mass energy $\sqrt{s} = 7$ TeV collected with the CMS detector in 2011, we evaluated the performance of the reconstruction of MET, which was based on a particle-flow algorithm, and in particular how it was affected by pile-up events. After false MET was removed, the MET spectrum was well described by MC simulation. While the MET response exhibited little dependence on the number of pile-up events, the MET resolutions became worse as the number of pile-up events increased. We have developed several techniques to mitigate the effect of the pile-up events. As the LHC luminosity is expected to keep rapidly increasing, it will be important to continue to develop and refine techniques to handle a large number of pile-up events.

In 2012, the CMS detector is collecting proton-proton collision data at the center-of-mass energy $\sqrt{s} = 8$ TeV at the increasing luminosity. MET will remain an important object to be reconstructed for a variety of physics analyses at the CMS experiment.

References

[1] CMS Collaboration 2008 *JINST* **3** S08004
[2] CMS Collaboration 2011 *JINST* **6** P09001 (*Preprint* 1106.5048)
[3] CMS Collaboration 2011 *CMS Detector Performance Summary* **CMS-DP-2011-010**
[4] CMS Collaboration 2012 *CMS Detector Performance Summary* **CMS-DP-2012-003**
[5] CMS Collaboration 2009 *CMS Physics Analysis Summary* **CMS-PAS-PFT-09-001**
[6] CMS Collaboration 2011 *JINST* **6** P11002 (*Preprint* 1107.4277)
[7] CMS Collaboration 2012 *Phys.Lett.* **B710** 91–113 (*Preprint* 1202.1489)

Performance of Missing Transverse Momentum Reconstruction in ATLAS with Proton-Proton Collisions at $\sqrt{s} = 7$ TeV

Irene Vichou on behalf of the ATLAS Collaboration
University of Illinois at Urbana-Champaign
E-mail: Irene.Vichou@cern.ch

Abstract. The performance of the missing transverse momentum reconstruction in the ATLAS detector is evaluated using data collected in proton-proton collisions at a centre-of-mass energy of 7 TeV in 2010. Evaluation of the systematic uncertainty on the missing transverse momentum scale is presented. A method to mitigate the effects arising from additional proton-proton interactions superimposed on the hard physics process is also discussed for the 2011 beam conditions.

1. Introduction

In a hadron collider event the missing transverse momentum is defined as the event momentum imbalance in the plane transverse to the beam axis, where momentum conservation is expected. Such an imbalance may signal the presence of undetectable particles, such as neutrinos or new stable, weakly-interacting particles. The vector momentum imbalance in the transverse plane is obtained from the negative vector sum of the momenta of all particles detected in a proton-proton (pp) collision and is denoted as missing transverse momentum, $\boldsymbol{E}_{\mathrm{T}}^{\mathrm{miss}}$. The symbol $E_{\mathrm{T}}^{\mathrm{miss}}$ is used for its magnitude.

An important requirement on the measurement of $\boldsymbol{E}_{\mathrm{T}}^{\mathrm{miss}}$ is the minimization of the impact of limited detector coverage, finite detector resolution, the presence of dead regions and different sources of noise, as well as cosmic-ray and beam-halo muons crossing the detector, that can produce fake $E_{\mathrm{T}}^{\mathrm{miss}}$. The ATLAS calorimeter [1] coverage extends to large pseudorapidities[1] to minimize the impact of high energy particles escaping in the very forward direction.

The $\boldsymbol{E}_{\mathrm{T}}^{\mathrm{miss}}$ measurement is significantly affected by the contributions of additional pp collisions superimposed on the hard physics process, referred to as pile-up in the following, so it is important to develop methods to suppress such contributions. Such a method is described in the last section of this note. The performance, in terms of resolution, scale and tails, of an optimised reconstruction and calibration of $\boldsymbol{E}_{\mathrm{T}}^{\mathrm{miss}}$ [2] developed by the ATLAS Collaboration is described. The event samples used to assess the quality of the $\boldsymbol{E}_{\mathrm{T}}^{\mathrm{miss}}$ reconstruction are minimum bias and

[1] ATLAS uses a right-handed coordinate system with its origin at the nominal interaction point (IP) in the centre of the detector and the z-axis coinciding with the axis of the beam pipe. The x-axis points from the IP to the centre of the LHC ring, and the y axis points upward. Cylindrical coordinates (r, ϕ) are used in the transverse plane, ϕ being the azimuthal angle around the beam pipe. The pseudorapidity is defined in terms of the polar angle θ as $\eta = -\ln\tan(\theta/2)$.

QCD di-jet events and events with leptonically decaying W and Z bosons. The performance in both data and Monte Carlo simulation as well as their comparison is studied. The data sample corresponds to an integrated luminosity of $36\,\mathrm{pb}^{-1}$ from 2010 and $4.2\,\mathrm{fb}^{-1}$ from 2011. The latter is used for the study of the perfromance in pile-up conditions, where the average number of extra interactions ranges from 6 to 15.

2. Calibration and Reconstruction

The $\boldsymbol{E}_{\mathrm{T}}^{\mathrm{miss}}$ reconstruction includes contributions from energy deposits in the calorimeters and muons reconstructed in the muon spectrometer. Tracks are added to recover the contribution from low-p_{T} particles which are missed in the calorimeters. Muons reconstructed from the inner detector are used to recover muons in regions not covered by the muon spectrometer. The $\boldsymbol{E}_{\mathrm{T}}^{\mathrm{miss}}$ reconstruction uses calorimeter cells calibrated according to the reconstructed physics object to which they are associated. Calorimeter cells are associated with a reconstructed and identified high-p_{T} parent object in a specific order: electrons, photons, hadronically decaying τ-leptons, jets and finally muons. Cells not associated with any such objects (referred to as "CellOut" below) are also taken into account in the $\boldsymbol{E}_{\mathrm{T}}^{\mathrm{miss}}$ calculation [2].

Once the cells in the calorimeters are associated with objects as described above, the $\boldsymbol{E}_{\mathrm{T}}^{\mathrm{miss}}$ is calculated as follows:

$$\begin{aligned} E_{\mathrm{x(y)}}^{\mathrm{miss}} &= E_{\mathrm{x(y)}}^{\mathrm{miss},e} + E_{\mathrm{x(y)}}^{\mathrm{miss},\gamma} + E_{\mathrm{x(y)}}^{\mathrm{miss},\tau} + E_{\mathrm{x(y)}}^{\mathrm{miss,jets}} \\ &+ E_{\mathrm{x(y)}}^{\mathrm{miss,softjets}} + (E_{\mathrm{x(y)}}^{\mathrm{miss,calo},\mu}) + E_{\mathrm{x(y)}}^{\mathrm{miss,CellOut}} + E_{\mathrm{x(y)}}^{\mathrm{miss},\mu}, \end{aligned} \tag{1}$$

where each term is calculated from the negative sum of calibrated cell energies, projected onto the x and y direction, inside the corresponding objects (within $|\eta| < 4.9$) and the $E_{\mathrm{x(y)}}^{\mathrm{miss},\mu}$ is calculated from the negative sum of the momenta of muon tracks reconstructed with $|\eta| < 2.7$. Note that the $E_{\mathrm{x(y)}}^{\mathrm{miss,calo},\mu}$ term, the enegy loss by a muon in the calorimeter, is only taken into account for muons isolated with respect to jets at the calorineter. Because of the high granularity of the calorimeter, it is important to suppress noise contributions. Therefore, cells used in the $\boldsymbol{E}_{\mathrm{T}}^{\mathrm{miss}}$ sum belong to three-dimensional topological clusters, referred to as topoclusters hereafter, with the exception of electrons and photons for which a different clustering algorithm is used.

The calibration scheme used is the one yielding the best performance in data, as described in detail in Ref. [2]. Electrons are calibrated with the default electron calibration, photons with the electromagnetic scale (EM) [2] and τ-jets with the local hadronic calibration (LCW). The LCW scheme calibrates clusters according to their classification as hadron-originated or not, and corrects for energy losses in the dead material. The jets are reconstructed with the anti-k_t algorithm [3] with distance parameter $R = 0.4$. They are calibrated with the LCW scheme if $10 < p_{\mathrm{T}} < 20\,\mathrm{GeV}$ (soft jets) and with the LCW+JES scheme, where JES is the jet energy scale (from calorimeter to particle level) if $p_{\mathrm{T}} > 20\,\mathrm{GeV}$ [4]. The contribution from topoclusters not associated to high-p_{T} objects (CellOut) is calculated with LCW calibration combined with tracking information. The value of $E_{\mathrm{T}}^{\mathrm{miss}}$ is calculated as:

$$E_{\mathrm{T}}^{\mathrm{miss}} = \sqrt{(E_{\mathrm{x}}^{\mathrm{miss}})^2 + \left(E_{\mathrm{y}}^{\mathrm{miss}}\right)^2}. \tag{2}$$

The total transverse energy in the calorimeters, $\sum E_{\mathrm{T}}$, is defined as the the scalar sum of transverse energy of cells associated to the different objects where cells are calibrated according to the scheme described above for $\boldsymbol{E}_{\mathrm{T}}^{\mathrm{miss}}$.

[2] The EM scale is the basic calorimeter signal scale for the ATLAS calorimeters. It provides the correct scale for energy deposited by electromagnetic showers. It does not correct for the lower energy hadron shower response nor for energy losses in the dead material.

3. Performance and resolution

The distributions of $\boldsymbol{E}_{\mathrm{T}}^{\mathrm{miss}}$ in minimum bias, di-jet, $Z \to \ell\ell$ and $W \to \ell\nu$ events from data are compared with the expected distributions from the MC samples. In the minimum bias, di-jet events and $Z \to \ell\ell$ events, apart from a small contribution from the semi-leptonic decay of heavy-flavour hadrons in jets, no genuine $E_{\mathrm{T}}^{\mathrm{miss}}$ is expected. Thus most of the $\boldsymbol{E}_{\mathrm{T}}^{\mathrm{miss}}$ reconstructed in these events is a direct result of imperfections in the reconstruction process or in the detector response. The comparisons show very good agreement and control of tails. Figure 1 shows the very good description of the data by the MC for $W \to e\nu$ events, where the $E_{\mathrm{T}}^{\mathrm{miss}}$ is non-zero due to the neutrino.

The $E_{\mathrm{T}}^{\mathrm{miss}}$ resolution is estimated from the width of the combined distribution of E_x^{miss} and E_y^{miss} in bins of $\sum E_{\mathrm{T}}$, expressed by the fitted σ of a Gaussian distribution over twice the expected resolution, as obtained from earlier studies. Figure 2 shows the resolution from data at $\sqrt{s} = 7$ TeV for $Z \to \ell\ell$ events, minimum bias and di-jet events as a function of the total transverse energy in the event, obtained by summing the p_{T} of muons and the $\sum E_{\mathrm{T}}$ in calorimeters. The resolution of the two $\boldsymbol{E}_{\mathrm{T}}^{\mathrm{miss}}$ components is fitted with a function $\sigma = k \cdot \sqrt{\Sigma E_{\mathrm{T}}}$ and the parameter k quantifies the $\boldsymbol{E}_{\mathrm{T}}^{\mathrm{miss}}$ resolution. There is reasonable agreement in the $\boldsymbol{E}_{\mathrm{T}}^{\mathrm{miss}}$ resolution in the different physics channels, as can be seen from the fit parameters k reported in the figure. They range from 0.42 GeV$^{1/2}$ for $Z \to \ell\ell$ events to 0.51 GeV$^{1/2}$ for di-jet events and agree with the k values obtained in simulated events. For all data samples the uncertainty on the fitted k value is small. The $\boldsymbol{E}_{\mathrm{T}}^{\mathrm{miss}}$ resolution is better in $Z \to \ell\ell$ events because the lepton momenta are measured with better precision than jets.

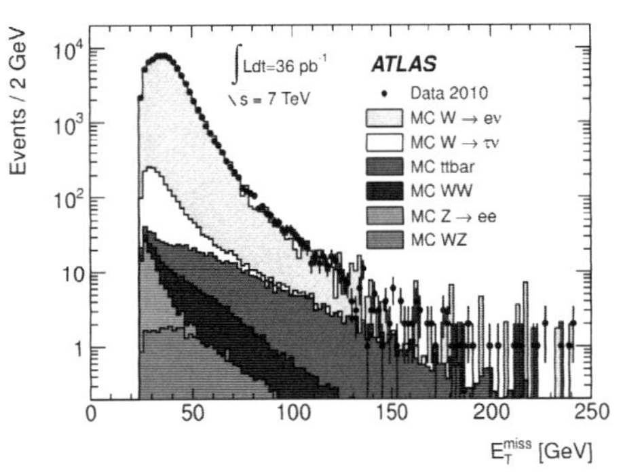

Figure 1. *$E_{\mathrm{T}}^{\mathrm{miss}}$ distribution in $W \to e\nu$ events compared to Monte Carlo simulation [2].*

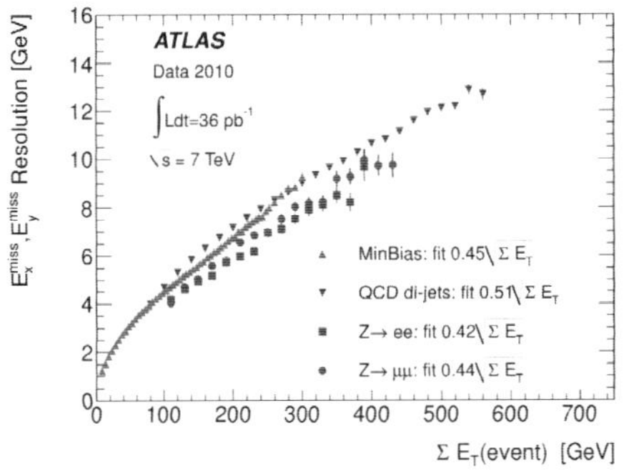

Figure 2. *E_x^{miss} and E_y^{miss} resolution as a function of the total transverse energy in the event. The fitted values of k, in GeV$^{1/2}$, (see text) are reported [2].*

4. Response and Scale

The in situ evaluation of the scale bias in events with no $\boldsymbol{E}_{\mathrm{T}}^{\mathrm{miss}}$ and the validation of the scale in events with $W \to \ell\nu$ decays is presented here.

In $Z \to \ell\ell$ events one can define an axis in the transverse plane such that the component of $\boldsymbol{E}_{\mathrm{T}}^{\mathrm{miss}}$ along this axis is sensitive to detector resolution and biases. This axis, $\boldsymbol{A_Z}$, is defined by the reconstructed momenta of the leptons as $\boldsymbol{A_Z} = (\boldsymbol{p_T}^{\ell^+} + \boldsymbol{p_T}^{\ell^-})/|\boldsymbol{p_T}^{\ell^+} + \boldsymbol{p_T}^{\ell^-}|$, where $\boldsymbol{p_T}^{\ell}$ are the vector transverse momenta of the lepton and anti-lepton and represents the direction of motion of the Z boson.

The mean value of the projection of $\boldsymbol{E}_{\mathrm{T}}^{\mathrm{miss}}$ onto the longitudinal axis, $\langle \boldsymbol{E}_{\mathrm{T}}^{\mathrm{miss}} \cdot \boldsymbol{A}_{\mathbf{Z}} \rangle$, is a measure of the $\boldsymbol{E}_{\mathrm{T}}^{\mathrm{miss}}$ scale, as this axis is sensitive to the balance between the leptons and the hadronic recoil. These mean values are used as a diagnostic to validate the $\boldsymbol{E}_{\mathrm{T}}^{\mathrm{miss}}$ reconstruction algorithms. If the leptons perfectly balanced the hadronic recoil the $\boldsymbol{E}_{\mathrm{T}}^{\mathrm{miss}} \cdot \boldsymbol{A}_{\mathbf{Z}}$ would be zero. Instead, $\langle \boldsymbol{E}_{\mathrm{T}}^{\mathrm{miss}} \cdot \boldsymbol{A}_{\mathbf{Z}} \rangle$ displays a small bias, as shown in Figure 3 as a function of p_{T}^{Z} separately for events with $Z \rightarrow \ell\ell$ produced in association with jets or not. The figure shows a negative bias in $\langle \boldsymbol{E}_{\mathrm{T}}^{\mathrm{miss}} \cdot \boldsymbol{A}_{\mathbf{Z}} \rangle$ for events with no jets, dominated by the $E_{\mathrm{T}}^{\mathrm{miss,CellOut}}$ term, which increases with p_{T}^{Z} up to 6% and confirms that the bias is due to mis-calibration of the soft hadronic recoil.

It is very important to validate in situ the $E_{\mathrm{T}}^{\mathrm{miss}}$ scale in events with genuine $E_{\mathrm{T}}^{\mathrm{miss}}$ from neutrinos. Two complementary methods to determine the absolute scale of $E_{\mathrm{T}}^{\mathrm{miss}}$ using $W \rightarrow \ell\nu$ events are exploited. The $E_{\mathrm{T}}^{\mathrm{miss}}$ range discussed here starts at $\approx 25\,\mathrm{GeV}$.

The first method uses a fit to the distribution of the transverse mass, m_{T}, of the lepton-$E_{\mathrm{T}}^{\mathrm{miss}}$ system, and is sensitive both to the scale and the resolution of $E_{\mathrm{T}}^{\mathrm{miss}}$.

The lepton transverse momentum, p_{T}^{ℓ}, and the $E_{\mathrm{T}}^{\mathrm{miss}}$ are used to calculate $m_T = \sqrt{2p_{\mathrm{T}}^{\ell} E_{\mathrm{T}}^{\mathrm{miss}}(1-\cos\phi)}$, where ϕ is the azimuthal angle between the lepton and $E_{\mathrm{T}}^{\mathrm{miss}}$ directions. Template histograms of the m_{T} distributions are generated by convolving the true transverse mass distribution with a Gaussian function $E_{x(y)}^{\mathrm{miss,smeared}} = \alpha\, E_{x(y)}^{\mathrm{miss,True}} * Gauss(0, k \cdot \sqrt{\Sigma E_{\mathrm{T}}})$, with parameters α and k being the $E_{\mathrm{T}}^{\mathrm{miss}}$ scale and resolution respectively.

These parameters are determined through a fit of the m_{T} distribution to data using a linear combination of signal and background m_{T} distributions obtained from simulation. The results are shown in Table 1, with the values of $\alpha - 1$ representing the bias. The results for the α and k using the m_{T} distribution of the simulated signal are also shown in Table 1, and they are in good agreement with the results from data. The results obtained with this method are compatible, at

Table 1. Results of m_{T} fit in $W \rightarrow \ell\nu$ events. The second and third columns show the scale and resolution parameters. The uncertainties take into account background subtraction uncertainties and correlations [2].

Channel	$\alpha - 1$ (%)	k	$\chi^2/ndof$
$W \rightarrow \mu\nu$ data	5.1 ± 0.8	0.52 ± 0.01	68/87
$W \rightarrow \mu\nu$ MC	5.5 ± 0.8	0.50 ± 0.01	70/78
$W \rightarrow e\nu$ data	-0.8 ± 1.6	0.49 ± 0.01	54/75
$W \rightarrow e\nu$ MC	1.8 ± 1.7	0.50 ± 0.01	38/54

the few percent level, with the ones from simulated W events on scale linearity and resolution (see figure 2). The uncertainty due to background subtraction is already included in the uncertainty reported in Table 1. The $\boldsymbol{E}_{\mathrm{T}}^{\mathrm{miss}}$ scale uncertainty after checking the dependence of $\alpha - 1$ when varying selections and generator, is about 1.5% and 2% for the $W \rightarrow e\nu$ and $W \rightarrow \mu\nu$ decay channels, respectively, while the bias is less than 5%

The second method uses the interdependence of the neutrino and lepton momenta in the $W \rightarrow e\nu$ channel, and the $E_{\mathrm{T}}^{\mathrm{miss}}$ scale is determined as a function of the reconstructed electron transverse momentum to determine the scale in $W \rightarrow e\nu$ events . The average bias is calculated to be $2.0 \pm 0.1 \pm 2.0\%$ in data, while the calculation on signal MC events alone results in $2.9 \pm 0.1\%$.

5. Systematic uncertainties

The $E_{\mathrm{T}}^{\mathrm{miss}}$uncertainty is evaluated from the individual uncertainties of the reconstructed objects used to build it. The overall systematic uncertainty on the $E_{\mathrm{T}}^{\mathrm{miss}}$ scale is then calculated by combining the uncertainties associated with each term. The relative impact of the uncertainty of the constituent terms on $E_{\mathrm{T}}^{\mathrm{miss}}$ depends on the final state being studied.

The uncertainty on the scale of the $E_{\mathrm{T}}^{\mathrm{miss,CellOut}}$ term, which is built from topoclusters with a correction based on tracks, can be calculated from the topocluster energy scale uncertainties. These uncertainties can be estimated from comparisons between data and MC simulation using the E/p response from single tracks, measured by summing the energies of all calorimeter clusters around a single isolated track. The effects of these uncertainties on the $E_{\mathrm{T}}^{\mathrm{miss,CellOut}}$ term can be evaluated by varying the energy scale of topoclusters that contribute to the $E_{\mathrm{T}}^{\mathrm{miss,CellOut}}$ term in $W \to e\nu$ MC samples where it is dominant.

The shift in the topocluster energy scale is applied by multiplying the topocluster energy by the function $1 \pm a \times (1 + b/p_T)$, with $a = 3(10)\%$ for $|\eta| < (>)3.2$, $b = 1.2$ GeV. The a parameter addresses the uncertainty on the cluster energy scale, obtained by comparing the ratio of the cluster energy and the measured track momentum, E/p, in data and MC simulation while the b parameter addresses the possible change in the clustering efficiency and scale in a non-isolated environment.

The value of the fractional $E_{\mathrm{T}}^{\mathrm{miss,CellOut}}$ uncertainty evaluated by shifting the energy of the topoclusters, is found to be approximately 13%, decreasing slightly with increasing $\sum E_{\mathrm{T}}{}^{\mathrm{CellOut}}$. This uncertainty is very conservative and much larger than the alternative estimation based on detector or reconstruction effects, shower modeling and generator uncertainties. Similarly, the uncertainty of the $E_{\mathrm{T}}^{\mathrm{miss,softjets}}$ term is evaluated to be about 10%.

Using as inputs the systematic uncertainties on the different reconstructed objects and on $E_{\mathrm{T}}^{\mathrm{miss,CellOut}}$ and $E_{\mathrm{T}}^{\mathrm{miss,softjets}}$, the overall $E_{\mathrm{T}}^{\mathrm{miss}}$ systematic uncertainty in $W \to e\nu$ and $W \to \mu\nu$ events is estimated.

Figure 4 shows the overall fractional $E_{\mathrm{T}}^{\mathrm{miss}}$ uncertainty together with the contributions of the individual terms uncertainties, as a function of the event $\sum E_{\mathrm{T}}$ in $W \to e\nu$ events. The uncertainties on $E_{\mathrm{T}}^{\mathrm{miss,softjets}}$ and $E_{\mathrm{T}}^{\mathrm{miss,CellOut}}$ are considered to be fully correlated. The average overall uncertainty on the $E_{\mathrm{T}}^{\mathrm{miss}}$ scale is estimated to be about 2.6% both for $W \to e\nu$ and $W \to \mu\nu$ events. The $E_{\mathrm{T}}^{\mathrm{miss}}$ scale uncertainty depends strongly on the event topology.

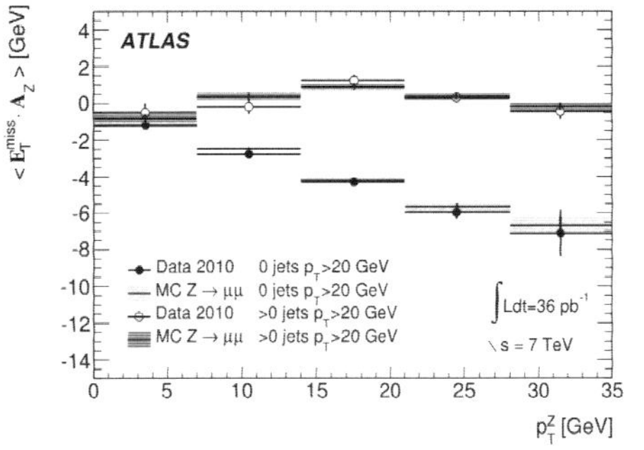

Figure 3. *Mean value of* $\boldsymbol{E}_{\mathrm{T}}^{\mathrm{miss}} \cdot \boldsymbol{A}_{\mathrm{Z}}$ *as a function of* p_{T}^{Z} *for* $Z \to \mu\mu$ *events in the presence of jets or not [2].*

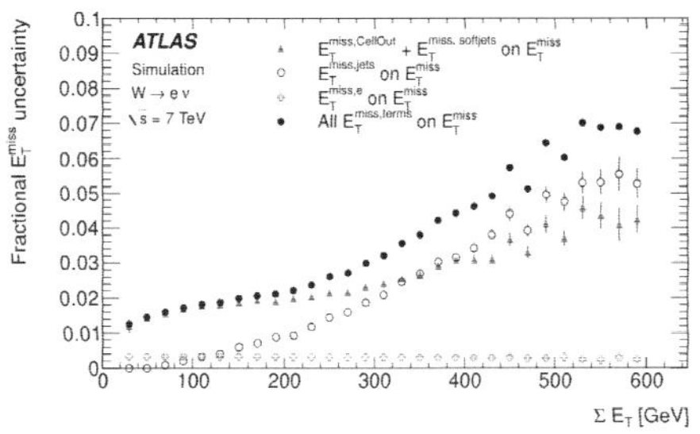

Figure 4. *The systematic overall fractional uncertainty on the* $E_{\mathrm{T}}^{\mathrm{miss}}$ *scale as a function of* $\sum E_{\mathrm{T}}$ *in MC* $W \to e\nu$ *events [2].*

6. Performance in pile-up conditions

A large deterioration of the resolution is observed in 2011 data and MC simulation with respect to that in 2010 due to the increased average number of pile-up interactions per event, as can be seen in Figure 5 which shows the ($E_{\mathrm{x}}^{\mathrm{miss}}$, $E_{\mathrm{y}}^{\mathrm{miss}}$) resolution as a function of the number of primary vertices in the event. The $\boldsymbol{E}_{\mathrm{T}}^{\mathrm{miss,SoftTerm}}$ term, the combination of $E_{\mathrm{T}}^{\mathrm{miss,CellOut}}$ and $E_{\mathrm{T}}^{\mathrm{miss,softjets}}$, is significantly affected by pile-up. This term cannot be ommited since it has an important contribution to the momentum balance in the event.

The method used here scales the $\boldsymbol{E}_{\mathrm{T}}^{\mathrm{miss,SoftTerm}}$ with the soft term vertex fraction (STVF), i.e. the fraction of tracks matched to the $\boldsymbol{E}_{\mathrm{T}}^{\mathrm{miss,SoftTerm}}$ which are associated with the hard scattering vertex and is calculated as: $\mathrm{STVF} = \sum_{\mathrm{track,PV}} p_{\mathrm{T}} / \sum_{\mathrm{track}} p_{\mathrm{T}}$, where the sums are taken over the tracks unmatched to physics objects and PV denotes the tracks associated to the main primary vertex[3]. Figure 5 shows that, after the correction of the $\boldsymbol{E}_{\mathrm{T}}^{\mathrm{miss,SoftTerm}}$ with STVF, the dependence of the $\boldsymbol{E}_{\mathrm{T}}^{\mathrm{miss}}$ resolution on N_{pv} is significantly reduced in MC $Z \to \mu\mu$ events with no jets with $p_{\mathrm{T}} > 20\,\mathrm{GeV}$. The dependence of the $\boldsymbol{E}_{\mathrm{T}}^{\mathrm{miss}}$ resolution on N_{pv} in inclusive $Z \to \mu\mu$ events is further reduced if each jet is scaled using the value of the jet vertex fraction, JVF, which is calculated in the same way as STVF using the tracks matched to the jet and has been used as an estimator of the probability of a jet to originate from pile-up. The pile-up suppression using STVF does not create additional tails in the $E_{\mathrm{T}}^{\mathrm{miss}}$ distribution; very few tails (< 1 per mil) are created by the pile-up suppression in jets using JVF. Figure 6 shows that the pile-up suppression in $E_{\mathrm{T}}^{\mathrm{miss}}$ restores the resolution to the level of events with average number of interactions per bunch crossing $\langle\mu\rangle = 0$.

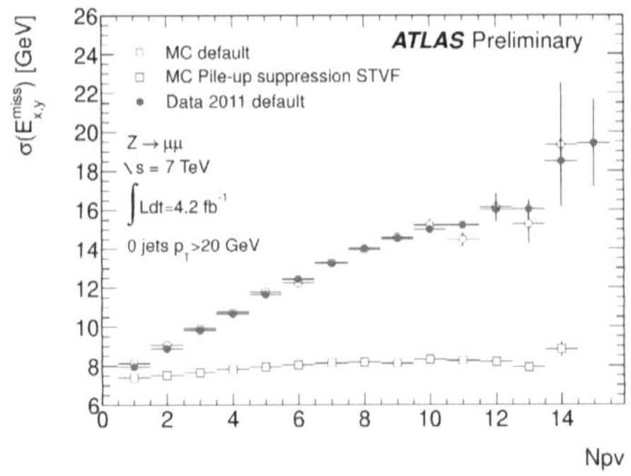

Figure 5. $E_{\mathrm{x}}^{\mathrm{miss}}$ *and* $E_{\mathrm{y}}^{\mathrm{miss}}$ *resolution vs* N_{pv} *in* $Z \to \mu\mu$ *events without jets of* $p_{\mathrm{T}} > 20\,\mathrm{GeV}$. *The pile-up suppressed* $\boldsymbol{E}_{\mathrm{T}}^{\mathrm{miss,SoftTerm}}$ *is shown for MC [5].*

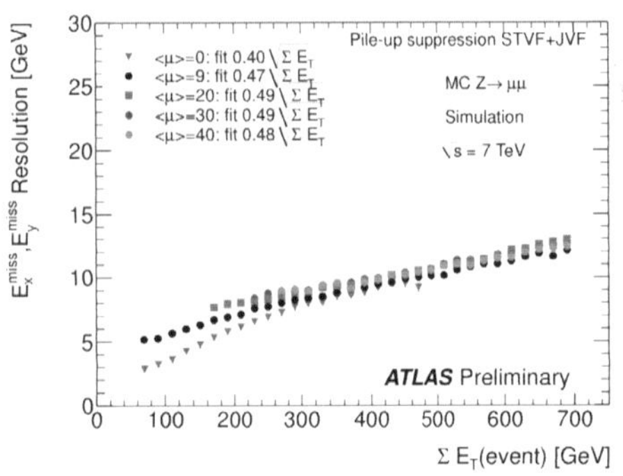

Figure 6. $E_{\mathrm{x}}^{\mathrm{miss}}$ *and* $E_{\mathrm{y}}^{\mathrm{miss}}$ *resolution vs the event* $\sum E_{\mathrm{T}}$ *in* $Z \to \mu\mu$ *MC events for different values of* μ *[5].*

7. References

[1] The ATLAS Collaboration, *The ATLAS Experiment at the CERN Large Hadron Collider*, 2008 *JINST* **3** S08003.
[2] The ATLAS Collaboration, *Performance of missing transverse momentum reconstruction in proton-proton collisions at 7 TeV in ATLAS*, Eur. Phys. J. C **72** (2012) 1844, `arXiv:1108.5602`.
[3] M. Cacciari, G. P. Salam and G. Soyez, *The anti-k_t jet clustering algorithm*, 2008 *JHEP* **5** 63
[4] The ATLAS Collaboration, *Jet energy measurement with the ATLAS detector in proton-proton collisions at 7 TeV*, submitted to Eur. Phys. J. C, `arXiv:1112.6426`.
[5] `https://twiki.cern.ch/twiki/bin/view/AtlasPublic/JetEtmissApproved2011EtmissWithPileup`

[3] The main primary vertex is the primary vertex with the largest track $\sum p_{\mathrm{T}}{}^2$ associated with it.

XVth International Conference on Calorimetry in High Energy Physics (CALOR2012) IOP Publishing
Journal of Physics: Conference Series **404** (2012) 012012 doi:10.1088/1742-6596/404/1/012012

Jet Energy Scale Uncertainties in ATLAS

Teresa Barillari on behalf of the ATLAS Collaboration
Max-Planck-Inst. für Physik, Foehringer Ring 6, 80805k, München, Germany
E-mail: barilla@mpp.mpg.de

Abstract. The first proton-proton collisions at a centre of mass energy of $\sqrt{s} = 7\,\text{TeV}$ have been used by the ATLAS experiment to achieve an accuracy of the jet energy measurement between 2% and 4% for jets transverse momenta between 20 GeV and 2 TeV and in the absolute pseudorapidity range up to 4.5. The jet energy scale uncertainty is derived from measurements *in situ* of the calorimeter single response to hadrons together with systematic variations in the Monte Carlo simulation. The transverse momentum balance between a central and a forward jet in events with two high transverse momenta jets is used to set the jet energy uncertainty in the forward region. The obtained uncertainty is confirmed by *in-situ* measurements. Jets in the TeV energy range have been tested using a system of well calibrated jets at low transverse momenta against high transverse momenta jets. A further reduction of the jet energy scale uncertainty between 1% and 2% for jets transverse momenta above 30 GeV has been achieved using data from the 2011 run based on an integrated luminosity of $5\,\text{fb}^{-1}$.

1. Introduction

Since Spring 2010, the ATLAS detector [1] has been collecting data from proton-proton (p-p) collisions at a center-of-mass energy $\sqrt{s} = 7\,\text{TeV}$ delivered by the Large Hadron Collider (LHC) at CERN. Understanding and measuring the performance of jets is crucial for the physics at LHC: the energy of jets (jet energy scale, or JES) is input to many analyses, and its uncertainty is the dominant experimental uncertainty for measurements such as the di-jet cross section, the top quark mass and new physics searches with jets and missing transverse momentum in the final state. In the following, the determination of the jet energy scale and its systematic uncertainty for inclusive jets in the ATLAS experiment is described. Sections 2-4 of this note describe the calorimeters, the Monte Carlo (MC) simulation framework, and the jet reconstruction. Sections 5-6 describe the technique used to calibrate the JES. Sections 7-8 describe the sources of systematic uncertainties for the JES and their derivation using simulated and collision data, the conclusions are given at the end.

2. The ATLAS Detector

The ATLAS detector [1] is a multipurpose particle physics apparatus with a forward-backward symmetric cylindrical geometry and nearly $4\,\pi$ coverage in solid angle. The layout of the detector features four superconducting magnet systems, which comprise a thin solenoid surrounding the inner tracking detectors and three large toroids supporting a large muon spectrometer. The calorimeters are of particular importance to the work here presented. In the pseudorapidity region $|\eta| < 3.2$, high-granularity liquid-argon (LAr) electromagnetic sampling calorimeters are used. An iron-scintillator tile calorimeter provides hadronic coverage over $|\eta| < 1.7$. The end-cap and forward regions, spanning $1.5 < |\eta| < 4.9$, are instrumented with LAr calorimetry for both electromagnetic and hadronic measurements.

3. Data and Monte Carlo simulation

The analysis here presented is based on data collected in the 2010 run at $\sqrt{s} = 7\,\text{TeV}$ for a total integrated luminosity of $38\,\text{pb}^{-1}$. Preliminary results obtained using the 2011 run for a total integrated luminosity of $5\,\text{fb}^{-}1$ are also briefly presented.

The MC simulated events (default MC) are produced with various event generators. The inclusive quantum chromodynamics (QCD) jet events are generated using PYTHIA [2] and ALPGEN [3].

PYTHIA is an event generator used to simulate non-diffractive p-p collisions. The parton distribution function (PDF) set used in PYTHIA is MRST LO* [4]. The tune of PYTHIA is done by a set of parameters denoted as ATLAS MC10 and is described elsewhere [5]. An independent tune of PYTHIA is PERUGIA2010 that better reproduces jet shapes and hadronic event shapes [6].

ALPGEN is a leading order matrix-element generator for hard multi-parton processes in hadronic collisions. It is interfaced to the HERWIG generator [7] to produce parton showers in leading-log approximation. Parton showers are matched to the matrix-element with the MLM matching scheme [8]. Soft multiple parton interactions are modelled using JIMMY [9] and the ATLAS MC09 tune [10]. The PDF set used in ALPGEN is CTEQ6L1 [11].

After all the MC events are generated, the GEANT4 software toolkit [12] is used to propagate the particles through the detector and to simulate their interactions with the detector material. The simulated detector signals are then reconstructed with the same reconstruction software used for the data.

4. Jet reconstruction

Jets are reconstructed [13] using the anti-$k_\perp$ algorithm [14] with values of the radius parameter R = 0.4 or $R = 0.6$. The four-momentum recombination scheme is used. Jet finding is done using the event-coordinates $y - \phi$, while jet corrections and performance studies are often done using the detector coordinates $\eta - \phi$. The threshold on the jet transverse momentum (p_T^{jet}) reconstruction is $p_T^{jet} > 7\,\text{GeV}$. The input to jets reconstructed in the calorimeter (called calorimeter jets) can be topological energy clusters (topo-clusters) [15] or towers [13] of calorimeter cells. In the following only results with jets built from topo-clusters are presented. In addition to calorimeter jets, track jets built from charged particle tracks originating from the primary hard scattering vertex are used to study close-by jet effects. These jets are less sensitive to effects of multiple p-p collisions within the same bunch crossing (pile-up). Finally MC simulation jets (called truth jets) are built from stable particles defined to have a proper lifetime longer than $10\,ps$, excluding muons and neutrinos.

5. Jet energy calibration scale

Reconstructed jets are first calibrated to the electromagnetic (EM) scale. The EM energy scale is established using test-beam measurements [16, 17, 18, 19, 20]. This EM energy scale accounts correctly for the energy of photons and electrons, but it does not correct for instrumental (detector) effects including calorimeter non-compensation, energy losses in inactive regions of the detector (dead material), particles which are not totally contained in the calorimeter (leakage), particles that fall out of the reconstructed jet but are included in the truth jet, and inefficiencies in calorimeter clustering and jet reconstruction. The goal of the JES calibration is to correct the energy and momentum of the jets measured in the calorimeter. The choice of JES calibration for the first ATLAS data is a jet by jet correction applied as a function of the p_T^{jet} and $|\eta|$ (called simply JES). Beside the above described EM calibrated jets, there are other more sophisticated jet calibration schemes developed in ATLAS like the local calibration weighting (LCW) scheme [21], and the global sequential calibration scheme [13]. All these methods require a final JES calibration, with the resulting final calibrated jets referred to as

EM+JES, or LCW+JES jets, etc. The more sophisticated jet calibration methods mentioned above are under commissioning and will be part of the default energy scale in future.

6. Jet energy calibration in the EM+JES scheme

The EM+JES calibration scheme applies p_T^{jet} and $|\eta|$ corrections to jets reconstructed at the EM scale. The additional energy due to pile-up is corrected before the hadronic energy scale is restored. The EM+JES calibration scheme consists of three main subsequent steps. As a first step the average additional energy due to pile-up is subtracted from the energy measured in the calorimeters using correction constants obtained from *in situ* measurements (pile-up correction). In the second step the direction of the jet is corrected such that the jet originates from the main primary vertex of the interaction instead of the geometrical centre of the detector (vertex correction). Finally in the third step the jet energy and direction as reconstructed in the calorimeters are corrected using constants derived from the comparison of the kinematic observables of reconstructed jets and those from truth jets (jet energy and direction correction). Once the jet origin and energy corrections are made, the origin-corrected jet η is further corrected for a bias due to poorly instrumented regions of the calorimeter [13]. This correction is in general small ($\Delta\,\eta < 0.01$) for most regions of the calorimeter but larger in the transition region between the barrel and end-cap calorimeters.

7. Jet energy scale uncertainties for the EM+JES scheme

The JES systematic uncertainty in the well-understood central barrel region is derived combining information from the single hadron response measured *in situ* and single pion test-beam measurements [13], uncertainties on the amount of material of the detector, the description of the electronic noise, and the MC modelling used in the event generation. Dedicated MC samples are generated with different conditions with respect to the standard MC sample. These variations are expected to provide an estimate of the systematic effects contributing to the JES uncertainty. The η bins used for the estimate of the JES uncertainty divide the detector in eight main η-regions [13]. The JES systematic uncertainty for all jets with $|\eta| > 0.8$ is determined using the JES uncertainty for the central barrel region ($0.3 < |\eta| < 0.8$) as a baseline. This choice is motivated by the good knowledge of the detector geometry in the central region, and by the use of pion response measurements in beam tests [16, 17, 18, 19, 20]. This section describes the sources of systematic uncertainties and their effect on the response of EM+JES calibrated jets.

7.1. Uncertainty in the JES calibration

After the jets are calibrated, the jet energy and the jet response still show at low p_T slight deviations from unity when the MC simulated jet is compared to its matched truth jet. Any deviation from unity in the jet energy or p_T^{jet} response after the application of the JES to the MC simulated sample implies that the kinematic observables of the calibrated calorimeter jet are not restored to that of the corresponding truth jet (non-closure). The non-closure is mainly due to the application of the same correction factor for jet energy and p_T^{jet}. The systematic uncertainty due to the non-closure of the JES calibration is taken as the larger deviation of the response in either jet energy or p_T^{jet} from unity.

7.2. Uncertainty on the calorimeter response

The response and corresponding uncertainties for single particles interacting in the calorimeters can be used to derive the JES uncertainty in the central calorimeter region as detailed in Ref [22]. In the simulation the true calorimeter energy deposits in each calorimeter cell can be traced to the particles generated in the collision. The uncertainty in the calorimeter response to jets can then be obtained from the response uncertainty of the individual particles constituting the jet. The *in situ* measurement of the single particle response detailed in Ref. [22] significantly reduces the uncertainty due to the limited knowledge of the detector geometry, in particular that due to the description of the dead material, and the modelling of the way particles interact in the detector.

7.3. Uncertainties due to the detector simulation

The topo-clusters used to build jets are constructed based on a signal-to-noise ratio of energy. Discrepancies between the simulated noise and the real noise in data can lead to differences in the cluster shapes and to the presence of fake topo-clusters. For data, the noise can change over time, while the noise RMS used in the simulation is fixed at the time of the production of the simulated data sets. These effects can lead to biases in the jet reconstruction and calibration. The effect of the calorimeter cell noise mis-modelling on the jet response is estimated by reconstructing topo-clusters in MC using the noise RMS measured from data. The response for jets reconstructed with modified noise thresholds are compared with the response for jets reconstructed in the same sample using the default MC noise thresholds. The maximal observed change in jet energy or p_T^{jet} response is used to estimate the uncertainty on the jet energy measurement due to the calorimeter cell noise modelling. The JES is also affected by possible deviations in the material description and in the geometry assumed for the detector as simulated in the default MC sample. Specific MC simulation samples have been produced using distorted geometries. The uncertainty contribution due to the description of the detector material is estimated by comparing the EM+JES jet response in the default MC simulation sample with the jet response in MC simulation samples with distorted geometries.

7.4. Uncertainties due to the event modelling in Monte Carlo generators

The contributions to the JES uncertainty from the modelling of the fragmentation, the underlying event and other choices in the event modelling of the MC event generator are obtained from samples based on ALPGEN+HERWIG+JIMMY and the PYTHIA PERUGIA2010 tune. By comparing the baseline PYTHIA sample to the PYTHIA PERUGIA2010 tune, the effects of soft physics modelling are tested. The ALPGEN MC uses different theoretical models for all steps of the event generation and it is used to estimate additional systematic variations. The ratios of the baseline MC simulated jet to that of the two above MC jet samples are used to estimate the systematic uncertainty to the JES.

7.5. In situ intercalibration using events with di-jet topologies

The response of the calorimeters to jets depends on the jet direction, due to the different calorimeter technology and to the varying amounts of material in front of the calorimeters. A calibration is therefore needed to ensure a uniform calorimeter response to jets. Since this is achieved by applying correction factors derived from MC simulations, such corrections need to be validated *in situ*. The relative jet calorimeter response and its uncertainty is studied by comparing the p_T^{jet} of a well-calibrated central jet and a jet in the forward region in events with only two high p_T^{jet} (di-jets). The JES uncertainty in the region $|\eta| < 0.8$ obtained using the single particle response and systematic variations of the MC simulations, is transferred to the forward regions using the results from the di-jet balance method (η-intercalibration method). These uncertainties are then included in the final uncertainty [13]. The uncertainty measurements are performed with average $20\,\text{GeV} < p_T^{jet} < 110\,\text{GeV}$. The uncertainty for $p_T^{jet} > 100\,\text{GeV}$ is taken as the uncertainty of the last available p_T^{jet}-bin. The uncertainties are evaluated separately for anti-$k_\perp$ jets with $R = 0.4$ and $R = 0.6$, and are found to be slightly larger for $R = 0.4$.

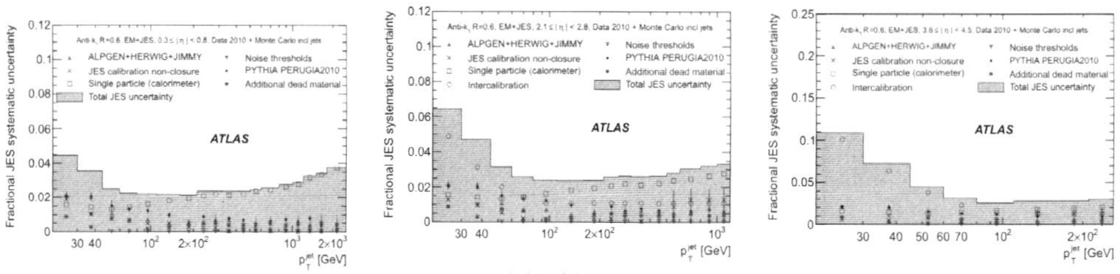

(a) $0.3 \leq |\eta| < 0.8$ (b) $2.1 \leq |\eta| < 2.8$ (c) $3.6 \leq |\eta| < 4.5$

Figure 1. Fractional JES systematic uncertainty as a function of p_T^{jet} for jets in the $0.3 < |\eta| < 0.8$ in the calorimeter barrel (a), $2.1 < |\eta| < 2.8$ in the calorimeter endcap (b), and in the forward pseudorapidity region $3.6 < |\eta| < 4.5$. The total uncertainty is shown as the solid light shaded area. The individual sources are also shown.

7.6. Uncertainties due to multiple proton-proton collisions

The uncertainty for the pile-up corrections can be obtained by studying the jet response with respect to the transverse momenta (p_T) of the track jets as a function of the number of primary vertices (NPV). In the case of NPV$=2$ the uncertainty due to pile-up for central jets with $p_T = 20$ GeV and $|\eta| < 0.8$ is about 1%, while it amounts to about 2% for jets with $2.1 < |\eta| < 2.8$ and to less than 2.5% for all jets with $|\eta| < 4.5$. In the case NPV$=3$ the pile-up uncertainty is approximately twice that of NPV$=2$, and with NPV$=4$ the uncertainty for central, endcap and forward jets is less than 3%, 6% and 8%, respectively. The pile-up uncertainty needs to be added separately to the estimate of the total JES uncertainty.

7.7. Summary of JES systematic uncertainties

The total JES uncertainty is derived by considering all the individual contributions described in the previous sections. In the central region ($|\eta| < 0.8$) for each (p_T^{jet}, η)-bin, the uncertainty contributions from the calorimeter, the jet non-closure, and systematic MC simulation variations are added in quadrature. For $|\eta| > 0.8$ the η-intercalibration contribution is estimated for each η bin in the endcap region. The η-intercalibration contribution is added in quadrature to the total JES uncertainty determined in the $0.3 < |\eta| < 0.8$ region to estimate the JES uncertainty for jets with $|\eta| > 0.8$, with the exception of the non-closure term that is taken from the specific η-region. For $p_T^{jet} < 45$ GeV, this choice leads to partially double counting the contribution from the dead material uncertainty, but it leads to a conservative estimate in a region where it is difficult to estimate the accuracy of the material description. The contribution to the uncertainty due to pile-up is added separately, depending on the NPV in the event. Figure 1 shows the final fractional JES systematic uncertainty and its individual contributions as a function of p_T^{jet} for anti-$k_\perp$ jets with $R = 0.6$ for three selected η regions. The fractional JES uncertainty in the central region amounts to 2% - 4% for $p_T^{jet} < 60$ GeV, and it is between 2% and 2.5% for 60 GeV$< p_T^{jet} < 800$ GeV. For jets with $p_T^{jet} > 800$ GeV, the uncertainty ranges from 2.5% to 4%. The uncertainty amounts to up to 7% and 3%, respectively, for $p_T^{jet} < 60$ GeV and $p_T^{jet} > 60$ GeV in the endcap region, where the central uncertainty is taken as a baseline and the uncertainty due to the intercalibration is added. In the forward region, a 13% uncertainty is assigned for $p_T^{jet} = 20$ GeV. The increased uncertainty is dominated by the modelling of the soft physics in the forward region that is accounted for in the η-intercalibration contribution. This uncertainty contribution is estimated conservatively. The same study has been repeated for anti-$k_\perp$ jets with $R = 0.4$ and the JES uncertainty for these jets is comparable to that obtained for anti-$k_\perp$

jets with $R = 0.6$, it is between $\approx 4\%$ (8%, 14%) at low p_T^{jet} and $\approx 2.5\%$ - 3% (2.5% - 3.5%, 5%) for jets with $p_T > 60$ GeV in the central (endcap, forward) region. These results have been confirmed by *in-situ* validation studies [22].

Recent studies done using the 2011 5 fb^{-1} data with increased pile-up conditions and the *in-situ* direct transverse momentum balance method described in Ref. [23], show for anti-$k_\perp$ jets with $R = 0.4$ JES uncertainties between 1% and 2%. The precision achieved with this *in situ* method is about 10% at low p_T^{jet} and 1% to 2% for $p_T^{jet} > 30$ GeV. Similar results are obtained using EM+JES or LCW+JES jets.

8. Conclusions

The JES calibration and its relative systematic uncertainty have been estimated for inclusive jets produced in p-p collisions at $\sqrt{s} = 7$ TeV. The jets are reconstructed with the anti-$k_\perp$ algorithm applied to topological energy clusters reconstructed in the ATLAS calorimeters. The method used to estimate the JES and its uncertainty relies on the MC simulation of inclusive QCD jets events. The procedure to calibrate the JES in different ranges of the jet kinematic variables has been described step-by-step for 20 GeV< $p_T^{jet} < 800$ GeV and $|\eta| < 4.5$. The contributions to the relative systematic uncertainty from the jet kinematic variables, the detector performance, the MC simulation model and the data taking conditions have been described and analyzed. For the low pile-up conditions of the 2010 data taking period, the JES relative uncertainty is on average $(3-4)\%$ and is maximum ($\sim$ 10%) for jets with p_T^{jet} of about 20 GeV in the forward region, $|\eta| > 3.2$. *In situ* studies have been performed using the much larger data sample collected in 2011 reaching a JES precision of about 10% at low p_T^{jet} and 1% to 2% for $p_T^{jet} > 30$ GeV.

References

[1] ATLAS Collaboration, 2008, *JINST* **3** S08003.
[2] Sjostrand T, Mrenna S, and Skands Z, 2006, *JHEP* **0605** 026.
[3] Mangano M L, Moretti M, Piccinini F, Pittau R., and Polosa A D, 2003, *JHEP* **07** 001.
[4] Sherstnev A and Thorne R S, 2008, *Eur. Phys. J.* **C55** 553575.
[5] ATLAS Collaboration, July 2010, *ATLAS-CONF-2010-046*.
[6] Skands P Z, 2010, *Phys. Rev.* **D 82** 074018.
[7] Corcella G et al., arXiv:hep-ph/0210213.
[8] Mangano M L, Moretti M, and Pittau R, 2002, *Nucl. Phys.* **B632** 343362.
[9] Butterworth J M, Forshaw J R, and Seymour M H, 1996, *Z. Phys.* **C72**, 637646.
[10] ATLAS Collaboration, March 2010, *ATLAS-PHYS-PUB-2010-002*.
[11] Pumplin J et al., 2002, *JHEP* **07** 012.
[12] GEANT4 Collaboration, Agostinelli S et al., 2003, *Nucl. Instrum. Meth.* **A 506** 250 - 303.
[13] ATLAS Collaboration, Dec. 2011, *CERN-PH-EP-2011-191*, 100pp.
[14] Cacciari M, Salam G P, and Soyez G, 2008, *JHEP* **04** 063.
[15] Lampl W et al., April 2008, *ATL-LARG-PUB-2008-002*.
[16] Abat E et al., 2010, *JINST* **5** P11006.
[17] Aharrouche M et al., 2010, *Nucl. Instrum. Meth.* **A 614** 400432.
[18] Colas J et al., 2007, *Nucl. Instrum. Meth.* **A 582** 429455.
Aharrouche M. et al., 2006, *Nucl. Instrum. Meth.* **A 568** 601623.
[19] Adragna P et al., 2009, *Nucl. Instrum. Meth.* **A 606** 362394.
[20] Pinfold J et al., 2008, it Nucl. Instrum. Meth. **A 593** 324342.
[21] Barillari T et al., Dec. 2008, *ATL-LARG-PUB-2009-001*.
[22] ATLAS Collaboration, March 2011, *ATLAS-CONF-2011-028*, March 2011, *ATLAS-CONF-2011-029*, May 2011, ATLAS-CONF-2012-053.
[23] ATLAS Collaboration, May 2012, *ATLAS-CONF-2012-053*.

XVth International Conference on Calorimetry in High Energy Physics (CALOR2012) IOP Publishing
Journal of Physics: Conference Series **404** (2012) 012013 doi:10.1088/1742-6596/404/1/012013

Determination of the Jet Energy Scale in CMS

Henning Kirschenmann, on behalf of the CMS collaboration
University of Hamburg, Institut für Experimentalphysik, Luruper Chaussee 149, 22761 Hamburg, Germany

E-mail: henning.kirschenmann@desy.de

Abstract. Measurements of the jet energy scale in CMS and the status of the jet energy corrections for 2011 analyses are presented. The measurements have been performed with a data sample collected in proton-proton collisions at a centre-of-mass energy of 7 TeV, corresponding to an integrated luminosity of $4.9\,\mathrm{fb}^{-1}$. Dijet and photon/Z+jets events are used to measure the jet energy response in the CMS detector. The results are presented for the Particle Flow approach, which attempts to reconstruct individually each particle in the event, prior to the jet clustering, based on information from all relevant subdetectors.

1. Introduction

Jets are the experimental signatures of quarks and gluons produced in high-energy processes such as hard scattering of partons in proton-proton collisions. The jet energy corrections relate the energy of reconstructed jets - on average - to the true particle level energy, which is independent of the detector response. The detailed understanding of the jet energy scale is of crucial importance for many physics analyses.

The measurements discussed here have been performed by the Compact Muon Solenoid (CMS) collaboration at the CERN Large Hadron Collider (LHC). The central feature of the CMS apparatus is a superconducting solenoid, of 6 m internal diameter, providing a field of 3.8 T. Within the field volume are a silicon pixel and strip tracker, a crystal electromagnetic calorimeter (ECAL) and a brass/scintillator hadron calorimeter (HCAL). Muons are measured in gas-ionization detectors embedded in the steel return yoke. A detailed account of the detector can be found in [1].

The techniques for the determination of the jet energy scale and transverse momentum resolution at CMS have been discussed in detail in Ref. [2]. In the following, we summarize the current status of jet energy corrections and present the 2011 results on the determination of the energy scale in data.

2. Jet-types used at CMS

Within CMS, three different methods to reconstruct jets have been commissioned (see Ref. [3]): a calorimeter-based approach, the "Jet-Plus-Track" approach, which improves the measurement of calorimeter jets by exploiting the associated tracks, and the Particle Flow (PF) approach, which attempts to reconstruct individually each particle in the event, prior to the jet clustering, based on information from all relevant subdetectors. The resulting jet energy resolutions from simulation are depicted in Fig. 1.

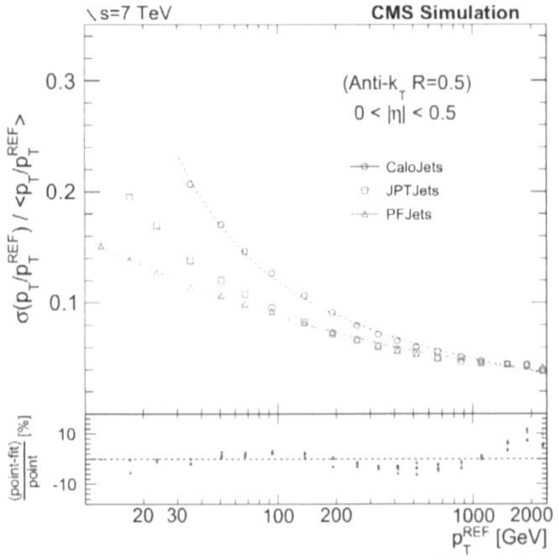

Figure 1. Jet energy resolution from simulation for the three types of reconstructed jets at CMS supplied with jet energy corrections: Calorimeter jets, Jet-Plus-Track jets, and Particle Flow jets. Taken from Ref. [4]

The improvement of the jet energy resolution and the more uniform and linear energy response of PF-objects indicate that the Particle Flow approach offers advantages with respect to the other jet types. As expected for the fragmentation of partons to jets, about 65% of the energy of PF-jets at central rapidity and $p_T \approx 100\,\mathrm{GeV}$ is carried by charged hadrons, 25% by photons, and 10% by neutral hadrons. For charged hadrons the tracker information is relevant ($\sigma_{tracker}(p_T)/p_T \approx 1\%$), while photons are reconstructed from isolated ECAL clusters ($\sigma_{ECAL}(E)/E \approx 1\%\sqrt{E} \oplus 0.3\%$) and only the remaining neutral hadrons are reconstructed from HCAL-information alone, leading to the aforementioned significant improvement of the jet energy resolution in comparison to the other jet-types. The study of the energy composition of PF-jets, i.e. the energy fractions of the different particle candidate classes, provides an important additional handle on the quality of the simulation of the detector and understanding of the PF algorithm. It is performed using a tag-and-probe in dijet events. The results in Fig. 2 show that the differences in the observed energy fractions between data and simulation are below 1% in the central detector region.

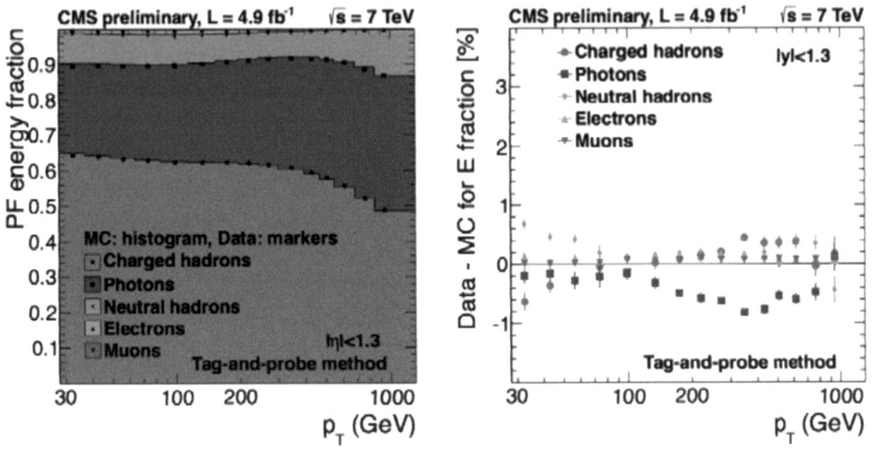

Figure 2. PF energy fractions in the central detector region. Taken from Ref. [5]

3. Factorized jet energy corrections

The success of the CMS simulation in describing the jet properties (see Ref. [4]) allows its effective use in a factorized approach to jet energy corrections (JEC) that CMS has adopted. After the first level correction, which subtracts the additional energy in the event induced by pileup effects (additional minimum bias events), the reconstructed jets are corrected to compensate for the non-linear response of the calorimeters (as a function of p_T) and variations of the response in η. These corrections are derived from simulation.

Subsequently, small residual corrections are applied which are based on measurements of the relative scale as a function of η from dijet events and the absolute scale in the central detector region ($|\eta| < 1.3$) from Z+jet and γ+jet events.

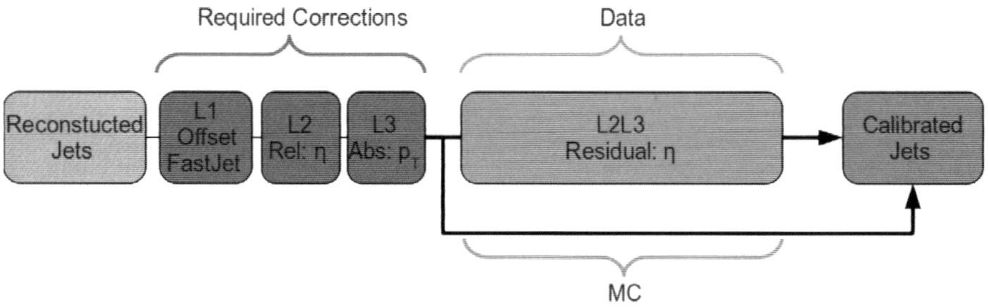

Figure 3. Sketch of the mandatory jet energy correction levels at CMS.

3.1. L1 - Pileup corrections

Corrections for pileup effects have become increasingly important during the LHC-running. The mean number of primary vertices in 2011 data has been $\langle N_{PV}\rangle \approx 7$ and increases to more than 20 in 2012. The average energy offset amounts to $\Delta p_T = 0.72\,\mathrm{GeV}/N_{PV}$ for central PF-jets. For each event, an average p_T density ρ per unit area is estimated which characterizes the soft jet activity and is a combination of the underlying event, the electronic noise and the pileup. Following the jet area approch described in [8, 9], ρ and the jet-area A are used to calculate an η-dependent correction factor to subtract the offset energy of individual jets on an event-by-event basis.

3.2. L2L3 - MC truth corrections

The L2L3-corrections are based on simulation and correct the energy of the reconstructed jets such that it equals - on average - the energy of the jets at particle level. Simulated jet events, generated with PYTHIA6 [6], tune Z2 and processed through the full, GEANT4 [7] detector simulation are used for the derivation of these corrections. The generated and reconstructed jets are matched spatially and the transverse momentum of the reconstructed jet p_T^{reco} and the response $p_T^{\mathrm{reco}}/p_T^{\mathrm{gen}}$ are determined in fine bins of p_T^{gen} and η^{gen}. The correction factor is then determined as the inverse of the mean response as a function of p_T^{reco} in fine η-bins.

4. Residual corrections from data-driven techniques

The corrections from simulated data are the foundation of the jet energy correction chain in CMS. To complete the chain, these corrections are applied to data and simulated data in order to validate the jet energy scale. Response estimators from the p_T balance and the MPF[1] method (used extensively at [10]) are used to determine the mean response in data and simulated data.

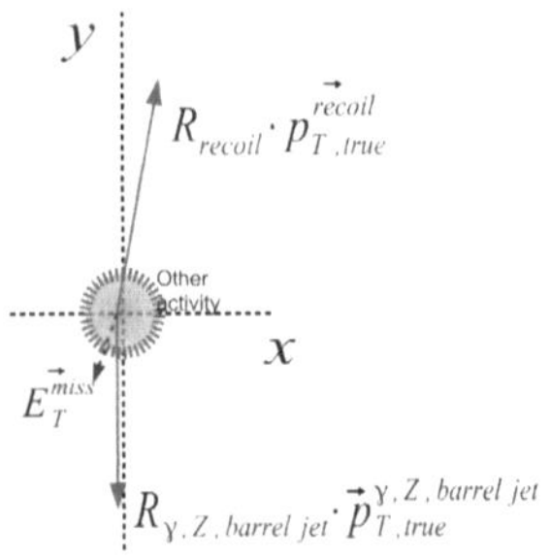

Figure 4. Sketch of MPF-method

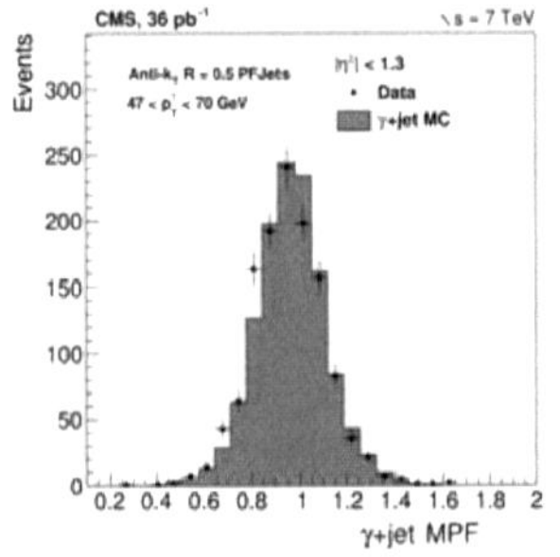

Figure 5. Sample MPF-distribution for γ+jet events

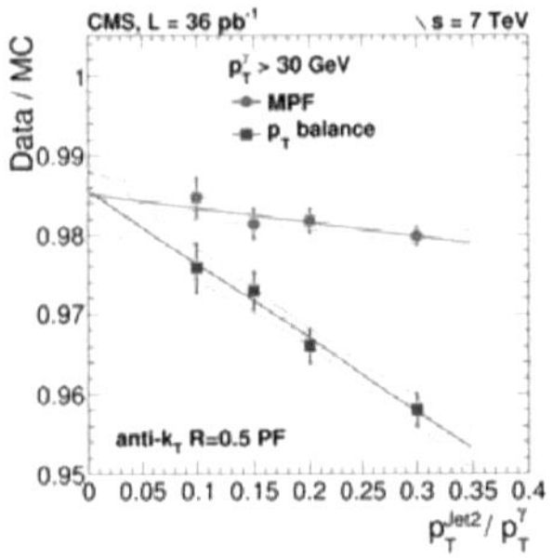

Figure 6. Extrapolation to zero additional event activity for γ+jet

4.1. Absolute scale from Z+jet and γ+jet

Z+jet and γ+jet events provide a very clean signature with a well understood and precisely measured reference object balancing the jet. The jets are selected to be in the barrel region ($|\eta| < 1.3$) and back to back to the reference object.

The response estimators are defined as

$$\mathcal{R}_{\text{balance}} = \frac{p_T^{jet}}{p_T^{\gamma,Z}} \tag{1}$$

for the p_T-balance method and as

$$\mathcal{R}_{\text{recoil}} = \mathcal{R}_{\gamma,\text{Z}} + \frac{\vec{\not{E}}_T \cdot \vec{p}_{\text{T}}^{\gamma,\text{Z}}}{(\vec{p}_{\text{T}}^{\gamma,\text{Z}})^2} \equiv R_{\text{MPF}} \equiv \mathcal{R}_{\text{probe}} \tag{2}$$

for the MPF-method, where $\vec{\not{E}}_T$ is the missing transverse energy and $\mathcal{R}_{\gamma,\text{Z}}$ the response of the reference photon or Z-boson. The idea underlying the MPF response estimator (see Fig. 4) is that there is no intrinsic $\vec{\not{E}}_T$ in such events and that the measured $\vec{\not{E}}_T$ is instead induced by mismeasurents of the hadronic recoil. The projection of $\vec{\not{E}}_T$ along the reference object axis can then be used to yield a MPF response estimator.

In order to determine the energy scale as a function of p_T, the distributions of the previously defined response estimators $\mathcal{R}_{\text{balance}}$ and $\mathcal{R}_{\text{MPF}}$ are evaluated. An example distribution is shown in Fig. 5. The mean of the estimated response is determined in bins of p_T for different values of the requirement on the maximum of the relative second jet p_T. The ratio of data to simulation of $\mathcal{R}_{\text{balance}}$ and $\mathcal{R}_{\text{MPF}}$ is extrapolated to zero additional event activity (corresponding to $p_T^{Jet2}/p_T^{\gamma} = 0$) to suppress the influence of soft radiation on the results. An example of such an

[1] Missing transverse energy projection fraction

extrapolation is shown in Fig. 6 and illustrates the significantly reduced dependence of $\mathcal{R}_{\mathrm{MPF}}$ on the radiation modeling in simulation with respect to the p_T balance method. The resulting ratio of data to simulation as a function of p_T is depicted in Fig. 7 and corresponds to a scale deviation of $\approx 1\%$ in the central detector region.

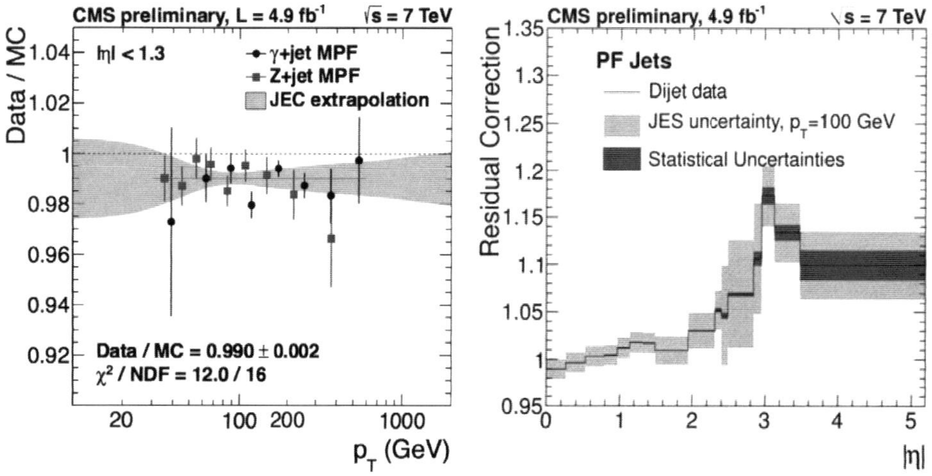

Figure 7. Absolute scale determined from Z+jet and γ+jet and the relative residual correction factor determined from dijet data. Taken from Ref. [5]

4.2. Relative scale from dijet events

With the absolute scale being verified in the central detector region, the missing component is the relative scale. This is determined using dijet events which provide a large number of events and a high p_T-reach. The two leading jets are required to be back-to-back in the azimuthal angle φ and at least one of the jets is required to be in the central ($|\eta| < 1.3$) detector region.

With this selection, the energy scale can be constrained relative to the central detector region by response estimators, i.e. the relative response

$$\mathcal{R}_{\mathrm{rel}}(\eta^{probe}, p_{\mathrm{T}}^{\mathrm{ave}}) = \frac{1 + \langle \mathcal{A} \rangle}{1 - \langle \mathcal{A} \rangle}, \text{ with } \mathcal{A} = \frac{p_T^{probe} - p_T^{barrel}}{p_T^{probe} + p_T^{barrel}} \tag{3}$$

and $\mathcal{R}_{\mathrm{MPF}}$ as before. As depicted in Fig. 7, these relative differences are very small in the central detector region and below 5% in the region up to $|\eta| < 2.5$. These small remaining differences observed in Z/γ+jet and dijet events are explicitly corrected for in data in the L2L3Residual correction step which completes the previously discussed jet energy correction chain. A significant advantage of this approach is that biases inherent to data-driven methods are canceled to first order by determining only the ratio of data to simulation.

5. Conclusions

As a result of the factorized calibration procedure outlined above, the energy scale is known to the percent level over a wide range of the available phase space (compare Fig. 8) and as precise as 1% at 200 GeV for a central jet. A better understanding of the scale uncertainties allows many physics analyses to improve on their JEC-related systematic uncertainties in turn,

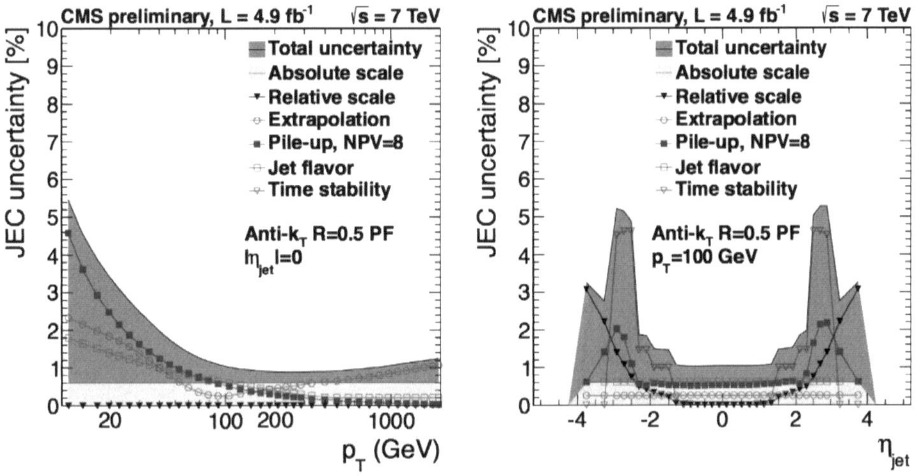

Figure 8. Jet energy scale uncertainties as a function of p_T and η. Taken from Ref. [5]

leading to more precise results in measurements involving jets. More details on pileup-mitigation techniques and on how to reduce the impact of jet energy scale uncertainties on measurements are covered in the corresponding contributions in these proceedings [11, 12].

References

[1] The CMS Collaboration 2008 The CMS experiment at the CERN LHC *JINST* **3** S08004 doi:10.1088/1748-0221/3/08/S08004

[2] The CMS collaboration 2011 Determination of jet energy calibration and transverse momentum resolution in CMS *JINST* **6** P11002 doi:10.1088/1748-0221/6/11/P11002

[3] The CMS Collaboration 2010 Commissioning of the Particle-flow Event Reconstruction with the first LHC collisions recorded in the CMS detector *CMS-PAS* PFT-10-001

[4] The CMS Collaboration 2010 Jet Performance in pp Collisions at 7 TeV *CMS-PAS* JME-10-003

[5] The CMS Collaboration 2012 Jet Energy Scale performance in 2011 *CMS Detector Performance Summaries* DP-2012-006

[6] T. Sjöstrand, S. Mrenna, and P. Skands 2007 PYTHIA 6.4 physics and manual *JHEP* **05:026**

[7] S. Agostinelli et al. 2003 Geant 4 A Simulation Toolkit *Nucl. Inst. Meth.* **A** 506 250303. doi:10.1016/S0168-9002(03)01368-8

[8] M. Cacciari and G.P. Salam 2008 Pileup subtraction using jet areas *Phys. Lett. B* **659** 119 [arXiv:0707.1378]

[9] M. Cacciari, G. P. Salam and G. Soyez 2012 *Eur. Phys. J. C* **72** 1896 [arXiv:1111.6097 [hep-ph]].

[10] D0 collaboration 1999 Determination of the absolute jet energy scale in the D0 calorimeters *Nucl. Instrum. Meth.* **A** 424 352

[11] Alexx Perloff, on behalf of the CMS collaboration 2012 Pile Up measurement and mitigation techniques in CMS *JPCS*

[12] Ricardo Eusebi, on behalf of the CMS collaboration 2012 Jet energy scale uncertainties and usage improvements in CMS *JPCS*

Journal of Physics: Conference Series **404** (2012) 012014 doi:10.1088/1742-6596/404/1/012014

Jet energy corrections and uncertainties in CMS: reducing their impact on physics measurements

Ricardo Eusebi, on behalf of the CMS collaboration.

Texas A&M University, Department of Physics and Astronomy,
College Station,Texas, 77843-4242. USA.

E-mail: eusebi@physics.tamu.edu

Abstract. The measurement of any physical quantity at CMS includes the estimation of a systematic error propagated from the uncertainties on the jet energy calibration. For some physical quantities the main systematic error results from this propagation. This talk will present the main uncertainties sources considered on the jet energy calibration, the effect these have on some particular analyses, and a novel way of handling uncertainties recently deployed at CMS to reduce the dependence on these uncertainties.

1. Introduction

Since the LHC began operations less than 2 years ago the CMS collaboration has shown an excellent understanding of the CMS detector as well as of its corresponding Monte Carlo simulations, it has produced accurate calibration constants that are the among the best in the world and published the first basic analysis rediscovering the known features of the Standard Model. But the current path to discovery requires now more sophisticated analysis techniques, as well as a more accurate set of calibration constants. The jet energy scale is a perfect example of the growing success achieved in the calibration area where the current jet energy scale uncertainties has been reduced to about 2%.

There are currently several studies being developed at CMS with varying degrees of complexity, and as detailed in the main text and despite the excellent accuracy of the jet energy calibrations many of these studies still report the jet energy scale as one of the largest source of uncertainties. This paper describes common methodologies to estimate the effect of these uncertainties on physical observables as well as novel procedures for reducing their overall impact.

Section 2 gives a brief overview of the jet energy corrections and uncertainties used at CMS, section 3 details standard methodologies for their application to most analyses, and section 4 details improvements in the application to reduce the overall impact of the uncertainties. The paper finished with conclusions on section 5.

2. Jet calibration at CMS

The detector response and its efficiency together with the details of the jet algorithm used for jet clustering result in measurement of jet energies that are different than the energy of the jet obtained at particle level. The jet energy is corrected, or calibrated, on average to the particle-level energy of the jet in the generic flavor mixture of the QCD sample. This is achieved by applying a sequence of corrections that individually deal with different aspects of the calibration. A brief overview of the correction level follows:

- Pile up: this is the first correction applied to jet. It subtracts the average extra energy deposited in the jet that comes from other vertices present in the event and is therefore not part of the hard jet. It is parameterized as a function of the median jet energy in the event as defined in [1], or alternatively on the number of primary vertices (N_{PV}). This function has an excellent linearity up to $N_{\mathrm{PV}} = 30$ and has very little η dependance within the tracker coverage. It is derived from zero-bias data and Monte Carlo simulations (MC). A more detailed discussion on pile up mitigation techniques can be found in [1].
- η dependance: it corrects for the non-uniform response of the detector in eta. The response of each position in eta is related to the detector response averaged over $|\eta| < 1.3$. Derived from Monte Carlo simulations shows an excellent performance correcting back to the 0.5% level for jet $p_{\mathrm{T}} > 30$ GeV. The response of different flavors is within the 1.5% of the QCD flavor mixture.
- p_{T} dependance: it corrects for the non-uniform response of the detector as a function of p_{T} . Obtained from Monte Carlo simulations.
- η and p_{T} residuals: applied to data only to correct for η and p_{T} residual differences between data and Monte Carlo. The η residual is derived from QCD dijet sample and the p_{T} residual from γ+jet and Z+jets samples in data.

A more detailed explanation of the correction levels and methodologies to derived them can be found in [2]. More discussion and a set of updated plots can be found in [3].

2.1. Jet energy uncertainties at CMS

Each one of these correction levels has uncertainties arising from many different sources. In general these sources can be categorized as being one of the following:

- Physics modeling in MC such as showering, modeling of underlying event, etc.
- MC modeling of true detector response and properties such as noise, zero suppression, etc.
- Potential biases in the methodologies used to estimate the corrections.

At CMS more than 16 sources of uncertainties have been identified from the above origins. Several of these uncertainty sources are related and can be combined into groups we label "Absolute scale", "Relative scale", "Extrapolation in p_{T} ", "Pile-up", "Jet flavor", and "Time stability". The figures 1 and 2 show the contribution of the jet energy correction uncertainty from the combined sources as a function of the jet p_{T} and jet η. At CMS the total uncertainty on the jet energy correction is computed as the quadrature sum of the uncertainty of each different source and it shown for reference in figures 1 and 2 in a solid grey color.

3. Application of jet energy correction uncertainties

The measurement of any physical quantity includes the estimation of an uncertainty on the quantity due to the uncertainty on jet energy calibrations. At CMS the most common practice consists of the evaluation of the change in the measured quantity when the jet energy is fluctuated up and down accor'ding to the total jabet energy uncertainty. The following examples show the effect of this application on three different analysis.

3.1. WW cross section in the lepton+jets channel

As an example application of let's consider the measurement of the WW cross section in the lepton+jets channel. Figure 3 shows a simultaneous fit to an average pseudodata to the W+jets background and the WW diboson event yields on the dijet invariant mass. In this fit the jet

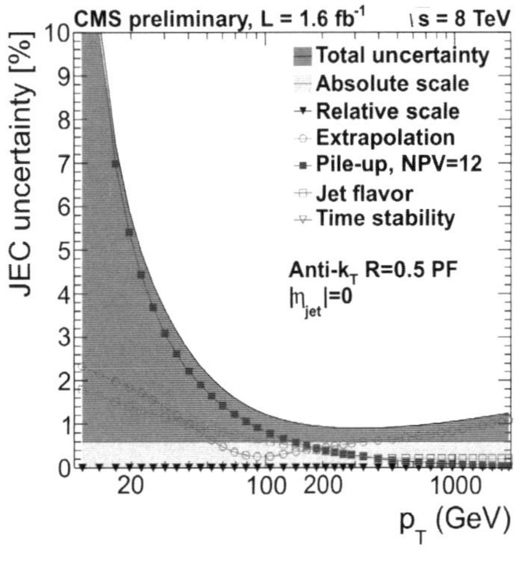

Figure 1: Jet energy uncertainty as a function of jet p_T for combined uncertainty sources.

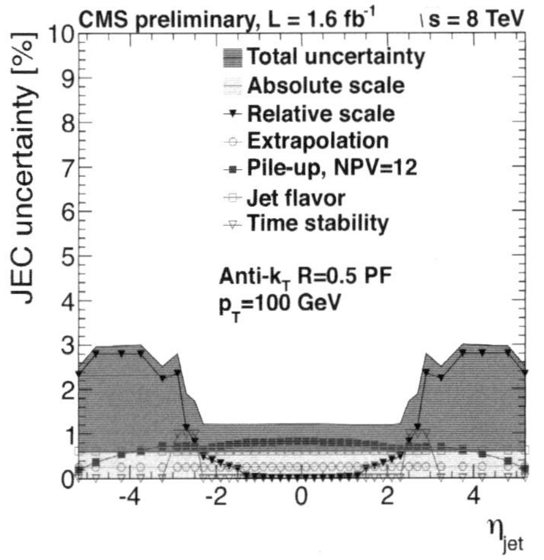

Figure 2: Jet energy uncertainty as a function of η for combined uncertainty sources.

energy uncertainty was not considered and the result of the fit is 322 ± 98 diboson events. However if the same fit is performed when the jets are scaled down (up) by their uncertainty the result is 363 ± 98 (279 ± 100) diboson events. This indicates that the error is obtained by simply scaling jets up and down by their uncertainty is about 43 events, or about 13% the diboson signal size.

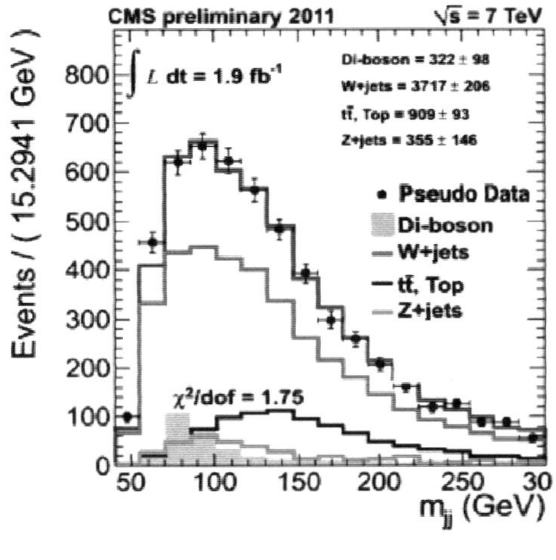

Figure 3: Simultaneous fit to an average pseudo-dataset expected of the W+jets background and the WW diboson event yields on the dijet invariant mass. Uncertainties on the jet energy corrections were not considered. The fit resulted in an expectation of 322 ± 98 diboson events and 3717 ± 206 W+jets events.

3.2. Top mass measurement in the lepton+jets channel

The measurement of the top mass is in general very sensitive to the jet energy corrections and uncertainties. Figure 4 shows a relative large effect of about 5 GeV on the fitted top mass when varying the jet energy scale about 4% as shown for example from [4]. To reduce this dependance the top mass analysis perform a simultaneous fit to the top mass and the jet energy scale as determined directly from the dataset. A summary table of the all the systematic errors on the top mass measurement after the jet energy fit is shown in table 1. From this table it can be seen that, even with a reduced jet energy dependance, the residual systematic error on the top mass due to jet energy uncertainties still amounts to about 23% of the total error on the top mass measurement.

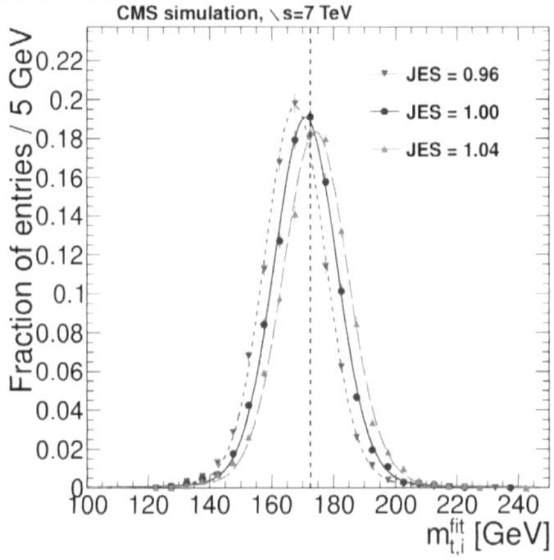

Figure 4: Fitted top invariant mass as a function of the jet energy scale for Monte Carlo simulations

	δ_{m_1} (GeV)	δ_{JES}
Calibration	0.15	0.001
b-tagging	0.17	0.002
b-JES	0.66	0.000
p_T and η dependent JES	0.23	0.003
Jet energy resolution	0.21	0.003
Missing transverse energy	0.08	0.001
Factorization scale	0.76	0.007
ME-PS matching threshold	0.25	0.007
Non-$t\bar{t}$ background	0.09	0.001
Pile-up	0.38	0.005
PDF	0.05	0.001
Total	1.18	0.012

Table 1: List of systematic uncertainties extracted from [4].

3.3. Inclusive jet cross section in pp collisions

The results of the measurement of the ratio of the jet differential cross section to the theoretical prediction as a function of p_T [5] is shown in figure 5. The shaded yellow band represents the total experimental systematic uncertainty which is dominated by jet energy uncertainties followed by a smaller contribution from uncertainty on luminosity estimations. The uncertainties from the proton parton distribution function are included as part of the uncertainty in the theory. Changes in the jet energy uncertainty propagate directly to this measurement.

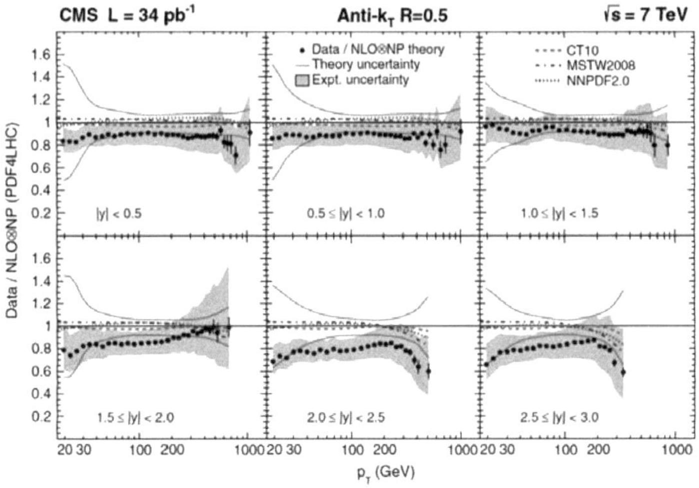

Figure 5: Ratio of the inclusive jet cross section of data over theoretical prediction as a function of jet p_T for different jet η regions. The yellow bands indicate the experimental uncertainty and the solid lines the theoretical uncertainty.

4. Reducing the effect of jet energy correction uncertainties

4.1. Better methodologies

The basic way to reduce the effect of jet energy uncertainties on a physical quantity is the utilization of better and smarter methodologies. Many analysis take advantage of other information on the events to produce an measurement of the jet energy corrections from their same dataset and often simultaneously measure the quantity of interest and the jet energy scale.This is most commonly done for top quark mass measurements, where a measurement of the correction scale is obtained from the hadronic decay of the W boson [4]. Furthermore, the measurement is not only done in the same dataset where the measurement is done, but an uncertainty is obtained "in-situ" in an event by event basis.

Another example is the study of dijet invariant mass spectrum in the W+2jet channel where the particular study is in the 125 to 180 GeV region in order to investigate an excess of event reported by the CDF collaboration [6]. This analysis fits the signal region to the excess seen at CDF but also reduces the dependance on jet energy uncertainty by performing a simultaneous fit to it. In addition, this dependance is further reduced by convoluting the fit in the signal region with an independent measurement of the systematic error due to jet energy uncertainties performed on a control region. This also require the consideration of the possible difference in jet-flavor between both regions. This methodology results in a very limited effect of the jet energy uncertainties on the final result.

4.2. Using full-fledged uncertainty correlations

There is another techniques that could help reduce the effect of uncertainties even more. At CMS we have provided not only up and down absolute uncertainties but also the up and down uncertainties as a function of the p_T and η of the jet for each different source. Each source represents a $\pm 1\sigma$ uncertainty on the jet energy. The plot on figure 6 shows some of the total uncertainty and the breakdown by sources that depend on p_T .

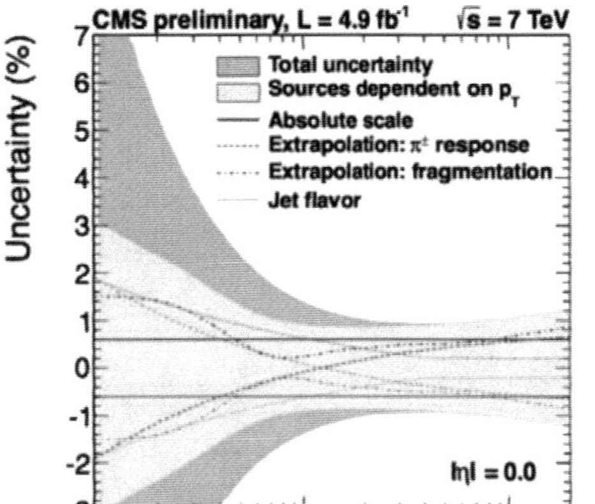

Figure 6: Sources of jet energy scale uncertainty. The yellow band represent the sources that depends on p_T . The grey band represent the total uncertainty and contain uncertainties that depend on eta.

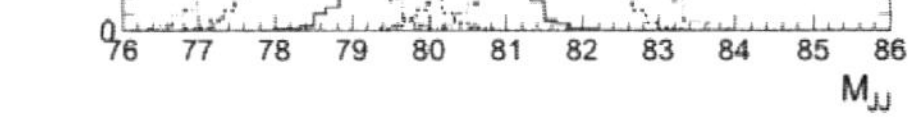

Figure 7: Difference between using absolute uncertainties and the uncertainty sources and their correlations on the dijet invariant mass distribution. Produced with toy Monte Carlo. Standard Up and down approach shifts the peak 1.85 GeV. Using the fully correlated uncertainty sources the shift is 1.18 GeV.

For any given jet these uncertainty sources are mutually uncorrelated, but each source is fully correlated with the same source of another jet, and this true whether the other jet is in the same event or not. When a physical quantity is derived from information that depends on one or more jets the proper uncertainty on this quantity can be obtained by varying each source independently and evaluating the systematic change in the physical quantity. The final error on the quantity can then computed as the sum in quadrature of the systematic error obtained from each source. This methodology is akin to the computation of the uncertainties due to the different eigenvalues of the parton distribution function.

These correlations can reduce the uncertainty in cases where there is a direct data-Monte Carlo comparison of shapes, regardless of the actual "position" of either data or MC in the distribution of interest. For example the Higgs boson production in association of vector boson is described in [7] and the results are obtained from a direct comparison of the dijet invariant mass of a Higgs boson produced with a given mass to data. In this case the correlation of the different uncertainty sources between data and Monte Carlo can play a beneficial role reducing the overall sensitivity to jet energy uncertainties.

In general let's consider the effect on the dijet invariant mass distribution produced with toy Monte Carlo and shown in figure 7. The peaked distribution could represent a quantity such as the background-subtracted distribution where a resonance signal is to be fit. The distribution is narrow just to emphasize the effect but the same results apply to any peaked distribution. When using the absolute energy uncertainty up and down the peak shifts about 1.85 GeV with respect the original peak. However, when using the fully correlated uncertainties the shift is only 1.18 GeV. It should be noted that taking the uncertainty correlation into account is always the right way of propagate the uncertainties in each source. However it is not guaranteed that the net effect on the physical quantity is the reduction of the systematic error when comparing with the error computed using the absolute total uncertainty.

5. Conclusions

The effect of jet energy uncertainties on physical observables can be significant and needs to be dealt with appropriately. Several analyses achieve a reduced sensitivity to these uncertainties with the incorporation of smarter methodologies such as the simultaneous fit of the uncertainties when possible or the further constraint of these uncertainties from control regions of the analysis.

The incorporation of the different uncertainty sources into the analysis is preferred against the simple use of total absolute uncertainties because of the proper consideration of uncertainty correlations between jets. Although not necessarily by construction, the usage of uncertainty sources can result in smaller uncertainties as shown in the example case discussed in the previous section.

References

[1] The CMS collaboration 2012 CALOR-2012 conference proceedings. "Pileup measurement and mitigation techniques in CMS"
[2] The CMS collaboration 2011 *JINST* **6** P11002
[3] The CMS collaboration CALOR-2012 conference proceedings. "Determination of the Jet Energy Scale in CMS"
[4] The CMS collaboration 2012 "Measurement of the top quark mass in the muon+jets channel" *CMS-PAS* TOP-11-015
[5] The CMS collaboration 2011 "Measurement of the Inclusive Jet Cross Section in pp Collisions at $\sqrt{s} = 7$ TeV" *Phys. Rev. Lett.* **107**, 132001
[6] The CMS collaboration 2012 "Study of the dijet mass spectrum in W+2 jets events" *CMS-PAS* EWK-11-017
[7] The CMS collaboration 2011 "Search for Higgs Boson in VH Production with H to $b\bar{b}$" *CMS-PAS* HIG-11-031.

Upgrade plans for the ATLAS Forward Calorimeter at the HL-LHC

John Rutherfoord on behalf of the ATLAS Liquid Argon Calorimeter Group

Physics Department, University of Arizona, Tucson, AZ 85721, USA

E-mail: rutherfo@physics.arizona.edu

Abstract. Although data-taking at CERN's Large Hadron Collider (LHC) is expected to continue for a number of years, plans are already being developed for operation of the LHC and associated detectors at an increased instantaneous luminosity about 5 times the original design value of 10^{34} cm^{-2} s^{-1}. The increased particle flux at this high luminosity (HL) will have an impact on many sub-systems of the ATLAS detector. In particular, in the liquid argon forward calorimeter (FCal), which was designed for operation at LHC luminosities, the associated increase in the ionization load at HL-LHC luminosities creates a number of problems which can degrade its performance. These include space-charge effects in the liquid argon gaps, excessive drop in potential across the gaps due to large HV supply currents through the protection resistors, and an increase in temperature which may cause the liquid argon to boil. One solution, which would require opening both End-Cap cryostats, is the construction and installation of new FCals with narrower liquid argon gaps, lowering the values of the protection resistors, and the addition of cooling loops. A second proposed solution, which does not require opening the cryostat cold volume, is the addition of a small, warm calorimeter in front of each existing FCal, resulting in a reduction of the particle flux to levels at which the existing FCal can operate normally.

1. Introduction

Each of the two identical ATLAS Forward Calorimeters (FCal) is designed to work well in the harsh radiation environment near the LHC beams where the energy and density of the debris from the copious min-bias events are highest. The calorimeter design [1] is quite simple, employing only a few radiation tolerant materials. The active material of these sampling calorimeters is liquid argon. The ionization electrons drifting in the liquid argon gaps form the signal. These gaps are unusually small (~ 0.27 mm in the first module) in order to avoid space-charge effects which would otherwise degrade the signal. Mechanical tolerances are held at the 1% level with a rod-tube electrode structure giving a good uniformity of response. Each FCal is made up of three modules, one behind the other, as seen from the interaction point (IP). The first, FCal1, is designed as an EM calorimeter with copper as the absorber material. Fig. 1 shows the structure of a unit cell of the FCal1 module. The second (FCal2) and the third (FCal3) are hadronic modules made of tungsten absorber in order to limit the longitudinal and transverse spread of the showers. Their unit cells look similar to the FCal1 unit cell.

The present ATLAS FCals perform very well at the luminosities reached to date ($\sim 6 \times 10^{33}$ cm^{-2} s^{-1}) and are expected to continue to work well up to luminosities at the LHC design value of 10^{34} cm^{-2} s^{-1} with a comfortable margin. But as the High Luminosity LHC (HL-LHC) phases in, several effects will lead to a degradation of the FCal response, first at the highest values of $|\eta|$ and then spreading to lower $|\eta|$ values as the luminosity approaches the new design value of 5×10^{34} cm^{-2} s^{-1}. The present CERN schedule projects this phase of the upgrade to come on-line after the long shutdown starting in 2022. Two R&D projects, in progress, are designed to project how the present FCal response will degrade at the HL-LHC luminosities. Each project is the subject of another talk at this conference [2][3].

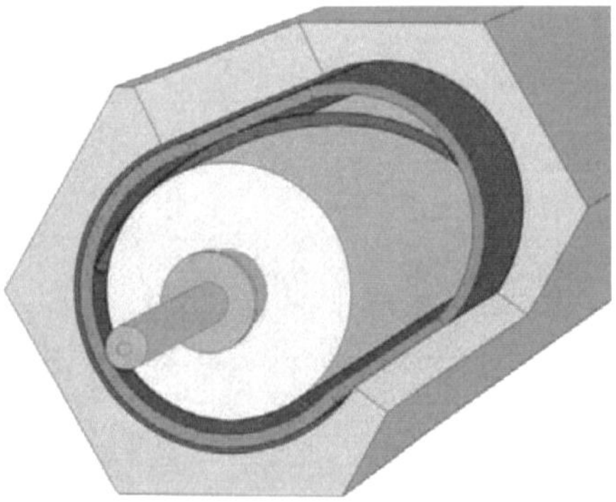

Figure 1. FCal1 unit cell. Each FCal1 module is made up of 12,260 unit cells, each parallel to the LHC beam line. The nearest neighbor spacing is 7.5 mm. The copper rod diameter is 4.712 mm while the copper tube inner diameter is 5.250 mm, making the liquid argon gap 0.269 mm. The depth of the module and of each unit cell is 445 mm. The insulating PEEK fiber of diameter 0.25 mm is helically wound around the rod to keep it centered within the tube. The copper absorber surrounding the tube is not physically separated from the absorber of the neighbor cells.

2. Luminosity limitations

Three possible limitations at HL-LHC luminosities have been identified so far. 1) Space-charge effects [4] will degrade the response. The slowly-drifting positive argon ions build up in the liquid argon and distort the electric field. At sufficiently high ionization rates the electric field falls almost to zero near the anode where electrons then also begin to accumulate. In this "dead region" of the gap electrons and positive argon ions recombine thereby degrading the signal. 2) The High Voltage protection resistors are too large for the d.c. currents expected at the HL-LHC luminosities. These resistors were designed to limit the current drawn from the supplies in case of arcing, thereby protecting the sensitive front-end electronics from large pulses. The potential drop across these resistors will cause the potential across the liquid argon gaps to sag to unacceptably low values at HL-LHC luminosities. 3) The heat generated in the calorimeter absorbers due to dE/dx energy losses (and, to a lesser extent, due to the Ohmic heating from the electrical currents in the liquid argon gaps) will grow to levels which threaten to boil the liquid argon. We expect that temperature measurements in the ATLAS End-Cap calorimeters at the higher luminosities achieved this year, coupled with our detailed heat-flow simulations, should allow reasonably accurate projections of this boiling threshold. From simulations alone it appears to be close to 10^{35} cm^{-2} s^{-1}, a bit beyond the presently projected HL-LHC luminosity.

Fig. 2 shows one estimate of how the signal will degrade due to the first two limitations. This estimate has large uncertainties which will be reduced as the two R&D projects are completed.

3. Upgrade options

Three possible paths have been identified so far to address the degradation of the FCal performance at HL-LHC luminosities. Either 1) replace the present FCals with new ones designed for the higher luminosities, or 2) insert a new, small module, a miniFCal, in front of each of the present FCals, or 3) do nothing.

The first path will be forced upon us in the event the cold electronics for the Hadronic End-Cap Calorimeter (HEC) must be replaced. This electronics was not designed for the

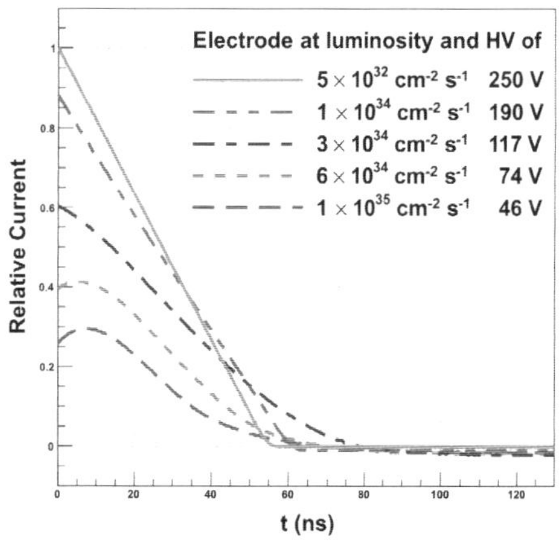

Figure 2. Simulated electrical signals at the FCal1 electrodes (before shaping by the electronics) near $|\eta| = 4.7$ where the ionization rate is highest. These signals are on top of a constant (i.e. no simulated fluctuations) d.c. background of ionization due to various rates of min-bias events. Various luminosities are shown. In the spirit of a safety margin, luminosities in excess of the expected values are included in this study. Note that the signal not only degrades in amplitude but also the signal shape changes in a way which depends on the poorly known electron-argon ion recombination rate.

HL-LHC integrated luminosities but the safety margin in effect at the design time may be sufficient to ensure their survivability. Another talk at this conference [5] will describe the present understanding of the future of the HEC cold electronics. In order to access the HEC cold electronics the FCal must first be removed from the cryostat. Because the FCal will be activated it cannot be re-installed due to excessive personnel exposure. Even if the present HEC cold electronics will survive, the replacement of the present FCal modules with new ones is still an option.

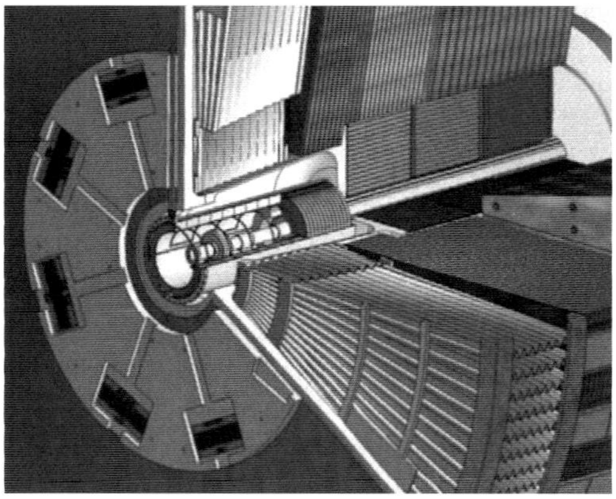

Figure 3. Inner portions of the ATLAS End-Cap Cryostat with parts of the Electromagnetic End-Cap Calorimeter, Hadronic End-Cap Calorimeter, and FCal calorimeter in view. Sections of the cryostat and of the calorimeter modules have been cut away to show more of the structure, represented schematically. Also shown is the proposed diamond wafer miniFCal at the far (from the IP) end of the cryostat alcove. The other miniFCal versions look similar on the scale of this drawing. The front-end electronics is seen mounted on the front face of the calorimeter at larger radius from the beams.

New FCal modules will have the same design as the present modules with the following exceptions. 1) The liquid argon gaps will be smaller (of order 0.10 mm for FCal1) to avoid degradation due to space-charge effects at the highest HL-LHC luminosities. Substantial R&D has shown that reliable gaps as small as 0.10 mm can be manufactured. 2) The protection resistors in the High Voltage distribution system (located inside the cryostat) will be reduced sufficiently so that the highest current draw will not significantly reduce the potential across the liquid argon gaps. The present protection resistor values are now understood to be overly conservative since not a single preamplifier has been lost due to arcing in the liquid argon gaps. 3) Liquid nitrogen cooling loops will be integrated into the outer periphery of the modules to better remove the heat due to the energy deposited in the modules.

In terms of the ATLAS calorimeter performance, this first path is the best option. But there are concerns about replacing the FCals in the case that the present HEC cold electronics is deemed able to survive throughout the HL-LHC running. Although the End-Cap cryostats were designed to be re-opened, it was found that the seals would not hold during the original closing of the cold covers and so these covers were welded shut. This increases the difficulty of opening and then re-closing these cryostats. The associated risk, cost, and schedule will figure into the decision.

The second path is very attractive from a risk and schedule point-of-view. The present End-Cap cryostats include a cylindrical "alcove", coaxial with the beam line, and extending from the front face of the cryostat to the cryostat walls just in front of the FCal1 module as can be seen in Fig. 3. This alcove is open to the air and has space for a cylindrical calorimeter of outer radius of about 175 mm. A warm sampling calorimeter of sufficient depth in this location ahead of the present FCal would reduce the energy deposited in the present FCal modules so that the three problems described above would be avoided. The challenge is to find a sensitive medium which will survive the radiation and yet make a calorimeter with adequate performance.

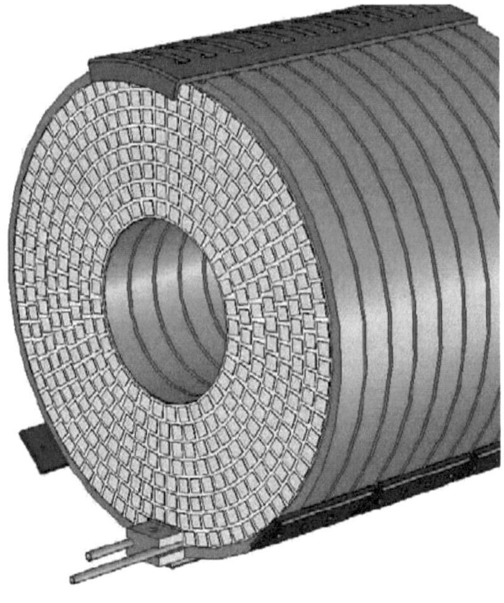

Figure 4. Drawing of miniFCal with diamond wafers. The first (of 12) copper plate is removed to show the first (of 11) ceramic disk with one version of tiling by diamond wafers. A complementary pattern of wafers is on the back side of the ceramic disk.

The miniFCal proponents have considered diamond wafers (of order 1 by 1 cm^2) as the sensitive material. See Fig. 4. The charge collected from the wafers would form the signal. While diamond is known for its radiation hardness, the environment near shower maximum in the forward region is particularly severe. Neutron fluxes of order 3×10^{16} $\mathrm{n/cm^2/yr}$ are predicted [6] at HL-LHC luminosities. Measurements [7] at TRIUMF show the response of diamond wafers near hadronic shower max dropping by about a factor 20 during an exposure of about 2×10^{17} $\mathrm{protons/cm^2}$.

More R&D would be necessary to develop the diamond wafer technology to the point of producing a reliable calorimeter for a working experiment. Questions to address are: 1) How linear is the diamond signal response to energy deposit? 2) Over what dynamic range is this response linear? 3) What is the in situ calibration method? 4) What energy resolution can be achieved? 4) Is the signal sufficiently stable with environmental conditions? 5) Is the sampling fraction large enough to minimize spurious signals? 6) Is the response sufficiently uniform over the face of the calorimeter? One of the strong features of the ATLAS liquid argon calorimeter system is that it has minimal transitions in η and ϕ. The miniFCal would introduce an awkward transition in an η region which presently has very uniform response. Finally there is worry that the cost of the diamond wafers is rather high.

Because of concerns about the degradation of the diamond wafer response with exposure to neutrons, two other technologies for a miniFCal have been considered. Ionization of xenon gas is likely to have stable response over the years of HL-LHC running. High pressures, of order 10 atmospheres, will help in reaching an acceptable sampling fraction. But the miniFCal must have sufficient density to limit the spread of showers in the longitudinal and transverse directions and this pushes the sampling fraction down to uncomfortable levels. Furthermore locating a pressure vessel in a semi-enclosed volume surrounded by the rather thin End-Cap cryostat warm wall and open to the ATLAS inner tracker raises safety concerns.

The other alternative is the well-understood liquid argon technology. In this case the miniFCal unit cell might be of similar or the same design as that proposed for the upgraded FCal1. This miniFCal might sit in its own cold vessel but share the warm vessel of the End-Cap cryostat. But the cold cryostat walls, the signal and HV cold feedthroughs, and the cryostat services will require additional space which will reduce the fiducial coverage of the miniFCal and add dead material which will degrade the overall calorimeter performance.

The miniFCal concept, regardless of technology, involves trade-offs which will lead to compromises in the calorimetric performance in the forward region. The "do nothing" path also leads to degraded calorimetric performance in this region as described above. Which is worse? The answer is not yet clear, mostly because the miniFCal technologies have not been developed far enough to allow reasonable projections. But the two liquid argon R&D studies at high ionization rates [2][3] must also be completed before the degradation of the present FCal will be well-understood. And even then we anticipate there will be aspects of the degradation which will remain uncertain.

4. Summary

In summary, the present ATLAS liquid argon FCal works well and is expected to continue to work well up to luminosities a little above the LHC design luminosity of 10^{34} cm^{-2} s^{-1}. At higher luminosities the performance of the present FCal will start to degrade due to space-charge effects, sagging of the potential across the liquid argon gaps, and possible boiling of the liquid argon. Three paths for dealing with the degradation have been identified. Either 1) replace the present FCals with ones designed for higher ionization rates, or 2) place a new miniFCal in front of each of the present FCals, or 3) do nothing and suffer the degraded performance of the present FCals. Paths 2) and 3) both lead to degraded performance. The magnitude of this degradation and its effects on ATLAS physics may be determined with further R&D.

References

[1] Artamonov A *et al.*, *The ATLAS Forward Calorimeter*, 2008 JINST 3 P02010.

[2] Rutherfoord J and Walker RB, *Luminosity limits for liquid argon calorimetry*, 2012 in this CALOR Conference Proceedings.

[3] Seifert F, *Liquid argon calorimeter performance at high rates*, 2012 in this CALOR Conference Proceedings.

[4] Rutherfoord J, *Signal degradation due to charge buildup in noble liquid ionization calorimeters*, 2002 Nucl. Instrum. Meth. A **482** 156.

[5] Nagel M *Neutron and proton tests of different technologies for the upgrade of cold readout electronics of the ATLAS hadronic endcap calorimeter* 2012 in this CALOR Conference Proceedings.

[6] Shupe M, http://atlas.physics.arizona.edu/~shupe/LHC_Upgrade_Studies/ATLAS_MINIFCAL/FLUXTEXTS_69008/Zoom_NeutFlux.txt. The table gives the flux of neutrons at an instantaneous luminosity of 10^{34} cm^{-2} s^{-1} predicted by a GEANT3/CALOR simulation of the ATLAS detector with its shielding in the cavern.

[7] Axen D *et al.*, *Diamond detector irradiation tests at TRIUMF*, 2011 JINST 6 P05011

Journal of Physics: Conference Series **404** (2012) 012016 doi:10.1088/1742-6596/404/1/012016

Luminosity limits for liquid argon calorimetry

Rutherfoord J and Walker RB

Physics Department, University of Arizona, Tucson, Arizona 85721

E-mail: rutherfo@physics.arizona.edu

Abstract. We have irradiated liquid argon ionization chambers with betas using high-activity Strontium-90 sources. The radiation environment is comparable to that in the liquid argon calorimeters which are part of the ATLAS detector installed at CERN's Large Hadron Collider. We measure the ionization current over a wide range of applied potential for two different source activities and for three different chamber gaps. These studies provide operating experience at exceptionally high ionization rates. We can operate these chambers either in the normal mode or in the space-charge limited regime and thereby determine the transition point between the two. From the transition point we indirectly extract the positive argon ion mobility.

1. Introduction

The liquid argon sampling calorimeters in the ATLAS detector [1] collect the ionization produced in the argon by penetrating, high energy charged particles. Most of these high energy charged particles within the calorimeter are in electromagnetic or hadronic showers produced by a higher energy particle incident on the face of the calorimeter. These incident particles typically come from one of the copious min-bias proton-proton collisions produced by CERN's Large Hadron Collider (LHC) [2]. At the LHC design energy and luminosity there are of order 100 particles produced in each min-bias proton-proton collision and these collisions come at a rate of almost 1 GHz. The number of charged particles above a kinetic energy of 1 MeV in a typical shower varies with depth in the calorimeter, reaching many tens at shower maximum. At small angles ($\sim 1^\circ$) to the incident proton beams at an instantaneous luminosity of 5×10^{34} cm^{-2} s^{-1} (as proposed for the high luminosity upgrade of the LHC, the HL-LHC) at about 5 m from the interaction point there will be about one gamma of average energy about 15 GeV per square centimeter per 25 ns crossing incident on the face of the calorimeter. The ionization rate in a liquid argon gap located at electromagnetic shower maximum is about 3×10^{11} /mm^3/s.

The liquid argon ionization chambers, distributed within a metal matrix, are made up of two electrodes separated by a gap of about 2 mm in most of the ATLAS calorimeters. But in the forward direction where the energy and density of particles incident on the face of the calorimeter are highest, the gaps are much smaller, approximately 0.27 mm at depths up to hadronic shower maximum. For either gap the applied electrical potential is adjusted to give an electric field within the gap of approximately 1000 V/mm. At this electric field the ionization electrons drift at a speed of about 4.3×10^6 mm/s and the positive argon ions drift in the opposite direction at about 100 mm/s. So for the 0.27 mm gap an ionization electron takes about 63 ns to drift from one electrode to the other while the positive argon ion takes about 2.7 ms. On the scale of the gap size this ionization is roughly spatially uniform. And on the scale of ion drift times the current drawn from the high voltage power supply due to this ionization is roughly constant in time.

2. Phenomenology

High energy charged particles passing through the liquid argon shake loose outer shell electrons from those argon atoms close to the particle track. Without an electric field, the ionization electron would likely recombine with its argon ion. But with an electric field the ionization electrons may escape and drift freely in the liquid argon. Measurements [3] give the average energy loss of the charged particle in liquid argon required to create one electron-ion pair as $W = 23.6^{+0.5}_{-0.3}$ eV at an electric field of 22 kV/cm. This gives about 9000 ionization electrons per mm of track.

Several forms of recombination deplete the ionization electrons. We will divide these forms into two classes. In the first category ionization electrons recombine with their corresponding ion or with other ions produced by the same particle track. This form of recombination is independent of the number of tracks in the chamber and dominates when the tracks are widely separated in space and/or time. In the second category the ionization electrons recombine with ions produced by a different particle track. This "bulk" recombination is proportional to the density of ionization electrons and ions and dominates at high ionization rates.

At such high ionization rates the slowly-drifting positive argon ions build up in the liquid argon gap enough to distort the electric field [4]. At the critical ionization rate D_c the distortion is so large that the electric field is almost completely screened next to the anode. In this case the normally fast-drifting electrons also slow so much that they accumulate in this region next to the anode. As the ionization rate increases above this critical value the screened region near the anode, with its accumulation of electrons, expands toward the cathode. The density of positive argon ions equals the density of ionization electrons in this region and they recombine at the same rate they are produced.

As is usual, we will assume the drift velocity of the positive argon ions is given by

$$v_+ = \mu_+ E \tag{1}$$

where μ_+, the positive ion mobility, is independent of electric field E. In this case the critical ionization rate is given by

$$D_c = \frac{4V_0^2 \epsilon \mu_+}{ea^4} \tag{2}$$

where V_0 is the electrical potential across the gap, ϵ is the permittivity of liquid argon, e is the magnitude of the electron charge, and a is the gap width, i.e. the distance from anode to cathode. The units of D_c are the number of electron-ion pairs produced in the liquid argon per unit volume per unit time. Note that D_c increases as the gap width decreases, allowing the chamber to operate normally at higher rates. We next define D_i as the actual ionization rate. This is proportional to the flux of high energy particles in the liquid argon gap. And finally we define the *relative* ionization rate

$$r = \frac{D_i}{D_c} \tag{3}$$

The value of the positive ion mobility μ_+ is not well known. As explained in reference [4] there are large inconsistencies in the many measurements of μ_+ reported in the literature.

3. The measurement

In order to determine the threshold ionization rate D_c separating the normal operating regime from the space-charge limited regime, we designed a liquid argon ionization chamber with an argon volume which is approximately uniformly illuminated by betas as shown in Fig. 1. Two

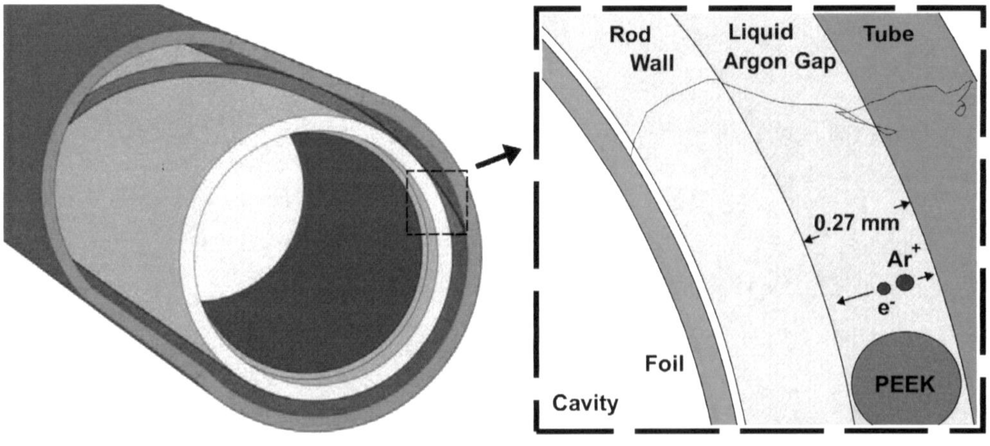

Figure 1. Liquid argon ionization chamber with Strontium-90 beta source. The perspective drawing on the left shows a cut-away of the two electrodes: a copper rod (yellow) within a thin-walled copper tube (green). The rod is held in place within the tube by an insulating PEEK fiber (red) wound helically around the rod. The gap between the rod and tube fills with liquid argon forming the ionization chamber. The solid copper rod of diameter 4.712 mm has a concentric cavity of diameter 4.212 mm machined out to a depth, along the axis of the rod, of 14 mm. Inserted into this cavity is a rolled copper foil (blue) of 10 mm length, coated on the outer surface with the Strontium-90 beta source. The open end of this rod is then welded to another solid copper rod (not shown), making one rod of longer length. The drawing on the right shows a cross-sectional close-up of this structure. Overlayed is the track of one simulated beta, from the Yittrium decay product of a Strontium-90 decay, leaving the source on the surface of the foil with kinetic energy of 1.48 MeV and passing through the liquid argon gap. An energetic delta ray, produced in the liquid argon, is represented by the red track.

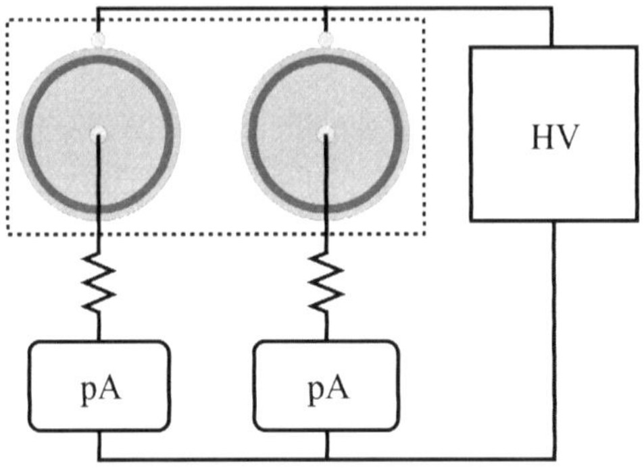

Figure 2. Schematic current readout circuit. Shown at the signal connection end are two ionization chambers with copper parts in yellow and liquid argon in red. HV is the high voltage power supply which places a potential between the rod and tube and each pA is a picoAmmeter. The resistors are 4 MΩ for the 50 mCi source and 10 MΩ for the 2 mCi source. The currents we measure are so small that the potential drop across each resistor is almost negligible.

rods containing beta sources were produced, one specified at 50 mCi and the other at 2 mCi. The ratio of activities turned out to be 21.37 rather than the expected value of 25. Tubes of three different inner radii were used at different times during this measurement to give liquid argon gaps of three different widths. An electric potential across the electrodes causes the charges in the intervening liquid argon to drift, inducing a current in an external circuit shown in Fig. 2.

A "data run" consisted of measurements of the currents from the 50 mCi and the 2 mCi

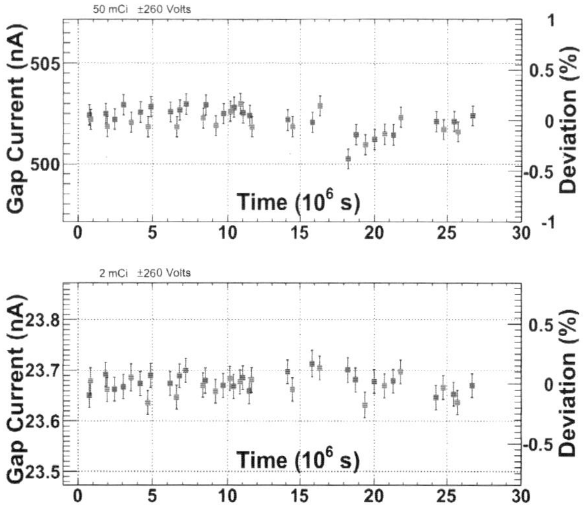

Figure 3. Currents in the 268 μm ionization chambers at the operating potential of 260 V as a function of time of the run, after correction for the decay of the beta sources. Red (green) data points are at positive (negative) polarity. This potential corresponds to the standard electric field of approximately 1000 V/mm used in the ATLAS liquid argon calorimeters.

chambers with the same gap width at a series of closely-spaced values of the potential across the electrodes. Each run lasted between 3 and 8 hours. Eight runs, each with a different HV power supply, were required to cover the full range of potentials at both polarities. A "fill" refers to the filling of our cryostat with liquid argon and operating it during a series of runs with the two chambers of a given gap width. Runs were repeated many times during a fill to check for reproducibility. These runs were interspersed with runs to test for various small effects like the very small leakage of radiation from the 50 mCi chamber into the 2 mCi chamber. There were three fills in all and between fills the tubes were changed to give different gap widths. Fills lasted several months, the longest being a little over 10 months.

Fig. 3 shows measurements of the chamber currents at a potential of 260 V when using 268 μm gaps during the long 10-month fill (about 2.7×10^7 seconds). For the 50 mCi chamber this is approximately equivalent to the exposure experienced by one of the ATLAS forward calorimeter ionization chambers at a depth in the calorimeter near electromagnetic shower maximum at an angle to the LHC proton beam of 1.04° for an integrated luminosity of about 500 fb^{-1}. Over this 10-month period the activity of the Strontium-90 sources (half life of about 29 years) dropped by about 2%. The currents shown here are corrected for this effect. The corrected currents are stable at a level of approximately 0.1%. This result suggests that the liquid argon technology as the sensitive medium of a sampling calorimeter is a good choice for experiments operating at very high intensities.

To map out the behavior of liquid argon ionization chambers as they are driven into the space-charge limited regime we would ideally vary the ionization rate by varying the activity of the beta sources. But with only two different source activities we were forced to adopt a different strategy. This alternate approach is based on the observation that the critical ionization rate D_c depends on the potential V_0 applied across the two electrodes. Rather than varying D_i (by varying the source activity) we, instead, vary D_c (by varying V_0). But in varying V_0 we not only change D_c we also change D_i, a potential complication.

However, if we divide the currents from the 50 mCi chamber by the currents from the 2 mCi chamber we cancel effects which are not rate-dependent and make manifest those effects which are rate-dependent, at least for potentials above the point where the 2 mCi chamber enters the space-charge limited regime.

4. Interpretation of the data

Fig. 4 shows the currents as a function of the potential applied across the gaps for one of the three gap widths. The currents from the 2 mCi chamber have been multiplied by 21.37 for comparison with the currents from the 50 mCi source. Such data is referred to as a High Voltage Plateau Curve because the current "plateaus" to a relatively constant value at higher values of potential. It is interesting that this data shows a very slow rise with potential over the full high end of the range we were able to cover. That is, there is no evidence that the current actually becomes constant at high potentials.

For potentials above about 100 V the currents from the 2 mCi and 50 mCi chambers have the same shape. This is the range in which both chambers are operating in the normal regime. Below about 100 V the 50 mCi chamber goes into the space-charge limited regime. This can be seen better in Fig. 5 where the ratios of the currents are shown for all three gap widths. At the higher values of potential both chambers, at a given gap width, are operating in the normal regime. Going down in potential, the 50 mCi chamber first enters the space-charge limited regime where the ratio falls off. Eventually the 2 mCi chamber also enters the space-charge limited regime where the ratio approximately levels off again. As predicted by Eq. 2 the potential of the critical ionization rate varies with the gap width. A simplified analytic approximation superimposed on the ratio data shows qualitative agreement. This simplified calculation predicts the ratio at low potentials will be the 3/4ths power of the ratio at the high potentials.

The simple predictions shift left and right depending on the value assumed for the positive

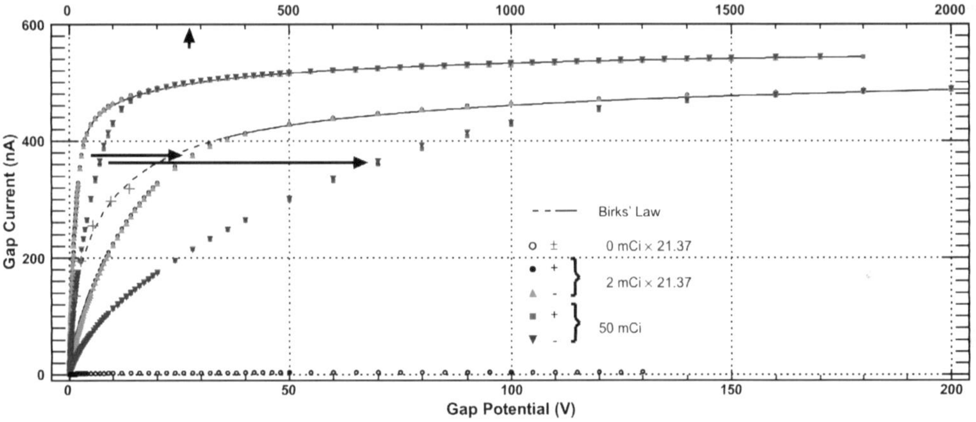

Figure 4. Currents from the ionization chambers with 268 μm gaps versus the applied potential. Both polarities are shown. Absolute values are taken for the negative currents and potentials. For potentials at and below 200 V the data points are plotted twice, once using the potential axis at the bottom of the plot and once using the potential axis at the top of the plot. Horizontal arrows connect two example pairs of identical data points. For potentials above 200 V data points are plotted only once using the axis at the top. The currents from the 2 mCi ionization chamber are multiplied by 21.37 to show how the currents agree in shape at higher potentials. The currents from the 0 mCi ionization chamber, multiplied by 21.37, are also shown. A Birks' Law [5] inspired parameterization is represented by the solid curve in the region of potential where the data was fit. This parameterization, extrapolated to lower potentials, is shown as a dashed line. A vertical arrow on the top axis shows the nominal operating potential in the ATLAS detector.

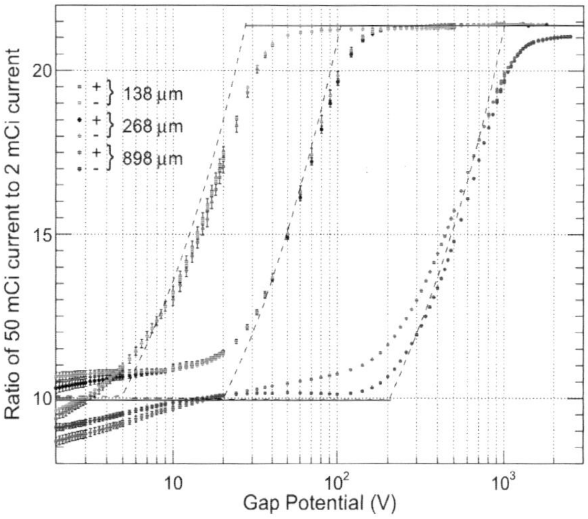

Figure 5. Ratio of the 50 mCi chamber currents to the 2 mCi chamber currents as a function of the potential across the ionization chamber for, from left to right, the 138 μm gap, the 268 μm gap, and the 898 μm gap. Data for the two polarities of potential are shown separately as indicated in the key. A simple analytic approximation is shown by dashed curves.

argon ion mobility μ_+. The value chosen for the plot in Fig. 5 is $\mu_+ = 0.10$ mm^2/V·s. At this time we assign an rms error of ± 0.02 mm^2/V·s dominated by uncertainties in the simple analytic approximation. We have written a detailed numerical simulation program with all the effects we can imagine might play a role. Many of the parameters governing these effects have not been measured at all or not measured over the range of electric fields covered by our data so we are trying to determine the sensitivity of the data to these unknown parameters and their correlations.

5. Summary

Using a strong-activity beta source distributed uniformly across a large-area foil we have exposed liquid argon chambers to ionization equivalent to the worst location in the ATLAS calorimeter system at luminosities approximately a factor two above the LHC design value. A weaker source acted as a control. We found that the chamber currents were stable at the 0.1% level during very long runs. This bodes well for prolonged running at the LHC design luminosity and a little above.

By varying the electrical potential across the liquid argon ionization chambers we were able to operate these chambers in the normal operating regime or in the space-charge limited regime. The ratio of currents from the 50 mCi chamber to currents from the 2 mCi chamber cancelled out effects which are rate independent leaving a measure of rate dependent effects. Using simple analytic approximations we estimate the potential which separates the normal regime from the space-charge limited regime and extract a value for the positive argon ion mobility of 0.10 ± 0.02 mm^2/V·s. The error includes an estimate of uncertainties introduced by the simple analytic approximation. Full simulations, when completed, should allow smaller uncertainties.

6. References

[1] ATLAS Collaboration, *The ATLAS Experiment at the CERN Large Hadron Collider*, 2008 JINST 3 S08003.
[2] Lyndon Evans and Philip Bryant (editors), *LHC Machine*, 2008 JINST 3 S08001.
[3] Miyajima M, Takahashi T, Konno S, Hamada T, Kubota S, Shibamura H, and Doke T, *Average energy expended per ion pair in liquid argon*, 1974 Phys. Rev. A 9 1438 and Erratum in 1974 Phys. Rev. A 10 1452.
[4] J.Rutherfoord, *Signal degradation due to charge buildup in noble liquid ionization calorimeters*, Nucl. Instr. and Meth. A482 (2002) 156 and references therein.
[5] J.Birks, *Theory and Practice of Scintillation Counting*, Pergamon Press, 1964.

Journal of Physics: Conference Series **404** (2012) 012017 doi:10.1088/1742-6596/404/1/012017

Liquid Argon Calorimeter Performance at High Rates

Frank Seifert

Institut für Kern- und Teilchenphysik, Technische Universität Dresden, D-01062 Dresden.

E-mail: F.Seifert@physik.tu-dresden.de

Abstract. The expected increase of luminosity at HL-LHC by a factor of ten with respect to LHC luminosities has serious consequences for the signal reconstruction, radiation hardness requirements and operations of the ATLAS liquid-argon calorimeters in the endcap, respectively forward region. Small modules of each type of calorimeter have been built and exposed to a high intensity proton beam of 50 GeV at IHEP/Protvino. The beam is extracted via the bent crystal technique, offering the unique opportunity to cover intensities ranging from 10^6 p/s up to $3 \cdot 10^{11}$ p/s. This exceeds the deposited energy per time expected at HL-LHC by more than a factor of 100. The correlation between beam intensity and the read-out signal has been studied. The data show clear indications of pulse shape distortion due to the high ionization build-up, in agreement with MC expectations. This is also confirmed from the dependence of the HV currents on beam intensity.

1. Introduction

The Large Hadron Collider (LHC) started operation in 2009 and delivered about $5\,\text{fb}^{-1}$ collision data to both multi-purpose experiments, ATLAS and CMS, at a center of mass energy of 7 TeV until the end of 2011. A data collection of around $20\,\text{fb}^{-1}$ at 8 TeV is expected until the end of 2012. The increase of the instantaneous luminosity of the LHC is planned in multiple steps of upgrades. Therefore several phases of shutdowns are foreseen, where machine upgrades as well as upgrades of the LHC detectors will take place.

The maximum instantaneous luminosity reached at the time of writing is about $6.5 \times 10^{33}\,\text{cm}^{-2}\text{s}^{-1}$. A first shutdown is planned in 2013 and 2014 with the main goal to reach a center of mass energy of 13 TeV or 14 TeV. Also a luminosity of $1\text{-}2 \times 10^{34}\,\text{cm}^{-2}\text{s}^{-1}$ is expected to be reached after the so-called Phase 0 upgrade from 2014 on. After some years of data taking, the Phase I upgrade is planned to take place in 2018 with a luminosity of up to $3 \times 10^{34}\,\text{cm}^{-2}\text{s}^{-1}$ at the end of the period. Finally, a third shutdown period is foreseen in 2022 implementing the Phase II upgrade, after which the instantaneous luminosity is planned to be at $5\text{-}7 \times 10^{34}\,\text{cm}^{-2}\text{s}^{-1}$. This long term schedule is under discussion and might be modified in the future.

A main challenge after the Phase II upgrade is the operation of the liquid-argon (LAr) calorimeters of the ATLAS detector in the high luminosity environment, especially the endcap and forward calorimeters [1]. In the following, testbeam studies with small modules of the HEC, the EMEC and the FCal to prove their performance in high intensity environments are presented. The testbeam setup, the calorimeter test modules and the obtained results are described in more detail.

2. The ATLAS liquid-argon calorimeter system

The LAr calorimeter system of ATLAS uses copper, tungsten or lead as absorber material and liquid-argon as active medium. It is used for the EM barrel calorimeter, the EM endcap calorimeter (EMEC), the hadronic endcap calorimeter (HEC) and the forward calorimeter (FCal). Pictures or layouts of the EMEC, the HEC and the FCal calorimeters can be seen in Figure 1. The EM barrel and the EMEC modules use an accordion shape structure of the absorbers and LAr layers to reach a full coverage in ϕ without any cracks. This is also shown in Figure 1 left.

Figure 1. The LAr endcap calorimeter parts. From left to right: Photograph of one part of an EMEC module, graphics of a part of a HEC module and schematic structure of the FCal1 module. More details can be found in Reference [2].

The FCal is included inside the endcap cryostat and consists of one EM module (FCal1) and two hadronic modules (FCal2, FCal3). The size of the LAr layers is typically 2.1 mm in the barrel, between 0.9 mm and 3.1 mm in the EMEC, 1.95 mm in the HEC, and 0.250 mm (0.375 mm, 0.50 mm) for the FCal1 (FCal2, FCal3). In the following, it is referred to the FCal1 module only as it is exposed to the highest particle intensities.

A Phase II upgrade scenario of the FCal for the highest luminosities includes the replacement of the FCal1 module with a new one. The layout would be similar to the current FCal1, but with 100 μm LAr gaps instead of the 250 μm ones. This so-called sFCal module is expected to show better performance during the high intensities after the Phase II upgrade.

3. Liquid-argon endcap calorimeter testbeam studies

In 2006 a testbeam project at the U-70 proton synchrotron in Protvino, Russia was set with the aim to test the current EMEC, HEC, FCal1 and the proposed upgraded sFCal module with the 100 μm LAr gaps in the high intensity environment of the high luminosity LHC (HL-LHC). A testbeam setup which simulates the particle flux distribution through the calorimeter test modules as it is in ATLAS over a wide intensity range was installed in beamline 23 at IHEP. Therefore, the RF bunch structure was kept and used to simulate well separated calorimeter input pulses. Every sixth RF bunch was filled with protons leading to one extracted bunch every 1 μs. One accelerator fill was extracted over roughly 1.2 s (spill) using a bent crystal extraction technique. The extracted beam intensity ranges from 10^6 p/spill to around 3×10^{11} p/spill. The spill cycle time was about 9.5 s.

A schematic overview of the experimental setup in beamline 23 can be seen in Figure 2 [3]. The beam instrumentation consists of a secondary emission chamber for beam profile measurement, six scintillation counters for beam intensity monitoring in the low intensity region, a Cherenkov monitor for bunch-based beam intensity measurement, a scintillator hodoscope for beam position- and intensity monitoring and an ionisation chamber for beam intensity measurement over the whole intensity range.

After the beam instrumentation a first iron absorber of 0.7 λ and the cryostat housing the FCal test module is installed. Then a second iron absorber of 1.8 λ follows to adjust the particle flux through the following test modules of the EMEC and the HEC. This setup was optimized with Monte-Carlo simulations using the 2D gaussian beam shape of about 10 mm size.

The FCal test module consists of two parts with four tube groups each (four readout channels and four HV channels), one with the 250 μm LAr gaps and one with the proposed 100 μm ones for the sFCal. Either of these parts can be centered to the beam. The EMEC module consists of four absorber plates and three copper electrodes with 2 mm LAr gaps in between. The HV is divided in three separate channels and the readout is split perpendicular to the beam axis into four readout channels. The HEC module includes four HV and four readout channels and the size of the test modules upright to the beam is $\approx 36\,\text{cm}^2$ with a sensitive area of nearly $25\,\text{cm}^2$. A schematic view of the modules is shown in Figure 3 and more details can be found in [3].

Charged particles passing the LAr gaps induce the typical triangular pulse. This is shaped by the front end electronics, similar to ATLAS, and the readout is done by 25 ns sampling ADCs. Two of them are used in normal- and delayed mode so that an effective sampling period of 12.5 ns is reached. Two gains, low and medium, are available and the understanding of the whole readout chain is very good [3]. Figure 4 shows the average signal pulse shape of the EMEC module for four different beam intensity ranges. One main goal was to study the signal degradation of the calorimeter pulses with increased beam intensities. This is already visible in Figure 4 where the beam intensity increases from the upper left plot to the lower right plot.

The reason for the signal degradation is the positive ion buildup in the gaps, which becomes critical when it equals the charge at the electrodes. This leads to high currents drawn over the protection resistors and reduced electric fields over the gaps, which are partly even zero above the critical intensity. In this case the drift of the produced electrons is reduced or even completely stopped, leading to a reduction of the induced signal. The absolute ionization rate, D, divided by the critical ionization rate, D_C, is defined as the relative ionization rate, r:

$$r = \frac{D}{D_C} \tag{1}$$

The amount of the signal degradation is then dependent on r and also on the recombination

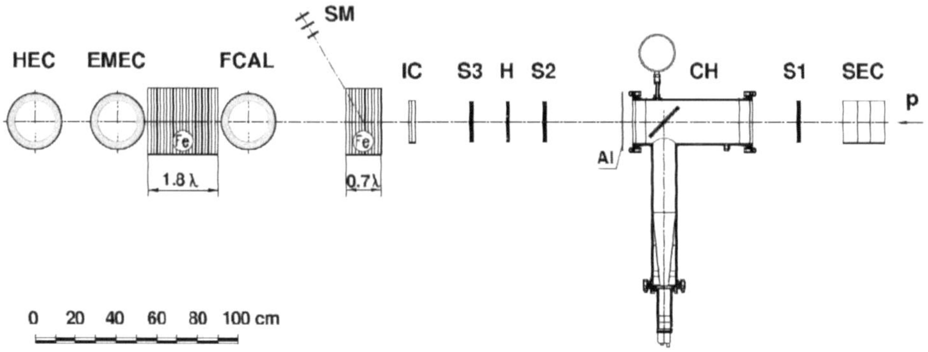

Figure 2. Schematic view of the experimental setup [3] in beamline 23 at IHEP, Protvino. The beam direction is from right to left. More details are described in the text.

Journal of Physics: Conference Series **404** (2012) 012017 doi:10.1088/1742-6596/404/1/012017

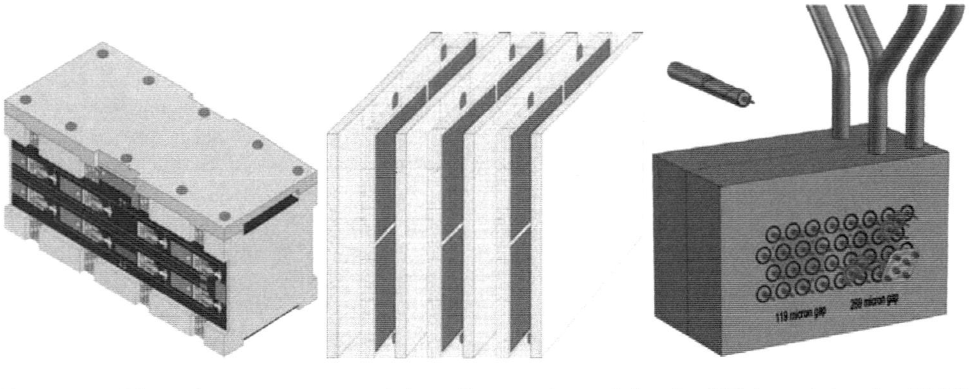

Figure 3. The calorimeter test modules. From right to left: the FCal module, the EMEC module and the HEC module. More details are given in the text and in Reference [3].

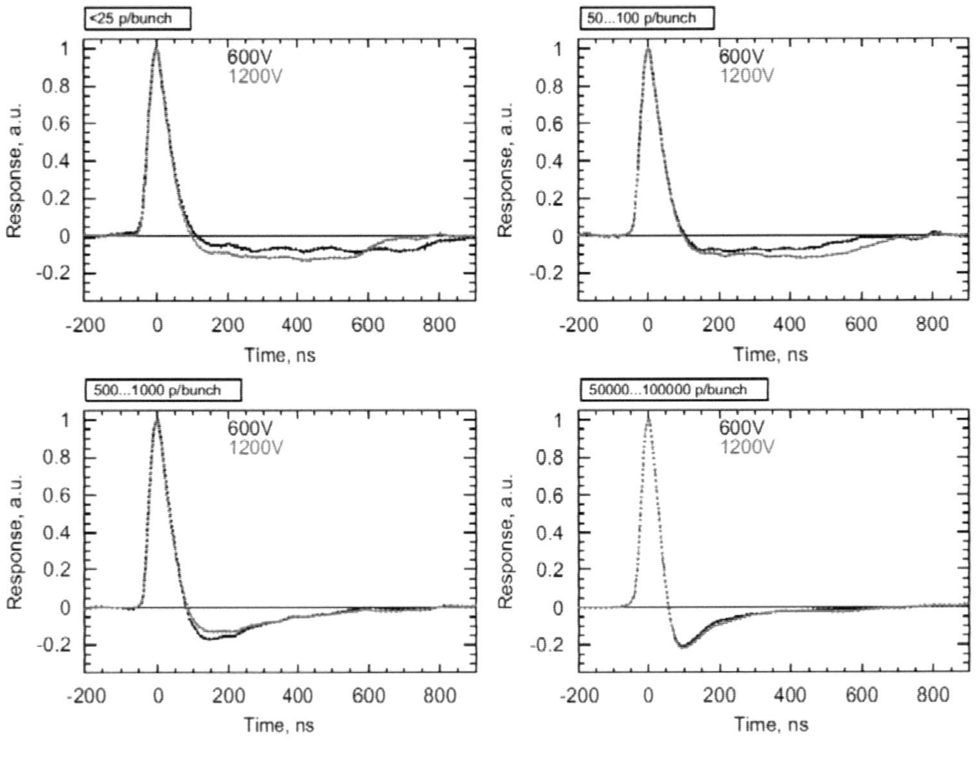

Figure 4. Average signal pulse of the EMEC calorimeter test module for four different intensity ranges, increasing from the upper left to the lower right [3].

rate w. The theoretical prediction for the dependence of the signal amplitude (normalized to the bunch intensity), s, on r is flat below the critical intensity and drops proportional to:

$$s \propto \frac{1}{r^{1/4}} \tag{2}$$

above the critical intensity [4]. More details can be found in Reference [4]. Therefore, the breakpoint in the behaviour can be used to determine the critical intensity in the operation for each calorimeter module.

The dependence of the calorimeter signal response per proton on the beam intensity was studied in detail for all calorimeter test modules during the testbeam runs between 2008 and 2010. A summary of the results can be seen in Figure 5 together with the fitted model function. The expected dependence with the flat plateau until the critical intensity and the decreasing behaviour is clearly visible. The beam intensity corresponding to the nominal LHC luminosity of $10^{34}\,\mathrm{cm}^{-2}\mathrm{s}^{-1}$ was calculated and compared to the obtained critical beam intensity for each calorimeter module (nominal voltage equivalent). The measured results are such that the critical intensity for the EMEC is at about 5 times the nominal LHC intensity, for the HEC at around 8 times and the FCal with the $250\,\mu$m LAr gaps becomes critical already at 1.5 times the nominal LHC intensity. This confirms the expectations and underlines the need to adapt the EM FCal part after the Phase II upgrade. The proposed sFCal design with the $100\,\mu$m LAr gaps shows no signal degradation below intensities of about 10 times the nominal LHC intensity, which

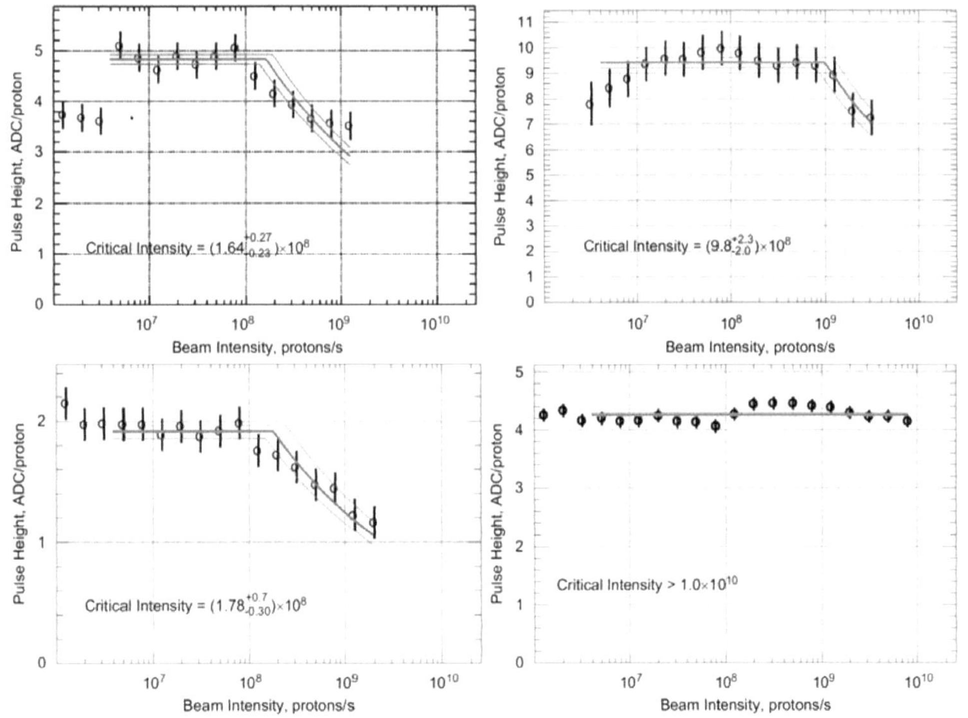

Figure 5. Response of the calorimeter signal amplitude (ADC counts per proton) in dependence of the beam intensity. Top left: EMEC module, bottom left: HEC module, top right: FCal module, bottom right: sFCal module. [3].

shows that it is suitable to run during the high luminosity phase. Due to substantial systematic effects at the highest beam intensities, the significance of the breakpoint at the critical intensity has to be taken with care [3].

In addition the HV return currents of the FCal and the EMEC calorimeter test modules have been measured with high accuracy during the testbeam runs and their dependence on the beam intensity was studied. For the FCal, it was possible to show that the currents with high precision depend linearly on the beam intensity (below the critical region, see Figure 6 left) and that, therefore, a relative luminosity measurement of the LHC with the EMEC and FCal return currents is possible [5]. This method has become one of the main methods for measuring relative luminosity changes in ATLAS.

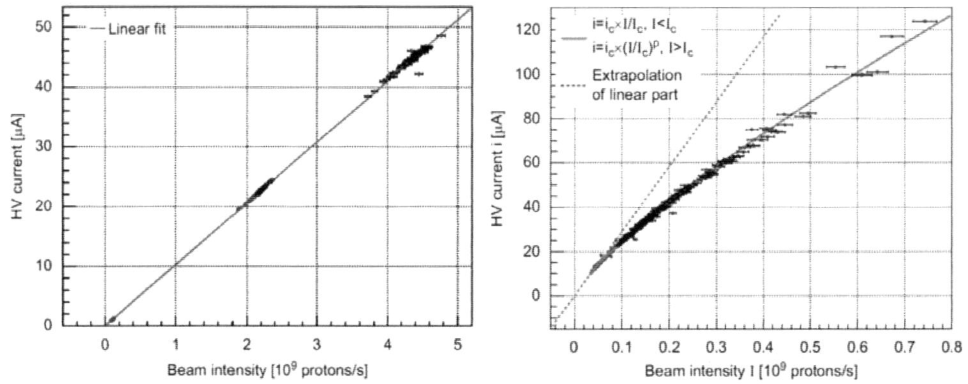

Figure 6. HV return currents in dependence on the beam intensity for the FCal test module (left) from data taken in 2008 and the EMEC (right) test modules from data taken in the 2010 run [3].

Further, the HV currents of the EMEC module were analysed to compare them with the predictions at and above the critical intensity. After a linear behaviour below the critical intensity a dependence on the beam intensity to the power $p = 0.75$ is predicted [4]. Figure 6 (right) shows the measured data together with the corresponding model fit with the critical intensity, I_C, the critical current, i_C, and the power, p, as free fit parameters. The result of the fit is a dependence above the critical intensity with a power $p = 0.76 \pm 0.03$ in agreement to the predictions [3].

Acknowledgments
The support of the IHEP staff in running the U-70 synchrotron is gratefully acknowledged. This work was financed by INTAS under the project INTAS-CERN 05-103-7555 and supported in part by the German Ministry of Education and Research (BMBF) and the Helmholtz Alliance "Physics at the Terascale".

4. References

[1] The ATLAS Collaboration, *Letter of Intent for the Phase-I Upgrade of the ATLAS Experiment.*, CERN-LHCC-2011-12, 2011.
[2] The ATLAS Collaboration, *The ATLAS Experiment at the CERN Large Hadron Collider*, JINST 3 S08003, 2008.
[3] Hilum ATLAS Liquid Argon Endcap Collaboration, *Liquid argon calorimeter performance at high rates.*, Nuclear Instruments and Methods in Physics A 669 (2012) 47-65.
[4] J. Rutherfoord, *Signal degradation due to charge buildup in noble liquid ionization calorimeters.*, Nuclear Instruments and Methods in Physics A 482 (2002) 156-178, 2002.
[5] The Hilum ATLAS Endcap Collaboration, *Relative luminosity measurement of the LHC with the ATLAS forward calorimeter.*, JINST 5 P05005, 2010.

Upgrade of the CMS Hadron Outer Calorimeter with SiPM sensors

Benjamin Lutz for the CMS collaboration
DESY, Notkestrasse 85, D-22603 Hamburg, Germany
E-mail: Benjamin.Lutz@desy.de

Abstract. The CMS Hadron Outer Calorimeter (HO) is undergoing an upgrade to replace the existing photo-detectors (HPDs) with Silicon Photo Multipliers (SiPMs). The chosen device is the Hamamatsu $3 \times 3\,\mathrm{mm}^2$, $50\,\mu\mathrm{m}$ pitch Multi Pixel Photon Counter (MPPC). A complete control system of bias voltage generation, leakage current monitoring, temperature monitoring, and temperature control using solid state Peltier coolers has been developed. The system is design to be a "drop-in" replacement of the HPDs. The complete system of more than 2800 channels has been produced and is currently burned-in. The installation of the SiPM based readout will be performed during the first LHC long shutdown scheduled for 2013.

1. Overview

The CMS central hadron calorimeter is composed of barrel (HB), end-cap (HE), and outer (HO) regions. While the barrel calorimeter is located inside the magnet bore, the Hadron Outer [1] component is installed just behind the magnet coil inside the magnet flux return yoke. It provides and additional calorimeter coverage of about 3λ thickness. The outer calorimeter is composed of two layers of scintillator in the central section of the detector and one layer otherwise. The scintillation light is collected with wavelength shifting fibers and transmitted over clear fibers to front end electronics placed close to the layers. Figure 1a shows the schematic layout of the calorimeters in CMS and shows the location of the HO scintillator layers. Figure 1b shows a photograph of the scintillator tiles with embedded wavelength shifting fibers. The tile size is chosen to match projective towers in the HB. Currently, hybrid photo diodes (HPD) are used as photo-sensors, but for performance and operational reasons it is desired to upgrade these with SiPMs.

The CMS HCAL group has developed a drop-in replacement for the current front end based on SiPMs as photo-sensor [2, 3]. The choice of SiPMs is driven by several properties beneficial for this application: They are inherently magnetic field insensitive which allows a straight-forward design of the front-end considering that it has to operate in the 0.2 T return magnetic field. The SiPMs have a relatively high photon-detection efficiency and high gain. Finally, the devices are available as SMD mounts and can easily fit into the limited available space. The possible limitations of SiPMs like radiation hardness, limited dynamic range, and limited aperture don't pose a problem in this case: The radiation levels are modest with a lifetime expected flux of less than 10^{11} neutrons/cm^2 ($E > 100\,\mathrm{KeV}$, $3\,\mathrm{ab}^{-1}$ LHC integrated luminosity). The energy flux into HO is small and the rate of larger energy depositions is low. The needed cross-sectional area is small due to the fiber based readout technique.

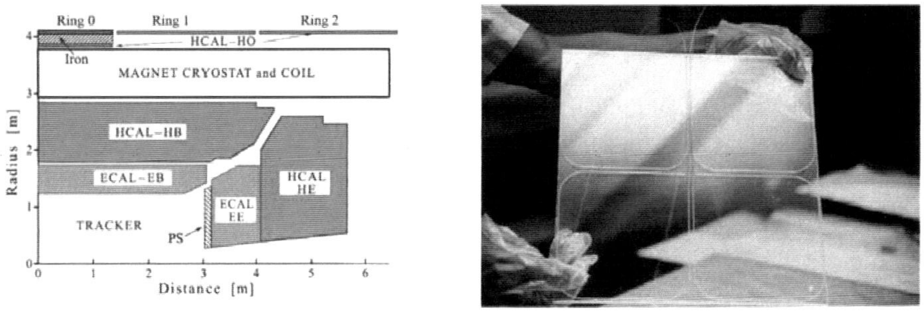

(a) Diagram of a quarter section of CMS showing the location of the HCAL HO

(b) An HO scintillator tile

Figure 1

2. SiPM choice

Following requirements have been chosen to select the right SiPM model for this application:

- Radiation tolerance up to 10^{11} neutrons ($> 100\,\mathrm{KeV}$)/cm^2
- Dynamic range sufficient for HO (2500 photo-electrons)
- Pulse recovery time short enough to accommodate HO rate-occupancy needs.
- Leakage current after radiation damage less than 40 μA
- Photon detection efficiency at least as good as the HPD (15% at 500 nm)
- Temperature dependence of gain small enough
- Source capacitance small enough not to significantly distort the pulse shape

The evaluation [4, 5] of different SiPM candidates lead to the choice of the Hamamatsu $3 \times 3\,\mathrm{mm}^2$ $50\,\mu$m pitch MPPC as the appropriate device. These devices have 3600 micro-pixels which is a good match for our required dynamic range of 2500 photo-electrons (pe). At the wavelength of interest, 500 nm, the MPPC has a photon detection efficiency (PDE) between 25% and 30%. The capacitance of the MPPC (at operating voltage) is about 300 pF. The gain of the MPPC is about 6×10^5 when operated at 1 volt over-voltage (voltage difference to the point where the gain becomes zero). For 2500 pe this corresponds to 240 pC of charge. The MPPC pulse width for our signal is roughly 50 ns with a recovery time of less than 50 ns.

The MPPC gain depends linearly on the temperature. At the foreseen operation point this corresponds to a maximal relative dependence of 8% gain shift per degree C. This temperature dependence requires to actively control the temperature of the SiPM. Therefore, we require the temperature of the SiPMs to be regulated better than 0.1° C.

Figure 2 shows properties of 2100 MPPCs from the total order of 2900. Figure 2a is the dark count in photo-electrons and Figure 2b is the variation in operating voltage.

3. System design

The photo-detector and front end electronics are housed in crates (RBX) that are installed inside the magnet return yoke. Electronics and photo-detectors are arranged inside the RBX into independent Readout Modules (RMs) currently containing one HPD each. The original HPD [6] has 18 optical pixels with individual readouts. The drop-in replacement needs to match this physical layout as well as the existing front end ADC. Additionally, the drop-in system has

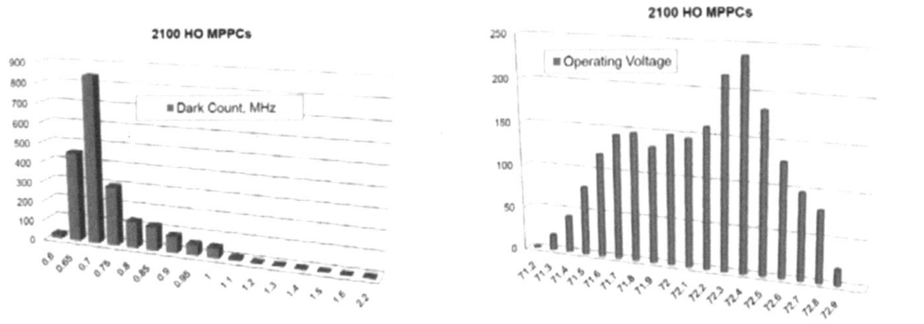

(a) Variation in dark count for 2100 MPPCs

(b) Variation in operating voltage for 2100 MPPCs

Figure 2

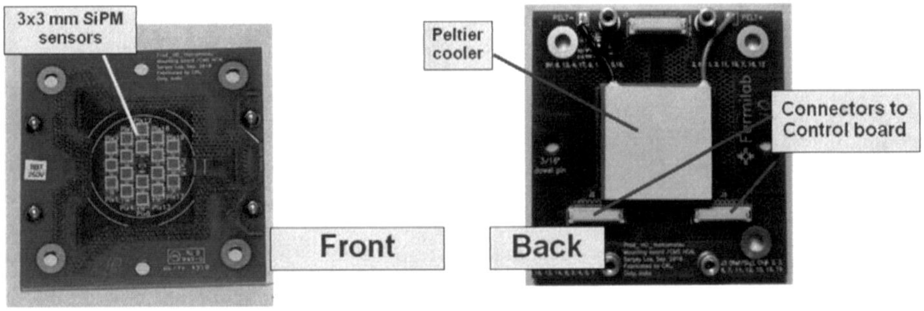

Figure 3: Photographs of the two sides of the Mounting Board (MB).

to supply a self-contained operating environment for the SiPM as only already existing supplies can be used.

The new system is designed to provide following functionality:

- Bias generation and regulation
- SiPM current monitoring
- SiPM temperature stabilization and monitoring
- Gain and shape matching to the existing ADC
- Matching of the detector optics

It consists of 3 circuit boards, the Mounting Board (MB), the Control Board (CB), and the Bias Board (BB). The array of 18 SiPMs is mounted on one side of the MB. On the other side a Peltier cooler is placed to regulate a constant temperature of the SiPM. The CB is connected on the Peltier side to the MB. The Both sides of the MB are shown in Figure 3. The BB is sitting as a piggy-back on the CB. Figure 4 shows details of the placement of the CB, MB, and BB in the RM unit of the RBX. To set the scale, the CB and MB are about 6.5 cm $\times$ 6.5 cm.

The control board shapes the SiPM signals and sends them to the 40 MHz flash front end ADC (QIE) [7, 8]. It also regulates the bias voltages, reads out the leakage currents, measures the

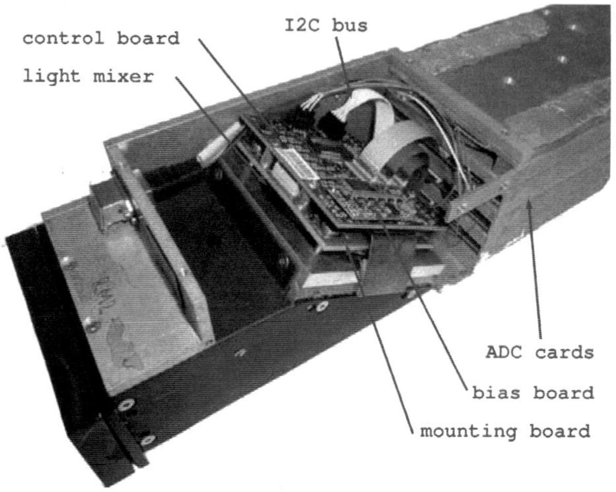

Figure 4: View of an opened readout-module where the mounting board, control board, and bias board can be seen. The ADCs (QIE) in the back are connected with three flat band cables. The optical system is coupled with a light mixer to the SiPMs.

temperatures and voltages of the system, and supplies the Peltier with power. Communication to the outside is done via pre-existing paths using the I2C protocol.

The SiPM is AC coupled through a 22 nF capacitor to the QIEs. The coupling network achieves signal charge attenuation through capacitive splitting with the resulting gain of 50,000. For the largest anticipated pulse this corresponds to about 20 pC, split approximately into 1/2 in one 25 ns sample, and the remaining 1/2 split into 2 trailing samples. The maximum charge measurement for the QIE is 10 pC which matches the attenuated maximum anticipated signal of 200 MIPs.

The SiPM bias voltage (BV) is generated on the BB from 6.5 V LV using a Cockcroft-Walton multiplier. The CB can regulate the 18 independent channels from 0 V to 100 V and supply 100 μA per channel. The BV is adjustable with a least count of 25 mV and is stable to about 5 mV. The SiPM leakage current can be read back for each channel, with a precision of 10 nA.

The temperature is stabilized using a Peltier cooler. About 1 watt of thermal power can be removed per RM. The mounting board has arrays of perforations around the SiPM area (Fig. 3) to provide a better thermal isolation. The temperature of the SiPM area is read out using a precision platinum resistor. The CMS slow control system reads out the temperature (through the CB) and generates a correction voltage to apply to the Peltier cooler. The control loop is executed about once per minute. Figure 5a shows a typical measurement of temperature vs time for the SiPM area. The least count of the readout is 0.018° C. We see that the temperature is stable to about 1 least count over a period of hours. The 2 transients are forced temperature excursions to demonstrate the temperature settling time. A hardware protection circuit is set to prevent the Peltier from driving the temperature below the local dew point. The trip temperature is 15° C and clamps off the Peltier power.

In order to ensure uniform illumination of the $3 \times 3\,\mathrm{mm}^2$ SiPM surface, a light mixer of a few millimeters thickness is placed between the input fiber bundle and the SiPM. These mixers prevent local SiPM saturation from light coming from the 1 mm diameter fibers from the scintillator tiles. Several different mixer designs have been tested to identify the optimal solution: Air-core mixers with cylindrical or conical reflective inner sides and mixers using optical fiber

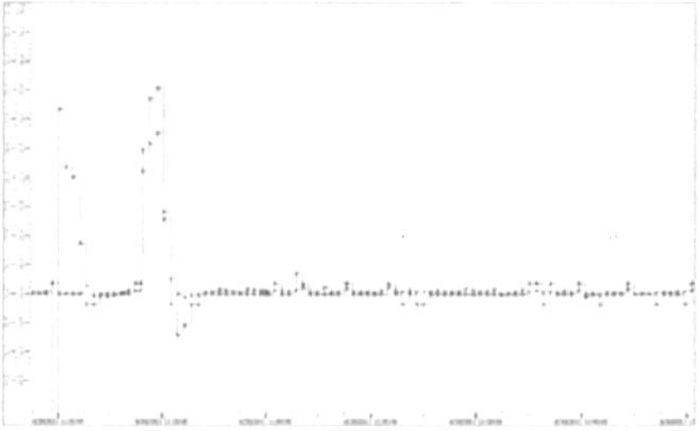

(a) Temperature measurement of the SiPM area vs time (Green curve)

(b) Light mixer to distribute the light homogeneously over the SiPM surface. The fiber stubs responsible for the mixing have been elevated for better display.

Figure 5

inserts. Several metallic reflectors (Al, Au, and Cr) have been compared in the tests of the air-core mixers. The mixer based on optical fibers was chosen due to its superior performance in light transmission and homogeneity. Figure 5b shows such a mixer which is a black plastic disk with fiber stubs inserted.

The different parameters of the electronics system are summaries in table 1.

Table 1: Control Board Parameters

Bias Voltage Range	0–100 V
BV Setting Resolution	25 mV
BV Current Limit (per channel)	100 mA
Leakage Current Full Scale	40 mA
Leakage Current Least Count	10 nA
Temperature Resolution	0.018 ° C

4. Qualification measurements and quality control

Once built into the detector, the system has to operate reliably and stably for many years without the possibility for repairs or modifications. A comprehensive and diverse program of laboratory, test-beam, and on detector tests have been performed to validate the design. 108 channels of a pre-production have been installed into the CMS detector and are successfully collecting data since more than 2 years. The light mixer performance has been studied in details at the test-beam and in laboratory tests.

The final production cards undergo an extensive quality control and burn-in program before they get declared fit for use. This process starts with the characterization of the SiPMs operational parameters at the manufacturer. A sub-sample is re-tested at CERN where also accelerated aging and radiation hardness are measured. It is followed by tests at the production sites for the different boards: Fermilab (USA) mounting board, control card and bias card at TIFR (India). Finally, the assembled card-packs undergo a full system test in a setup similar to the detector readout at CERN for at least one week. The SiPMs and electronic cards are tracked during the full system production process, from the assembly to the final burn-in. Figure 6 shows

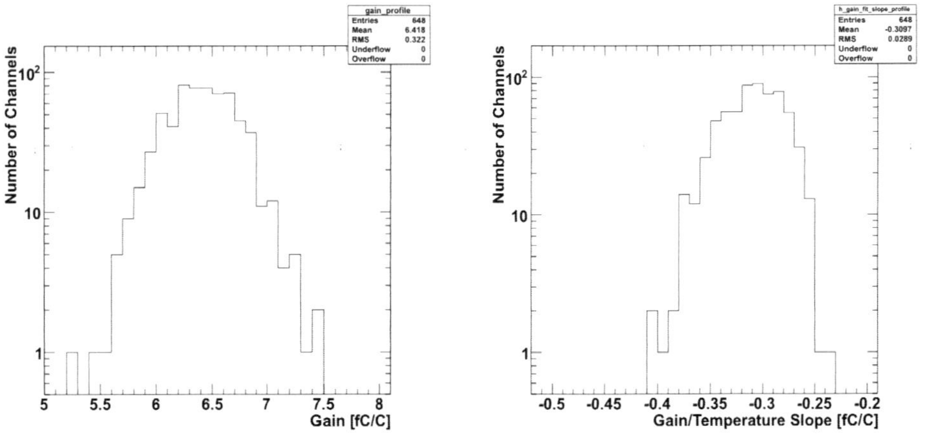

(a) Gain distribution of 648 SiPMs that have passed the quality control. The measurement was performed in the final system configuration.

(b) Temperature dependence of 648 SiPMs that have passed the quality control. The measurement was performed in the final system configuration.

Figure 6

the gain and the gain temperature dependence from already qualified card-packs. These are only two out of the many different parameters monitored during the quality control.

5. Schedule

During the first long LHC shut-down (scheduled to start Feb. 2013) all HCAL Outer HPDs, will be replaced with the new system. A complex choreography of detector access preparation, RM extraction, RM refurbishing, and quality control is currently developed to ensure an efficient use of the limited time during LHC shutdown. The removed HPDs will be reused as spare devices for the barrel and end-cap hadron calorimeters.

6. Summary

The CMS HCAL collaboration is replacing the existing photo-detectors (HPDs) in the Hadron Outer Calorimeter with SiPMs. A complete control system including bias voltage generation, leakage current monitoring, and temperature stabilization has been developed. The system is design to be a "drop-in" replacement of the HPDs with no need for new supplies or mechanical modifications. The full system of more than 2800 channels has been produced and the electronics is currently undergoing an extended quality check. The installation of the SiPM based readout will be performed during the first LHC long shutdown scheduled for 2013.

References

[1] Abdullin S *et al.* (CMS HCAL Collaborations) 2008 *Eur.Phys.J.* **C57** 653–663
[2] Freeman J 2010 *Nucl. Instrum. Meth.* A **617** 393–395
[3] Freeman J 2010 *Int. J. Mod. Phys.* A **25** 2421–2436
[4] Heering A *et al.* 2007 *2007 IEEE Nuclear Science Symposium Conference Record* **Vol. 2** 1545
[5] Heering A *et al.* (USCMS) 2007 *Nucl.Instrum.Meth.* **A576** 341–349
[6] P B Cushman A H H 2002 *ICFA Instrum. Bull.* **25** 1–16
[7] Shaw T *et al.* (CMS collaboration) 2002 *IEEE 2002 NSS*
[8] Zimmerman T and Hoff J R 2004 *IEEE J.Solid State Circuits* **39** 895–905

Upgrade of the CMS hadron calorimeter for an upgraded LHC

Jacob Anderson, for the CMS Hcal Collaboration
Fermi National Accelerator Laboratory, Batavia, IL 60510, USA
E-mail: andersj@fnal.gov

Abstract. The CMS barrel and endcap hadron calorimeters (Hcal) upgrading the current photo-sensors are hybrid photodiodes (HPDs) to meet the demands of the upgraded luminosity of the LHC. A key aspect of the Hcal upgrade is to add longitudinal segmentation to improve background rejection, energy resolution, and electron isolation at L1 trigger. The increased segmentation can be achieved by replacing the HPD's with multi-pixel Geiger-mode avalanche photodiodes. The upgraded electronics are required to operate in a harsh environment and are constrained by the existing infrastructure. The proposed solutions span from chip level to system level. They include the development of a new ADC ASIC, the design and testing of higher speed transmitters to handle the increased data volume, the evaluation and use of circuits from other developments, evaluation of commercial FPGAs, better thermal design and improvements in the overall architecture.

1. Introduction

The Large Hadron Collider (LHC) at CERN is planning to upgrade to higher instantaneous luminosities ($\sim 5 \times 10^{34}\,\mathrm{cm}^2/\mathrm{s}$). In order to improve the performance of its hadron calorimeter (Hcal), the Compact Muon Solenoid (CMS) Collaboration foresees upgrading the Hcal front-end electronics including the photo-sensors. The CMS Hcal is exploring replacing the existing hybrid photodiodes (HPD) with multi-pixel Geiger-mode avalanche photodiodes (also known as silicon photomultipliers or SiPMs). The new sensors and capabilities of the Hcal will require new ADC and data transmission ASICs.

The CMS hadron calorimeter is composed of barrel (HB), endcap (HE) and tail-catcher (HO) subsystems visualized in Fig 1. There is also a forward (HF) subsystem not shown. Reference [1] contains a detailed description of the CMS detector and its calorimeters. The HB and HE are composed of scintillating tiles interleaved with brass absorber layers and lie within the 3.8 T CMS solenoid. The scintillator tiles are read out via wavelength-shifting fiber (WLS) using HPDs. The calorimeter energy deposits are digitized on the front-end before being transmitted off of the detector via optical fiber. This paper is a report on some of the ongoing research and development to upgrade the front-end readout system of the HB and HE subsystems. It includes the upgrade goals, the SiPM sensor development, new ADC ASIC development, foreseen improvements to the digital readout electronics and some results from prototypes used in beam tests.

The goals are to improve the performance and reliability of the Hcal even under the more demanding luminosity environment of the upgraded LHC. These goals include: lowering instrumental noise, improving the calorimeter's ability to separate physics signals from the

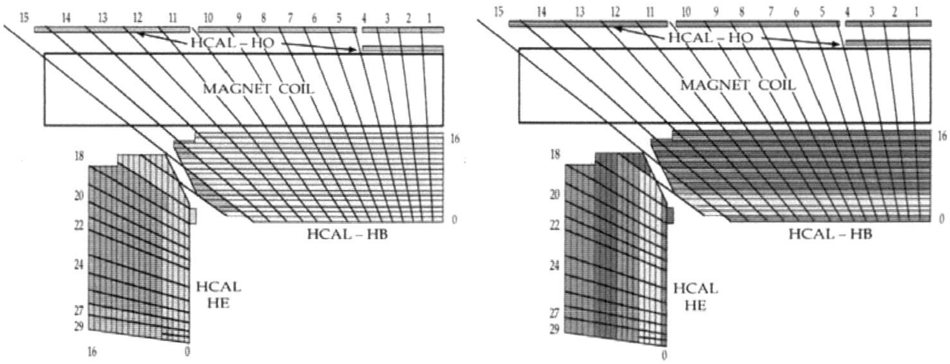

Figure 1. Diagram of the longitudinal segmentation of the HB and HE for the current detector (left) and a proposed upgrade (right).

multiple proton-proton interactions that occur each LHC bunch crossing, rejecting backgrounds from out-of-time sources, improved robustness against component failure. Along with these goals, there are also constraints on the reuse of service infrastructure such as cooling, cables, fibers, etc. The development of the SiPM sensor has the potential to allow CMS to achieve it goals while staying within its present constraints.

Because the SiPM is a compact sensor it has allowed CMS to explore increasing the longitudinal segmentation of the Hcal, in particular the HB. Figure 1 shows the current segmentation of the Hcal and one proposal for how to increase the segmentation. The increased segmentation would provide additional handles to deal with pile-up from multiple proton-proton interactions and to maintain calorimeter performance after higher radiation doses. The SiPM is also well suited to achieve the Hcal goals because of its large gain, its insensitivity to magnetic fields, its fast signals and its radiation tolerance.

2. SiPM research and development

To the end of specifying a SiPM sensor for the CMS Hcal, we have developed a simplified device simulation which models most of the essential characteristics of the device, namely its dynamic range, pixel recovery time, cross-talk and temperature dependence. Figure 2 shows some of the output of such a model. We have also cross-checked the model with measurements of devices from various vendors. Using this model we have been able to find for a light pulse which is distributed in time, like that from our WLS fiber, a SiPM with a fast pixel recovery time can see a significantly extended dynamic range as seen in Fig. 3. We observe that we could get $\sim 3\times$ boost in the dynamic range for a SiPM with fast pixel recovery. Using our models and measurements from the test bench and test beams we have been able to create a list of requirements for a SiPM for the HB/HE subsystems shown in Table 1.

3. SiPM radiation hardness

We have also investigated the radiation hardness of the SiPM sensors. The neutron dose at the location of the front-end electronics is expect to be around 2×10^{12} n/cm^2. Using the irradiation facilities of the CERN PS, we observe deterioration of photon-detection efficiency (PDE) and gain and an increase in the leakage current through the diodes. The results for several devices can be seen in Fig. 4. In addition to these gradual degradations with dose, we look for radiation induced anamalous signals from the sensors. We do observe such signals in some of the diodes,

Table 1. Key requirements for a SiPM device to meet the needs of the CMS Hcal HB/HE.

Parameter	Spec value	Parameter	Spec value
size	$2.2\,\text{mm}^2$	gain	6×10^4
effective pixels/device	> 20k	pixel recover time	$\tau_{RC} < 10\,\text{ns}$
photo-detection efficiency (515 nm)	> 15%	excess noise factor	< 1.4
leakage current after $2 \times 10^{12}\ n$	$< 200\,\mu\text{A}$	optical cross-talk	< 15%
neutron sensitivity	no	bias voltage	< 100 V
operating temperature	22° C	temerature dependence	< 5%/°C

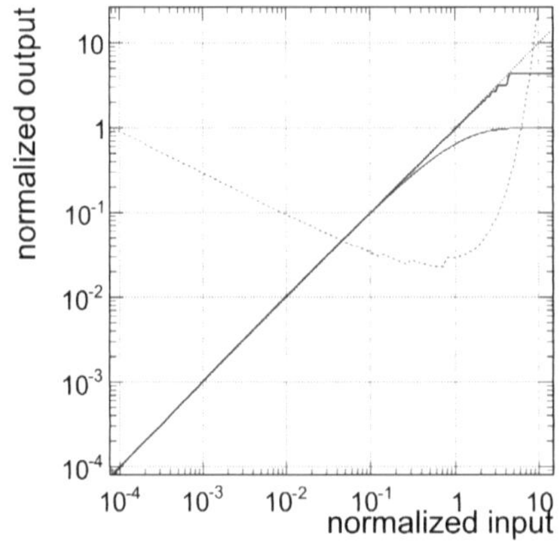

Figure 2. Simulation of a SiPM device the input and output are normalized to the total number of micro-pixels on the device. The blue curve is the raw output of of the device after QIE ADC simulation. The black curve is the output after correction for saturation. The red curve shows the total fractional error on the correction and includes contributions from the SiPM itself, Poisson statistics of the photons and uncertainty due to the correction.

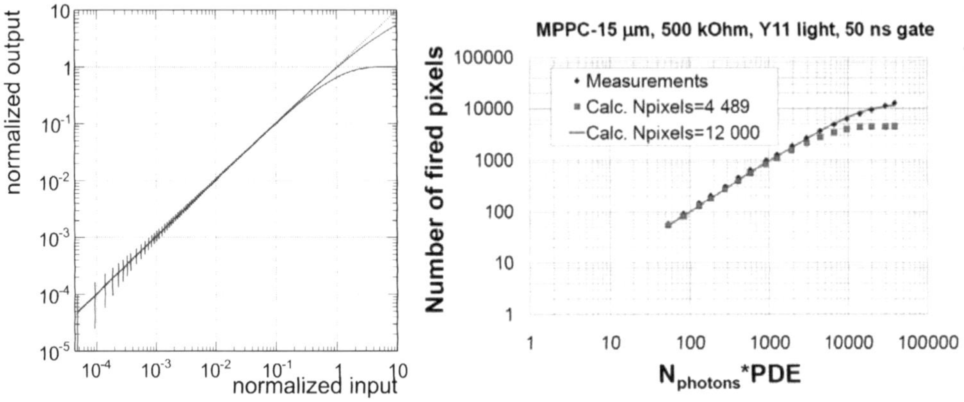

Figure 3. Increase in dynamic range of a fast SiPM device for a pulse from a Y11 WLS fiber. The left plot is from simulation where the purple curve is for the fast SiPM and the blue is for a slow one. The right plot is a measurement for a device with ~4500 pixels and shows a measurement of the same behavior.

but at a rate low enough not to produce significant spurious energy deposits in CMS. We are further working with vendors to further reduce such signals.

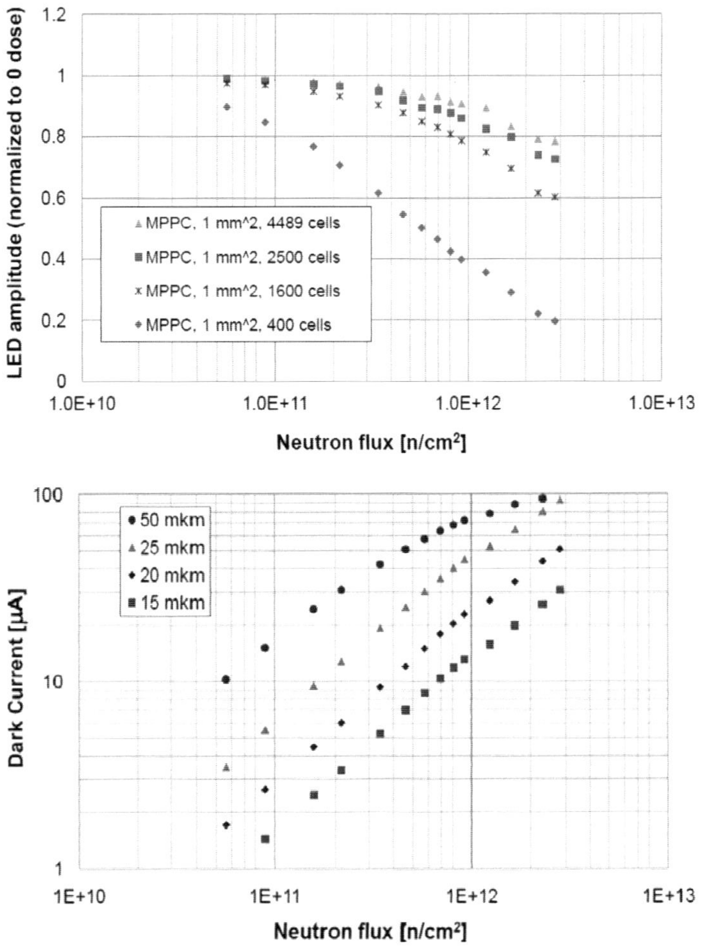

Figure 4. Effects of radiation damage on SiPM sensors from Hamamatsu as a function of neurton fluence. The top plot shows the degradation of gain × PDE. The bottom plot shows the increase in leakage current. There is no bias correction applied in either case.

4. ADC and readout development

The added gain of the SiPM device requires an improved ADC over the current QIE ASIC used by the CMS Hcal. In a joint project between the CMS and ATLAS collaborations, Fermilab is developing a new version of their QIE ADC ASIC. This new ASIC takes a similar approach as the previous QIE designs compressing the number of bits needed to cover a large dynamic range by increasing the distance between codes. Figure 5 shows that the quantization error of the QIE10 will be ~ 1%, much lower than the detector resolution. This new QIE10 chip will have a significantly increased dynamic range compared to the QIE8 chip currently used. It will also include TDC functionality to provide precision timing of the signal within the 40 MHz integration frequency. CMS expects to have a full prototype of the ASIC in the latter portion of 2012.

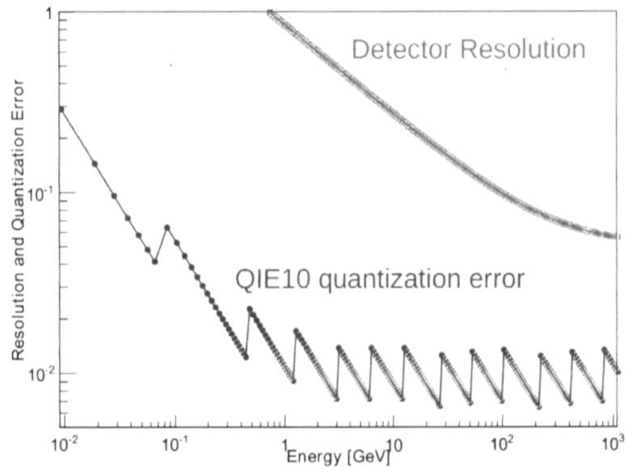

Figure 5. A comparison of the QIE quantization error and the detector resolution. The blue QIE resolution is significantly smaller than the overall detector resolution in red.

In addition to a new ADC, the data links from the CMS detector to the back-end electronics will be upgraded. We need additional bandwidth to accomodate the increased data payload of the QIE10 ADC+TDC information and the added longitudinal segmentation. We are evaluating the GBT chipset being developed by CERN[2, 3]. In addition to transmitting data off of the detector the same links can be used to distribute the LHC clock to the read-out electronics.

5. Beam test results

We have received and evaluated high pixel-density sensors from Hamamatsu[4], FBK[5], Zecotek[6] and Ketek[7]. The Zecotek diodes have a pixel recovery time which is too slow to be a viable option for the CMS Hcal. The others are much faster in particular the Hamamatsu pixel recovery time is ~ 3.3 ns. The Ketek and FBK sensors have the smallest anomolous response to neutrons. Each of these has been further tested in the CMS Hcal calorimeter setup at the CERN SPS H2 test beam line.

We can observe muon signals in the calorimeter and confirm the improved signal to noise ratio. The MIP signal from only the first depth in the HB, as seen in Fig. 6, has a separation from pedestal similar to that observed for the entire HB with an HPD.

We use pion beams to look at the hadronic resolution of the calorimeter. A representative distribution is shown in Fig. 7. We observe comparable resolution with new SiPM sensors and 4-depth longitudinal segmentation to that from previous beam tests using HPD sensors [8].

6. Summary

The CMS Hcal has embarked on an advanced study to replace its existing photo sensors with SiPMs. This has identified characteristics that are required for a device to meet the needs of the Hcal and perform in the challenging LHC environment. In conjunction with the new sensors, we are improving our digitization and read-out chain to increase dynamic range, add precision timing and increase bandwidth. The development of these systems is getting very advanced and prototype tests have been done in the lab and in test beams. The Hcal plans to upgrade its HB an HE front-end electronics systems during the LHC long shutdown two currently forseen in 2017 and 2018.

References

[1] Chatrchyan S *et al.* (CMS) 2008 *JINST* **3** S08004

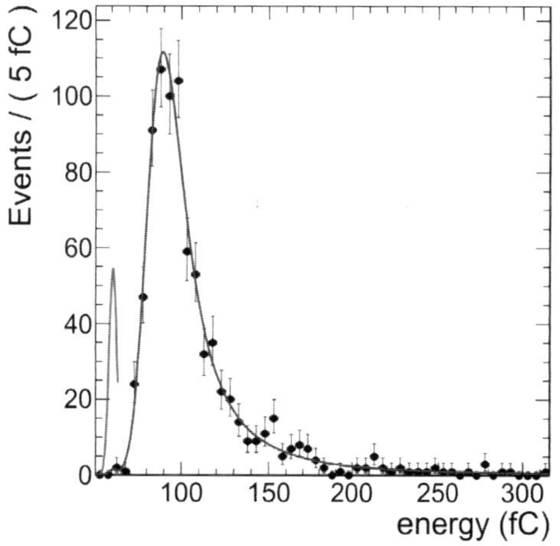

Figure 6. The distribution of 150 GeV muons in the first depth of the CMS Hcal in the H2 test beam. The blue curve is a fit to a Landau shape and the red curve is the pedestal distribution.

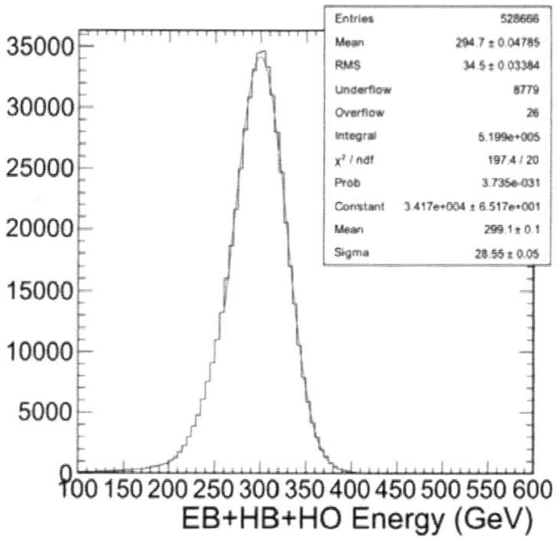

Figure 7. The distribution of the energy sum in the calorimeters for a 300 GeV pion beam. The Hcal has four longitudinal depths, read out via SiPMs. The red curve is a simple fit with a Gaussian shape.

[2] Moreira P *et al.* 2009 The GBT Project *Topical Workshop on Electronics for Particle Physics (TWEPP 09)*
[3] Moreira P 2007 The GBT, a proposed architecture for multi-Gbps data transmission in high energy physics *Topical Workshop on Electronics for Particle Physics (TWEPP 07)* p 71
[4] `http://www.hamamatsu.com`
[5] `http://cmm.fbk.eu/en/home`
[6] `http://www.zecotek.com/503/1107/`
[7] `http://www.ketek.net/products/sipm/`
[8] Abdullin S *et al.* (CMS HCAL/ECAL) 2009 *Eur.Phys.J.* **C60** 359–373

EM Calorimeters for SoLID at Jefferson Lab

Z W Zhao[1], J Huang[2], M Meziane[3], X Zheng[1], P E Reimer[4], D Armstrong[5], T Averett[5], W Deconinck[5] and the SoLID Collaboration

[1]Department of Physics, University of Virginia, 382 Mccormick Rd, Charlottesville, VA 22904, USA
[2]Los Alamos National Laboratory, Box 1663, MS H846, Los Alamos NM 87545, USA
[3]Department of Physics, Duke University, Physics Bldg., Science Dr., Box 90305, Durham, NC 27708, USA
[4]Physics Division, Argonne National Laboratory, 9700 S. Cass Avenue, Argonne, IL 60439, USA
[5]Department of Physics, College of William and Mary, P.O. Box 8795, Williamsburg, VA 23187-8795, USA

E-mail: zwzhao@jlab.org

Abstract. Several approved experiments at Jefferson Lab for the 12 GeV era will use the proposed Solenoid Large Intensity Device (SoLID) spectrometer. Two EM calorimeters with a total area of 15 square meters are required for electron identification and electron-pion separation. The challenge is to build calorimeters that can withstand high radiation doses in high magnetic field region and bring photon signals to low field region for readout. Several types of calorimeters were considered and we are favoring Shashlyk type as a result of balancing performance and cost. Our preliminary design and simulation of SoLID EM calorimeters are presented.

1. Introduction

The Solenoidal Large Intensity Device (SoLID) is a versatile spectrometer based on a large solenoid magnet. Presently, there are two distinct series of measurement that have been proposed for this spectrometer at Jefferson Lab in the 12 GeV era. In one configuration, the spectrometer will be used to measure parity violating deep inelastic scattering (PVDIS) [1]. The experiment will study lepton-quark neutral current interactions by recording collisions of polarized electrons with cryogenic hydrogen and deuterium targets. Measuring the polarization asymmetry due to the interference of the photon and the Z boson over a wide range of kinematics will determine of the axial-quark, vector-electron couplings while at the same time providing a measurement of charge symmetry violation at the quark level. Measurement of PVDIS with a hydrogen target can provide a determination of the d/u ratio in the proton without the need for nuclear corrections. The second configuration will be used for studies of semi-inclusive deep inelastic scattering (SIDIS) [2] with nearly 2π azimuthal-angle coverage and momentum transfers up to 9 GeV^2. The SIDIS experiment will measure semi-inclusive electroproduction of charged pions from transversely/longitudinally polarized ^3He and from transversely polarized proton. With the SoLID spectrometer, the single target spin and double beam-target spin azimuthal asymmetries can be measured with high precision and within a large 4-dimension (x,Q^2,z,P_T) kinematic space, which will lead to a more comprehensive understanding of

Transverse Momentum dependent parton Distribution functions (TMD) for the neutron and proton in the valence quark region.

The PVDIS and SIDIS setups will share the same magnet yoke and many of the same detector elements, but configured in different ways. The magnet will be a large bore solenoid of 2-3 meter diameter and 1.5T field. Re-use of an existing large superconducting solenoidal magnet (such as BaBar [3] and CLEO-II [4]) is being pursued. For PVDIS setup, the detection system requires excellent electron identification that is achieved by a gas Cerenkov counters and an electromagnetic calorimeter at forward angles. Gas Electron Multiplier (GEM) chambers are used for tracking at high rates. A series of baffles would be deployed to block soft electrons and line of sight photons' trajectories. For SIDIS setup, the electron-hadron separation would be accomplished with a light gas Cerenkov and an electromagnetic calorimeter at forward angles. At large angles, the electron-hadron separation will be accomplished by an electromagnetic calorimeter alone. A heavy gas Cerenkov and a layer of multi-gap resistant plate chamber (MRPC) as time of flight will be used for pion identification. GEMs will be also used for tracking. Both setups (with the CLEO-II magnet) are shown in Figure 1.

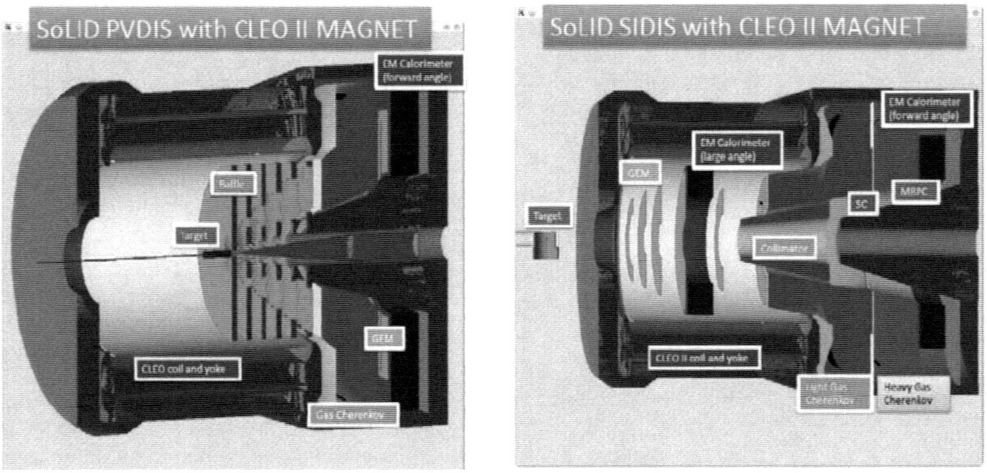

Figure 1. SoLID PVDIS setup (left) and SIDIS setup (right).

SoLID Electromagnetic Calorimeter (EC) is a key detector, used in both PVDIS and SIDIS experiments to provide electron-hadron separation. The EC modules need to be designed to fit the needs of both PVDIS and SIDIS experiments and be able to be reconfigured for the different setups. The challenge is to build the calorimeters that can withstand high radiation dose in high magnetic field region and bring photon signals to low field region for readout. The EC's main characteristics are determined by both the physics goals and the designed running conditions of the experiments. We are favoring Shashlyk type calorimeters [5] as a result of balancing the performance and cost.

2. Prelimnary Design

A Shashlyk calorimeter is a sampling calorimeter constructed from alternating layers of scintillator and heavy absorber e.g. lead. Scintillation light is absorbed, re-emitted and transported to the photon detector by wavelength shifting (WLS) optical fibers penetrating through the calorimeter modules longitudinally.

Our Shashlyk calorimeter module design is based on the COMPASS module [6] as shown in Figure 2. A Geant4 simulation was used to study the key specifications for optimal physics results and to lower the cost.

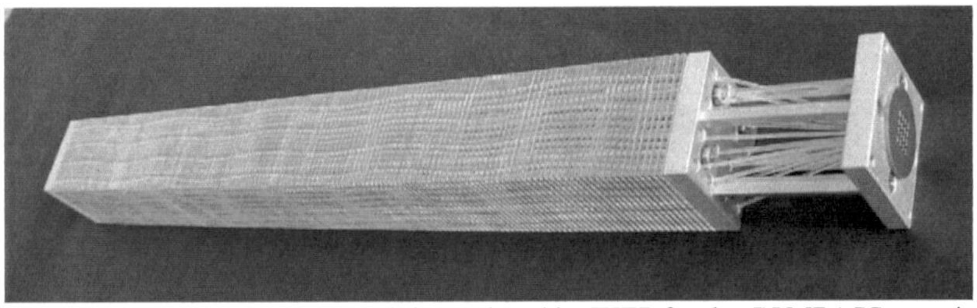

Figure 2. A Shashlyk calorimeter module produced by IHEP for the COMPASS experiment.

The main features of our current design are the following

- Scintillator and absorber ratio:

Each layer consists of a 1.5 mm-thick scintillator plate and an absorber plate made of lead. The Pb absorber thickness of 0.6 mm or less is favored to provide a fine sampling, and therefore an energy resolution of 5%. For the 0.6 mm Pb configuration, effective radiation length is about 21 mm.

- Longitudinal size:

A total length about 20 radiation length is sufficient to contain to 95% of the electromagnetic shower and short enough to maximize the difference in energy deposition between electrons and pions. A calorimeter module of 20 radiation length us approximately 42 cm composed of 183 layers.

- Longitudinal segmentation:

Having both preshower and shower segments in EC is essential to reach the required pion rejection factor of between 10 and 100 depending on the particle's momentum. The preshower thickness of 3-5 radiation lengths can maximize the difference in energy deposition between electrons and pions at the momentum ranges of 2-7 GeV for SoLID. The proposed design is to build the Shashlyk calorimeter in one unified structure, but separating the preshower and shower parts by different readout fibers. The preshower fibers must be optically isolated from shower part.

- Lateral size:

The lateral size of each module determines the position resolution and the background level of the reconstructed shower hits. It is also the most relevant factor to the overall cost of the calorimeter. The Monte Carlo study showed that a lateral size of about 10 cm would provide a good balance between position resolution and cost as shown in Figure 3.

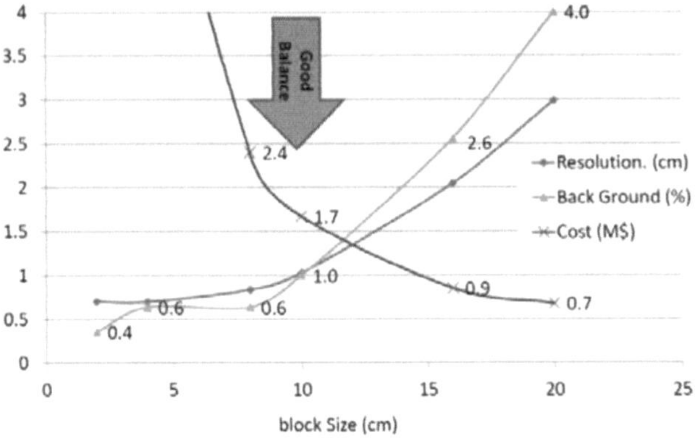

Figure 3. Lateral size dependence of the position resolution, background level from simulation and the nominal cost of the SoLID calorimeter if using COMPASS type module.

- Radiation hardness:
  At the designed luminosity of 10^{39}/cm^2/s for PVDIS and 10^{36}/cm^2/s for SIDIS, the radiation level at EC reaches 60 krad/year in the PVDIS configuration and 400krad/year in the SIDIS configuration (the difference comes from luminosities, targets, acceptance angles and the use of baffles for PVDIS). The COMPASS module has been tested up to 500 krad. Increasing the thickness of the first lead layer and using more radiation resistant scintillator and fibers can further improve the radiation hardness.
- Fiber connection:
  Because the high rate and high magnetic field environment, the photon sensors need to stay outside of the SoLID yoke structure. We plan to combine and connect the WSL fibers into clear fibers and lead them outside of the yoke to where the photon sensors will be located.
- Module layout:
  The modules need to be easily swapped between PVDIS and SIDIS settings. And the layout needs to preserve the reflection symmetry in the spectrometer. Modules with a square cross section can be stacked in the layout shown in Figure 4 to meet this criteria.

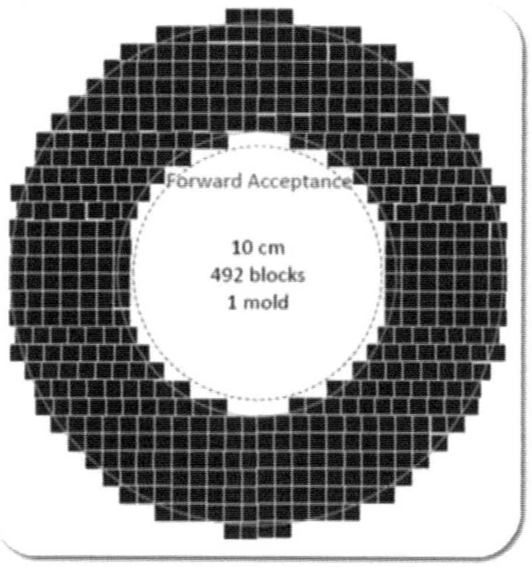

Figure 4. The layout of SIDIS large angle calorimeter with square modules.

3. Summary

SoLID at Jefferson lab will open doors to exciting physics through precise measurements of PVDIS and SIDIS. EC is one of key detectors of SoLID. Its primary role is the identification and separation of pions and electrons. This must be done at the trigger level for the proposed PVDIS measurements. The required features of the EC system were outlined, and these features appear to be available in Shashlyk-type calorimeter modules, similar to those used in COMPASS.

References

[1] Jefferson lab 12 GeV SoLID PVDIS proposal E12-10-007
http://www.jlab.org/exp_prog/PACpage/PAC37/proposals/Proposals/Previously%20Approved/E12-10-007.pdf
Jefferson lab 6 GeV Hall A experiement E08-011
http://hallaweb.jlab.org/experiment/PVDIS/pac33/PR_pvdis3_submitted.pdf

[2] Jefferson lab 12 GeV SoLID SIDIS proposal E12-10-006, E12-11-007,PR12-10-108
http://wwwold.jlab.org/exp_prog/PACpage/PAC38/proposals/Previously_Approved/E12-10-

006-update.pdf
http://wwwold.jlab.org/exp_prog/PACpage/PAC38/proposals/Previously_Approved/E12-11-007_Update.pdf
http://wwwold.jlab.org/exp_prog/proposals/11/PR12-11-108.pdf
Jefferson lab 6 GeV Hall A experiement E06-010, E06-011
Qian X *et al.* 2001 *Phys.Rev.Lett.* **107** 072003
Huang J *et al.* 2012 *Phys. Rev. Lett.* **108** 052001
[3] Aubert B *et al.* 2002 *Nucl. Instr. and Meth. A* **479** 1
[4] Kubota Y *et al.* 1992 *Nucl. Instr. and Meth. A* **320** 66
[5] Atoian G *et al.* 2004 *Nucl. Instr. and Meth. A* **531** 467
Atoian G et al. 2008 Nucl. Instr. and Meth. A 584 291
[6] Private communication with the Institute for High Energy Physics of Russia (IHEP).

Journal of Physics: Conference Series **404** (2012) 012021 doi:10.1088/1742-6596/404/1/012021

Secondary Emission Calorimeter Sensor Development

David R Winn[1] and Yasar Onel[2]

[1] Department of Physics, Fairfield University
Fairfield, CT 06880 USA
[2] Department of Physics, University of Iowa
Iowa City, Iowa USA

E-mail: winn@fairfield.edu

Abstract: In a Secondary Emission electron(SEe) detector module, Secondary Emission electrons (SEe) are generated from an SE surface/cathode, when charged hadronic or electromagnetic particles, particularly shower particles, penetrate an SE sampling module placed between absorber materials (Fe, Cu , Pb, W etc) in calorimeters. The SE cathode is a thin (10-50 nm thick) film (simple metal-oxides, or other higher yield materials) on the surface of a metal plate, which serves as the entrance "window" to a compact vacuum vessel (metal or metal-ceramic); this SE film cathode is analogous to a photocathode, and the SEe are similar to p.e., which are then amplified by dynodes, also is in a PMT. SE sensor modules can make use of electrochemically etched/machined or laser-cut metal mesh dynode sheets, as large as ~30 cm square, to amplify the Secondary Emission Electrons (SEe), much like those that compact metal mesh or mesh dynode PMT's use to amplify p.e.'s. The construction requirements easier than a PMT, since the entire final assembly can be done in air; there are no critical controlled thin film depositions, cesiation or other oxygen-excluded processes or other required vacuum activation, and consequently bake-out can be a refractory temperatures; the module is sealed by normal vacuum techniques (welding or brazing or other high temperature joinings), with a simple final heated vacuum pump-out and tip-off. The modules envisioned are compact, high gain, high speed, exceptionally radiation damage resistant, rugged, and cost effective, and can be fabricated in arbitrary tileable shapes. The SE sensor module anodes can be segmented transversely to sizes appropriate to reconstruct electromagnetic cores with high precision. The GEANT4 and existing calorimeter data estimated calorimeter response performance is between 35-50 Secondary Emission electrons per GeV, in a 1 cm thick Cu absorber calorimeter, with a gain per SEe $>10^5$ per SEe, and an e/pi<1.2. The calorimeter pulse width is estimated to be <15 ns. With fine mesh sampling only (no thick absorbers) the resolution is ~25 MeV at 1 GeV.

1 Introduction

The initial purpose of secondary emission calorimeters is particle (e, gamma, n, p, meson, nuclei) energy measurements in extreme radiation environments and very high rates. Other benefits include fine segmentation and close packing, operation in harsh environments (space environs, high temperature/pressure environs e.g. boreholes>1 km deep), imaging detectors in medicine and industry, national security, and beyond. Calorimetry, crucial for new discoveries in Nuclear-High Energy Physics experiments often must: a) work in very high radiation environments for many years, and b) operate very high rates and/or with sub-ns time resolution; and c) can be configurable in compact, dense and homogeneous units for "energy flow" calorimetry; d) are sufficiently inexpensive to enable large absorber volumes; e) have energy resolution capable of distinguishing signals over a large dynamic range. Additional examples of novel calorimetry include experiment operations remotely, where the detection mechanisms are both rugged and operate without need for repair, service, recalibration or resupply for many years, such as in satellites, in remote or hostile environment locations on earth, in surveillance such as homeland security, in medical or industrial imaging, and in complex physics detector systems, where the calorimeter in the assembled detector is relatively inaccessible, densely tiled, and may need to operate for many years without disassembly. One extreme example might be operation at temperatures ~200C in deep (20,000 feet) boreholes for geophysical exploration.

2 Secondary Emission in Existing Detectors

Secondary Emission (SE) of electrons from surfaces[1] which have ionizing radiation absorbed on them is commonly used in nuclear and medium energy physics. Examples include the dynode gain mechanism used in photomultiplier tubes, and in accelerator beam rate and position monitors.

2.1 Photomultiplier Tubes

It is estimated that the SE emission oxide surfaces of the last dynodes of PMT operate in excess of 100GigaRads from 100-300 eV electron bombardment in the 10-50 nm thick surface layer that secondary emits. Figures 1a shows sheet-like dynodes, quite appropriate for inclusion in a sampling calorimeter ionization SE electron sensor, from Hamamatsu. The etched metal sheets are used in the Hamamatsu compact PMT, and can be made in large sheets up to ~0.5m x 0.5m. A stack with a gain exceeding 500,000 per electron is 6 mm thick. These dynode sheets are similar in function but more efficient than fine metal mesh dynodes, as also shown below in 2 types in Figures 1b. Modern micromachining techniques and tooling developed for flat-panel HDTV enable complex metal structures up to ~1 m sizes.

Figs. 1a: Machined/ Etched Metal Mesh Hamamatsu etched metal mesh dynode examples. The "tube" at left is 12.5 cm (5") across –right shows a closeup of the metal mesh. 2" (52mm) square H10966-type PMT. Such metal mesh dynodes can be made in large sheets and could be used to amplify secondary electrons generated in a thin film cathode from the particles in calorimeter showers.

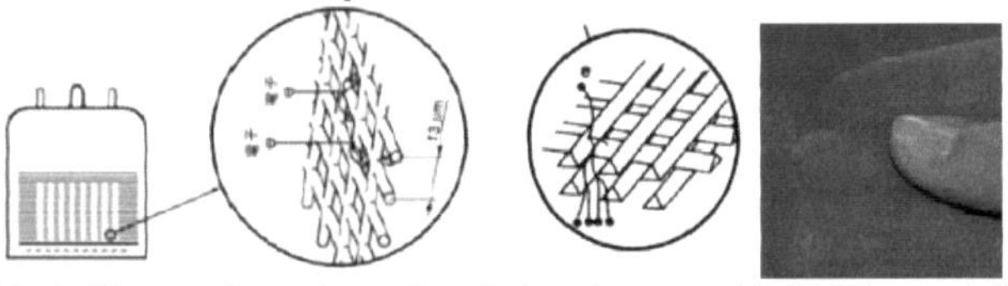

Figs. 1b: Fine Wire Mesh: Two configurations of mesh dynodes as used in PMT, especially for operation in magnetic fields; ~15 dynodes yield a gain of 5 x 105. The mesh-mesh spacings are between 0.5-1 mm, as designed c1985. The compact tube at right has 19 mesh dynodes in <1cm. It operates in 10^5 gain in 1.25 T magnetic fields. Modern micromachined meshes can be spaced at 0.3 mm.A commercial metal mesh similar to that used in the mesh tubes: CuBe Mesh 37% transparent 75 µm apertures, cost <$20/m^2$.

The risetime of such mesh dynode tubes is typically less than 1.5 ns, transit times < 8ns, with bandwidths exceeding 300 MHz, reflecting the speed and rate capability (when properly powered) of SE dynodes as a gain mechanism, which we propose to apply in large areas to amplify SEe which are produced by showers crossing SE surfaces in calorimeters.

2.2 Secondary Emission Accelerator Beam Monitors

As an example, the secondary emission BLMS (Beam Loss Monitor System) used at CERN has a quoted life of 80 MGy per year, for 20 years[2]. The SE Yield was found to be constant up to 10^{20} p/cm^2 of integrated dose (!), which far exceeds many calorimeter requirements. Additionally, the risetime of these detectors are sub-nsec. This BLM is shown in Fig.2. Based on these proven technologies, SE electrons that could be produced by charged particles in sampling calorimeter showers are a candidate for use in high radiation, high rate experiments.

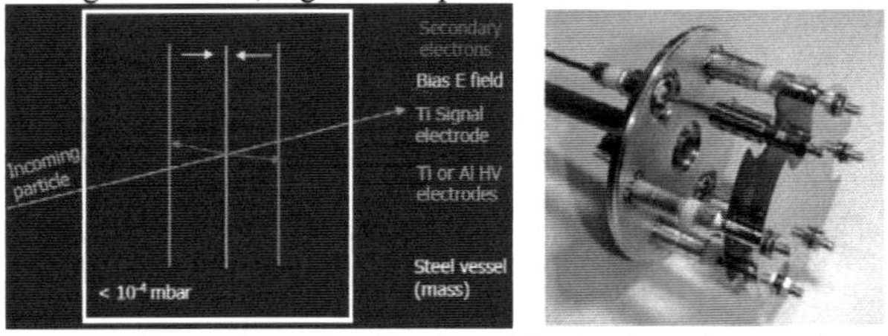

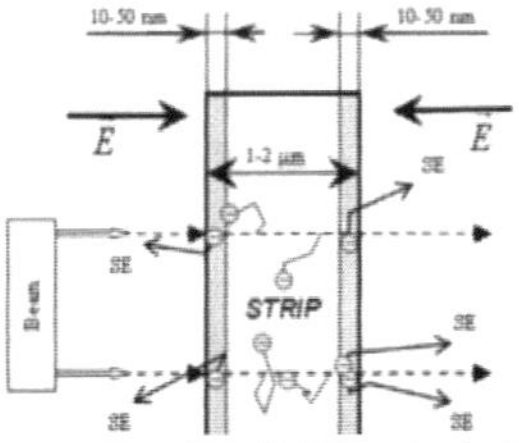

Fig. 2: Cartoon of operation and realization of the SEe beam loss monitor element used at CERN. A similar device with mesh-dynode-like gain stages would be used for the SEe calorimeter sensors.

Fig 3 (R): Cartoon of Secondary emission as used in an accelerator beam monitor, and as would be used in a calorimeter. Electrons freed by ionization random walk into the surface oxide, a thickness typically 10-50 nm thick, over typically 1-2 μm in the underlying metal. A few escape into the vacuum, enhanced by an applied electric field, becoming "secondary electrons".

3 Secondary Emission Precis for Calorimeter Sensors

The calorimeter sensor herein proposed is based on the Secondary Electron (SE) emission from solids. When a charged particle passes through the signal cathode, it can excite conduction band or inner shell electrons. These so called "True Secondary Electrons" can diffuse only several nm, as they usually have energies lower than 50 eV independent of the primary particle's energy and type[3] in contrast to the "knock-on" delta electrons. The material from which SE electrons escape is only from a thin surface layer of the traversed material, and are subsequently drifted into vacuum, usually helped by a bias electric field. The Secondary Electron Emission Yield (SEY) is proportional to the electronic energy loss of the particle in the surface layer of the signal electrode. A cartoon of this process is shown in Figure 3. It is not straightforward to simulate the SE in GEANT4 as it has no corresponding process defined. A modified semi-empirical formula of Sternglass[4] (the contribution of δ electrons to the true SEY could also be included) can be used to calculate the SEY for metal oxide (TiO2, Al2O3, MgO, BeO etc) surfaces.

$$SEY = 0.01 C_F L_S \, dE/dx \; ; \; L_S = (0.23 N \sigma_g)^{-1}. \qquad (1)$$

Where dE/dx stands for ionization energy loss, L_S for effective penetration distance of secondary electrons to the surface, N for number of atoms per unit volume and $\sigma_g = 1.6 Z^{1/3} 10^{-16}$ cm^{-2}. The calibration factor $C_F = 0.8$ is used in order to match the experimental data for *Al2O3* [7] and *TiO2* [8]. The SE yield curves for Alumina, Titania, Al and Ti are shown in Figure 4 for protons at normal incidence between 10 KeV and 5 TeV. The emission yields for other charged particles (pions, electrons, etc) scale as the momentum, according to the Bethe-Bloch formulae for dE/dx.

The universal curve for the secondary emission yields normalized to the peak yield vs the energy ratioed by the peak yield energy is shown in Figure 5a. Briefly, the yield rises from zero incident energy to a maximum yield at a particular material dependent energy, and then falls, since the specific ionization per track length falls, and ionized electrons formed deeper in the SE surface cannot escape to the surface. Fig. 5b shows the yield for MgO vs energy in KeV. At the present state of this proposal, the angular distributions of the incident calorimeter shower particles on the SE surface and that of the secondary electrons are not considered. Suffice it to say, this serves as a lower limit, since the secondary yield is a strong function of incident particle angle, typically increasing by x1.5 at 45 degrees and doubling at 60 degrees incident. Diamond thin film yield is shown below in Fig. 5c. Nucleating diamond crystalites on a W thin film surface could create many SE electrons, as shown below – a prelude for subsequent research.

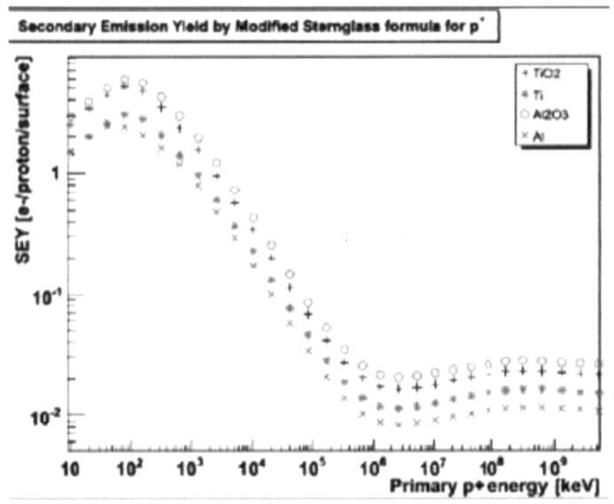

$$\gamma_b = \Lambda \frac{dE}{dx}_{el} \left(1 + \frac{1}{1 + \frac{E_p}{0.1836\, A_p}} \right)$$

Fig. 4a: The Sternglass formula SE vs proton energy in KeV - for the secondary electron yield from 4 different surfaces Afor protons of normal incidence, as used for the design of the CERN BLM. SEY for other particles scale as gamma-beta. The DESY Beam Monitors, similar, yield about 0.05-0.07 SEe per mip.

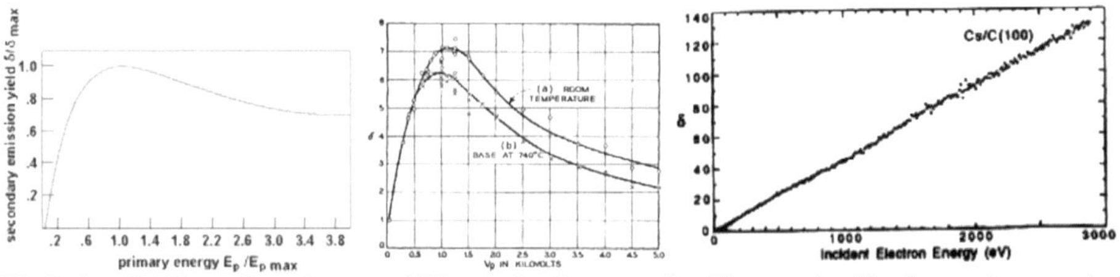

Fig.5a,b, c (L) The universal curve of SE emission from metal oxides as ratioed by the maximum emission vs the ratioed particle energy to that of the maximum emission. (M) The emission from MgO as an example. For example, Cu-BeO peaks at a yield of 8 SEe per electrom at 800 V. (R) Synthetic Diamond terminated with Cs SE emission, exceeding 120 per strike at 3 KeV electrons – remarkably, the Cs dipole bound to the C SE surface is stable in air. Since tungsten is a pseudo lattice match, W foils or plates overcoated with 10-100 nm crystallites would make a unity detection efficiency e-m module SEcathode, and possibly an e-m preradiator/isolated charged particle detector (future work beyond this proposal).

4 Generation of Secondary Emission Electrons by E-M and Hadronic Shower Particles

We now turn to predicting or estimating the number of secondary emission electrons that could be generated from showers in calorimeters – i.e. the number of ionizing charged particles that are produced and could traverse SE cathode surfaces in a reasonably configured/sampled sampling hadron or electron calorimeter. As can be seen in Figure 6, a basic first approximation of the average SE yield is between 0.05-0.1 electrons per particle for essentially all particles (e, pi being the most important for calorimeters) above 40-50 MeV.

4.1 Estimates from Existing Data

Typically the response of a near-compensated hadron calorimeter with 5cm iron/scintillator sampling is the equivalent to the response of about 16 MIP equivalents/GeV[5] (Holder et al., 1978) (dE/dx Fe ~ 2 GeV/meter). Scaling to an increase in sampling from 5 cm to 1 cm absorbers would produce ~100 MIP equivs per GeV x 0.05 SEe/MIP = 5 SEe/GeV. We note that a MIP is the SE lowest yield (above ~100 of eV incident particle energy), and thus 5 SEe/GeV represents a *minimum* likely signal. A rough equivalent might be a response of 5 p.e./GeV for a scintillator calorimeter, assuming the SEe can be amplified like a p.e. to a detectable level (see next section). In a more direct measurement of particles, the CALICE collaboration[6] measures (corrected data for containment) ~12 hits per GeV in a purely digital (i.e. each RPC readout pad is either hit or not, with no proportional energy or count of particles within a pad) energy flow/tracking hadron calorimeter consisting of 1 cm square RPC readout pad arrays between 2 cm thick steel plates (but only about 4 Lint thick). This again sets a *lower* limit on a SE signal in 1 cm thick sampling of about 24 hits/GeV x 0.05 SEe/hit = 1.2 SEe per GeV. A more realistic signal estimate for SEe is considerably higher, since a yes/no hit pad may have several particles through it, and many of the particles have a higher SE yield than 0.05 SEe per hit as at minimum ionizing (as shown in the SE yield curves vs energy shown above).

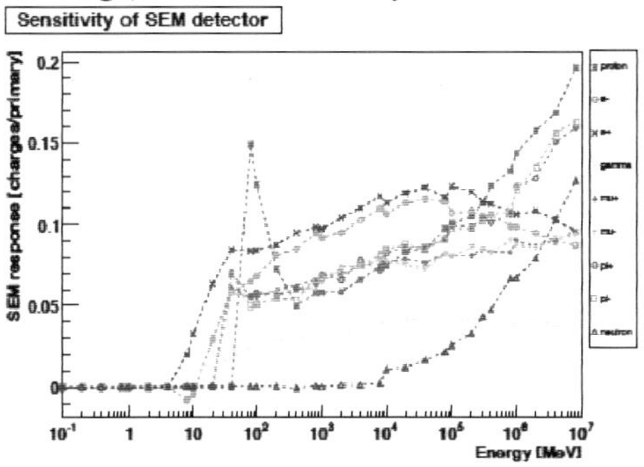

Fig. 6: The number of SE electrons per primary collected in the CERN BLMS system cell (alumina) vs primary energy- p(red square), e-, e+(green-solid, blue-x), μ+, μ-(purple, magenta), pi+, pi-(red-circle, green-open box), n(blue triangle), gamma(yellow) primaries. The high yields above ~ 10 GeV are speculated to be from delta-rays. From Ref [7]: B. Dehning et al. The energy bins < ~10 MeV have a threshold from the thick metal entrance window. (NB: negative signal at 8 MeV is absorption of e- inside the signal electrode).

4.2 Estimates from Monte Carlo

To obtain a more direct estimate of the response we use the GEANT4 Monte Carlo to estimate calorimeter configuration yields.

4.2.1 Sampling Calorimeter Simulation

To obtain a more direct estimate of an SE calorimeter signal, we have made a simple GEANT4 simulation, simply tabulating the numbers of charged particles crossing each 1 cm thick absorber sampling gap in Cu, 30 x 30 cm wide x 1.6m (160 gaps) long in several energy bins. Note that GEANT4 is not accurate at calculating shower particle energies below ~1 KeV, where the secondary yield is maximal for e, pi,p secondary shower particles. Figures 7 show 100 GeV pions and 10 GeV electrons in the 1cm sampling Cu block, with the number of shower particles crossing from a 1 cm absorber per KeV incident per longitudinal distance vs energy of incident shower particles and vs longitudinal distance. We averaged 100 particles of 10, 100 GeV x electrons, pions. GEANT4 tabulates ~740 particles per GeV for pions crossing those gaps in 1 cm layers of Cu, and ~900 for electrons (and is the same within errors at both 10 & 100 GeV). Assuming a very conservative minimal signal of 0.05 SEe per particle as entering an SEe module in the calorimeter gaps, we predict ~37 SEe per GeV for pion showers, and ~45 for electron showers, with an e/pi ~ 1.2, a reasonable first guess. A caveat is that the spectra of incident particle energies generating SEe is similar in electron and pion initiated showers, as seen above, but not identical, and with different particle types – so e/pi will be different when properly weighted by the energy & particle dependence This forms the basis for a very good signal, *provided that the SEe can be amplified by subsequent dynodes to a level consistent with fast electronic measurement* – a gain typically of ~10,000 would be a minimal requirement to stand out from electronic noise. We remark in passing that the number of SE electrons in terms of calorimeter performance are comparable to the photoelectrons in a scintillator or Cerenkov calorimeter. In the optical/scintillator case, ~10-100 thousands of photons are generated in a sampling calorimeter per GeV. but typically only 0.001-0.0001 are collected and converted to p.e., whereas in the SEe case, the produced SEe are far fewer, but essentially all are collected and amplified, giving a similar performance to a possible SE calorimeters.

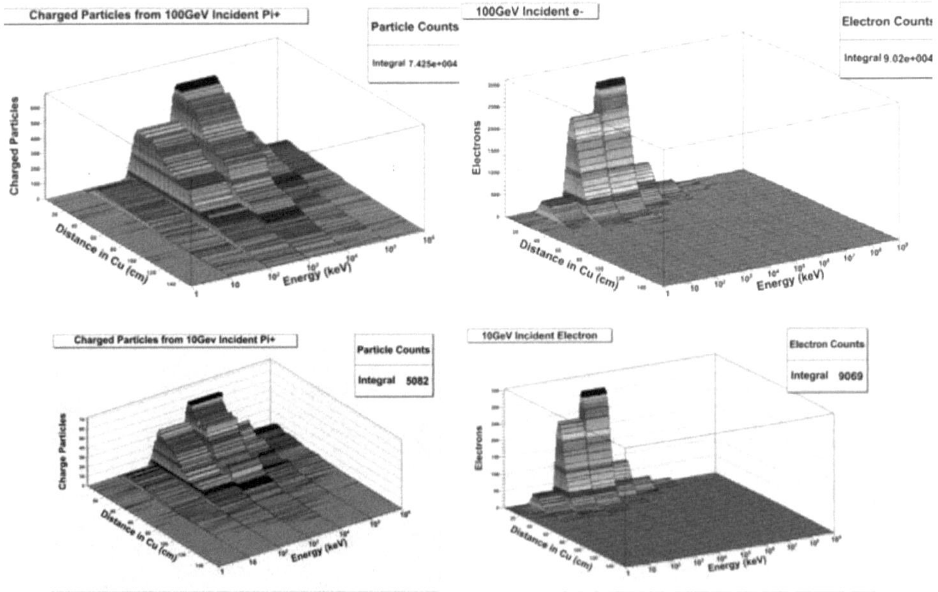

Fig. 7: GEANT4 MC of 100 GeV pions (top) and 10 GeV electrons (bottom) in a 1cm sampling of a Cu block. Number of shower particles crossing from a 1 cm absorber per KeV incident per longitudinal distance vs energy of incident shower particles and vs longitudinal distance into Cu.

4.2.2 Fine Quasi-Homogeneous Electromagnetic Calorimeter Simulation

We have run a detailed simulation using 20 micron Cu meshes spaced at 20 microns, as the entire calorimeter (Note: 100 Volts mesh-mesh voltage/20 microns = 5 volts/micron < 10 V/μm, the acceptable electric field in vacuum). Gammas of 1 GeV generated ~8,000 shower electrons of energy >100 eV, crossing the mesh gaps. This generated an average of 1,600 SE electrons(SEe)/GeV, which when amplified by the meshes, would be a resolution as good as +/-25 MeV at 1 GeV. Even finer meshes are available (10 micron meshes) of W, and the 50/50 metal/vacuum ratio means that a calorimeter with ~half the density of tungsten is possible.

5 Amplification of the SE electron signal

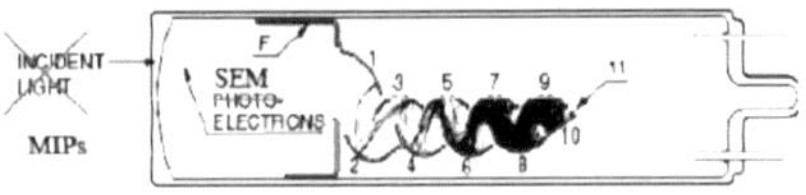

Fig. 8: A PMT modified as an SE beam monitor:photocathode replaced by an aluminum film SE cathode.

For practical device with 20:1 S/N at ~100 MHz requires a gain of >10,000, or 10k electrons as a least count for 1 SEe (similar to 1 p.e. in optical calorimeters). This is possible with dynode-like amplification: since the SEe are generated in vacuum, this is possible by making a vacuum box with a metal "window" with an SE surface on the inside, and using planar proximity collected SEe from the entrance metal oxide cathode – using the mesh-type dynodes as shown above. A detector somewhat similar to this has already been fabricated by EMI as a beam monitor. It used a standard Thorn PMT, simply replacing the photocathode with an Aluminum film SE cathode! Figure 8 shows the data sheet from Thorn-EMI of a beam monitor used by DESY.

6 Initial Tests of SE Calorimetry

We have tested this concept at a CERN test beam using metal mesh PMT as shown in Fig 10 below. The photocathode was disabled by using a +HV base, operating the anode at ~ +2KV, D1 at ground, and the photocathode at small positive voltages or connected to ground through 400kOhms. Blue LED showed that the photocathode was insensitized. The tube was operated mainly with particles entering from the base side, with the photocathode window blackened. The tube was exposed to beams of 225 GeV muons and 100 GeV electrons entering a 3cm radiator, to simulate shower max.

The muon signal was 74% efficient, consistent with 3 pe, consistent with MIP delta rays/SE electrons>50 eV from the photocathode surface, and SE from the 19 dynodes. The electron shower max was fully efficient, with a mean number of SEe of ~30-40, roughly what would be expected for a sample of a 100 GeV shower at 5.5 Lrad and the dynode stack diameter. This implies with homogeneous mesh stack calorimeter (no other absorber) that >600 SEe per GeV are possible with this mesh, fairly consistent with the MC in section 4, and a resolution of ~4%/√E.

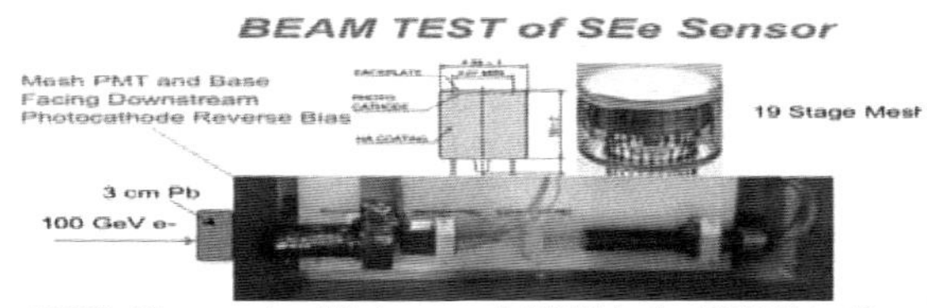

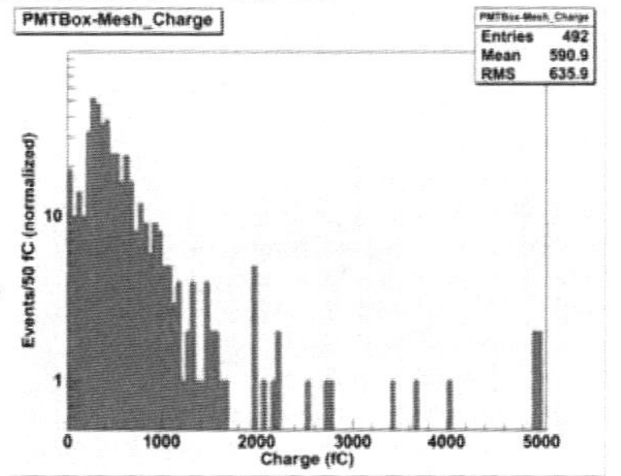

Fig. 10(L): Hamamatsu 19 stage mesh PMT in the CERN test beamline, in the phototube test box. 225 GeV μ+ or 100 GeV e^- hitting 3 cm of Pb radiator incident from left. The photocathode was disabled using a +HV base, anode at +2.1KV, Gain=$8x10^4$, and D1 grounded. Photocathode: +10V or ground through 400k Ohms.
Fig 11(R): Beam test results of Mesh SEe "sensor": 100 GeV electrons into 3 cm Pb (as shown in 10), with one SE sampling "cell". Peak in mesh PMT signal corresponds to 41 SEelectrons (mesh gain ~10^5) at about shower max. The large width is essentially entirely consistent with shower fluctuations on a single shower sampling after 3 cm Pb. This implies > 600 SEe/GeV possible - the same as ~600 pe in a scintillator calorimeter.

7. Conclusions

The properties of Secondary Emission generated electrons as a calorimeter sensing signal has been outlined, and shown to be promising in initial test, consistent with MC and calorimetric data estimates. Very high rad-hardness, and speeds nearly cotemporal with shower development times are evident, and thereby applications to LHC or NLC forward calorimetry at η>3 are of high interest. With a homogeneous mesh-dynode stack SE sensor - effectively a very high sampling fraction - precision calorimeters at the ~2%/√E are indicated by MC, depending on the fine-ness of the mesh (or other) dynodes. While an SEe is similar to a p.e. in terms of signal, SE amplifiers can be constructed in air and taken to high temperatures, as contrasted with PMT. Construction techniques and practical examples are explored in subsequent publications.

Acknowledgements: We thank colleagues at Fairfield & Iowa, especially Grekim Jennings (FU), and B. Bilki, T. Yetkin, E. Albayrak & W. Clarida (U.I), and support from DOE, CMS HCal, and US-CMS.

REFERENCES:

[1] H. Bruining, Physics and Applications of Secondary Electron Emission. 1954. Pergamon Press.
[2] G. Ferioli and R. Jung, Evolution of the Secondary Emission Efficiencies of Various Materials Measured in the CERN SPS Secondary Beam Lines, CERN-SL-97-071-BI, (1997)
[3] D. Hasselkamp et al., Particle Induced Electron Emission II, Springer-Verlag, (1992)
[4] E.J. Sternglass, Phys. Rev. **108** (1957) 1.
[5] Diddens et al. CDHS, Nuclear Instruments and Methods, Vol. **178**, 1 Dec (1980) 27
[6] CALICE Collaboration publications 2007-9: http://llr.in2p3.fr/activites/physique/flc/calice.html
[7] B. Dehning et al., LHC Beam Loss Detector Design: Simulation and Measurements, FRPMN072 Proceedings of PAC07 (2007)

The large-angle photon veto system for the NA62 experiment at the CERN SPS

F Ambrosino[1], B Angelucci[2], A Antonelli[3], F Costantini[2], G D'Agostini[4], D Di Filippo[1], R Fantechi[2], S Gallorini[2], S Giudici[2], E Leonardi[4], I Mannelli[2], P Massarotti[1], M Moulson[3,*], M Napolitano[1], V Palladino[4], F Rafaelli[2], M Raggi[3], G Saracino[1], M Serra[4], T Spadaro[3], P Valente[4], S Venditti[2]

[1]Dipartimento di Scienze Fisiche dell'Università and Sezione INFN, Napoli, Italy
[2]Dipartimento di Fisica dell'Università and Sezione INFN, Napoli, Italy
[3]Laboratori Nazionali di Frascati dell'INFN, Frascati, Italy
[4]Dipartimento di Scienze Fisiche dell'Università and Sezione INFN, Napoli, Italy

[*]E-mail (speaker): `Matthew.Moulson@lnf.infn.it`

Abstract. The branching ratio (BR) for the decay $K^+ \to \pi^+\nu\nu$ is a sensitive probe for new physics. The NA62 experiment at the CERN SPS will measure this BR to within about 10%. To reject the background from dominant kaon decays with final state photons, the large-angle photon vetoes (LAVs) must detect photons of energy as low as 200 MeV with an inefficiency of less than 10^{-4}. The LAV detectors make use of lead glass blocks recycled from the OPAL electromagnetic calorimeter barrel. We describe the mechanical design and challenges faced during construction, the characterization of the lead glass blocks and solutions adopted for monitoring their performance, and the development of front-end electronics to allow simultaneous time and energy measurements over an extended dynamic range using the time-over-threshold technique. Our results are based on test-beam data and are reproduced by a detailed Monte Carlo simulation that includes the readout chain.

1. The NA62 Experiment

The decays $K^+ \to \pi^+\nu\bar{\nu}$ and $K_L \to \pi^0\nu\bar{\nu}$ are flavor-changing neutral-current processes for which the rates are highly suppressed in the Standard Model (SM). At the same time, largely because the hadronic matrix elements can be obtained from experimental data on $K_{\ell 3}$ decays, the SM branching ratios (BRs) for these decays can be predicted with minimal intrinsic uncertainty (see Ref. 1 for a recent review). The $K \to \pi\nu\bar{\nu}$ decays are therefore a sensitive probe of the SM flavor sector and provide constraints on the CKM unitarity triangle that are complementary to those from measurements of B-meson decays. On the other hand, the tiny BRs for these decays are notoriously difficult to measure, not least because of the three-body kinematics with two undetectable neutrinos in the final state. At present, the experimental value of the BR for the decay $K^+ \to \pi^+\nu\bar{\nu}$ is $1.73^{+1.15}_{-1.05} \times 10^{-10}$ on the basis of seven detected candidate events [2]. The goal of NA62, an experiment at the CERN SPS, is to detect ~100 $K^+ \to \pi^+\nu\bar{\nu}$ decays with a S/B ratio of 10:1 in two years of data taking beginning in 2014. The experiment is fully described in Ref. 3. The experimental layout is illustrated in Fig. 1.

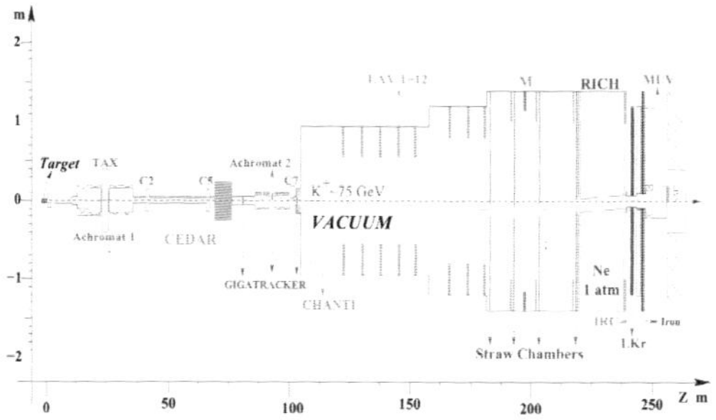

Figure 1. The NA62 experimental layout.

NA62 will make use of a 75-GeV unseparated positive secondary beam with a total rate of nearly 800 MHz, of which ~50 MHz is K^+'s. All 800 MHz of beam particles are tracked by three silicon-pixel tracking detectors just upstream of the vacuum decay volume, providing measurement of the K^+ trajectory and momentum. The vacuum decay volume (10^{-6} mbar) begins ~100 m downstream of the production target. 5 MHz of kaon decays are observed in the 65-m long vacuum decay region. Large-angle photon vetoes (LAVs) are placed at 11 stations along the decay region and provide full coverage for decay photons with 8.5 mrad $< \theta <$ 50 mrad. The last 35 m of the decay region hosts a dipole spectrometer with four straw-tracker stations operated in vacuum. The NA48 liquid-krypton calorimeter [4] is used to veto high-energy photons at small angle. Additional detectors further downstream extend the coverage of the photon veto systems (e.g. for particles traveling in the beam pipe).

The experiment must be able to reject background from dominant decays such as $K^+ \to \pi^+\pi^0$ at the level of 10^{12}. Cuts on the K^+ and π^+ momenta provide a rejection factor of 10^4 and ensure that the photons from the π^0 have 40 GeV of energy. Forward photons that are intercepted by the LKr calorimeter and small-angle vetoes have much higher energies than the photons intercepted by the LAVs. Nevertheless, photons in the LAVs from decays such as $K^+ \to \pi^+\pi^0$ may have energies from a few tens of MeV to several GeV. In order to detect the π^0 with an inefficiency of $\leq 10^{-8}$, the maximum tolerable inefficiency in the LAV detectors for photons with energies as low as 200 MeV is 10^{-4}. In addition, the large-angle vetoes must have good energy and time resolution (10% and 1 ns for 1 GeV photons) and must be compatible with operation in vacuum.

2. The Large-Angle Veto System

The NA62 LAV detectors make creative reuse of lead glass blocks recycled from the OPAL electromagnetic calorimeter barrel [5], which became available in 2007 when various technologies were under consideration for the construction of the LAV detectors. Other solutions considered included a lead/scintillating tile design originally proposed for use in the (later canceled) CKM experiment at Fermilab, and a lead/scintillating-fiber design based on the electromagnetic calorimeter for the KLOE experiment. Prototype instruments based on each of the three technologies were obtained or constructed, and tested with the electron beam at the Frascati Beam-Test Facility. These tests demonstrated that all three technologies are suitable for use in NA62 [6]. In particular, the inefficiency for the detection of single, tagged electrons with the OPAL lead glass modules was measured to be $1.2^{+0.9}_{-0.8} \times 10^{-4}$ at 203 MeV and $1.1^{+1.9}_{-0.7} \times 10^{-5}$ at 483 MeV. Basing the construction of the LAV system on the OPAL lead glass modules provides significant economic advantages.

The modules from the central part of the OPAL electromagnetic calorimeter barrel consist of blocks of Schott SF57 lead glass. This material is about 75% lead oxide by weight and has a

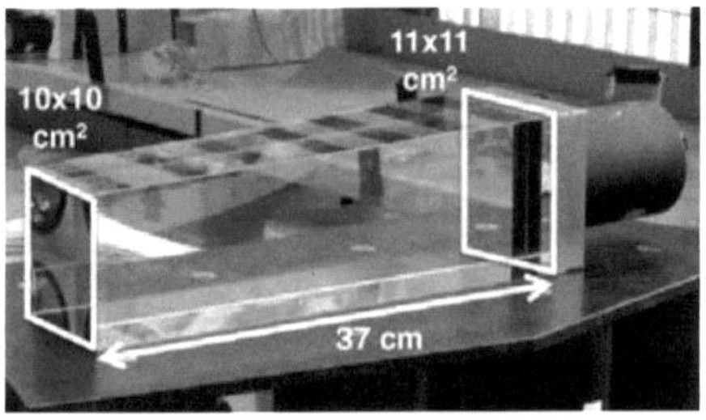

Figure 2. A module from the OPAL calorimeter, without wrapping and with reinforcement plates at the interface between the glass and the steel flange.

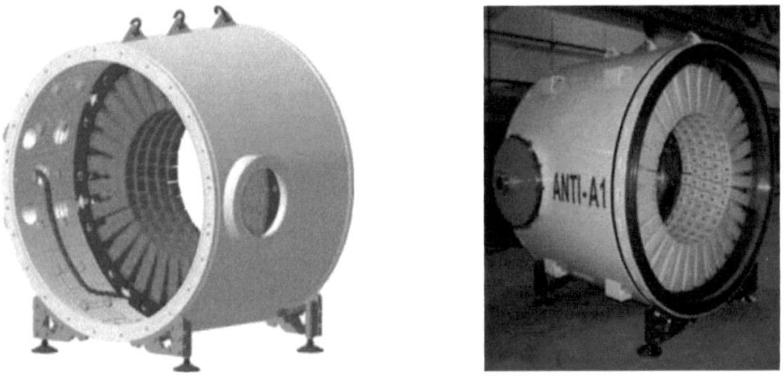

Figure 3. Design study (left) and completed prototype A1 veto station (right).

density $\rho = 5.5$ g/cm^3 and a radiation length $X_0 = 1.50$ cm; its index of refraction is $n \approx 1.85$ at $\lambda = 550$ nm and $n \approx 1.91$ at $\lambda = 400$ nm. The Cerenkov light produced by electromagnetic shower particles in the lead glass is read out at the back side of the block by a Hamamatsu R2238 76-mm PMT coupled via a 4-cm long cylindrical light guide of SF57 of the same diameter as the PMT. The rear face of the glass block is glued to a 1-cm thick stainless steel flange. A mu-metal shield surrounding the PMT and light guide is also glued to the flange. Figure 2 shows a picture of a complete module.

A LAV station is made by arranging these blocks around the inside of a segment of vacuum tank, with the blocks aligned radially to form an inward-facing ring. Multiple rings are used in each station. The blocks in successive rings are staggered in azimuth; the rings are spaced longitudinally by about 1 cm. The LAV system consists of a total of 12 stations. Stations A1–A5 are 217 cm in diameter and contain 160 blocks arranged in 5 layers, stations A6–A8 are 266 cm in diameter and contain 240 blocks arranged in 5 layers, and stations A9–A11 are 306 cm in diameter and contain 240 blocks arranged in 4 layers. A12 is operated in air; its design is somewhat different from that of the other stations. As a result of the staggering scheme, particles incident on any station are intercepted by blocks in at least three rings, for a total minimum effective depth of 21 X_0. Most incident particles are intercepted by four or more blocks (27 X_0).

Station A1 was constructed as a prototype during the first half of 2009. It was installed in the NA62 beamline at CERN and tested in October 2009. The design study and completed station are shown in Fig. 3. Various improvements to the design were made on the basis of the results from the test, and station A2 was constructed and tested in the T9 beamline at the CERN PS in August 2010. A1 was then rebuilt to incorporate the design improvements, and construction of the remaining stations was commenced.

Figure 4. Left: Test stand for characterization of detector modules. The top and bottom rows of blocks form a telescope for the selection of vertical cosmic rays. During operation the front panel is closed. Right: Installation of modules into the steel vacuum vessel.

3. LAV Construction

The OPAL detector modules (lead glass block plus PMT) were manufactured by Hamamatsu during the mid-1980s. Their recycling requires substantial care throughout the assembly procedure.

After the storage area in which the modules were kept was inundated during a flash flood in spring 2008, the modules were subject to an extensive sorting and clean-up effort carried out at CERN by an industrial recovery firm. They were subsequently shipped to Frascati, where the LAVs are assembled.

The glass at the interface with the stainless steel flange is fragile and is found to be fractured in a few percent of the modules upon first examination. This is attributed to thermally induced stress. While damaged modules are discarded, the first step in the processing of intact modules is to reinforce the interface. Using epoxy resin, 20 cm^2 $\times$ 0.5-mm thick stainless steel plates are attached across the glass-steel interface on all four sides of the block. Calculations indicate and static tests confirm that the reinforced bond is several times stronger than the original bond. After reinforcement, the blocks are unwrapped and cleaned with isopropanol or acetone. They are then wrapped with a new laser-cut and heat-welded Tyvek cover.

During the A1 test beam, ringing of the analog signal was observed to lead to errors in charge reconstruction using the time-over-threshold technique discussed below. This problem was traced to a small parasitic inductance in the PMT dynodes and solved by replacing the original OPAL HV dividers with new dividers of our own design. The new divider features additional resistors on the last three dynodes and anode to damp out the oscillation, storage capacitors for the last three dynode stages to improve response linearity for large signals, and a decoupling resistor between the HV and signal grounds to decrease noise.

After the divider is replaced, the blocks are tested and characterized 12 at a time using a test station featuring an LED pulser and cosmic-ray telescope (Fig. 4). The PMT gains are measured first, by varying the HV settings and mapping out the response for each block. Using the gain curves so obtained, the PMTs are set to a reference value of the gain (1×10^6), and the response to cosmic rays selected by the telescope is measured; the photoelectron yield for the block (p.e./MeV) is then obtained assuming that vertically incident cosmic rays leave 77 MeV in each block. Yields of 0.34 p.e./MeV are typical. Finally, using the gain curves and the measurements of photoelectron yield, the PMT voltages are set to the values expected to produce a common output charge level of 4.5 pC for cosmic-ray events. The response is measured and the HV setting is validated. Thus, at the end of a 12-hour cycle, which is fully automated using LabView, we have PMT gain and photoelectron yield measurements as well as the operational HV settings for 12 modules. Additional data (current-draw measurements, dark-count rates) are also collected using the test station.

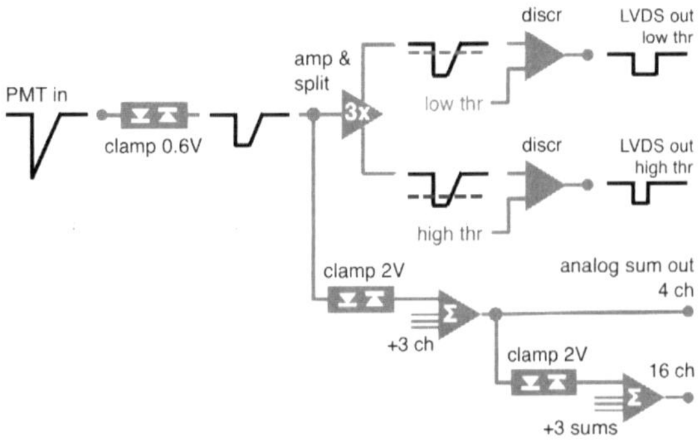

Figure 5. Conceptual schematic of the ToT discriminator board.

The OPAL design features an optical port at the base of each module. Blue LEDs are installed in these optical ports as part of the calibration and monitoring system, and will allow monitoring of the operational status and relative timing for each block. A low-capacitance LED was chosen to minimize the rise and fall times of the light pulse; this is important for use with the time-over-threshold-based readout system discussed in the following section. In principle, the LED system should allow in-situ gain calibration as well.

After testing and characterization, the blocks are arranged in groups of four in an aluminum mounting bracket. For the installation, the vacuum vessel is turned on end. The aluminum mounting bracket with four modules is lowered into the upended vessel by overhead crane and bolted to the wall, as illustrated in Fig. 4, right.

4. Front-End Electronics

While the amplitude of the PMT signal from a minimum ionizing particle is about 20 mV, signals from 20-GeV showers may be as large as 10 V. Our readout scheme is based on the time-over-threshold (ToT) technique and furnishes time and energy measurements over a large range of incident photon energies. The scheme is implemented using a dedicated front-end ToT discriminator board of our own design [7] and a digital readout board (TEL62) used by various NA62 detector subsystems [8].

The ToT discriminator converts the analog signals from the detector to low-voltage differential signal (LVDS) pulses, with width equal to the duration of the analog signal from the detector above a specified threshold. After a clamping stage for protection, the signal from each PMT is passively split. One copy of the clamped analog signal is summed with the signals from adjacent channels and made available via a front-panel connector for diagnostic purposes. The other copy is amplified $\times 3$ and used as input to a two different comparators, corresponding to a low threshold and a high threshold. Each comparator generates a separate LVDS output. We expect to use thresholds of about 5 and 50 mV against a noise level of about 2 mV; the threshold values are programmable. A conceptual schematic of the ToT discriminator board is presented in Fig. 5. The production model has 32 input channels and 64 output channels.

The TEL62 is based on the design of the TELL1 readout board developed for the LHCb experiment. On the TEL62, TDC mezzanines measure the leading and trailing times of the LVDS pulses. Using a time-to-charge calibration parameterization, the FPGA on board the TEL62 calculates the time corrected for slewing, as well the charge for each hit as reconstructed from the pulse width above threshold. This information is sent to the subsequent DAQ stages. Level-0 trigger primitives are also calculated on board the TEL62 and sent to the level-0 trigger processor.

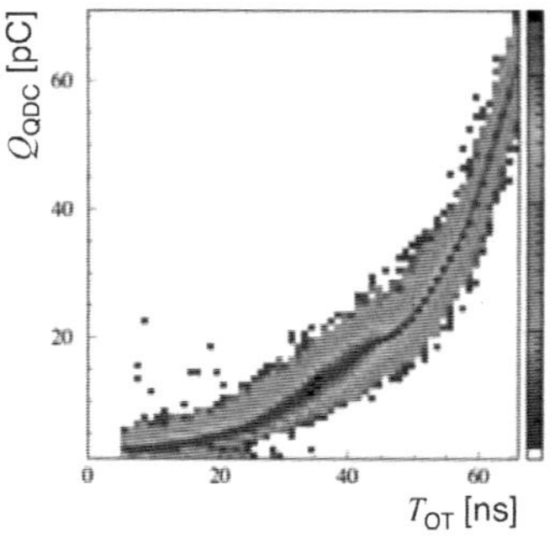

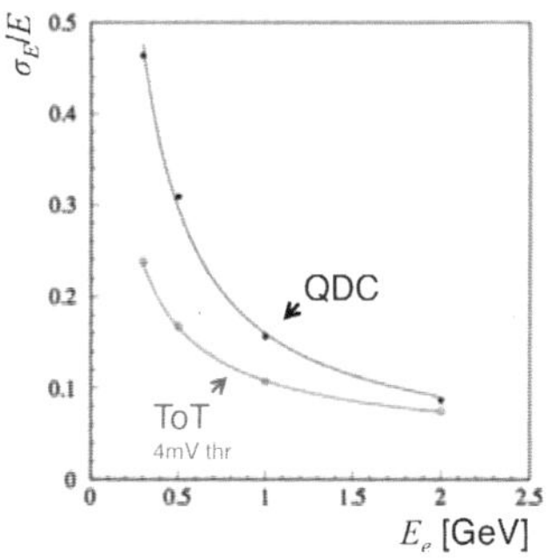

Figure 6. Left: Scatter plot of signal charge measured using QDCs vs. signal ToT, for electrons of various energies. Right: Energy resolution obtained using ToT compared with that obtained using QDCs, as a function of incident electron energy.

5. Test-Beam Performance

LAV A2 was tested using the positive secondary beam in the T9 area at the CERN PS in August 2010. Data were collected at various beam momenta over the interval 0.3–10 GeV. The composition of the T9 beam changes as a function of the momentum setting. At 0.3 GeV, the beam is roughly 70% e^+ and 30% π^+ (including decay μ^+), while above 6 GeV the e^+ component drops off sharply. Two threshold CO_2 Cerenkov counters in the beamline allowed samples enriched e^+ and μ^+ to be selected. The beam focus coincided approximately with A2's first layer of lead glass blocks. Two crossed scintillator paddles were placed at the beam entrance, and one larger paddle was placed at the beam exit. A prototype of the ToT discriminator board was used for the test, mainly in conjunction with commercial QDCs and TDCs.

Figure 6 shows a scatter plot of the signal charge measured using the QDCs vs. the signal time over a threshold of 4 mV, for electrons. The data are summed for all blocks on which the beam was incident and over runs of different energies. For small signals, a small increase in the integrated charge corresponds to a large increase in the time over threshold; the ToT measurement provides more sensitivity for small signals than does the QDC measurement. This is a desirable property for the LAV detectors, since high detection efficiency is required for low-energy photons. Parameterizations of the type illustrated in Fig. 6 are used to convert ToT to an effective charge measurement. The energy resolution obtained using the ToT technique is compared with that obtained using the QDCs in Fig. 6. As expected from the form of the curve in Fig. 6, at low energies, the resolution obtained with the ToT technique is better than that obtained with the QDCs. The fits in Fig. 6 give

$$\begin{aligned} \text{QDC}: &\quad \sigma_E/E = 8.6\%/\sqrt{E\ [\text{GeV}]} \oplus 13\%/E, \\ \text{ToT}: &\quad \sigma_E/E = 9.2\%/\sqrt{E\ [\text{GeV}]} \oplus 5\%/E \oplus 2.5\%. \end{aligned}$$

While, as expected, the statistical contribution to the energy resolution is about the same with either readout scheme, the contribution from noise (term proportional to $1/E$) appears to be significantly smaller with the ToT technique. The presence of the constant term with the ToT technique may be due to small differences in the charge vs. ToT curves from block to block.

The width of the signal time distribution in slices of charge gives a measurement of the time resolution. A fit to the measurements of σ_t vs. charge from ToT (4 mV threshold) for a single block gives $\sigma_t = 220\ \text{ps}/\sqrt{E\ [\text{GeV}]} \oplus 140$ ps, where the constant term is assumed to be due to trigger jitter. The intrinsic time resolution of the detector can be estimated from the width of the distribution of signal time differences for two successive blocks. Assuming that the two

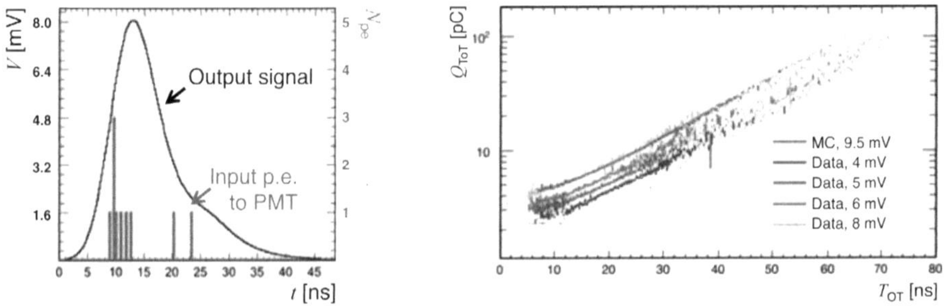

Figure 7. Left: Simulation of PMT signals. Number of photoelectrons produced in bins of time (red, scale at right) and resulting PMT signal (black, scale at left). Right: Comparison of charge vs. ToT curves for MC (9.5 mV threshold) and data (various thresholds).

blocks have the same intrinsic time response and that there is no common-mode contribution to the resolution, $\sigma_{t_1} = p/\sqrt{E_1}$ and $\sigma_{t_2} = p/\sqrt{E_2}$, so that we expect $\sigma_{\Delta t} = p/[E_1 E_2/(E_1 + E_2)]^{1/2}$, with no constant term. Our measurements are consistent with $p = 210$ ps and no constant term.

The Geant4-based NA62 Monte Carlo (MC) simulation includes a detailed description of the LAV geometry and materials. The arrival times of the Cerenkov photons at the photocathode are simulated as shown by the histogram in red in Fig. 7. A complete simulation of the PMT uses this information to generate an output signal, taking into account the PMT gain and transit-time fluctuations, the capacitance of the PMT, and dispersion in the readout cable. The resulting signal, which is illustrated by the black curve in Fig. 7, is input to a full simulation of the ToT discriminator.

Figure 7 shows charge vs. ToT curves for test-beam muons for four different threshold settings (4, 5, 6, and 8 mV). The red curve is from the simulation, with a threshold at 9.5 mV and all other adjustable parameters set to typical or measured values. The simulation with a threshold of 9.5 mV accurately reproduces the shape of the measured charge vs. ToT curve with threshold 8 mV. This slight discrepancy is easily attributed to the accuracy of the manual threshold adjustment on the prototype ToT discriminator board. These results thus confirm our detailed understanding of the detector and the readout chain.

6. Status and Outlook

As of July 2012, LAVs A1 through A8 have been completed and installed in the beamline. An additional station, A11, is nearly complete at Frascati. Serial production of the front-end electronics boards is underway. As the front-end electronics become available during summer 2012, LAVs A1 though A8 will be read out without beam. They will be fully tested in an NA62 technical run scheduled for November 2012. NA62 data taking with all detectors installed is planned for early 2014.

References

[1] Cirigliano V *et al.* 2011 In press, arXiv:1107.6001
[2] E949 Collaboration, Artamonov A *et al.* 2008 *Phys. Rev. Lett.* **101** 191802
[3] Hahn (ed) F *et al.* 2010 NA62 technical design document NA62 Document 10-07
[4] NA48 Collaboration, Fanti V *et al.* 2007 *Nucl. Instrum. Meth. A* **574** 433
[5] OPAL Collaboration, Ahmet K *et al.* 1991 *Nucl. Instrum. Meth. A* **305** 275
[6] Ambrosino F *et al.* 2007 *2007 IEEE Nuclear Science Symposium Conf. Record* (Honolulu) pp N05–6 arXiv:0711.3398
[7] Antonelli A *et al.* 2011 *JINST* **7** C01097
[8] Ambrosino F *et al.* 2011 *2011 IEEE Nuclear Science Symposium Conf. Record* (Valencia, Spain) pp NP2.S–105

Journal of Physics: Conference Series **404** (2012) 012023 doi:10.1088/1742-6596/404/1/012023

Results of R&D on a new construction technique for W/ScFi Calorimeters

O D Tsai[1], L E Dunkelberger[1], C A Gagliardi[2], S Heppelmann[3], H Z Huang[1], G Igo[1], K Landry[1], Y X Pan[1], S Trentalange[1], W Xu[1] and Q Zhang[1]

[1] University of California, Los Angeles, California 90095
[2] Texas A&M University, College Station, Texas 77843
[3] Pennsylvania State University, University Park, Pennsylvania 16802

E-mail: tsai@physics.ucla.edu

Abstract. We report on results of an R&D program to develop new, simple and cost effective techniques to build compact sampling calorimeters utilizing tungsten powder and scintillating fibers. Such calorimeter detectors are under consideration for experiments at the planned Electron Ion Collider and the future upgrade of the STAR experiment at RHIC (BNL). In the first year of this R&D project we built two prototypes of very compact electromagnetic calorimeters and tested them at FNAL test beam T1018 in January 2012. Details of the construction technique, results of the test run and future plan will be presented.

1. Introduction

A high-energy polarized Electron-Ion Collider (EIC) will present new scientific opportunities to advance the understanding of cold nuclear matter and the spin structure of the proton in QCD with unprecedented precision [1]. Calorimetry is a critical component of a dedicated EIC detector. The requirements for calorimeters for this detector (end caps, barrel, very forward) differ significantly in terms of resolution, granularity and radiation hardness. All these calorimeters need to be compact and hermetic, in addition, they must be able to work in the presence of a magnetic field of about 2 T. Detailed requirements for these detectors are presently under investigation.

In addition, current RHIC experiments, both STAR and PHENIX, are planning for major upgrades to explore new physics opportunities along with the accelerator upgrade. The STAR experiment, for example, is working on a forward instrumentation upgrade to elucidate the dynamics that underlie the observed large transverse single-spin asymmetries in polarized $p+p$ collisions and explore the onset of gluon saturation in $p+A$ collisions. Important components of this program will require the ability to measure identified hadrons (π^0, η, Λ...) at large rapidity, charged hadrons, direct photons, $e^+ e^-$ pairs from Drell-Yan and J/Ψ production, and jets, as well as di-hadron and γ+hadron correlations. A compact, highly granulated, compensated Forward Calorimeter System (FCS) is being pursued by the STAR collaboration for a future upgrade.

A scintillating fiber calorimeter technique [2, 3, 4] seems to be a suitable option for some part of the dedicated EIC calorimeter system and for the upgrade of the STAR forward calorimeter system. This technique allows construction of versatile compact and fine-grained sampling electromagnetic and hadronic calorimeters with good energy resolution, hermeticity, homogenity, timing and position resolutions [5, 6, 7]. We proposed to investigate new methods of construction of fiber calorimeters for possible application for EIC and STAR under the generic EIC detector R&D program sponsored by the DOE nuclear physics and BNL. In the first year of this program we aimed at demonstrating the 'proof-of-principle' with construction and beam tests of small prototypes. The targeted electromagnetic energy resolution was chosen somewhere in between preliminary requirements for EIC and STAR, and is close to 12%/√E. In addition, the sampling fraction f_s was kept close to 2% in order to satisfy the requirement of compensation for the STAR FCS system. The sampling term for energy resolution for such a detector is expected to be close to $2.7\%/\sqrt{(d/f_s)}$ where d is the diameter of the scintillating fibers in mm [3].

2. Construction technique

Despite the simplicity of fiber calorimeters, building them is not a trivial process. Detector components have to be produced with very high tolerances to keep the detector homogeneous. In the past, extrusion, machining or rolling techniques were employed to produce the absorption plates. These processes were not simple. Usually construction of special machines and tooling to produce absorber plates were needed. Assembling these calorimeters is a labour intensive process, because multiple elements of the detector had to be handed essentially one by one. These factors contributed to higher costs for fiber calorimeters compared to scintillation plate detectors. Traditional techniques are also prone to limitations with respect to increased sampling frequency, because thinner absorber layers and thinner fibers become more and more difficult to produce and handle. Construction and assembly techniques for JETSET, KLOE and H1 (figure 1) calorimeters are described in [5, 6, 7].

A succinct description of our technique is the following: first we form a matrix of fibers and then the absorber is poured into this matrix. The main difference from previous techniques is that the individual elements of the calorimeter do not need to be handled separately.

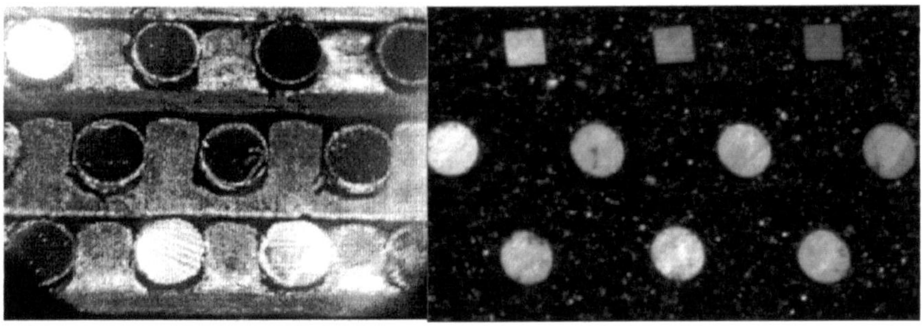

Figure 1. Fiber/absorber structure of typical ScFi calorimeters [8] (left); fibers 0.5 mm in diameter, fiber spacing 0.9 mm. (right) Fiber/tungsten powder structure of one of our mechanical prototypes: square fibers 0.25 mm x 0.25mm, lower rows: round fiber 0.33 mm diameter, spacing 0.88 mm.

The first key element in our technique is the utilization of tungsten powder as an absorber. We used tungsten powder produced by Tungsten Heavy Powder, Inc. The properties of this powder (Technon® Ultra) are: particle size distribution 90% between 40 and 150 microns, bulk density 18.5 g/cm^3, tap density 11.25 g/cm^3. The chemical composition is: W > 99.3%, Fe < 0.05%, Ni < 0.05%, O_2 < 0.5%, others (Co, Cr, Mo and Cu) < 0.5%. This tungsten powder also has very good fluidity, an important property in our application. The only operation with the absorber material that needs to be performed is to measure the right amount of powder before pouring it into the fiber matrix assembly. The cost of this tungsten powder by weight is about 30% of the cost of 1 mm thick tungsten sheets. The powder is produced from scrap tungsten metal.

The second key element in our technique is a simple method of forming a matrix of scintillating fibers. The matrix is defined by a set of precision brass meshes produced by photo-etching. Typical mechanical tolerances for these meshes are 30 microns on overall dimensions for 300 micron thick meshes and about 15 microns on center-to-center distances between the holes for the scintillation fibers. Scintillation fibers were cut to desired length by a thermo-cutter. This causes the end of the fibers to melt, forming a small drop which works as a stopper to prevent the fibers from slipping through the openings in the mesh. With all the meshes stacked together, the scintillation fibers are fed through by simply dropping about 500 fibers at a time into the container holding the meshes. With a slight tapping on the holding container all the fibers will flow through the set of meshes in a matter of a few seconds. For our latest prototype the whole process of forming a matrix consisting of about 1500 fibers took about 30 minutes. A short video clip explaining assembly process is available on-line [9].

The idea of using W powder as an absorber was briefly investigated by the UCLA group in 2003 with the construction and test of a small electromagnetic prototype at SLAC. The mechanical structure of the towers required a thin walled brass container to hold dry powder and fibers in place. It was later found that the assembly technique at that time was not perfect, leading to large transverse non-uniformities in the response of the detector due to variations of sampling fraction along the depth of the towers and possible displacement of the fibers during packing. A compact calorimeter requires very strict mechanical tolerances and very uniform internal structure to achieve theoretical resolution. To solve problems with non-homogeneity in the composition of the towers and at the tower boundaries we introduced intermediate meshes to hold fibers along the towers and developed a vacuum-assisted method to infuse optical epoxy into the tungsten powder/fiber assembly. Since the whole assembly becomes rigid and stable there is no need for any type of external container. This eliminates dead material in the tower assembly. These additional meshes also allow arrangement of fibers in an accordion configuration, a mechanical structure almost impossible to build with traditional construction techniques. The use of precision meshes and precise molding forms to produce sub-assemblies assured good mechanical tolerances and uniform spacing between fibers as well as uniform sampling fraction along the towers and from one tower to another tower. A few mechanical samples were examined for homogeneity. The variation of density along the towers was found to be within 1%.

Using this technique we constructed two small prototypes of electromagnetic calorimeters.

Figure 2. Matrix of fibers in the molding forms prior filling with tungsten powder (a). Finished block of 'spacal' type prototype before light guides were glued in place, as seen from the photo-detector side (b). Different stages of the gluing of the 'spacordion' prototypes from small sub-assembly units (c).

Each detector consists of sixteen towers. The main difference between these two detectors is the arrangement of the scintillation fibers. In the first detector (called a 'spacordion') we used Bicron BCF 12 [10] 0.33 mm diameter fibers spaced by 0.88 mm and arranged in an accordion shape as seen in figure 2 (a). In the second detector (called a 'spacal') we used KURARAY SCSF 78 [11] 0.47 mm diameter fibers spaced by 1 mm and placed parallel to long axis of the tower. The parameters of the 'spacal' prototypes are: final density of 10.17 g/cm^3 , radiation length – 7 mm, Molere radius – 23 mm, sampling fraction for electrons 2%. Each 'spacal' super module was glued from two subassemblies. Dimensions of the super modules are: 166 x 53.3 x 53.3 mm^3. The average weight of super modules is 4636 grams. The difference in the weight of different super modules is about 0.5%. Each super module holds 3120 fibers. The composition of the towers at the front (first 2.5 mm) and back sides (last 5 mm) is epoxy and fibers only. These regions are free from tungsten powder. That was made to achieve good quality optical surfaces by machining the ends of the super module (otherwise particles of tungsten powder scratch the fibers). For the test run the light from each 'spacal' super module was collected by 20 cm long tapered trapezoidal acrylic light guides enclosed in an Enhanced Specular Reflector (ESR) mirror pipe into a single photo-multiplier (Electron Tube 9125B). The front face mirror was made from ESR film and coupled to the front face of the super module with optical grease. PMTs were coupled to the light guides by optical grease. To verify the uniformity of light collection one super module was scanned across the front face of the detector with a 350 nm LED. Photographs of these prototypes are given in figure 2.

3. Results of the Test Run

The goal for the first year of R&D was to demonstrate 'proof-of-principle' with a beam test of the prototypes. The most important parameters for future developments are: energy resolution, more precisely, how close is the measured resolution to a GEANT4 prediction, and the absolute light yield.

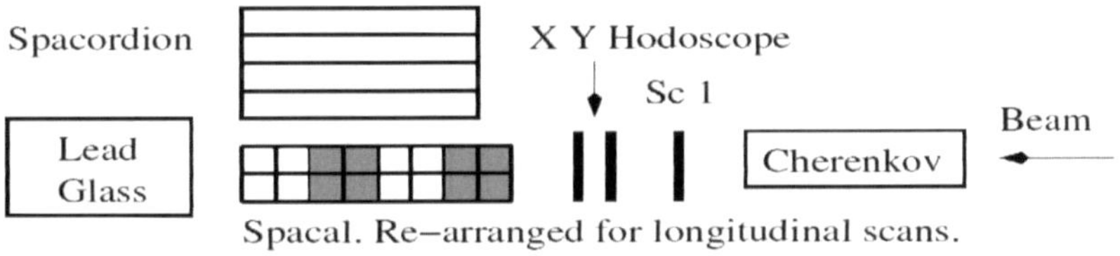

Figure 3. T1018 test setup at the FNAL Meson Test Beam Facility.

The FNAL test beam setup T1018 in the Muon Test Beam Facility (MTBF) is given in figure 3. The absolute energy scale of the beam was set with a high resolution large lead glass block provided by the MTBF. The momentum spread of the beam below 4 GeV was estimated by previous T1005 test beams to be about 2.7% [12] which is consistent with our measurements with the lead glass calorimeter. At 8 GeV we estimated the momentum spread of the beam to be 2.3%. Electrons were identified with a threshold Cherenkov counter (standard equipment at the MTBF). Impact position was defined by a scintillation XY hodoscope (4.9 mm wide scintillation square rods readout by SiPMTs). Calorimeter prototypes were mounted on a movable platform at an angle of 3° with respect to the beam. The lead glass calorimeter was located just behind the detectors under test. To measure attenuation lengths and homogeneity along the towers all 'spacal' modules were re-arranged as shown in figure 3, the 'spacordion' was rotated 90°. In some tests two of the 'spacal' modules had their front face mirrors replaced with black absorber tape. To measure energy resolution of the 'spacal' the impact points were restricted to an area 2x2 cm^2 at the center of the spacal super module (corner of four towers, which represents the worst possible case for non-uniformities). The energy resolution for the 'spacordion' is derived from the energy measured by 3x3 towers clusters. To identify minimum ionizing particles (MIP) during longitudinal scans, signals in all four 'spacal' super modules (total depth $\sim 30X_0$) and signal in the lead glass calorimeter located behind them, were all required to be within MIP peak. To verify the uniformity of response in the transverse direction the impact points in the central area of 'spacal' was selected by the hodoscope. Within 2x2 cm^2 the transverse non-uniformity measured with 4 GeV electrons is 1.4%. The light yield from the 'spacal' measured with a front mirror is close to 2000 photo electrons (pe) per GeV (PMTs were preselected and calibrated prior to the test run with single photo-electron response). The ESR mirror at the front face adds about 70% more light. In any case (with or without mirror) the contribution of the photo-statistics to the energy resolution is very small. The light yield from the 'spacordion' was measured to be close to 1000 pe/GeV. The beam settings at 4 GeV were somewhat questionable. It is possible that beam settings during calibration runs with the lead glass and during test measurements were different. The resolution results are presented as a raw (without any corrections) and with beam momentum spread subtracted. This correction is small at low energy. No other corrections were applied.

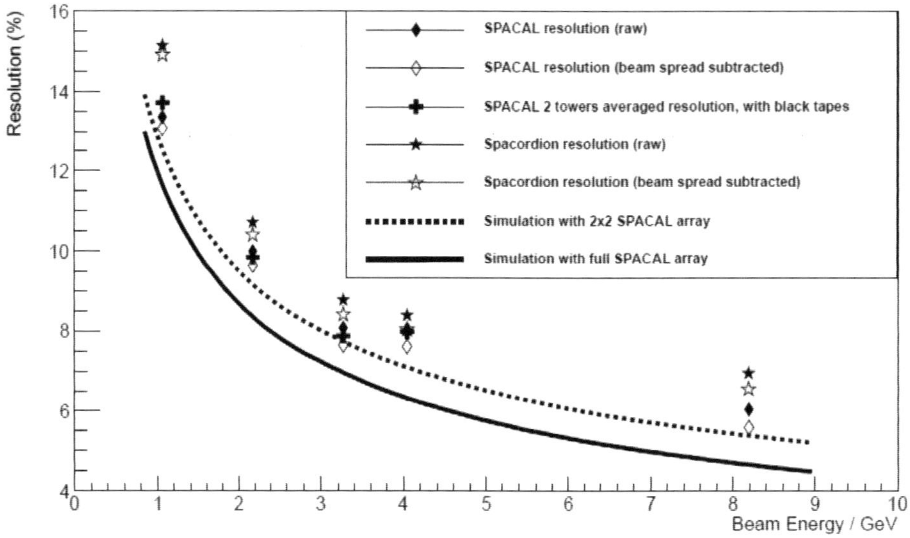

Figure 4. Measured energy resolution of 'spacal' and 'spacordion' prototypes compared with GEANT4 prediction.

The measured resolution for 'spacal' is very close to GEANT4 prediction (dotted line in figure 4). The energy resolution of a single 'spacal' super module is slightly worse than theoretical limit (solid line in figure 4) due to transversal energy leakage. Prior to the test run the expectation was that the energy resolution for the 'spacordion' should be somewhat better compared to the 'spacal' due to increased sampling frequency with about the same sampling fraction. That was not confirmed by the test.

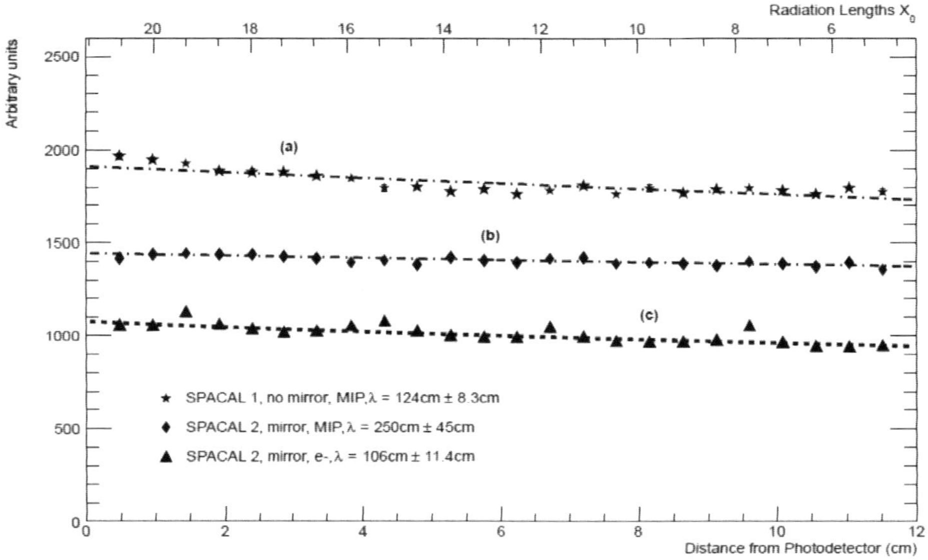

Figure 5. Attenuation length for 'spacal' measured with MIPs and electrons.

The longitudinal scans with 8 GeV beam along the towers were performed to measure the uniformity of composition along the length of the towers and to measure the attenuation of light in the fibers. From a single data sample both MIP and electrons can be cleanly identified for the 'spacal'. For the 'spacordion,' the position of the MIP peak was too close to pedestal, thus only electrons were used to measure the light attenuation length. The uniformity of composition was found to be better than 5%, (figure 5, data points (c)). Four points above the fit were excluded because they correspond to the 300 micron thick brass meshes which effectively shift the development of the shower. 'Spacal' 2 was located second downstream in the beam line as shown in figure 3. The attenuation lengths measured using MIPs with and without a front face mirror were close to what was expected. The same light attenuation length measured with electrons was much shorter, 106 cm vs 250 cm. The only explanation for this at this point is the difference in the spectrum of light produced by MIPs and particles in the electromagnetic shower, mainly the proportion of Cherenkov light. The attenuation length for the 'spacordion' was found to be close to 50 cm. The much shorter attenuation length of 'spacordion' partially explains the worse-than-expected energy resolution for this detector, which was late confirmed with GEANT4 simulations.

4. Conclusions

The 'proof-of-principle' for a new simple and cost effective method of building compact scintillation fiber calorimeters has been demonstrated. In principle, very high sampling frequencies and non-traditional arrangements of fibers can also be achieved with this method. We also learned that thin fibers (diameters smaller than 0.5 mm) will limit resolution due to a very short attenuation length. Wiggling of scintillation fibers significantly complicates assembly of the detector; in addition it probably caused some decrease in the attenuation length. We will not pursue these directions in our future R&D effort. There are a lot of questions beyond 'proof-of-principle' which need to be answered: mechanical integrity under load, light response under load, and challenging question of compact readout. An extension of this technique for wedge-shaped towers is also under consideration. We plan to address these questions in the near future.

Acknowledgments

We thank FNAL for giving us opportunity to test our detectors with the beams and A. Soha and E. Ramberg for their help during the test run. This study was supported by the United States Department of Energy.

References

[1] Boer D *et al.* 2011, Gluons and the quark sea at high energies: distributions, polarization, tomography, arXiv:1108.1713v2
[2] Burmeister H *et al.* 1984, Nucl. Instr. and Meth. **A225**, 530
[3] Livan M *et al.* 1995, *Scintillating-fiber calorimetry*, CERN-95-02
[4] Acosta D *et al.* 1991, Nucl. Instr. and Meth. **A308**, 481
[5] Hertzog D *et al.* 1990, Nucl. Instr. and Meth. **A294**, 446
[6] Nicholls T *et al.* 1996, Nucl. Instr. and Meth. **A374**, 149
[7] Lee-Franzini J *et al.* 1995, Nucl. Instr. and Meth. **A360**, 201
[8] Appuhn R-D *et al.* 1996, *The H1 Lead/Scintillating-Fiber Calorimeter*, DESY 96-171
[9] Tsai O, http://www.youtube.com/watch?v=MceBdr-4Ijo&feature=youtu.be number)' style
[10] Saint-Gobain Crystals and Detectors, 12345 Kinsman Road, Newbury, OH 44065.
[11] KURARAY CO.,LTD., 1-1-3 Otemachi, Chiyoda-ku, Tokyo 100-8115, Japan.
[12] McNabb R *et al.* 2009, Nucl. Instr. and Meth. **A602**, 396

XVth International Conference on Calorimetry in High Energy Physics (CALOR2012) IOP Publishing
Journal of Physics: Conference Series **404** (2012) 012024 doi:10.1088/1742-6596/404/1/012024

The Calorimeter Systems for the sPHENIX Experiment at RHIC

C Woody and E Kistenev, for the PHENIX Collaboration

Physics Department, Building 510C
Brookhaven National Lab
Upton, NY 11973

E-mail: woody@bnl.gov, kistenev@bnl.gov

Abstract. A major upgrade is being planned for the PHENIX experiment that will have greatly enhanced physics capabilities to measure jets in relativistic heavy ion collisions at RHIC, as well as in polarized proton interactions, and eventually electron ion collisions at an Electron Ion Collider. This upgrade, sPHENIX, will include two new calorimeter systems. One will be a hadronic calorimeter, which will be the first hadronic calorimeter ever used in an experiment at RHIC, and another will be a new compact electromagnetic calorimeter. Both calorimeters will cover a region of ± 1.1 in pseudorapidity and 2π in phi. The hadron calorimeter will be based on scintillator plates interspersed between steel absorber plates and read out with wavelength shifting fibers. The electromagnetic calorimeter will be an accordion design that will utilize scintillating fibers embedded in a matrix consisting of tungsten plates, tungsten powder and epoxy. The readout for both calorimeters will use silicon photomultipliers. The overall design of these two calorimeter systems is described along with the R&D efforts currently being pursued to develop them along with their readout.

1. Introduction

The PHENIX Experiment is planning a major new upgrade that will greatly enhance its physic capabilities for measuring jets in heavy ion collisions at RHIC, and well as in polarized proton interactions, and eventually in electron ion collisions at an Electron Ion Collider. This upgrade, called sPHENIX [1], will involve removing the current PHENIX central magnet and two central spectrometer arms and replacing them with a new superconducting solenoid magnet and two new calorimeter systems, one electromagnetic and the other hadronic, that will cover approximately $\pm$ 1.1 units in pseudorapidity and 2π in phi. The hadronic calorimeter will be the first hadronic calorimeter ever used in an experiment at RHIC and will, for the first time, provide the ability to measure the total energy of jets produced in heavy ion collisions, and allow a detailed study of the Quark Gluon Plasma near the region of its critical temperature. The electromagnetic calorimeter will be located just outside the solenoid magnet at a radius of 95 cm. These calorimeters, along with the VTX silicon tracking detector from the current PHENIX experiment, will form the basis of the new central spectrometer for sPHENIX. Figure 1 shows this new configuration with the present PHENIX North and South Muon Arms still in place. However, there are also additional plans to upgrade both of these arms in the future to enhance the capabilities of sPHENIX for pA, polarized proton and EIC physics.

Journal of Physics: Conference Series **404** (2012) 012024 doi:10.1088/1742-6596/404/1/012024

Figure 1. New central detectors for the sPHENIX Experiment with the present PHENIX North and South Muon Arms still in place. The yellow and red objects are the inner and outer sections of the hadron calorimeter and the dark grey cylinder is the electromagnetic calorimeter. The white cylinder is the superconducting solenoid and the small inner cylinder near the beam pipe is the current PHENIX VTX detector.

2. Detector requirements

The design of the two calorimeter systems is driven by the detector requirements to perform the physics measurements in the sPHENIX experiment. The primary goal is to measure jets in heavy ion collisions, which places demands on the energy resolution and segmentation of both the EMCAL and the HCAL. However, for events containing jets in heavy ion collisions, there is a significant amount of energy deposited in the calorimeter from the large number of particles produced in the underlying event of accompanying soft collisions. Various sophisticated algorithms have been developed to subtract this background energy from the total energy measured for the jet, but ultimately, the fluctuations on this energy limits the jet energy resolution. As a result, the calorimeter energy resolution need not be better than the limitations imposed on the jet energy resolution by the fluctuations in the energy of the underlying event. This leads to a requirement of ~ $100\%/\sqrt{E}$ for the HCAL and ~ $15\%/\sqrt{E}$ for the EMCAL for measuring the jet energy. Other physics measurements, such as measuring γ-jet correlations and direct photons at high p_T ($p_T > 10$ GeV/c) are also consistent with these requirements. The requirements on the segmentation are $\Delta\eta\times\Delta\phi \sim 0.1\times0.1$ for the HCAL and $\Delta\eta\times\Delta\phi \sim 0.025\times0.025$ for the EMCAL, which are mainly driven again by the high particle multicity in heavy ion collisions. For other physics measurements, such as measuring heavy quarkonia, tagged heavy flavor jets and high-z fragmentation functions, an additional tracking system that would add additional layers of silicon tracking to the current VTX detector would be required, and is being pursued as a future upgrade option. In addition, a future preshower detector would be added that would enable the measurement of high p_T π^0's (p_T up to ~ 40 GeV/c) and also provide the capability of resolving the high mass Y states.

We have performed a full GEANT4 simulation of the sPHENIX detector and used this to study the expected performance of the two calorimeter systems. The EMCAL was simulated using a simple geometry consisting of cylindrical layers of absorber and scintillator, as opposed to the actual geometry as described below, but with the same sampling fraction as the real calorimeter will have. The hadronic calorimeter was simulated using the actual geometry of the proposed design.

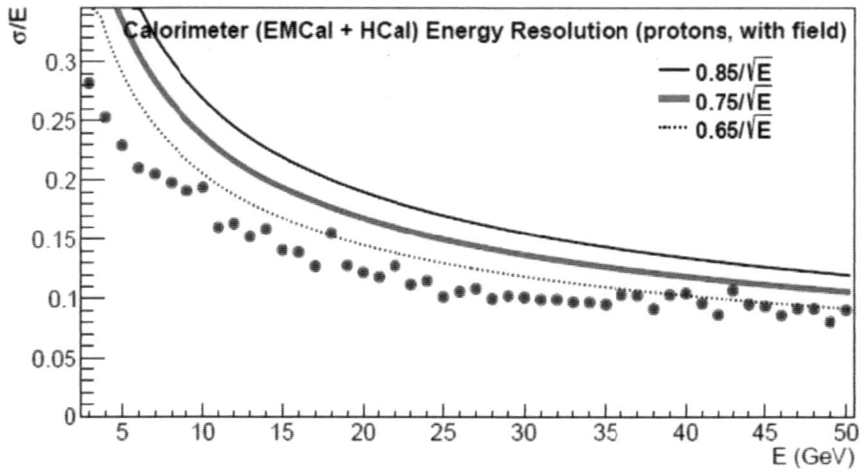

Figure 2. Energy resolution of the combined EMCAL and HCAL for protons in the full sPHENIX detector. Blue points are the simulated results compared with three resolution curves. The red curve at 75%/√E is the single particle design requirement.

Figure 2 shows the energy resolution of the combined EMCAL and HCAL systems for a single particle (proton) as a function of energy obtained from a full GEANT4 simulation of the sPHENIX detector. The calorimeters sit behind the solenoid magnet and VTX detector and include the effect of the magnetic field. The blue points give the simulated results, and the curves show three different values for the energy resolution. The design requirement for the single particle energy resolution is 75%/√E, and the results therefore indicate that the current design exceeds this requirement.

3. The Hadronic Calorimeter

The hadronic calorimeter will consist of alternating layers of steel and scintillator plates oriented parallel to the beam direction in a fin-like arrangement as shown in Fig. 3. The steel plates will be tapered such that their thickness increases with radius with uniform thickness (~7 mm) scintillator plates interspersed in between. Since the absorber thickness varies with the depth inside the calorimeter, the sampling fraction for the hadronic shower will also change with depth. However, the calorimeter will be divided into two longitudinal sections, as shown in Fig. 3a, which will allow a correction to the measured energy using the weighted average of the longitudinal shower position obtained from the two sections of the calorimeter. The total depth will be ~ 5 λ_{abs} (1 m), with approximately 1.5 λ_{abs} in the front section and 3.5 λ_{abs} in the back section. The electromagnetic calorimeter in front will add an additional 1 λ_{abs} to the total hadronic absorption length. It will be divided into 22 towers in η and 64 towers in ϕ for both the inner and outer sections for a total of 2816 towers. The plates will be slightly tilted (~ ± 5°) in the ϕ direction in order to avoid channeling of particles passing through the scintillator alone. The scintillator plates will be read out with wavelength shifting fibers placed in a serpentine groove, as shown in Fig. 3c, which in turn will be read out using silicon photomultipliers (SiPMs), similar to the readout used for the T2K experiment. We therefore expect a light yield similar to what T2K obtained (~ 12 p.e./MIP/tile, giving ~ 400 p.e./GeV) [3]. The front section of the calorimeter will be read out at the inner radius and the rear section will be read out at the outer radius. The steel plates will also serve as the flux return for the solenoid magnet.

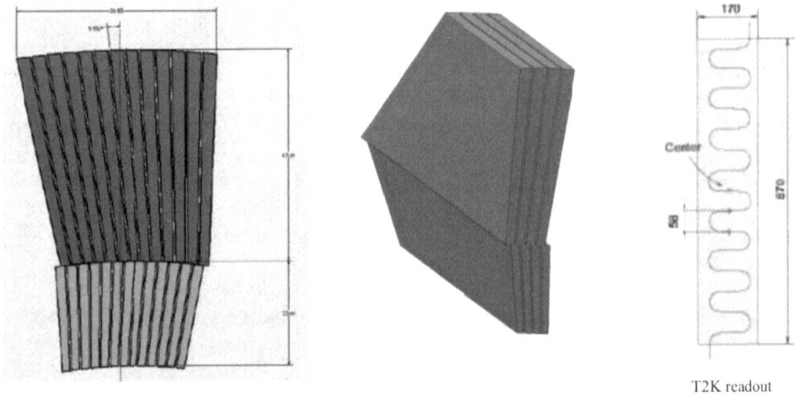

Figure 3. a) Orientation of the plates in the front (orange) and rear (grey) compartments of the hadronic calorimeter b) Steel plates joined together to form one section c) Readout of a scintillating tile using a wavelength shifting fiber in a serpentine groove used by the T2K experiment. A similar readout will be used for this calorimeter.

4. The Electromagnetic Calorimeter

The electromagnetic calorimeter will be an accordion design consisting of layers of scintillating fibers and tungsten plates held together with epoxy containing tungsten powder. The resulting matrix will have a very high density (~ 17 g/cm^3), a small Moliere radius (R_M = 1.5 cm) and a short radiation length (X_0 = 5 mm), making the calorimeter very compact, and allowing it to be placed at a small radial distance from the beam (~ 95 cm) and still provide good separation of nearby showers. The total depth will be ~ 17 X_0 and will extend only ~ 10 cm in the radial direction. It will have sampling frequency of ~ 0.6 X_0, a sampling fraction ~ 4%, and an energy resolution ~ 15%/$\sqrt{E}$. It will be projective in the transverse plane (perpendicular to the beam) and approximately projective in pseudorapidity (only approximate projectivity is required in η since the interaction verticies are spread over ~ ± 30 cm along the beam direction). The calorimeter will be divided into 96 towers in η and 256 towers in ϕ, each with a dimension of ~ 2x2 cm^2, for a total of 24,576 towers.

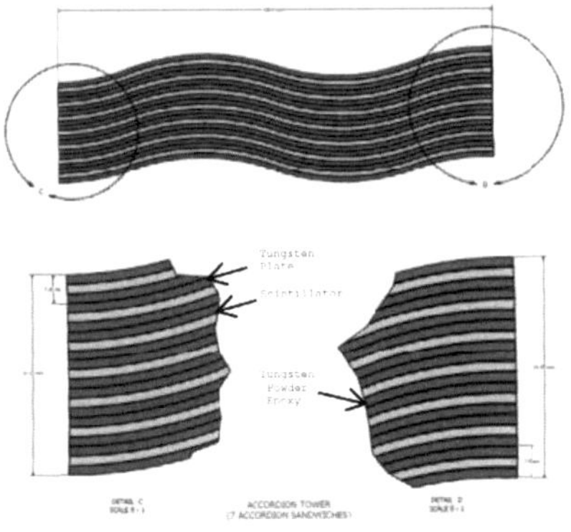

Figure 4. Layered structure of the tungsten-scintillating fiber accordion.

Fig 4 shows the layered structure of the tungsten scintillating fiber accordion design. It consists of 1 mm scintillating fibers embedded in a matrix of tungsten powder and epoxy that is glued between two thin (~ 1 mm) accordion shaped tungsten plates that have a length of ~ 1.4 m (half the length of the calorimeter) along the beam direction. The thickness of the tungsten-epoxy layer increases with depth, thus providing a projective geometry in the r-ϕ plane. The fibers are placed in a fan-like arrangement along the beam direction such that they point back towards the intersection region. Seven of the layer structures are glued together to form towers, then several towers are glued together to form modules, which are then assembled together in a ring to form the complete calorimeter.

Figure 5 shows an example of a tungsten powder epoxy layer containing several scintillating fibers that was produced by Tungsten Heavy Powder [2], which is a company that we are working with to develop the various calorimeter components. The results so far have been extremely encouraging in terms of producing thin tungsten sheets with embedded layers of scintillating fibers and tungsten powder epoxy that can be formed an accordion shape. During the coming year, we plan to build a prototype calorimeter module consisting of a 5x5 array of ~ 2x2 cm^2 towers that will be placed in a test beam in order to carry out a detailed study of its properties.

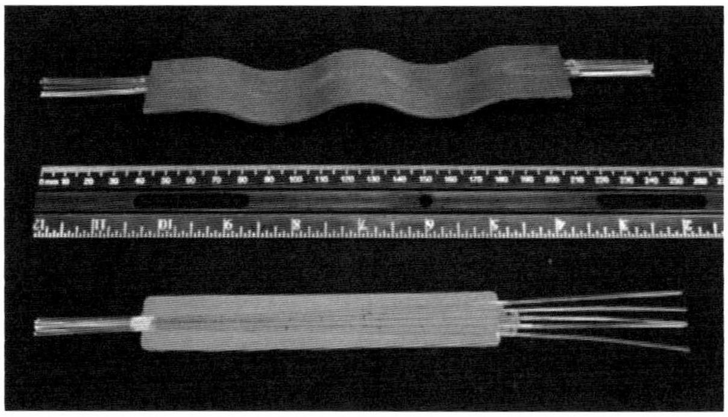

Figure 5. Accordion layer (top) containing tungsten powder epoxy and embedded scintillating fibers.

In the accordion configuration, the light output of the fibers is of primary concern, since it is important to have sufficient light output to achieve good energy resolution. We have studied the light output from various types of fibers embedded in different types of epoxies in a geometry similar to that of the accordion calorimeter. We find that the light output from the fibers is significantly reduced when they are embedded in the epoxy, mainly due to the loss of the cladding light, and the resulting light yield is ~ 100 photoelectrons per MeV of energy deposit in the scintillator when the fibers are read out directly with a photomultiplier tube. With a sampling fraction of ~ 4% in our current design, this corresponds to ~ 4000 p.e./GeV of energy deposit in the calorimeter. However, we will need to randomize and collect the light from approximately 150 fibers from a roughly 2x2 cm^2 tower onto a single 3x3 mm^2 SiPM within a very short distance (~ 2-3 cm). We plan to do this with either a small reflecting cavity or a wavelength shifting block at the back of the calorimeter tower. Assuming only a geometrical factor of 2% for the light collection (which we expect will be much higher with either the reflecting cavity or the WLS block), and a factor of 2.5 higher photon detection efficiency for the SiPM relative to the PMT, we should be able achieve a light yield ~ 200 p.e./GeV for the calorimeter, which is sufficient to not affect the overall energy resolution.

5. Electronics and Readout

Both the EMCAL and the HCAL will use silicon photomultipliers for their readout. We are currently investigating devices from different manufacturers, but a device such as the 3x3 mm^2 Hamamatsu S10362-33-025C would be suitable. It contains 14.4K 25 μm micropixels, which would provide a dynamic range of a few times 10^3, and a photon detection efficiency of ~ 25%. Each SiPM will have its own biasing circuit, temperature sensor and preamp. The temperature sensor will be used to measure the temperature locally and apply a small correction voltage to the bias in order to achieve gain equalization and stability. We are considering two different readout systems at the present time, one which is based on an existing PHENIX ADC system (previously used for the PHENIX Hadron Blind Detector) and another that would use a new custom designed ASIC (currently being developed by PHENIX collaborators at Oak Ridge National Lab) that would provide a preamp for the SiPMs which would be read out using the Beetle chip as an analog buffer and the CERN Scalable Readout System (SRS). Both readout systems are currently being investigated and a complete readout chain will be tested during the coming months.

6. Summary and Conclusions

The PHENIX Experiment at RHIC is planning a major upgrade that will include two new calorimeter systems. One will be a hadronic calorimeter that will provide a measurement of the total energy of jets produced in heavy ion collisions for the first time at RHIC, and a new electromagnetic calorimeter that will provide an independent measurement of the electromagnetic energy. Both calorimeters will cover a region of ± 1.1 units in pseudorapidity and 2π in phi. The requirements on the energy resolution for the two calorimeters is not particularly stringent due to the large background in measuring the jet energy from the fluctuations in the energy deposited in the calorimeter from the underlying event. The energy resolution requirement for measuring jets in the hadron calorimeter is ~ $100\%/\sqrt{E}$ and ~ $15\%/\sqrt{E}$ for the electromagnetic calorimeter. A preliminary design of both calorimeters has been completed and is included in the sPHENIX upgrade proposal that was submitted to Brookhaven National Laboratory in July of 2012, and will later be submitted as a Major Instrumentation and Equipment (MIE) proposal to the Nuclear Physics Office of the U.S. Department of Energy. R&D and further development on current design is proceeding and we expect to build and test prototypes of both calorimeters during the coming year.

7. References

[1] Aidala C 2012 *sPHENIX: An Upgrade Concept from the PHENIX Collaboration*, arXiv:1207.6378v1 [nucl-ex].
[2] Tungsten Heavy Powder, San Diego, CA, http://www.tungstenheavypowder.com/
[3] Izmaylov A, et.al., *Scintillator Counters with WLS fiber/MPPC Readout for the Side Muon Range Detector (SMD) for the T2K Experiment*, Nucl. Inst. Meth. A623 (2010) 382-384.

Journal of Physics: Conference Series **404** (2012) 012025 doi:10.1088/1742-6596/404/1/012025

A Study on Radiation Damage in BGO and PWO-II Crystals

Fan Yang, Rihua Mao, Liyuan Zhang and Ren-Yuan Zhu
256-48, HEP, California Institute of Technology, Pasadena, CA 91125, USA
E-mail: zhu@hep.caltech.edu

Abstract. We report in this paper an investigation on radiation damage in 20 cm long BGO crystals produced at SIC as well as 20 cm long PWO-II crystals produced at BTCP for the proposed Panda experiment at GSI. The optical and scintillation properties of BGO and PWO-II crystals, such as transmittance, light output and light response uniformity, were measured before and after γ-ray irradiations at different dose rates. BGO crystals produced recently are compared to crystals used in the L3 experiment at LEP which was produced in early eighties. PWO-II crystals are compared to crystals used in the CMS experiment. Progresses are observed in the radiation hardness for both crystals. Their use for crystal calorimeters in future high energy physics experiments is discussed. To maintain crystal precision, a precision light monitoring system is mandatory to trace variations of crystal transparency *in situ* for a BGO or PWO crystal calorimeter in radiation environment.

1. Introduction

Because of their high density (7.13 g/cm^3 and 8.3 g/cm^3), short radiation length (1.12 cm and 0.89 cm), $Bi_4Ge_3O_{12}$ (BGO) and $PbWO_4$ (PWO) crystals have been used to construct electromagnetic calorimeters (ECAL) for high energy physics (HEP) experiments, such as the L3 experiment at LEP [1] and the CMS experiment at LHC [2]. They are also being considered for future HEP experiments, where precision measurements of photons and electrons are crucial for their physics mission. The use of large size scintillation crystals in future HEP experiment, however, faces a challenge from the severe radiation environment expected in these experiments. Recently, BGO and PWO crystals with improved radiation hardness were grown at Shanghai Institute of Ceramics (SIC) and Bogoroditsk Techno-Chemical Plant (BTCP). In this paper, we present a study of the radiation hardness for large size BGO and PWO-II crystals grown recently, and compare them to the crystals used for the L3 and CMS experiment respectively.

Figure 1. BGO samples used in this study

Figure 2. PWO samples used in this study

Fig.1 shows the BGO samples investigated in this study. The two rectangular samples were grown in 2004 and 2011 at SIC, and have a dimension of 25×25×200 mm^3. The tapered BGO sample was grown at SIC in 1980 for the L3 experiment with a dimension of 20×20 mm^2 at the small end, 30×30 mm^2 at the large end and 240 mm in length. Fig.2 shows PWO samples investigated. Two shorter samples are PWO-II crystals grown recently at BTCP for the proposed Panda experiment [3] with a dimension of 24.5×24.5×200 mm^3. Two longer samples are typical PWO crystals grown at SIC and BTCP for the CMS experiment. A total of 52 CMS PWO samples were studied in our previous study [4], in which 21 from BTCP and 32 from SIC. All 21 BTCP samples are of the CMS endcap size with a slightly tapered shape: 28.5×28.5 mm^2 at the small end, 30×30 mm^2 at the large end and 220 mm long. Among the 32 samples from SIC, 20 samples have a rectangular shape of 22×22×230 mm^3 and the other 12 are of the CMS endcap size. All six faces of these BGO and PWO samples are polished.

Longitudinal transmittance (LT) was measured using a PerkinElmer LAMBDA 950 UV/Vis spectrophotometer with double beam, double monochromator and a large sample compartment equipped with an integrating sphere. The systematic uncertainty in repeated measurements was about 0.15%. Scintillation light output (LO) was measured before and after irradiations by using a Hamamatsu R1306 PMT with a bi-alkali photo-cathode and a borosilicate glass window for BGO, and a Hamamatsu R2059 PMT with a bi-alkali photo-cathode and a quartz window for PWO. For the LO measurement, the large (tapered samples) or one (rectangular samples) end of the samples was coupled to the corresponding PMT with Dow Corning 200 fluid, while all other faces of the sample were wrapped with Tyvek paper. A collimated ^{137}Cs source was used to excite the samples. The γ-ray peak positions were obtained by a simple Gaussian fit.

The γ-ray irradiations were carried out in two facilities at Caltech: a 50 curie ^{60}Co source for irradiations at low dose rates (2, 8 and 30 rad/h) and a 7,000 curie ^{137}Cs source for high dose rate at 7,160 rad/h. The dose rates were calibrated. The main consequence of γ-ray induced radiation damage in BGO and PWO crystals is radiation induced absorption which causes degradation of transmittance and light output [5,6]. It may also affect crystal's light response uniformity (LRU) and cause a degradation of the energy resolution if the light attenuation length (LAL) is reduced significantly [7,8]. The γ-ray induced absorption may also spontaneously recover, leading to a dose rate dependent radiation damage [5].

2. Gamma ray induced radiation damage in BGO crystals

2.1 Initial properties

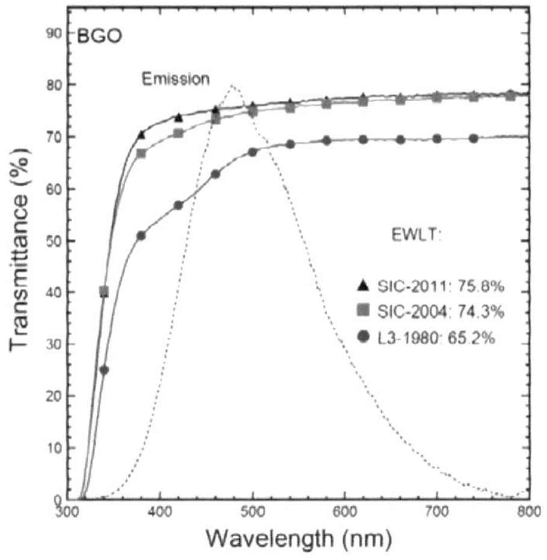

Figure 3. Initial LT and EWLT of BGO samples

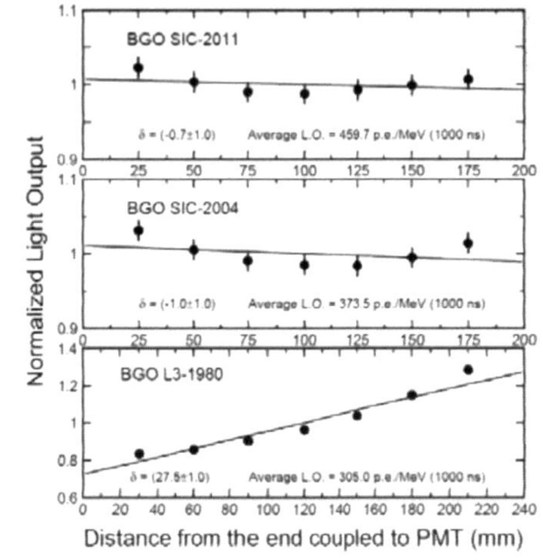

Figure 4. Initial LO and LRU of BGO samples

All three BGO crystals were annealed at 200°C for 200 minutes and then irradiated by γ-ray at 2, 8, and 30 rad/h step by step until reaching equilibrium, following which recovery measurement was carried out.

Fig.3 shows longitudinal transmittance and the emission for all samples before irradiations. Also shown in the figure is the numerical value of the emission weighted longitudinal transmittance (EWLT), which is defined as:

$$EWLT = \frac{\int LT(\lambda)Em(\lambda)d\lambda}{\int Em(\lambda)d\lambda} \qquad (1)$$

The EWLT is a good representation of crystal's transparency better than the transmittance at the emission peak. Comparing with other BGO samples, sample SIC-2011 shows the best LT and the highest EWLT, indicating a clear progress in crystal's optical property.

Fig. 4 shows the LO and LRU before irradiations for all samples. The LRU was measured by moving a collimated γ-ray source along the longitudinal axis of the long sample at seven evenly distributed points. The response (y) was fit to a linear function:

$$\frac{y}{y_{mid}} = 1 + \delta\left(\frac{x}{x_{mid}} - 1\right) \qquad (2)$$

where y_{mid} represents the light output at the middle of the crystal, δ represents the shape of the LRU; x is the distance from one end of the crystal. It is clear that sample SIC-2011 has the best LO among three samples. While both rectangular crystals have a flat response, the response of the L3 BGO at the far (small) end is significantly higher than that at the near (large) end which is coupled to the PMT. This is due to a well-known optical focusing effect caused by the tapered geometry of this sample.

2.2 γ-ray induced radiation damage and its recovery

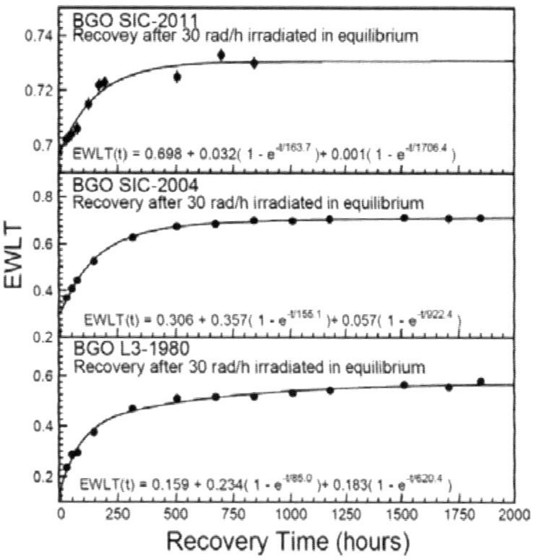

Figure 5. Recovery of EWLT for all samples after reaching equilibrium at 30 rad/h

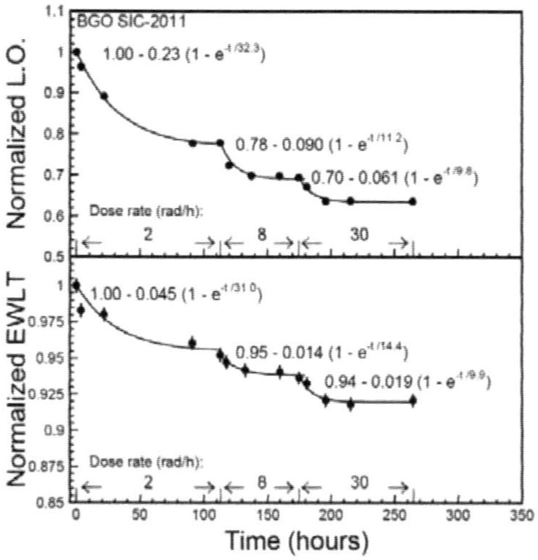

Figure 6. Normalized LO and EWLT for sample SIC-2011 under γ-ray irradiations

Fig. 5 shows the γ-ray induced radiation damage (EWLT) recovers under room temperature. All three crystals were irradiated at 2, 8, and 30 rad/h step by step until reaching equilibrium, following which recovery was measured. Two time constants were determined by an exponential fit to describe the recovery process, indicating at least two color centers at different depth. Comparing to the initial

EWLT (75.8%, 74.3% and 65.2% respectively for sample SIC-2011, SIC-2004 and L3-1980), the EWLT values measured after a few thousands hours recovery is significant lower, indicating additional deep color centers in these crystals.

Because of the recovery, the γ-ray induced radiation damage in BGO is dose rate dependent. The degradation of both LT and LO under a certain dose rate would reach an equilibrium, when the speed of the color center formation (damage) equals to the speed of the color center annihilation (recovery) [5]. Fig. 5 shows the normalized LO and EWLT for sample SIC-2011 during γ-ray irradiations in three steps at dose rates of 2, 8 and 30 rad/h. Both EWLT and LO reached equilibrium during irradiations under a defined dose rate, showing a clear dose rate dependent damage. Consistent time constants were determined by exponential fits for both normalized LO and EWLT.

2.3 Radiation damage as a function of the dose rate

Fig. 7 shows the normalized LO and EWLT measured in equilibrium as a function of the dose rate for all three samples. The corresponding data are listed in Table 1. It is clear that progress has been made in the radiation hardness of sample SIC-2011 as compared to other two BGO crystals produced early.

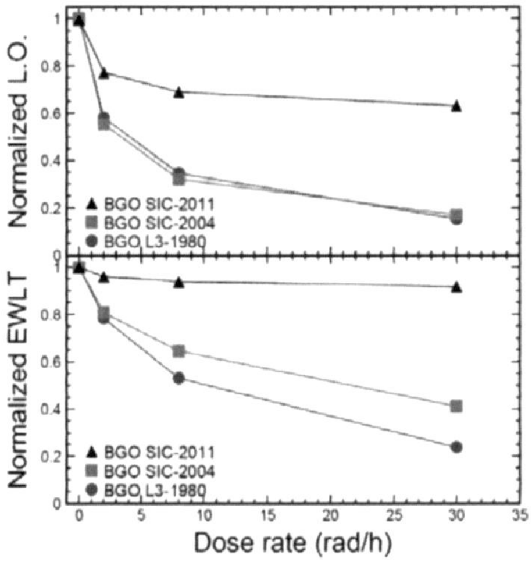

Figure 7. Normalized LO and EWLT as a function of the dose rate

Table 1. LO and EWLT loss of BGO samples in equilibrium under different dose rates

Samples		SIC-2011	SIC-2004	L3-1980
Initial LO (p.e./MeV)		460	373	305
Initial EWLT (%)		75.8	74.3	65.2
LO Loss (%)	2rad/h	22	45	42
	8rad/h	31	68	66
	30rad/h	36	83	85
EWLT Loss (%)	2 rad/h	4	19	22
	8 rad/h	6	36	47
	30rad/h	8	59	76

3. Gamma ray induced radiation damage in PWO-II crystals

Two PWO-II crystals were annealed at 200°C for 200 minutes to remove residual radiation damage and then irradiated at 2, 8, 30 and 7,160 rad/h step by step until reaching equilibrium in each step. It is well known that the γ-ray induced radiation damage in PWO crystals recovers under room temperature, leading to dose rate dependent radiation damage [6,7]. Their recovery was measured after 30 rad/h in equilibrium. The radiation damage data of these two PWO-II samples are compared to that of CMS PWO crystals [4].

3.1 Initial properties and recovery of radiation induced damage

Fig. 8 shows correlations between the numerical values of the initial LO and the initial LT at 360 nm for PWO crystals grown at SIC (top) and BTCP (bottom). Also shown in these figures are the linear fits to the data and the linear correlation coefficients (CC), defined as:

$$CC = \frac{\sum(x-\bar{x})(y-\bar{y})}{\sqrt{\sum(x-\bar{x})^2 \sum(y-\bar{y})^2}} \qquad (3)$$

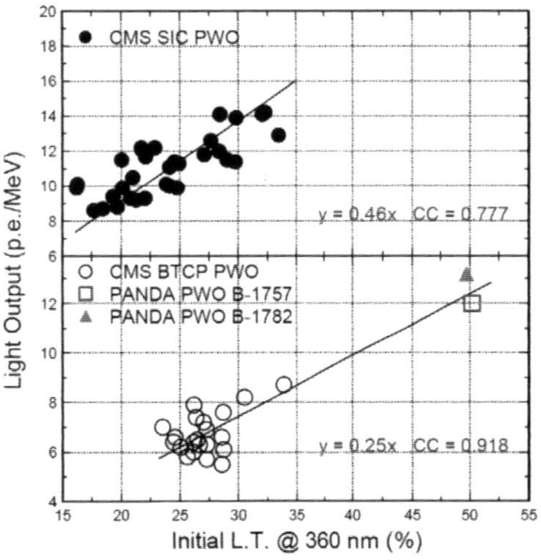

Figure 8. Correlations between the initial LO and the initial LT at 360 nm

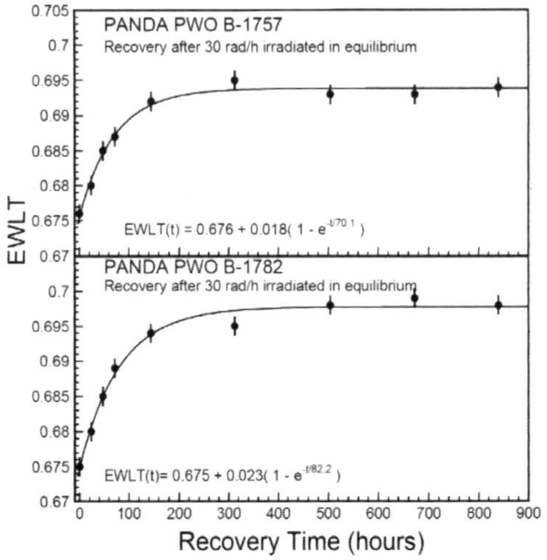

Figure 9. Recovery of EWLT as irradiated at 30 rad/h in equilibrium

The LT values of PWO-II crystals at 360 nm are significantly higher than CMS crystals. Their light output is also higher than the CMS PWO crystals grown at BTCP, but is compatible with the CMS PWO crystals grown at SIC. We note that the BTCP samples have a higher longitudinal transmittance values at 360 nm as compared to the SIC samples. This is partly due to the birefringent nature of the PWO crystals. While the BTCP PWO crystals were grown along the "a" axis the SIC crystals were grown along the "c" axis [4]. Because of the birefringence PWO crystals grown along the "a" axis should have higher theoretical limit of transmittance than that grown along the "c" axis. The linear fits also show different slopes for the SIC and BTCP samples, which may also be caused by their different crystal orientations. The nature of this correlation is understood due to the self-absorption effect: the very blue part of the PWO emission spectrum is self-absorbed because of the cross-over of the transmittance and emission spectra at 360 nm [4].

Both PWO-II crystals were irradiated at 2, 8, 30 and 7,160 rad/h step by step until reaching equilibrium. Fig. 9 shows the recovery of EWLT under room temperature for two PWO-II samples after 30 rad/h irradiations in equilibrium. Consistent recovery time constants were determined by exponential fits. The initial EWLT values (70.0% and 70.1% for sample B-1757 and B-1782), however, are not reached after nine hundreds hours' recovery, indicating deep color centers in these PWO crystals.

3.2 Radiation damage as a function of dose rate

Fig. 10 shows normalized LO (top) and EWLT (bottom) as a function of the dose rate for two PWO-II samples in equilibrium. The corresponding data are listed in the Table 2. The LO loss of PWO-II crystals is about 45% at 7,160 rad/h in equilibrium and corresponding EWLT loss is about 10%. These two PWO-II crystals have consistent radiation harness up to 7,160 rad/h.

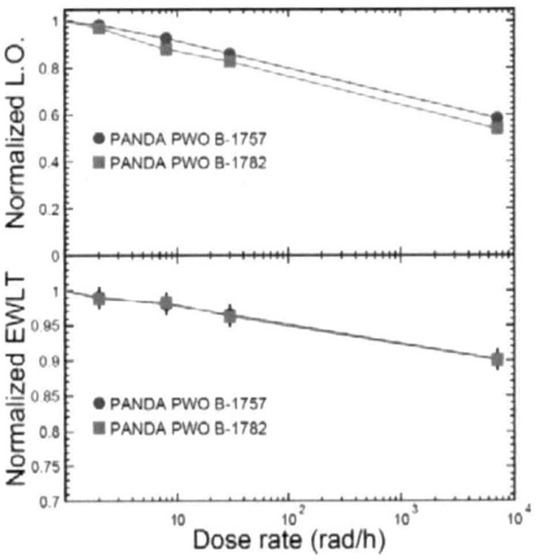

Figure 10. Normalized LO and EWLT as a function of the dose rate for PWO-II crystals

Table 2 LO and EWLT loss of PWO samples in equilibrium under different dose rates

Samples		PWO B-1757	PWO B-1782
Initial LO (p.e./MeV)		12.0	13.2
Initial EWLT (%)		70.0	70.1
LO Loss (%)	2rad/h	2	3
	8rad/h	8	12
	30rad/h	14	17
	7160rad/h	42	46
EWLT Loss (%)	2 rad/h	0.8	1.1
	8 rad/h	2.0	1.7
	30rad/h	3.4	3.7
	7160rad/h	9.9	10.0

3.3 Comparison of radiation hardness for PANDA PWO-II and CMS PWO

The radiation induced absorption coefficient and the emission weighted radiation induced absorption coefficient (EWRIAC) can be extracted from the LT data measured before and after irradiations [9,10]. The EWRIAC is a measure of the radiation damage, which is defined as

$$EWRIAC = \frac{\int RIAC(\lambda)Em(\lambda)d\lambda}{\int Em(\lambda)d\lambda}, \qquad (4)$$

where *RIAC* is the radiation induced absorption coefficient or the radiation induced color center density, and *Em* is the scintillation emission as a function of wavelength. The *RIAC* can be calculated as the difference of the color centers represented as the inverse of the light attenuation length (LAL):

$$RIAC = \frac{1}{LAL_{equilibrium}} - \frac{1}{LAL_{before}}, \qquad (5)$$

where "equilibrium" refers to "in equilibrium", and "before" refers to "before irradiation". The light attenuation length was calculated by using the longitudinal transmittance data according to:

$$\mathrm{LAL} = \frac{l}{\ln\left\{[T(1-T_S)^2]/\left[\sqrt{4T_S^4+T^2(1-T_S)^2-2T_S^2}\right]\right\}}, \qquad (6)$$

where T is the transmittance measured along crystal length l and T_S is the theoretical transmittance without internal absorption.

Fig. 11 shows correlations between the relative LO losses versus the relative LT losses at 440 nm (top) and the corresponding EWRIAC values (bottom) for all PWO samples in the equilibrium. Positive correlations are observed with the linear correlation coefficient at 50% level. The data of PWO-II crystals groups well with the data of CMS PWO crystals. The good correlation between the LO losses versus the LT losses at 440 nm indicates that the LT at 440 nm is a good measure of the radiation damage. This result is consistent with our previous investigation where 440 nm was chosen for the monitoring wavelength because of its best linearity [11].

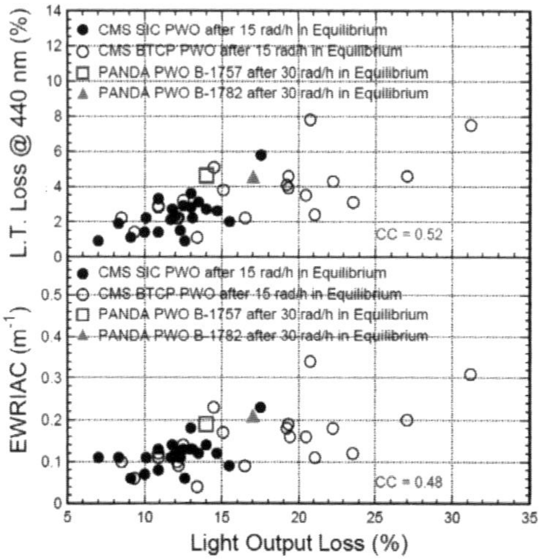

Figure 11. Correlations between the LO loss versus the LT loss at 440 nm (top plot) and the EWRIAC (bottom plot) for all PWO crystals

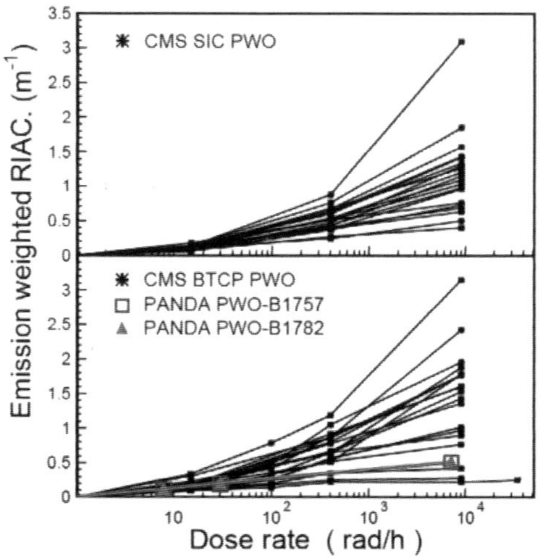

Figure 12. Distributions of the EWRIAC as a function of the dose rate

Fig. 12 shows distributions of the EWRIAC values as a function of the γ-ray dose rate. It is clear that two PWO-II crystals are in the better portion of the CMS PWO crystals. The EWRIAC values of PWO-II crystals are less than 0.5 m^{-1} up to 7,160 rad/h, indicating no damage to the light response uniformity for PWO-II crystals under such dose rate

4. Summary

Significant progress is observed in both BGO crystals grown at SIC and PWO crystals grown at BTCP. The BGO sample SIC-2011 has the better initial LT and LO as well as a better radiation hardness than previous BGO samples. It is radiation hard up to 30 rad/h. Investigation will continue for these BGO samples at high dose rate. PWO-II samples have a better LT at 360 nm and LO as compared to CMS PWO crystals. Their radiation hardness is in the better portion of the CMS PWO crystals. They are radiation hard up to 7,160 rad/h. Correlations are confirmed between: (1) LO and LT at 360 nm, (2) LO loss versus LT loss at 440 nm, and (3) LO loss versus EWRIAC.

Radiation damage in BGO and PWO crystals recovers under room temperature, leading to a dose rate dependent damage level. Because of the recovery, a precision light monitoring system is mandatory for a BGO or PWO crystal calorimeter in a radiation environment to trace variations of crystal transparency and provide crucial monitoring corrections for inter-calibration *in situ*.

Acknowledgments

The authors would like to thank Prof. Shaohua Wang of SIC and Prof. R. Novotny of the Giessen University for providing BGO and PWO-II samples respectively used in this study. This work was supported in part by the US Department of Energy Grant DE-FG03-92_ER40701.

References

[1] U. Micke 1987 *Helv Phys Acta* **60**, 685.

[2] E. Auffray 2001 *Nucl Instrum Meth A* **461**, 365.

[3] R. W. Novotny 2004 *IEEE T Nucl Sci* **51**, 3076.

[4] J. M. Chen, R. H. Mao, L. Y. Zhang, and R. Y. Zhu 2007 *IEEE T Nucl Sci* **54**, 375.

[5] R. Y. Zhu 1998 *Nucl Instrum Meth A* **413**, 297.

[6] R. Y. Zhu, H. Stone, H. Newman, T. Q. Zhou, H. R. Tan, and C. F. He 1991 *Nucl Instrum Meth A* **302**, 69.

[7] R. Y. Zhu 1997 *IEEE T Nucl Sci* **44**, 468.

[8] D. A. Ma and R. Y. Zhu 1993 *Nucl Instrum Meth A* **332**, 113.

[9] R. H. Mao, L. Y. Zhang, and R. Y. Zhu 2004 *IEEE T Nucl Sci* **51**, 1777.

[10] D. A. Ma and R. Y. Zhu 1993 *Nucl Instrum Meth A* **333**, 422.

[11] X. D. Qu, L. Y. Zhang, and R. Y. Zhu 2000 *IEEE T Nucl Sci* **47**, 1741.

XVth International Conference on Calorimetry in High Energy Physics (CALOR2012) IOP Publishing
Journal of Physics: Conference Series **404** (2012) 012026 doi:10.1088/1742-6596/404/1/012026

Quality of Long LSO/LYSO Crystals

Ren-Yuan Zhu
256-48, HEP, Caltech, Pasadena, CA 91125, USA
E-mail: zhu@hep.caltech.edu

Abstract. Because of their high stopping power (X_0 = 1.14 cm) and fast (t = 40 ns) and bright (4 times of BGO) scintillation cerium doped silicate based heavy crystal scintillators (LSO and LYSO) have attracted a broad interest in the high energy physics community pursuing precision electromagnetic calorimeter for future high energy physics experiments. Their excellent radiation hardness against γ-rays, neutrons and charged hadrons also makes them a preferred material for calorimeters to be operated in a severe radiation environment, such as the HL-LHC. The optical and scintillation properties and its radiation hardness against γ-ray irradiations up to 1 Mrad are presented for the first 2.5 × 2.5 × 28 cm LYSO sample. An absorption band was found at the seed end of this sample and three other long samples, which was traced back to a bad seed crystal used in the corresponding crystal growth process. Significant progresses in optical and scintillation properties were achieved for large size LYSO crystals after eliminating this absorption band. Their application in future HEP experiments at HL-LHC are discussed.

1. Introduction

In the last two decades, cerium doped silicate based heavy crystal scintillators have been developed for the medical industry. As of today, mass production capabilities of lutetium oxyorthosilicate (LSO) [1] and lutetium-yttrium oxyorthosilicate (LYSO) [2, 3] are established. Because of their high stopping power (ρ = 7.4 g/cm^3, X_0 = 1.14 cm and $R_{Moliere}$ = 2.07 cm), high light yield (about 4 times of BGO) and fast decay time (τ = ~40 ns) these crystals have also attracted a broad interest in the physics community for future high energy physics experiments [4], such as the proposed SuperB forward endcap calorimeter [5], the KLOE experiment [6] and the Mu2e experiment [7]. Investigations on large size (2.5 × 2.5 × 20 cm) LSO/LYSO crystals also show that this new generation of inorganic crystal scintillators has excellent radiation hardness against γ-rays [8], neutrons [9] and charged hadrons [10]. This material thus may also be considered for calorimeters to be operated in a severe radiation environment, such as the HL-LHC, and seems the only suitable scintillator for which mass-production capability is presently available.

The main obstacles of using this material in the experimental physics are two fold: the availability of high quality crystals in sufficiently large size and the high cost associated with their raw material cost and high melting point (>2,000^{o}C). An effort has been made at Sichuan Institute of Piezoelectric and Acousto-optic Technology (SIPAT), Chongqing, China, to grow large size (up to 28 cm or 25 X_0 long) LSO and LYSO crystals for high energy physics applications. In this work, the first 2.5 × 2.5 × 28 cm LYSO sample from SIPAT is evaluated, including its optical and scintillation properties and its radiation hardness against γ-rays up to 1 Mrad. Poor longitudinal transmittance was observed in this sample, which was found to be caused by an absorption band at the seed end. After removing this absorption band both the

Published under licence by IOP Publishing Ltd

optical and scintillation properties are improved significantly for large size samples. This report is a part of an on-going R&D program to further develop this material to be of practical use at the HL-LHC.

2. 20 cm long LSO/LYSO Crystal Crystals

Large size LSO and LYSO crystals are routinely grown in industry [11]. Figure 1 shows four long crystal samples of 2.5 × 2.5 × 20 cm^3 from CTI Molecular Imaging (CTI), Crystal Photonics, Inc. (CPI), Saint-Gobain Ceramics & Plastics, Inc. (Saint-Gobain) and Sichuan Institute of Piezoelectric and Acousto-optic Technology (SIPAT). Figure 2 shows the spectra of 0.51 MeV γ-rays from a ^{22}Na source observed with coincidence triggers [11]. The readout devices used are a Hamamatsu R1306 PMT (Left) and two Hamamatsu S8664-55 APDs (Right). The FWHM resolution for the 0.51 MeV γ-ray is about 12% for the PMT readout, which can be compared to 15% for the BGO sample. With the APD readout, the γ-ray peaks are clearly visible. The energy equivalent readout noise was less than 40 keV for these long LSO and LYSO samples.

Figure 1. A photo shows four long crystal samples with dimension of 2.5 × 2.5 × 20 cm^3.

LSO/LYSO crystals is also found to be much more radiation hard than other crystals commonly used in high energy and nuclear physics experiment, such as BGO, CsI(Tl) and $PbWO_4$ [12]. Their scintillation mechanism is not damaged by γ-ray irradiation. Radiation damage in LSO and LYSO crystals recovers very slow under room temperature but can be completely cured by thermal annealing at 300°C for ten hours. The energy equivalent γ-ray induced readout noise was estimated to be about 0.2 MeV and 1 MeV respectively in a radiation environment of 15 rad/h and 500 rad/h for LSO and LYSO samples of 2.5 × 2.5 × 20 cm^3 [8]. Figure 3 (Left) shows an expanded view of the longitudinal transmittance spectra for three samples before and after several steps of the γ-ray irradiation with integrated dose of 10^2, 10^4 and 10^6 rad. Also shown in the figure is the corresponding numerical values of the photo-luminescence weighted longitudinal transmittance ($EWLT$). Figure 3 (Right) shows the normalized average light output as a function of integrated dose for five long samples from

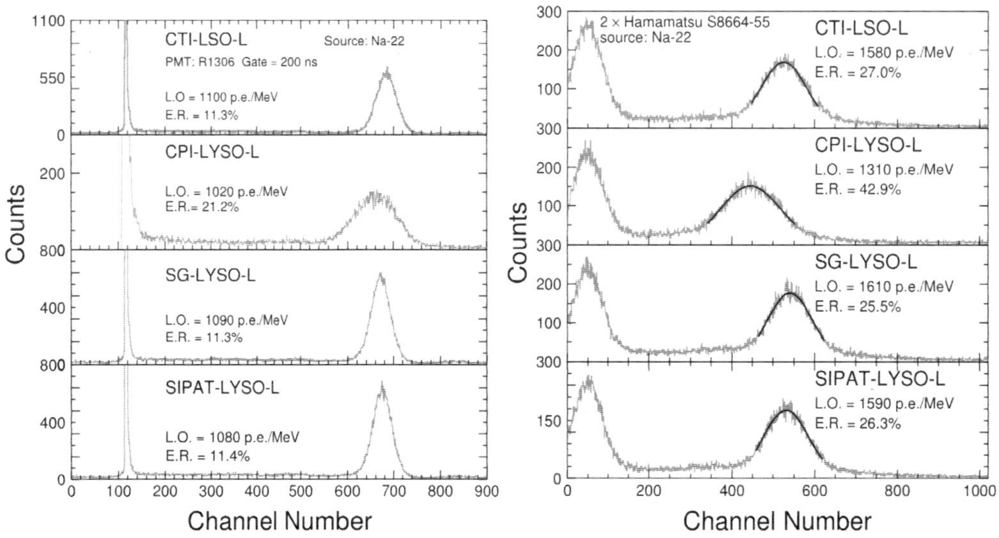

Figure 2. The spectra of 0.511 MeV γ-rays from a ^{22}Na source, measured by a Hamamatsu R1306 PMT (Left) and two Hamamatsu S8664-55 APDs (Right), with a coincidence trigger for four long LSO and LYSO samples from CTI, CPI, Saint-Gobain and SIPAT.

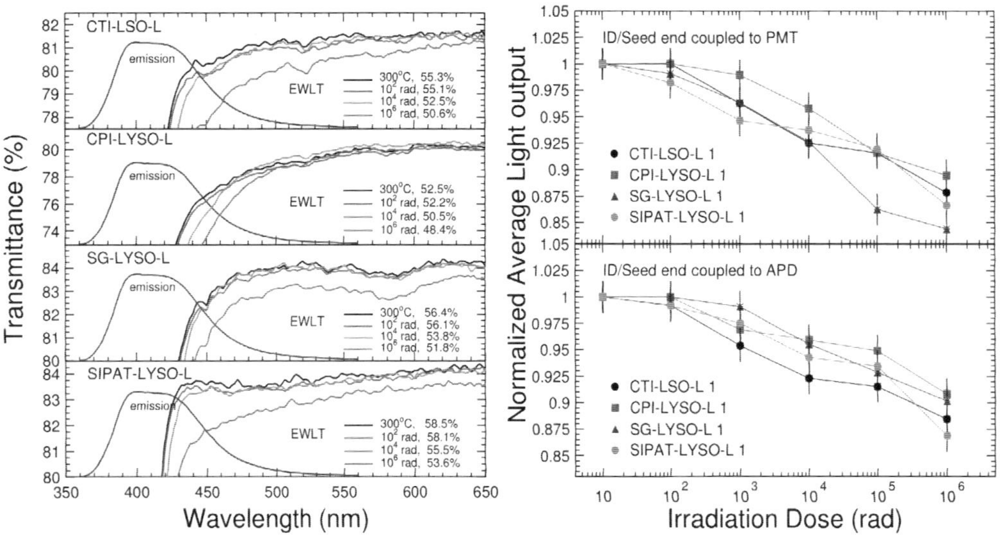

Figure 3. Left: The longitudinal transmittance spectra are shown as a function of wavelength in an expanded scale together with the photo-luminescence spectra for four LSO and LYSO samples before and after the irradiation with integrated doses of 10^2, 10^4 and 10^6 rad. Right: The normalized light output is shown as a function of the integration dose for four long LSO and LYSO samples with PMT (top) and APD (bottom) as the readout devices.

various vendors. It is interesting to note that all samples show consistent radiation resistance with degradations of both the light output and transmittance at 10 to 15% level after γ-ray irradiation with an integrated dose of 1 Mrad.

3. The First 28 cm long LYSO Sample

Figure 4 is a photo showing a LYSO ingot (top) and the first 2.5 × 2.5 × 28 cm LYSO sample (bottom) grown at SIPAT. The diameter of the ingot is about 60 mm and the length of the constant diameter is about 310 mm. From this ingot the first 28 cm long LYSO sample (SIPAT-LYSO-L7) was obtained.

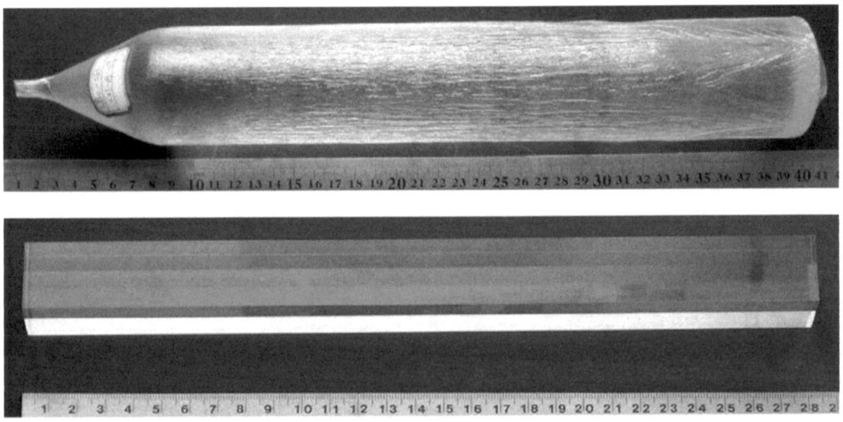

Figure 4. Top: A LYSO ingot grown by SIPAT with constant diameter of 60 mm and 310 mm long (top). Bottom: The first 25 X_0 long LYSO sample of 25 × 25 × 280 mm cut from the ingot.

All six surfaces of the sample are polished. Its dimension is found to be within 100 μm tolerance to the nominal values. The sample is colorless, crack free, without any visible inclusion. No thermal treatment was applied before the initial measurements. According to the manufacturer, the yttrium concentration is about 7%. The nominal cerium doping level is about 0.2%. The actual cerium concentration in the sample, however, would be less than the nominal value because of the cerium segregation. The distribution of the cerium concentration along the long sample's axis is not uniform.

The UV excited photo-luminescence spectra at two ends of the sample were measured by a Hitachi F-4500 fluorescence spectrophotometer. The angle between the sample normal and the excitation beam is set to be 10° to avoid internal absorption of the crystal [13]. The Left plot of Figure 5 shows the excitation (dashed lines) and photo-luminescence (solid lines) spectra measured at the seed (top) and tail (bottom) end of SIPAT-LYSO-L7. Both the excitation and emission spectra are consistent between these two ends, indicating no deviation of scintillation along the crystal.

The longitudinal and transverse transmittance spectra were measured using a Perkin Elmer Lambda 950 spectrophotometer equipped with double beam, double monochromator and a general purpose optical bench (GPOB) with light path up to 40 cm. The systematic uncertainty in repeated measurements is about 0.15%.

The right plot of Figure 5 shows the longitudinal transmittance spectrum (black line) for SIPAT-LYSO-L7. Also shown in this figure are the theoretical limits of the transmittance

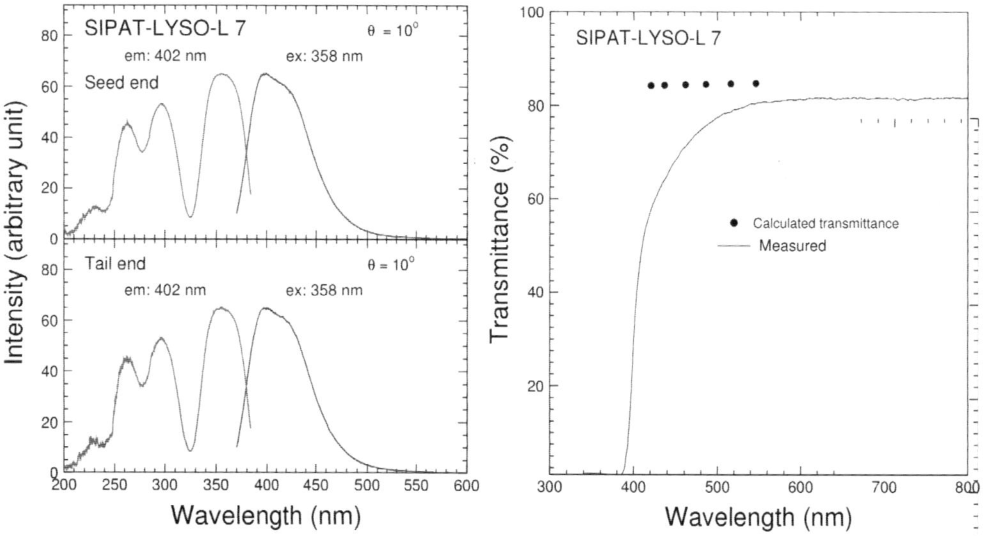

Figure 5. Left: The excitation (dashes lines) and photo-luminescence (solid lines) spectra are shown as a function of wavelength for SIPAT-LYSO-L7. Right: The longitudinal transmittance spectrum (black lines) is shown as a function of wavelength for SIPAT-LYSO-L7 together with the theoretical limits (black dots).

(black dots) calculated by using the refractive index of LYSO [13]. Taking into account multiple bouncing between two end surfaces the theoretical limits of the transmittance without internal absorption can be calculated as [14]

$$\begin{aligned} T_s &= (1-R)^2 + R^2(1-R)^2 + ... \\ &= (1-R)/(1+R), \end{aligned} \tag{1}$$

where

$$R = \frac{(n_{crystal} - n_{air})^2}{(n_{crystal} + n_{air})^2}, \tag{2}$$

where $n_{crystal}$ and n_{air} are the refractive index of crystal and air respectively, which is a function of wavelength.

A comparison of the measured optical transmittance and its theoretical limit reveals sample's overall optical quality. Poor transmittance was found between 400 nm and 600 nm. The left plot of Figure 6 shows the transverse transmittance spectra measured every 2 cm from the seed (curve-1) to the tail (curve-13) end. A large divergence of the transverse transmittance spectra between 400 and 600 nm is observed. The insert shows an expanded view between 420 nm and 450 nm, illustrating clearly the degradation of transmittance from the tail end to the seed end. While the transverse transmittance measured at 2 cm from the tail end (curve-13) is very good, that measured at 2 cm from the seed end is rather poor. From these spectra alone one can not judge if the poor transmittance observed at the seed end is caused by internal absorption or scattering. Also shown in this figure is the cut-off wavelength for the transverse transmittance. It is defined as the wavelength at which the transmittance value is 50% of the theoretical value at 600 nm. From the numerical values of the cut-off wavelength one may extract the cerium concentration by using an established correlation. This will be discussed in our future publication.

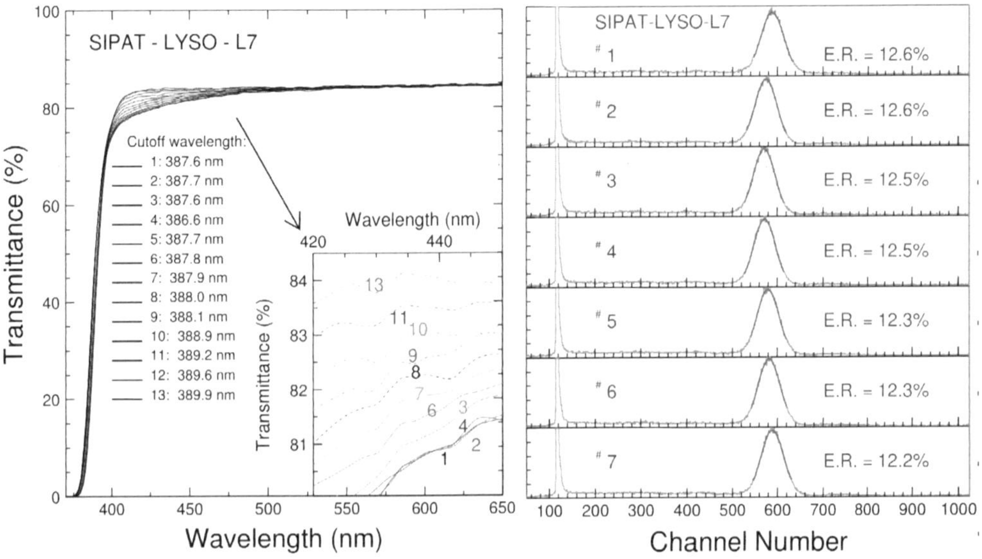

Figure 6. Left: The transverse transmittance spectra measured every 2 cm from the seed (curve-1) to the tail (curve-13) end are shown as a function of wavelength for SIPAT-LYSO-L7. The inset is an expanded view between 420 nm and 450 nm. Right:The pulse height spectra of 0.511 MeV γ-ray peaks (solid lines) and corresponding Gaussian fits measured by a Hamamatsu R1306 PMT are shown at seven points evenly distributed along SIPAT-LYSO-L7. Also shown are the numerical values of the FWHM energy resolutions (E.R.).

Despite the poor transmittance observed at the seed end this 28 cm long LYSO sample provides adequate light output and energy resolution. Two photodetectors, a Hamamatsu R1306 PMT and a Hamamatsu S8664-1010 APD with 1 cm^2 area, were used in the light output measurement. In these measurements the seed end of the sample was coupled to the PMT via Dow Corning 200 fluid, while all other faces of the sample were wrapped with two layers of Tyvek paper. To reduce the effect of the intrinsic natural radioactivity, a collimated ^{22}Na source was used with a coincidence trigger provided by a BaF_2 crystal. The detail of the setup used in these measurements is discussed in reference [4]. The γ-ray peak position was determined by a simple Gaussian fit. The right plot of Figure 6 shows the pulse height spectra measured by the Hamamatsu R1306 PMT at seven points evenly distributed along SIPAT-LYSO-L7. The FWHM resolutions obtained for 0.511 MeV γ-rays from the ^{22}Na source are about 12.5%, which is quite good for crystals of such length.

The γ-ray peak positions obtained by sending a collimated beam of γ-ray at seven points evenly distributed along the crystal were used to extract the light response uniformity of the crystal. A linear fit is used to fit the normalized response

$$\frac{LO}{LO_{mid}} = 1 + \delta\left(\frac{x}{x_{mid}} - 1\right), \tag{3}$$

where LO_{mid} represents the average light output at the middle of the sample, x is the distance from the end coupled to the readout device and δ represents the deviation of the light response uniformity.

This sample was measured with both PMT and APD readout. The average light output is 1,380 p.e./MeV and 1,510 p.e./MeV for the PMT and APD readout respectively. The corresponding light response uniformities (δ values) are 0.7 ± 1% and 2.8 ± 1.5%. Despite its poor transmittance observed at the seed end this 28 cm long LYSO sample shows rather good longitudinal light response uniformity.

4. Radiation Resistance of the First 28 cm long sample against γ-ray Irradiations

Radiation damage of this sample against γ-rays was investigated. A ^{137}Cs source was used for the irradiations with a dose rate of 7,500 rad/h. The irradiation was carried out step by step to integrated doses of 10^2, 10^4 and 10^6 rad. Since radiation damage in LYSO crystals does not recover, and is not dose rate dependent [8], the total integrated dose is used to represent the level of the radiation applied to this sample in this study.

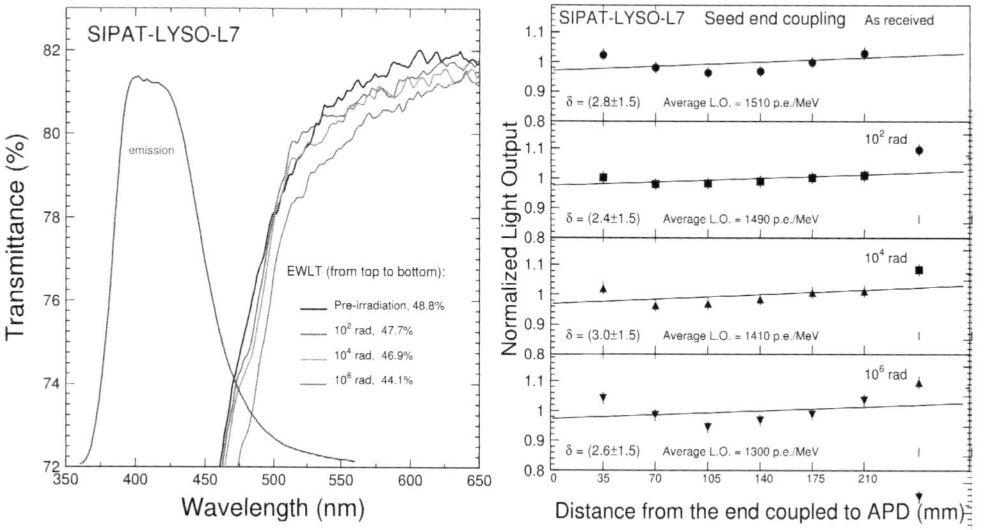

Figure 7. Left: An expanded view of the longitudinal transmittance spectra measured before and after γ-ray irradiations in several steps up to 1 Mrad is shown for SIPAT-LYSO-L7. Also shown is the emission spectrum and the values of EWLT defined in the text. Right: Normalized light output and light response uniformity measured by Hamamatsu S8664-1010 APD before and after γ-ray irradiations in several steps up to 1 Mrad are shown for SIPAT-LYSO-L7.

The left plot of Figure 7 shows an expanded view of the longitudinal transmittance spectra before and after the γ-ray irradiations with an integrated dose of 10^2, 10^4 and 10^6 rad. Also shown in the figure is the emission spectrum and the values of the emission weighted longitudinal transmittance (EWLT), which is defined as

$$EWLT = \frac{\int LT(\lambda)Em(\lambda)d\lambda}{\int Em(\lambda)d\lambda} \quad (4)$$

The numerical value of EWLT represents how transparent the crystal is to the scintillation light so is a good measure of its transparency. Its degradation represents the radiation damage effect on transparency. For SIPAT-LYSO-L7 its radiation damage in EWLT is 9.6% after 1 Mrad irradiations.

The right plot of Figure 7 shows the normalized light output and response uniformity measured by the Hamamatsu APD before and after γ-ray irradiations with an integrated dose of

10^2, 10^4 and 10^6 rad. The damage on the light output was found to be about 13% after 1 Mrad irradiations. This is much better than 26% light output loss measured for typical lead tungstate (PWO) crystals after 10 krad with a dose rate of 400 rad/h [15]. We also note that the light response uniformity of SIPAT-LYSO-L7 does not change, indicating that its energy resolution may be maintained even after 1 Mrad irradiations. Despite the poor transmittance at the seed end the radiation resistance of this 28 cm long LYSO sample SIPAT-LYSO-L7 is quite good.

5. Progress on Optical and Scintillation Properties in Long LYSO Crystals

Since the poor transmittance between 400 nm and 600 nm observed in SIPAT-LYSO-L7 reduces its light output it therefore should be eliminated. Similar poor transmittance was also found in other three LYSO samples of 20 cm long grown at SIPAT for a SuperB beam test: SIPAT-LYSO-L8, L9 and L10. Their longitudinal and transverse spectra were found to be similar to that shown in figures 5 and 6 respectively with poor transmittance located at the seed end. Poor

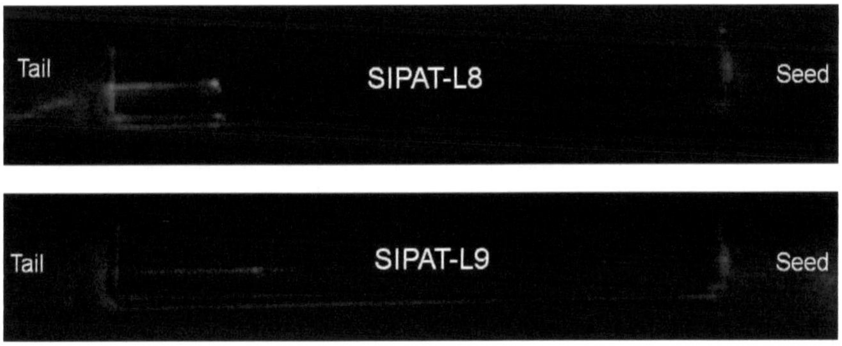

Figure 8. A picture showing scattering centers exists at the tail end of SIPAT-LYSO-L8 and SIPAT-LYSO-L9.

transmittance may be caused by point defects which absorb light in the crystal. It may also be caused by scattering centers introduced by macroscopic defects, such as cracking and inclusions etc. Scattering centers in a crystal may be observed by shooting a He-Ne red laser beam through the crystal. Figure 8 shows two photos for SIPAT-LYSO-L8 and SIPAT-LYSO-L9 with the laser beam entering longitudinally from the seed end (right) of the crystals. Scattering centers are clearly visible at the tail end of these crystals. Observations also show scattering centers at the tail end for other crystals with poor longitudinal transmittance. Since the poor transmittance was found at the seed end this observation ruled out macroscopic defects at the seed end. We thus conclude that the poor transmittance in these samples are caused by point defects at the seed end which absorb light.

This issue of the absorption band at the seed end found in SIPAT-LYSO-L7, L8, L9 and L10 was brought to the attention of SIPAT, and was discussed in a site visit. The problem was traced back to a bad seed crystal used in the growth process. After a rigorous quality control for the seed crystals the absorption band at the seed end was eliminated. Figure 9 show the longitudinal transmittance spectra for eleven LYSO crystals: ten from SIPAT and one from Saint-Gobain. No absorption was found between 400 nm and 800 nm for SIPAT-LYSO-L11 to L16, which were grown with seed crystals of high quality, and the Saint-Gobain sample. A specification of 75% at 420 nm was introduced to reject crystals with poor longitudinal transmittance. Because of the improvement in optical transmittance the light output of 20 cm long LYSO crystals from SIPAT is 30% higher than that with the absorption problem as shown in Table 1. The average

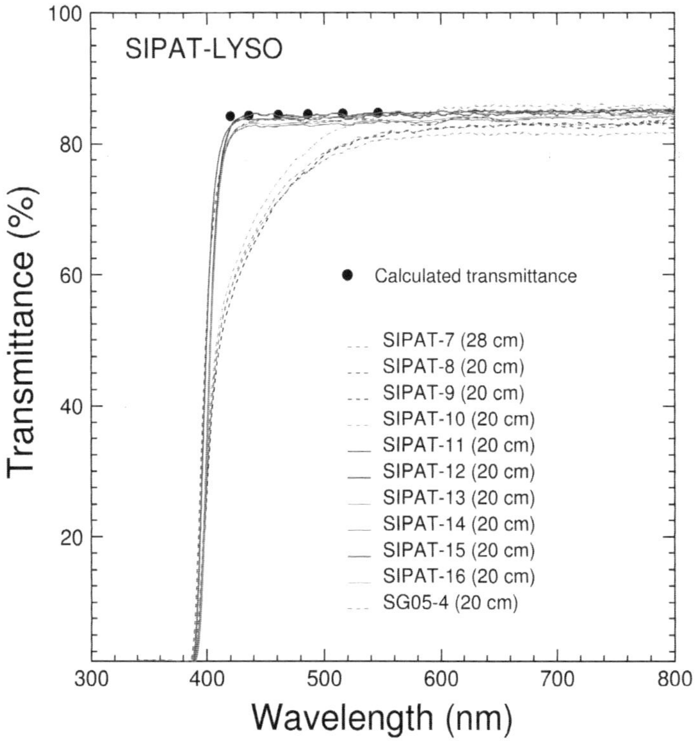

Figure 9. Longitudinal transmittance spectra are shown as a function of wavelength for eleven LYSO crystals: ten from SIPAT and one from Saint-Gobain.

FWHM resolution measured by the Hamamatsu R1306 PMT with 0.511 MeV γ-rays from a ^{22}Na source for long LYSO samples have also been improved from >12% to better than 11% as shown in Table 1.

Table 1. Light Output Measured with Hamamatsu R1306 PMT for SIPAT LYSO Samples

ID	Light Output (p.e./MeV)	Energy Resolution FWHM (%)
SIPAT-7	1380	12.2
SIPAT-8	1330	13.1
SIPAT-9	1450	12.3
SIPAT-10	1490	12.3
SIPAT-11	2010	10.7
SIPAT-12	1970	10.4
SIPAT-13	2050	11.5
SIPAT-14	2100	10.9
SIPAT-15	2040	10.5
SIPAT-16	2050	10.1

6. Summary

In a brief summary, LSO/LYSO crystals are an excellent material for a total absorption electromagnetic calorimeter for a future high-energy physics experiment in a severe radiation environment. The first 2.5×2.5×28 cm (25 X_0) LYSO sample was successfully grown at SIPAT. It has consistent emission, adequate light response uniformity and good radiation resistance against γ-rays up to 1 Mrad. This sample and three other large size samples from SIPAT, however, showed poor longitudinal transmittance between 400 nm and 600 nm as well as poor transverse transmittance at the seed end. This poor transmittance at the seed end was understood as being caused by point defects which absorb light, and was traced back to a bad seed crystal used in their growth. With rigorous quality control on seed crystals recently grown LYSO crystals show no absorption band at the seed end and have a light output of 30% more than those with this problem. The corresponding average FWHM resolution measured by using 0.511 MeV γ-rays from a ^{22}Na source is also improved from >12% to better than 11%. Because of the crystals excellent radiation hardness, a LSO/LYSO crystal calorimeter is capable of making precision measurement for electrons, photons and jets and thus provides a great physics discovery potential in a severe radiation environment, like the HL-LHC.

Acknowledgments

This work is partially supported by the U.S. Department of Energy Grant No. DE-FG03-92-ER40701 and the U.S. National Science Foundation Award PHY-0612805 and PHY-0516857.

References

[1] C. Melcher and J. Schweitzer, "Cerium-doped lutetium oxyorthosilicate: a fast, efficient new scintillator," *IEEE Trans. Nucl. Sci.* **39** (1992) 502–505.
[2] D.W. Cooke, K.J. McClellan, B.L. Bennett, J.M. Roper, M.T. Whittaker and R.E. Muenchausen, "Crystal growth and optical characterization of cerium-doped $Lu_{1.8}Y_{0.2}SiO_5$," *J. Appl. Phys.* **88** (2000) 7360–7362.
[3] T. Kimble, M Chou and B.H.T. Chai, "Scintillation properties of LYSO crystals," in *NSS conference Record* 2002 IEEE, Vol 3, 1434-1437.
[4] J.M. Chen, R.H. Mao, L.Y. Zhang and R.Y. Zhu., "Large size LSO and LYSO crystal for future high energy and nuclear physics experiments," *IEEE Trans. Nucl. Sci.* **54** (2007) 718–724.
[5] C. Cecchi, A LYSO Calorimeter for the super B factory, 2011 J. Phys.: Conf. Ser. 293 012066.
[6] F. Happacher , *CCALT: Crystal Calorimeter at KLOE2*, 2011 J. Phys.: Conf. Ser.
[7] The Mu2e Experiment, see http://mu2e.fnal.gov/.
[8] J.M. Chen, R.H. Mao, L.Y. Zhang and R.Y. Zhu, "Gamma-ray induced radiation damage in large size LSO and LYSO crystal samples," *IEEE Trans. Nucl. Sci.*, **54** (2007) 1319–1326.
[9] Rihua Mao, Liyuan Zhang and Ren-yuan Zhu, "Effects of neutron irradiations in various crystal samples of large size for future crystal calorimeter," *2009 IEEE NUCLEAR SCIENCE SYMPOSIUM CONFERENCE RECORD*, VOLS 1-5 (2009) 2041-2044.
[10] F. Nessi-Tedaldi, G. Dissertori, P. Lecomte, D. Luckey and F. Pauss, "Studies of Cerium Fluoride, LYSO and Lead Tungstate Crystals Exposed to High Hadron Fluences",Paper N32-3, 2009 IEEE NSS Conference.
[11] J.M. Chen, R.H. Mao, L.Y. Zhang and R.-Y. Zhu, *IEEE Trans. Nucl. Sci.* **54** (2007) 718.
[12] J.M. Chen, R.H. Mao, L.Y. Zhang and R.-Y. Zhu, *IEEE Trans. Nucl. Sci.* **54** (2007) 1319.
[13] Rihua Mao, Liyuan Zhang and Ren-yuan Zhu, "Optical and Scintillation Properties of Inorganic Scintillators in High Energy Physics," *IEEE Trans. Nucl. Sci.* **55** (2008) 2425–2431.
[14] D. A. Ma and R. Y. Zhu, "Light attenuation length of barium fluoride crystals," *Nucl. Instr. and Meth.* **A333**, (1993) 422–424.
[15] Rihua Mao, Liyuan Zhang and Ren-yuan Zhu, "Gamma-ray induced radiation damage in PWO and LSO/LYSO crystals," *2009 IEEE NUCLEAR SCIENCE SYMPOSIUM CONFERENCE RECORD*, VOLS 1-5 (2009) 2045-2049.
[16] R.H. Mao, L.Y. Zhang and R.-Y. Zhu, *Optical and Scintillation Properties of Inorganic Scintillators in High Energy Physics*, *IEEE Trans. Nucl. Sci.* **NS-55** (2008) 2425–2431.
[17] D.A. Ma and R.-Y. Zhu, *Nucl. Instr. and Meth.* **A333** (1993) 422.

Test and Simulation of a LYSO+APD matrix with a tagged Photon Beam from 40 to 300 MeV

M. Cordelli[1], S. Giovannella[1], F. Happacher[1], A. Luca'[1], S. Miscetti[1], A. Saputi[1], I. Sarra[1], G. Pileggi[1], B. Ponzio[1], P. Otte[2], A. Thomas[2], E. J. Downie[3]

1) Laboratori Nazionali di Frascati, INFN, Frascati, Italy
2) Johannes Gutenberg Universitat Mainz, Germany
3) The George Washington University, Washington DC, USA

E-mail: `ivano.sarra@lnf.infn.it`

Abstract. Understanding the energy resolution terms for LYSO based calorimeters with APD readout at low energy ($<$ 500 MeV) is relevant both for the completion of the KLOE-2 experiment, at DAΦNE, and for the design of the Mu2e calorimeter. In this work, we present a dedicated comparison between experimental data, taken in 2011 at the MAMI tagged photon beam facility with a crystal matrix prototype, and a full Geant-4 simulation of this detector. The crystal prototype matrix consisted of 9 2×2 × 15 cm^3 LYSO crystals read-out by 10x10 mm^2 Hamamatsu avalanche photodiodes (APD) surrounded by 8 $PbWO_4$ crystals read-out by Bialkali photomultipliers for outer leakage recovery granting a total transverse coverage of 3 Rm. An energy resolution of ~5.4% at 100 MeV has been achieved. A fit to the energy dependence of the resolution provides the following parametrization: $\sigma_E/\mathrm{E} = 2.1\%/\mathrm{E}^{1/4} \oplus 3.6\%$, where the sum is in quadrature. The noise term is completely negligible as expected by the reduced level of the electronic noise achieved. The constant term is still leakage dominated.

Introduction

To understand the resolution terms for LYSO based calorimetry is important for many experiments. Our interests were motivated by the need of improving the calorimeter angular coverage of the KLOE-2 experiment at DAΦNE in Frascati and to design the calorimeter system for the Mu2e experiment at Fermilab. In Frascati we are realizing the CCALT calorimeters, fig. 1, that will increase the acceptance to low energy photons of the central calorimeter from 18 to 10 degrees, covering the region where inner quadrupoles are located. This will improve the signal/background discrimination for rare K_s and η, η' rare decays. The main requirements, for the CCALT, are to get an accurate timing resolution (300-400 ps at 20 MeV) to reject the background from machine background events. The calorimeter must be compact due to space constrains and to have a very high efficiency for 20-300 MeV photons. Finally it must be able to work inside 0.5 T magnetic field. We are also designing the electromagnetic calorimeter for the Mu2e experiment, fig. 2, where instead the main requirement is to obtain an energy resolution $\leq$ 2% for 100 MeV energy electrons. Indeed the calorimeter will be used to confirm that a reconstructed track is a well measured, well identified conversion electron candidate and was not created by a spurious combination of hits in the tracker. The calorimeter must be radiation hard (80 Gy/y) and able to work inside 1 T magnetic field. In both cases a crystal calorimeter based on LYSO is an excellent choice.

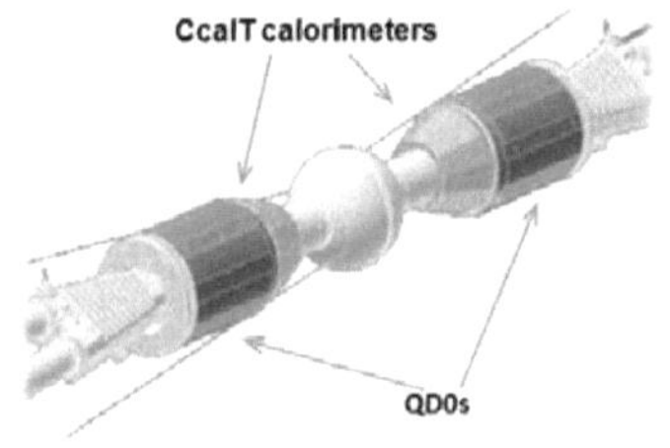

Figure 1. CCALT: a Crystal Calorimeter with Timing for the KLOE-2 upgrade.

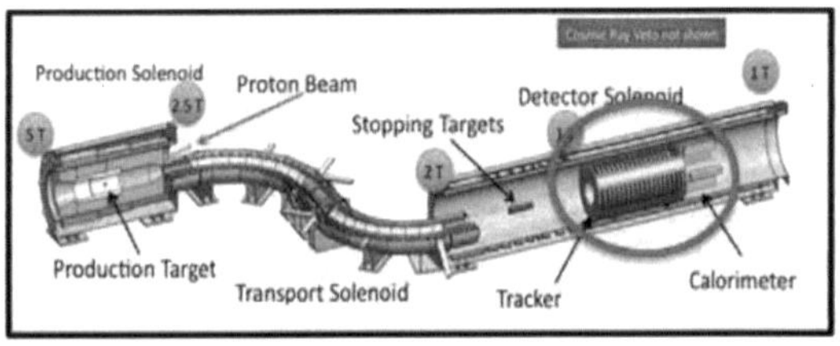

Figure 2. Overall design of the Mu2e experiment at Fermilab.

1. A LYSO Crystal Calorimeter Prototype

After a first test done in 2009 at BTF (Beam Test Facility of LNF) with a smaller size prototype [1], we have built a larger size matrix prototype to test it to a tagged photon beam. This second matrix was built In February 2011 to reach transverse dimensions corresponding to ~2.8 R_M and longitudinal dimensions corresponding to 11 ÷ 12 X_0. The prototype consists of an inner matrix of 9 LYSO crystals by SICCAS (20×20 × 150 mm^3) readout by APDs S8664 (10×10 mm^2) produced by Hamamatsu. Due to budget constraint we surrounded it with a cheaper outer matrix, for leakage recovery, composed of 8 $PbWO_4$ crystals of mixed dimensions readout by standard Hamamatsu Bialcali photo multipliers of 1 inch diameter, as shown in Figure 3.

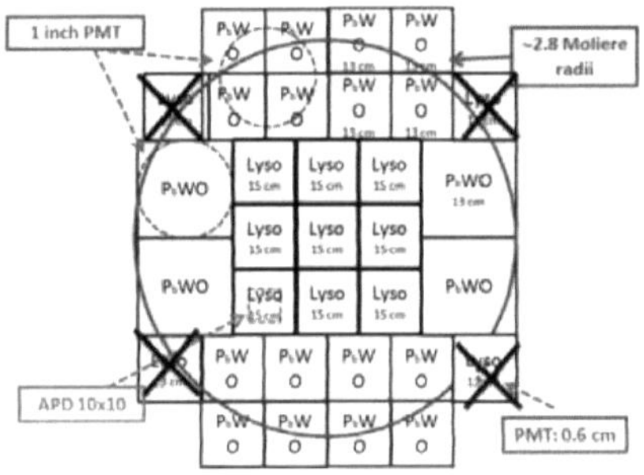

Figure 3. Layout of the improved matrix.

Each crystal was first wrapped with a 150 μm thick Tyvek sheet, which diffusely reflects the scintillation light escaping the crystal with high efficiency, and helps to increase the light collection of photosensors, in optical contact to the rear surface of the crystal by means of BICRON grease. The LYSO crystals are then wrapped with 1 mm thick black tape, to avoid cross-talk in the outer matrix which has a much higher photo sensor gain (due to the phototubes readout). The entire matrix was installed in a light shielded box, with the front layer made of Copper and the lateral and back layers made of Aluminium.

2. The MAMI accelerator facility

The MAMI (Mainz Microtron) electron beam facility produces up to a 1.5 GeV high quality ~ 100% duty factor electron beam. In the facility hall A2 the electron beam is converted to an intense beam of real photons through bremsstrahlung in a thin metal foil radiator. The scattered electrons in this process are momentum analyzed by a plastic scintillator spectrometer which provides a determination of the energy of the associated bremsstrahlung photon with a resolution of few per mil. The tagged photon beam is excellent, having Δp(FWHM) = 1 MeV and a cross section on the calorimeter front face of about 8 mm diameter. The calorimeter prototype was installed over a movable table that allowed to adjust the position of the matrix with respect to the photon beam, Fig. 4, with a precision better than 0.2 mm.

Figure 4. The calorimeter prototype is installed over a movable table in front to the photon's beam.

We have triggered using a coincidence between the OR of the discriminated signals from the inner matrix and the selected strip signal from the beam tagging system. The calorimeter analog sum discriminator (LeCroy 4419) threshold was set as low as possible ($\sim \sim$ 15 mV) and corresponded to an energy threshold of $\sim$ 10 MeV. We acquired data with a CAMAC system, reading out LeCroy ADC and TDC boards with a sensitivity of 250 fC/count and 100 ps/count, respectively. The proton beam had an average rate of 10-20 kHz. The data acquisition was writing on disk at sim 10 Hz. The temperature of the experimental hall ($\sim$ $24\,^oC$) was continuously monitored with thermo sensors attached to the electronics and preamplifiers and was stable at the level of $\pm$ 0.5 ^{o}C.

3. Data Collection

We have taken data for four days in March 2011. All of data were collected at eleven different settings of the photon-beam energy, in the range from 40 to 300 MeV. About 10k events were collected in each run. For most of the runs, the photon beam was impinging at the center of the inner crystal matrix. The same photon energy (100 MeV) was used for the position scan. Pedestal and test pulse runs were performed once every few hours during the beam time to check the system stability. Cosmic Rays (CR) were used for cross calibration of the gains. We have taken CR calibration data during the off-beam periods by means of an external trigger provided by a coincidence of a pair of scintillator counters. The average number of collected CR events per run was about 100k and was taken in about 10 hours.

We summed the charge of the 9 LYSO + APD crystals of the inner matrix and the charge of the 8 PBWO + PMT of the external matrix. The overall noise for LYSO inner matrix was of $\sim$ 3.2 counts. We use the linearity relation among energy and MIP peak to estimate the noise in MeV as follows:

$$\sigma(\text{noise})_{\text{tot}} = (3.2\text{ count} \times 9.6\text{ MeV/cm} \times 2\text{ cm})/120\text{ count} \approx 500\text{ keV} \qquad (1)$$

Similarly, we estimated that the noise contribution of each channel is 150 keV as seen from the sigma of the pedestal distribution. Thus we observe a reasonable incoherent noise in the detector as proven by the scaling law with the number of channels,N_{ch}: $\sqrt{\text{N}_{\text{ch}} * \sigma(\text{noise})_{\text{i}}^2} = 450$ keV . For the external matrix, the overall $PbWO_4$ + PM noise was of $\sim$ 2 counts. Thus, we get a total of 3.6 counts, 720 keV of noise for the whole calorimeter.

4. Linearity of the response

The total response of the detector is defined as:

$$Q_{TOT} = \sum_i (Q_i - P_i) \cdot 1/M_i \quad (2)$$

where Q_i and P_i are the collected charge and the pedestal of the i-th channel and M_i is the peak for a minimum ionizing particle. The energy distribution of Q_{TOT}, for different photon beam energies is shown in Figure 5. Each energy distribution has been fit with a Log Gaussian function and the peak value and the sigma has been extracted to estimate linearity and resolution.

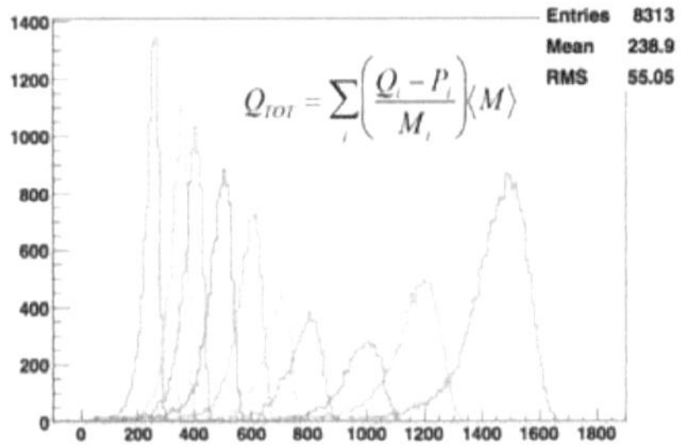

Figure 5. Distribution of Q_{tot} for the 11 selected beam energies between 40 and 300 MeV.

5. Test results and comparison with MC simulation

Test beam results have been compared with Monte Carlo studies. We ran a complete Geant-4 simulation respecting all the construction features of the matrix: dimensions, positioning, photosensors, wrapping. In Figure 6, the simulated prototype is shown. The detector has been modeled with the same geometry as the actual beam-test setup. The yellow line represents an impinging photon.

Figure 6. An event display of an 100 MeV photon (yellow line), impinging on the simulated calorimeter prototype, simulation done using Geant-4. The shower development inside the crystals is shown.

Figure 7 shows a comparison between real data and MC results, for the energy deposit of each crystal in the central row of the inner matrix. The black circles are the experimental data, the red distribution is the G4-simulation result. We are forced to add a 4% intrinsic fluctuation term to each crystal. After having inserted this correction, an excellent reproducibility of the energy distributions both in the integral and in the shower shape in the transversal direction is observed. This result has been achieved both for the inner and for the recovery matrix.
We have finally simulated and studied also the effects related to the non-linearity in response of the LYSO and to the longitudinal non-uniformity of response along the crystal. We find that the first contribution is negligible, while to explain our ad-hoc correction term we need to have a longitudinal non-uniformity of 26%. Measurement of this effect is in progress on the used crystals. After the ad-hoc correction, an excellent data-MC agreement is obtained both for the

inner and the whole matrix. In Figure 8 the superposition between data and MC is shown for the whole matrix.

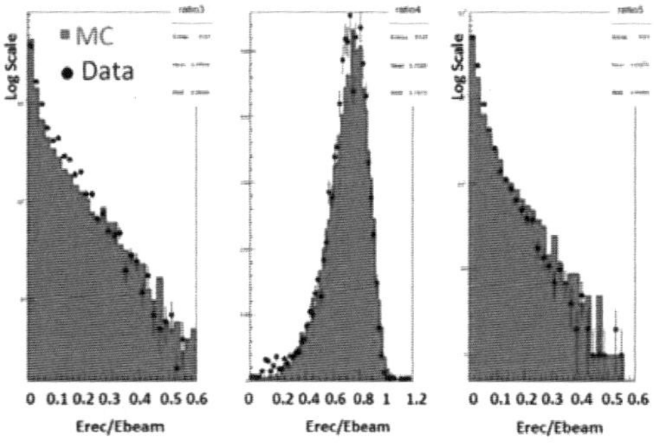

Figure 7. Comparison between real data (black circles) and MC results (red distribution), for the energy deposit in the central row of the inner matrix.

In order to evaluate the energy resolution, we have selected 11 energy values between 50 and 300 MeV. We have fit each energy distribution with a logarithmic gaussian function [2]. We used a automatic fit procedure to evaluate the systematic error of the fit. Starting from the position of the maximum in the energy distribution, we fit the histogram automatically changing the fit ranges as follows.

The fit uses as input the maximum of the histogram and its RMS as standard deviation of the gaussian; then, using an iterative procedure the left end side and the right end side of the fit range are varied by a fraction of the RMS value independently. For each energy distribution we performed several fits using about 100 different ranges. Only the fit results with χ^2/ndf between 0.5 and 2, $\Delta\sigma/\sigma_E$ and $\Delta E_{peak}/E_{peak}$ less than 0.05 are taken into account. The parameters, surviving the selection, are collected. Then we used their mean value and the RMS to evaluate the systematic error.

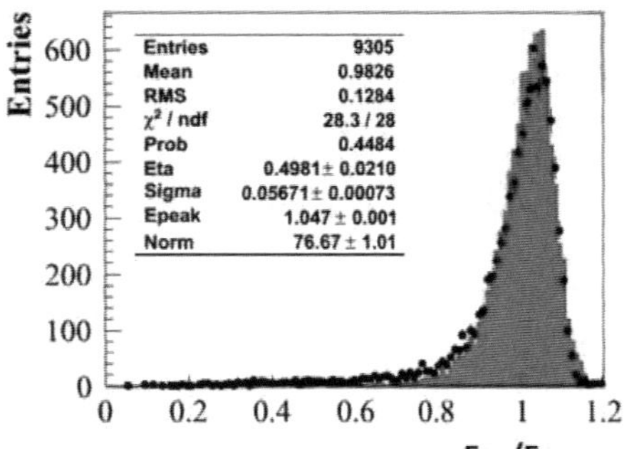

Figure 8. Data MC Comparison, sum of the whole crystals matrix. Data in black, MC in red.

In Figure 9, we show the energy dependence of the energy resolution for the whole matrix. MC is shown with and without the additional 4% spread per crystal (referred as full/ideal in the following). The dependence of the energy resolution has been fit with the standard equation:

$$\sigma_E/E = a/\sqrt[4]{E[GeV]} \oplus b/E[GeV] \oplus c \tag{3}$$

where the sum is done in quadrature. The noise term is expected to be 0.06% and is negligible in the fit procedure. when fitting the MC distribution for the whole matrix for the ideal case we get a c term of 3.0 % and an a term of 1.4 %. Since the contribution of the photoelectron statistics is negligible the a term represents the energy dependence of the leakage fluctuation. The full simulation closely follows the data distribution. The a term increase is related to the non uniformity of the crystals.

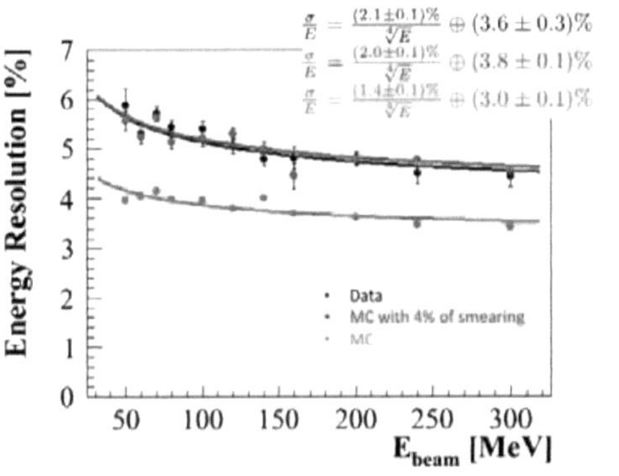

Figure 9. Dependence of the energy resolution on beam momentum, for all crystals. Black points data, red (pink) points MC with (without) the 4% correction. We get a stochastic term, a =2.1±0.1% and a constant term, c, =3.6±0.3% for data in good agreement with MC expectation. The noise term b is practically negligible.

5.1. *Timing resolution*

We could not determine the timing resolution at MAMI, due to the large time jitter of our trigger (1 ns). We determined it at BTF of Frascati, with a test performed on March 2009 with a smaller matrix prototype. We obtained σ_t = 250 (49) ps at 500 MeV, 291 (120) ps at 100 MeV without (with) correction for trigger jitter [3].

5.2. *Position resolution results*

We have also determined the position resolution with centroid method defined as $X_{pos} = \sum Q_i \cdot X_i/Q_{tot}$. We observe a position resolution of $\sim$ 3.4 mm RMS for 100 MeV photon impinging directly the calorimeter center.

6. Conclusions

A test beam study of a prototype LYSO calorimeter for KLOE-2 and Mu2e detector has been tested at a photon beam with energy from 40 to 300 MeV at MAMI. The achieved energy resolution of 5.4% is dominated by leakage and can be improved enlarging the crystals array. Results obtained at the beam test are well reproduced by the Geant-4 simulation where we have investigated the effect on the resolution related to intrinsic non-linearity in response and longitudinal non-uniformity of the crystals. We realized that the longitudinal non-uniformity is the most important contribution and we were forced to add a 4% spread to each crystal to obtain a fair data-MC agreement. To conclude the R&D for the Mu2e experiment we aim to get an energy resolution of O(2%) and we therefore need to get an energy resolution below 1% related to the longitudinal non-uniformity which has to be contained below 5 %. We are planning to build a new 5×5 LYSO Crystals (3×3×13 cm^3) matrix with more uniform crystals and better electronics to test it again at MAMI with the goal of reaching $\sigma_E/E \sim$ 2-3% at 100 MeV.

References

[1] M Cordelli et al. 2011 *J. Phys.* **293** 012010
[2] H. Ikeda et al. 2000 *NIM in Physics Research A* B **441** 401-426
[3] M Cordelli et al. 2010 *NIM A* **671** 109-112

Why a homogeneous dual readout calorimeter won't work

Don Groom
Lawrence Berkeley National Laboratory, Berkeley CA 94720
E-mail: degroom@lbl.gov

Abstract. If the response to a hadronic shower in a semi-infinite uniform calorimeter structure is S relative to the electronic response, then $S/E = [f_{em}+(1-f_{em})(h/e)]$, where E is the incident hadron energy, f_{em} is the electronic shower fraction, and h/e is the hadron/electron response ratio. In conventional calorimeters the resolution is dominated by the stochastic variable f_{em}, whose broad, skewed pdf has an energy-dependent mean. The slow increase of the mean with E is responsible for response nonlinearity and the skewness results in a non-Gaussian response. If the cascade is observed in two channels with different values of h/e (typically scintillator(S) and Cherenkov (C)), f_{em} can be eliminated. An energy estimator, linear in C and S, is obtained which is proportional to the incident hadron's energy. The resolution depends upon the contrast in h/e between the two channels. The Cherenkov h/e will be 0.20–0.25. In sampling calorimeters, h/e can be increased to about 0.7 by arranging for preferential absorption of the electromagnetic (EM) shower energy in the absorber (decreasing e) and using a hydrogenous detector (organic scintillator) to enhance h through the contribution of recoil protons in n–p scattering. *Neither mechanism is available in a homogeneous crystal or glass scintillator*, where h/e is expected to be in the vicinity of 0.4 because of invisible hadronic energy loss and other effects. The h/e contrast is very likely too small to provide the needed energy resolution. We support this conclusion with simple Monte Carlo simulations.

1. Introduction

A homogeneous dual readout hadron calorimeter has been proposed for the future linear collider. The machine will probably be an e^+e^- collider with a long bunch spacing (100's of ns), so that detectors with time constants in this range can be used. Discrimination between the Cherenkov signal (C) and scintillator (S) optical signals is expected to be made using a combination of timing and color. For the best candidate materials (inorganic crystals and glasses), $\lambda_I \geq 20$ cm. A hadron calorimeter $\geq 5\lambda_I$ thick lying outside the tracking region and (probably) outside of a large solenoid will have a volume of several hundred cubic meters, setting stringent price as well as properties requirements on the material.

For the purposes of this analysis we will assume that corrections for leakage, cracks, etc. have been made correctly, and consider the energy resolution of a semi-infinite calorimeter with uniform structure that is either fine-sampling or homogeneous.

Mostly via $\pi^0 \to 2\gamma$ in a sequence of high-energy hadronic collisions, a fraction f_{em} of the incident energy is deposited by EM showers. The mean, $\langle f_{em}\rangle$, is ≈ 0.5 for 100–150 GeV incident pions. The hadronic response S to an incident hadron with energy E is

$$S = E[\,f_{em} + (1 - f_{em})(h/e)]\ . \tag{1}$$

Published under licence by IOP Publishing Ltd

(Here and elsewhere the energy E is normalized to the electron response.) The EM deposit is detected with relative efficiency e, and the hadronic energy deposit with relative efficiency h. Both vary from event to event, but the variance of h is much larger than the variance of e. It makes sense to treat h/e as the stochastic variable, since the distribution of the more usual ratio e/h is fairly pathological.

The energy-independent ratio $\langle h/e \rangle$ depends upon calorimeter composition and structure, and also depends upon readout—an organic scintillator readout is somewhat sensitive to the otherwise-invisible neutron content of the cascade, and a Cherenkov readout is relatively blind to the hadronic content. $\langle \eta_C \rangle$ and $\langle \eta_S \rangle$[1] can be found from fits to the average π^-/e response at a sequence of test-beam energies.

If $\langle h/e \rangle$ ($\equiv \langle \eta \rangle$) is not unity, then the broad, skewed f_{em} probability distribution function (p.d.f.) significantly degrades and skews the distribution of S. It in fact usually dominates, producing the familiar wide, non-Gaussian energy distributions, as well as the non-linear response via its energy-dependent mean. If f_{em} could be *measured* for each event, then an energy estimator distribution could be obtained that has a much narrower Gaussian distribution and a mean proportional to the energy. This was realized more than 3 decades ago, when the physics summarized in Eq. 1 was first understood.

One approach is to obtain two signals for each event, one preferentially sensitive to the EM content (invariably Cherenkov readout) and one as sensitive as possible to the hadronic content (scintillator). This could be done either by readout of separate elements, as so successfully demonstrated by the DREAM collaboration[1, 2], or by separation via color and timing information in a single material[3].

This study is motivated by renewed interest in the latter approach, in particular a homogeneous dual-readout heavy crystal (or glass) calorimeter for use at a future linear collider. The collider's bunch spacing is to be in the 100–500 μs range, a good match to the properties of many high-$\langle Z \rangle$ inorganic crystals and glasses.

While the desired signal separation has already been demonstrated by Akchurin, et al.[4, 5], at least for BGO and $PbWO_4$, we are concerned that low h/e for the scintillator component (η_S) will prevent adequate hadronic energy resolution. In this paper we explore the likely resolution as a function of energy and η_S using resolution contributions based on published homogeneous crystal and sampling calorimeter performance. Simple, transparent Monte Carlo's (MC's) are used by preference, to make the physics more transparent than if a sophisticated MC such as GEANT4 were used. The probability distribution function (p.d.f.) of f_{em} is approximated with some care, while other resolution contributions are taken to be Gaussian.

2. $\langle h/e \rangle$ in a high-density crystal or glass scintillator

In an EM cascade relativistic electrons deposit most of the energy. The result is a response very nearly linear in the incident electron or photon energy.

Hadronic interactions deposit energy in a variety of ways (See Table 1 (by Gabriel and Schmidt) in Ref. [6]; detailed discussions can be found in Refs. [7, 8] and other recent reviews). 20–40% of the energy goes to nuclear dissociation and is "invisible." Much or most of the neutron energy is lost. Low-energy protons and charged fission fragments produce saturated signals in scintillator[9]. All of these factors result in low visible response to the hadronic component of the cascade relative to response to the EM component.

Detection of recoil protons in neutron scattering in hydrogenic detectors increases h. A disproportionate fraction of the EM energy is deposited in the higher-Z absorber, decreasing e. Both of these effects increase h/e. In practical sampling calorimeters $\langle h/e \rangle$ is typically 0.7, and can be made to approach unity with careful design.

[1] We introduce the less cumbersome notation $\eta = h/e$.

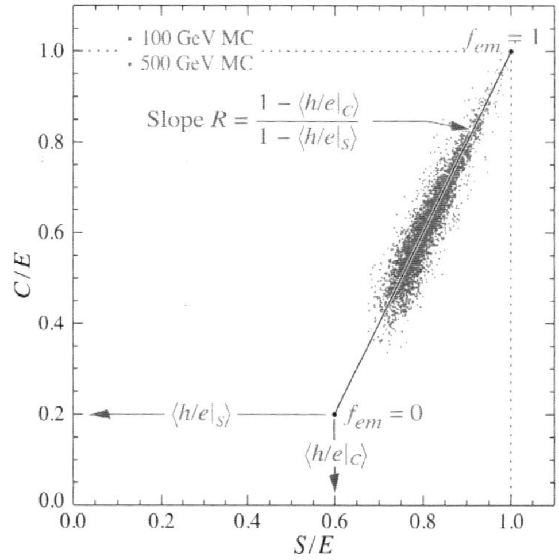

Figure 1. Energy-independent event locus in the C/E–S/E plane. With increased energy, resolution improves and the mean moves upward along the locus.

Neither mechanism for increasing h/e is available to a high-density homogenous calorimeter.
The resolution is dependent on the "h/e contrast," the difference between the values of h/e for the Cherenkov (η_C) and scintillation(η_S) readouts.

Based on experience with quartz-fiber readout calorimeters, $\langle \eta_C \rangle = 0.20$–$0.25$[1, 2, 10]. There are few data concerning $\langle h/e|_S \rangle$ in a homogeneous calorimeter, but there is no way to hide EM energy in absorber and there is very little neutron sensitivity. We might expect $\sim 40\%$ of the hadron energy to be expended on nuclear dissociation and therefore invisible, and 15%–20% to be carried by neutrons. These alone would result in $\langle \eta_S \rangle \approx 0.5$. There are other effects, such as incomplete Cherenkov-scintillator separation and saturated response to highly ionizing particles, so we might expect $\langle \eta_S \rangle = 0.35$–$0.5$. This is corroborated by a measurement reported in Ref. [5]: "The e/h value of ECAL [$PbWO_4$] as a scintillation device is much larger than for the Cu/plastic sampling structure in DREAM: 2.4 vs. 1.3." ($h/e = 0.43$ vs 0.7).

3. Dual-readout hadron calorimetry

The proposed scenario is a Cherenkov readout (C) that is fairly blind to hadronic activity and a scintillator readout (S) with optimized hadronic response. Eq. 1 is replaced by

$$S = E[f_{em} + (1 - f_{em})(\eta_S)] \; ; \qquad C = E[f_{em} + (1 - f_{em})(\eta_C)] \; . \tag{2}$$

In parametric form, Eqs. (2) describe a straight line-segment event locus in the (C–S) (or (C/E–S/E)) plane, as illustrated in Fig. 1. If the cascade is "all electomagnetic" ($f_{em} = 1$), then $S/E = C/E = 1$. If the cascade is "all hadronic" ($f_{em} = 0$), then $S/E = \eta_S$ and $C/E = \eta_C$.

Equations (2)are linear in $1/E$ and f_{em}, with solutions[11, 12]

$$E = \frac{S(1-\eta_C) - C(1-\eta_S)}{\eta_S - \eta_C} \; ; \qquad f_{em} = \frac{C\,\eta_S - S\,\eta_C}{S\,(1-\eta_C) - C(1-\eta_S)} \; . \tag{3}$$

There is an important difference between Eqs. 2 and Eqs. 3: Eqs. 2 give estimators of the response, given the incident energy, f_{em}, and η_S, or η_C. In contrast, Eq. 3 provides estimators of the energy and f_{em} given S, C, η_S and η_C. Here η_C and η_S are the values *peculiar to that event*. These are unknown—and unknowable, until some way of tagging the hadronic content becomes available. But in an experimental situation, an estimator of the energy must be established for a given event. There is little choice but to replace η_C and η_S by their means, $\langle\eta_C\rangle$ and $\langle\eta_S\rangle$, for this purpose.

It should be evident from all of these relationships, especially Eq. 3, that the estimator E is exceedingly sensitive to the difference between η_S and η_C, both of which are subject to large event-to-event fluctuations.

4. Electromagnetic fraction

The f_{em} p.d.f. has an energy-dependent mean (near 0.5 at 100–150 GeV and approaching unity as $E \to \infty$). Its standard deviation is about 11%, depending only weakly on energy, and it is skewed to the large-f_{em} side[7, 11]. For the present study, the mean and fractional standard deviation of f_{em} as parameterized in Ref. [11] for cascades in a large lead cylinder are used. A skewness $\gamma_1 = \mu_3/\sigma^3 = 0.6$ was assumed. In Ref. [11] it was also pointed out that the Beta p.d.f., $B(x;\alpha,\beta) \propto x^{\alpha-1}(1-x)^{\beta-1}$, is very similar to f_{em}'s p.d.f. with proper choice of α and β, except for the position of the mean. To construct the p.d.f. for f_{em}, α and β are found which give the desired variance and skewness. The distribution is then displaced so that the mean is at $\langle f_{em}\rangle$.

5. Resolution contributions

Except in a few compensated sampling calorimeter cases, the f_{em} distribution dominates single-readout calorimeter resolution. In a dual-readout calorimeter f_{em} has been elevated from a stochastic quantity to a measured quantity.

The C and S distributions are broadened by a variety of instrumental effects, which we assume are correctable. Dominant remaining stochastic contributions include photoelectron (p.e.) statistics and intrinsic fluctuations that contribute differently to the two channels:

(i) Cherenkov p.e.: Based upon the best Pb glass detectors (e.g. OPAL[13]), the contribution is assumed to be 10–20%$/\sqrt{C}$.

(ii) Scintillator p.e.: For good signal separation, the scintillator response cannot be very large compared with the Cherenkov signal. We assume a small multiple of the Cherenkov detector's p.e. contribution, scaling as $1/\sqrt{S}$.

(iii) Cherenkov intrinsic fluctuations, presumably dominated by fluctuations in the number of relativistic pions: Following Wigmans's estimates in Ref. [7], $\sigma_{h,C} = 170\%/\sqrt{E}$ is assumed. (This has a small effect because η_C is small.)

(iv) Scintillator intrinsic fluctuations: Based on best compensated sampling calorimeter results coupled with separation of sampling and intrinsic contributions by Drews et al.[14], $\sigma_{h,S} = 11\%/\sqrt{E}$ is assumed.

Except possibly for $\sigma_{\text{p.e.}}$, all of the standard deviations described in this section are highly uncertain, and can be used only as guides. Since the energy distributions reported for compensating calorimeters are all consistent with Gaussian distributions, it is evidently valid to consider these as Gaussian as well.

6. Simulation of the energy estimator distribution

Equation 3 describes the estimator of incident pion energy as obtained from the observed scintillation and Cherenkov signals. Using the above estimates, our elementary simulation proceeds as follows:

(i) Choose the incident energy E and the detection efficiency ratios $\langle\eta_S\rangle$ and $\langle\eta_C\rangle$.
(ii) Choose the resolution parameters, as estimated above.
(iii) Generate N values of f_{em} chosen from the displaced Beta distribution for energy E.
(iv) Generate N values of η_S from a normal distribution with mean $\langle\eta_S\rangle$ and fractional standard deviation $\sigma_{h,S}$. Similarly, generate N values of η_C from a normal distribution with mean $\langle\eta_C\rangle$ and standard deviation $\sigma_{h,C}$.
(v) Construct the corresponding S and C arrays via Eqs. 2.
(vi) Replace each S and C as calculated in step 4 with values chosen from normal distributions with means S and C and standard deviations $\sigma_{\text{p.e.},S}\sqrt{S}$ and $\xi\sigma_{\text{p.e.},S}\sqrt{C}$, respectively. The resulting S and C include p.e. statistics, and are used for subsequent "data analysis."
(vii) Find the N energy estimators via Eq. 3, using the mean values of η_C an η_S.

The results of four representative simulations are shown in Fig. 2, at 75 and 200 GeV using realistic (0.4) and optimistic (0.6) values of η_S. In every case the mean value of the estimator of E scaled by the beam energy is 1.00, and the distribution is Gaussian, as shown by the smooth curves drawn over the E histograms. In nearly all cases $|\gamma_1| \leq 0.05$ for 10000 event samples, where γ_1 is the third moment about the mean divided by σ^3.

The fractional standard deviation of the energy estimator scales as $1/\sqrt{E}$. The coefficient (resolution at 1 GeV) as a function of $h/e|_S$ is shown in Fig. 3 for two values of $h/e|_C$ that bracket the range found in SiO_2 calorimeters.

7. Discussion and conclusions

While the input variances are subject to question, the overall conclusion remains: in a non-hydrogeous homogeneous hadron calorimeter, the h/e contrast between the Cherenkov and scintillator readouts is insufficient to obtain the needed resolution. On the other hand, the mean of the energy estimator distribution is the beam energy, and the distribution is Gaussian.

Acknowledgments

This work was supported by the U.S. Department of Energy under Contract No. DE-AC02-05CH11231.

References

[1] R. Wigmans, "First results of the DREAM project," Proc. 11th Inter. Conf. on on Calorimetry in Particle Physics, Perugia, Italy, 29 March–02 April, 2004, ed. C. Cecchi, P. Cenci, P. Lumbrano, & M. Pepe, World Scientific (2005) 241.
[2] N. Akchurin, et al., Nucl. Instr. and Meth. A 537 (2005) 537.
[3] D. R. Winn and W. A. Worstell, IEEE Trans. Nuc. Sci 36, (1989) 334.
[4] Akchurin, et al., Nucl. Instr. and Meth. A 550 (2005) 185; Nucl. Instr. and Meth. A 595 (2008) 359; Nucl. Instr. and Meth. A 598 (2009) 710; Nucl. Instr. and Meth. A 604 (2009) 512. Nucl. Instr. and Meth. A 621 (2010) 212.
[5] Akchurin, et al., Nucl. Instr. and Meth. A 584 (2008) 273.
[6] C.W. Fabjan, et al., Nucl. Instr. and Meth. 141 (1977) 61; T. A. Gabriel & W. Schmidt, ORNL/TM-5105 (1975).
[7] R. Wigmans, Calorimetry: Energy Measurement in Particle Physics, International Series of Monographs on Physics, vol. 107, Oxford University Press (2000).
[8] C. Leroy & P.-G. Rancoita, Radiation Interaction in Matter and Detection, World Scietific (2004).
[9] See fig 3.25, p. 145 in Ref. [7].
[10] N. Akchurin, et al., Nucl. Instr. and Meth. A 399 (1997) 202.
[11] D.E. Groom, Nucl. Instr. and Meth. A 572 (2007) 633.
[12] J. Beringer, et al. (Particle Data Group), Phys. Rev. D86, 010001 (2012).
[13] OPAL collaboration, Nucl. Instr. and Meth. A 305 (1991) 275.
[14] G. Drews, et al., Nucl. Instr. and Meth. A 335 (1990) 335.

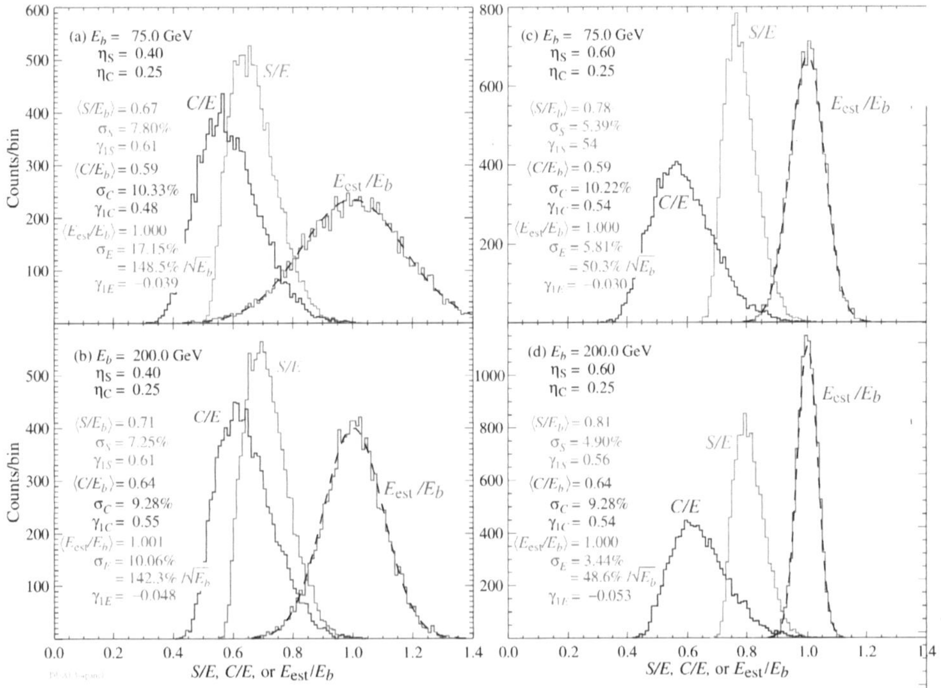

Figure 2. Monte Carlo distributions of C, S, and the estimator of E. (a) and (c) are for beam energies of 75 GeV; for (b) and (c) the beam energy is 200 GeV. For (a) and (b) $\eta_S = 0.40$, while for (c) and (d) $\eta_S = 0.60$. In all cases $\eta_C = 0.25$. Gaussians with the "measured" σ_E and mean relative to beam energy are shown as dotted black lines.

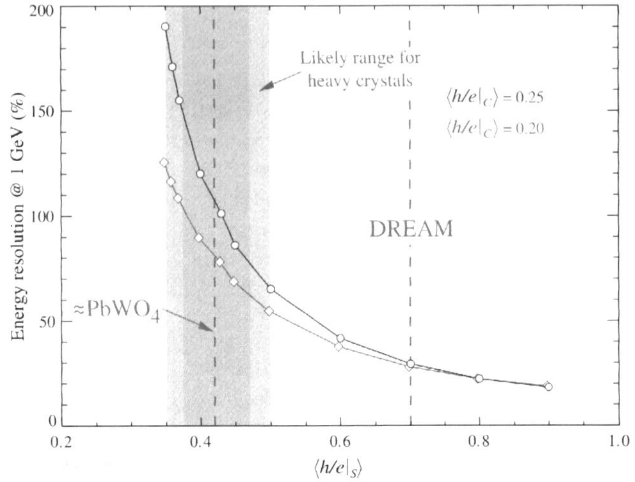

Figure 3. The coefficient of $1/\sqrt{E}$ upon $\eta_S \equiv h/e|_S$ at two values of $\eta_C \equiv h/e|_C$. The results for $\eta_C = 0.20$ are essentially those for $\eta_C = 0.25$ displaced to the left by 0.05; it is the *contrast* between η_C and η_S that determines the resolution.

Crystals for the HHCAL Detector Concept

Rihua Mao, Liyuan Zhang and Ren-Yuan Zhu
256-48, HEP, Caltech, Pasadena, CA 91125, USA
E-mail: zhu@hep.caltech.edu

Abstract. Crystal calorimeter has traditionally played an important role in precision measurement of electrons and photons in high energy physics experiments. Recent interest in calorimeter technology extends its application to measurement of hadrons and jets with dual readout for both Cherenkov and scintillation light. Optical and scintillation properties of crystal scintillators commonly used in particle physics experiments are reviewed. Technologies to discriminate Cherenkov and scintillation light is elaborated. Candidate crystals for the homogeneous hadronic calorimeter detector concept and their recent development are discussed.

1. Introduction

Total absorption shower counters made of inorganic crystal scintillators have been known for decades for their superb energy resolution and detection efficiency for electrons and photons [1]. In high energy and nuclear physics, crystal calorimeters have been constructed, and their use has been a key factor in the successful physics programs of many experiments. The physics discovery potential of crystal calorimeter was early demonstrated by the Crystal Ball experiment through its study of radiative transitions and decays of the Charmonium family [2]. With proper calibration and monitoring, crystal calorimeters usually deliver their designed resolution *in situ* [3].

Table 1 summarizes parameters of past and present crystal electromagnetic calorimeters in high energy physics experiments [1]. One notes that each of these calorimeters requires several cubic meters of high quality crystals. The most ambitious crystal calorimeter is presumably the CMS calorimeter which uses 11 m^3 $PbWO_4$ crystals [4]. Recent interest in the homogeneous hadronic calorimeter (HHCAL) detector concept extends the application of crystals to the measurement of hadrons and jets with high resolution [5]. This HHCAL detector concept adapts dual readout for both Cherenkov and scintillation light, which has been extensively studied recently by the Dream collaboration [6].

Section 2 of this paper describes the optical and scintillation properties of heavy crystal scintillators commonly used in particle physics experiments. Discrimination between Cherenkov and scintillation light is discussed in Section 3. Section 4 discusses candidate crystals for the HHCAL detector concept. A brief summary is given in Section 5.

2. Properties of Crystal Scintillators

Table 2 [7] lists the basic properties of heavy crystals with mass production capability: NaI(Tl), CsI(Tl), BaF_2, bismuth gemanade ($Bi_4Ge_3O_{12}$ or BGO), lead tungstate ($PbWO_4$ or PWO), LSO [8]/LYSO [9] and PbF_2. All, except PbF_2, are scintillators with the characteristics of their scintillation light listed. All have either been used in, or actively being pursued for, high energy

Table 1. Crystal Calorimeter in High Energy Physics: Past and Present

Experiment	C. Ball	L3	CLEO II	KTeV	*BaBar*	BELLE	CMS
Accelerator	SPEAR	LEP	CESR	Tevatron	PEP II	KEK	LHC
Date	75-85	80-00	80-00	90-10	94-10	94-10	95-20
Crystal Type	NaI(Tl)	BGO	CsI(Tl)	CsI	CsI(Tl)	CsI(Tl)	$PbWO_4$
B-Field (Tesla)	-	0.5	1.5	-	1.5	1.0	4.0
Inner Radius (m)	0.254	0.55	1.0	-	1.0	1.25	1.29
Number of Crystals	672	11,400	7,800	3,300	6,580	8,800	76,000
Crystal Depth (X_0)	16	22	16	27	16 to 17.5	16.2	25
Crystal Volume (m^3)	1	1.5	7	2	5.9	9.5	11
L. Yield (p.e./MeV)	350	1,400	5,000	40	5,000	5,000	2
Photo-sensor	PMT	Si PD	Si PD	PMT	Si PD	Si PD	APD‡
Photo-sensor Gain	Large	1	1	4,000	1	1	50
Noise/Can. (MeV)	0.05	0.8	0.5	Small	0.15	0.2	30
Dynamic Range	10^4	10^5	10^4	10^4	10^4	10^4	10^5

† Wavelength Shifter.
‡ Avalanche photo-diode.

and nuclear physics experiments, which are also listed in the table. The experiment name in bold indicates possible future crystal calorimeters in the next decade. LSO and LYSO crystals are also widely used in the medical industry. Mass production capabilities exist for all these crystals.

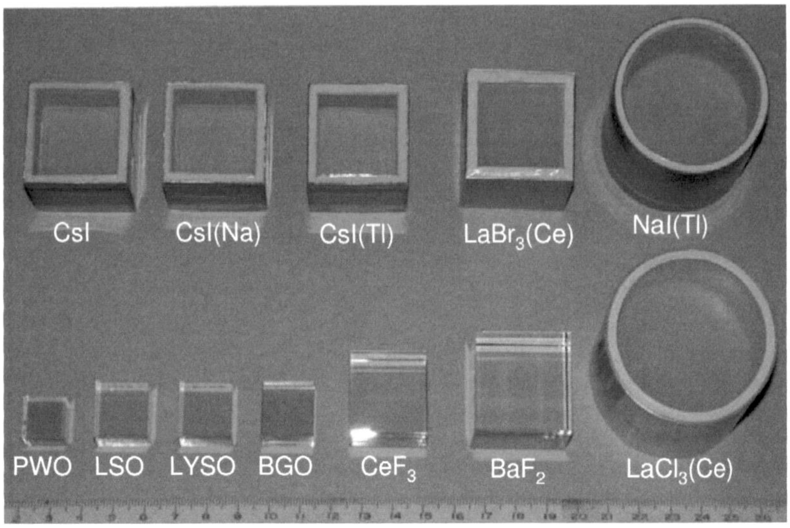

Figure 1. A photo shows twelve crystal scintillators with dimension of 1.5 X_0.

Figure 1 is a photo showing twelve crystal scintillator samples. In addition to samples listed in Table 2 pure CsI, CsI(Na), LYSO as well as $LaCl_3$ and $LaBr_3$ [10] are also shown in this photo although the last two are not yet in a mass production stage. Samples are arranged in

Table 2. Properties of Heavy Crystal with Mass Production Capability

Crystal	NaI(Tl)	CsI(Tl)	CsI	BaF_2	BGO	$PbWO_4$	LSO	PbF_2
Density (g/cm^3)	3.67	4.51	4.51	4.89	7.13	8.3	7.40	7.77
Melting Point (°C)	651	621	621	1280	1050	1123	2050	824
Radiation Length (cm)	2.59	1.86	1.86	2.03	1.12	0.89	1.14	0.93
Molière Radius (cm)	4.13	3.57	3.57	3.10	2.23	2.00	2.07	2.21
Interaction Length (cm)	42.9	39.3	39.3	30.7	22.7	20.7	20.9	21.0
Refractive Index[a]	1.85	1.79	1.95	1.50	2.15	2.20	1.82	1.82
Hygroscopicity	Yes	Slight	Slight	No	No	No	No	No
Luminescence[b] (nm) (at Peak)	410	560	420 310	300 220	480	425 420	420	?
Decay Time[b] (ns)	245	1220	30 6	650 0.9	300	30 10	40	?
Light Yield[b,c]	100	165	3.6 1.1	36 4.1	21	0.30 0.077	85	?
d(LY)/dT[b,d] (%/°C)	-0.2	0.4	-1.4	-1.9 0.1	-0.9	-2.5	-0.2	?
Experiment	Crystal Ball	CLEO *BaBar* BELLE BES	KTeV	TAPS	L3 Belle	CMS ALICE PrimEx Panda	Mu2e SuperB	A4

a At the wavelength of the emission maximum.
b Top line: slow component, bottom line: fast component.
c Relative light yield of samples of 1.5 X_0 and with the PMT quantum efficiency taken out.
d At room temperature.

an order of their density, or radiation length. All non-hygroscopic samples are wrapped with white Tyvek paper as reflector. Hygroscopic NaI, CsI, $LaBr_3$ and $LaCl_3$ are sealed in a package with two ends made of quartz windows of 3 or 5 mm thickness to avoid surface degradation. To minimize uncertainties in the light output measurement caused by the sample size dependence all samples have a cubic shape of $1.5 \times 1.5 \times 1.5\ X_0^3$, except NaI(Tl) and $LaCl_3$ which are a cylinder with a length of 1.5 X_0 and areas at two ends equaling to $1.5 \times 1.5\ X_0^2$ to match the 2 inch diameter of the PMT cathode.

Figure 2 shows a comparison of the transmittance, photo-luminescence and excitation spectra for ten samples. The solid black dots in these plots are the theoretical limit of the transmittance, which is calculated using the corresponding refractive index as a function of wavelength taking into account multiple reflections between the two parallel end surfaces and assuming no internal absorption [11]. Most samples, except $LaBr_3$ and $LaCl_3$, have their transmittance approaching the theoretical limits, indicating negligible internal absorption. The poor transmittance measured for $LaBr_3$ and $LaCl_3$ samples is probably due to scattering centers inside these samples. It is interesting to note that BaF_2, BGO, NaI(Tl), CsI(Tl) and $PbWO_4$ have their emission spectra well within the transparent region showing no obvious self-absorption effect. The UV absorption edge in the transmittance spectra of LSO, LYSO, CeF_3, $LaBr_3$ and $LaCl_3$, however, cuts into the emission spectra and thus affects crystal's light output. This self-absorption effect is more serious in long crystal samples used in high energy and nuclear physics experiment as extensively discussed for LSO and LYSO crystals [12, 13]. We also note that the values of the cut-off wavelength, at which the transmittance data show 50% of that at 800 nm, are 140 nm, 280 nm, 293 nm, 315 nm, 318 nm, 342 nm,358 nm, 365 nm and 390 nm

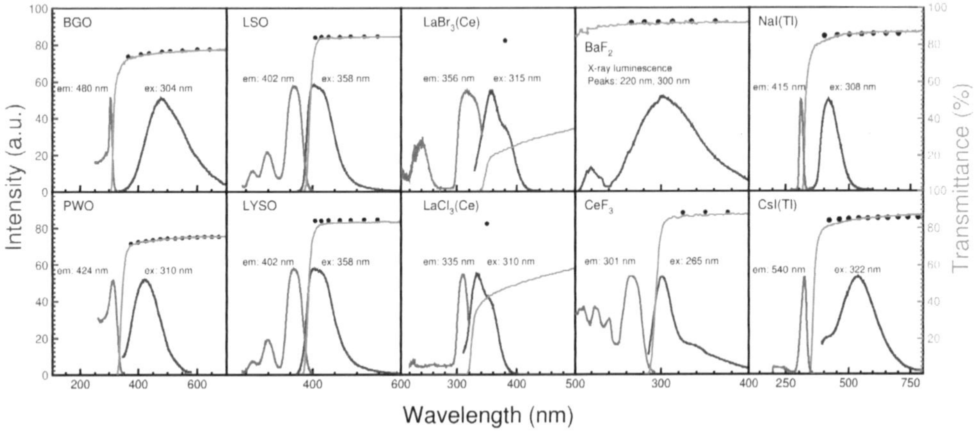

Figure 2. The excitation (red) and emission (blue) spectra (left scale) and the transmittance (green) spectra (right scale) are shown as a function of wavelength for ten crystal scintillators. The solid black dots are the theoretical limit of the transmittance.

for BaF_2, CsI, CeF_3, BGO, CsI(Na), PWO, CsI(Tl), NaI(Tl) and LSO/LYSO respectively. On the other hand it is 250 nm for PbF_2, indicating that it is a good Cherenkov radiator.

Figure 3 shows the ^{137}Cs γ-ray pulse hight spectra measured by a Hamamatsu R1306 PMT with a bi-alkali cathode for twelve crystal samples. Also shown in these figures are the corresponding FWHM energy resolution (E.R.), which is affected by the quantum efficiency of the readout device. γ-ray spectroscopy with a few percents resolution is required to identify isotopes for the homeland security applications. It is clear that only $LaBr_3$ approaches this requirement. All other crystals do not provide sufficient energy resolution at low energies.

Figure 4 shows light output in photo-electrons per MeV energy deposition as a function of the integration time, measured using a Photonis XP2254b PMT with a multi-alkali photo cathode, for six fast crystal scintillators (Left): $LaBr_3$, LSO, LYSO, CeF_3, un-doped CsI and $PbWO_4$

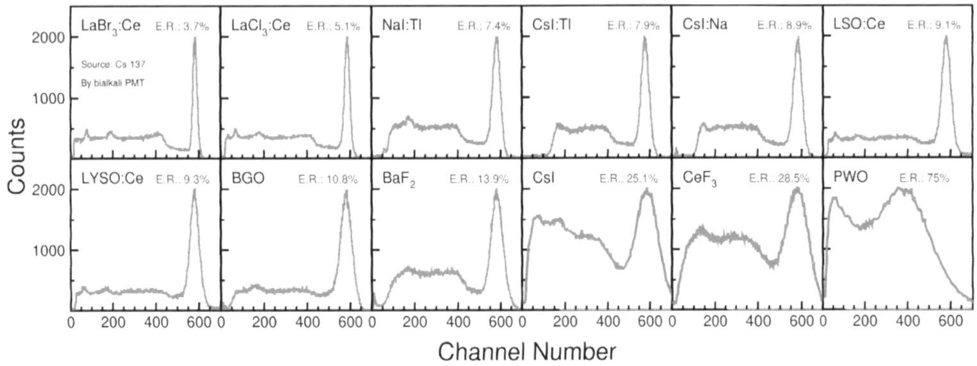

Figure 3. ^{137}Cs γ-ray pulse hight spectra measured by a Hamamatsu R1306 PMT are shown for twelve crystal samples. The numerical values of the FWHM resolution (E.R.) are also shown in the figure.

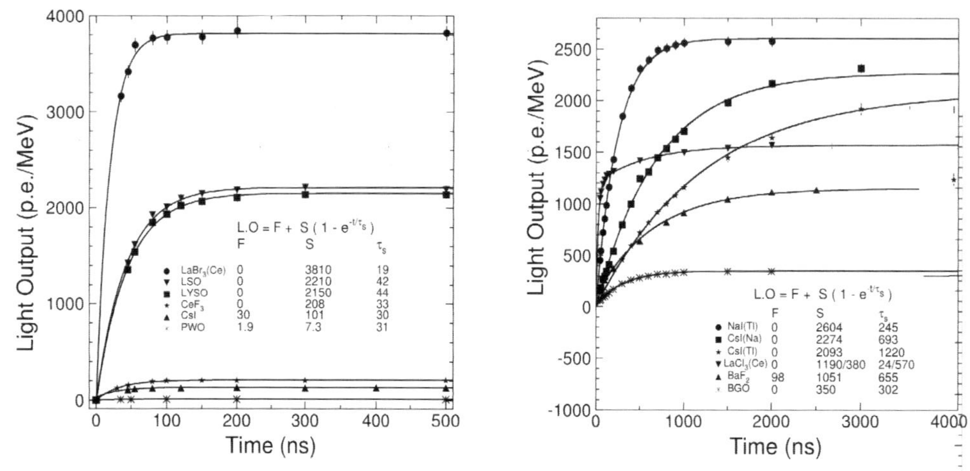

Figure 4. Light output measured using a XP2254b PMT is shown as a function of integration time for six fast (Left) and slow (Right) crystal scintillators.

and six slow crystal scintillators (Right): NaI(Tl), CsI(Na), CsI(Tl), $LaCl_3$, BaF_2 and BGO. The corresponding fits to the exponentials and their numerical results are also shown in these figures. The un-doped CsI, $PbWO_4$, $LaCl_3$ and BaF_2 crystals are observed to have two decay components. Despite its poor transmittance the cerium doped $LaBr_3$ is distinguished by its bright fast scintillation, leading to the excellent energy resolution for the γ-ray spectroscopic

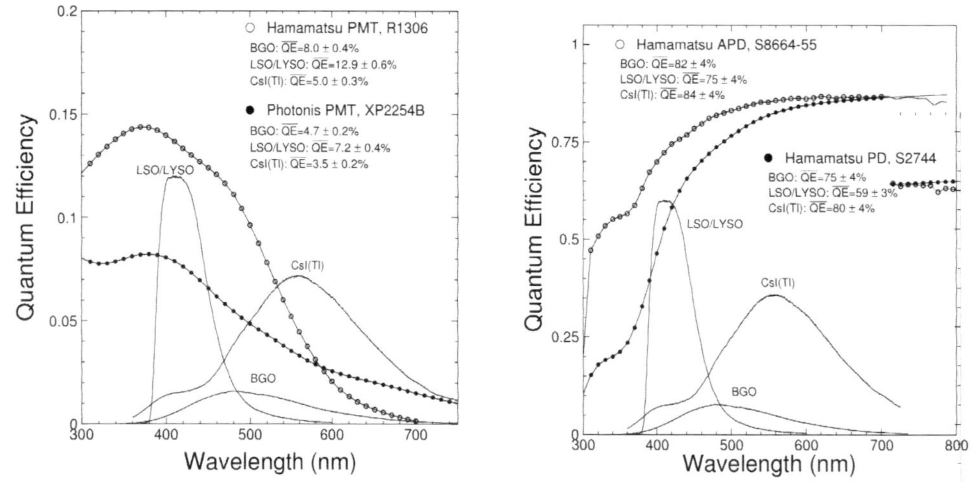

Figure 5. Left: Quantum efficiencies of a Hamamatsu 1306 PMT with a bi-alkali cathode (open circles) and a Photonis 2254B PMT with a multi-alkali cathode (solid dots) are shown as a function of wavelength together with the emission spectra of the LSO/LYSO, BGO and CsI(Tl) samples, where the area under the emission curves is proportional to their corresponding absolute light output. Right: The same for a Hamamatsu S8664 Si APD (open circles) and a Hamamatsu S2744 Si PIN diode (solid dots).

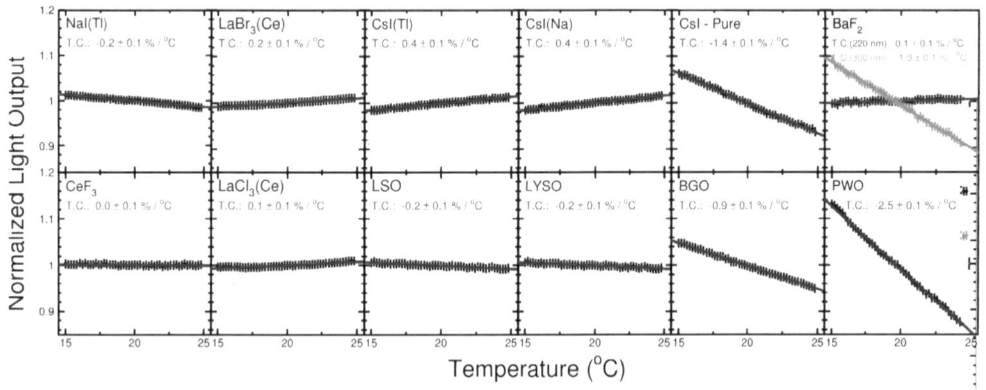

Figure 6. Light output temperature coefficient obtained from linear fits between 15°C and 25°C for twelve crystal scintillators.

applications. The LSO and LYSO samples have a consistent fast decay time (~40 ns) and photo-electron yield, which is 6 and 230 times of BGO and $PbWO_4$ respectively.

Since the quantum efficiency of the PMT used for the light output measurement is a function of wavelength, it should be corrected for to directly compare the light output between different crystals. Figures 5 shows the typical quantum efficiency as a function of wavelength for a PMT with a bi-alkali cathode (Hamamatsu R1306) and a PMT with a multi-alkali cathode (Photonis 2254B) (Left), and a Si APD (Hamamatsu S8664) and a Si PIN PD (Hamamatsu S2744) (Right). The emission spectra of LSO/LYSO, BGO and CsI(Tl) crystals are also shown in these figures. The light output values listed in Table 2 are corrected for the PMT quantum efficiency. The light output of LSO and LYSO crystals is a factor of 4 and 200 of that of BGO and $PbWO_4$ respectively.

The scintillation light yield of crystals may also depends on temperature. Fig 6 shows the light output variations for twelve crystal samples between 15°C and 25°C. The corresponding temperature coefficients obtained from linear fits are also listed in the figure. The numerical result of these fits is also listed in Table 2.

3. Discrimination between Cherenkov and Scintillation Light

Crystals have recently been proposed to construct a homogeneous calorimeter, including both electromagnetic and hadronic components [5]. This HHCAL detector concept removes the traditional boundary between ECAL and HCAL, thus eliminating the effect of dead materials in the middle of the hadronic shower development. It takes advantage of the recently implemented dual readout approach to measure both Cherenkov and scintillation light to achieve good energy resolution for hadronic jet measurements [6]. Because of the un-precedent volume (70 to 100 m^3) foreseen for such a calorimeter [5], the crystal material must be dense (to reduce the volume), UV transparent (to effectively collect the Cherenkov light) and allows a clear discrimination between the Cherenkov and scintillation light.

Figure 7 shows samples of three $5 \times 5 \times 5$ cm^3 crystal samples: PbF_2, BGO and PWO. Crystals of this size can be seen as typical building block for a crystal hadronic calorimeter. All material are dense with a nuclear interaction length in a range from 21 cm (PWO and PbF_2) to 23 cm (BGO).

Figure 8 (Left) shows the transmittance spectra of PbF_2 (green), BGO (blue), PWO (red) and a UG11 filter (black) as a function of wavelength together with the Cherenkov emission

Figure 7. A photo shows three crystal samples of $5 \times 5 \times 5$ cm^3 investigated for the homogeneous hadronic calorimeter concept.

spectrum (dashed blue). The UG 11 filter can be used to select the Cherenkov light with little or no scintillation contamination. Also shown in this figure is the normalized figure of merit for the Cherenkov measurement using the UG11 filter, which is defined as the transmittance weighted Cherenkov emission spectrum (TWEM). Their numerical values are 1.0:0.53:0.21, which would be 1.0:0.82:0.75 without the UG11 filter. Among these materials PbF_2 is the most efficient in collecting the Cherenkov light because of its good UV transmission.

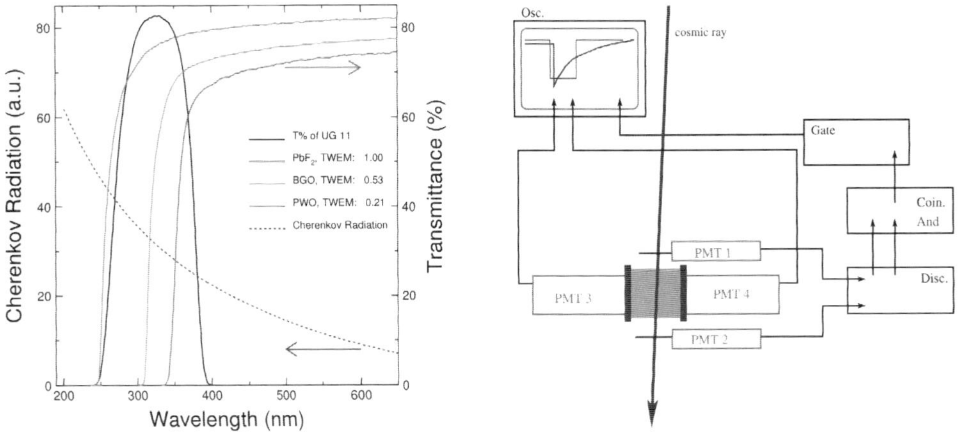

Figure 8. Left: The transmittance spectra of PbF_2 (green), BGO (blue), PWO (red) and UG11 (black) are shown as a function of wavelength. Also shown in this figure are the Cherenkov emission spectrum (dashed blue) and the normalized figure of merit for the Cherenkov light measurement with the UG11 filter. Right: A schematic showing a simple set-up used to measure cosmic-ray generated Cherenkov and scintillation light simultaneously using two Hamamatsu R2059 PMTs.

An effective technology to discriminate Cherenkov and scintillation light can be realized using an optical filter or pulse timing. Figure 8 (Right) shows a set-up used to investigate Cherenkov light collection and its discrimination from the scintillation light. Samples shown in Figure 7 were

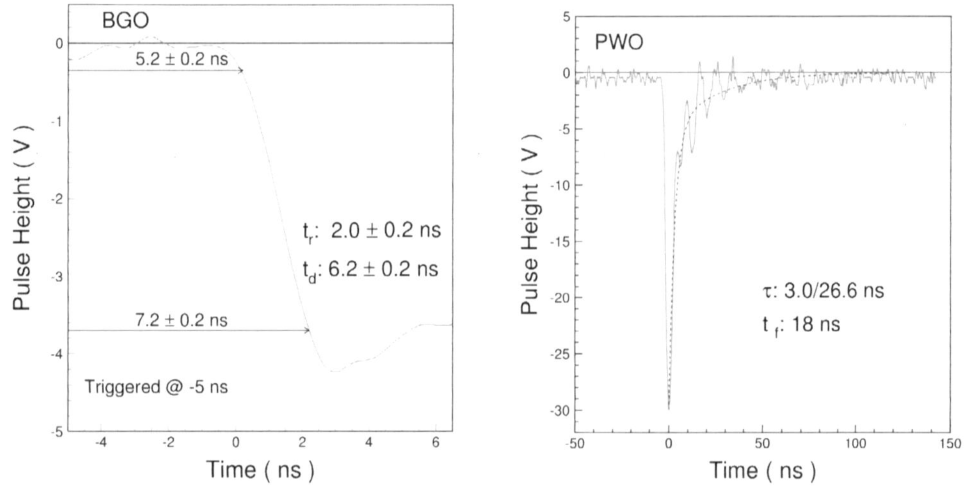

Figure 9. The light pulses are recorded by an Agilent 6052A digital scope. Digital scope traces of the scintillation light front edge measured by a Hamamatsu R2059 PMT with GG400 filter for the BGO (Left) and PWO (Right) samples.

used in this investigation. To avoid the multiple light reflection effect all crystals were wrapped with black paper. Cosmic-rays were triggered by two finger counters with coincidence. The Cherenkov and scintillation light pulses generated by cosmic-rays were measured simultaneously by two Hamamatsu R2059 PMT coupled to the sample through optical filters UG11 and GG400. GG400 is a low-pass filter with cut-off at 400 nm. The UG11 filter is used to select the Cherenkov light as shown in Figure 8. The GG400 filter is used to select the scintillation light with small contamination of the Cherenkov light. The output of these two PMTs were digitized by an Agilent 6052A digital scope.

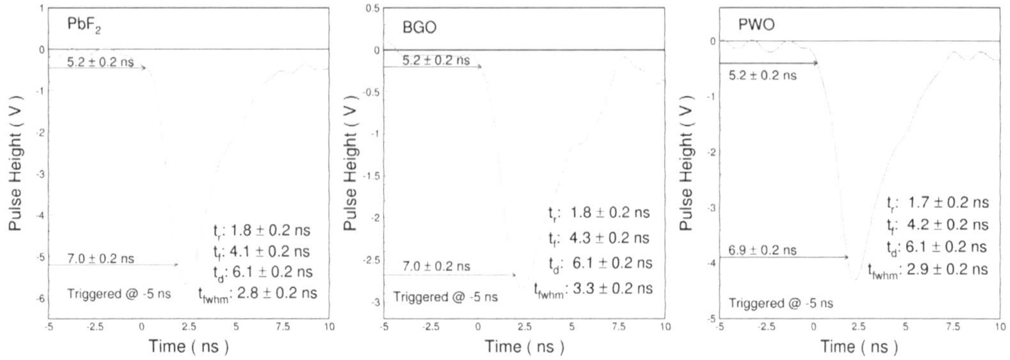

Figure 10. Digital scope traces of Cherenkov light pulse measured by a Hamamatsu R2059 PMT with UG11 filter for the PbF_2 (Left), BGO (Middle) and PWO (Right) samples.

Figure 9 shows the front edge of the scintillation light pulse from BGO and PWO, observed through the GG400 filter. Their delay from the trigger (t_d) and rise time (t_r) are identical with numerical values of 6.2 ns and 1.9 ns respectively with the later dominated by the rise time of

DSO (0.7 ns) and PMT (1.3 ns). Figure 10 shows the Cherenkov light pulse shape observed for PbF_2 (Left), BGO (Middle) and PWO (Right) through the UG11 filter. All pulses have a consistent time structure in the delay (6.1 ns), the rise time (1.8 ns), the fall time (4.2 ns) and the FWHM width (3.0 ns). It is interesting to note that there is actually no difference observed in the delay and rise time between the Cherenkov and scintillation light, indicating that only the light pulse width and fall time are useful for the discrimination between the Cherenkov and scintillation light. A slow scintillator may actually help this discrimination.

Because of the difference in the spectrum and direction between the Cherenkov and scintillation light the multiple reflections inside crystals may cause a difference in the measured rise time between Cherenkov and scintillation light. Figure 11 shows DSO traces of scintillation light for the BGO sample with different wrappings (green: Tyvek and blue: black). The multiple reflections introduced by white Tyvek paper wrapping clearly increase the measured scintillation decay time (left) and rise time (right).

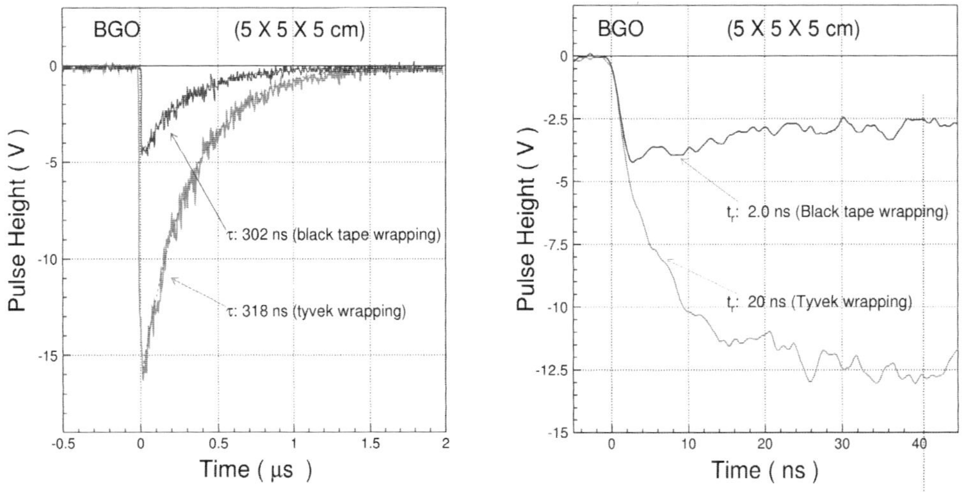

Figure 11. Agilent 6052A digital scope traces of the scintillation light for the BGO sample with Tyvek (green) and black (blue) wrappings are shown with long (left) and short (right) time scales.

The ratio of Cherenkov versus scintillation light was measured to be 1.55% and 22% for BGO and PWO respectively. These values are consistent with the scintillation light yield shown in Table 2, the emission weighed quantum efficiency of the bi-alkali cathode of the Hamamatsu R2059 PMT shown in Figure 5 and the TWEM values shown in Figure 8.

4. Crystals for the HHCAL Detector Concept

Based upon our experience accumulated in building electromagnetic calorimeters, an initial detector design with a pointing geometry was proposed. It may provide better resolutions for both energy and position measurements, and thus a good jet mass reconstruction. Figure 12 shows a schematic of a typical HHCAL cell with a pointing geometry [14]. It is similar to a typical calorimeter cell of a crystal ECAL, but has several longitudinal segments with a total length of about 1 m. By using dense active materials such detector depth would provide about 5 nuclear interaction lengths, adequate for hadronic jet energy reconstruction. The readout devices are mounted on the side faces of these crystal segments. Due to the recent development

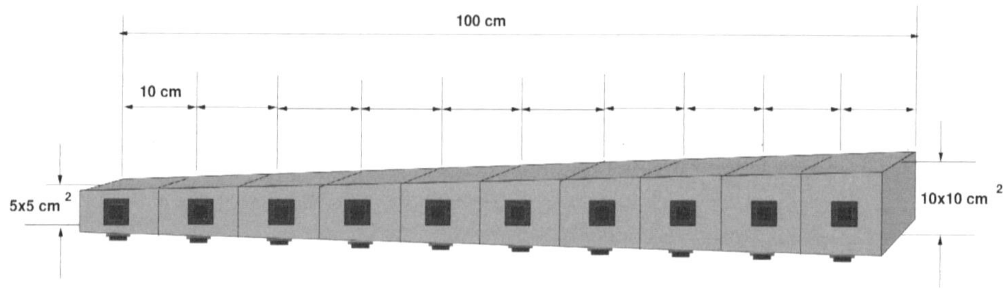

Figure 12. A schematic showing a typical cell for the HHCAL detector concept with a pointing geometry [14].

in compact solid state readout devices, e.g., silicon PMTs, such a readout scheme is now feasible.

Because of the huge volume required for an HHCAL detector, development of cost-effective materials is crucial. Table 3 summarized the basic properties of candidate crystals being considered for this detector concept, where two crystal cost drivers, melting point and raw material cost, are also listed. While BGO may be the best material to be used for such a calorimeter its mass production cost is prohibitively high. While PWO, PbF_2, PbFCl and BSO are under investigation, PbF_2 and PbFCl are preferred because of their low melting point and raw material cost.

R&D is actively being pursued by the high energy physics community in collaboration with the material science community to search for adequate materials for the HHCAL detector

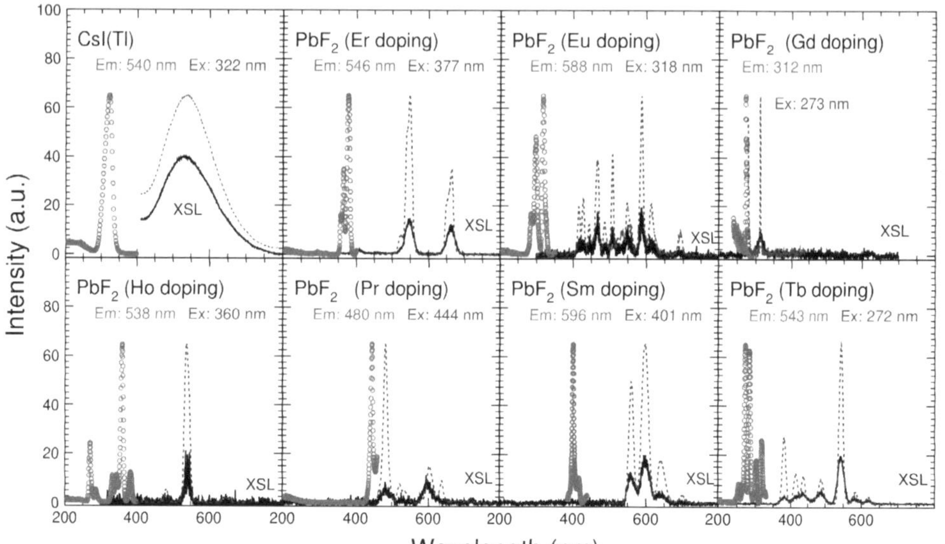

Figure 13. The excitation (red dots) and Photo- (blue dashes) and X- (black lines) luminescence spectra are shown as a function of wavelength for the PbF_2 samples doped with Er, Eu, Gd, Ho, Pr, Sm and Tb as well as a reference CsI(Tl) sample.

Table 3. Candidate Crystals for the HHCAL Detector Concept

Parameters	BGO	PWO	PbF_2	PbFCl	BSO
Density (g/cm^3)	7.13	8.29	7.77	7.11	6.8
λ_I (cm)	22.8	20.7	21.0	24.3	23.1
n @ λ_{max}	2.15	2.20	1.82	2.15	2.06
τ_{decay} (ns)	300	30/10	?	30	100
λ_{max} (nm)	480	425/420	?	420	470
Cut-Off λ (nm)	310	350	250	280	300
Light Output (%)	100	1.4/0.37	?	17	20
Melting Point (°C)	1050	1123	842	608	1030
Raw Material Cost (%)	100	49	29	29	47

concept. One approach is to develop scintillating PbF_2 crystals by selective doping. Observations of fast scintillation in Gd doped PbF_2 crystals were reported early by Shen and Woody [15, 16]. Our investigation shows that rare earth doping introduces scintillation in PbF_2, but not at the level can be measured by using a γ-ray source [17]. Figure 13 shows the excitation, photo-luminescence and x-luminescence spectra for Er, Eu, Gd, Ho, Pr, Sm and Tb doped PbF_2 crystal samples. It is noted that some of this scintillation light is between 500 to 600 nm, which is desirable for Cherenkov/scintillation discrimination.

The photo-luminescence decay time constant of these doped PbF_2 samples was measured using a pulsed laser as the excitation source. Figure 14 shows the photo-luminescence pulse shape (blue circles), the corresponding exponential fit (red lines) and the decay time constant for the PbF_2 samples doped with Er, Ho, Eu, Sm and Tb as well as a reference CsI(Tl) sample.

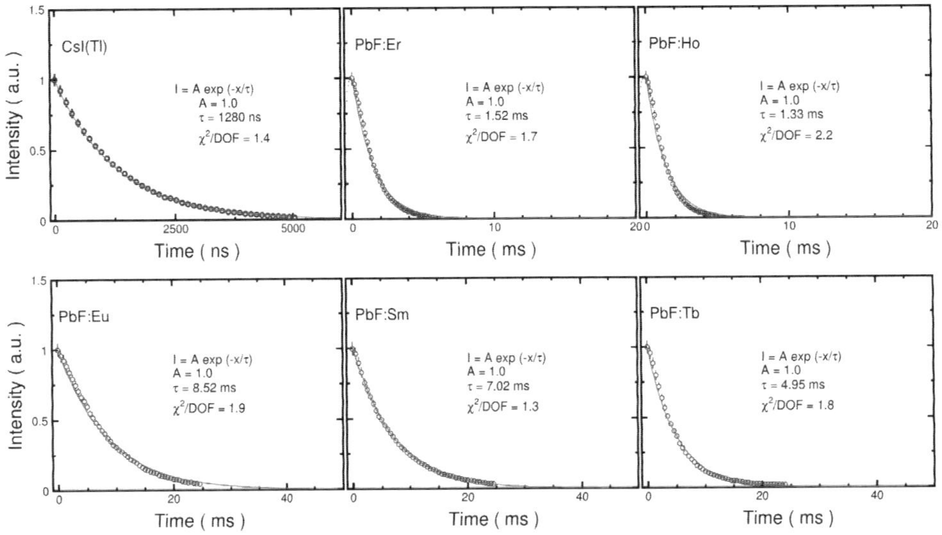

Figure 14. The photo-luminescence pulse shape (blue circles), corresponding fit to an exponential (red lines) and the decay time constant are shown for the PbF_2 samples doped with Er, Ho, Eu, Sm and Tb as well as a reference CsI(Tl) sample.

The photoluminescence intensity of PbF_2 samples doped with Pr and Gd are too weak to be useful to extract the decay time constant. Table 4 summarizes the decay time constants for the PbF_2 samples doped with Er, Ho, Eu, Sm and Tb. They were found at a millisecond scale as expected from the f-f transition of these rare earth elements [18]. These time constants are too long to be useful for high energy physics experiments.

Table 4. Decay Time constant for Doped Lead Fluoride Crystals

Dopant	Er	Eu	Ho	Sm	Tb
Decay time constant (ms)	1.5	8.5	1.3	7.0	5.0

In a brief summary, the combination of potential low cost, good UV transmittance and mass production capability makes PbF_2 crystals a favored material for the HHCAL detector concept. Our search for scintillation found consistent photo- and x-ray luminescence spectra in PbF_2 samples doped with Er, Eu, Gd, Ho, Pr, Sm and Tb with decay time at a millisecond scale. While some Eu doped samples show γ-ray induced anode photo-current larger than the undoped samples, the numerical result indicates that the scintillation light, if any, is less than 3 p.e./MeV measured by a PMT with a bi-alkali photo-cathode. Their ^{137}Cs γ-ray excited pulse height spectra measured with up to a 10 μs integration time were also found identical to that of the undoped sample, confirming that their luminescence is too weak to show a peak. Investigation will continue to search for scintillation in doped PbF_2 phases and mixtures.

On the other hand, R&D is also actively being pursued by the material science community [19]. One other approach worthwhile to mention is to develop PWO crystals with a slow scintillation emission. Green (560 nm) and slow emission with a few μsec decay time was observed by selective doping in PWO crystals [20]. Such crystals were reported to have a factor of ten more light than yttrium doped PWO crystals used in high energy physics experiment. This slow and green scintillation would be desirable for this application. Additional approaches are also being pursues, such as PbFCl, BSO, BSO glasses and various ceramics [21].

5. Summary

Precision crystal electromagnetic calorimeters have been an important part of high energy physics detectors. The availability of mass production capability of large size LSO and LYSO crystals provides an opportunity to build a LSO/LYSO crystal electromagnetic calorimeter with good energy resolution over a large dynamic range down to the MeV level. Such a calorimeter, if built, would greatly enhance the physics discovery potential for high energy and nuclear physics experiments in the next decade.

Recent interest in the high energy physics community to pursue a homogeneous hadronic calorimeter with dual readout opens a new area of crystal calorimetry to achieve good energy resolution for hadronic jet measurements in the next decade. The main challenge for this concept is to develop cost effective heavy scintillators with good UV transmission and excellent Cherenkov/scintillation discrimination. Dense crystals, scintillating glasses and ceramics offer a very attractive implementation for this detector concept.

Acknowledgments

The crystals samples used in this investigation are provided by Professors Guohao Ren, Shaohua Wang and Hui Yuan of Shanghai Institute of Ceramics, and Prof. Dingzhong Shen of Scintibow. This work is partially supported by the U.S. Department of Energy Grant No. DE-FG03-92-ER40701.

References

[1] G. Gratta, H. Newman and R.-Y. Zhu, "Crystal calorimeters in Particle Physics," *Annu. Rev. Nucl. Part. Sci.* **44** 453–500 (1994).

[2] E. Bloom and C, Peck, *Ann. Rev. Nucl. Part. Sci.* **33** (1983) 143–197.

[3] R.-Y. Zhu, "Calibration and monitoring for crystal calorimetry," *Nucl. Instr. and Meth.* **A537** (2005) 344–348.

[4] *The CMS Electromagnetic Calorimeter Project*, CERN/LHCC 97-33 (1997).

[5] A. Para, "Very High Resolution Hadron Calorimetry," FERMILAB-CONF-11-519-CD, and H. Wenzel *et al.*, "Simulation of a Totally active dual read out calorimeter for future Lepton Colliders," FERMILAB-PUB-11-531-CD-E.

[6] N. Akchurin *et al.*, "Separation of crystal signals into scintillation and Cherenkov components", *Nucl. Instr. and Meth.* **A537** (2005) 537.

[7] R.H. Mao, L.Y. Zhang and R.-Y. Zhu, "Optical and Scintillation Properties of Inorganic Scintillators in High Energy Physics," *IEEE Trans. Nucl. Sci.* **NS-55** (2008) 2425-2431.

[8] C. Melcher, V.S. Patent 4958080 (1990) and 5025151 (1991).

[9] D.W. Cooke, K.J. McClellan, B.L. Bennett, J.M. Roper, M.T. Whittaker and R.E. Muenchausen, "Crystal growth and optical characterization of cerium-doped $Lu_{1.8}Y_{0.2}SiO_5$," *J. Appl. Phys.* **88** (2000) 7360 and T. Kimble, M Chou and B.H.T. Chai, "it Scintillation properties of LYSO crystals," *2002 IEEE NSS Conference Record.*

[10] E. Loef, P. Dorenbos, E. Eijk, K. Kraemer and H. Guedel, "Scintillation Properties of $LaBr_3$:Ce^{3+} crystals: Fast Efficient and high-energy resolution scintillators," *Nucl. Instr. and Meth.* **A486** (2002) 254.

[11] D.A. Ma and R.-Y. Zhu, "Light attenuation length of barium fluoride crystals," *Nucl. Instr. and Meth.* **A333** (1993) 422–424.

[12] J.M. Chen, R.H. Mao, L.Y. Zhang and R.-Y. Zhu, "Large Size LSO and LYSO Crystals for Future High Energy Physics Experiments," *IEEE Trans. Nucl. Sci.* **54** (2007) 718–724.

[13] J.M. Chen, R.H. Mao, L.Y. Zhang and R.-Y. Zhu, "Gamma-Ray Induced Radiation Damage in Large Size LSO and LYSO Crystal Samples," *IEEE Trans. Nucl. Sci.* **54** (2007) 1319–1326.

[14] R.-Y. Zhu, talk presented in International Linear Collider Workshop 2008 at Chicago. http://www.hep.caltech.edu/~zhu/talks/ ryz_081118_lcws.pdf.

[15] D. Shen *et al.*, *Jour. Inor. Mater.* **Vol 101** (1995) 11.

[16] C. Woody *et al.*, in *Proceedings of SCINT95*, Delft University Press, Delft, The Netherlands, (1996).

[17] R.H. Mao, L.Y. Zhang and R.-Y. Zhu, "A Search for Scintillation in Doped Cubic Lead Fluoride Crystals," *IEEE Trans. Nucl. Sci.* **57** (2010) 3841–3845.

[18] D.F. Anderson M. Kobayashi, Y. Yoshimura, C.L. Woody, "Lead Fluoride: An Ultracompact Cherenkov Radiator for EM Calorimetry," *Nucl. Instr. and Meth.* **A290** 385-389, and C. Woody *et al.*, "A Study on the Use of Lead Fluoride for Electromagnetic Calorimetry," *IEEE Trans. Nucl. Sci.* **NS-40** (1993) 546-551.

[19] See, for example, presentations in the 2nd Workshop for Material Development for the HHCAL Detector Concept, Beijing, May 9, 2010, See http://indico.ihep.ac.cn/sessionDisplay.py?sessionId=2&slotId=0&confId=1470#2010-05-09.

[20] R.H. Mao *et al.*, "Precision Crystal Calorimeters in High Energy Physics: Past Present and Future," in *Proceedings of IX International Conference on Calorimetry in Particle Physics*, Ed. B. Aubert *et al.*, Franscati Physics Series Vol. XXI (2000) 709–720.

[21] See presentations and papers in this conference.

Preliminary Results from a Test Beam of ADRIANO Prototype

C Gatto[1], V. Di Benedetto[2] and A. Mazzacane[3]
On behalf of T1015 Collaboration

[1] INFN, Sezione di Napoli, via Cinthia, 80126 Napoli, Italy
[2] INFN, Sezione di Napoli and Università del Salento - Lecce, Italy
[3] Fermilab, Batavia (IL), USA

E-mail: corrado.gatto@infn.it

Abstract. The physics program at future colliders demands an energy resolution of the calorimetric component of detectors at the limits of traditional techniques. The ADRIANO technology (*A Dual-readout Integrally Active Non-segmented Option*) is under development with an expected excellent performance. Results from detailed Monte Carlo studies on performance with respect to energy resolution, linear response and transverse containment and a preliminary optimization of the layout are presented. A baseline configuration is chosen with an estimated energy resolution of $\sigma(E)/E \approx 30\%/\sqrt{E}$, to support an extensive *R&D* program recently started by *T1015* Collaboration at Fermilab. Preliminary results from a test beam at the *Fermilab Test Beam Facility* (FTBF) of a $\sim 1\lambda_I$ prototype are presented, along with simulation studies. Future prospects with ultra-heavy glass are, also, summarized.

1. Introduction

The physics program envisaged for the experiments at future lepton and hadron colliders will be dominated by studies of processes involving multi-jets event. In such an environment, calorimeters will play a fundamental role as particle detector. A broad based *R&D* and Monte Carlo simulation activity is already in progress within the lepton colliders communities[1].
A consensus has been established on the fact that the minimum hadronic energy resolution of the calorimetric systems needed to successfully distinguish the W from the Z signal is $\sigma(E)/E \approx 30\%/\sqrt{E}$. Such a resolution is unprecedented and it is being reached only by massive compensating calorimeters with very small volume ratio between passive and active materials[2]. However, the large size needed to contain the showers, because of their relatively low density, make them impractical to build in experiments with colliding beams. As a consequence, it would be challenging to achieve at the same time effective compensation and shower containment. Furthermore, the resolution of traditional calorimeters is limited, among other sources, by the fluctuation in the electromagnetic (EM) content of the hadronic shower and by the unequal response of such devices to the EM and hadronic components of the shower itself[3]. In recent years, an alternative technique has been developed in order to cope with such effects: the dual-readout calorimetry[7], based on an event-by-event measurement of the EM fraction of the shower. Such a technique is based on the simultaneous measurement of signals generated by different shower production mechanisms, thus providing complementary information on the composition of the hadronic shower. When applied to a realistic detector layout expected at an

experiment at a high energy collider, such a technique appears capable of meeting the necessary requirements[5].

The dual read-out calorimetry falls under two broad categories: sampling and integrally active. Sampling dual-readout techniques are currently investigated by the DREAM[7] and the 4th Concept Collaborations[5]. Test beams and extensive simulations have proved that these techniques provide excellent energy resolution. However they introduce two important sources of fluctuations: Poisson fluctuations in the Čerenkov signal, induced by the low photo-electron statistics and sampling fluctuations. These fluctuations do not only impair the energy resolution for hadronic showers, but have also detrimental consequences on the detection quality of photons and electrons. Consequently, the detection of high energy jets where a large contribution by EM particles occurs is similarly affected. An obvious solution to the latter problem would be to design a detector with two distinguished regions: a front EM section and a rear hadronic section. However, it is well known that such a configuration, most often consisting of media with very different properties, is sub-optimal in terms of energy measurement of hadronic particles due to the extra fluctuations introduced by the development of the shower into two different sections [3].

In this article we propose a novel Dual-Readout, Integrally Active and Non-homogeneous Option (ADRIANO), based on signals produced in high transmittance optical glasses and scintillating fibers that has the potential to overcome the above limitations.

2. Description of ADRIANO technique

The new detection technique named ADRIANO (*A Dual-readout Integrally Active Non-segmented Option*) is a part of an extensive *R&D* program recently started by *T1015*[4] Collaboration. ADRIANO prototypes recently built all have a modular structure, with the base unit consisting of an individual cell of parallelepiped shape with 40×40 mm^2 cross-section and either 15 cm or 25 cm length. The cell consists of a sandwich of scintillating fibers and high density, optical grade heavy glass. The glass behaves as an absorber and as an active medium at the same time, generating almost exclusively Čerenkov light. The scintillation and Čerenkov sections of ADRIANO are optically separated. Therefore, the two generated lights are well separated, with minimal chance of cross-talk. For the results presented in this report, we have considered various techniques to optically separate the two regions: white and silver coating of the glass, white coating, silver coating and aluminum sputtering of the scintillating fibers and finally, a thin layer of Teflon between the glass plates and the array of scintillating fibers.

The scintillating fibers are either accommodated in grooves in the glass itself or in white plastic trays sandwiched among plates of glass. They run parallel to the longitudinal axis of the cell and are responsible for the generation of the scintillation component of the dual readout calorimeter. The pitch between nearby fiber is sufficiently small compared to the nuclear hadronic interaction length of the detector so that the shower sampling fluctuations are small.
The Čerenkov light generated in the glass plates is collected by WLS fibers running inside grooves parallel to the scintillating fibers and optically coupled to the glass by especially formulated optical compounds. The two light components are read out at the back of each cell with two distinct photodetectors. In some sense, ADRIANO is a spaghetti calorimeter with the passive absorber replaced by an active, transparent absorber made of heavy glass. Another advantage of ADRIANO relies on the fact that the heavy glass absorber can be used to detect electromagnetic showers in exactly the same ways as it has been done in the past with lead glass based electromagnetic calorimeters. Therefore, ADRIANO does not require a front electromagnetic section.

Several heavy glasses (mostly lead and bismuth based) have been tested, with the intent of comparing the Čerenkov light yield and propagation. Their refractive index ranges from 1.85 through 2.24 while the densities range from $5.5gr/cm^3$ through $7.5gr/cm^3$. Several constructions techniques have been considered: diamond machining, precision molding, glass

Figure 1. *Molding fabrication technology.*

melting, laser drilling and photo-etching. However, only the former two have been considered for the production of the eleven ADRIANO prototypes presented in this report. A picture of a $8mm$ thick glass slice obtained with the precision molding technique is shown in Fig.1 . A detailed description of the layouts of the eleven prototype and their corresponding construction techniques will follow in an upcoming article.

3. ADRIANO readout system

The scintillating and WLS fibers from each ADRIANO cell were bundled and routed each to a photodetector. In order to compare the performance of the light collection system in variuos situations we used three different photomultipliers (R647 and H3165 from Hamamatsu and P30CW5 from Sensetech) and two different SiPM ($4x4mm^2$ square and $\O 2.7mm$ round from FBK) for WLS fibers and only one type of SiPM ($4x4mm^2$ square from FBK) for the scintillating fibers. When PMT were used, the fibers were routed through a plastice fixture and coupled to the photosensor window with a custom made optical grease. In the case of SiPM, we used either acrylic ligth concentrators (designed and produced by INFN Trieste) in direct contact with the fibers on one side and spaced $0.1m\mu$ from the active SiPM surface or we routed the fibers directly to the SiPM up to $0.1m\mu$ from the active SiPM surface. A picture of a INFN-Trieste light concentrators is shown in top left of Fig.2. The output of the SiPM and PMT used was digitized by the TB4 DAQ system developed at Fermilab. Among the features of the DAQ, the most relevant for our application are:

- 50Ω inputs
- 14 bit ADC;
- $\sim$ 30 MHz bandwidth;
- $\sim$ 212 MSPS digitizer;
- Up to 16 channels per Motherboard;
- Bipolar, so both positive (from SiPM) and negative (from PMT's) signals can be acquired simultaneously;
- Slow-control over USB, readout over 100 Mbit Ethernet.

4. Preliminary Results From Test Beams

From March 2011 through April 2012 four test beam have been completed at FTBF Facility of Fermilab (Batavia, US). A total of eleven ADRIANO cell prototypes (with different dimensions)

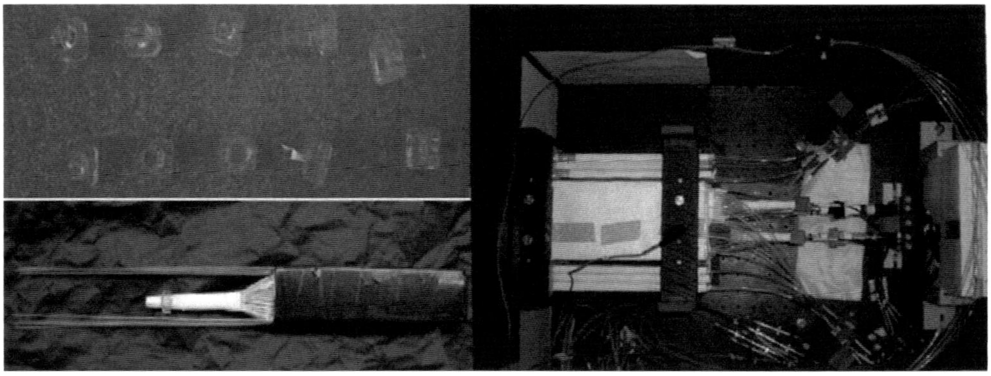

Figure 2. *INFN-Trieste light concentrators, a typical ADRIANO prototype, with scintillating fibers bundled together and WLS unbundled on the left and 2012 test beam setup at FTBF on the right.*

in different configurations and with different construction techniques have been illuminated by the secondary beam available at FTBF. A picture of a typical ADRIANO prototype, with scintillating fibers bundled together and WLS unbundled is shown in the bottom left of Fig.2 In summary we have tested:

- 4 glass type: lead and bismuth based.
- 2 glass coatings: TiO2, Silver paint.
- 3 WLS fibers: Y11 (1.2 mm) & BCF92 (1.0 mm, 1.2 mm).
- 1 Scintillating fiber: SCSF81.
- 4 scifi coating: TiO2, BasO4, Silver paint, Al sputter.
- Several optical glues (mostly custom made).
- 5 photodetectors: 2 SiPMs (ø$2.7mm$ round and $4x4mm^2$ square) & 3 PMTs (P30CW5 , R647, H3165)
- 4 light coupling systems: direct glass + direct WLS + 4 light concentrators

The goal of these experiments were to optimize the construction technique and the materials used in order to maximize the Čerenkov light yield. We also measured some of the parameters that were needed for Monte Carlo simulations with ILCroot framework [8]. Due to the limited size of the cells, we do not expect to be able to to test the dual-readout concept nor to evaluate the energy resolution of ADRIANO.

Typical waveforms for several beam configurations, as obtained by TB4 DAQ, the Čerenkov spectrum various prototypes when exposed to a 5 GeV beam at FTBF and the uniformity of Čerenkov response across the cell are shown in Fig.4 (right side).

Finally, the Čerenkov light yield from the eleven prototypes tested at FTBF is summarized in Table 1.

5. Summary of results

Since the start of ADRIANO *R&D* program, we were able to constantly improve the Čerenkov light yield by refining the construction techniques and the materials employed. The current limit for a $40x40x250mm^3$ is roughly 160 pe/GeV. We found that the Čerenkov light attenuation length inside the glass absorber critically depends on coating type and on the surface finishing of

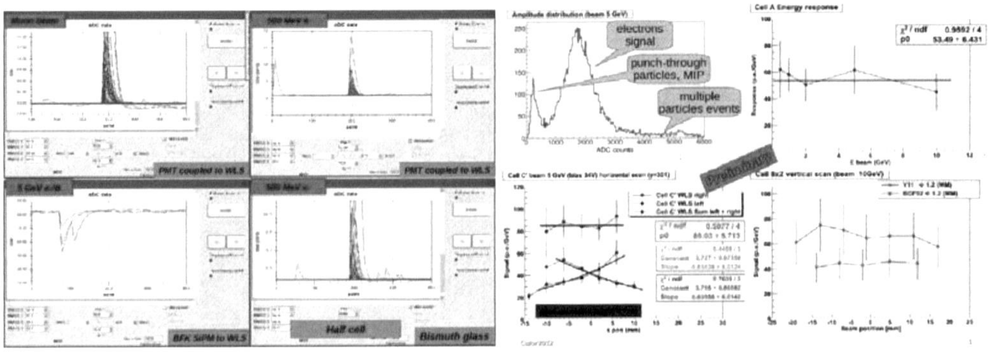

Figure 3. *Typical waveforms,the Čerenkov spectrum from various prototypes and the uniformity of Čerenkov response.*

Prototype #	Layout	Glass	gr/cm^3	L.Y	Notes
1	5 slices, machine grooved, unpolished, white	Schott SF57HHT	5.6	82	SiPM readout
2	5 slices, machine grooved, unpolished, white, v2	Schott SF57HHT	5.6	84	SiPM readout
3	5 slices, precision molded, unpolished, coated	Schott SF57HHT	5.6	55	15 cm long
4	2 slices, ungrooved, unpolished	Ohara BBH1	6.6	65	Bismuth glass
5	5 slices, scifi silver coated, grooved, clear, unpolished	Schott SF57HHT	5.6	64	15 cm long
6	5 slices, scifi silver coated, grooved, clear, unpolished	Schott SF57HHT	5.6	120	improoved version
7	10 slices, white, ungrooved, polished	Ohara PBH56	5.4	> 30	DAQ problems
8	10 slices, white, ungrooved, polished	Schott SF57HHT	5.6	76	
9	5 slices, wifi Al sputter, grooved, clear, polished	Schott SF57HHT	5.6	30	2 wls/groove
10	5 slices, ungrooved, polished	Schott SF57HHT	5.6	158	Small wls groove
11	2 slices, plain	Ohara experimental	7.5	–	DAQ problems

Table 1. Summary of light yields for ADRIANO modules for 2011-2012 test beams.

the glass slice. Coupling of fibers to SiPM's is critical as well; any thin air gap between the light concentrator and the active surface of an SiPM more than halves the light yield. Kuraray Y11 fibers produced almost 50% more light than Bicron BCF92 with the same diameter. Different optical glues, used to couple the glass to WLS fibers produced up to a factor of 2 in variation of light yield. This result should not surprise since we are attempting to collect light from an optical medium with considerably large refractive index $(1.8-2.2)$ by transferring the Čerenkov light to plastic fiber with far lower refractive index (typical refractive index of the external cladding of WLS fibers is 1.45). Cold vs hot glass construction methods make no appreciable difference in Čerenkov light yield. Direct reading of Čerenkov light from glass with an acrylic light guide coupled to the back of a cell yields consistently less light than when reading using the WLS fibers coupled longitudinally to the cell. This is a clear indication of the fact that, with thin layers of heavy glass having an absorption length of few centimeters, it is preferable to use light collection mechanisms based on WLS fibers distributed across the glass where the short path traveled by Čerenkov photons more than compensates the lower light collection efficiency. SiPM and PMT produce comparable signals, when their respective gains are taken into account. However, large noise from present generation of SiPM make them hard to use in low energy applications. This situation should quickly improve with the latest, low noise, generations of silicon devices.

6. Future Prospects

The first two year of ADRIANO *R&D* have already produced clear directions. Precision molding technique is, at present, the preferred fabrication technique since it has the potential of making quick (less than 30 minutes) glass slices with optical surface finish and with appropriate grooves. Nonetheless, we will keep exploiting other fabrication techniques as they might have other potential advantages when compared to precision molding. Photo-etching techniques, for example, are expected to be equally fast and cheap, although they present quite severe chemical hazards.

A great boost in ADRIANO development is currently obtained with Ohara sponsorship/partnership as they provided bismuth glass strips of commercial optical glass ($6.6gr/cm3$, $nd = 2.0$) and strips of an even denser experimental glass with density of $7.54gr/cm^3$ and refractive index of 2.24.

7. Conclusions

In this report we have presented the novel ADRIANO dual-readout technique along with preliminary results from several test beam at FTBF of eleven ADRIANO prototypes. The Čerenkov light yield we have obtained across the board is more than adequate for an hadronic calorimeter with an energy resolution of 30%/sqrt(E) or better. Our goal is to increase the light yield to a level that ADRIANO can be used as an electromagnetic and hadronic calorimeter at the same time. Hardware *R&D* and Monte Carlo simulations are in advanced state. More generally, current results from several test beams proved that Čerenkov light readout from heavy glasses with WLS collection system is feasible and provides equal or better results than traditional methods employing a large area PMT directly coupled to the glass. Correctly matching calorimetric techniques with SiPM and Front End Electronics is also crucial for a good performance of the detector. T1015 Collaboration will address these issues in the future and it will exploit new calorimetric techniques based on heavy glass, including scintillating heavy glasses.

8. Acknowledgments

We would like to thank Fermilab for its on-going support and for providing all the infrastructure for the construction and testing of ADRIANO prototype. We also would like to thanks Ohara for providing us with free samples of their BBH-1 bismuth based glass and of a new super-dense (about $7.5gr/cm^3$) experimental glass which have been used for the construction of ADRIANO prototype #6 and #11.

References

[1] *http : //www.linearcollider.org/ http : //clic − study.web.cern.ch/clic − study/*
[2] S. Buontempo et al., "Construction and test of calorimeter modules for the CHORUS experiment", Nucl. Instr. and Meth. A 349 (1994), p. 250.
[3] R. Wigmans, Calorimetry energy measurement in particle physics, in: International Series of Monographs on Physics, vol. 107, Oxford University Press, Oxford, 2000.
[4] *wwwppd.fnal.gov/FTBF/MOUpDF/T1015_mou.pdf*
[5] 4th Concept Collaboration - Letter of Intent from the Fourth Detector ("'4th"') Collaboration at the International Linear Collider, 2009 also at: http://www.4thconcept.org/4LoI.pdf
[6] N. Akchurin, et. al, "New Crystals for Dual-Readout Calorimetry", Nucl. Instr. Meth. A604(2007)512-526
[7] All DREAM papers are accessible at http://www.phys.ttu.edu/dream, and also http://highenergy.phys.-ttu.edu.particle physics, in: International Series of Monographs on Physics, vol. 107, Oxford University Press, Oxford, 2000.
[8] *http : //www.dmf.unisalento.it// danieleb/IlcRoot/*

XVth International Conference on Calorimetry in High Energy Physics (CALOR2012) IOP Publishing
Journal of Physics: Conference Series **404** (2012) 012031 doi:10.1088/1742-6596/404/1/012031

Development of a Gas Electron Multiplier -based Digital Hadron Calorimeter

A.White[1], J.Yu, S. Park

Physics Department, University of Texas at Arlington,
Arlington, Texas 76019, USA

awhite@uta.edu

Abstract. We describe the development of large-area (1m x1m), very thin, and high sensitivity gas detector based on gas electron multiplier (GEM) technology as a sensitive medium for sampling calorimeters for a use in future collider experiments.

1. Introduction

Experiments at future high energy colliders aim for high precision measurements of physics processes and exploration of new physics areas – electroweak symmetry breaking, the Higgs boson and its properties, the origin of mass, and searches for super-symmetry, extra space-time dimensions, and the emergence of unexpected new particles and forces. Much of this ambitious physics program requires high-precision measurements of jet energies and jet-jet invariant masses. This paper reports on the development of large area Gas Electron Multiplier (GEM) [1] technology as an important option for implementing calorimetry for future collider experiments.

A promising new approach to the calorimetric measurement of jet energies, the Particle Flow Algorithm (PFA) approach, is based on the observation that over 60% of the particles in jets are charged particles; therefore their momenta can be measured with high precision in the detector's tracking system. Recently, the PFA approach to jet-energy measurement has received considerable attention; it has been the subject of development by CMS [2] at the LHC and two of the major ILC detector design groups [3,4] and for CLIC[5].

Using small calorimeter cells, typically a few cm^2, makes it possible to track charged particles in the calorimeter and associate their energy deposits with the corresponding tracks from the tracking system. After removal of this charged-particle associated energy from the calorimeter information, the remaining energy is measured using the residual hits in the calorimeter, namely those without associated incoming charged particle tracks. Our studies have shown that there is an approximately linear relationship between the number of digital hits, associated with a track producing a calorimeter cluster, and the cluster's energy [6] as shown in Fig.1. Using digital hadron calorimetry, in which a cell is either on or off, but no

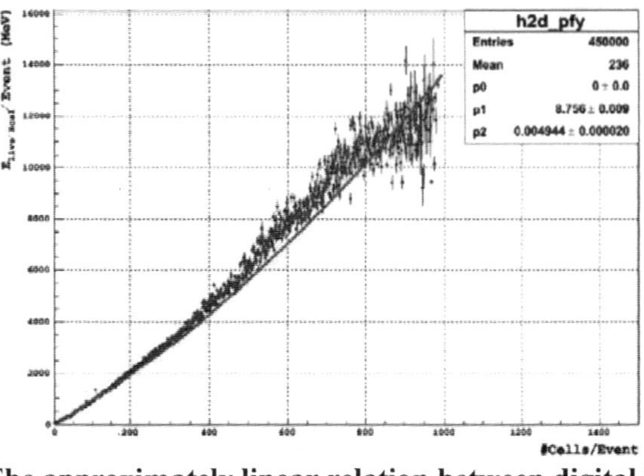

Figure 1 The approximately linear relation between digital hits and energy for a simulated GEM-based digital hadron calorimeter.

analog signal measurement is made, along with this new PFA approach, an overall hadronic (jet) energy resolution of about 3 – 4% for a range of jet energies up to 500 GeV should be achievable. We note that for 100 GeV this corresponds to $30\%/\sqrt{E}$.

Our proposed digital hadron calorimeter comprises a stack of steel absorbers, of sufficient thickness to contain hadronic showers, interlaced with active gaseous sampling elements. A primary concern is to keep the active layer thickness small to minimize lateral spread of the showers and as it is repeated typically 40 times through the limited depth of the calorimeter stack which could be located inside a high-cost superconducting magnetic coil in lepton collider experiments. In order to take full advantage of the PFA approach, it is of critical importance to minimize contributions to the energy resolution due to confusion in track-energy cluster matching. Accomplishing this requires a calorimeter that can provide excellent matching between energy clusters and tracks whose momenta are measured in the tracking detector. This means a small cell size and fine readout granularity for which GEM offers a promising solution. The basic layout of a GEM-based digital hadron calorimeter is shown in Fig. 2, which also shows details of our double-GEM structure.

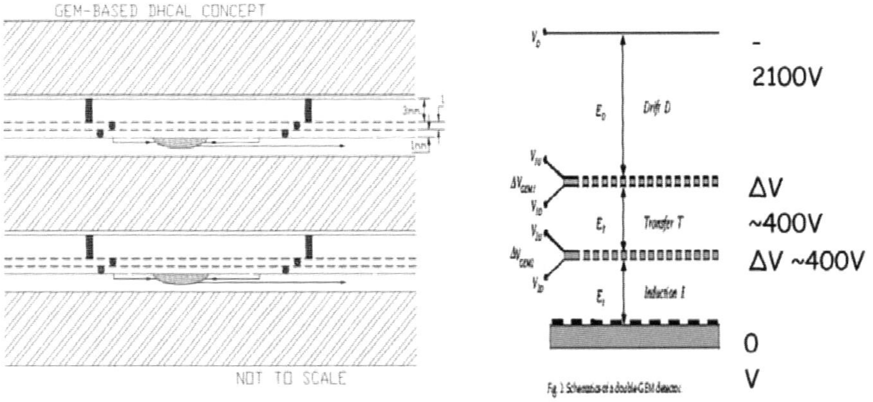

Figure 2 (left) Principle of GEM DHCAL implementation using GEM layers, (right) Details of double-GEM structure used in each DHCAL layer

2. Prototype Development

We have built and tested a series of GEM chamber prototypes, using 30cm x 30cm foils first from 3M Corporation, and later from CERN. Figure 3 shows a prototype chamber with an 8 x 8 array of 1 cm^2 anode pads. We use an 80/20 Argon/CO_2 gas mixture, which gives a gain of up to 11,000 from the double foil arrangement. Figure 3 also shows the results of exposing the chamber to an ^{55}Fe source. The characteristic double-peak spectrum can clearly be seen for each cell.

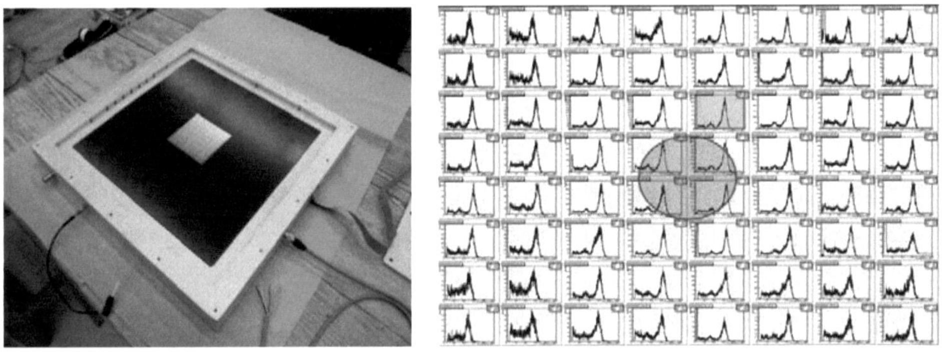

Figure 3. (left) GEM-DHCAL prototype chamber with 8x8 anode pad array, (right) ^{55}Fe spectrum in each channel – red circle indicates source position.

For the readout of the pad signals we have used three approaches. In early tests, and to establish a baseline for each new chamber, we have used a single channel analog chip. For multi-pad readout we have used two chips, the analog KPiX chip [7] from SLAC, and the digital. DCAL chip [8] from ANL. One anode pad is connected to one channel on a chip. The KPiX chip has a four-deep pipeline, on-board DAC charge injection calibration, and a Wilkinson 13-bit ADC for each channel. The digitized signals are readout via an FPGA then sent to a PC via a USB connection. The DCAL chip outputs a digital result by comparing the signal from a pad with a pre-downloaded threshold. The collected digital results are read out via a data concentrator to a VME system. Since a final digital hadron calorimeter system would have about 50 million pixels, there is a critical need to have a high-density readout scheme.

In the process of constructing prototype chambers using the $30cm \times 30cm$ foils from CERN GDD, we encountered gains smaller than anticipated. We have learned that the standard foil production technique at present focuses more on the stability of the detector than higher gains. Since for DHCAL, it is necessary to maintain the detector as thin as possible, we use a double-GEM layer configuration instead of triple-GEMs. This requires the gains from each of the GEM layer to be as high as possible without damaging the foil surfaces from frequent discharges. To remedy this issue, we requested CERN to produce GEM foils with higher gain and we measured the gains from detectors with the two different sets of GEM foils. Since the two sets of GEM foils show dramatic differences in gains, we decided to construct $30cm \times 30cm$ prototype detectors for beam test using the higher gain GEM foils.

3. Test beam results

We constructed one $30\text{cm} \times 30\text{cm}$ chamber with the 13-bit KPiX9 readout board with 64 active channels in the center and three chambers with DCAL boards that have a center $16\text{cm} \times 16\text{cm}$ active area. We exposed these four chambers to particle beams in August 2011 at the Fermilab Beam Test Facility (FTBF)[9]. We took a total of 7 million events with 32 GeV muons and pions, as well as 120 GeV protons. The data were taken to measure chamber responses, gains as a function of high voltage across each GEM foil, position dependence of the response and threshold dependences of efficiencies. In addition, we took a 32 GeV pion shower run with 8-inch steel bricks causing the pions to shower in front of the detector array.

Figure 4 (left) shows a photograph of our test beam set up with the stack of 8-inch steel bricks in front of the detector array that causes pion showers. Beam was incident from the left hand-side of the photo.

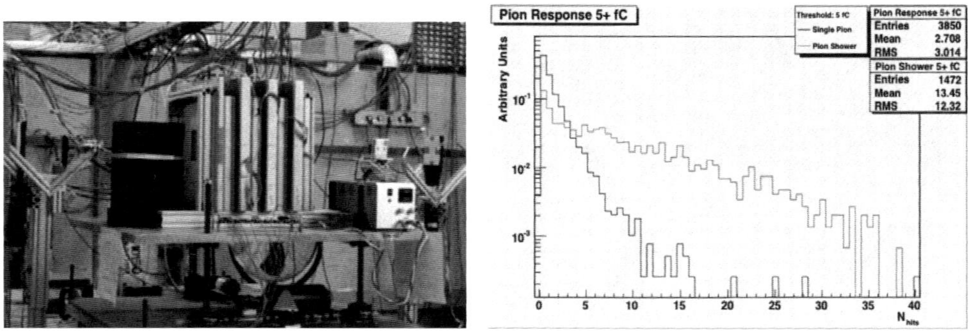

Figure 4. Four GEM chamber setup at Fermilab Test Beam Facility.

Figure 4 (right) shows the distributions of the numbers of hits from the single pions (black) and the pion shower (red) above 5fC measured in the 13 bit KPiX chamber. One can clearly see the dramatic difference between these two sets of data. This indicates that the chambers and the electronics can handle multiple simultaneous hits. A similar trend is seen for the three DCAL chambers as well, depicting progressively broadened hit patterns in the chambers as the shower propagates downstream.

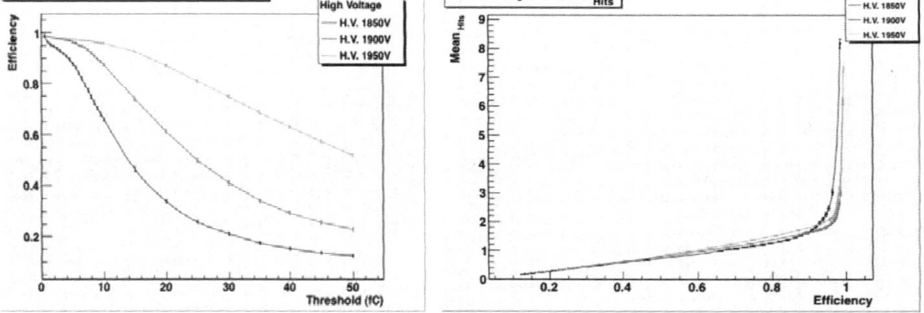

Figure 5. (left) Measured GEM chamber efficiency above threshold for various high voltages. (right) Chamber efficiency vs. number of cells hit.

We measure a chamber gain of about 11,000 at an operating voltage of 390V across each GEM foil, consistent with other previous results. Figure 5 (left) shows the chamber efficiency as a function of threshold for three different high voltages across the chamber. The preliminary chamber efficiency is measured to be 98% with 2fC threshold.

Figure 5. (right) shows how the number of hit cells for one incident charged muon correlates with chamber efficiency. The figure indicates that for an efficiency of ~95% there is on average only a 50% chance of an extra hit. This would be a good operating point for a calorimeter system designed for particle flow.

4. Development of Large Area Chambers

The next phase in our work is the development, characterization and construction of full scale $1m^2$ GEM chambers for DHCAL. We are developing the mechanical structure, the electronic readout board schemes and the schemes for integrating the three unit chambers ($33cm \times 100cm$) into one $1m \times 1m$ plane. We plan to construct two of these $33cm \times 100cm$ unit chambers in the next year. The team has been working with the CERN GDD workshop [10] in developing $33cm \times 100cm$ GEM foils. The foils have been made using a new single-sided-etching approach, which eliminates some of the manufacturing problems associated with side-to-side registration in the standard double-sided-etching process. The first set of five foils has been produced and delivered to UTA. Testing and qualification of these foils were carried out in a systematic study by UTA students. In order to test and qualify these foils, the resistance of each of the 31 independent HV strips on each foil was measured using the Keithley6145 electrometer. The instrument measures the resistance by applying constant amount of current 1nA and measure the voltage difference across the top and bottom of the foil. The saturation voltage corresponds to a resistance of 260GΩ. A foil is considered qualified if all strips have the saturation voltage, indicating the resistance of the strips greater than 260GΩ.

We have completed the design of the first $33cm \times 100cm$ chamber whose schematic diagram is shown in Fig. 6 (left). In order to avoid thick side walls for the chamber resulting from tensioning the large foils, we will use a 2mm steel strong-back on which to mount the chamber. In a final calorimeter configuration, this would be part of the absorber structure between two active layers. The required separation of the cathode, foils and anode layers is achieved by the use of thin spacers, also made for us by the CERN GDD Workshop.

Fig. 6 (right) shows a 33cm x 100m chamber under construction at UTA. Once we have assembled, tested, and successfully operated a $33cm \times 100cm$ chamber, we will proceed to construct and test a full $1m \times 1m$ plane. A schematic view of such a plane is also shown in Fig. 6 (right).

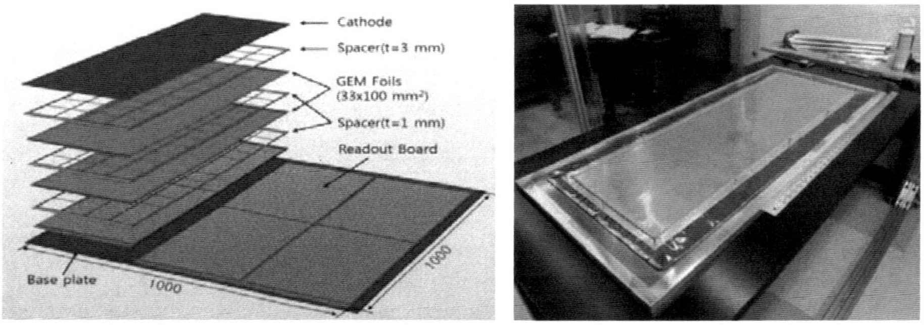

Figure 6. (left) Schematic view of a 33cm x 100cm chamber and extension to a $1m^2$ chamber, (right) a 33cm x 100cm chamber under assembly.

We will build 6 additional $33cm \times 100cm$ unit chambers for two $1m \times 1m$ planes and characterize using long-term cosmic rays tests. We will then integrate three $33cm \times 100cm$ chambers with DCAL chip based anode boards for assembly of each final $1m \times 1m$ DHCAL sensitive gap plane. As funding allows, we will construct nine more $33cm \times 100cm$ unit chambers for three additional $1m \times 1m$ chamber planes using DCAL readout boards. In the long term these chambers will allow us to construct a total of fifteen $33cm \times 100cm$ unit chambers for five DCAL $1m \times 1m$ chamber planes. These five planes will then be inserted into the existing CALICE calorimeter stack and run together with the Si/W ECAL and 35 RPC planes in the HCAL.

References

[1] F. Sauli, GEM: A new concept for electron amplification in gas detectors, Nucl. Inst. Meth. A386, 531 (1997); R. Bouclier, *et al.*, The Gas Electron Multiplier (GEM), IEEE Trans. Nucl. Sci. NS-44, 646 (1997).

[2] CMS Experiment, Compact Muon Solenoid Exp., Spokesperson Dr. J. Incandela, http://cms.cern.ch/

[3] SiD concept collaboration, H. Weerts, J. Jaros and A. White (Spokespersons) http://www-sid.slac.stanford.edu/; SiD Detector LOI, https://confluence.slac.stanford.edu/display/SiD/ LOI

[4] International Large Detector (ILD) concept collaboration, http://ilcild,org; ILD Lettor of Intent, http://ilcild.org/documents/ild-letter-of-intent.

[5] CLIC Linear Collider Detector (LCD) group, https://twiki.cern.ch/twiki/bin/view/CLIC/ Detector

[6] Venkat Kaushik, Masters thesis, UTA 2008, unpublished.

[7] D. Freytag et al. SLAC-PUB-13462, Dec 11, (2008).

[8] G.Drake et al., Nucl.Instr. and Method. A578 (88) 2007; Conceptual design of the readout system for the Linear Collider Digital Hadron calorimeter prototype, http://www.hep.anl.gov/repond/ dcal_sys_desc_113.pdf

[9] Fermilab Test Beam Facility, http://www-ppd.fnal.gov/FTBF/

[10] CERN Gas Detector Development workshop, http://gdd.web.cern.ch/GDD/

Journal of Physics: Conference Series **404** (2012) 012032 doi:10.1088/1742-6596/404/1/012032

Belle2c Calorimeter Upgrade (Status and Plan)

Isamu Nakamura (中村 勇)
KEK, 1-1 Oho, Tsukuba, Ibaraki 305-0801, Japan
E-mail: isamu.nakamura@kek.jp

Abstract. Upgrade of KEKB Accelerator and Belle2 Detector are on going, aiming for starting physics data taking with 50 times higher instantaneous luminosity in FY2015. The belle electromagnetic calorimeter consists of CsI(Tl) crystals read out by two photodiodes, covering polar angle of 12-155 degrees. A total of 8736 counters are installed in a barrel and two endcap modules. The Belle2 upgrade of electromagnetic calorimeter is in two steps. For the beginning of Belle2, all readout electronics will be replaced to cope with higher event rate. A waveform sampling will be introduced to reduce an effect from severe accelerator background. Replacement of CsI(Tl) crystals to faster undoped CsI crystals are also planned. In the talk, detail of Belle2 upgrade will be presented as well as status of preparation and plans for crystal replacement.

1. SuperKEKB and Belle2

After successful operation, the KEKB asymmetric e^+e^- collider and the Belle Detector [1] accumulated data corresponding to an integrated luminosity of 1 ab^{-1}. The experiment stopped data taking by the end of June 2010 and started to improve its performance. The SuperKEKB, successor of KEKB accelerator aims for 50 times more luminosity by squeezing the beam to about 1/20 of current KEKB and by putting twice more current. With SuperKEKB accelerator, trigger rate of Belle experiment is expected to reach as high as 20–30 kHz. Beam related background is also expected to increase. Requirement for the Belle2 detector is to have same or better performance compared to the Belle detector under 10–20 times more background, though it is very difficult to make an accurate estimation of beam background without actual accelerator operation. In the Belle2 detector upgrade, completely new tracking detector consists of two layers of DEPFET Pixel detector, four layers of Double-sided Silicon Strip detector and small cell drift chamber will be installed. Particle ID devices will also be completely new. For barrel region, Time-Of-Propagation Čerenkov detector replaces current TOF and Aerogel Čerenkov detector while proximity focusing Aerogel-RICH detector will be installed in the forward endcap. For calorimeter crystals and preamplifiers are kept but all readout electronics will be replaced. The solenoid magnet as well as the barrel muon and K_L^0 detector(KLM) will be kept. The RPC of the endcap KLM will be replaced with Scintillator strip detector. All DAQ electronics and trigger system will also be renewed. The detail of SuperKEKB/Belle2 upgrade can be found in elsewhere [2, 3].

2. The Belle Calorimeter

The main purpose of the electromagnetic calorimeter is to detect γ and π^0 from B meson decay. Energy of such γ ranges from MeV to GeV. Energy resolution is very important because for some analyses one has to separate signal and background events only with mass distribution. To separate two photons from high momentum π^0, fine granularity is required. Hermeticity is also very important to study final states with neutrino. The Belle electromagnetic calorimeter consists of 8736 CsI(Tl) crystals with typical dimension of $6\times6\times30$ cm^3. The length corresponds to about 16 X_0. Figure 1 shows the schematic view of the calorimeter. The calorimeter has pointing geometry and covers polar angle from 12° to 155°. The scintillation light of the CsI(Tl) crystal has λ = 560 nm with rather slow decay constant of 1.3 μsec but abundant light output of about 50000 p.e./MeV. The light is read out by two independent sets of silicon PIN photodiodes (Hamamatsu S2744-08) with sensitive area of 10×20 mm^2 connected to charge sensitive preamplifier. The signals are sent to the front-end electronics with cable of typically 10 m long. Two signals from preamplifiers are first added and then processed with two shaper with 0.2(fast) and 1μsec(slow) shaping time. The fast signal are used to generate trigger signal while the slow signal is then sent to LeCroy MQT300A charge-to-time converter. The MQT converts 3 range/12 bit charge information into 15 bit timing information which corresponds to 18 bit dynamic range. The time encoded energy information is then recorded with LeCroy 1877S Fastbus TDC. The readout system consists of about 100 TDC boards in five crates read out by three VME systems. It takes about 30 μsec to read out one physics event. The trigger rate is about 300 Hz with instantaneous luminosity of 10^{34} cm^{-2}sec^{-1}, which corresponds to about 1% readout deadtime. With 10 kHz trigger rate deadtime is 30% that is not acceptable. Although the signal is recorded with TDC, only energy is recorded. No event timing information is available.

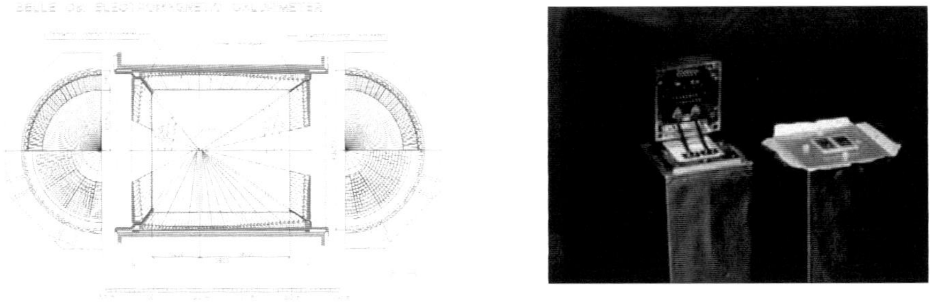

Figure 1. (Left) The Belle Electromagnetic Calorimeter. (Right) A counter.

3. Belle2 Calorimeter Upgrade

In the Belle2 experiment trigger rate will be as high as 30 kHz and background will be 10 times as high as Belle. To cope with such situation waveform sampling together with pipelined readout will be introduced. The former enables to use timing information to discriminate off-timing hits while the latter can eliminate effectively readout deadtime. Figure 2 shows simulated timing distribution.

To realise above readout electronics are replaced to new one. Figure 3 shows the schematic diagram of new readout system. Hart of the system is Shaper Digitizer board which receives preamplified signals from 16 crystals. Two preamplifier signals from one crystal are summed and processed with shaper with time constant of 0.5μs which is shortened from 1.0μs. Reducing shaping time further is not efficient as the time constant of scintillation is 1.3 μs. Signal after shaping is digitized by 18 bit ADC with 1.8 MHz sampling rate and sent to pipeline buffer in FPGA which is 512(290μs) deep. Once trigger is issued, waveform fitting is performed in FPGA

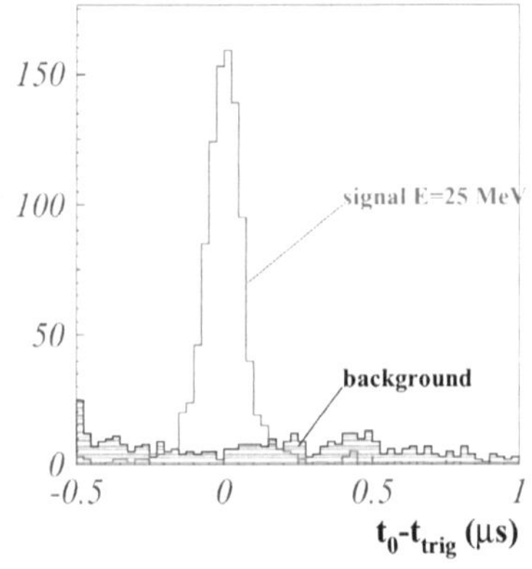

Figure 2. Expected timing distribution with new electronics.

using 16 samples to extract timing and amplitude. According to simulation study a factor of seven reduction of fake cluster is expected with upgraded readout system.

To readout all 8736 crystals, total of 576 Shaper Digitizer boards are placed in 52 VME crates. In each VME crate a Collector board will be installed. The Collector board receives trigger signal, distribute it to Shaper Digitizer boards in the crate, receives fitted timing and amplitude and sends them to backend DAQ system via Belle2 standard optical link. Introducing Collector board greatly reduce the number of cables from detector to electronics hut. The Collector board can also supply calibration signal.

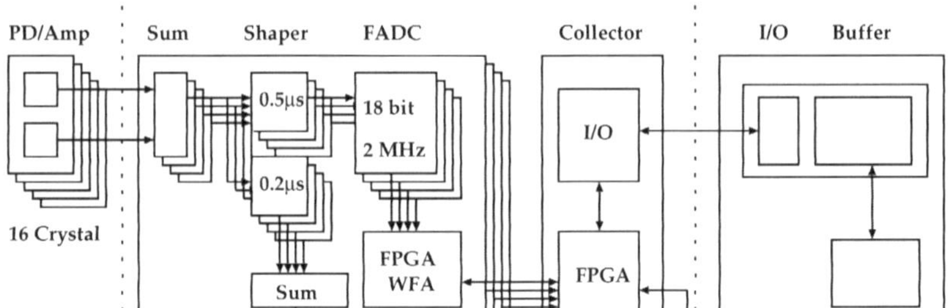

Figure 3. Schematic Diagram of new DAQ system.

4. Status

4.1. Detector

After the runend, Belle detector was rolled out from the accelerator and was dismantled for the preparation of upgrade. All barrel sub-detectors inside barrel calorimeter as well as the endcap calorimeters(Figure 4) were taken out. After dismantling, the big earthquake happened on Mar.11. Since the detectors were already taken apart, there's no major damages visible from outside. Since the crystals, photodiodes and preamplifier circuit will be used in Belle2, test was performed to make sure the counters are functional. Since the old readout system were already removed from the detector, an older prototype shaper amplifier module(Figure 4) was used. The module has the same shapers and preamplifiers as the belle readout system, can send calibration signals to the preamplifiers as well as driving pulse to LED flasher installed in the counter. Using the module about 900 counters corresponding to one eighth of all barrel channels were tested and found to be all functional. The test of all barrel channels and endcap channels will be performed after a part of new readout system become ready.

Figure 4. (Left) Backward endcap calorimeter taken out from the Belle. (Right) The old prototype readout module.

4.2. Readout System

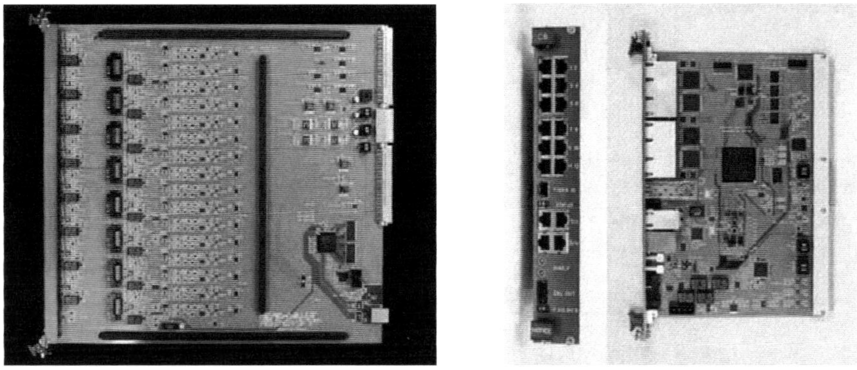

Figure 5. (Left) Shaper Digitizer prototype. (Right) Collector prototype.

Recently 12 "final" prototype shaper digitizer boards(Figure 5 Left) and a second prototype collector module(Figure 5 Right) were delivered. It enable us to perform "one crate test" with configuration similar to the actual experiment. These prototype shaper digitizer boards are tested using either calibration pulse distributed from the prototype collector module or cosmic ray signals from the counters. Figure 6(Left) shows the distribution of amplitude from pedestals. Noise of the read out were measured to be ~300 keV and ~20 keV for incoherent and coherent noise, respectively. Figure 6(Center) shows the correlation of input charge and measured amplitude and Figure 6(Right) shows non-linearity. The non-linearity of ~0.3% was obtained. Both noise and non-linearity measured in the prototype modules fulfill our requirement.

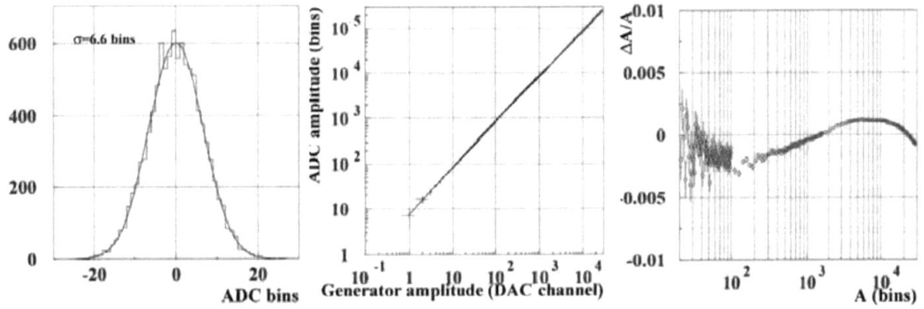

Figure 6. (Left) Pedestal distribution. (Center) Measured amplitude as a function of input charge. (Right) Non-linearity of the board.

Currently 100 barrel shaper digitizer boards are being produced and will be delivered on March 2013. Our plan is to produce all 432 barrel boards and all 52 collector boards by the end of March 2014, and to start installation in 2014. The endcap part will be ready by early 2015.

References

[1] A. Abashian *et. al.*, Nuclear Instruments and Methods A479 117(2002)
[2] SuperKEKB Letter of Intent, KEK Report 04-4
[3] Belle II Technical Design Report in preperation.

Gamma rays, electrons and positrons up to 3 TeV with the Fermi Gamma-ray Space Telescope

Philippe Bruel, on behalf of the Fermi-LAT collaboration
Laboratoire Leprince-Ringuet, Ecole polytechnique, CNRS/IN2P3, Palaiseau, France
E-mail: Philippe.Bruel@llr.in2p3.fr

Abstract. The Fermi Gamma-ray Space Telescope (formerly known as Gamma-ray Large Area Space Telescope, GLAST) was successfully launched on June 11 2008. Its main instrument is the Large Area Telescope (LAT), which detects gamma rays from 20 MeV to more than 300 GeV. It is a pair-conversion telescope with 16 identical towers (tracker and calorimeter), covered by an anti-coincidence detector to reject charged particles. The calorimeter is a hodoscopic array of CsI(Tl) crystals, arranged in 8 alternating orthogonal layers, with a total thickness of 8.6 radiation lengths. In this paper we will present the performance of the LAT, with special attention to the calorimeter, which provides a good energy measurement up to 3 TeV. We will also review some of its scientific results after 4 years of operation, focusing on measurements which extend up to very high energy, such as the spectrum of the extragalactic diffuse emission, the spectrum of cosmic electrons and the positron fraction.

1. Introduction

The main scientific goal of the Fermi Large Area Telescope [1] is to study the gamma-ray sources from 20 MeV up to more than 300 GeV. There is a large variety of such gamma-ray sources, from galactic sources (pulsars, pulsar wind nebulae, supernova remnants, microquasars) to extragalactic sources (active galactic nuclei and gamma-ray bursts). After 3 years of operation, the LAT had detected more than 1800 gamma-ray sources. On top of individual sources, we also seek to measure the galactic diffuse emission and the extragalactic diffuse emission.

The point source sensitivity is mainly the result of two characteristics of the telescope: its angular resolution and its total acceptance (the effective area times the field of view). The angular resolution drives the ability of finding a source on top of a relatively flat background. The total acceptance determines the number of gamma rays that can be detected from a given source. A large effective area is important, but a large field of view is also very important: it allows the telescope to observe a large portion of the sky at the same time, so that the total amount of time during which a given source is seen, is a large fraction of the total observational time of the telescope. Achieving the best angular resolution and the largest total acceptance were the main goals of the design of the LAT.

Regarding the energy measurement, the LAT was optimized to have the largest energy range: down to about 20 MeV and up to 300 GeV and more. In this energy range, gamma-ray sources generally have a smooth power-law energy spectrum with a photon index about -2. Thus measuring their energy spectra only requires a moderate energy resolution. But, since the spectra fall strongly with energy, any significant migration of reconstructed energy toward higher energies would easily distort the energy spectrum measurement. So the energy reconstruction must minimize the risks of overestimating the energy. Though most of the gamma-ray sources have a smooth energy spectrum, some, like pulsars and active galactic nuclei, have energy cutoffs or breaks that we want to measure as precisely as possible. We also want to reach the best sensitivity to gamma-ray lines that could result from dark matter annihilation. As a consequence, achieving the best energy resolution is also important for the LAT.

2. From EGRET to the LAT

In order to detect gamma rays above 10 MeV and measure their directions and energies, a space-based detector needs three sub-detectors: a tracker, in which gamma rays convert to pairs, to measure their directions, a calorimeter to measure their energies and an anti-coincidence detector (ACD) to reject charged particles. The tracker is made of some passive material in which the conversion occurs and some active material that tracks the e^+e^- pair. The effective area depends on the fraction of converted gamma rays, which increases with the amount of converter. But below $\sim$ 10 GeV, the angular resolution is limited by multiple scattering, which also increases with the amount of converter. So a compromise must be found between increasing the effective area and degrading the angular resolution.

The predecessor of the LAT, EGRET [2], which operated between 1991 and 2000, had a tracker made of spark chamber modules interleaved with tantalum foils. Its volume was $80 \times 80 \times 45$ cm^3 and the total on-axis depth was 0.54 radiation lengths, leading to an on-axis conversion efficiency of about 30%. Between the tracker and the calorimeter, there was a time of flight coincidence system. So they were separated by $\sim$ 80 cm. As a consequence, the field of view of EGRET was limited to about $\sim 25^o$. The calorimeter of EGRET was a $77 \times 77 \times 20$ cm^3 block of NaI, without segmentation. The total on-axis depth was about 8 radiation lengths. Its energy resolution was $\sim$ 9% from 100 MeV to 10 GeV. The ACD of EGRET was a monolithic dome-shaped scintillator. At high energy, the electromagnetic showers in the calorimeter produce backsplash, which can produce some signal in the ACD, which mimics charged cosmic rays. So the maximum energy reached by EGRET was limited to about 20 GeV for two reasons: because the calorimeter was not segmented (preventing a good energy measurement) and because the ACD was not segmented (no ability to reject false vetos from backsplash).

2.1. The Large Area Telescope

The LAT was designed to overcome EGRET's limitations. It is composed of 16 identical towers arranged in a 4×4 array, with each tower made of a tracker module on top of a calorimeter module. The overall surface of the LAT is 150×150 cm^2. A tracker module consists of 18 silicon (x,y) planes. The first 16 planes are interleaved with a tungsten foil, which thickness is 0.03 radiation lengths for the first 12 planes and 0.18 radiation lengths for the following 4 planes. The total on-axis depth is 1.5 radiation lengths, leading to an on-axis conversion efficiency of about 70%. The height of the tracker is 60 cm and there are 5 cm between the tracker and the calorimeter. So the overall aspect ratio of the LAT is much smaller than that of EGRET and its field of view is about 60°. Each calorimeter module is an hodoscopic array of 8×12 CsI(Tl) crystals, arranged in 8 layers, each layer being 19.9 mm thick. The total on-axis depth is 8.6 radiation lengths. Despite of its modest thickness, we will see in section 3 that the calorimeter segmentation allows us to measure gamma rays up to 3 TeV with good resolution. The ACD is segmented in 89 plastic scintillators tiles in order to minimize its sensitivity to backsplash. Thanks to its calorimeter segmentation and its ACD segmentation, the LAT is able to detect and measure gamma rays up 3 TeV, much higher than EGRET.

2.2. The LAT calorimeter

Each calorimeter module is made of 8 layers with 12 crystals per layer. Each crystal is a $326 \times 26.7 \times 19.9$ mm^3 block of CsI(Tl). The orientation of the crystals depends on the layer:

along x for the first layer, along y for the second, and so forth. This geometry allows for full imaging of gamma-ray showers. The crystal height is 1.07 radiation lengths and its width is 0.75 Moliere radii. This thin segmentation gives 2 coordinates of the energy deposition in a crystal. The third one, the longitudinal position along the crystal, is given by the asymmetry between the signals read out at each end.

In order to have a large energy range, each crystal end is readout by two PIN photodiodes (25 mm^2 and 147 mm^2) with two gains per diode, leading to four ranges (2 MeV to 100 MeV, 2 MeV to 1 GeV, 30 MeV to 7 GeV, 30 MeV to 70 MeV). During normal data taking, the lowest non-saturated range is chosen. This very large dynamic range (2 MeV to 70 GeV) allows us to measure very low energy deposits (low energy gamma rays, minimum ionizing protons or ground muons) while avoiding crystal saturation for gamma rays up to $\sim$ 1 TeV. After 4 years of operation, there is only one dead diode out of 6144.

The in-flight calibration of the crystals is done in several steps. We use a charge-injection system to correct for electronics non-linearities. The lowest energy range is calibrated with the minimum ionizing cosmic proton peak at 11 MeV. The inter-range calibration is performed using non-interacting cosmic protons and heavy nuclei. The latter are also used to calibrate the asymmetry between the two ends, with the help of the tracker from which the reconstructed direction can be extrapolated to the crystal to give the longitudinal position reference. The observed crystal light yield attenuation due to radiation damage is about 1% per year.

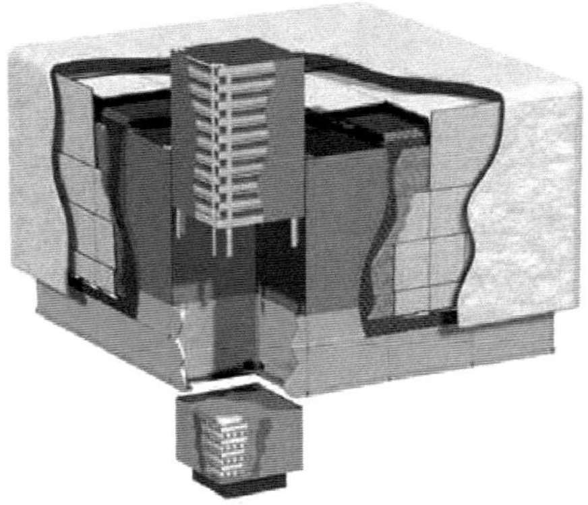

Figure 1. Schematic diagram of the Large Area Telescope.

Figure 2. Exploded view of a calorimeter module showing the hodoscopic 12×8 array of CsI detectors in the carbon fiber structure.

2.3. Event reconstruction and selection

The goal of the LAT is to reconstruct and select good photon candidates. We use the tracker to find the direction of the event and the conversion point. The energy measurement relies on both the tracker and calorimeter. Very low energy gamma rays do not reach the calorimeter and their energy is determined from the number of hits in the tracker. For higher energies, we correct the energy deposited in the calorimeter for the various losses by using the position of the shower maximum or by performing a fit of the longitudinal profile of the shower, as fully described in section 3.

The direction information is used to extrapolate the trajectory of the particle back to the ACD in order to reject charge particle background. This information, in addition to the topology of the shower in the tracker and in the calorimeter, is used for the selection of the gamma-ray candidates. The reconstruction and selection are based on classification trees and have been optimized using a detailed GEANT4 simulation. The performance of the instrument has been carefully checked in beam tests and on orbit [3]. The on-axis effective area is 0.25 m^2 at 100 MeV, 0.7 m^2 at 1 GeV and 0.8 m^2 at 100 GeV.

The rate of cosmic-ray background passing through the LAT can be as high as 10 kHz. Since the characteristic signal shaping times of the tracker, calorimeter and ACD are respectivelly 10, 3.5 and 4 μs, such a rate implies that the probability that particles pass through the LAT a few μs before a gamma ray triggers the detector, can be higher than 10%. These out-of-time particles can leave sizable signals in all the three subdetectors. These spurious signals can cause serious errors in the event analysis and thus must be cleaned by the event reconstruction. In the calorimeter this is done by running a clustering algorithm and finding the clusters related to the real event.

3. Energy reconstruction from 3 GeV up to 3 TeV

The LAT phase space is very large: energies from 20 MeV up to more than 300 GeV and angles up to 60 deg with respect to the vertical axis of the detector. Because of the geometry of the instrument, the fraction of energy deposited in the calorimeter varies very much throughout this phase space. For normally incident photons, the fraction of energy deposited in the calorimeter is maximum at 1 GeV and is about 80%. Above 1 GeV, leakage from the calorimeter starts to be the main cause of energy loss. Another cause of energy loss is the 4 cm gaps between the towers. Here we describe an energy reconstruction method that works for photons above ~ 3 GeV, when the energy deposited in the tracker is small compared to that in the calorimeter, and where the leakage from the calorimeter can be large.

The longitudinal segmentation of the calorimeter allows a fit of the shower longitudinal profile, which is a good way to correct for energy loss when the shower is not fully contained in the calorimeter. In order to be able to perform such a fit, we must predict the energy deposited in each layer for any longitudinal profile. At normal incidence, the layers are perpendicular to the shower axis and the prediction is obvious. But at large incidence angle, the energy contained in a slice of the shower at a given age is shared among several layers or part of it escapes the calorimeter. As a consequence, for a given shower, we have to model accurately its development through the layers of the calorimeter, taking into account the gaps between towers.

The crystal electronic signals saturate at about 70 GeV. This saturation, which occurs for photons above 1 TeV, must be taken into account by the shower fit. When a crystal is saturated, we do not know how much energy has been deposited in it, so it cannot be used in the shower fit. Since the saturated crystals lie along the core of the shower, the only information we have comes from the transverse tails of the shower, which contain less than 10% of the energy. That means that, in order to provide a useful energy resolution above 1 TeV, we have to be able to model very accurately the development of the shower through the calorimeter at the crystal level, and paying close attention to those crystals that lie at more than one Moliere radius from the shower axis.

As a consequence, the shower longitudinal profile fit is based on the modelling of the longitudinal and transverse profiles of electromagnetic showers and on the modelling of the development of the showers through the LAT calorimeter.

3.1. Shower longitudinal profile parameterization

The average longitudinal shower profile can be described by a gamma function [4]:

$$\left\langle \frac{dE(t)}{dt} \right\rangle = P_{(E,\alpha,\beta)}(t) = E \times \frac{(\beta t)^{\alpha-1}\beta e^{-\beta t}}{\Gamma(\alpha)} \tag{1}$$

where t is the longitudinal shower depth in units of radiation length, α the shape parameter and β the scaling parameter. The position of the maximum of the profile is $T = (\alpha - 1)\beta$. As shown in [5], the same function can be used to describe individual profiles. In order to model the longitudinal profile, we have to determine how the distributions of α and β vary with energy.

In order to do so, simulations of photons in a very large CsI calorimeter with GEANT4 at various energies have been performed. The calorimeter is segmented in 400 layers of 1.85 mm = 0.1 radiation length. Its total depth is 40 radiation lengths so photons are fully contained. For each event, the conversion point is used to define the shower starting point $t = 0$. In order to smooth the longitudinal profile, 0.5 radiation length bins were used to construct the histogram of the longitudinal profile, which is fitted with equation (1).

Fig 3 shows the distributions of α and β and their logarithm for 10 GeV photons. The distribution of $\ln\alpha$ is more gaussian than α, whereas the distribution of β is more gaussian than $\ln\beta$. That is why we decide to use $\ln\alpha$ and β.

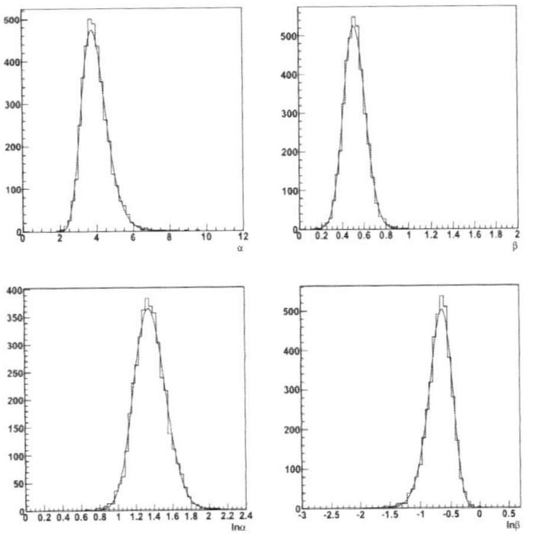

Figure 3. Distributions of α and β and their logarithms for 10 GeV photons.

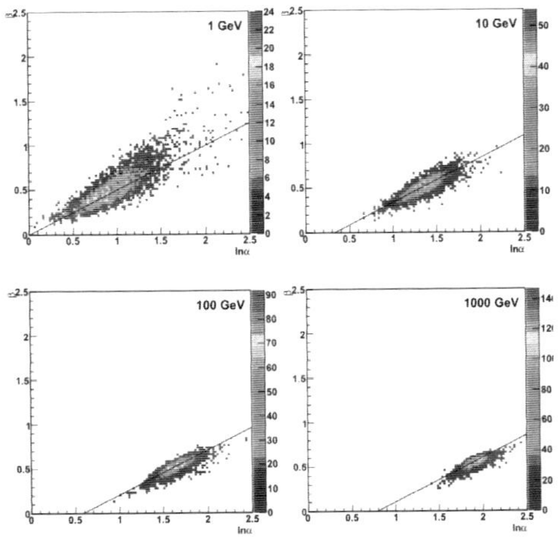

Figure 4. Correlation between $\ln\alpha$ and β for 1,10,100 and 1000 GeV photons.

Fig 4 shows the correlation of these two variables for various energies. The correlation is independent of energy with a slope of $\tan\theta_c = 0.5$. As it is preferable to use uncorrelated variables, we use the following variables:

$$\begin{aligned} S_0 &= \ln\alpha\cos\theta_c + \beta\sin\theta_c \\ S_1 &= -\ln\alpha\sin\theta_c + \beta\cos\theta_c \end{aligned}$$

Fig 5 shows the distributions of S_0 and S_1 for various photon energies. The last step of the description of the shower parameters S_0 and S_1 is the parameterization of the variation of their mean and RMS with logE, as shown in Fig 6. The dependence of the mean can be simply parameterized with a quadratic polynomial. The dependence of the RMS is quadratic below ~ 30 GeV and linear above.

In order to determine the error of the modeling of the shower profile, that is to say the error that will be used in the calculation of the χ^2 during the fit, we use the residuals of the shower profile fit as a function of the shower depth t. The RMS of these residuals are almost flat with a broad maximum at $0.5 < t/T < 1.5$. As a consequence we assume that the error does not depend on the shower depth and we use the same modeling error for all layers. We define ϵ_T as the RMS of the residuals at shower maximum divided by the energy deposited at shower maximum. Again ϵ_T varies with energy. Its variation can be parameterized by $\epsilon_T(E) = 0.17e^{-\log E/1.38}$, with E in GeV. During the fit, the error used in the χ^2 calculation is $\delta e = \epsilon_T(E) \times \max(E_i)$, where E_i is the energy deposited in layer i.

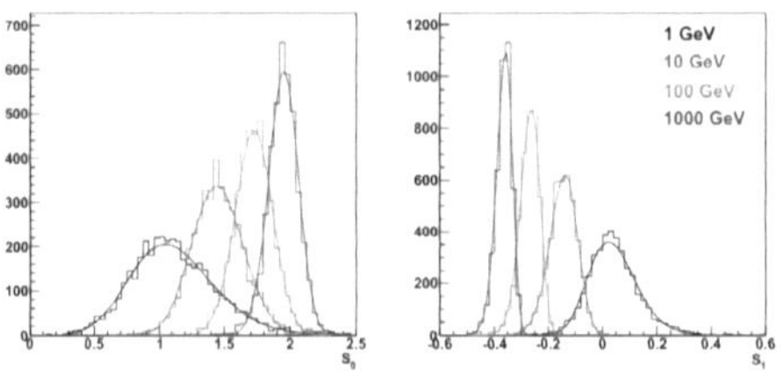

Figure 5. Distributions of the shower parameters S_0 and S_1 for 1, 10, 100 and 1000 GeV photons.

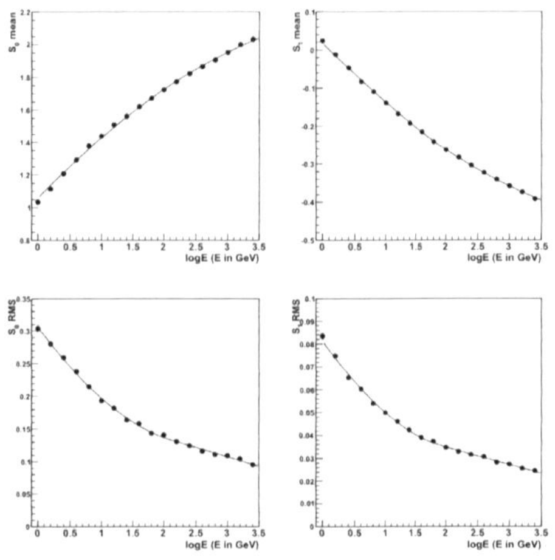

Figure 6. Parameterization of the mean and RMS of S_0 and S_1 as function of logE.

3.2. Shower transverse profile parameterization

The average transverse energy profile can be described by two components, a core and a tail. As in [5], we use the following function:

$$f(t/T, r) = \frac{1}{dE(t)}\frac{dE(t,r)}{dr} = p\frac{2rR_C^2}{(r^2+R_C^2)^2} + (1-p)\frac{2rR_T^2}{(r^2+R_T^2)^2}$$

where R_C and R_T are the medians of the core and the tail components (in units of Moliere radius), and p is the relative weight of the core component ($0 < p < 1$). This transverse profile depends on the shower depth t/T. In order to study the dependence of R_C, R_T and p on t/T, simulations were performed with `GEANT4`. In this case, the CsI calorimeter was divided longitudinally in slices of 0.1 radiation length and radially in rings with a width of 0.1 Moliere radius, which is 3.5 cm for CsI. For each event, we perform the longitudinal profile fit in order to determine T, the position of the shower maximum.

We compute the average fraction of energy contained in a cylinder as a function of the cylinder radius for various intervals of the shower depth t/T. Fig. 7(left) shows some examples of these cumulative transverse profiles for 100 GeV photons. The fit of these profiles allows the determination of R_C, R_T and p. Fig. 7(right) shows how these parameters vary with t/T. The variation of the average transverse profile with energy was found to be very weak, so we neglect it and use the 100 GeV profile.

3.3. Constraining the longitudinal fluctuations during the fit

The fit of the shower longitudinal profile works well when the shower maximum is well contained in the calorimeter. Since the LAT calorimeter is composed of 8 layers, this is the case when there is significantly less energy in the last layer than in the seventh layer. For incidence angles smaller than 25 deg, the fraction of photons for which the shower maximum is not well contained is $\sim$ 25% for photons at 100 GeV. The situation above 1 TeV is even worse since almost no photons within 45 deg of the detector axis have a well contained shower maximum.

In order to be able to reconstruct the energy of these photons, we have to use more information than the deposited energies in the layers. The idea is to use the knowledge on the fit parameters S_0 and S_1 (especially the dependence of their mean and RMS with E) by adding to the χ^2 a term that constrains the parameters to be close to their expected values. Since S_0 and S_1 are gaussian and uncorrelated, we use $s_0^2 + s_1^2$, where $s_i(E) = (S_i - \mu_{S_i}(E))/\sigma_{S_i}(E)$, i.e. the distance of S_i to its mean value and divided by its RMS. The shower fit χ^2 becomes:

$$\chi^2(S_0, S_1, E) = \sum_{i=1}^{8} \frac{(e_{m,i} - e_{p,i}(E))^2}{\delta e^2(E)} + c\left(s_0^2(E) + s_1^2(E)\right)$$

where $e_{m,i}$ and $e_{p,i}$ are respectively the measured and predicted energies in layer i. The parameter constraint term is multiplied by a factor c. We varied c and found that $c = 1$ was optimal. Fig. 8 compares the distributions of $E_{\text{fit}}/E_{\text{true}}$ with $c = 0$ and $c = 1$, for 100 and 1000 GeV photons at normal incidence. When $c = 0$, the energy can be overestimated by a factor greater than 2, whereas setting $c = 1$ allows us to remove entirely the high energy tail, even when the shower maximum is not well contained.

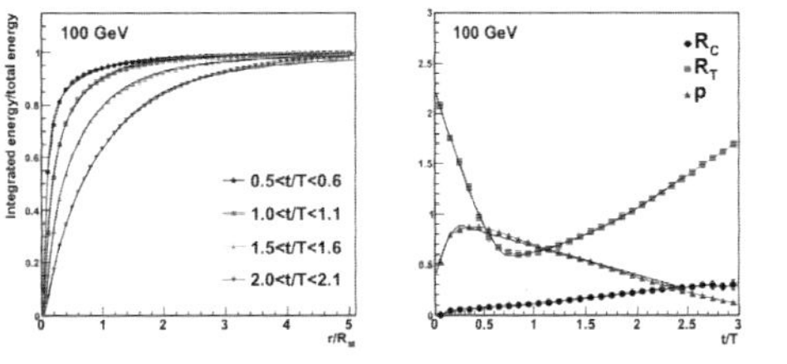

Figure 7. Left: average transverse profiles at 100 GeV. Right: variation of the parameters of the transverse profile with the position along the shower at 100 GeV.

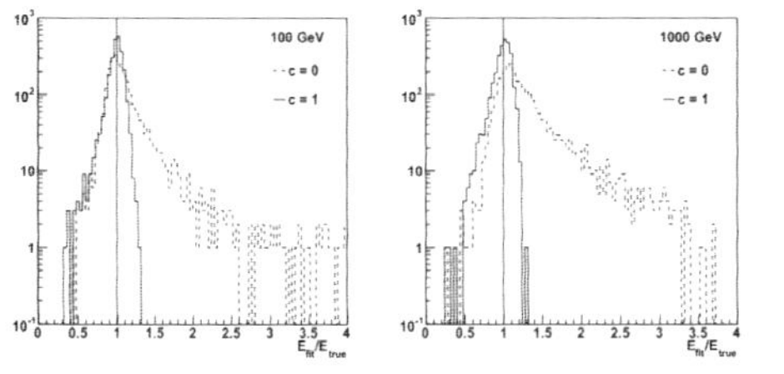

Figure 8. Comparison of the distributions of $E_{\text{fit}}/E_{\text{true}}$ with and without the parameter constraining term at 100 GeV (left) and 1 TeV (right).

3.4. The development of showers through the LAT calorimeter

For a given shower longitudinal profile $P_{(E,\alpha,\beta)}$, the predicted energy deposited in layer i is:

$$e_{p,i} = \int_0^\infty f_i(t) P_{(E,\alpha,\beta)}(t) \mathrm{d}t$$

where $f_i(t)$ is the fraction of energy deposited in layer i by the shower slice between t and $t + \mathrm{d}t$. For normal incidence photons, $f_i(t)$ only depends on the position of the shower slice t. But when the incidence angle is greater than 0, the layers are no longer perpendicular to the axis of the shower and so the transverse profile must be taken into account. Since the transverse profile is a function of t/T, the energy fraction $f_i(t)$ also depends on T.

The computation of $f_i(t)$ depends on the photon trajectory. If the photon converted in the tracker, we use the results of the tracking algorithm to get the direction and the conversion point, corresponding to the starting point of the shower. We then compute the number of radiation lengths crossed by the shower between the conversion point and the extrapolated entry point of the photon in the calorimeter. If no track has been found, we use the results of the moments analysis of the shower in the calorimeter to get the direction and we assume that the conversion point corresponds to the entry point of the photon in the calorimeter.

Using this trajectory information, the computation of $f_i(t)$ is done as follows (illustrated in Fig. 9):

- divide the trajectory into $\sim$ 1.85 mm steps, starting 100 cm ahead of the entry point of the photon in the calorimeter;
- at each step along the trajectory, at a distance x in mm along the trajectory from the starting point of the shower, corresponding to the shower depth t:
  - define a disk centered on the step point, perpendicular to the trajectory, whose radius is 3 Moliere radii = 95 mm (the red line in Fig. 9);
  - define a grid of test points on the disk, with a step of 0.05 Moliere radius;
  - the test points are used to compute the fraction of energy deposited in the layers at the shower depth t, using the transverse profile parameterization depending on t/T (represented in Fig. 9 by the red area);
  - compute $\mathrm{d}t$, the number of radiation lengths corresponding to the longitudinal step, taking into account the active (CsI cystals) and passive (gaps between calorimeter modules) material seen by the shower, and add it to t.

While developing through the calorimeter, the shower becomes wider. At some point, two different parts of the shower at the same shower depth t evolve independently. In other words, if one part of the shower escapes the calorimeter, the rest of the shower evolves as if the whole shower was inside the calorimeter. This is taken into account when computing $\mathrm{d}t$. At this stage of the computation, we have the energy fractions as a function of x as well as the shower depth t as a function of x. We then deduce the energy fractions as a function of t: $f_i(t)$.

Examples of $f_i(t)$, the fractions of energy deposited in the layers, are shown in Fig. 10, for a shower maximum T equal to 8 radiation lengths. The fraction of energy deposited in layer i at normal incidence is very simple, being 1 when the shower is inside layer i, and 0 otherwise. For an incidence angle of 45 degrees, the situation is more complex, because of the sharing of the energy between layers, which increases with t as the shower becomes wider.

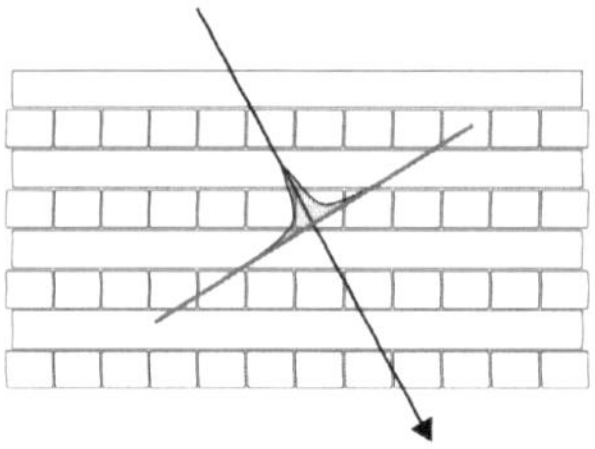

Figure 9. One step of the computation of the development of an individual shower inside a calorimeter module. The arrow corresponds to the photon trajectory. The red disk represents a shower slice at a given shower depth for which we compute the fraction of energy deposited in each layer.

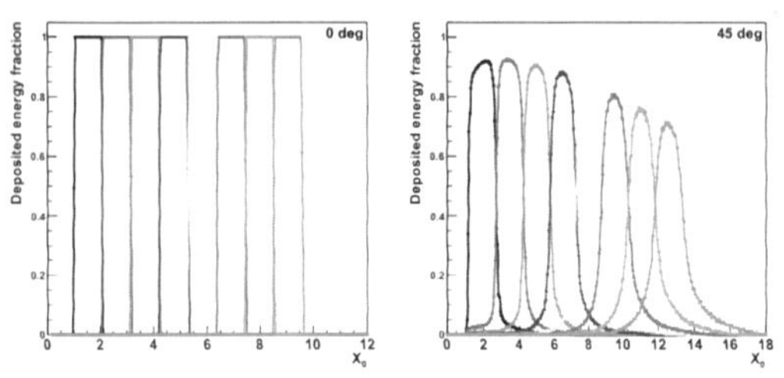

Figure 10. Predicted fraction of energy deposited in the layers of the calorimeter for a photon at 0° (left) and 45° (right) as a function of the shower depth. Each curve corresponds to one layer: from the first layer on the left to the last layer on the right.

3.5. Crystal saturation handling

Due to the readout limitation, the signal of a crystal saturates when the deposited energy is greater than $\sim$ 70 GeV. It occurs for photons with $E > 1$ TeV. In order to take this saturation effect into account in the shower fit, we consider the saturated crystals separately, and especially take out their energy from the layer energies. The computation of the shower development through the calorimeter (presented in section 3.4) is used to determine the fraction of energy deposited in the saturated crystals as a function of t. The standard part of the χ^2 becomes:

$$\chi^2 = \sum_{\text{layers}} (e_{m,i} - e_{p,i})^2 / \delta e^2 + \sum_{\text{sat.crystals}} (\max(0, e_{m,i} - e_{p,i}))^2 / \delta e^2$$

where the layer energies are summed over the non saturated crystals and where we introduced the term $\max(0, e_{m,i} - e_{p,i})$ in order that the saturated crystal contribution vanishes as soon as the predicted energy is greater than the measured energy.

3.6. Results

The performance of the shower profile fit is summarized in Fig 11. The bias is less than 3% and is easily corrected for. The energy resolution varies with energy and incidence angle. At normal incidence, it degrades when the energy increases. This is because the shower containment decreases with energy, as shown in Fig 11(right). At 45°, the resolution improves slightly from 3 GeV to 30 GeV because the fraction of energy deposited in the tracker decreases. Above 30 GeV, the resolution starts to degrade because the shower containement decreases. It must be noted that at 3 TeV, the shower containement is only $\sim$ 10% and the mean number of saturated crystals is $\sim$ 6. But still we have a decent energy resolution, especially for off-axis photons.

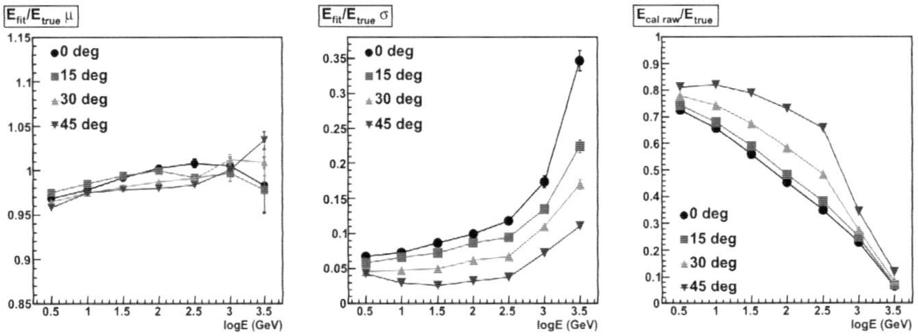

Figure 11. Performance of the shower profile fit: bias (left), resolution (center), shower containment (right).

4. Some Fermi results at high energy

It is impossible to summarize all the results obtained by Fermi after 4 years of operation. We only focus on results at high energy for which the measurement of the energy is important.

4.1. In-flight absolute energy scale

There is no gamma-ray source with a sharp and well known spectral feature for absolute energy scale calibration. But we can use the cutoff of the electron spectrum due to the Earth's magnetic field to check our absolute energy scale [6]. As explained in section 4.6, Fermi can detect electrons and positrons as efficiently as gamma rays. The geomagnetic cutoff rigidity depends on the geographic longitude and latitude. The altitude of the Fermi orbit is 565 km and its inclination is 26.5^{o}. So the Fermi spacecraft spans a range of geomagnetic cutoff that goes from 6 to 14 GeV. This range is divided into 6 bins. In each geomagnetic cutoff bin, we first determine the expected primary electron+positron spectrum. This is done by using a tracer code that computes the trajectories in the reverse direction (from the spacecracft to space) in order to reject secondary particles that are produced in the atmosphere. The measured primary spectrum is obtained after subtraction of the secondary fraction which is derived using the azimuthal distribution. The ratio of the cutoffs of the expected and measured spectra is shown in Fig. 14. From this comparison, we find that, in the energy range [6,14 GeV], the energy scale is known within 5%.

4.2. Gamma-ray sources spectral features

By detecting more than 1800 sources, Fermi is able to peform detailed population studies. Concerning active galactic nuclei, the most numerous sources for Fermi, spectral breaks have been observed to be common in Flat Spectrum Radio Quasars (the break energy ranging from 1 to 10 GeV in the source frame for the brightest sources), and present also in some bright low-synchroton peaked BL Lacs [7]. These breaks represent challenges for theoretical models aiming at describing the blazar phenomenon. Another important class of objects are pulsars: Fermi has currently detected more than 100 pulsars, whereas EGRET only detected 6. Pulsars have a power-law spectrum with a cutoff, which position and shape depends on the location of the gamma-ray emission region in the pulsar magnetosphere. The cutoffs of all pulsars measured by Fermi [8] are consistent with exponential cutoffs at a couple of GeV. This result rules out the low-altitude emission models which predict a sharper cutoffs.

4.3. Gamma-ray bursts and Lorentz invariance violation

Quantum gravity may cause the violation of Lorentz invariance and the speed of light could then depend on energy. It would imply that photons with different energies emitted simultaneously by a very distant source would not arrive at the same time. Fermi is able to constrain the violation of Lorentz invariance by looking at the highest energy events of gamma-ray bursts. GRB090510 [9] was a short gamma-ray burst with a redshift $z = 0.9$. The highest energy event has 31 GeV and is detected 0.86 ms after the start of the emission below 1 MeV (measured by the Gamma-ray Burst Monitor, the second instrument on-board Fermi). In the framework of linear Lorentz invariance violation, the delay between two photons is proportional to the energy difference divided by M_{QG}, the quantum gravity mass scale. Assuming that the high energy emission did not start before the burst, this 31 GeV event sets the limit: $M_{\mathrm{QG}}/M_{Planck} > 1.2$.

4.4. Earth limb spectrum

Gamma rays from the Earth limb are the result of the interaction of cosmic rays in the atmosphere. Above 10 GeV, the emission that is detected by Fermi is only due to grazing incidence cosmic rays in the upper atmosphere. The altitude range of the emission is rather thin: cosmic-rays passing too high do not interact enough and the gamma-rays produced by cosmic-rays that pass too low are attenuated by interacting in the atmosphere. The measured Earth limb spectrum [10] is a power law with spectral index $= 2.79 \pm 0.06$, in agreement with the spectral index of cosmic-ray proton $= 2.75$. The spectrum is very smooth up to 1 TeV. The absence of any significant sharp feature is an additional proof that our photon selection is well under control.

4.5. Extragalactic diffuse emission

The largest source of Fermi photons is the galactic diffuse emission, due to the interaction of cosmic-rays with the matter in the Galaxy. But there is a much fainter diffuse emission that has an extragalactic origin and that represents the history of the non-thermal universe. Thanks to a careful background subtraction, we measured the extragalactic diffuse emission up to 100 GeV [11] and expanding this measurement up to 1 TeV is under way. Currently, unresolved sources like blazars, radio galaxies and star-forming regions account for 50% to 80% of this emission.

4.6. Cosmic-ray electron and positron spectrum

Since Fermi's primary goal is to detect gamma-rays, the main trigger configuration rejects a large fraction of charged particles by requiring that there is no activity in the ACD above the tracker tower that triggered. But, in order to be fully efficient at high energy, all events that deposit more than 20 GeV in the calorimeter are sent to the ground, which is the case for high energy electrons and positrons. Since they produce electromagnetic showers very similar to photon induced showers, Fermi is able to reconstruct and select electrons and positrons a well as gamma-rays. The fraction of residual background is estimated to vary smoothly from 4% at 20 GeV to 20% at 1 TeV. Since measuring such high energies with a 8.6 radiation length thick calorimeter is not an easy task, we checked that our results did not change when selecting off-axis events that pass through more than 12 radiation lengths in the calorimeter. The measured spectrum [12] rules out the excess around 700 GeV seen by the ATIC experiment [13] and can be explained by a conventional model with an adjusted injection spectrum or with an additional source, which could be a nearby pulsar.

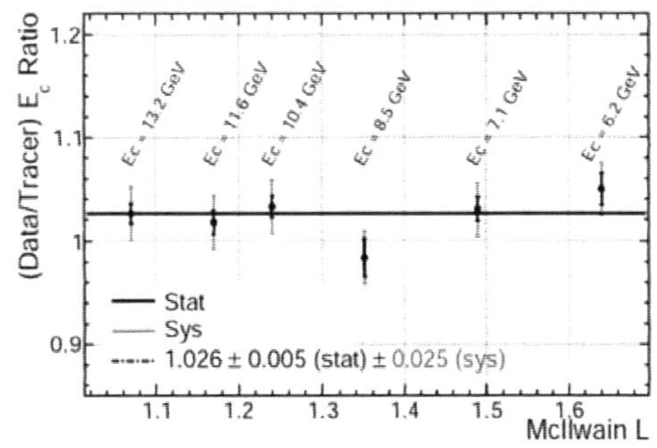

Figure 12. Distribution of cutoff energy ratios as a function of McIlwain L (with corresponding cutoff energies indicated). The statistical errors are represented by the black error bars, and the systematic errors are in gray. From [6]

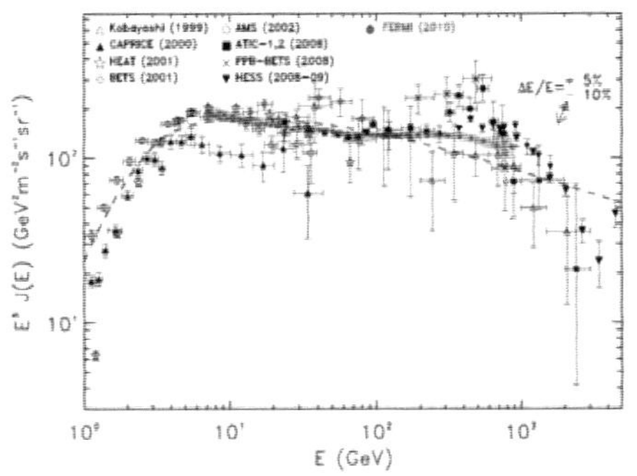

Figure 13. Cosmic-ray electron spectrum as measured by Fermi LAT for 1 yr of observations along with other recent high-energy results. Systematic errors are shown by the gray band. From [12].

4.7. Cosmic-ray positron fraction

Without any magnet on-board, Fermi is not able to discriminate between electrons and positrons. But because of the geomagnetic field, there is a region of the Earth's horizon from which only electrons can come and another region from which only positrons can come, as shown in Fig. 14. So Fermi is able to measure separately the spectra of electrons and positrons between 20 GeV and 200 GeV (above 200 GeV these regions become too small). From these spectra we can deduce the positron fraction, which is shown in Fig. 15. Our measurement [14], which was the first above 100 GeV, confirms the increase of the fraction with energy as reported by PAMELA [15].

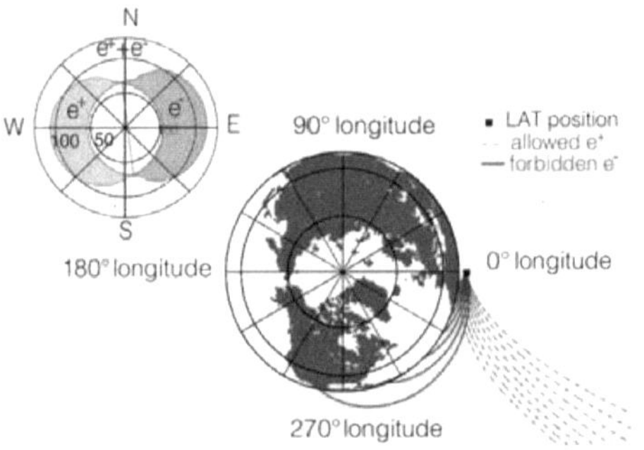

Figure 14. Examples of calculated electron (red) and positron (blue) trajectories arriving at the detector, for 28 GeV particles arriving within the Equatorial plane (viewed from the North pole). Forbidden trajectories are solid and allowed trajectories are dashed. From [14].

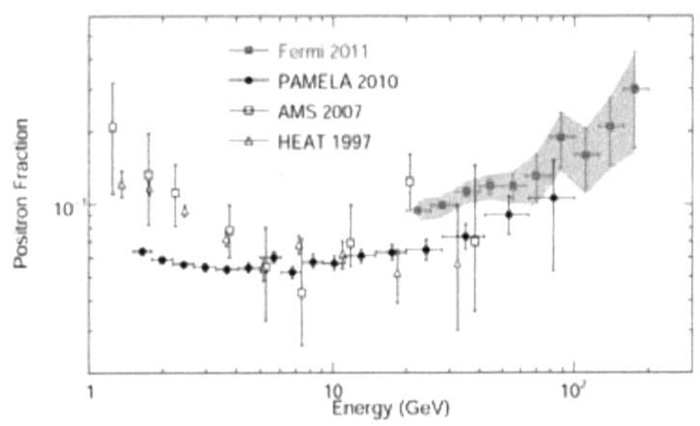

Figure 15. Positron fraction measured by the Fermi LAT and by other experiments. The Fermi statistical uncertainty is shown with error bars and the total (statistical plus systematic uncertainty) is shown as a shaded band. From [14].

5. Conclusion

After 4 years in orbit, Fermi has proved to be an excellent gamma-ray telescope. Thanks to its large effective area and field of view, to its good angular and energy resolution, Fermi has already unravelled the gamma-ray sky with much more precision than its predecessor EGRET. By detecting more than 1800 sources, it allows population studies of pulsars, supernova remnants, active galactic nuclei and gamma-ray bursts. The diffuse emission of the Galaxy and the extragalctic diffuse emission are also known much better. Fermi has proved to be a very good detector of electrons and positrons as well, and, by using the geomagnetic field, it is even able to discriminate between them. Part of the success of Fermi is due to its large energy range. Thanks to the fine segmentation of its calorimeter, we are able to measure and select gamma rays, electrons and positrons up to 3 TeV, despite the modest thickness of the calorimeter. Taking benefit from our experience after 4 years in orbit, we are currently optimizing the event reconstruction and selection. With a better sensitivity, Fermi will certainly continue to improve our understanding of the high energy sky.

References

[1] Atwood W B *et al.* 2009 *Astrophys. J.* **697** 1071–1102 (*Preprint* astro-ph/0902.1089)
[2] Thompson D J *et al.* 1993 *Astrophys. J. Suppl.* **86** 629
[3] Ackermann M *et al.* 2012 *submitted to Astrophys. J. Suppl.* (*Preprint* astro-ph/1206.1896)
[4] Longo E and Sestili I 1975 *Nucl. Instrum. Methods* **128** 283
[5] Grindhammer G and Peters S 1993 (*Preprint* hep-ex/0001020)
[6] Ackermann M *et al.* 2012 *Astropart. Phys.* **35** 346–353 (*Preprint* astro-ph/1108.0201)
[7] Ackermann M *et al.* 2011 *Astrophys. J.* **743** (*Preprint* astro-ph/1108.1420)
[8] Abdo A A *et al.* 2010 *Astrophys. J. Suppl.* **187** 460–494 (*Preprint* astro-ph/0910.1608)
[9] Abdo A A *et al.* 2009 *Nature (London)* **462** 331–334 (*Preprint* astro-ph/0908.1832)
[10] Abdo A A *et al.* 2009 *Phys. Rev. D.* **80** (*Preprint* astro-ph/0912.1868)
[11] Abdo A A *et al.* 2010 *Phys. Rev. Lett.* **104** (*Preprint* astro-ph/1002.3603)
[12] Ackermann M *et al.* 2010 *Phys. Rev. D.* **82** (*Preprint* astro-ph/1008.3999)
[13] Chang J *et al.* 2008 *Nature (London)* **456** 362–365
[14] Ackermann M *et al.* 2012 *Phys. Rev. Lett.* **108** (*Preprint* astro-ph/1109.0521)
[15] Adriani O *et al.* 2010 *Astropart. Phys.* **34** (*Preprint* astro-ph/1001.3522)

Journal of Physics: Conference Series **404** (2012) 012034 doi:10.1088/1742-6596/404/1/012034

Performance of the AMS02 Electromagnetic Calorimeter in space

Sylvie Rosier-Lees on behalf of the AMS02 collaboration
LAPP IN2P3-CNRS and Université de Savoie, 74940 Annecy le Vieux, France
E-mail: rosier@lapp.in2p3.fr

Abstract.
The Alpha Magnetic Spectrometer (AMS-02) is a general purpose high energy particle detector which was successfully deployed on the International Space Station (ISS) on May 19, 2011 to conduct a long duration mission of fundamental physics research in space. After one year of operation, AMS-02 has collected more than 17 billions of events. The main scientific goals of the experiment are the searches for antimatter and dark matter, the high precision measurement of charged cosmic ray spectra and fluxes and the study of gamma rays, in the GeV to TeV energy range. In AMS-02, the Electromagnetic Calorimeter (ECAL) is required to measure e+, e- and gamma spectra and to discriminate electromagnetic showers from hadronic cascades. To fulfill these requirements the ECAL is based on a lead/scintillating fiber sandwich, providing a 3D imaging reconstruction of the showers. AMS-02 has been tested during Summer 2010 in a test beam at CERN, using 10 to 250 GeV electron and positron beams and proton beam at 400 GeV. After a summary of the AMS02 performance in space, results on the measurements of ECAL parameters and performance will be reviewed.

1. Introduction

AMS-02 is a general purpose particle detector capable of identifying and measuring simultaneously all cosmic ray particle species: photons, electrons, protons and nuclei as well as all corresponding anti-particles. This feature becomes very important to distinguish signals from new phenomena and background processes, given a significant uncertainty in the background calculations related to the modeling of the standard processes and subsequent propagation.

AMS will measure spectra for nuclei in the energy range from 0.5 GeV/nucl to 2 TeV/nucl with 1% accuracy over the 11-year solar cycle. The scientific goals of AMS are to reach a sensitivity of antimatter search of10^{-10} (ratio of anti-helium to helium), an e+/p rejection of 1/ 10^6 and to measure the composition and spectra of charged particles with of an accuracy of 1%.

Looking at the recent measurements of $e^+/(e^+ + e^-)$ by AMS-01, HEAT, PAMELA and Fermi [1] which indicate a large deviation of this ratio with respect to the model expectations (that include only secondary species produced in the cosmic ray collisions), AMS-02 detector is expected to extend the energy domain and the statistics allowing to perform more precise measurements and provide more informations to understand the nature of this deviation. Undoubtedly, the calorimeter will play a leading role in this challenging search.

2. The AMS Detector

AMS-02 is a general purpose detector to study primordial cosmic ray particles in the energy range from 0.5 to 2000 GeV. In order to ensure that technologies used in the detector construction work reliably in space, a scaled down detector (AMS-01) was built and flown in 1998 on board the STS-91 mission for 10 days [2] Layout of the AMS-02 detector is presented in Figure 1. It consists of six complementary sub-detectors, providing measurements of the energy, the mass and the charge leading to an unambiguous identification of the cosmic rays.

It consists of: a Transition Radiation Detector participating to the e/p rejection and the charge measurement via dE/dx ; four planes of Time of Flight counters , the key detector for the trigger and measuring the timing, the velocity and the charge via dE/dx; a Permanent Magnet important to determine the sign of the charge combined with a precision silicon Tracker consisting of 9 layers, out of which 7 are in the magnetic field (Inner tracker). This combined detector, provides the rigidity and the charge via dE/dx measurements and participates to the e/p rejection by comparing the rigidity with the energy measured in the calorimeter; an array of Veto Counters included in the trigger system, surrounding the Tracker; a Ring Image Cerenkov detector providing the charge and the velocity measurements and the Electromagnetic Calorimeter which contributes to the trigger, the e^+, e^-, γ identification and the energy as well as the charge via dE/dx measurements . The maximal AMS acceptance given by the TRD and the inner Tracker, amounts to 0.5 m^2.str.

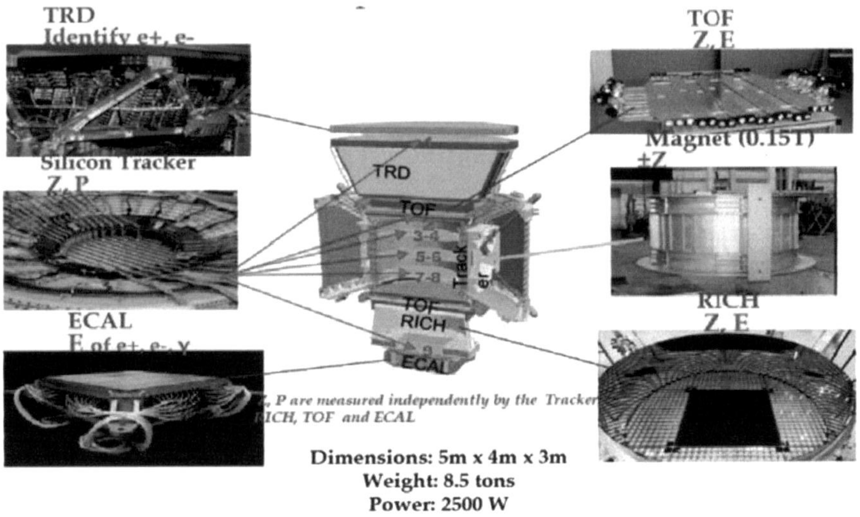

Figure 1. AMS layout with the different sub detectors

3. AMS Operation on ISS

AMS was successfully installed on the International Space Station on May 19, 2011 and since that date the detector is collecting data at an average rate of 10 Mega bits per seconds as presented at ICRC2011 [3]. Particle rates over one ISS orbit vary between 200 Hz near the equator to about 2000 Hz near the Earth magnetic poles. Data acquisition efficiency is on average 85% (it reaches 96% near the equator and 65% near the poles) resulting in an average event acquisition rate of 700 Hz. Over 17 billion events have been collected during the first year of operations in space. Over its lifetime of approximately 20 years AMS will collect 300 billions of triggers. Typical Positron and Boron events are displayed in Figures 2 and 3 illustrating the complementary responses of the sub-detectors; the different charges of both candidate can be observed qualitatively through the amplitudes of the signals in the detectors.

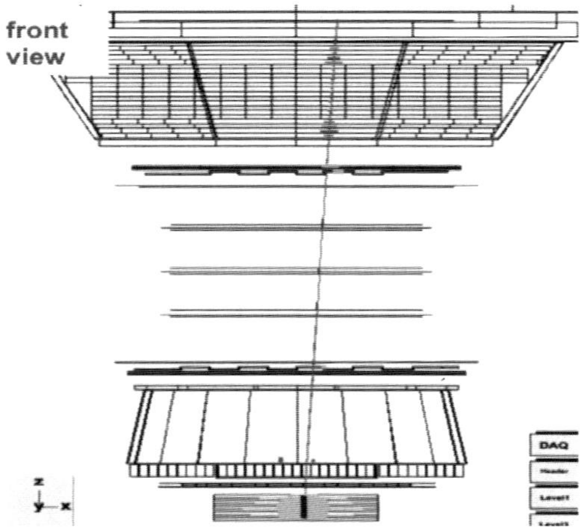

Figure 2. Positron candidate of 464 GeV

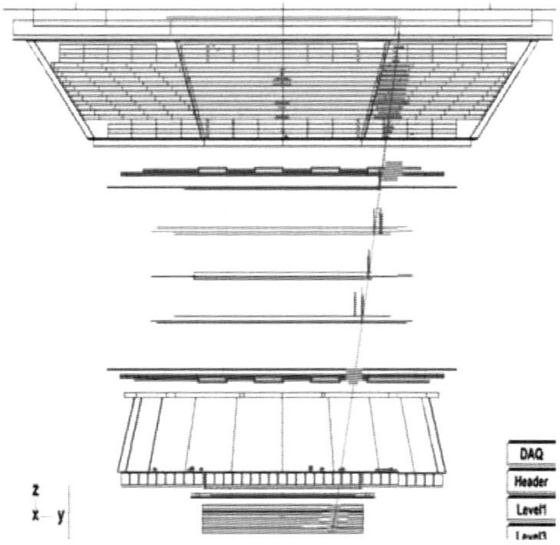

Figure 3. Boron candidate, seen at the minimum ionizing in the calorimeter

The status of all AMS subsystems is constantly monitored in the AMS Payload Operations and Control Center, POCC, located at CERN. Dynamically changing running parameters (data downlink bandwidth, distribution of available electrical power, rotation of ISS solar panels and radiators near AMS,) are followed by shifters, who are also in permanent contact with the NASA ground staff on voice loops. Commands are sent to AMS from POCC in response to changing conditions, as necessary. All AMS subsystems are fully operational with the performance expected from ground measurements, including the gas leak rate of the TRD which is found to be identical to the one measured on the ground. Variations of ambient conditions (temperature in first place) are studied and will be accounted for with proper calibrations and alignments. Therefore, calibration work for all detector systems is going on now in order to maximize the accuracy of the measurements

4. The Electromagnetic Calorimeter of AMS

The role of the calorimeter is threefold:

- Measure precisely the energy of the electromagnetic particles on a wide energy range from 0.5 GeV to 1 TeV: the largest depth in radiation length(x_0) is mandatory.
- Identify photons, electrons, positrons, thus be able to reject protons. Since the expected secondary positron flux is 1000 to 10000 lower than those of protons, a rejection factor of 10^5 is needed. This is achieved with a high granularity in the longitudinal and lateral views. In addition, the ratio between the interaction length (λ_I) and x_0 has to be at least 22.
- Trigger on the non converted photons. Since 75% of the photons are not converted in the TRD, a standalone Ecal trigger is needed, at the first level of the trigger to accept the gamma events. In addition such redundant trigger is mandatory to increase the electron efficiency at high energy that could generate high multiplicity hits in the Veto counters with the backsplash.

On top of that physics requirements, one had to cope with space constraints and environment and in particular with the limitation on weight, on power consumption and on number of channels.

4.1. 3D imaging Calorimeter

The 3D imaging electromagnetic calorimeter (ECAL), consists of 9 modules (called superlayer) made of a sandwich of grooved lead foils and of layers of scintillating fibers glued together [4], [5], representing an active area of 648x648 mm^2 and a thickness of 166.5 mm. The ECAL acceptance amounts to 0.06 m^2.str. /

Each superlayer is 18.5 mm thick and made of 11 grooved (1 mm thick) lead foils interleaved with 10 layers of 1 mm diameter scintillating fibers. In each superlayer, the fibers run in one direction only and are alternatively read at one end . Each anode covers an active area of 9x9 mm2, corresponding to 35 fibers, defined as a cell as illustrated in Figure 4. In total the ECAL is subdivided into 1296 cells (324 PMTs).Thus the calorimeter consists of 18 longitudinal independent samplings leading to a segmentation in the longitudinal view (vertical axis) of 0.95 radiation length and in each lateral view (X and Y) of 0.5 Moliere radius (R_M) .

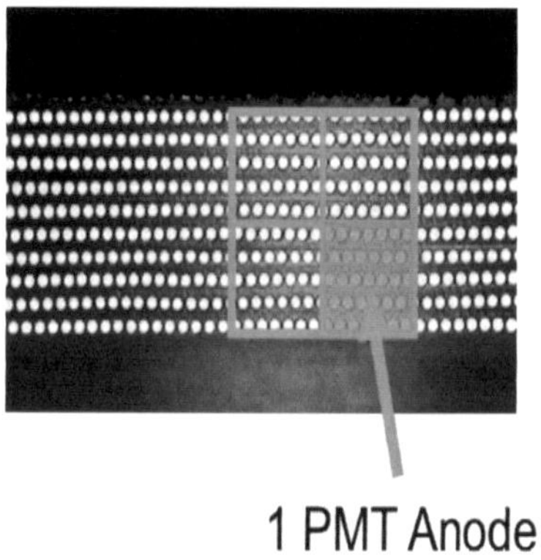

Figure 4. One of the 9 superlayers of the calorimeter with the fibers lightened on the back, with the footprint of the PMT superimposed.

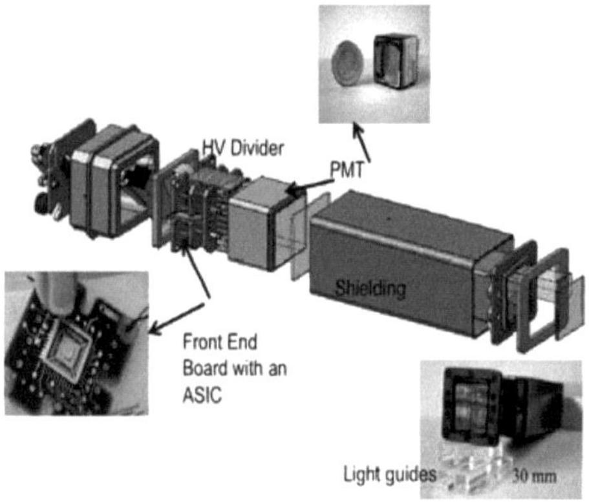

Figure 5. Exploded view of the light collection system of the calorimeter

Two calorimeters, the engineering Model (IM: 2001-2005) and the flight model (FM 2005 to 2010) have been built and extensively tested from 2001 to 2010, passing successfully the functional and flight qualification tests(thermal, vibration, electromagnetic interference and radiation hardness). The ECAL performance of the flight model have been measured in different beam tests at CERN : both for an ECAL standalone mode (2007) and with the complete AMS detector (2010).

4.2. Light Collection and readout system

The light collection system, detailed in Figure 5, consists of four 30 mm wrapped long light guides, the PMT and the front-end electronics inserted into a polycarbonate support tube, this latter being inserted itself into a magnetic shielding iron tube. The front-end electronics was implemented in a dedicated ASIC chip [6] in BiCMOS 0.8 microns technology with very low power consumption (21 mW in total). The signals in the 9 channels of the chip (2x4 anodes + 1 dynode) are shaped to give a peaking time around 2.2 μ sec. and are hold until they are read

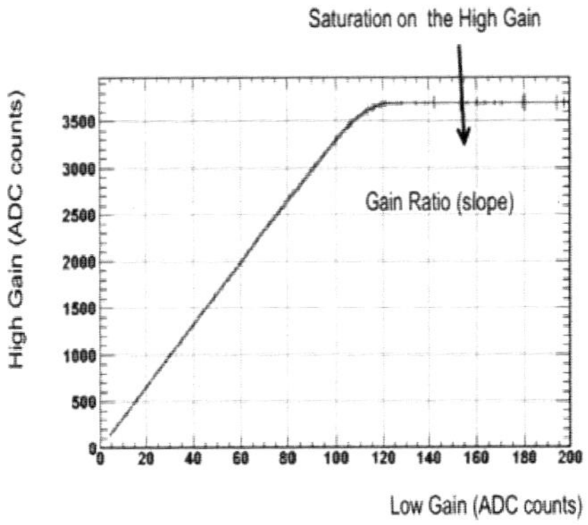

Figure 6. Example of a gain ratio (slope) value for one channel, 3 runs on ISS (~1 hour)

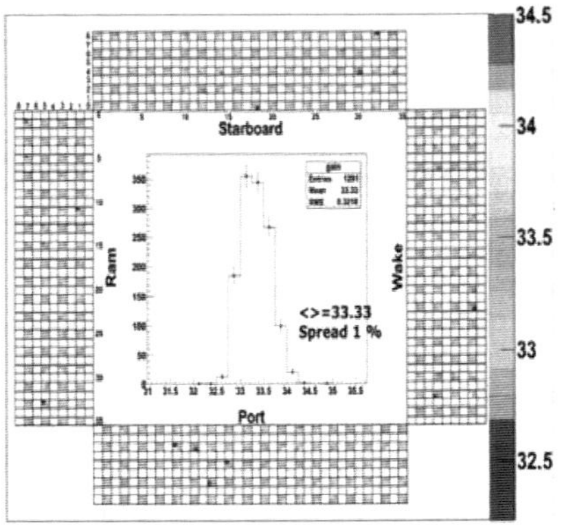

Figure 7. Gain ratio map for each anode, measured on ISS. The vertical color scale represents the gain values.

via a multiplexing circuit and sent sequentially to a serial 12 bits ADC (AD7476A, also used by other sub-detectors). The High Voltage dividers are implemented on two 26 × 26 mm2 boards. The 3 boards (Front end and HV dividers) are assembled together and mounted behind each PMT. The HV dividers and the rear of the PMT are completely potted to avoid corona effect while the Front-End board is only coated to save weight.

About 8 photoelectrons per cell are expected for a minimum ionizing particle while 48000 photoelectrons are produced in one cell by the electromagnetic showering of a 1 TeV electron. In addition the signal-to-noise MIP (minimum ionizing particle) ratio should be greater than 10, leading to a total dynamic range of 60000. To insure this, for each anode, the front end electronics is designed with two gains (high and low) with a ratio high to low of 33.33 on average as illustrated in Figure 6. Figure 7 shows the gain distribution map as measured on the ISS, during one run. The gain dispersion is of the order of 1% , as expected. The pedestal values in space and on ground are fully consistent, the rms-noise both on the high and low gains are of the order of 0.4 ADC counts leading to a signal-to-noise ratio for the MIP of 30 (with the present High Voltage settings). The saturation has been also measured for each anode on both gains and the saturation is reached on the low gain for electromagnetic shower of 1 TeV; Thus, there is some room to reach 2 TeV by simply decreasing the High Voltage supplies by 50 volts, the signal-to-noise for the MIP becoming 15.

The EDR (Ecal Reduction Board) receives the digitized signals from 27 PMTs (3 EIB) over Low Voltage Differential Signal (LVDS) lines. The data reduction, thought the DSP of the EDR, computes pedestals and its rms for each channel, every 2 runs (i.e 42 minutes, at the equator) then subtracts pedestals, suppresses zeros and sends the results throughout the backplane to the central Data Acquisition chain [5]. The EDR also acts as a bridge for the Front End power supply (+3.5, -2.5 VDC) and for the control of EIBs using single ended TTL (5 V) signaling. The 12 EDR boards are distributed into 2 crates. Each EDR is fully redundant. Each half of the EDR is protected by a solid state circuit breaker (SSF).

The electronics was designed following stringent requirements on mechanical and thermal stability, power consumption, radiation hardness and double redundancy. The full system has successfully gone through the space qualification tests [6]. After one year of data taking on the ISS, all the front end boards are operating, as well as all EDRs or EIBs. In total, since 2005, 2 PMTs have been lost. Less than 0.2 % of the anodes are noisy time to time, nominal behavior being recovered when performing HV power cycles. In term of gain, rms-noise, pedestal, saturation, noise, the electronics behaves nominally, the environment in term of noise being better in space with respect to the ground.

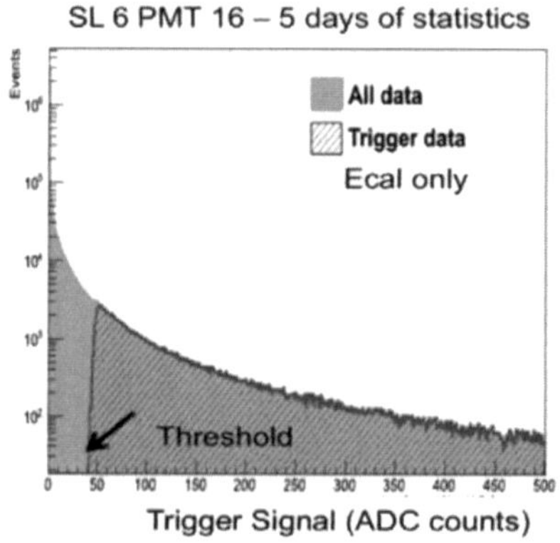

Figure 8. The digitized dynode signal events from the EFE are compared to those containing in addition a trigger (light blue)

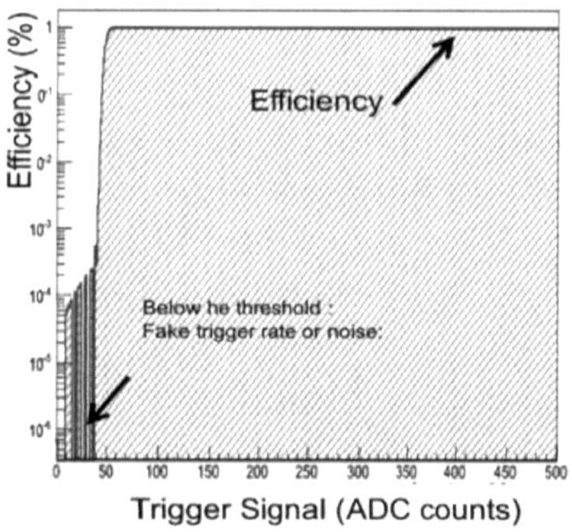

Figure 9. Trigger efficiency resulting from the division of the blue histogram by the green histogram of figure 8

4.3. Trigger data

The ECAL is included in the global trigger for detecting γ and to identify electrons. The constraints are the following: the decision of the fast ECAL trigger should be given within 180 ns, the additional trigger rate should be less than 50 Hz (compared to the maximal value of 2 kHz) and the power consumption below 10 Watts. During the 2002 test beam, it was demonstrated that the last dynode signal after amplification can be used in a fast standalone trigger for photons with energies above 2 GeV. The analog signal has to be amplified by a factor 10, in order to get a signal amplitude of 25 mV for an energy deposition of 100 MeV.Thus, the signal of the last dynode of the PMT is read out and simultaneously sent to a low gain electronic channel in the chip for redundancy purpose and amplified(by a factor of 10) and sent to a comparator for the analog gamma-trigger.

The trigger decision is build into 2 steps: a fast decision, available within 180 ns, given by the count of PMTs above threshold in the 6 central superlayers of the calorimeter; a level 1 trigger decision, well before 1 μs, obtained with a fast reconstruction of the particle direction [5].

In order to be able to trigger on gamma of 1 GeV, thresholds as low as 20 mV have been set corresponding to a signal of 2 mV before amplification. To this end EMI and grounding issues have been carefully studied. Performance of the analog part are detailed in [6]. Results concerning the comparison of the events with or without trigger, for a threshold of 33 mV, are illustrated in Figure 8 and 9. The efficiency has been measured for each threshold and is close to 100 % for 98.7 % of the trigger channels (2 channels exhibit some inefficiency since 2007 and one trigger channel has been lost since the launch). The rising time is of the level of 5 % of the threshold, as expected. The random trigger rate has been also measured and be found to be less than 0.1 % on ground and seems to a factor 10 lower in space [6].

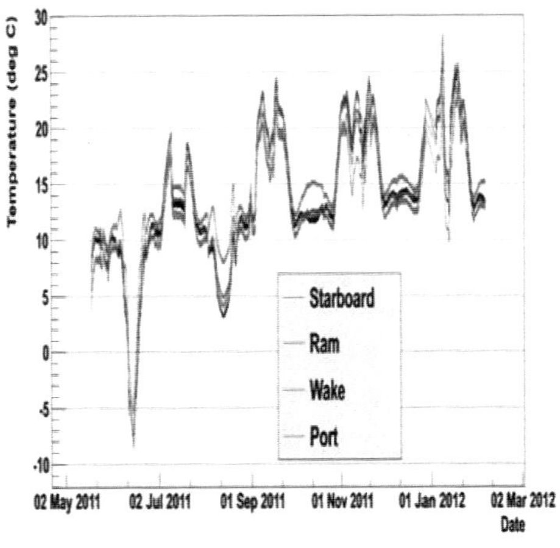

Figure 10. Temperatures measured on the four faces of the calorimeter as a function of time during the 9 first months of operation on the ISS

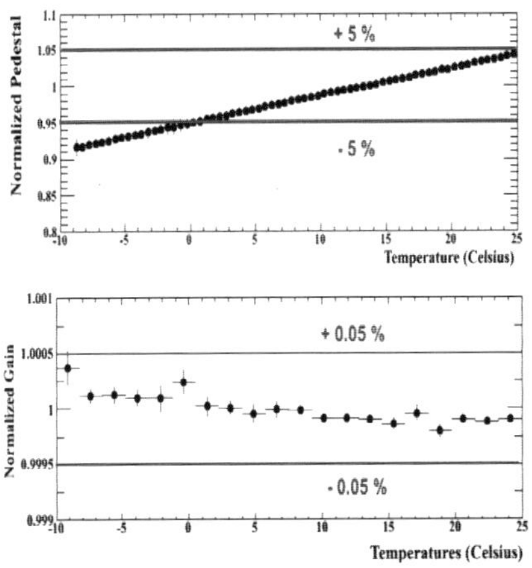

Figure 11. pedestal (top) and gain relative variation as a function of the temperature. Behaviors are linear (correlated or anti correlated) and variations are measured with a high precision, at the level of 10^{-5}

4.4. Slow Control

The calorimeter must be able to stand extreme temperatures without irreversible damages, the reference temperatures being -30 ^{o}C and + 50°C; thus the temperature variations have been and are the main concerns. Each part of the detector readout has been tested under vacuum individually in the laboratories. In addition, at ESTEC in Holland, thermal vacuum tests (TVT) have been performed with the complete assembly detector in Spring 2010. The TVT took 20 days, and were very useful to verify the performance of the complete detector under vacuum and in a wide range of the temperatures close to those of those on the space station.

40 thermal sensors are used to measure the temperatures of the calorimeter (6 on each face), the crates and on the HV power supplies and the mechanical structure (4 sensors per corner). Due to the position of the ISS with respect to the sun and its orientation, temperatures may vary over a large range, typical variations as a function of time for the nine first months of operation are shown in Figure 10. The temperature dependence of the pedestals, the pulse heights, the saturations and gains have been carefully measured since the beginning of data taking. Relative dependence with a sensitivity as low as 10^{-5} have been measured . Examples are given in Figure 11. Everything is behaving similarly in space and on the Earth.

5. Energy reconstruction

5.1. Corrections at the pixel level: Equalization and attenuation

The first set of corrections is meant to equalize the response of each anode, to correct from the light attenuation . The equalization is performed using the minimum ionizing particle (MIP) signal obtained in each cell from a scan along the X and Y axis in the middle of the calorimeter. Each distribution is fitted by a Landau function convoluted by a Gaussian function plus an exponential background. The MIPs are distributed with an average of 15 ADC counts and a sigma of 2.3 ADC counts. MIP measurements have been measured on the ground (test beam and during the commissioning at Kennedy Space Center) and in space. An anti correlated variation of the MIP with the temperature is observed, as illustrated in Figure 12, and a global correction is applied run per run.

The signal is attenuated along the fibers and needs to be corrected depending upon the original position along the fibers. For each layer, attenuation curves have been fitted simultaneously by the following function $f \times exp(-x/\lambda_{fast}) + (1-f) \times exp(-x/\lambda_{slow})$ letting free an individual amplitude per cell; an example of the fit is given in Figure 13[7]; Test beam, cosmics data and ISS data agree well as depicted in Figure ??. The correction is applied such that there is no correction in the center of the calorimeter.

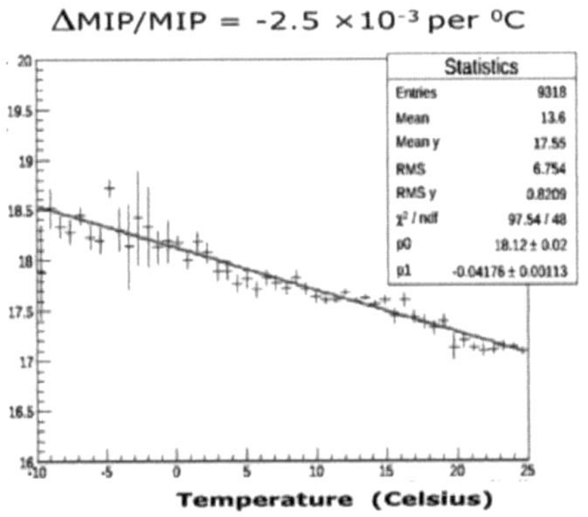

Figure 12. Variation of the MIP pulse height, averaged over all the channels, with temperature

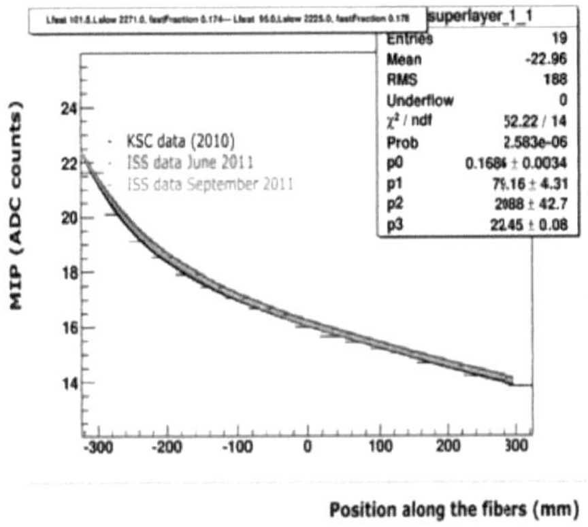

Figure 13. MIP signal attenuation along the fiber average over one layer, for three diffrent periods . The attenuation on the total length is about 30 %, for 16 layers and about 38 % for 2 layers

5.2. Global corrections : Impact and Rear leakage corrections

To take account of the light loss at the edges of the light collection guides, a standalone method correction was developed. The signal loss is maximal for particle with normal incidence impinging between two pixels, i.e a dead zone and is then of the order of 10 %. This impact correction relies on the fraction of energy deposited in the central cell (S1) of the electromagnetic shower relative to the one deposited in the neighbouring ones (S3). Thus, the ratio S1/S3 is estimated ones per layer and summed over all layers in each direction (X or Y). The dependence of the deposited energy as a function of this variable is nicely linear; it was checked with test beam date that the amplitude of the correction does not depend on the energy [7]. The effect of the correction is illustrated in Figure 14 where the deposited energy before and after impact correction are compared as a function of the impact position as measured with the center of gravity.

The last correction will correct for the energy rear leakage. The relative leakage energy is approximated by a linear function of the fraction of energy measured in the latest layers of the calorimeter. This linear relation scales with energy, as observed with test beam data. Therefore this dependence was accounted for, using the energy divided by the longitudinal barycenter which is roughly proportional to the incident energy. Figure 15 shows the principle of the correction. The rear leakage has been also estimated by fitting the longitudinal profile of the shower, well parametrized by a Gamma function; Both methods give similar performance, the latter one including all the layers.

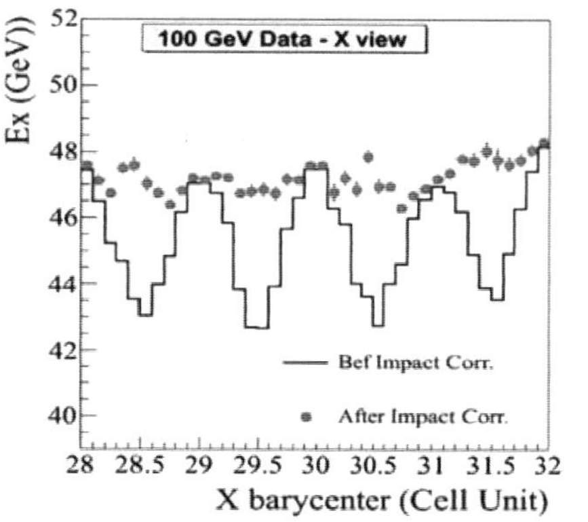

Figure 14. Illustration of the impact correction, as a function of the position of the shower, in the Y view

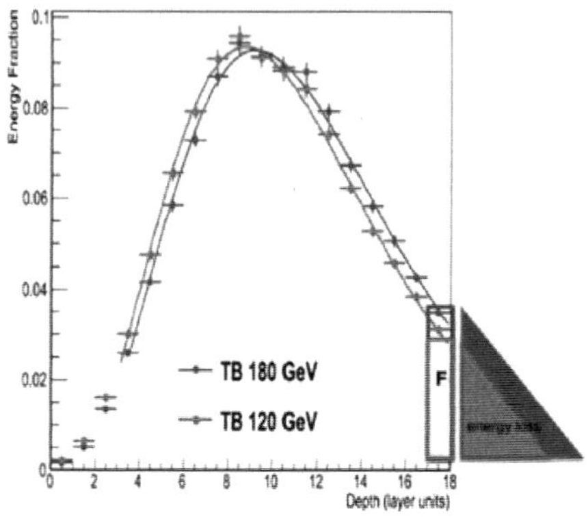

Figure 15. Fits of the longitudinal profile for two different energies (120 and 180 GeV) as measured during 2010 test beam. The fraction, F, of deposit energy in the 2 last layers is proportional to the rear leakage, represented by the full triangles.

5.3. Performance on the Ground (test beam)

During test beams in 2007, the properties of the calorimeter have been extensively measured, such as the total radiation length X_0 , by extracting first the position of the maximum of the longitudinal profile, then fitting the linear dependence of the maximum with the logarithm of the energy. Thus, X_0 has been measured to be 1.05(0.02) layer unit. Therefore, the calorimeter consists of 17.1 radiation lengths ([5],[7]). Thus, the expected rear leakage for 1 TeV electrons is on average, 16 % and the maximum of the shower is expected to be in layer 11. The critical energy has been also deduced from the radiation length measurement and the fit according to equation 1 propagating all the relevant uncertainties. The measurements are compared to the expectation (Rossi formula $E_c = \frac{610MeV}{Z+1.24}$ [8]). A critical energy of 7.6 ± 0.2 MeV is obtained. Finally, since the Moliere radius is directly linked to the critical energy according to [8], typical values of $R_M = 2.5 - 2.6X_0$ have been also derived.

$$zmax = X_0 \times ln(\frac{Ebeam}{E_c}) + X_0 \times C_e \quad C_e = -0.5 \tag{1}$$

In 2010, the AMS detector was fully assembled and the performance in term of linearity and energy resolution, with less energy points are similar to what was previously measured [5], despite the ~ 0.65 expected radiation length in front of the calorimeter. Results on the energy resolution and the linearity are shown respectively in Figures 16 and 17. The resolution was fitted to be

$$\frac{\sigma_E}{E} = \frac{10.4\%}{\sqrt{E}} \oplus 1.4\%$$

while the linearity is kept within 1%. Using the same complete set of corrections, the linearity and the resolution are not degraded when the incidence angle of the beam is 15 degrees.

6. Charge Measurement

The ionizing energy loss is well described by the Bethe Bloch equation which shows that in a given material, for a given velocity, the energy loss is proportional to the square of charge of the

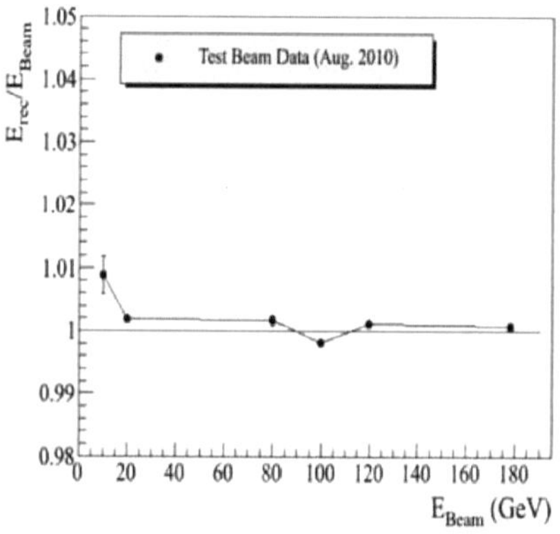

Figure 16. The linearity is better than 1%, after applying all the corrections

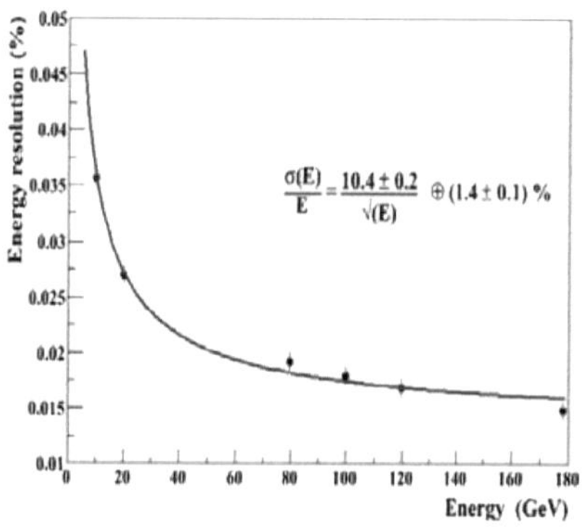

Figure 17. Energy resolution as measured in 2010, on electron and positron beams. The resolution is better than 2% for energies greater than 80 GeV

crossing particle. So the charge of the particle can be discriminated by the response of ECAL. Identification of the nuclei has been performed with the ECAL for the first time by analyzing the ISS data by measuring their deposited energy in ECAL. The non-interacting particles in ECAL are selected by requiring a low hit multiplicity layer by layer, thanks to the low noise level in the calorimeter. The deposited energy in the layer should be concentrated in one cell (i.e representing at least of 99 %) . Doing so, it is possible to identify the starting point of the shower (called hereafter apex) and determine the path length of the nucleus remaining at the minimum ionizing. The layers selected at the MIP, i.e before the apex, only contribute to the charge measurement of the nuclei. The energy loss dependence on the rigidity is shown in Figure 18. The total deposited energy of the same charge follows a Gaussian-like distribution from proton to oxygen nuclei as depicted in Figure 19. The charge determined this way is consistent with what is measured by the other specific charge-measuring sub-detectors and will be included in the AMS global charge estimator.

7. Electron Identification

As introduced previously, the imaging calorimeter of AMS has been designed to precisely reconstruct the longitudinal and lateral profiles of interactions and to measure the deposited energy. The longitudinal and lateral segmentation of the calorimeter, combined with the measurement of the particle energy loss, allows a very high identification power for electromagnetic showers. In particular, the high longitudinal granularity allows to reconstruct the apex position and then reject most of the proton (~80 %) starting an hadronic shower after the three first layers. Figure 20 exemplifies the different hadronic showers pattern of interacting proton, preselected with the TRD and the tracker; The probability of interaction for proton as a function of the calorimeter depths is shown in Figure 20; it turns out that, on average, 53 % of the protons crosses the calorimeter without developing a shower, leading to an interaction length of 28 layers (25.75 cm, or 26.6 X_0). The calorimeter depth is then equivalent to 0.64 λ_{int}.

After having applied topological cuts related to the starting point of the shower, specific selection cuts more based on the electromagnetic shower properties are applied. Global variables both in the longitudinal and lateral views are build. Figure 21 shows two typical discriminant variables, in the Y view (lateral) the fraction of energy deposited in 1.5 Moliere radius and also

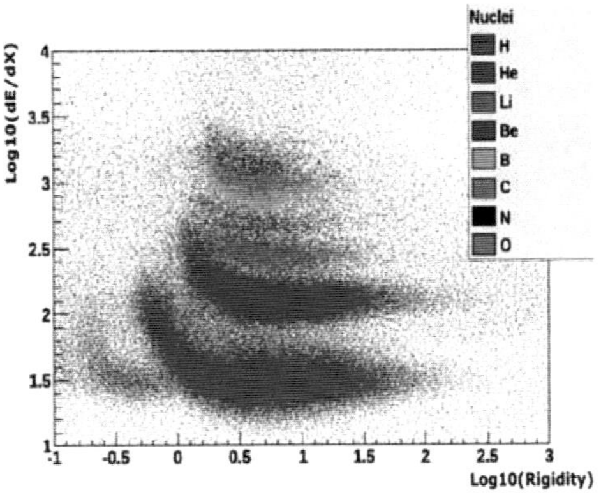

Figure 18. dE/dx measurement in the calorimeter as a function of the Rigidity measured in the Tracker, for Hydrogen to Oxygen nuclei

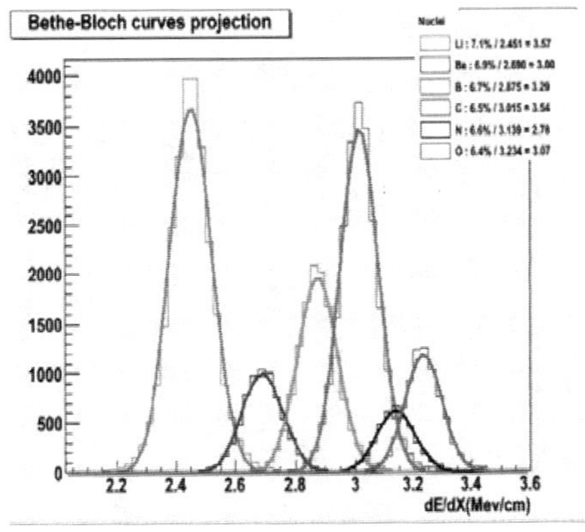

Figure 19. Charge measurement in the calorimeter (Log(dE/dx)) for the Lithium to the Oxygen Nuclei

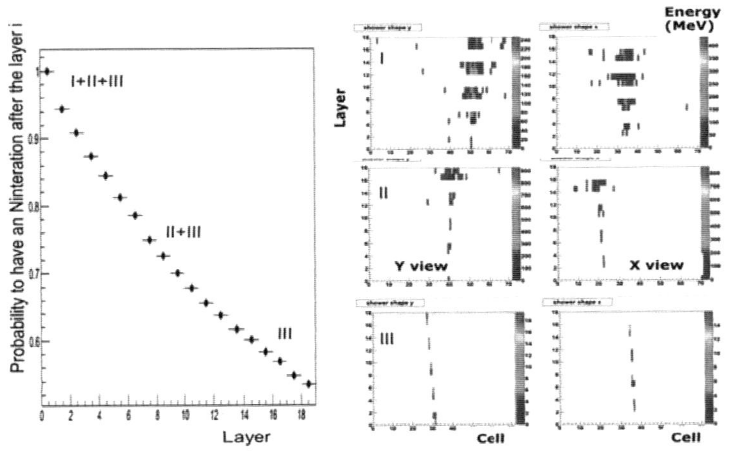

Figure 20. The typical interacting proton patterns in the calorimeter (right). Fraction of protons developing a shower as a function of the calorimeter depth(left)

the chi square of the longitudinal fit, for 400 GeV protons and 180 GeV electrons. The separation is clear and the method based on simple cuts allow to achieve a proton rejection factor ranging from 50 to 100 with an efficiency of 80-90 %. This is not enough to reach the rejection goal even after having included the TRD and the tracker, maintaining an overall efficiency of 90 %.

A multi variate analysis, based on the TMVA package [9] has been then proposed exploring the different techniques (Boost Decision Tree, neural network , Likelihood,..). One of the main issue is the selection of the training samples (background and signal) to avoid any overtraining or any bias that could be introduced due to the lack of statistics or a wrong simulation model for example. In what follows, trainings were performed on data only. For what concerns ISS data subsample data are selected using the TRD and the tracker, while for test beam data both electron and proton beams have been used. In total, 56 discriminant variables have been combined including global variables but also the fraction of the deposit energy and the width of the shower in each of the 18 layers. Analysis based on boosted decision trees has been studied, handling the correlations between the variables. The boosted decision trees have been trained in several energy bands between 2 GeV and 1 TeV. This leads to a large improvement of the analysis sensitivity, with respect to cut based analysis, reaching a rejection factor greater than 4000 for an efficiency of 90 %. The signal over background ratio is dramatically increased compared to the original cut based methods, as shown in Figure 22. Combined with the other detectors (TRD and Energy and momentum from the tracker matching) an overall rejection factor of 10^6 is achieved, from 2 GeV to 100 GeV. Analyses probing higher energies are ongoing.

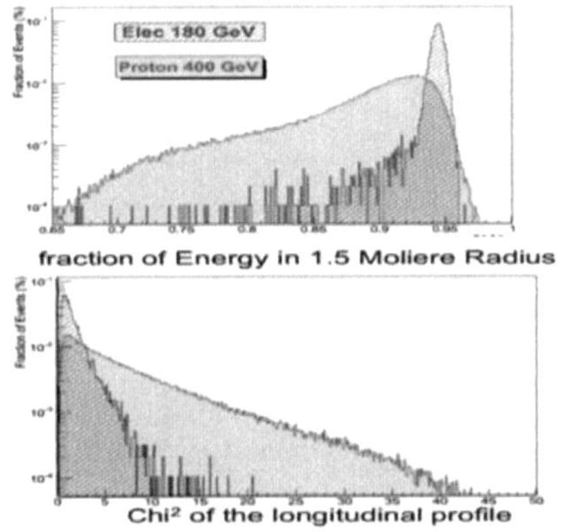

Figure 21. Distribution of two discriminant variables based on the lateral (top) and longitudinal (bottom) electromagnetic shower properties obtained with 2010 test beam data

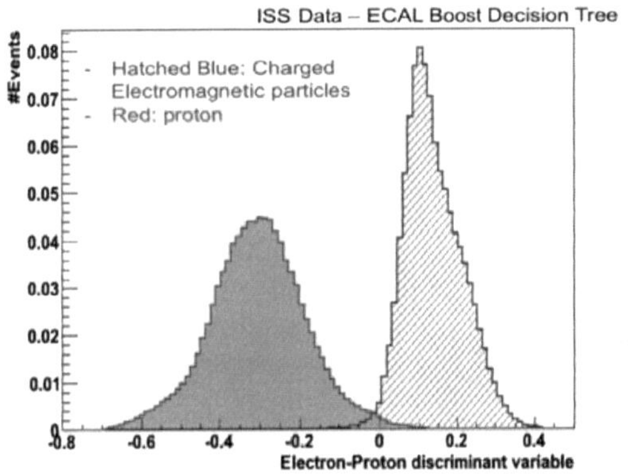

Figure 22. Ecal Boost Decision Tree output variable as obtained with ISS data.

8. Conclusions

The AMS experiment on the ISS has collected cosmic rays for more than one year. Throughout this period, the imaging calorimeter of the apparatus has been performing nominally. In particular, all functional parameters (pedestals, noise, signal, gain,attenuation) are nominal and temperature variations well monitored. The physics performance are consistent with the expectations and the test beam results. During this first year of operation in space, no failures nor loss/degradation of performance have been observed and the calorimeter proved to be able to fulfill the requirements needed to achieve the scientific goals of AMS.

9. References

[1] O.Adriani et al Nature 458, **607-609** (2009) and A. Abdo et al., Phys. Rev. Lett. **102**, (2009)
[2] E. Choumilov et al., Nucl. Instr. and Method A **426** (1999) 625.
[3] A.Kounine, *Proceedings of the 32nd International Cosmic Ray Conference* , Vol.12.
[4] F. Cervelli et al., Nucl. Instr. and Method A **490** (2002) 132.
[5] F. Cervelli et al., submitted to Nucl. Instr. and Method A September 2012.
[6] L.Basara et al."Front end electronics of the AMS-02 ECAL: Description and Performance", to be submitted to NIMA.
[7] C.Goy *Proceedings of Calor2008* J. Phys Conf Serie 160 012041 (2009) and references herein.
[8] *Particle Data Book, Particle interaction in the matter* PDG
[9] *Tools for Multi Variate Analysis* TMVA 4,CERN-OPEN-2007-007.

The NOνA Neutrino Calorimeter

Stephen Magill

Argonne National Laboratory

Abstract. The NOνA experiment is a long baseline neutrino detector designed to 1) observe oscillations of muon neutrinos to electron neutrinos, 2) determine the ordering of the neutrino mass states, and 3) observe CP violation in neutrinos if it exists. To accomplish this, the NOνA detector is a unique low-Z, high sampling fraction calorimeter capable of precise measurements of the particles produced in a neutrino interaction while also being able to reject particles from background cosmic rays. Some experience has already been obtained with the operation of a prototype near detector on the Fermilab site, and construction of the far detector is just beginning in northern Minnesota. The calorimetric properties of the NOνA detector will be described with emphasis on relevance to the overall experimental goals.

1. The NOνA Long Baseline Neutrino Experiment

The NOνA Collaboration is composed of over 150 Physicists representing more than 26 institutions from 5 countries. It will consist of a Near Detector (ND) on the Fermilab campus and a Far Detector (FD) ~800 km from Fermilab in northern Minnesota. The detectors are situated 14 mr off-axis of the NUMI neutrino beam which results in a narrow neutrino energy distribution of ~2 GeV on average.

The primary goal of NOνA is to measure the probability of neutrino oscillations from the ν_μs produced in the NUMI beam to ν_es detected in the FD. This will allow NOνA to measure the mixing angle θ_{13}, to determine the neutrino mass hierarchy, to study the phase parameter for CP violation in the lepton sector, and to determine the θ_{23} octant. With ν_μs, NOνA can make precision measurements of the parameters θ_{23} and Δm^2_{32}. Other topics are searching for sterile neutrinos, measurement of neutrino neutral current cross sections at 2 GeV, and detection of neutrinos from a galactic supernova.

2. Neutrino Oscillation Physics at NOνA

Recent results from reactor neutrino experiments have shown that the mixing angle θ_{13} is non-zero and large – the world average of all measurements is now $\sin^2(2\theta_{13}) = 0.092 \pm 0.012$ [1]. The $\nu_\mu \rightarrow \nu_e$ oscillation probability is therefore large enough for NOνA to be sensitive to the neutrino mass hierarchy in a comparison of the oscillation probabilities of neutrinos and anti-neutrinos. Figure 1 shows possible NOνA measurements of P(anti-ν_e) vs. P(ν_e).

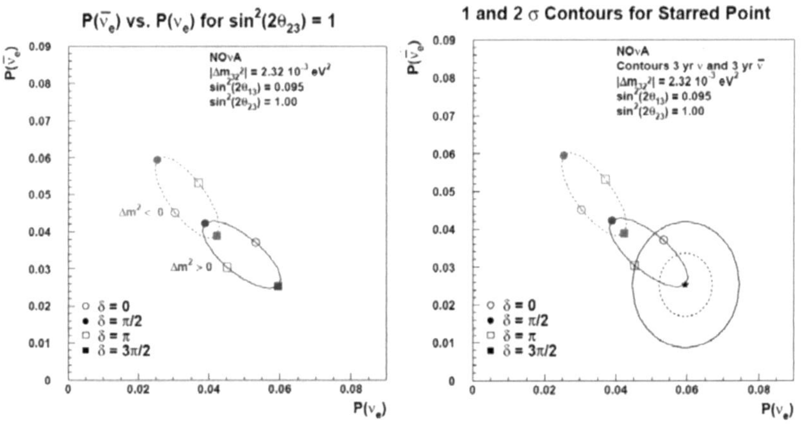

Figure 1. (left) Possible NOνA measurements of oscillation probabilities for all δ_{CP} and both mass hierarchies, and (right) 1- and 2σ contours for a measurement example if $\delta_{CP}=3\pi/2$.

The ellipses represent measurements at all possible values of the CP violation parameter δ_{CP} for each of the mass hierarchies, normal ($\Delta m^2>0$) and inverted ($\Delta m^2<0$). For a possible CP parameter $\delta_{CP}=3\pi/2$, the starred point with 1- and 2σ contours represents a NOνA measurement example with 3 years each of ν and anti-ν running. At (and near) the example measured point, the inverted mass hierarchy is excluded at >2σ. Figure 2 shows the significance of mass hierarchy resolution for all possible measurements of both hierarchy possibilities.

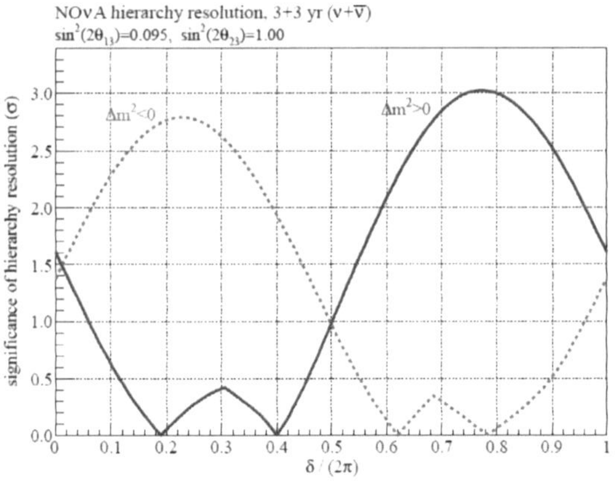

Figure 2. Significance of mass hierarchy resolution for both possibilities and for all δ_{CP}.

Since the ellipses overlap, there are 2 points for each hierarchy where there is no hiersrchy resolution significance. If, however, NOνA and T2K [2] (shorter baseline than NOνA) results are eventually combined, the difference in sensitivities of the 2 experiments resolves the ambiguity seen in the NOνA-only plot where the normal and inverted hierarchy ellipses overlap. A combined measurement would then result in at least 0.5σ sensitivity for both hierarchies over the full range of δ_{CP}.

NOνA can also resolve the current ambiguity in the value of θ_{23} – whether that mixing angle is less than or greater than 45 degrees. This determines whether the third neutrino mass state, ν_3, is comprised mostly of ν_μ or ν_τ.

3. The NOνA Detectors

The NOνA detectors are made from extruded PVC with cell sizes of 4 cm × 6 cm × 15.6 m long in the FD and × 3.9 m long in the ND. There are 16 cells in each PVC extrusion, 2 of which are glued together to form a 32-cell wide module. Each cell contains a looped WLS fiber, is filled with liquid scintillator, and is read out by an Avalanche Photodiode (APD). Figure 3 shows modules being stacked at the module factory (Univ. of Minn.).

Figure 3. Stack of NOνA modules ready for shipment to the FD.

The WLS fibers are routed through a manifold attached to one end of the module, ending at a connector for the APD. Each layer of the FD has 384 cells in 12 modules, while a ND layer consists of 96 cells in 3 modules. The full 14 kTon FD has 928 layers of PVC cells in alternating vertical and horizontal orientations per layer. The ND is 192 layers of alternating vertical and horizontal cell layers with the addition of a muon catcher block with steel absorber plates between PVC layers at the downstream end. In all, the NOνA detectors are comprised of >370K PVC cells, ~12K 32-pixel APDs, >11 million meters of WLS fiber, and over 3 million gallons of liquid scintillator. When complete, the NOνA FD will be the largest plastic structure in the world.

4. The NOνA Calorimeter

The NOνA detectors are calorimeters designed to detect and identify neutrino interactions. Produced as ν_μs in the NUMI beam at Fermilab, the neutrinos propagate towards the NOνA FD as a mixture of neutrino mass states. Due to differences in the propagation speed of the different mass states, the neutrino detected in the FD has a non-negligible probability to be in the ν_e flavor state – the phenomena known as neutrino oscillations. The length of the baseline – the distance from Fermilab to the NOνA FD site – was chosen such that the magnitude of the oscillation is near its maximum. The flavor of an interacting neutrino is identified by its final state configuration. Figure 4 shows simulated neutrino interaction signatures in the NOνA detector.

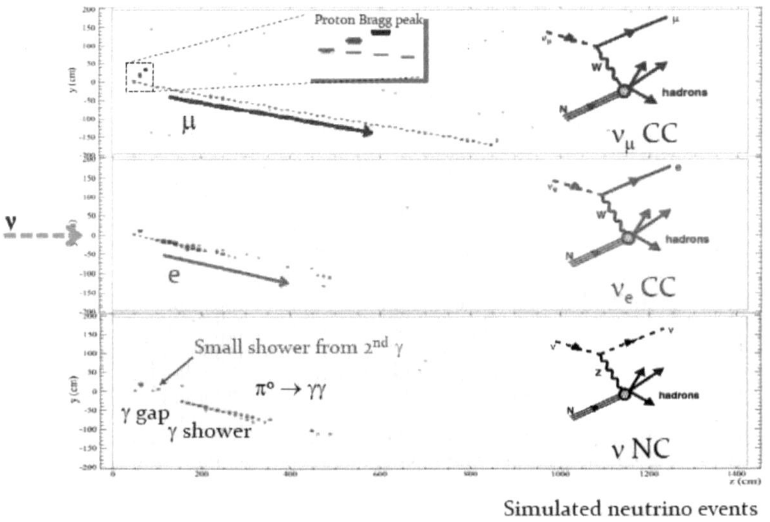

Figure 4. Neutrino interaction processes in the NOνA detectors; (top) a ν_μ charged current event with long muon track and a recoil proton, (middle) a ν_e charged current event with a typical electron shower, and (bottom) a ν neutral current event with a π^0 in the final state.

To measure the $\nu_\mu \rightarrow \nu_e$ oscillation probability, identification of an electron in the final state of a neutrino interaction must be made. Backgrounds from ν_μ CC events can be relatively easily identified by the presence of a muon in the final state as shown. A NC event containing a π^0 in which the 2 photons can not be resolved is a particularly difficult background to reject. The design of the NOνA experiment is optimized to detect and characterize electromagnetic showers, distinguishing photon showers (mainly from π^0s) from electron showers. The NOνA detectors are low Z sampling calorimeters with an active volume of ~77% (~66% active mass). Some parameters of interest are the critical energy E_C = 73 MeV, the radiation length, X_0 = 38 g/cm^2 → 37.5 cm along the axis of the detector, and the Moliere radius of ~10 cm (2.5 cell widths). A single NOνA layer is 6.63 cm thick so electromagnetic showers are sampled at ~0.18 X_0 per layer. Due to the relatively high E_C of 73 MeV (compared to ~10 MeV for the ZEUS U/Scintillator calorimeter), electromagnetic showers are short in X_0, being almost entirely contained in 10 X_0 as opposed to ~25 X_0 for ZEUS. However, in spatial dimensions, these showers are ~4 m long in NOνA compared to ~20 cm for ZEUS. Therefore, the high sampling rate per X_0 and the large spatial extent of electromagnetic showers will help in distinguishing electrons from photons and π_0s.

The longitudinal and transverse shower shapes of electromagnetic showers can be calculated using parameters of the detection medium. This is possible since the energy deposits in an electromagnetic shower are highly correlated with each other. The longitudinal profile of electron showers for energies of interest is calculated and shown in Figure 5.

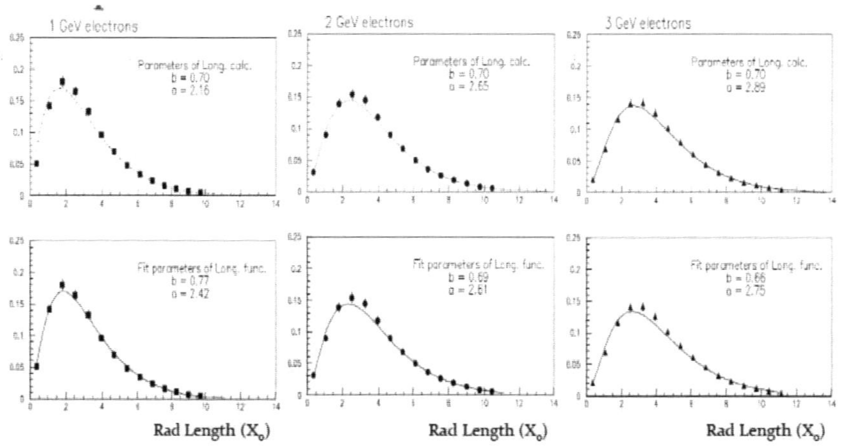

Figure 5. Longitudinal profile plots of electron showers in NOνA for (top row) calculation using detector parameters (lines) compared to simulated data (points), and (bottom row) fits of the longitudinal profile function (lines) compared to the simulated data (points).

At all electron energies, the parameters of the calculation (shown on the top row plots) are very similar to the parameters obtained from fits to the simulated data (shown on the bottom row plots). While the "b" parameter was chosen to be a constant in the calculations (an approximation), in the fits it exhibits a variation with energy which is expected.

Transverse shower distributions can also be calculated – one model being a double Gaussian formula. Figure 6 shows a double gaussian fit to the simulated data and the combined shower shape distribution using both longitudinal and transverse fits.

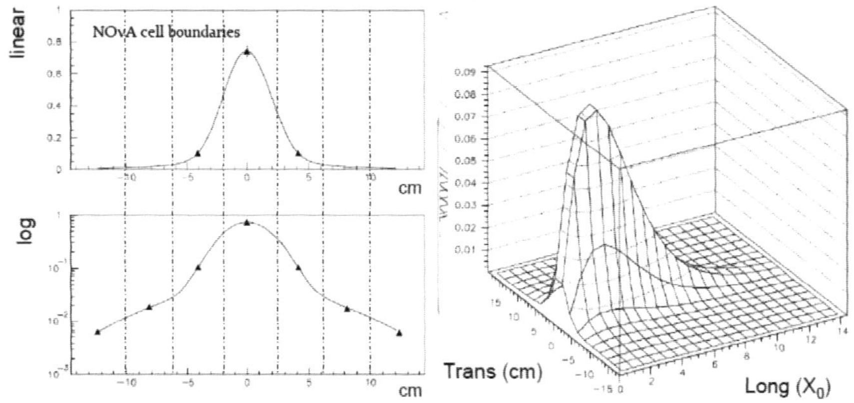

Figure 6. (left) Transverse profile fit (line) to simulated data shown on linear and log scale with cell boundaries overlaid(dashed lines), and (right) the combined transverse and longitudinal electron shower distribution.

The relative fractions of the narrow and wide Gaussians were allowed to float in the fit and the result yielded ~94% in the narrow term. The full longitudinal extent of the shower is contained in ~10 X_0, or ~4 m in the NOνA detector while near longitudinal shower max, ~7 cells contain the shower in the transverse dimension (full width of ~28 cm).

5. Summary

The NOvA detectors are unique low-Z calorimeters optimized for:

1) detection of neutrino oscillations in the NUMI beam at a long baseline → 14kTon Far Detector.

2) identification of the detected neutrino flavor by reconstructing final state particles including electromagnetic showers from electrons → highly segmented and granular calorimeter with high sampling fraction.

With the recent reactor results indicating a relatively large value for the mixing parameter θ_{13}, NOvA will contribute to understanding of mixing in the lepton sector – in particular towards resolving the hierarchy of neutrino mass states, further constraining the flavor content of the ν_3 mass state, and determining if CP violation in the lepton sector occurs.

References

[1] Double Chooz: Y Abe et al., Phys. Rev. Lett. 108, 131801 (2012).
DayaBay: F. P. An et al., arXiv:1203.1669.
RENO: J. K. Ahn et al., arXiv:1204.0626.

[2] T2K: K. Abe et al., Phys. Rev. Lett. 107, 041801 (2011).

Journal of Physics: Conference Series **404** (2012) 012036 doi:10.1088/1742-6596/404/1/012036

The LBNE near detector

Elena Guardincerri

Los Alamos National Laboratory, Los Alamos, NM, USA

Abstract. The Long-Baseline Neutrino Experiment (LBNE) is a next generation neutrino oscillation experiment currently proposed for construction in the United States with the main goal of studying muon (anti)neutrinos oscillations into electron (anti)neutrinos over a distance greater than 1000 km and over a wide range of neutrino energies. Its main physics results will be the precision measurement of δ_{CP} and all three mixing angles. A value of δ_{CP} different from 0 or π will appear in the LBNE data as a small (0% to 40%) asymmetry on the oscillation probability of neutrinos versus antineutrinos. Such a small effect requires a very good control of the systematic uncertainties affecting the measurement and this can only been accomplished using a near detector. I will describe the LBNE near detector complex focusing primarily on its reference design, which includes a set of muon detectors to monitor the beam and a magnetized liquid argon TPC surrounded by a muon identifier detector to measure neutrino interactions at the near site.

1. Introduction

The recent results by atmospheric[1], solar[2], and reactor[3] neutrino experiments estabilished that neutrinos oscillate, that is neutrinos produced in a given flavor state have a non zero probability of being detected in a different flavor state. The only consistent explanation of all the data collected so far is that neutrino are not massless, and that mass eigenstates (i.e. eigenstates of the free Hamiltonian) are different from weak eigenstates (i.e. eigenstates of the full Hamiltonian, including the interaction terms).

The oscillation probability in vacuum depends on the energy E of neutrinos, on the distance L between the source and the detector, on the differences between the neutrino mass eigenvalues Δm^2_{ij} and on the elements of the so called *neutrino mixing matrix* U, relating neutrino mass eigenstates ν_i with neutrino flavor eigenstates ν_α.

In the case of n neutrino flavors and n massive neutrinos the unitary $n \times n$ matrix U:

$$\nu_\alpha = U_{\alpha,i}\nu_i \quad \alpha = e, \mu, \tau, ..., n \quad i = 1, 2, 3, ..., n \tag{1}$$

can be parametrised by $n(n-1)/2$ Euler angles and $n(n+1)/2$ phases [4].

The measurement of the decay width of the Z^0 boson into neutrinos proves that the number of light ($m_\nu \leq m_Z/2$) active neutrinos is exactly three, therefore a fourth neutrino must not couple to the standard electroweak currents, that is, it must be sterile.

At present almost all neutrino data are consistent with the assumption that there are only three neutrino mass eigenstates and three flavor eigenstates, and therefore in this paper I shall consider the "standard" scenario where the three light active neutrino flavor states (ν_e, ν_μ, ν_τ) and the three known mass eigenstates (ν_1, ν_2, ν_3) participate in the oscillations. Then, U is a 3×3 matrix analogous to the CKM matrix for the quarks, and it can be parametrized by three mixing angles $\theta_{12}, \theta_{13}, \theta_{23}$ and three complex phases $(\delta_{CP}, \eta_1, \eta_2)$, two of which (η_1, η_2) are physical only if neutrinos are Majorana fermions [5] and do not affect neutrino oscillations [6] [7].

In order to relate the elements of the mixing matrix to experimental observables, one needs to order the neutrino mass eigenvalues. This is done by requiring that: $m_2^2 > m_1^2$ and $\Delta m_{21}^2 < |\Delta m_{31}^2|$. In this case, there are three mass related oscillation observables: Δm_{21}^2, $|\Delta m_{31}^2|$, and the sign of Δm_{31}^2. A positive(negative) sign for Δm_{31}^2 implies $m_3^2 > m_2^2$ ($m_3^2 < m_2^2$) and characterizes the so-called normal (inverted) neutrino mass hierarchy.

Our current knowledge of neutrino oscillation parameters is summarized by [8] [9] [10]:

$$\Delta m_{21}^2 = 7.59^{+0.20}_{-0.18} \times 10^{-5} eV^2, \quad \Delta m_{31}^2 = 2.50^{+0.09}_{-0.16} \times 10^{-3} eV^2 \quad (-2.40^{+0.08}_{-0.09} \times 10^{-3} eV^2), \quad (2)$$

$$\sin^2\theta_{12} = 0.312^{+0.017}_{-0.015}, \quad \sin^2\theta_{23} = 0.52 \pm 0.06, \quad \sin^2\theta_{13} = 0.092 \pm 0.017 \quad (3)$$

There is no information, at present, about the value of δ_{CP} and on the sign of Δm_{31}^2. The measurement of δ_{CP} and the determination of the correct ordering of neutrino masses are two main goals of the Long-Baseline Neutrino Experiment (LBNE), a next generation neutrino oscillation experiment to be constructed in the United States.

In its reference design LBNE will consist of three main facilities; a new high intensity *neutrino beam* at the Fermi National Accelerator Laboratory (Fermilab) aimed at the Sanford Underground Laboratory at Homestake (Sanford Laboratory), 1300 km away, where a massive *neutrino detector* will be built and, a near detector complex located at Fermilab downstream of the target. With a baseline L of 1300 km and a neutrino beam energy peaking between 1 and 6 GeV LBNE will achieve an unprecedented sensitivity for the determination of δ_{CP} and of the mass hierarchy.

The oscillation channel investigated by LBNE will be the $\nu_\mu \to \nu_e$ appearance and its antineutrino counterpart $\bar{\nu}_\mu \to \bar{\nu}_e$ and the measurement will rely on the MSW effect [11].

In the hypothesis that the density of the matter along the path of neutrinos is constant, a reasonable hypothesis for the baseline in question, the oscillation of $\nu_\mu \to \nu_e$ in matter in a 3×3 mixing scenario is approximately given by [12]:

$$\begin{aligned} P(\nu_\mu \to \nu_e) \approx & \sin^2\theta_{23} \frac{\sin^2 2\theta_{13}}{(\hat{A}-1)^2} \sin^2((\hat{A}-1)\Delta) + \\ \pm \quad & \alpha \frac{\sin\delta_{CP} \cos\theta_{13} \sin 2\theta_{12} \sin 2\theta_{13} \sin 2\theta_{23}}{\hat{A}(1-\hat{A})} \sin(\Delta) \sin(\hat{A}\Delta) \sin((1-\hat{A})\Delta) \\ + \quad & \alpha \frac{\cos\delta_{CP} \cos\theta_{13} \sin 2\theta_{12} \sin 2\theta_{13} \sin 2\theta_{23}}{\hat{A}(1-\hat{A})} \cos(\Delta) \sin(\hat{A}\Delta) \sin((1-\hat{A})\Delta) \\ + \quad & \alpha^2 \frac{\cos^2\theta_{23} \sin^2 2\theta_{12}}{\hat{A}^2)} \sin^2(\hat{A}\Delta) \end{aligned} \quad (4)$$

where $\alpha = \Delta m_{21}^2 / \Delta m_{31}^2$, $\Delta = \Delta m_{31}^2 L / 4E$, $\hat{A} = 2VE/\Delta m_{31}^2$, $V = \pm\sqrt{2} G_F n_e$ is the matter potential, G_F is the Fermi coupling constant, n_e is the density of electrons in the matter. The $+$ sign in the second term of equation 4 and in the matter potential refers to neutrinos, the $-$ sign to antineutrinos. The second and third terms in equation 4 control the sensitivity to CP violation through the parameter δ_{CP}, and the first three terms predict a matter induced enhancement for the normal mass ordering ($\Delta m_{31}^2 > 0$) and a suppression for the reversed mass ordering ($\Delta m_{31}^2 < 0$) through α, Δ and $\hat{A}$. At a baseline of 1300 km the oscillation probability will be in the 5-10% range at the first oscillation maximum between 2 - 3 GeV, and at lower energies both the size of the probability and the effect of the CP phase will be larger.

2. Motivation for a Near Detector complex

LBNE will study neutrino oscillations by measuring event rates at the far detector (FD) with high statistics. In order to extract the oscillation probabilities from the measurement above it is paramount to precisely predict such rates, and therefore to gain a detailed enough knowledge of both the signal and backgrounds.

In the energy range where the first and second oscillation maxima are expected at the LBNE FD the neutrino cross sections and the momenta of the outgoing particles are poorly known (pion production in the resonance regime is an example), and it is therefore crucial to measure the neutrino fluxes and interaction channels at a near site, before the fluxes have been affected significantly by neutrino oscillations. This task will be performed by the LBNE near detector complex, a suite of detectors with the purpose of making all those measurements required to mitigate the effect of the large systematic uncertainties affecting neutrino fluxes, cross sections and background in the region of interest in such a way that the ultimate sensitivity of LBNE is not limited by them.

Given the energy range where the $\nu_\mu \to \nu_e$ oscillation probability peaks, it is necessary, in particular, to focus on $E_\nu < 8$ GeV, as well as on higher neutrino energies that produce background in the region of interest for the oscillation signal. In this region it is critical to identify and measure processes that can mimic oscillation signals at the far detector; among these the intrinsic ν_e (and in its antineutrino counterpart when running the accelerator in antineutrino mode) background in the beam is of particular interest when investigating the $\nu_\mu \to \nu_e$ oscillation channel, and it has to be known with sufficient precision to be subtracted at the far site.

ν_es ($\bar{\nu}_e$s) are detected through their charged current (CC) interactions, and in particular through the measurement of the outgoing charged electron (anti-electron).

Another dangerous background for these interactions is represented by neutral current (NC) events containing π^0s, the dominant source of background from neutrino NC interactions.

π^0s decay to two gammas which, after conversion to e^+e^- pairs, initiate an electromagnetic shower that can be difficult to distinguish from an electron shower.

It is paramount to be able to separate muons from electrons in order to distinguish μ and e CC interactions.

The other important ingredient necessary to predict the event rates at the FD is the knowledge of the neutrino flux; since the hadron production models used to simulate hadron propagation and interaction in the target and horn materials are not well constrained, measurements of particles in the decay region as well as external measurements of hadron production and propagation in the target and horn materials are highly desirable.

3. The near detector complex

In what follows I shall describe the reference design for the near detector complex as described in the Conceptual Design Report document, phase 1; LBNE is currently undergoing a process of revision and the complex described here might not be the actual one.

The near detector complex will be located at Fermilab, downstream of the beamline, and will consist in a set of beam line measurements (BLM) and a neutrino detection (ND) system.

The ND will be placed underground in the Near Detector Hall 450 m downstream of the target, with the purpose of measuring or constraining neutrino fluxes and spectra and to measure the neutrino-interaction channels important for predicting the signal and backgrounds at the far site.

The BLM will be placed in the region of the absorber at the downstream end of the decay region. Its goal is to determine the neutrino fluxes and spectra and to monitor the beam profile on a spill-by-spill basis, and this will be achieved by measuring the muons exiting the decay volume. In fact, the two-body decays of pions and kaons that produce neutrinos also result in the creation of a daughter muon. Since muons and neutrinos come from the same parent pion and kaon decays, a measurement of the absolute muon flux in conjunction with the energy spectrum seen in the muon monitors can confirm the absolute neutrino flux.

LBNE will also rely on a set of *external* (performed, for example, using the MIPP or NA61 detectors) *hadron-production measurements*, some of which will be performed as a part of the project itself. The simulation of the hadronic cascades in the target, horn and decay tunnel require an accurate knowledge of the phase-space distribution of the initial proton beam, of the geometry and material of all the elements in the beamline, of the hadron-production cross sections and of the meson-to-neutrino decay rates. All these items can be simulated accurately except for the hadronic cascades, whose prediction relies on models based on hadron-production measurements. In order to keep the uncertainty in the near/far event-rate ratio from being limited by systematic uncertainties in the flux, the LBNE flux simulation must be accurate at the 4-5% level.

3.1. Muon Measurement Facilities

Figure 1 shows a conceptual layout of the muon measurement systems.

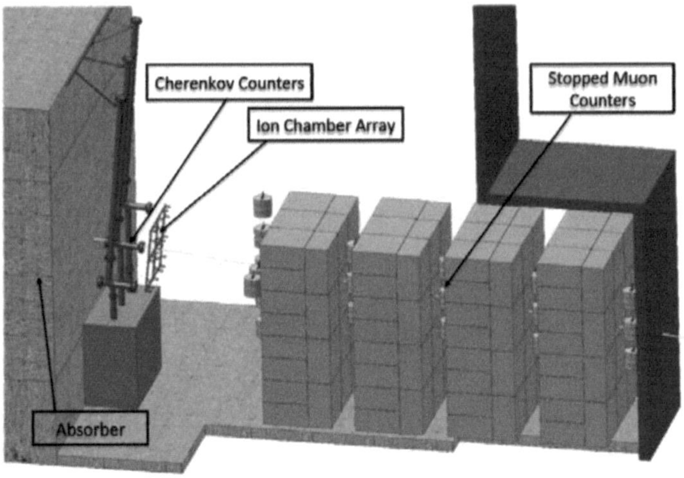

Figure 1. default

The first set of muon-measurement devices, from left to right, is the *ion-chamber array* which is mounted directly to the rear wall of the absorber. Following that, a set of three variable-pressure *gas Cherenkov counters*, and finally a set of *stopped-muon counters* which are interspersed between walls of steel blocks. The blocks are there to provide several depths at which to monitor the stopped muons as they range out in the material.

The ion-chamber array has the purpose of monitoring the stability of the beam direction over time; even a tiny misalignment of the beam can in fact result in a significant change of the flux at the FD. Because the muon monitors will be located approximately 275 m from the beam target, this requires a measurement of the muons to an accuracy of approximately 5 cm. To keep the change in the neutrino beam less than 1% in all energy bins, the beam direction must be known to a precision of approximately 0.2 mrad.

The muon monitors must also be capable of operating in a high-radiation environment; the rate of muons crossing the monitors will in fact be approximately equal to 50 million muons per cm^2 for a pulse of 10^{14} protons-on-target. Due to their high radiation tolerance, sealed ionization counters are the default technology, and the conceptual design is based on commercial ion chambers by LND, Inc. model 50343. The chambers will be mounted in a 5×5 grid, four meters by four meters and arranged as shown in figure 1.

One disadvantage to ionization counters is that they measure the total ionization deposited from all particle species (including the delta-ray electrons produced by the muons), making it challenging to convert the ionization signal into an absolute muon flux. The LBNE ND complex plans to use the ionization counters to monitor the beam stability, direction and shape, but not to determine the absolute flux of muons or to determine the muon-energy spectrum, for which, respectively, the gas Cherenkov counters and the stopped-muon counters will be used.

The method exploited by the stopped-muon counters is to stop muons in a material with significant carbon content and, via muon capture, to produce ^{12}B that will in turn undergo β decay. The high-carbon material, in this case graphite, surrounds a Cherenkov radiator material which is sensitive to electrons from muon decay or high-energy beta decays. μ^+ will be identified based on their 2.2 μs decay time. The detectors would only operate in the lower-rate environment that is present many microseconds after the beam pulse is over.

As mentioned above, the muon flux will be measured in the near detector complex using Cherenkov counters. They will be deployed downstream of the absorber and will not image individual Cherenkov rings, but rather see the integrated signal from many muons due to the very large instantaneous flux. In addition, by varying the radiator gas pressure, and hence the Cherenkov threshold, the systems index of refraction would vary, allowing it to map out the muon momentum distribution. The conceptual design is based on a traditional beamline Cherenkov counter, where a gas radiator is contained in a pressurized tube.

3.2. Neutrino Detector

The reference design for the LBNE FD is a massive Liquid Argon Time Projection Chamber (LAr-TPC); to avoid including unnecessary uncertainties from extrapolation between different nuclei, the neutrino detector at the near site will include the same nuclear target as the Far detector.

Its reference design is a magnetized LAr-TPC consisting of the following components:

- a cryostat to contain the cryogenic liquid-argon volume
- a time projection chamber (TPC) immersed in the LAr volume
- read-out electronics and data acquisition for the TPC
- a 0.4-T dipole magnet surrounding the cryostat
- a muon-identifier detector (MuID) on the sides and ends of the magnet

The total active mass of the TPC will be 18 t: with a 120-GeV proton beam, assuming 0.5×10^{14} protons per beam spill, this corresponds to 3 events per spill in the active volume.

The vertex resolution if the LAr-TPC will be sufficiently good (< 1 cm) to determine whether a track comes from the vertex (in the case of an electron) or is displaced from the vertex (in the case of a gamma).

The detector will also allow the reconstruction of the incident neutrino energy for CCQE events with good resolution in order to measure the neutrino fluxes as a function of energy and the reconstruction of neutral and charged pions produced in NC and CC scattering events. This is important because π^0 events constitute the largest background to ν_e and $\bar{\nu}_e$ appearance and π^+ and π^- events constitute the largest background to ν_μ and $\bar{\nu}_\mu$ disappearance (this last channel is used to make precision measurements of θ_{23} and Δm_{32}).

The presence of a magnetic field within the detection volume is required to measure the momentum and charge-sign of the reaction products from NC, CC and CCQE interactions to allow the determination of the beam flux at the near site and the measurement of ν_e appearance backgrounds. In this respect, separating μ^+ and μ^- is particularly important in antineutrino running, when the contamination of wrong sign neutrinos in the beam is larger. A magnetic field of 0.4 T over the length of the tracker is sufficient to allow for the determination of the charge-sign of the muons. The magnetic field will be generated by a dipole magnet modeled on the large-aperture UA1 magnet.

The MuID will consist of inexpensive scintillator planes interspersed between the thick steel plates of the dipole magnet. This detector is only meant to provide the identification of the muon, the muon momentum will be measured by the TPC inside the magnetic field. The motivation for it is that πs and μs have similar energy depositions, and πs misidentified as μs could be erroneously included in the CC ν_μs spectra.

Besides the reference design presented here a more expensive alternative design based on a Straw Tube Tracked has also been prepared, and the details about both designs can be found in the Conceptual Design Report.

I thank the LBNE collaboration and the DOE for support through grant #HEP-KA/NS03/9/ARRA-2 Rev.0.

References

[1] Fukuda S *et al.* (Super-Kamiokande) 2002 *Phys. Lett.* **B539** 179–187 (*Preprint* hep-ex/0205075)
[2] Ahmad Q R *et al.* (SNO) 2002 *Phys. Rev. Lett.* **89** 011301 (*Preprint* nucl-ex/0204008)
[3] Abe S *et al.* (KamLAND) 2008 *Phys. Rev. Lett.* **100** 221803 (*Preprint* 0801.4589)
[4] Eidelman S *et al.* (Particle Data Group) 2004 *Phys.Lett.* **B592** 1
[5] Gonzalez-Garcia M and Nir Y 2003 *Rev.Mod.Phys.* **75** 345–402 (*Preprint* hep-ph/0202058)
[6] Bilenky S M, Hosek J and Petcov S 1980 *Phys.Lett.* **B94** 495
[7] Langacker P, Petcov S, Steigman G and Toshev S 1987 *Nucl.Phys.* **B282** 589
[8] Schwetz T, Tortola M and Valle J 2011 *New J.Phys.* **13** 109401 (*Preprint* 1108.1376)
[9] Machado P, Minakata H, Nunokawa H and Funchal R Z 2012 *JHEP* **1205** 023 (*Preprint* 1111.3330)
[10] An F *et al.* (DAYA-BAY Collaboration) 2012 *Phys.Rev.Lett.* **108** 171803 (*Preprint* 1203.1669)
[11] Wolfenstein L 1978 *Phys.Rev.* **D17** 2369–2374
[12] Freund M, Huber P and Lindner M 2001 *Nucl.Phys.* **B615** 331–357 (*Preprint* hep-ph/0105071)

Cosmic Ray in the Northern Hemisphere: Results from the Telescope Array Experiment

C. C. H. Jui[†], for the Telescope Array Collaboration
[†]Department of Physics and Astronomy, University of Utah, Salt Lake City, UT, USA

Abstract. The Telescope Array Experiment the largest extensive air-shower cosmic-ray detector in the northern hemisphere. TA combines three fluorescence detectors and a scintillation counter surface detector. The physics goals of TA are to measure the spectrum, composition, and arrival direction anisotropy of ultrahigh energy cosmic rays. We will present the latest results from Telescope Array.

1. Introduction

The Telescope Array (TA) experiment [1] is located in the desert of Millard County, Utah. The physical layout of the experiment is shown in figure 1. Three fluorescence detectors (FDs) shown by the green squares, are sited at the edge of a central ground array 507 scintillation surface detectors (SDs) marked by the Black squares. The ground array is arranged in a square grid pattern with a nearest-neighbor separation of 1.2 km, covering a total of 700 km^2. Each FD station views approximately 30° in elevation and 110° in azimuth, inward onto the surface array. Overall, the telescope array detector systems constitute a very large, homogenous, sampling calorimeter. The total mass of the active volume is of the order of 10^{13} kg. It is a non-compensating calorimeter with an e/h ratio of about 1.1.

2. TA Surface Detectors

The active component of a TA SD counter is made of two layers of plastic scintillators, with a total collection area for particles of 3.0 m^2. Extruded grooves in the plastic carry wavelength-shifting optical fibers that collect and concentrate the scintillation light into two photo-multiplier tubes (PMTs), one for each layer. Each SD is powered entirely by a solar-panel battery power unit, and exchanges data over a 2.4 GHz wireless network. The SD counters are self-calibrated at ten-minute intervals using distribution of pulse heights from minimum-ionizing cosmic muons. Each PMT is continuously sampled at 40 million samples for second (40 MSPS). Data is stored at a counter for pulses corresponding to at least 0.3 vertical equivalent muon (VEM) over background. An event trigger occurs when a minimum of three adjacent counters are fired with at least 3 VEM each. A typical event is shown in figure 2.

In figure 2(a), the triggered counters and the their measured densities are shown by the circles and their area, respectively. The location, signal size, and arrival time detected by each counter are used to find the core location and arrival direction of the shower, using a modified Linsley time delay function [2]. The result of this fit is illustrated in figure 2(b). The particle densities from the participating SDs are then fitted to their distance from the shower core to the AGASA lateral density function (LDF) [3], and the density at 800 meters (S800) is interpolated from

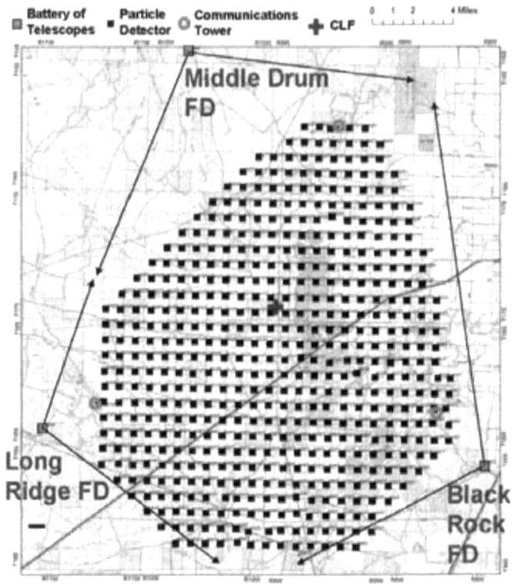

Figure 1. The physical layout of the Telescope Array experiment. Black squares show the location of the 507 SDs. Green squares mark the FD stations.

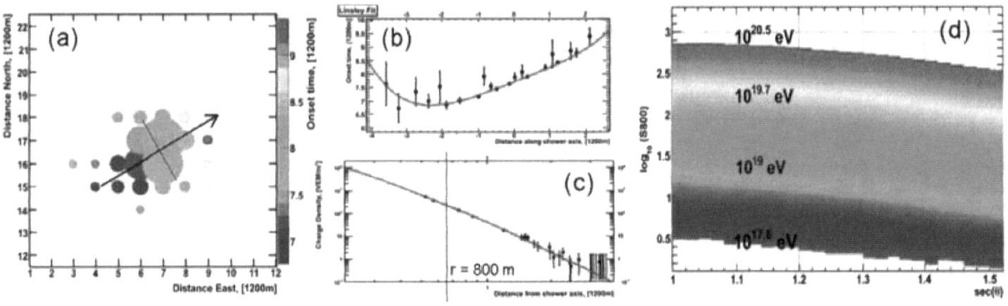

Figure 2. (a) Display of a typical air shower event captured by the TA surface detector array. Each circle represents a triggered SD unit. (b)Shower arrival times at hit SD units fitted to determine the shower arrival direction. (c) Bottom Right: Particle density vs. distance from the shower axis fitted and interpolated to obtain the density at 800 m (S800). (d) Variation of S800 with the energy and zenith angle of simulated air showers.

the fit, as shown in figure 2(c). The S800 value is then compared to the average from simulated events, shown in figure 2(d) and the energy is interpolated according to the zenith angle.

3. TA Fluorescence Detectors

A total of 38 fluorescence telescopes are deployed over three sites. The first two sites were constructed on Black Rock (BR) Mesa and Long Ridge (LR) at the southern flanks of the SD array. These 24 telescopes were built in Japan, based on the same optical design and 256-pixel segmentation as the HiRes experiment, but with 30% larger mirrors of 6.8 m^2 area. A typical event captured by the FD station at Black Rock is shown in figure 3(a). Signals are recorded by each PMT pixel at 10 MHz, those containing a three sigma excess over night-sky background are designated hit pixels, and displayed as circles in the figure. The area of each circle is proportional to the integrated pulse area.

The pointing directions of the hit pixels are used to fit for a shower-detector plane (SDP),

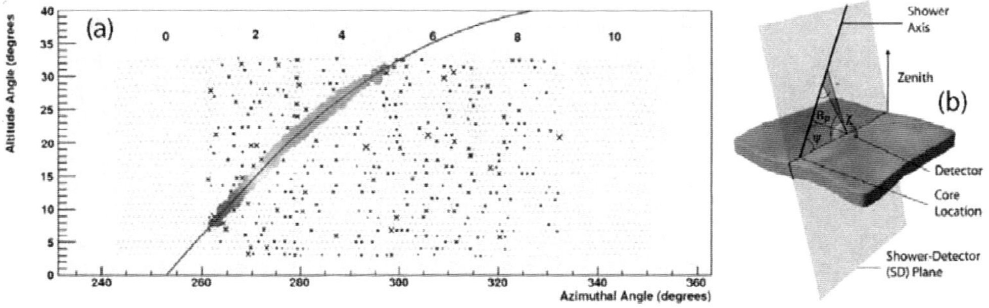

Figure 3. (a) Display of a typical downward air shower event captured by the FD station at Black Rock. The circles correspond to channels with pulses in excess of three sigma over background, and the area of each circle represents the integrated pulse area. (b) Illustration of the FD shower-detector plane, and the time-fit to determine the shower trajectory.

which appears as a curve in the cylindrical projection of figure 3(a). The colors of pixels indicate time progression: blue represents the earliest arrival times at the top of the event, and red represents the latest at the bottom. The event depicted was clearly a downward going air shower. Next, the shower axis is found by fitting the pixel arrival times to

$$t_i = t_0 + \frac{R_P}{c} \tan\left(\frac{\pi - \psi - \chi_i}{2}\right)$$

As illustrated in figure 3(b), The output of the fit are: (1) R_P, nearest distance of approach of the shower axis to the FD, (2) ψ, the angle made by the shower axis to the line of intersection between the SDP and the ground, and (3) t_0, physically corresponds to the time at which the shower passes the point of nearest approach. The inputs are the measured times t_i and the angles χ_i made between the PMT pointing direction (projected onto the SDP) and the ground, measured within the shower detector plane.

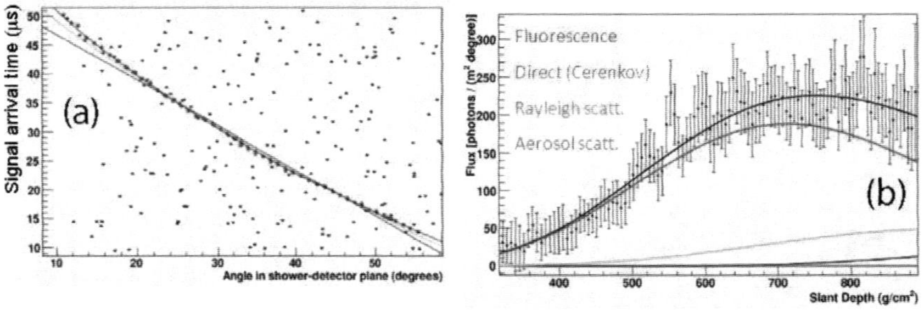

Figure 4. (a) Time-fit to determine R_P and ψ. (b) Profile-fit to determine the energy and X_{max} for the shower shown originally in figure 3(a)

Figure 4(a) shows the time-fit described above for the event shown in figure 3(a). Next, the pointing directions of the PMTs are converted to slant depth and the PMT signals are then fitted to a parametric Gaisser-Hillas form [4] for the shower size vs. depth. This fit includes scattered and direct Cherenkov light in addition to the fluorescence signal. The shower-profile fit for this same event is shown in figure 4(b). The Energy is extracted from the overall area of the curve, and the depth of the shower maximum, X_{max}, from the longitudinal extent of this curve. Over many showers, the X_{max} values give a statistical measure of the composition.

4. TA Energy Spectrum Measurement

A priority objective for TA was to resolve the controversy in the Greisen-Zatsepin-K'uzmin [5] cut-off. Using the fluorescence technique alone, HiRes reported the first observation of the GZK cut-off in 2008 [6], whereas earlier spectrum measurements by AGASA, using a scintillation ground array alone reported a continuing spectrum [7].

The MD FD station redeployed 14 telescopes from HiRes-1, the latter having provided most of the statistical significance for the GZK cut-off. This commonality between HiRes and TA allows us to compare the results of the two experiments directly. A variant of the monocular reconstruction was used that combined the timing and profile-fits of figure 4. This profile-constrained fit (PCF) was developed to overcome the very short tracks seen by the HiRes-1 detector, which only viewed up to 17° in elevation. To make a direct comparison, this analysis of the MD monocular FD data used exactly the same simulation and PCF reconstruction codes as was used for HiRes-1. The only changes made were updates of pointing geometry of the detectors, and lowering the trigger threshold in the simulation to reflect the reduced ambient background light at the TA site. Figure 5(a) shows the monocular spectrum from the Middle Drum FD station from its first three years of observation [8], overlaid with the HiRes monocular spectrum [6]. The two results agree both in the shape and normalization. The MD monocular spectrum is also consistent with the GZK cutoff.

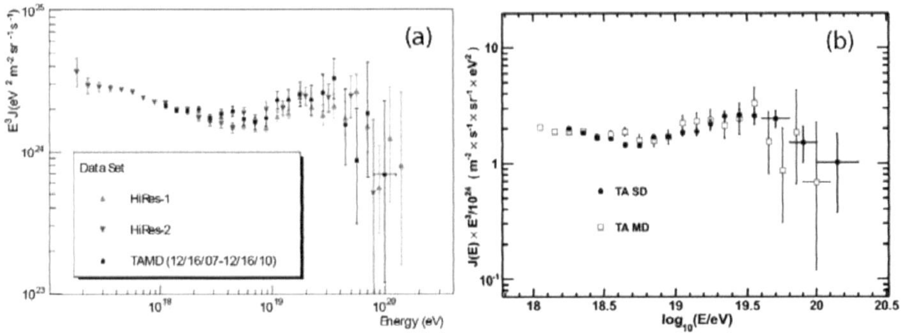

Figure 5. (a) TA monocular spectrum from the Middle Drum FD station from its first three years of observation, overlaid with the monocular spectra from HiRes. (b) TA surface detector spectrum with event energies scaled down by a factor of 1.27, overlaid with the monocular FD spectrum from Middle Drum.

Figure 5(b) shows the cosmic ray energy spectrum from the first three years of TA SD data, overlaid with the monocular MD FD spectrum. The FD and SD spectra are in excellent agreement, and the SD spectrum clearly shows a suppression at the expected GZK cut-off. Note that the energies of the SD events used to make Figure 5(b) were divided by a scaling factor of 1.27 obtained by comparing the reconstructed energies of hybrid events seen by both FD and SD. Figure 6(a) shows a scatter-plot of log FD energy vs. log SD energy for the coincident events, after applying the 1.27 correction factor to the SD. This plot shows a linear relationship between the two energy measurements over the 1.5 decades of energy above 3×10^{18} eV.

The hybrid event set can be analyzed using another variant of the monocular reconstruction, where times from the SD counters can be added to those of the FD for the fit to equation 3.

The benefit of this technique is that it provides much improved resolutions over that obtained my monocular reconstruction: angular resolution of $\sim 0.6°$ for hybrid compared to $\sim 5-8°$ for monocular, and X_{max} resolutions of $\sim$ 30 g/cm^2 for hybrid over that of $\sim 50-60$ g/cm^2 for monocular. Figure 6(b) shows a preliminary hybrid spectrum from the Middle Drum FD data overlaid with the SD spectrum. Again the two are in excellent agreement.

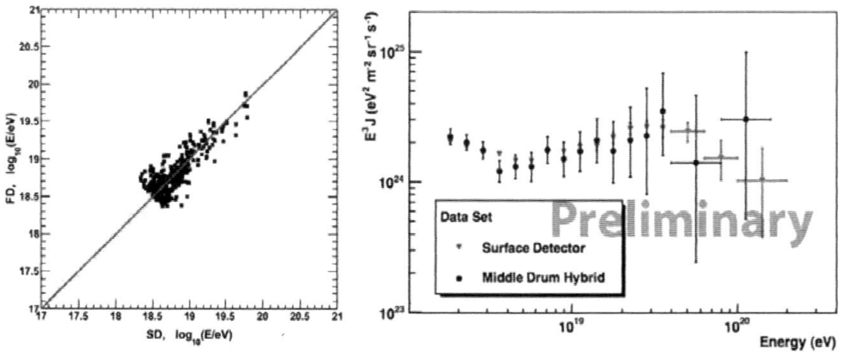

Figure 6. (a) Scatter-plot of the log FD energy vs. log SD energy (rescaled by 1.27) for hybrid events above 3×10^{18} eV. (b) The hybrid FD spectrum from Middle Drum overlaid with the (energy-rescaled) SD spectrum.

5. Composition and Anisotropy

Since 2009, there has been a discrepancy in the X_{max}-based composition results between the AUGER and HiRes collaborations. AUGER claims to see a trend toward heavier composition at above 10^{19} eV [9], whereas HiRes results are entirely consistent with proton composition [10]. Figure 7 shows the first TA composition result based on X_{max} from stereo events, where the shower trajectory can be determined from the intersection of the two SDPs In Figure 7(a), the distribution of X_{max} for TA stereo events is compared to those of iron and proton events simulated with CORSIKA using the QGSJET-II hadronic model. Figure 7(b) shows the plot of mean X_{max} vs. log energy for the same stereo data set. The accompanying curves show the predictions (folding in detector response and trigger selection) of CORSIKA simulations with different hadronic interaction models. The TA data, like that for HiRes, is again consistent with a predominately protonic composition, especially when compared to QGSJET models. Composition studies based on the width of the X_{max} distributions, and on the width of the shower profiles as well as those using hybrid events are nearing completion.

The anisotropy searches in TA are based on the SD data. After the first three years, our data is entirely consistent with isotropy. A search was made in TA data against the claim made by the AUGER collaboration in the 2007 Science article [11], where 8 of 13 AUGER events above 5.7×10^{19} eV were seen to be within 3.1° of Active galactic nuclei in the Veron-Cetty catalog [12] with $z < 0.018$. For the northern sky, the prediction for TA was 15 correlations out of 20 TA events above 5.7×10^{19} eV, with five accidental background coincidences. Of the 20 events, eight were seen to be in coincidence with AGNs. This result is not a particularly significant departure ($p = 0.13$) from the null hypothesis. More TA data are needed.

6. Acknowledgments

The Telescope Array experiment is supported by the Japan Society for the Promotion of Science through Grants-in-Aid for Scientific Research on Specially Promoted Research (21000002)

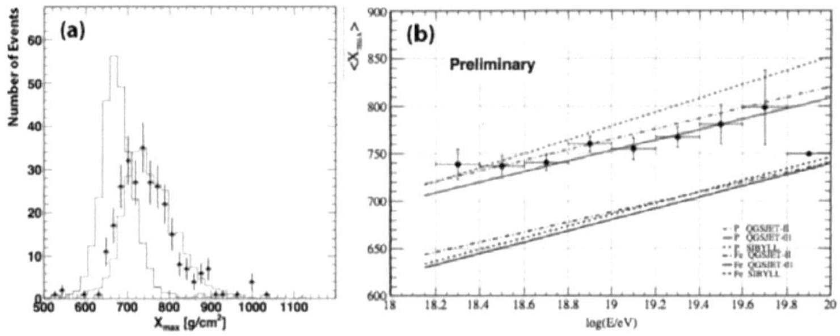

Figure 7. (a) Distribution of shower maximum depth (X_{max}) of TA stereo data compared to CORSIKA simulation for proton and iron, based on the QGSJET-II hadronic interaction model. (b) Plot of mean X_{max} vs. log energy for the same TA stereo data set.

"Extreme Phenomena in the Universe Explored by Highest Energy Cosmic Rays", and the Inter-University Research Program of the Institute for Cosmic Ray Research; by the U.S. National Science Foundation awards PHY-0307098, PHY-0601915, PHY-0703893, PHY-0758342, and PHY-0848320 (Utah) and PHY-0649681 (Rutgers); by the National Research Foundation of Korea (2006-0050031, 2007-0056005, 2007-0093860, 2010-0011378, 2010-0028071, R32-10130); by the Russian Academy of Sciences, RFBR grants 10-02-01406a and 11-02-01528a (INR), IISN project No. 4.4509.10 and Belgian Science Policy under IUAP VI/11 (ULB). The foundations of Dr. Ezekiel R. and Edna Wattis Dumke, Willard L. Eccles and the George S. and Dolores Dore Eccles all helped with generous donations. The State of Utah supported the project through its Economic Development Board, and the University of Utah through the Office of the Vice President for Research. The experimental site became available through the cooperation of the Utah School and Institutional Trust Lands Administration (SITLA), U.S. Bureau of Land Management and the U.S. Air Force. We also wish to thank the people and the officials of Millard County, Utah, for their steadfast and warm support. We gratefully acknowledge the contributions from the technical staffs of our home institutions and the University of Utah Center for High Performance Computing (CHPC).

[1] Fukushima, M., Institute for Cosmic Ray Research Mid-term (2004 - 2009) Maintenance Plan Proposal Book "Cosmic Ray Telescope Project". 2002, Tokyo University. http://www.telescopearray.org
[2] J. Linsely, J. Phys. G: Nucl. Phys., **12**, 51-57, 1986; D. Ikeda *et al.* Astrophys. Space Sci. Trans., **7**, 257263, (2011).
[3] S. Yoshida, N. Hayashida, K. Honda *et al.* J. Phys. G: Nucl. Phys., **20**, 651664, (1994).
[4] T. Gaisser and A. M. Hillas, Proc. 15th ICRC **8**, p. 353, (1977).
[5] K. Greisen, Phys. Rev. Lett. **16**, 48 (1968); T. Zatsepin and V. A K'uzmin, Pis'ma Zh. Eksp. Teor. Fiz., **4**, 114 (1966).
[6] R. U. Abbasi *et al.* Phys. Rev. Lett. **92**, 151101 (2008).
[7] M. Takeda *et al.*, Astropart. Phys. **19**, 447 (2003).
[8] T. Abu-Zayyad *et al.*, Astropart. Phys., in press (2012). http://arxiv.org/abs/1202.5141.
[9] J. Abraham *et al.*, Phys. Rev. Lett. **104** 091101 (2010).
[10] R. U. Abbasi *et al.*, Phys. Rev. Lett. **104** 161101 (2010).
[11] The Auger Collaboration, Science, 318 p. 938, (2007).
[12] M.-P. Cetty and P. Veron, Astronomy and Astrophysics, **455**(2), p. 773, (2006).

Status of RadioWave Neutrino Detection

Dave Besson

KU Physics Dept., 1082 Malott Hall, Lawrence KS 66045

E-mail: zedlam@ku.edu

Abstract. As of this writing, there are three dedicated experiments, all based in Antarctica, which seek first-ever measurement of the ultra-high energy neutrino flux at Earth. All three (ANITA, ARA and ARIANNA) exploit the Askaryan Effect to detect the so-called cosmogenic neutrinos which should result from interactions of ultra-high energy baryons with the Cosmic Microwave Background Radiation (CMBR). Photoproduction of those neutrinos, via $N\gamma \rightarrow \Delta \rightarrow N\pi^{\pm}$, with subsequent weak decays of those pions resulting in neutrinos. In-ice weak and neutral scattering of those neutrinos off ice molecules can yield in a detectable pulse of coherent, radio-frequency radiation. We summarize the three experiments, and discuss prospects.

History

Radio Detection of UHE Neutrinos, Theory

Coherent radio Cherenkov studies go back to Jelly, who first studied radio signals from cosmic ray air showers[1, 2]. Askaryan[3] subsequently predicted a net charge imbalance in dense media and coherent radio power scaling like the energy of the shower squared, providing impetus for continued progress in UHE air shower studies. The basic idea of coherent detection of radio-frequency Čerenkov radiation emanating from a neutrino induced electromagnetic shower is straightforward. The number of particles N in a shower of energy E is proportional to E. The shower has an approximately fixed ratio of electrons to positrons ($\simeq 1.25$) arising primarily because of scattering of atomic electrons into the developing shower; the shower will thereby build up an effective negative electrical charge proportional to its energy. Čerenkov radiation, which is intrinsically broadband, will be produced by this relativistically moving charge. If the electromagnetic shower develops in a dense, but transparent medium, the energy in the shower will be concentrated on a Čerenkov front which can travel a long distance with little attenuation. The radiation will be emitted incoherently for very small wavelengths relative to the lateral size of the shower, but coherently for wavelengths greater than the lateral size of the shower ($\sim 10\,\mathrm{cm}$ in ice). At such radio-frequencies, the power emission will be proportional to the shower energy squared. This rapid increase in power with energy substantially compensates for the decreasing neutrino flux, which has an integral spectrum falling roughly as $1/E^2$. A test beam experiment at SLAC by Gorham and Saltzberg has given the most direct confirmation of the Askaryan effect[10].

Russian collaborators performed the first experimental studies, and laid the initial groundwork for development of a radiowave-based neutrino detection experiment. That previous work includes the definitive measurements at Vostok of the temperature profile of the ice down to 2 km (1.5 km above bedrock), as well as measurement of the frequency and temperature

Published under licence by IOP Publishing Ltd

dependence of the ice absorption of radio waves. A pilot experiment ("RAMAND", for Radio wave Antarctic Muon And Neutrino Detector) tested many aspects of the neutrino radio detection idea at Vostok between 1985-1990, including critical initial measurements in Antarctica of natural and man-made radio impulse backgrounds[4, 5, 7, 8, 9]. Unfortunately, that nascent effort was abruptly terminated in 1991, when the Soviet political (as well as scientific) infrastructure collapsed.

In 1994, George Smoot and collaborators at the Lawrence Berkeley Lab characterized the radio-frequency environment at South Pole ("RAND"), however, that project also lacked follow-up. The RICE experiment, which first deployed in-ice antennas in 1995, represented a prototype of the in-ice neutrino detection technique and can perhaps be credited with first demonstrating the operational feasibility of such an effort.

Simulations of the signal-generation mechanism are, in principle, straightforward. For definiteness, consider a ν_e undergoing a charged current interaction, $\nu_e + N \to e + N'$. The primary UHE neutrino transfers most of its energy to the electron, which quickly builds an exponentially increasing shower of e^+e^- pairs. The number of pairs N_e scales like the primary energy. In the most populated region of the shower, at the "bottom" of its energy range, a charge imbalance develops as positrons drop out and atomic electrons scatter in. Detailed Monte Carlo calculations by Zas, Halzen and Stanev (ZHS)[11] and confirmed by later GEANT simulations find that the net charge of the shower is about 20% of N_e. The electric field produced by this relativistic pancake is dominated by coherent Cherenkov radiation for wavelengths in the radio frequency region. Equivalently, for wavelengths large comparable to the transverse size of the shower ($\sim 2r_{Moliere}$, or $\sim$10cm in ice), the relativistic pancake can be treated as a single, extended, radiating charge. (Clearly, in the limit $\lambda \to \infty$, the radiating region approaches a point charge.)

Cold polar ice offers several advantages which make it an extremely attractive candidate target medium for neutrino detection. In addition to comprising a huge, uniform target, measurements to date indicate that cold ice ($\leq -50°$C) has extremely long field attenuation lengths, of order $\geq$1-3 km for 100 MHz - 1 GHz radio signals[12], whereas optical photons in ice have absorption and scattering lengths which are typically an order of magnitude smaller.

Ultra-high energy (UHE) neutrinos point back to sources of high energy cosmic rays, providing an identification of the source as well as a powerful test of models for the acceleration mechanism. Detection of UHE neutrino fluxes simultaneous with gamma-ray bursts would provide essential information on the nature of these extraordinarily luminous sources[13]. At even higher energies, "GZK" neutrinos may help identify more exotic sources such as topological defects[14]. In the realm of particle physics, detection of UHE neutrinos from cosmological distances, if accompanied by flavor identification, may permit measurement of neutrino oscillation parameters over an unprecedented range of Δm^2[15, 16], or detect ν_τ via "double-bang" signatures[17]. The angular distribution of upward-coming neutrino events could be used to measure weak cross-sections at energies unreachable by man-made accelerators[18]. Alternately, if the high energy weak cross-sections are known, they can be used to test Earth composition models along an arbitrary cross-section (so-called 'neutrino tomography')[19]. Most exciting of all, as the history of astronomy has demonstrated, the implementation of each new technology has been rewarded with spectacular, yet unanticipated, discoveries.

Radio Detection of Neutrinos, Experiment

The ANITA Experiment

Initiated in 2003, the Antarctic Impulsive Transient Antenna (ANITA) is a balloon-borne antenna array primarily designed to detect radio wave pulses caused by neutrino collisions with matter, specifically ice. The basic instrument consists of a suite of 40 quad-ridged horn antennas, optimized over the frequency range 200-1200 MHz, with separate outputs for vertically vs. horizontally incident radio frequency signals, mounted to a high-altitude balloon. From an elevation of ~38 km, the balloon scans the Antarctic continent in a circumpolar trajectory. Details on the ANITA hardware, as well as the triggering scheme crucial to the analysis described herein, are provided elsewhere[20].

Two one-month long missions (ANITA-1; Dec. 2006-Jan. 2007 and ANITA-II; Dec. 2008-Jan. 2009) have yielded world's-best limits to UHE neutrino flux in the energy range to which ANITA is sensitive[21, 22]. A recent analysis of the ANITA-1 data sample has also provided a statistically large (16 events) sample of radio frequency signals attributed to geosynchrotron radiation associated with cosmic-ray induced extensive air showers (EAS)[23]. That analysis not only demonstrated that radio wave detectors can independently trigger on EAS, but also established the air-borne strategy as an efficient EAS detection technique, and further affirmed the broad range of science accessible to ANITA.

The heirarchical ANITA-II trigger allows a relatively low neutrino detection trigger threshold, while maintaining a tolerable (~Hz) thermal noise data rate written to disk. In contrast to ANITA-I, only signals from the VPol channel of the dual-polarization horn antennas contribute to the ANITA-II trigger. Following the antenna, signals routed through the 'trigger' (vs. 'digitization') path are tested for their spectral power in four frequency bands, approximately covering the intervals (in MHz) 200→350, 330→600, 630→1100 and 150→1240, respectively. This partitioning is performed in order to provide rejection power against narrow-bandwidth anthropogenic backgrounds, while retaining high efficiency for sharp duration (temporally), large-bandwidth neutrino signals. The frequency-banded signals are then passed through a tunnel-diode, which integrates roughly 7-ns units of data and provides a unipolar (negative) output pulse. The lowest-level trigger requires signal in one of the four frequency bands at a level exceeding approximately $2.3\sigma_V$, with σ_V the rms of the typical tunnel-diode output voltage at this point.

The heirarchical trigger and small occupancy allows ANITA to collect data very close to the intrinsic thermal noise floor, resulting in excellent sensitivity. Although no neutrinos have been found thus far, ANITA has, as mentioned previously, nevertheless demonstrated the ability to self-trigger on radio emissions from air showers and is, to date, the only existing experiment to have successfully done so. That observation is an important consideration in trigger and hardware modifications planned for ANITA-3, which is slated to fly in Dec., 2013.

The ARA Experiment

ARA stations are designed to detect a highly linearly-polarized RF impulse arriving from any direction, either as an approximate plane wave (for distant sources) or a spherical wave for closer sources. The individual antennas that comprise a station must have response that is preferably dipole-like in angular extent, but the antenna array must also be sensitive in two orthogonal polarizations to avoid any bias against detecting arbitrary planes of polarization, since the RF impulses that arise from the Askaryan effect may be observed at an arbitrary angle with respect to the plane of polarization. Because of practical limitations on drilling, antennas should fit within a 15 cm diameter borehole, and also be below the ~150 m thick firn, the top layer of the ice sheet made of compacted snow of density lower than that of solid ice. Due to the refractive index gradient, signals within the firn are subject to ray-bending which complicates triggering and reconstruction, and also leads to an effective inverted horizon, cutting off the detectability of more distance sources in the ice. Since South Polar ice has its highest RF clarity in the coldest ice, the antenna array response should be optimized for detection in the upper 1 to 1.5 km of ice, although deeper events will also be seen.

The prototype 'testbed' station deployed in 2010-11 at 20–30 m ice depths[24], with electronics and antennas very similar to the full-fledged ARA01 (2011-12) and ARA02/ARA03 (2012-13) stations to be deployed to 200 m depth. The ARA testbed installation was completed in the third week of January 2011, at a location approximately 1.8 km grid East of the IceCube detector, and has been operating nearly continuously since then.

The projected science performance of ARA-37 is summarized in Figure 1.

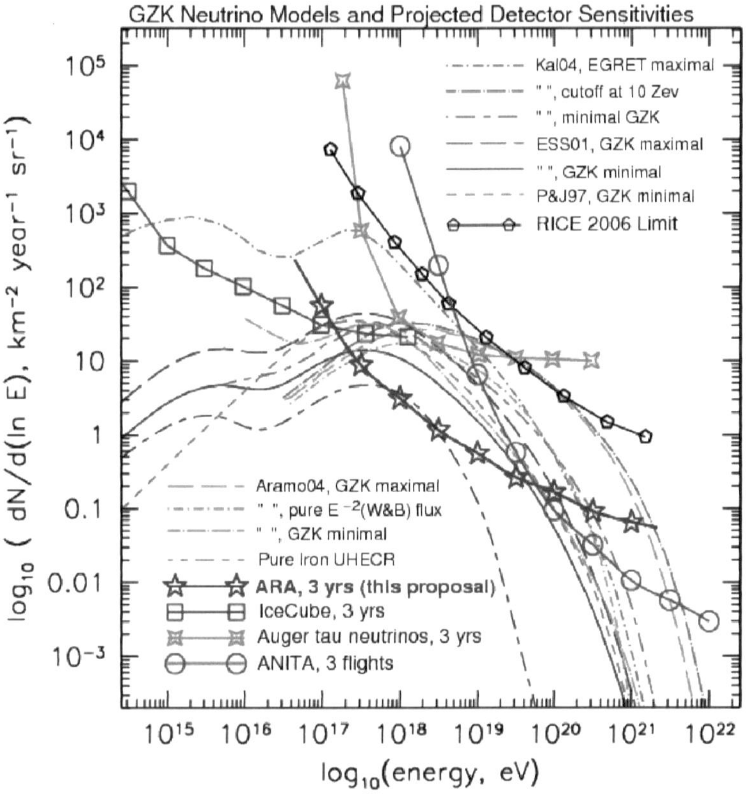

Figure 1. *Projected neutrino senstivity of the ARA-37 array.*

The ARIANNA Experiment

In the TeV-PeV energy range, Antares/Baikal and IceCube represent two complementary approaches to neutrino detection, namely in-water vs. in-ice. Carrying the analogy further, the difference between ARIANNA and ARA can be likened to the distinction between Antares and the Lake Baikal experiments, which are themselves based in salt vs. fresh water. Whereas ARA is sited at South Pole, and draws power off the South Polar grid, ARIANNA is sited at Minnabluff, just below the snow surface, on the Ross Ice Shelf, in a region shielded by local mountains from the radio background presented by McMurdo Station. As a result, the ARIANNA hardware design targets minimal power consumption, but requires none of the narrow-band notch filtering required by ARA.

Relative to ANITA, ARA and ARIANNA obtain a lower experimental energy threshold by locating their antennas 'close' to a prospective neutrino interaction point. The two experiments are compared, in tabular form, in Table 2.

	ARA	ARIANNA
Locale/geometry	South Pole/37 x 20-channel stations; 2 km spacing between stations	MinnaBluff/931 9-channel stations; 1 km spacing
Trigger Rate (current)	5 Hz	0.1 Hz
Trigger threshold:	4 x $4\sigma_{kT}$ over 400 ns	4 x $4\sigma_{kT}$ over 60 ns
Power Draw/(nRx+nTx)	100 W/(20 Rx + 2 Tx)	10 W/(8 Rx + 1 Tx)
Frequency band	4 chs x 25-700 MHz (surface) + 16 chs x 150-900 MHz (200 m)	8 chs x 110-900 MHz (surface)
Polarization response	Hpol/Vpol	Hpol
Attenuation length (E-field)	1500 m	400 m
$\Delta\theta/\delta\phi$ resolution	0.5°/0.1°	2.5°/2.5°
Air shower response	Yes, via RASTARA (surface Rx)	Studies in progress
Firn ray tracing effects	Small (for 200 m deployment)	Yes
Tx calibration	1 Hz	0.002 Hz
Neutrino Event Rate	50/yr for ARA37	50/yr for ARIANNA (6-mos/yr)
Current funding	**$4M OPP proposal for ARA01-ARA07 stations**	**$1 M MRI for first 7 ARIANNA stations**

Figure 2. *Mano-a-mano comparison of ARIANNA and ARA projects.*

Conclusions and Prospects

Prospects for detection of ultra-high energy neutrinos have recently brightened with the announced observation of two ultra-high energy neutrino candidates by the IceCube experiment. Although their origin has yet to be established, this observation nevertheless may, at long last, usher in the era of the 'neutrino telescope'. Correspondingly, the ARA and ARIANNA projects are now in the prototyping phase, in order to establish the hardware and deployment procedures that are anticipated for ultimate arrays 100-1000x larger than their current scales.

The ANITA experiment, although forced to an experimental threshold 1-2 orders of magnitude above the expected peak of the cosmogenic neutrino energy spectrum and therefore perhaps less likely (given our current understanding) to establish the UHE neutrino flux than the in-ice projects, nevertheless has the distinction of being, thus far, the only experiment to self-trigger on the radiowave signal resulting from a developing, in-air cosmic ray shower. The trigger and electronics for the ANITA-3 mission, scheduled for 2013-14 will enhance the direct CR detection rate, relative to ANITA-2, by at least two orders of magnitude, and, by the conclusion of it's flight, will have an ultra-high energy cosmic ray sample numerically competitive with Auger, albeit with somewhat degraded resolution.

[1] H. R. Allan, Progress in Elementary Particles and Cosmic Ray Physics, North-Holland Publishing Company, Amsterdam, 1971.
[2] J. V. Jelly *et al*, Nu. Cim. 46, 649 (1966)
[3] Askaryan, G. A., Zh. Eksp. Teor. Fiz. **41**,616 (1961) [Soviet Physics JETP **14**, 441 (1962)].
[4] G.A. Gusev and I.M. Zheleznykh, Piz'ma Zh. Exp. Teor. Fiz. 38 (1983) 505; M.A. Markov and I.M. Zheleznykh, Nucl. Inst. Methods A248 (1986) 242.
[5] A.L. Provorov for the RAMAND collaboration, in: Proc. 3rd Int. Workshop on Neutrino Telescopes, ed. Milla Baldo Ceolin (Venice, 1991) p. 337.

[6] A.L. Provorov and I.M. Zheleznykh, Astroparticle Physics 4 (1995) 55.
[7] R.D. Dagkesamansky and I.M. Zheleznykh, Radio astronomy method of super high energy neutrino detection; Piz'ma Zh. Exp. Teor. Fiz. (Sov. JETP Lett.) 50 (1989) pp. 233–235.
[8] Large-scale Cherenkov detectors in ocean, atmosphere and ice; Nucl. Inst. Meth. A248 (1986) pp. 242–251.
[9] A.V. Butkevich,..., I.M. Zheleznykh (5 co-authors), Ultra-high neutrino-nucleon and neutrino-electron cross-sections in the Standard Model, in supersymmetric and superstring model; Z. Phys. C39 (1988) pp. 241–250.
[10] P. W. Gorham, D. P. Saltzberg, P. Schoessow, et al., "Radio-frequency measurements of coherent transition and Cherenkov radiation: Implications for high-energy neutrino detection," Phys. Rev. E 62, 8590, (2000); hep-ex/0004007;
D. Saltzberg, P. Gorham, D. Walz, et al., "Observation of the Askaryan Effect: Coherent Mi- crowave Cherenkov Emission from Charge Asymmetry in High Energy Particle Cascades," Phys. Rev. Lett. 86, 2802 (2001); hep-ex/0011001.
"Accelerator Measurements of the Askaryan Effect in Rock Salt: Towards Teraton Neutrino Detectors", P. W. Gorham, D. Saltzberg, C. Field, G. Guillian, D. Walz, E. Zas, J. Alvarez- Muniz, R. Milincic, C. J. Naudet, and D. Williams, in prep. for Physical Review D, 2004.
[11] E. Zas, F. Halzen, and T. Stanev, Phys. Lett. **B257**, 432 (1991); E. Zas, F. Halzen, and T. Stanev, Phys. Rev. **D45**, 362 (1992).
[12] V.V. Bogorodsky, C.R. Bentley and P.E. Gudmandsen, *Radioglaciology*, Riedel, Dordrecht, Holland (1985).
[13] D. Guetta *et al.*, astro-ph/0302524
[14] V. Berezinsky, hep-ph/0303091, Talk given at XXI Texas Symposium (Florence)
[15] F. Halzen and D. Saltzberg, hep-ph/9804354, and Phys.Rev.Lett. 81 (1998) 4305-4308
[16] S. Pakvasa, hep-ph/0305317, presented at the Tenth International Conference on Neutrino Telescopes, Mar 11-14, 2003; Venice, Italy
[17] J. G. Learned and S. Paksava, U. of Hawaii preprint UH-511-799-94, Astropart. Phys. 3, 267 (1995).
[18] Yu. Andreev, V. Berezinsky and A. Smirnov, Phys. Lett. B, 84, 247 (1979); D. W. McKay and J. P. Ralston, Phys. Lett. B167, 103, 1986; M. H. Reno and C. Quigg, Phys. Rev. D37, 657 (1987); G. M. Frichter, D. W. McKay and J. P. Ralston, Phys. Rev. Lett. 74, 1508-1511 (1995); J. P. Ralston, D. W. McKay, and G. M. Frichter, in International Workshop on Neutrino Telescopes, (Venice, Italy, 1996), edited by M. Baldo-Ceolin; astro-ph/9606007; R. Gandhi, C. Quigg, M. H. Reno, I. Sarcevic, Phys. Rev. D58, 093009, (1998).
[19] See, e.g., P. Jain, J. P. Ralston and G. Frichter, Astropart. Phys. 12, 193 (1999). See also J. P. Ralston, in 26th International Cosmic Ray Conference Proc. (Salt lake City 1999) edited by D. Kieda and B. L. Dingus (IUPAP 1999).
[20] P. W. Gorham *et al.* (ANITA Collaboration), Astropart. Phys. **32**, 10 (2009).
[21] P. W. Gorham *et al.* (ANITA Collaboration), Phys. Rev. Lett **103**, 051103 (2009).
[22] P. W. Gorham *et al.* (ANITA Collaboration), Phys. Rev. **D82**, 022004 (2010), and *erratum* (arXiv:1011.5004).
[23] S. Hoover *et al.* (ANITA Collaboration), arXiv:1003.2961, submitted to Phys. Rev. Lett. (2010)
[24] P. Allison *et al.* (ARA Collaboration)

Long term performance of the MINOS calibration procedure and stability of the MINOS detectors

M Mathis, for the MINOS Collaboration

Department of Physics, College of William & Mary, Williamsburg, Virginia 23187, USA

E-mail: mmathis@fnal.gov

Abstract.
The MINOS detectors are steel-scintillator sampling tracking calorimeters and are calibrated using an in-situ light-injection system and cosmic ray muons. The MINOS Near and Far Detectors have been operating almost continuously since 2003 and 2005, providing opportunity to quantify the behavior of the various detector components, many of which are used in the next generation neutrino oscillation experiments, under long-term experimental operation. We report on the calibration procedure and its stability, as well as the time and temperature dependencies of the scintillator, wavelength-shifting fibers and photo-multiplier tubes.

The MINOS [1, 2] experiment is a two-detector long-baseline experiment. The 0.98 kT Near Detector (ND) is located approximately 1 km from the NuMI [3] target and is used to measure the neutrino energy spectrum and composition of the beam. The 5.4 kT Far Detector (FD) is located 734 km further down stream in the Soudan Underground mine. By comparing the neutrino energy spectrum observed at the Far Detector with that observed at the Near Detector neutrino oscillation parameters can be determined [4, 5, 6, 7]. A robust calibration technique has been developed [2, 8] and is employed to remove the spatial and temporal variations in the detector response which would otherwise degrade the calorimetric energy resolution and therefore the ability to measure the oscillation parameters.

The two MINOS detectors have been designed to be as similar as possible to minimize systematic errors due to uncertainties in the beam-flux and neutrino interaction cross-sections. Each detector is comprised of torroidally magnetized one-inch-thick steel layers interleaved with scintillator. The scintillator is made of Dow STYRON 664 polystyrene doped with 1% PPO by weight, plus 0.03% POPOP fluors [9]. Each scintillating layer is comprised of 4.1 cm wide, 1.0 cm thick strips of varying length. The strips in each alternate scintillator layer are rotated 90° to allow for three-dimensional track reconstruction. The scintillating strips are contained in a co-extruded TiO_2 cladding to maximize internal reflection. The scintillation light is read out by a 1.2 mm Kuraray "non-S" type wavelength shifting (WLS) fiber, containing the Y-11 fluor at 175 ppm, embedded in center of the wide face of each strip. The light is transmitted from the end of the scintillation module by clear acrylic fibers to Hamamatsu multi-anode photomultiplier tubes (PMTs) [10, 11, 12].

The Near Detector is required to read out several events per 10 μs spill, and is therefore created for continuous readout. The Near detector readout [13] integrates PMT charge in intervals of 19 ns, continuously reading out the pulse-shape via a QIE chip for each PMT pixel. Successive samples come from four charge-holding capacitors that are read out with a combination of an 8-bit ADC and a 4-bit range. The response function of this device is therefore very nonlinear, and is calibrated in hardware using a reference DAC to calibrate and linearize each channel. During normal data-taking, each sample is passed through a look-up table to determine the linear charge for that sample. The readout system is regularly re-calibrated, particularly when hardware is swapped, and therefore is known to be much more linear and stable than the PMTs. During beam spills, the detector reads out continually. Outside of beam spills (e.g., cosmic-ray events) the detector reads out a set of 8 samples from each PMT pixel when triggered by a dynode tap on that PMT.

In contrast, the Far Detector readout [14] is constructed to service many more channels of electronics at low rate. Each PMT is read out only when triggered by the PMT dynode tap. Charge from each pixel is stored on a capacitor and then each capacitor is read out sequentially through a common line shared among 3 PMTs. Pedestals are subtracted by the readout system, but the resulting charge is highly nonlinear due to a roll-off effect in the ADC at large charges. This nonlinearity is eventually calibrated out along with the PMT nonlinearity.

Because of the differences in detector conditions, the PMT readout electronics differ between the Near and Far Detectors. Because this is one of the few differences between Near and Far Detectors, it has been studied extensively in a test-beam environment [15].

1. Long Term Detector Trends

We have tracked the long term behavior of the detector response and the response of the PMT gains in both the Near and Far Detectors in the period from January 2005 to February 2011. During this time the detector and PMT response was frequently monitored and a uniform calibration scheme was applied. This data was used to measure the stability of the response of the MINOS detectors over time. We also quantified the effects of hall conditions such as temperature fluctuation. Since the detector response is a product of the PMT response and the amount of light incident on the PMT photo-cathodes, we can measure the changes in light level as a function of time by calculating the ratio of the detector response to the PMT gains.

The detector response, or drift, is monitored daily using cosmic ray muons. The response in both the MINOS Near and Far Detectors has decreased gradually over the life of the MINOS experiment. For the period between January 2005 and February 2011, the raw response of the detector has decreased by roughly 8%. This decrease is shown in Fig. 1. In this figure, the drift is shown as a percentage decrease from the first data for each detector. The start date for the Near and Far Detectors is slightly different, reflecting the fact that the Near Detector turned on after the Far was already running in a stable manner.

1.1. PMT Behavior

The MINOS Near Detector uses R5900-00-M64 multi-anode PMTs, while the Far uses R5900-00-M16 PMTs; the low occupancy of hits at the Far Detector allows an 8-fold multiplexing of non-neighboring scintillating strips on each of the PMT pixels. In total the Near (Far) Detector has 192 (1452) PMTs. The characteristics of these PMTs have been extensively evaluated [11, 12]. To determine the gain each Near (Far) Detector strip-end is pulsed 1000 (300) times per hour [2] by the LI system at approximately 50 PE. The data are then summed over a three day interval and the gains evaluated using photon statistics [8]. The gain determined is therefore an average over a 3 day period, averaging out daily fluctuations.

Figure 2 plots the Near and Far Detector gains as a function of time. Since November 2004, the gains have increased by 19% and 16% at the ND and FD, respectively. The ND data indicates that the PMTs experience an initial rapid increase in gain in their first year of operation. After this initial increase, the rate of the increase slows with time.

The short term fluctuations in gain are well correlated with temperature and are more pronounced in the ND data as the temperature variations are larger. The temperature of the FD

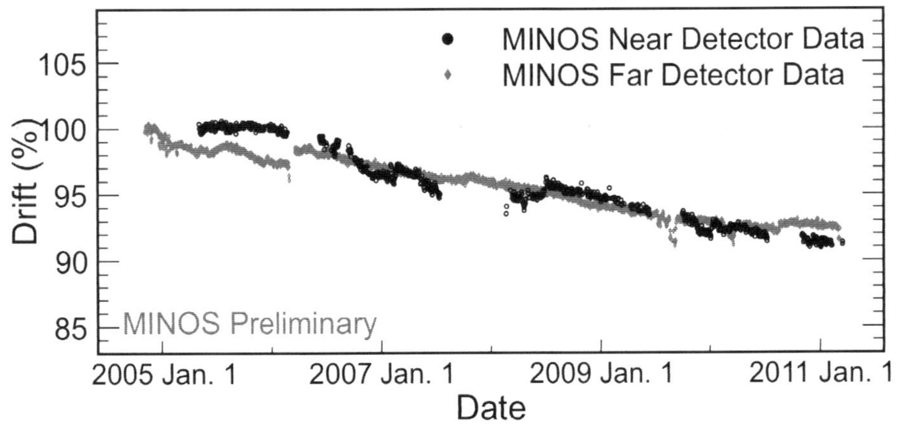

Figure 1. The daily change in detector response of the MINOS Near and Far Detectors.

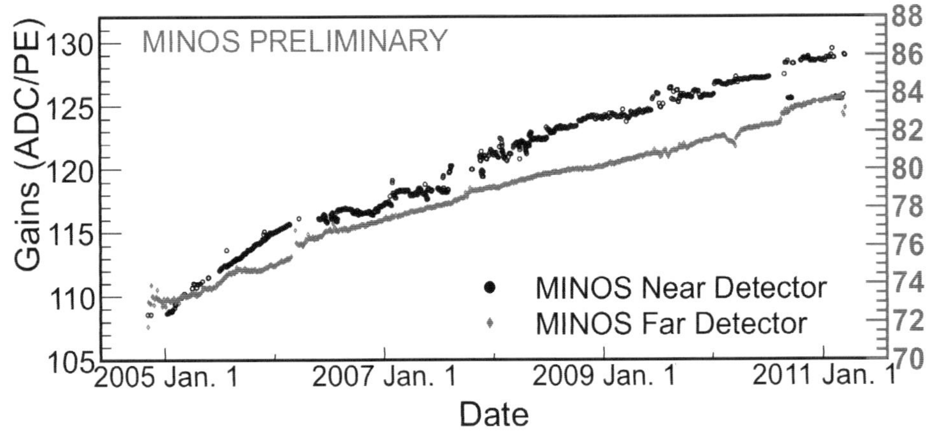

Figure 2. The measured Near and Far Detector gains as a function of time. The short-term fluctuations away from a smooth curve correlate well with temperature. The small discontinuity in the FD gains in 2006 is due to a re-tuning of the PMT high voltages.

PMTs closely tracks the cavern temperature. To identify the effect that variation in temperature has on the gain, short periods of time with large variations in temperature ($\sim 2°$C) have been identified. In these instances it was observed that the PMT gain changes with temperature at a rate of $(-0.58 \pm 0.05)\%/°$C for the FD and $(-0.232 \pm 0.019)\%/°$C for the ND.

The gain curves fit well to an exponential of the form $1 - Ae^{-t/\tau}$ with a time constant of 10.57 ± 0.05 years at the ND and 20.2 ± 1.8 years at the FD. The Near Detector has two distinct exponential periods, and we report the result for the latter and longer one, which covers June 1, 2006 through March 19, 2011. Moreover, individual PMTs and PMT pixels are aging at different rates increasing the spread in the gains by 4.6 % in the ND and 8.9 % in the FD.

1.2. Detector Light Yield

The amount of light incident on the PMTs can be defined by dividing the observed detector response by the average gain. This measurement is affected by the scintillator light yield, the attenuation in the WLS and clear fibers and the optical transparency of the readout chain. Figure 3 plots the light level for both detectors from November 2004 to February 2011. In this period of time, the overall light level in both detectors has dropped approximately 20% with an average rate of 3.5% per year in the ND and 3.0% per year in the FD.

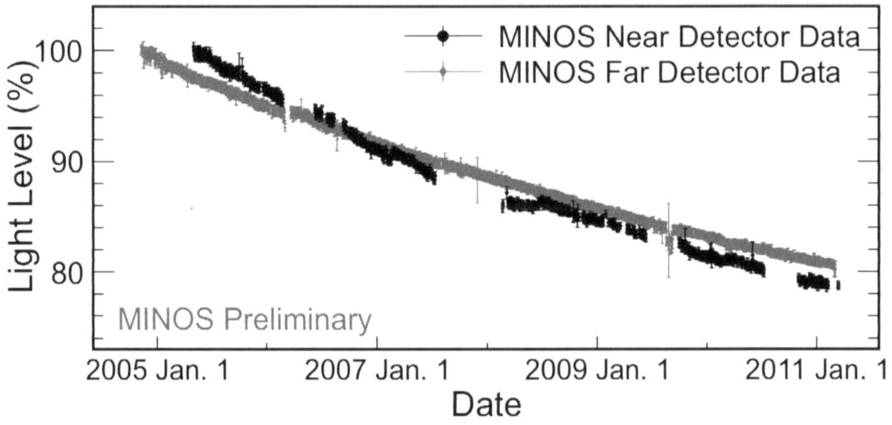

Figure 3. The relative light level of the MINOS Near and Far Detectors over time. The light level is a smoother curve than drift as it removes the large variations due to PMT gain.

As with the gain measurement, the rate at which the light level is dropping decreases with time. The data was fit to an exponential function of the form $(100 - A)e^{-t-t_0}/\tau + A$. The time constant is found to be τ=8.2 years for the Far Detector and τ=5.4 years for the Near Detector, and the parameter A is found to be A=68% for the Far Detector and A=64% for the Near Detector. This suggests that, ignoring the influence of more catastrophic types of aging such as large temperature changes, the scintillator in the MINOS detectors is aging much more slowly in 2011 than in 2004 and the light level will not decrease below 60% of its original value.

In addition to the long-term decrease in light level, short-term variations in the light level are observed. These short-term variations are due to a variety of sources including detector and hardware instabilities and temperature variations. For both detectors, the scintillator temperature tracks that of the detector hall. We identified several regions where the Near and Far hall temperature exhibits large, smooth changes in short regions of time, and fit these periods to determine the decrease in light level as a function of temperature. We find that the light level decreases by 0.2±0.06 $\%/^{\circ}C$ in the ND and 0.4±0.07 $\%/^{\circ}C$ in the FD.

1.3. Calibration Stability

The raw and calibrated detector response as a function of position is shown in Fig. 4. The 30% variation in response that is observed across the face of the detector is effectively removed by the calibration chain. Figure 5 plots the calibrated response of the Far Detector as a function of time; the response is stable over the complete run period. The response of the Near Detector is more stable than that of the Far Detector.

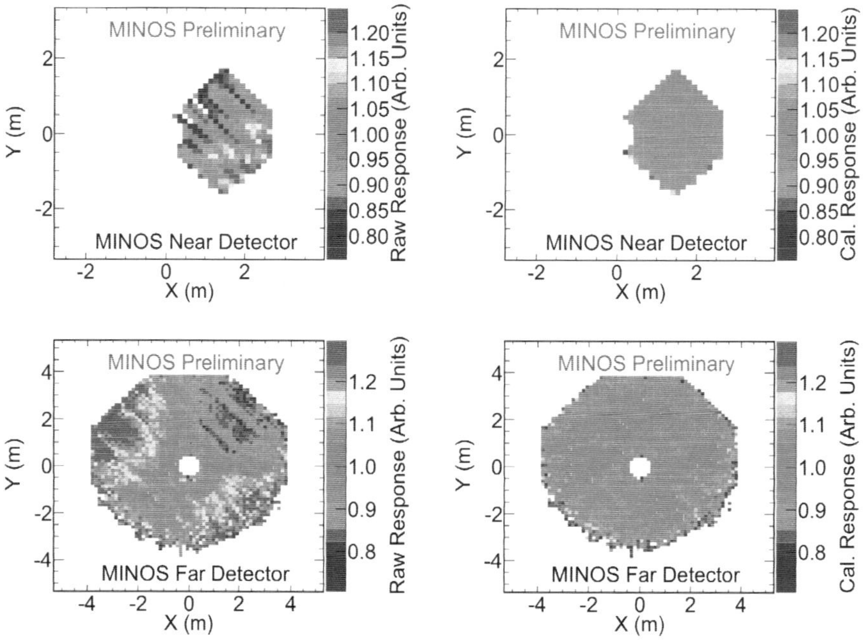

Figure 4. (left) The raw response of the detector as a function of XY coordinate. (right) The response of the detector after all the calibrations have been applied. The top row is the Near Detector response and the bottom row is the Far Detector response.

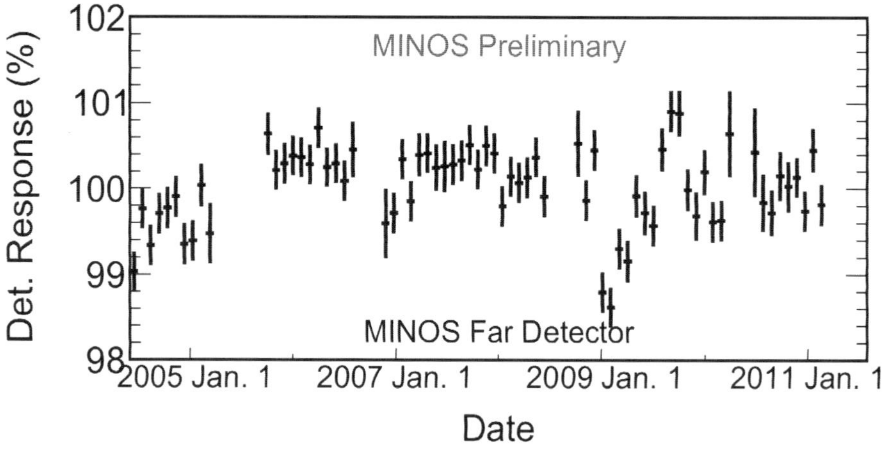

Figure 5. The calibrated response of the Far Detector as a function of time.

2. Absolute Energy Calibration

The MINOS Calibration Detector (CalDet) [8] was a scaled down, functionally identical version of the Near and Far Detectors. It was exposed to various test beams at CERN between 2001 and

2003 to determine the detector response to GeV energy protons, pions, electrons, and muons; the latter being important as stopping muons are used to determine equivalence of all three detectors. Using CalDet, it was found that the uncertainty on the absolute energy scale is 5.6%. Extrapolating this to the Near and Far Detectors, it becomes 5.7%. Taking into account differences between Near Detector data and Monte Carlo simulations, the error on the absolute energy scale of the Near Detector is 6.0% and the Near-to-Far relative energy scale is 2.1%.

3. Summary

During their operation, the responses of the Near and Far Detector have decreased by approximately 20%, which is better than the design specifications. We have demonstrated that the MINOS detector calibration method has been demonstrated to yield a uniform response as a function of position within the detector and in time. We measure the absolute energy scale to 5.7$ and relative energy scales to be 1.9% and 0.9% at the Near and Far Detectors, respectively.

Acknowledgments

This work was supported by the US DOE; the UK STFC; the US NSF; the State and University of Minnesota; the University of Athens, Greece; and Brazils FAPESP, CNPq, and CAPES. We gratefully acknowledge the Minnesota DNR, the crew of the Soudan Underground Laboratory, and the personnel of Fermilab for their contribution to this effort.

References

[1] Ambats I *et al.* (MINOS) NUMI-L-337
[2] Michael D G *et al.* (MINOS) 2008 *Nucl. Instrum. Meth.* **A596** 190–228 (*Preprint* 0805.3170)
[3] Crane D A *et al.* (NuMI Beam Group) 1995 FERMILAB-TM-1946
[4] Adamson P *et al.* (MINOS) 2008 *Phys. Rev. Lett.* **101** 131802 (*Preprint* 0806.2237)
[5] Adamson P *et al.* (MINOS) 2011 (*Preprint* 1104.0344)
[6] Adamson P *et al.* (The MINOS) 2010 *Phys. Rev.* **D81** 052004 (*Preprint* 1001.0336)
[7] Adamson P *et al.* (MINOS) 2009 *Phys. Rev. Lett.* **103** 261802 (*Preprint* 0909.4996)
[8] Adamson P *et al.* 2006 *Nucl. Instrum. Meth.* **A556** 119–133
[9] Pla-Dalmau A (MINOS Scintillator Group) 2001 *Frascati Phys. Ser.* **21** 513–522
[10] Barker M A *et al.* 2002 (*Preprint* hep-ex/0203001)
[11] Lang K *et al.* 2005 *Nucl. Instrum. Meth.* **A545** 852–871
[12] Tagg N *et al.* 2005 *Nucl. Instrum. Meth.* **A539** 668–678 (*Preprint* physics/0408055)
[13] Cundiff T *et al.* 2006 *IEEE Trans. Nucl. Sci.* **53** 1347–1355
[14] Oliver J, Felt N, Feldman G, Lebedev A and Lee R 2004 *IEEE Trans. Nucl. Sci.* **51** 2193–2195
[15] Cabrera A *et al.* (MINOS Collaboration) 2009 *Nucl.Instrum.Meth.* **A609** 106–113 (*Preprint* 0902.1116)
[16] Kordosky M A FERMILAB-THESIS-2004-34
[17] Dytman S, Gallagher H and Kordosky M 2008 (*Preprint* 0806.2119)
[18] Hartnell J J FERMILAB-THESIS-2005-51

Journal of Physics: Conference Series **404** (2012) 012040 doi:10.1088/1742-6596/404/1/012040

Current status and performance of the BESIII electromagnetic calorimeter

Jian Fang , Zhigang Wang

Institute of High Energy Physics, Chinese Academy of Science, Beijing 100049,China

wangzhg@ihep.ac.cn

Abstract. The design and construction of the BESIII electromagnetic calorimeter is introduced briefly. Radiation dose of CsI(Tl) crystals is monitored and history graph of integral dose of crystals is showed. LED-fiber system is used for monitoring the EMC light output, and large decrease of light output of several crystals is discussed. BESIII electromagnetic calorimeter works very well and its performance reach the design value.

1. Introduction

Beijing Electron Positron ColliderII and Beijing SpectroscopyIII (BEPCII/BESIII) [1] is a major upgrade of BEPC/BESII. The BESIII detector is equipped with a CsI(Tl) crystal based electromagnetic calorimeter (EMC) [2] which plays an important role in the BESIII physics. In this paper, we present an introduction of the design of the BESIII EMC, the work status and performance of it.

2. BESIII Electromagnetic Calorimeter

BESIII EMC is based on CsI(Tl) crystals, with a designed energy resolution of 2.5% and a position resolution of 6 mm for 1 GeV showers. The EMC consist of barrel and endcap parts as shown in Figure 1. In the barrel, there are 120 sectors in phi direction, each consist of 44 crystals in z direction. The number of barrel crystals is 5280 in total. The crystals are 28 cm long, and typical size of the crystal is 5.2 cm by 5.2 cm in the front surface and 6.5 cm by 6.5 cm in the rear face. In the endcap, the crystals with similar dimensions as those in the barrel are arranged in 6 rings in radial direction. There are total of 960 crystals in two endcaps. In order to minimize the dead material and gap between the crystals, the crystals are suspended from the support girder.

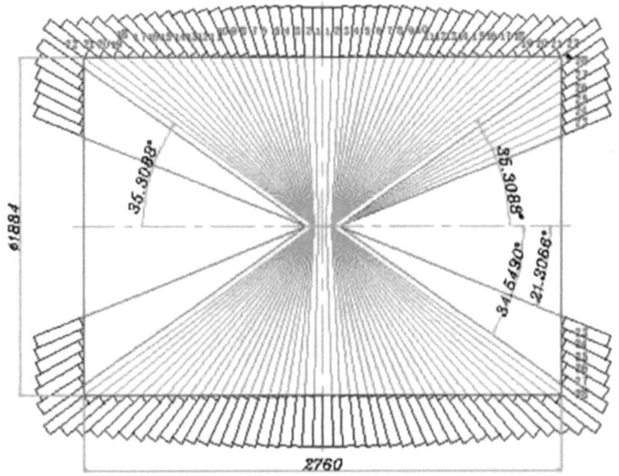

Figure 1. The structure of the BESIII EMC

Two photodiodes are attached to the back face of every crystals, the photon sensors are Hamamatsu S2744-08 (1 cm × 2 cm) photodiodes. Each diode is readout by one preamplifier [3]. The signals from the preamplifier are set to the main amplifier, the shaping time of the amplifier is 1 us. Then the signals are digitized in a Q module. The noise of the electronics is required to be less than 1100 electrons, equivalent to shower energy of 220keV. The readout units are housed in an Al box. The crystals are held by 4 screws drilled into the crystal, then mounted to the support structure. There are fibers emitting light from LED to each crystal for monitoring purpose.

3. Operation Experience

In order to ensure the safety of the crystals from the radiation, the radiation dose is carefully monitored by RadFET [4]. RadFET is an acronym for RADIATION-SENSING FIELD-EFFECT TRANSISTOR based on the metal-oxide-silicion p-channel structure. It is an integrating dosimeter which measures dose by virtue of the field effect caused by space charge trapped in an inorganic insulator. The RadFET probes are distributed at a radius close to crystal position, in the barrel 10 probes are positioned along the Z direction at one phi angle. There are 4 rows of probes at phi of 0, 90, 180 and 270 degrees, and there are 40 probes in barrel. In the endcap, each side placed 4 probes along the radius direction at one phi, and probes are placed in phi direction of 0, 90, 180 and 270 degrees, there are 32 probes in total in the endcap. The integral dose accumulated from beginning of collision to March 2012 is showed in Figure 2. The integral dose in the center of barrel is large than that at two ends. The integral dose at west endcap is larger than that at east. Integral dose accumulated from August 2008 to March 2012 is calculated. According to the design, the allowed radiation dose per year should be less than 200rads at crystals.

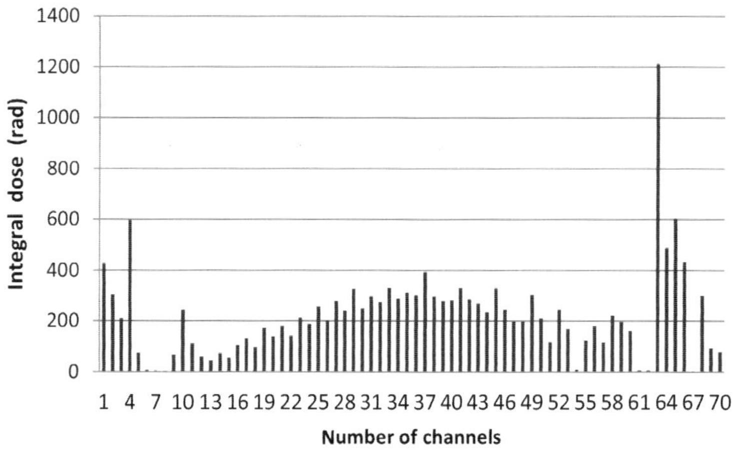

Figure 2. The integral dose of RadFET over four years

There is a LED-fiber system used to monitor the crystals light output during the EMC assembling and during data taking. Each crystal is connected to one fiber, it is used to check modules quality, monitor radiation hardness and EMC performance. The LED-fiber unit is composed of LED, light mixer and fiber bundle. The light intensity can be adjusted to simulate the energy deposited in crystal from 10 MeV to 2.5 GeV. There are two independent reference to monitor the light variances of each LED. The instability of the LED-fiber system is less than 1% in a run.

There are 76 LED-fiber units. According to the structure of EMC, 60 LED-fiber units in barrel, each serves 88 crystals, 16 units in endcap, each serves 60 crystals. Before assembly of super module, each cell is examined by LED-fiber system, if light output < 80% of PMT data, it will opened to check PD-crystal gluing and preamplifiers and so on. During data taking, the system serves to find out bad channels and monitor the performance of the EMC, including the gain reduction by radiation in each crystal. Relative decrease of light output measured by LED-fiber monitor is shown in Figure 3. In most of the crystal the light output decrease is less than 10%. There are 26 crystals light output decrease exceed 10%, among them 11 crystals exceed 20%. The result is agreement with the offline calibration by Bhabha data. Suddenly decrease of light output of some crystals has been found. This case has appeared before installation, and was confirmed that it was caused by loose gluing of PD to crystal. There is a crystal, it's light output decreased to only 20%, this may be caused by total drop of PDs from crystal. All these crystal damages can be recovered by offline calibration. So far, there is no dead channel in EMC.

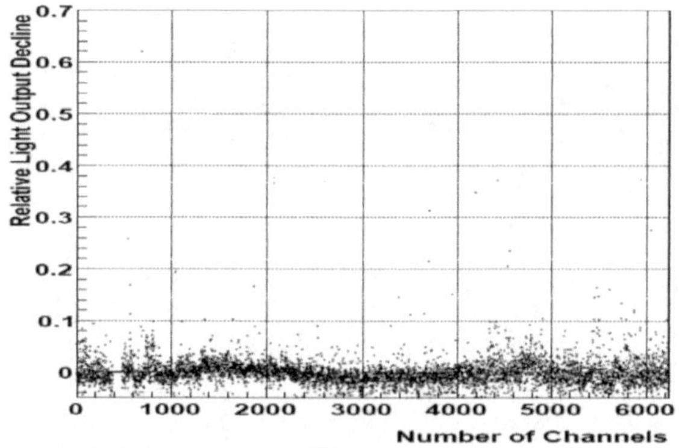

Figure 3. Relative decrease of light output measured by LED monitor

Every crystal is read out by two photodiodes. The two preamplifier outputs are received by two differential amplifiers and then go through a switching stage with two selections. Normally the selection A+B that sums the two photodiode outputs is used. If one photodiode or preamplifier lost, the selection 2A or 2B that boosts the signals by a factor of two is used. There are 10 crystals read out by one PD.

There is a water flow system used to control the temperature of the calorimeter in a suitable range. Water flow system of barrel operate normally and the maximum, minimum and average temperature of the barrel calorimeter over six months is showed in Figure 4. As a precaution of leakage, water flow system of endcap is stopped in 2009, the maximum temperature is blow 29 degree.

Gas-filled system was used to keep the calorimeter in a dry environment. It prevent the CsI(Tl) crystal from deliquescence and maintain the humidity in a suitable range. The gas of the system was changed from dry Nitrogen to freeze-dried compressed air and the humidity of crystals over six months is showed in Figure 5.

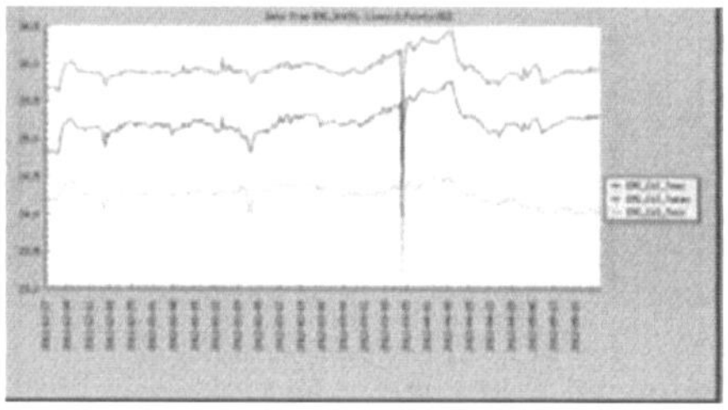
Figure 4. Temperature of crystals over six months

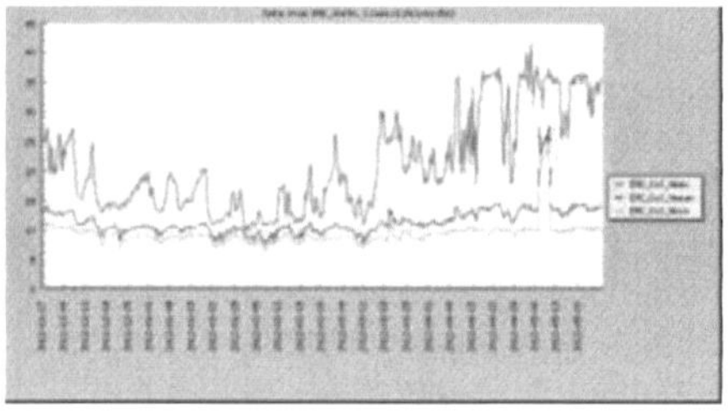
Figure 5. Humidity of crystals over six months

4. EMC Performance

In BESIII, Bhabha events are used to calibrate the gain of each crystal. From the Bhabha data, we can get energy peak and resolution of each crystal channel at 1.5 GeV. Those checks are shown in Figure 6 and Figure 7, respectively. It can be seen that data and MC results agree very well after the calibration with Bhabha. The energy resolution for $e^+e^- \rightarrow e^+e^-$ is ~2.3% at the center-of-mass energy 3.097 GeV in barrel and 4.1% in endcap. At the same time the performance of the calorimeter has no degradation after 3 years running.

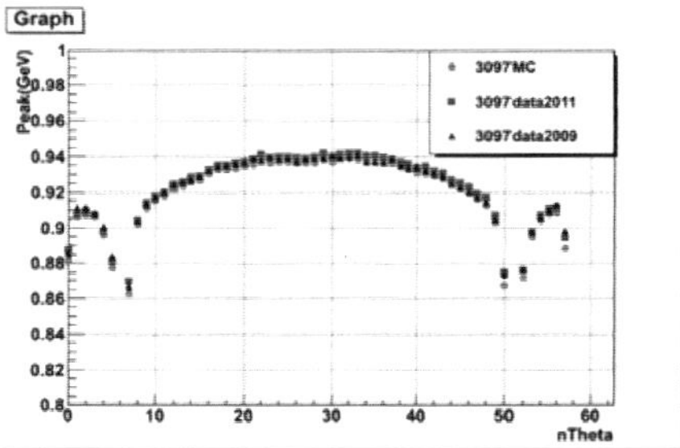

Figure 6. Energy peak at 1.5 GeV

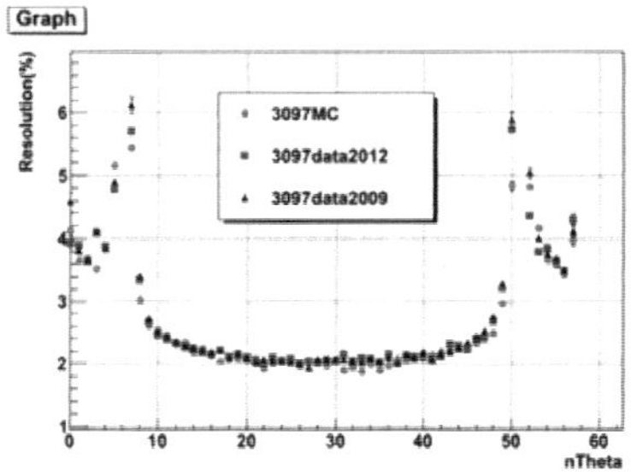

Figure 7. Energy resolution at 1.5 GeV

5. Conclusion

The performance of BESIII CsI(Tl) calorimeter is very good. All the 6240 channals are work well. The light output of the crystals is monitored by the LED-fiber system. The temperature and humidity control system are running normally. The performance observed meets the design requirements, and has no degradation after 3 years running.

References

[1] M. Ablikim, et al. (BESIII Collaboration), "Design and construction of the BESIII detector " Phys. Rev. D81,052005(2001).

[2] DONG Ming-Yi, et al. "Design and assembly of the CsI(Tl) crystal module of the BESIII electro-magnetic calorimeter", Chinese Physics C. Vol.32, No.11, Nov., 2008.

[3] CHANG Jin-fan, et al. "The design of BESIII EMC charge measurement with three ranges", Nuclear Electronics and Detection Technology. Vol.29, No.4, Jul. 2009.

[4] GONG Hui, et al. "Online measurement of the BEPCII background using RadFET dosimeters", Chinese Physics C. Vol.33, No. 8, Aug., 2009.

Monitoring and Correcting for Response Changes in the CMS Lead-tungstate Electromagnetic Calorimeter

Federico Ferri

CEA/Saclay Irfu/SPP, 91190 Gif-sur-Yvette, France

E-mail: federico.ferri@cern.ch

Abstract. The CMS Electromagnetic Calorimeter (ECAL) comprises 75848 lead-tungstate scintillating crystals. Changes in the ECAL response, due to crystal radiation damage or changes in photo-detector output, are monitored in real time with a sophisticated system of lasers to allow corrections to the energy measurements to be calculated and used. The excellent intrinsic resolution of the CMS ECAL requires the monitoring system itself to be calibrated to a high precision and its stability to be controlled and understood. The components of the CMS ECAL monitoring system, and how it has evolved to include modern solid-state lasers, are described. Several physics channels are exploited to normalise the ECAL response to the changes measured by the monitoring system. These include low energy diphoton resonances, electrons from W and Z decays (using shower energy versus track momentum measurements), and the azimuthal symmetry of low energy deposits in minimum bias events. This paper describes how the monitoring system is operated, how the corrections are obtained, and the resulting ECAL performance.

1. Introduction

The Compact Muon Solenoid (CMS) detector [1] is a general purpose detector installed at the CERN Large Hadron Collider (LHC). It consists of a silicon pixel and strip tracker surrounded by a crystal electromagnetic calorimeter (ECAL) and a brass/scintillator sampling hadron calorimeter (HCAL), all in an axial 3.8 T magnetic field provided by a superconducting solenoid of 6 m internal diameter. The muon system is composed of gas-ionization detectors embedded in the steel return yoke of the magnet. In addition to the barrel and endcap detectors, CMS has an extensive forward calorimetry system.

The conception of the electromagnetic calorimeter has been driven by the search for the Higgs boson via its electromagnetic decay $H \to \gamma\gamma$, appearing as a very narrow resonance on a large background of events with a pair of photon candidates, coming mainly from QCD processes. The electromagnetic calorimeter consists of 75848 lead tungstate ($PbWO_4$) crystals. It is divided into a central part (barrel) covering the region $|\eta| < 1.48$[1] and forward parts (endcaps) extending the coverage up to $|\eta| < 3$ for a particle originating from the nominal interaction point. The crystals are arranged in a projective geometry with a granularity of 0.0174 in both the η and φ directions

[1] In the CMS coordinate system, θ designate the azimuthal angle with respect to the counterclockwise beam direction, and the pseudorapidity is defined as $\eta = -\ln\left[\tan\frac{\theta}{2}\right]$.

in the barrel, and increasing with $|\eta|$ from 0.021 to 0.050 in the endcaps. The scintillation light is read out by avalanche photo-diodes (APDs) in the barrel and vacuum photo-triodes (VPTs) in the endcaps.
A preshower detector, consisting of two planes of silicon sensors interleaved with 3 radiation lengths of lead, is placed in front of the endcaps to cover the pseudorapidity region $1.65 < |\eta| < 2.6$.

The factors impacting the ECAL energy resolution are multiple. First of all the uniformity of the response and its time-stability, both monitored via a laser-based system. Secondly the channel-to-channel calibration, ultimately based on physics events such as $\pi^0/\eta \to \gamma\gamma$, $W \to e\nu$, $Z \to ee$, and on the azimuthal symmetry of the energy distribution in minimum bias events [2]. Finally the energy scale and linearity of the response to electrons and photons, both depending on effects such as bremsstrahlung and electron-positron pair production taking place in the tracker material in front of the calorimeter, determined with physics events such as $Z \to ee$ and $Z \to \mu\mu\gamma$.
We focus here on the first of these aspects, describing how a laser system is used to monitor and correct for changes in the ECAL response, mainly induced by radiation-dependent effects. For an overview of the calibration of the electromagnetic calorimeter and of the overall impact of its performance on the search for the Higgs boson see Ref. [3].

2. Changes to the ECAL response

There are three mains sources of change to the ECAL response: radiation due to the particles produced during collisions, which affect the crystal transparency via a dose-rate dependent effect and the response of the VPTs via the cumulated charge at the cathode [4]; temperature variations, affecting the light yield as $\Delta T/T \sim -2\%/^\circ$C at 18 °C and the gain of the APDs as $\Delta\mathcal{G}/\mathcal{G} \sim -2\%/^\circ$C; variations in the high-voltage bias applied to the APDs, as $\Delta\mathcal{G}/\mathcal{G} \sim 3\%$/V. The last two of these three sources as well as others minor contributions from, for example, the low voltage system and the readout electronics, are controlled to a very precise level and are very well within specifications [5].
The radiation-induced effects, on the other hand, are continuously evolving thus requiring a continuous monitoring and prompt deployment of corrections to compensate for any changes. A precision of better than 0.3% in the correction to the response is required in order to be negligible compared to other contributions to the energy resolution and achieve the best ECAL potential.

The passage of particles through a $PbWO_4$ crystal loses its transparency because of the formation of so called "colour centres". When the crystal is not subject to irradiation, it spontaneously recovers some of its transparency due to thermal annealing of shallow colour centres. The characteristic timescales of these two antagonistic processes of loss and recovery are different but are of the order of hours in both cases. The effect of the LHC operations on the relative response of the crystals is shown in Fig. 2, as probed by the laser monitoring system described in the Section 3. Rapid loss and recovery of the optical transmission corresponding to LHC cycles are clearly visible, as well as steady recovery during LHC technical stops and the low-irradiation Heavy-Ion collision period towards the end of the 2011 run. The red curve corresponds to the average variation in the ECAL barrel, while the other colours refer to the average variations in different ranges of $|\eta|$. Due to the tracker coverage, electron and photon in CMS are not reconstructed beyond $|\eta| = 2.5$.
The losses are consistent with expectations and with the rejection at construction time of crystals experiencing a response loss of more than 6% at a dose rate of 0.15 Gy/h, which is not exceeded up to $|\eta| \sim 2$ with the 2011 LHC peak luminosity ($3.5 \cdot 10^{33}$ cm^{-2}s^{-1}).
For the regions of the ECAL endcaps, Fig. 2 is *de facto* showing the convolution of the

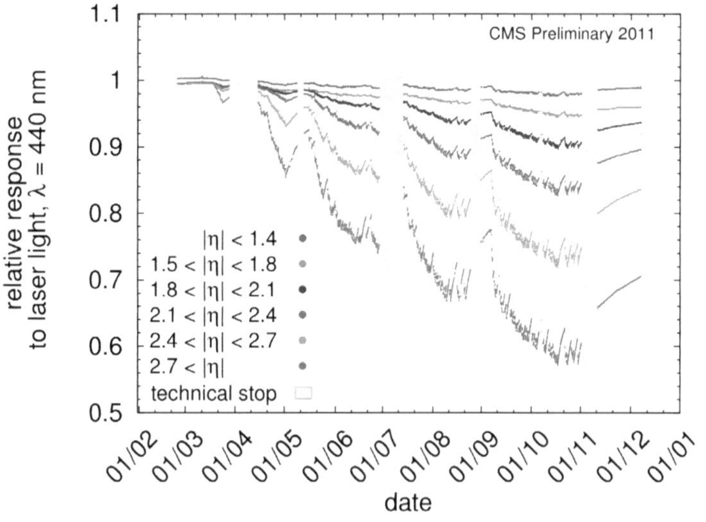

Figure 1. Relative response variation measured by the ECAL laser monitoring system in 2011. The red curve corresponds to the average variation in the ECAL barrel, while the other colours refer to the average variations in the ranges of $|\eta|$ listed in the legend. Technical stops without monitoring data are shaded. The LHC luminosity varied from 10^{33} cm^{-2}s^{-1} in April to $3.5 \cdot 10^{33}$ cm^{-2}s^{-1} in October. Heavy-Ion collisions took place in November.

transparency variations with the loss in response of the VPTs due to the accumulated charge on the cathode. Dedicated laboratory studies have been performed over six months on a limited number of VPTs immersed in a 3.8 T magnetic field. A load pulse has been used to induce an average current as expected at nominal LHC luminosity. Cycles corresponding to periods with/without beam of 8/4 and 16/20 hours have been simulated. By monitoring the VPT response with blue LED light it has been shown that the VPT average loss has been 12%, with a spread of 10% and a tendency towards a non-zero plateau after 5-20 mC of integrated charge, while in 2011 about 0.5 mC of charge have been integrated at $|\eta| = 2.1$. More precise assessments of these effects are ongoing. Currently it has not been possible to disentangle the VPT loss and transparency loss *in situ*.

3. The CMS ECAL monitoring system

The evolution of the crystal response is continuously measured with a dedicated laser system [6], operating at 100 Hz during about 1% of the LHC beam abort gaps, a period of 3 μs at the end of each 89 μs beam cycle. This mode of operation allows the monitoring of the response, along with other quantities such as the electronic noise and the electronic stability, without interfering with the normal data taking.

Laser pulses are injected via a system of optical fibres and diffusing spheres into each of the 75848 ECAL crystals in groups of a few hundred crystals (700 to 900, depending on the region). A relative response measurement is obtained by normalising the crystal response to reference silicon PN photo-diodes, which are associated to groups of 100 to 200 crystals and give the fraction of laser light received by the specific group. The laser signal is also sampled at 1 GHz at the source by a fast CAEN digitizer, model V1729.

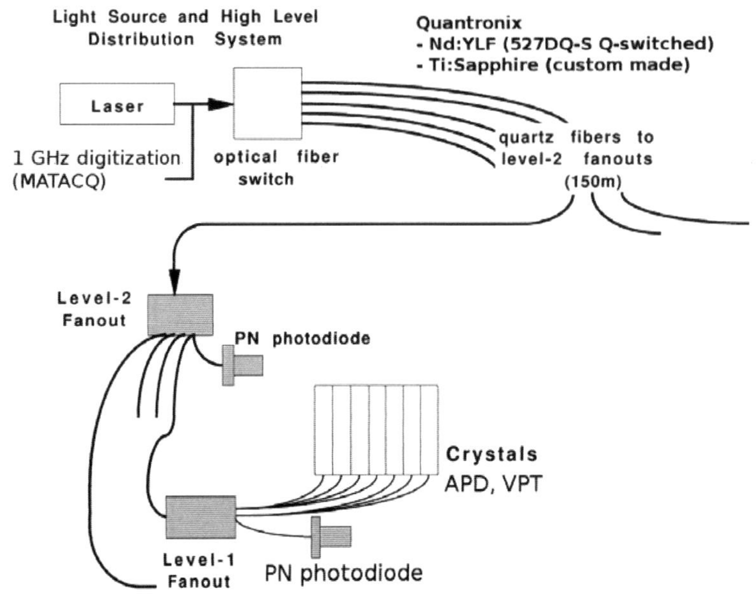

Figure 2. Schema of the laser monitoring system. The light is sampled at 1 GHz at the output of the source and injected into the crystals via a two level distribution systems. At the output of the Level-1 fanout, PN photo-diodes provide a reference to measure relative response changes.

More than one laser can be operated, to ensure adequate redundancy in the system and the possibility to probe the transparency with different wavelengths.
Because of the different optical paths within a crystal, it has been determined at test-beams that the response to scintillation light from an electromagnetic shower is related to the response measured via the laser light by

$$\frac{S}{S_0} = \left(\frac{L}{L_0}\right)^{\alpha} \tag{1}$$

where $\frac{S}{S_0}$ is the relative response to scintillation light, and $\frac{L}{L_0}$ is the relative response to laser light. The empirical parameter α has been studied in Test-Beam campaigns [7] for a limited number of crystals, from different production batches and manufacturers. An average value of 1.52 (1.00) has been found for BCTP[2] (SIC[3]) crystals, with a 10% rms spread within the crystals from the same manufacturer. A determination of the average α using physics events has tuned the value of BCTP crystals in EE to 1.16, which is also effectively correcting for the average VPT response variations not disentangled from the transparency changes by the current monitoring system. With the response variations measured in 2011, the spread of α is expected to contribute to the spread of the single crystal response by order of 0.1% in the barrel and 1% in the endcaps. Studies are ongoing to determine α on a crystal-by-crystal basis using physics events.

A transparency measurement for each of the ECAL crystal is performed in about 40

[2] Bogoroditsk Plant of Technochemical Products, 43, per. Vyazovsky, Bogoroditsk, Tula region, 301801, Russia
[3] Shanghai Institute of Ceramics, Chinese Academy of Sciences 1295 Dingxi Road Shanghai, 200050, P.R.China

minutes. Corrections to physics events are obtained by interpolating the response between each measurement and are delivered within 48h for prompt reconstruction of CMS data.
The main laser used throughout the whole 2010 and 2011 data taking period was manufactured by Quantronix (Nd:YLF 527DQ-S Q-switched) and provides light at $\lambda = 440$ nm, close to the scintillation emission peak. Residual instabilities in the pulse width and timing of the light source as well as other second order effects such as non-linearities of the silicon PN photo-diodes, ballistic deficit between the APD (VPT) and the PN photo-diodes are corrected for. In particular, at the beginning of the 2011 data taking the whole monitoring chain was re-calibrated with dedicated measurements spanning the range of laser parameters, and the single pulse response of each individual channel has been re-measured to optimise the quality of the corrections. To ease the maintenance and operation, an upgrade of the main laser was foreseen and took place at the beginning of 2012 [8].

4. Performance of the monitoring system

Tools are available in order to estimate the intrinsic precision of the monitoring system as well as the stability and precision in time of the response corrections. They make use of the response measurements themselves and of standard physics channels, and are briefly summarised in the following paragraphs.

4.1. Precision of the response measurement

It is possible to evaluate the precision of a single measurement of the response by using the data themselves, in periods subjected to low irradiation, thus having small response changes. In 2011 this has been the case during the Heavy-Ion period at the end of the data taking, Fig. 3-left. One can consider three consecutive response measurements p_{n-1}, p_n, and p_{n+1}, taken at time t_{n-1}, t_n, and t_{n+1}. The response, $i(t_n)$, is the interpolation between $(p,t)_{n-1}$ and $(p,t)_{n+1}$ evaluated at time t_n. The variance of the difference $\delta_n = i(t_n) - p_n$ computed every two measurements in the considered period gives an estimate of the intrinsic precision on the single measurement. It is worth noting that the method would give correct results even in the case of quadratic drifts. Since the typical distribution of δ_n for one channel is well described by a Gaussian curve, the precision has been estimated from the σ parameter of a fit to the distribution. The results of this analysis for the barrel and the two endcaps are shown in Fig. 3, and show a precision on the single response measurement of about $3 \cdot 10^{-4}$, very well within the requirements.

4.2. Performance of the response measurement

The invariant masses of π^0, η and Z, reconstructed from their electromagnetic decays, give an absolute reference to evaluate the stability of the corrected response. In particular, π^0 and η are selected by CMS at the High Level Trigger level via a dedicated stream with a rate of a few kHz, allowing a fine grain monitoring of the correction with a few points per hour, as shown in Fig. 4.

In contrast, Z bosons require more time to accumulate enough data, but the clean signal also provides information about the resolution of the calorimeter, via the width of the invariant mass distribution (Fig. 5).
In addition to that, it is possible to use isolated electrons and positrons from the decays of W and Z bosons as additional probes of the correction quality. The ratio E/p between the energy measured by the ECAL and the momentum measured by the tracker gives an estimate of the stability of the response corrections, under the verified assumption that the error on p is negligible with respect to the precision on E (Fig. 6).

The overall picture is that the ECAL monitoring system tracks response changes in ECAL with a stability of about 0.12% in EB and 0.35% in EE. The resolution is stable throughout the

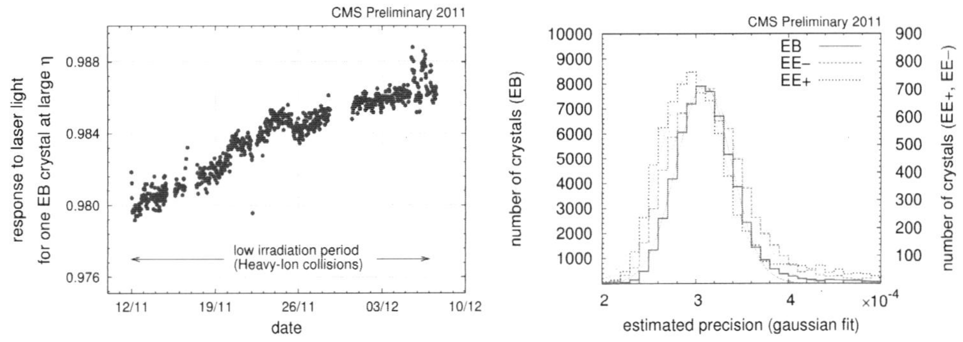

Figure 3. Left: example of a history plot for the relative response of one typical channel of EB at large $|\eta|$ during low irradiation period (Heavy-Ion run). Right: barrel and endcap distributions of the precision on the single response measurement estimated as explained in the text. The peaks at $3 \cdot 10^{-4}$ are much better than the required precision of $2 \cdot 10^{-3}$.

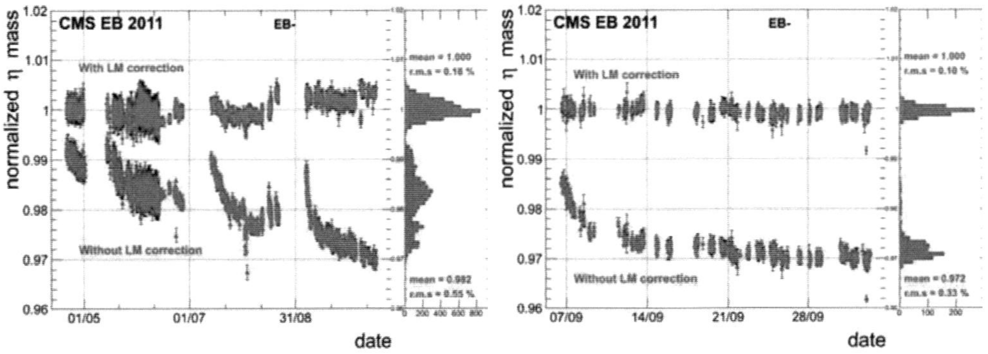

Figure 4. Stability of the energy response in 2011 as measured from the invariant diphoton mass from $\eta \to \gamma\gamma$ decays as a function of time in the barrel, before (red points) and after (green points) response corrections are applied. The right plot shows in greater detail the period of September 2011.

2011 data taking period in the barrel. In the endcaps, where the radiation-induced variations are larger and other effects such as *pileup* have a greater impact, a worsening of the resolution of about 1.5% in quadrature is observed. Refined measurements of crystal properties such as α at the single channel level and optimisation of the handling of additional effects such as the pile-up are being carried out, and can bring further improvements especially in the endcaps.

5. Conclusions

Operating the CMS electromagnetic calorimeter to its full potential in the LHC radiation environment is a very challenging exercise.

The excellent stability of environmental conditions has been such that response changes have depended to a large extent only on radiation-induced effect in the $PbWO_4$ crystals and in

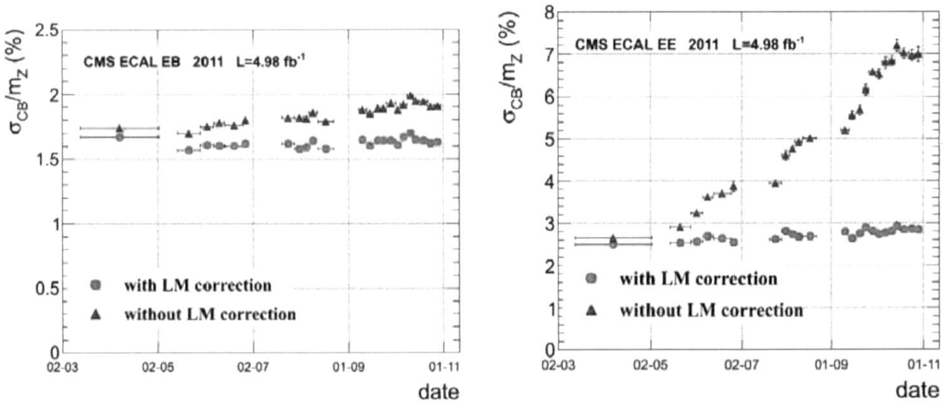

Figure 5. Stability of the mass resolution estimated from $Z \to e^+e^-$ for the barrel (left) and the endcaps (right), before (red points) and after (green points) response corrections are applied.

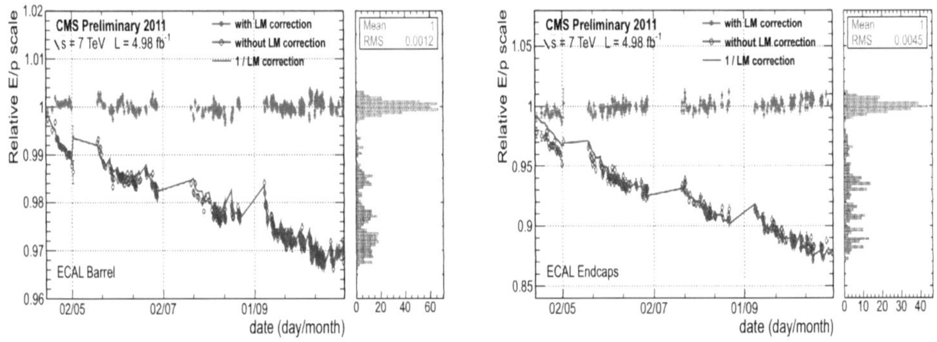

Figure 6. Stability of the energy response of the barrel (left) and endcap (right) from the ratio of electron energy E, measured in the ECAL to the electron momentum p measured in the tracker, before (red points) and after (green points) response corrections are applied.

the readout photo-detectors in the endcaps. The ECAL laser-based monitoring system has successfully operated during more than three years of LHC data taking campaigns. Along with a precise calibration of the calorimeter [2] this has put CMS in an excellent position to be able to observe or exclude the Standard Model Higgs boson through the $H \to \gamma\gamma$ decay [3]. At the time of writing these Proceedings, CMS has observed a new boson with a mass near 125 GeV using data samples of 5.1 fb^{-1} at $\sqrt{s} = 7$ TeV (2011) and of 5.3 fb^{-1} at $\sqrt{s} = 8$ TeV (2012), with a local significance in the $H \to \gamma\gamma$ alone of 4.1 σ [9].

References

[1] Chatrchyan S *et al.*, *The CMS experiment at the CERN LHC*, JINST 3 (2008) S08004, http://dx.doi.org/10.1088/1748-0221/3/08/S08004
[2] CMS Collaboration, *Electromagnetic calorimeter calibration with 7 TeV data*, CMS PAS-EGM-2010-003 (2010)

[3] Tabarelli de Fatis T for the CMS collaboration, *Role of the CMS electromagnetic calorimeter in the hunt for the Higgs boson in the two-gamma channel* , CMS CR-2012/251 (2012) and these Proceedings
[4] Leslie D E for the CMS collaboration, *The effect of pulse rate on VPT response and the use of an LED light to improve stability*, CMS-CR-2009-284 (2009)
[5] CMS Collaboration, *ECAL Detector Performance, 2011 Data*, CERN-CMS-DP-2012-007 (2012)
[6] Anfreville M *et al.*, *Laser monitoring system for the CMS lead tungstate crystal calorimeter*, Nucl. Instr. and Meth. A 594 (2008) 292-320
[7] Ghezzi A *et al.*, *Analysis of the response evolution of the CMS electromagnetic calorimeter under electron and pion irradiation*, CMS Note 2006/038 (2006)
[8] Zhang L for the CMS collaboration, *A Diode-Pumped DP2-447 Blue Laser for Monitoring CMS Lead Tungstate Crystals Calorimeter at the LHC*, CMS CR-2012/230 (2012) and these Proceedings
[9] Chatrchyan S *et al.*, *Observation of a new boson at a mass of 125 GeV with the CMS experiment at the LHC*, Phys.Lett. B 716 (2012) 30-61

Journal of Physics: Conference Series **404** (2012) 012042 doi:10.1088/1742-6596/404/1/012042

A Diode-Pumped Solid State Blue Laser for Monitoring the CMS Lead Tungstate Crystal Calorimeter at the LHC

Liyuan Zhang
On behalf of the CMS ECAL Collaboration
256-48, HEP, California Institute of Technology, Pasadena, CA 91125, USA
E-mail: liyuan@hep.caltech.edu

Abstract. Monitoring the changes in channel response is crucial to maintain the energy resolution of the CMS electromagnetic calorimeter. A new diode-pumped solid state blue laser was commissioned and installed at CERN for 2012 operation. This laser has a simple structure and is more reliable than the existing lamp-pumped lasers used by the monitoring system. The stability of critical parameters, such as the pulse energy, width and centre timing, is much better than that of the lamp-pumped lasers. The characteristics of this blue laser are described. Its performance *in-situ* at LHC is elaborated. The prospects for improving the light monitoring precision are discussed.

1. Introduction

The intrinsic width of the light Higgs is very narrow, so its discovery potential in the $\gamma\gamma$ decay mode is directly related to the energy resolution of the electromagnetic calorimeter (ECAL) [1]. The 75,848 lead tungstate ($PbWO_4$ or PWO) crystals in the CMS ECAL suffer from dose rate dependent radiation damage. Precision calibration thus is crucial for the Higgs discovery. During the time needed to accumulate sufficient statistics for the inter-calibration by using physics events, PWO crystals experience radiation damage during beam on and recovery during beam off. In addition there are response losses in the Vacuum Phototriodes (VPTs) in the ECAL Endcaps ($|\eta| > 1.48$). These effects are shown in Fig. 1 [2]. The changes of channel response *in situ* are corrected by using a light monitoring system which measures variations of crystal transparency and photodetector response at 440 nm [3].

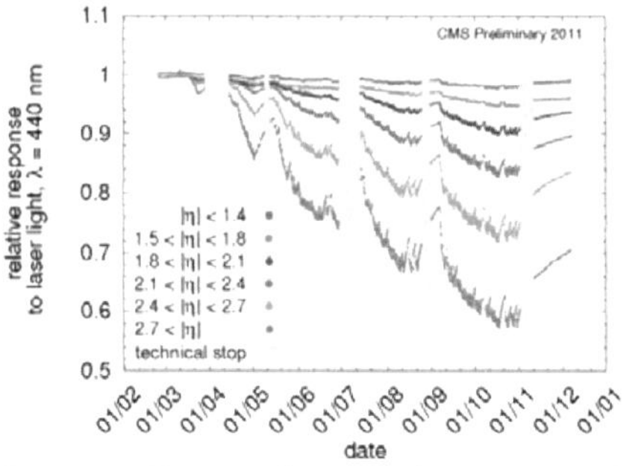

Figure 1. History of monitoring response in 2011 for channels at different rapidity.

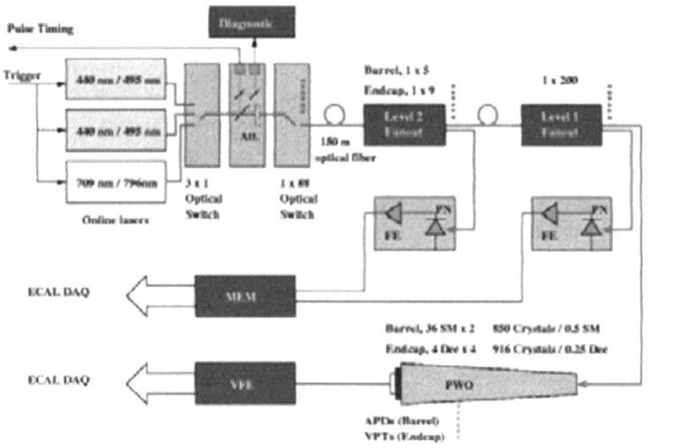

Figure 2. A schematic showing the CMS ECAL laser system in 2011

A light monitoring system was designed and constructed by the Caltech and Saclay groups for the CMS PWO crystal ECAL [4]. Fig. 2 shows that the monitoring laser pulses of different color selected by a 3 × 1 fiber optical switch are distributed via a 1× 88 switch to one of 88 calorimeter elements. A two stage distribution system mounted on each calorimeter element delivers monitoring laser pulses to each individual crystal. In the Barrel ECAL ($|\eta| < 1.48$) the light is detected by Avalanche Photodiodes (APDs), in the Endcaps by the VPTs. The laser pulse energy measured by ECAL readout is normalized to that measured by reference PN diodes. The APD/PN, or VPT/PN, ratio is the monitoring signal, which is a measure of the channel response. The monitoring signal must be measured to a precision of 0.2% to achieve a 0.5% inter-calibration precision.

The CMS ECAL monitoring light source uses three Quantronix laser systems. Each laser system contains an Nd:YLF pump laser and a tunable Ti:S laser. The pump lasers provide frequency-doubled pulse at 527 nm with intensity up to 20 mJ. The tunable Ti:S lasers provide pulse intensity up to 1mJ at two wavelengths. Two monitoring laser systems provide four wavelengths: 440, 495, 709 and 796 nm. The third laser system (440 and 495 nm) is used as a spare to guarantee 100% availability of the 440 nm. All three laser systems were installed at CERN in 2001to 2003 for beam tests, and were moved to the underground CMS cavern in 2007 and 2008 for the LHC data taking.

All three lasers are pumped by DC Kr lamps, which degrade when aging. To address this issue a software feedback was introduced in 2006, which compensates the lamp aging effect by trimming the Nd:YLF laser pumping current [5]. Excellent stability at 3% level for the laser pulse energy, 4% for the laser pulse width and 2 ns jitter for the laser pulse timing were observed. However, additional issues came up during the LHC operation. Quantronix discontinued lamp pumped Nd:YLF laser in 2005. Since then replacement parts for the laser system are no longer available. (Quantronix is actually merged into Continuum in 2012 and left the ns laser market.) Laser intervention (monthly lamp changing and tuning) caused steps in monitoring responses. A stable blue laser with no need for frequent maintenance intervention was therefore required. A decision was made to procure a diode-pumped solid state (DPSS) blue laser system for the 2012 LHC run with the following specifications.

1) Pulse intensity: 1 mJ/pulse at 440 nm, equivalent to 1.3 TeV in dynamic range.
2) Pulse intensity instability: < 3%.
3) Pulse FWHM: < 30 ns to match ECAL readout.
4) Pulse width instability: < 5%.
5) Pulse jitter: < 3 ns for synchronization with LHC.
6) Pulse repetition rate: 0-100 Hz, scan of full ECAL in 20 minutes.
7) Immune to stray B field of 30 Gauss.

Following a market survey and vendor visits, Photonics DP2-447 DPSS blue laser was selected in November, 2011, and was delivered to Caltech in February, 2012. After a short period of integration at Caltech it was successfully commissioned at CERN in March, 2012.

2. DP2-447 laser at Caltech

Figure 3. DP2-447 laser and its ancillary optics.

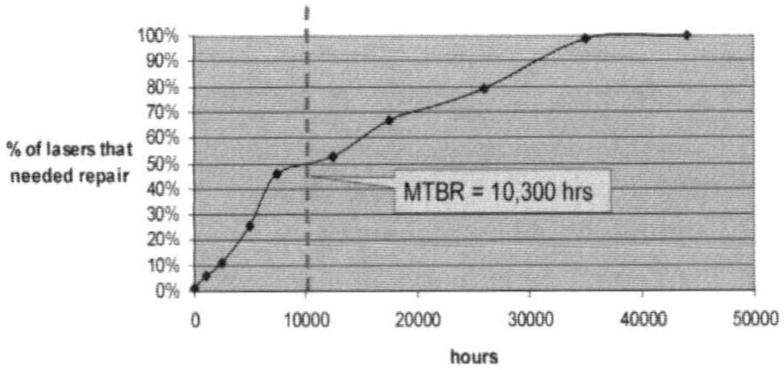

Figure 4. The expected reliability of the DPSS lasers. Courtesy of Photonics, Inc.

Photonics DP2-447 is a proprietary DPSS laser based on intra-cavity third harmonic generation (THG) technology. It is designed to be robust and compact with no user alignment required. It consists of a hermitically sealed laser head, a diode and power supply unit, a chiller and a delay unit which provides two external triggers for the pump diode and the Q-switch. Fig. 3 shows the laser and some of its ancillary equipment at Caltech. The expected mean time between failures (MTBF) for the DPSS laser is shown in Fig.4. It is longer than 1 year for 24/7 operations. This long MTBF indicates that maintenance related interventions, which cause steps in the monitoring response, could be avoided.

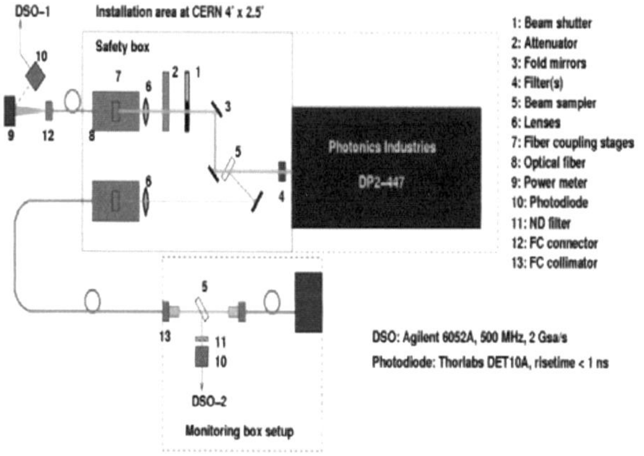

Figure 5. The set-up used at Caltech to evaluate the laser performance.

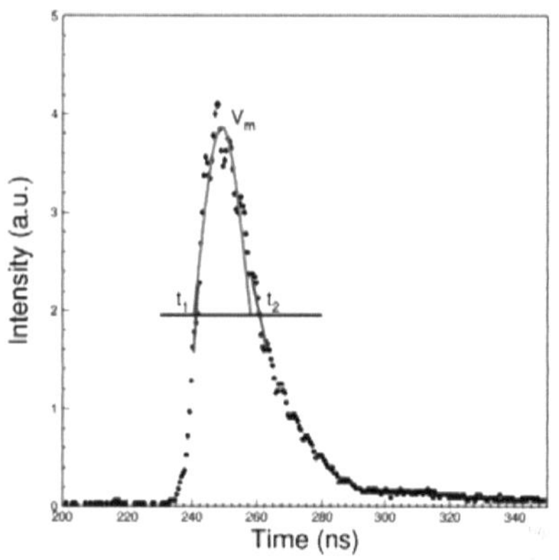

Figure 6. DP-2 laser pulse shape and its reconstruction.

Fig. 5 shows a schematic for the set-up used for laser evaluation at Caltech. The laser pulse was measured by a photodiode (labeled 10 in Fig. 5) which was read out by a digital oscilloscope DSO-2 (Agilent 6052A, 2GS/s). A typical laser pulse shape is shown in Fig. 6. The vertical jitter on the pulse shape, or intrinsic noise, is caused by the oscillation of the pulse inside the laser cavity. The algorithm used for the laser pulse reconstruction is as follows.

1) Find peak V_m,
2) Find time at $V_m/2$: t_1 and t_2,
3) Gaussian fit between (t_1, t_2),
4) Calculate pulse intensity: Σy_i between (-4σ, 8σ),
5) Calculate pulse center timing: $\Sigma t_i y_i / \Sigma y_i$ between (-4σ, 8σ),
6) Calculate pulse width by a linear fit of 5 points (2 before and 2 after) at t_1 and t_2 to find t_{1f} and t_{2f} at $V_m/2$ as well as FWHM = $t_{2f} - t_{1f}$.

The calculated pulse energy is calibrated by using a power detector inserted between the blue filter (4) and the beam sampler (5). All measurements were carried out with the default trigger setting: the pump diode is on for 88μs and the Q-switch is turned on at 86 μs.

Fig. 7 shows the pulse energy, width (FWHM) and center timing as a function of the diode pump current. Their instability (RMS) as a function of the diode pump current is shown in Fig. 8. While the pulse energy reaches the 1 mJ at 60A, the DP2-447 laser meets all other specifications with its diode pump current as low as 35A. This indicates that this laser can be run at a low pump current to prolong the lifetime of the diodes and other optical components if 1 mJ is not needed.

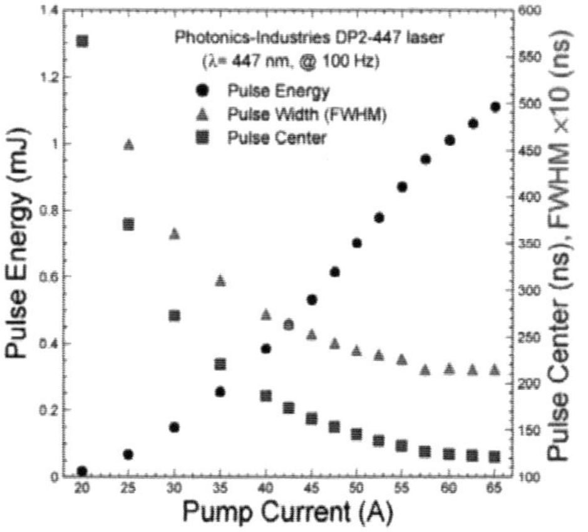

Figure 7. Pulse energy, width and center timing are shown as a function of the diode pump current.

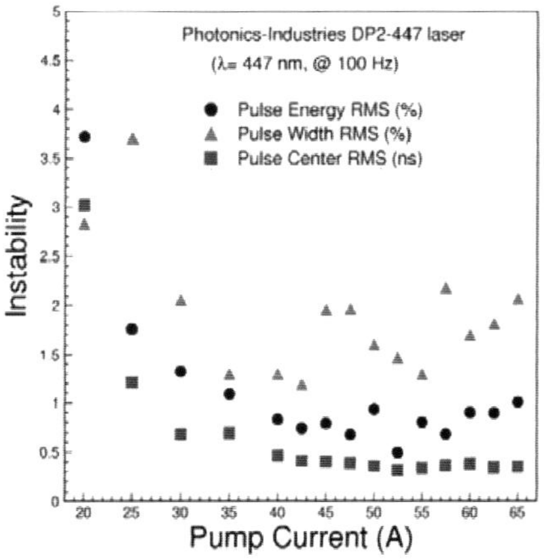

Figure 8. Instabilities of the pulse energy, width and center timing are shown as a function of the diode pump current.

Fig. 9 shows variations of the laser pulse energy and the room temperature in a test run of about 60 hours. The corresponding temperature dependence was determined to be about 2.8%/°C, as shown in Fig. 10. This means that the room temperature in the laser barracks needs to be stabilized to 1°C to maintain the pulse energy stability at a level of 3%.

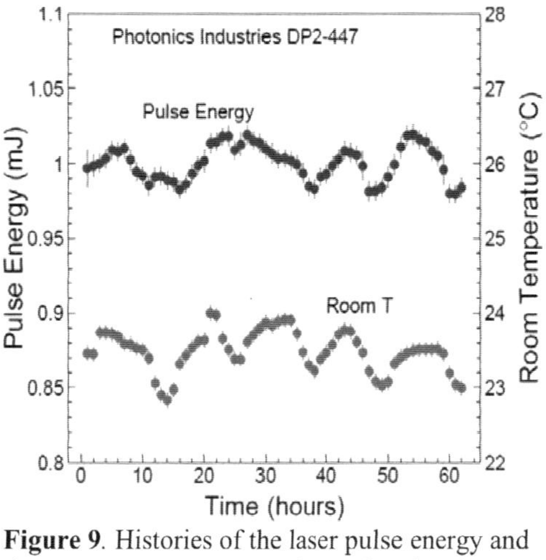

Figure 9. Histories of the laser pulse energy and the room temperature are shown for sixty hours.

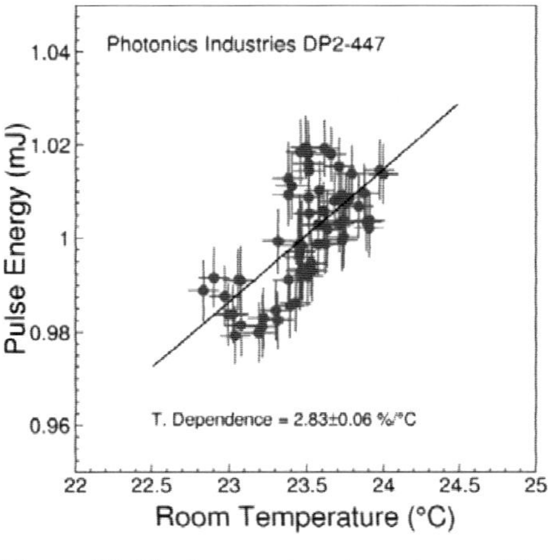

Figure 10. The temperature dependence of the laser pulse energy.

The result of a long run for 85 hours is shown in Fig.11. The observed stabilities are 0.7%, 2% and 0.3 ns respectively for the pulse energy, width and centre timing. They are much better than the 3%, 4% and 2 ns observed for the lamp-pumped Quantronix lasers [5]. All parameters measured are well within the specifications.

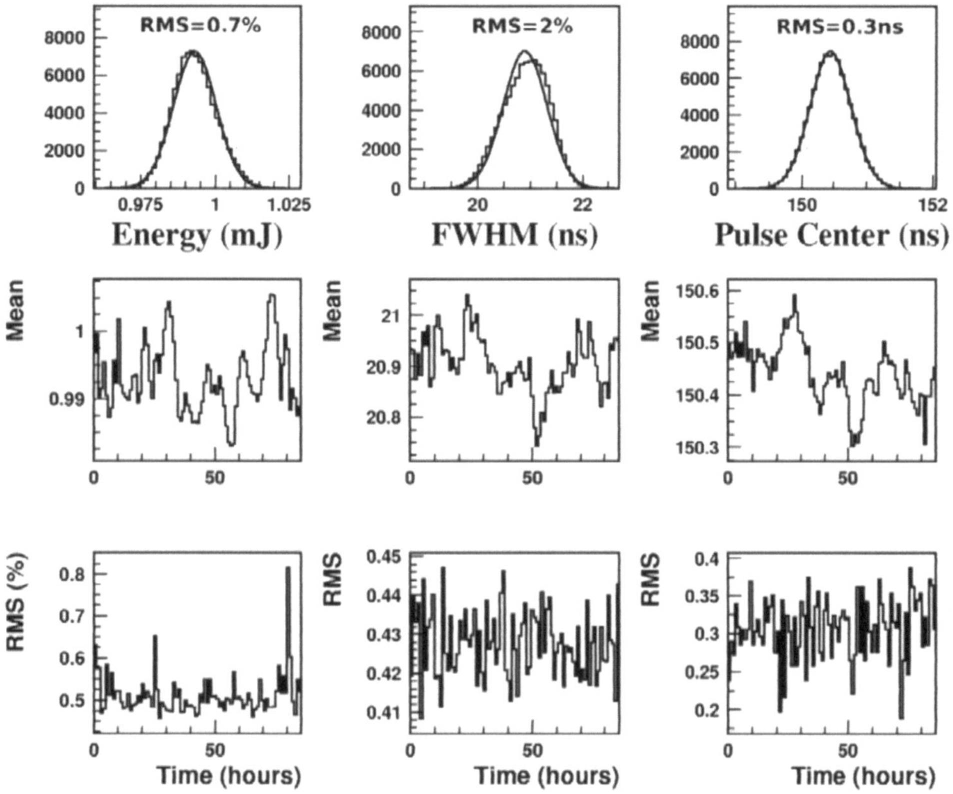

Figure 11. History of laser pulse energy, width and center timing are shown for a run of 85 hours.

3. DP2-447 laser *in situ* at LHC

Figure 12. DP2-447 laser installed in CMS

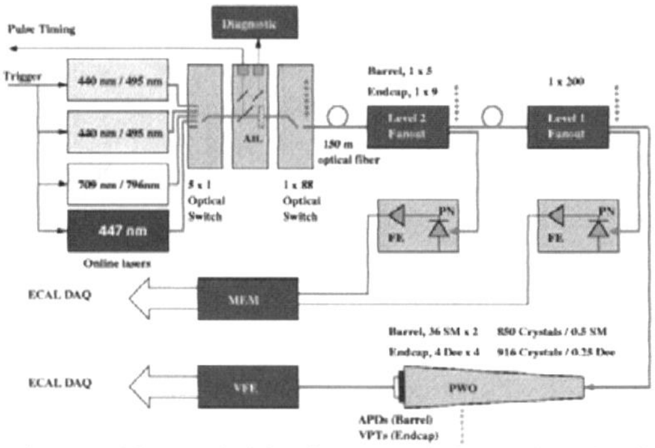

Figure 13. DP2-447 integrated into the CMS monitoring laser system.

The DP2-447 laser system arrived at CERN on March 20, 2012, and was installed in the CMS underground cavern on March 21. Its power was found to be consistent with that measured at Caltech, indicating no effect from the residual B field at a level of 30 Gauss. Fig.12 is a photo showing DP2-447 laser in the CMS underground cavern on March 21 for initial test. Following that, the ancillary optics and monitoring fibers were installed. The DP2-447 laser system was finally integrated into the existing monitoring system as shown in Fig. 13. As with the Quantronix laser systems, the DP2-447 has its own slow monitoring system, consisting of a beam sampler, a slow monitoring fiber, a photo-diode, a spectrometer and a digital scope. Its pulse is delivered to the ECAL detector through one of the channels in the 5 × 1 optical switch. The laser and its ancillary optics are controlled by a stand-alone PC which communicates with the PC controlling the Quantronix lasers via Ethernet.

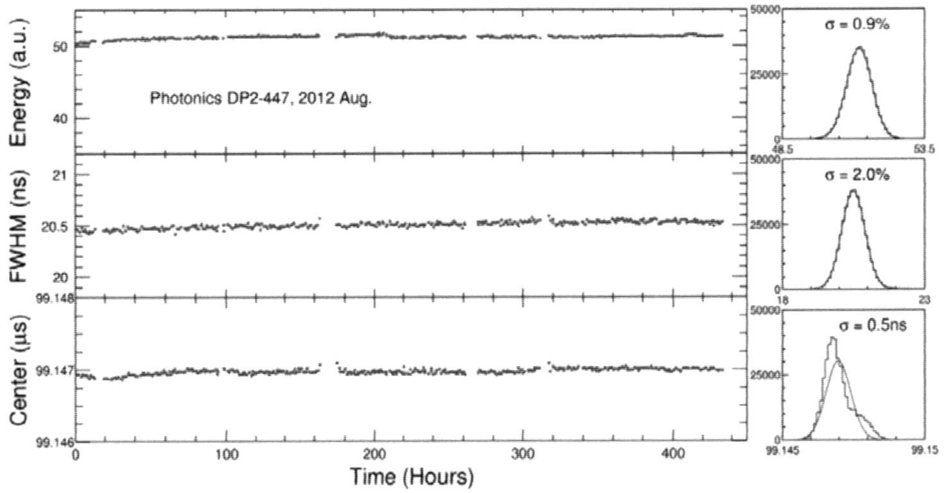

Figure 14. History of the DP2 laser pulse energy, width and timing measured *in situ* at LHC.

Fig.14 shows a typical history plot for the DP2-447 pulse energy, width and center timing *in situ* at LHC and their stabilities over 400 hours. The results show that the stability of the laser pulse energy, width and jitter are at 1%, 2% and 0.8 ns respectively, exceeding the 3%, 5% and 3 ns specifications. The comparison between responses of Quantronix and Photonics lasers is shown in Fig. 15. The overall consistency between these two lasers is excellent.

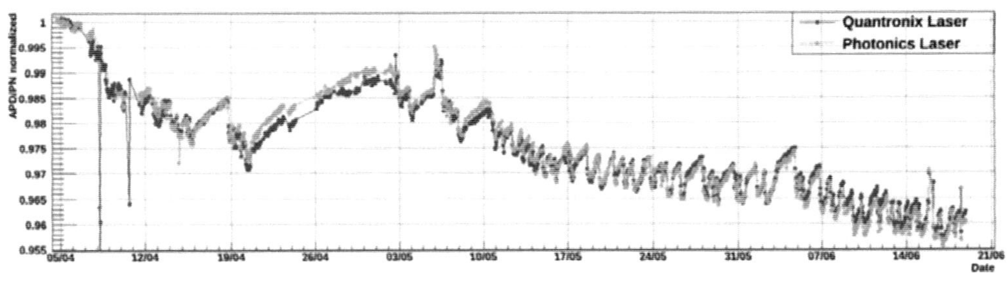

Figure 15. Comparison of the laser monitoring responses obtained with Quantronix and Photonics lasers in situ at LHC.

4. Summary

A new DPSS DP2-447 blue laser was successfully commissioned at CERN for the 2012 operation. This laser uses an Nd:YVO_4 crystal and a proprietary intra-cavity frequency triple technology. It has a simple structure and is more reliable than the existing lamp-pumped lasers. Measurements at Caltech and *in situ* at LHC show that this DP2-447 laser has good stabilities of 1%, 2% and 1 ns for the laser pulse energy, width and centre timing respectively. This new blue laser system provides a good foundation for precision monitoring of the 75,848 channels in the CMS ECAL *in situ* at LHC.

Acknowledgments

The authors would like to thank CMS ECAL colleagues for the collaborative effort on the work presented in this paper. This work was supported in part by the US Department of Energy Grant DE-FG03-92-ER40701.

References

[1] The CMS Collaboaration, **CMS-HIG-12-028-004;CERN-PH-EP-2012-220.**
[2] See, for example, T. Tabarelli in these proceedings.
[3] X. Qu, L. Zhang and R.-Y. Zhu, **2000** *IEEE Trans. Nucl Sci* 47, 1741.
[4] M. Anfreville *et al.*, **2008** *Nucl Inst. And Meth* A594, 292.
[5] L. Zhang, K. Zhu, D. Bailleux, A. Bornheim and R.-Y. Zhu, **2008** *IEEE T Nucl Sci* 55, 637.

Mitigation of Anomalous APD signals in the CMS Electromagnetic Calorimeter

D A Petyt, for the CMS collaboration
STFC Rutherford Appleton Laboratory, Didcot OX11 0QX, UK
E-mail: petyt@cern.ch

Abstract. Anomalous large signals are observed in the CMS Electromagnetic Calorimeter during proton-proton collisions at the Large Hadron Collider. They are ascribed to direct energy deposition by particles in the Avalanche Photodiodes (APDs) used for the light readout. Laboratory and test beam studies, as well as Monte Carlo simulations, have been used to understand their origin. The methods that have been employed to reject these signals in the online trigger of CMS and in the reconstruction of physics data are also presented.

1. The CMS Electromagnetic calorimeter

The Compact Muon Solenoid (CMS) [1] is a large general-purpose detector operating at the Large Hadron Collider (LHC) [2] at CERN. The principal goal of CMS is to search for Physics beyond the Standard Model at the TeV energy scale.

The main component of CMS to detect and measure the energies of electrons and photons is the Electromagnetic Calorimeter (ECAL) [3, 4]. The CMS ECAL consists of 75848 lead tungstate ($PbWO_4$) crystals, organised into a barrel and two endcap detectors and providing coverage in the pseudorapidity range $-3.0 < \eta < 3.0$. Two silicon preshower detectors are placed in front of the endcaps, covering the pseudorapidity range $1.65 < |\eta| < 2.6$.

Scintillation light emitted by the lead tungstate crystals is converted to electrical signals by photodetectors glued to the rear face of the crystals. Avalanche Photodiodes (APDs) are used in the ECAL Barrel, and Vacuum Phototriodes (VPTs) are used in the ECAL Endcaps.

The APD photodetectors are Hamamatsu type S8148 reverse structure avalanche photodiodes specially developed for the CMS ECAL [5]. The internal structure of the APD may be represented for the purpose of this discussion as having a 5 μm thick high-gain silicon layer (gain=50) and a 45 μm thick low-gain silicon layer (gain=1.4). Each APD has an active area of 5×5 mm^2 and a pair is mounted on each crystal and read out in parallel. A protective epoxy layer of about 400 μm thickness covers the front of the APD. Each pair is mounted in a capsule, a moulded receptacle (see Figure 1), which is then glued to the rear face of each crystal. The main properties of the APDs at gain 50 and 18°C are listed in [1].

2. Anomalous signals: ECAL spikes

Anomalous signals, consisting of isolated large signals with equivalent energies that can exceed 100 GeV in CMS, have been observed in the ECAL Barrel during LHC proton-proton (pp) collisions. An example event is shown in Figure 2. The isolated deposit in the top-right of this figure corresponds to an equivalent transverse energy of 690 GeV in the ECAL. These deposits,

Figure 1. Photograph of two CMS APDs (active area 5×5 mm^2) mounted in a capsule.

termed ECAL "spikes", are observed to occur at a rate that is proportional to the collision rate of the proton beams[1]. As a consequence, they present issues for triggering CMS at high luminosity and must be eliminated from physics analyses in order to prevent large biases in the energies of reconstructed electrons, photons and jets and in the computation of missing transverse energy (MET).

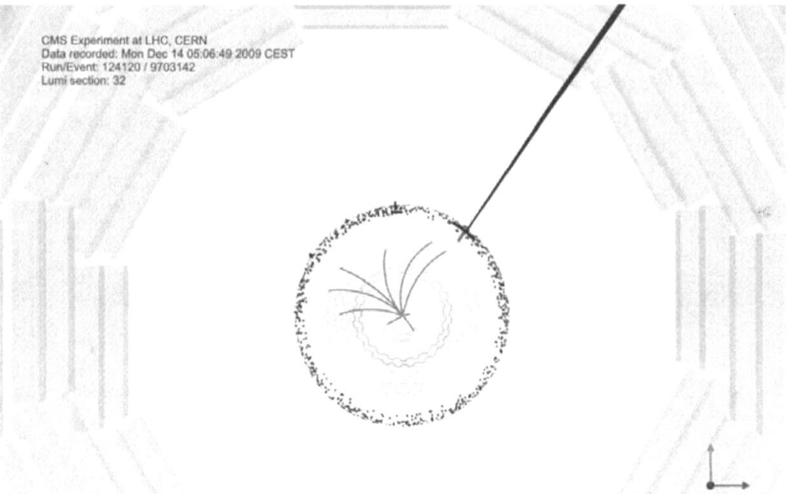

Figure 2. CMS Event Display of a pp collision event, showing an isolated ECAL spike (top-right) corresponding to a 690 GeV transverse energy deposit.

The spikes discussed in this note are understood to be associated with particles (produced in pp collisions) striking the APDs and very occasionally interacting to produce secondaries

[1] On average, one spike is observed per 370 minimum-bias triggers in CMS at $\sqrt{s} = 7$ TeV.

that cause large anomalous signals through direct ionization of the silicon. This hypothesis has been checked by studying the APD response via laboratory and test beam studies, and the development of Monte Carlo simulations where the APDs are treated as active volumes.

A model of the APD structure has been implemented in the CMS Monte Carlo simulation [6]. This simulation, which uses the GEANT 4 [7] package for particle transport, is employed to further understand the origin of spikes and their rates in CMS. The simulations show that direct ionization of the APD by protons/ions produced in pp collisions can produce large apparent energy signals after amplification (direct ionization produces $\sim 2.8 \times 10^5$ e/MeV, c.f. ~ 4.5 p.e./MeV for scintillation light in $PbWO_4$). The simulation results and laboratory measurements [8] indicate that a significant fraction of the anomalous signals are produced by np scattering in the protective epoxy coating of the APD, and the resulting proton directly ionizing the APD active volume.

3. Online rejection of ECAL spikes

Since the characteristics of ECAL spikes are localised high energy signals, they will often satisfy the conditions for triggering electrons and photons in CMS. A detailed description of the trigger algorithm can be found elsewhere [9]. If untreated, the rate of spikes would be a dominant component of the 100 kHz CMS Level-1 trigger rate bandwidth. Eliminating the spikes online is vital to maintain the lowest possible unprescaled electron and photon triggers that are required for CMS physics measurements.

Spike-like energy deposits are prevented from triggering CMS by exploiting additional functionality of the ECAL front-end electronics. The amplitude recorded in each channel in a "strip" (5×1 crystal array) is compared to a configurable threshold. The resulting 5-bit word, which records how many channels in the strip are above threshold, is used as the index in a lookup table, which outputs one bit per strip. It is configured to return zero if only one channel in the strip is above threshold (spike-like) and one if more than one channel is above threshold (corresponding to a typical electromagnetic (EM) shower). The resulting five "strip bits" for a given trigger tower (5x5 crystal array, corresponding to a single readout unit of the ECAL front-end electronics) are then ORed together to give a single bit per tower, the strip Fine-Grained Veto Bit (sFGVB). This will return zero for an isolated energy deposit (a single channel above threshold) or one for an EM-like energy deposit (multiple channels above threshold). Figure 3 shows the operation of the sFGVB on typical EM and spike-like energy deposits.

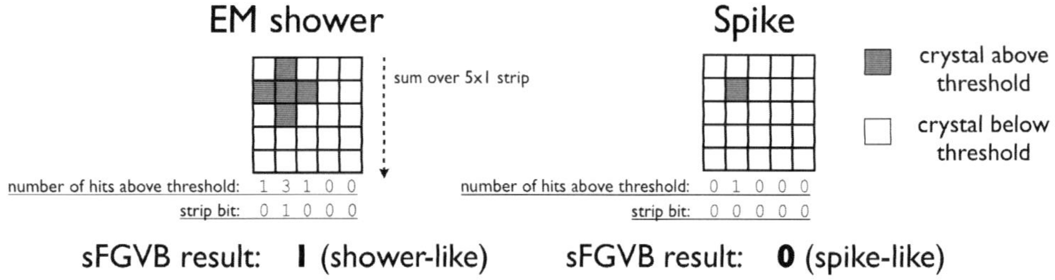

Figure 3. Operation of the strip Fine-Grained Veto Bit (sFGVB) on an electromagnetic shower (left) and a spike-like energy deposit (right).

If the sFGVB is set to zero, and the trigger tower transverse energy is greater than 8 GeV (12 GeV in 2012), the energy deposition is considered spike-like. The trigger tower energy is set to zero and the tower will not contribute to the triggering of CMS for the corresponding event.

The sFGVB is now implemented in CMS and has been operational since April 2011. It has been measured using data to reject > 95% of spikes with transverse energy greater than 8 GeV, with only a small (0.4%) effect on the efficiency for triggering well-identified electrons with transverse energy greater than 20 GeV. In 2012, the sFGVB was re-optimised for conditions with high event pile-up (mean of 20-30 interactions per LHC bunch crossing), achieving the same performance as for 2011 data.

4. Offline rejection of ECAL spikes

Spikes, which generally deposit energy in a single channel, have an anomalous pattern of energy sharing between crystals. A discriminating variable is constructed (see Figure 4), based on the ratio of the energy deposited in the central (highest energy) crystal in an ECAL energy deposit, $E1$, to the summed energy in the four adjacent crystals, $E4$. An EM shower, well-centered on an ECAL crystal, will typically contain $\sim 80\%$ of its energy in the central crystal and $\sim 20\%$ of its energy in the neighbouring crystals. Figure 5 shows the distribution of the "Swiss-cross" variable, $(1 - E4/E1)$, for CMS data and Monte Carlo (simulating EM energy deposits only). There is a clear additional peak around 1.0 in the data, due to the presence of ECAL spikes. A cut of $(1 - E4/E1) > 0.95$ rejects more than 99% of spikes with transverse energy greater than 10 GeV, with a negligible impact on the efficiency of selecting EM showers [10].

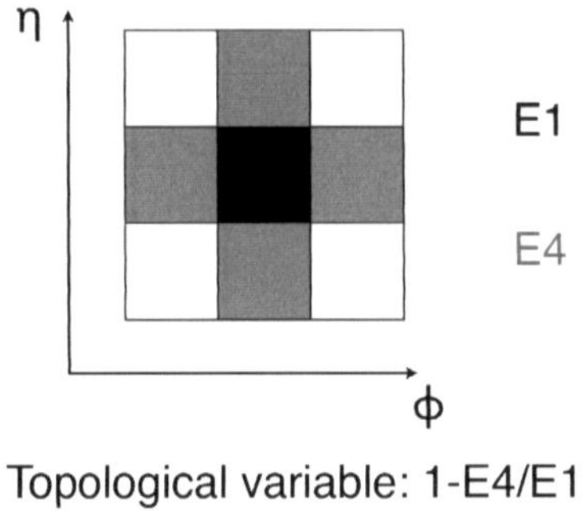

Figure 4. Definition of the "Swiss-Cross" topological variable $(1 - E4/E1)$.

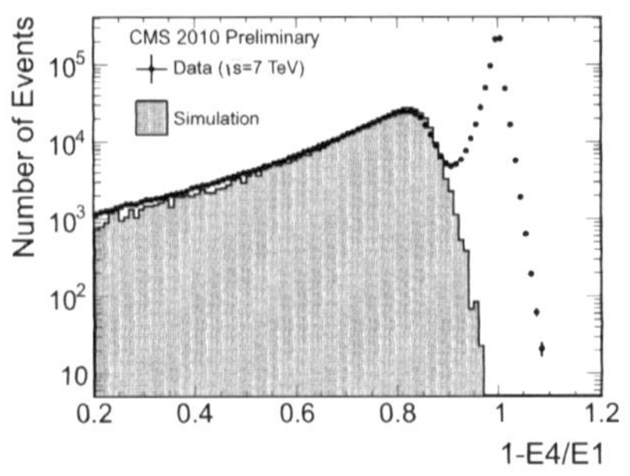

Figure 5. Distribution of the Swiss-Cross variable for the highest energy deposit in each event, for data (points) and simulation ($\sqrt{s} = 7$ TeV, filled histogram). Only events with an energy deposit with transverse energy > 3 GeV are plotted. The two distributions are normalized to the same total number of minimum-bias events with $(1 - E4/E1) < 0.9$.

Spikes and electromagnetic (EM) energy deposits have different signal pulse shapes. Since a spike is produced by a particle directly hitting the APD, the resulting signal does not include the response time of the scintillation light. Figure 6 shows representative pulse shapes from spikes and EM energy deposits measured from CMS data. When the pulse is fitted to extract the timing of the signal, the spikes generally appear "early", due to the faster rise time of the spike pulse. Figure 7 shows the reconstructed timing distribution for ECAL energy deposits with $E_T > 3$ GeV in both data and simulation[2]. The timing distribution from simulation

[2] The timing calibration is defined such that a relativistic particle originating from the nominal collisions vertex

(which does not include the anomalous signals) is centered at zero. The timing distribution from data shows a significant component of out-of-time signals, due to spikes[3]. A cut on the signal timing of ± 3 ns is $> 90\%$ efficient at removing spike pulses, with a negligible impact on EM energy deposits[4] [10]. By combining the topological and timing cuts, which rely on independent variables, a spike rejection factor of significantly better than 10^2 is achieved.

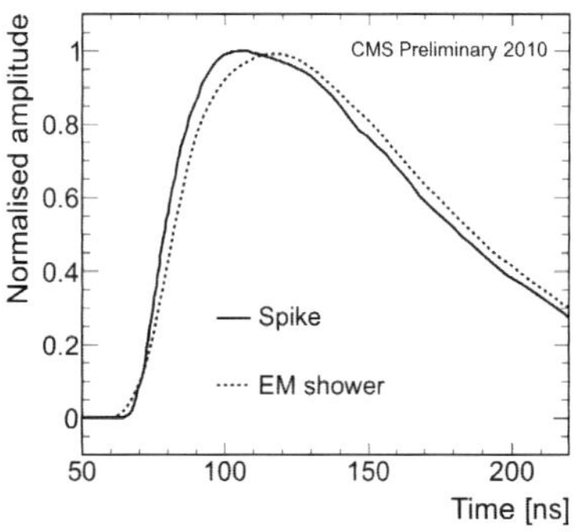

Figure 6. Average pulse shape (normalised signal amplitude versus time) for spike (solid line) and electromagnetic (dashed line) energy deposits, measured from CMS data.

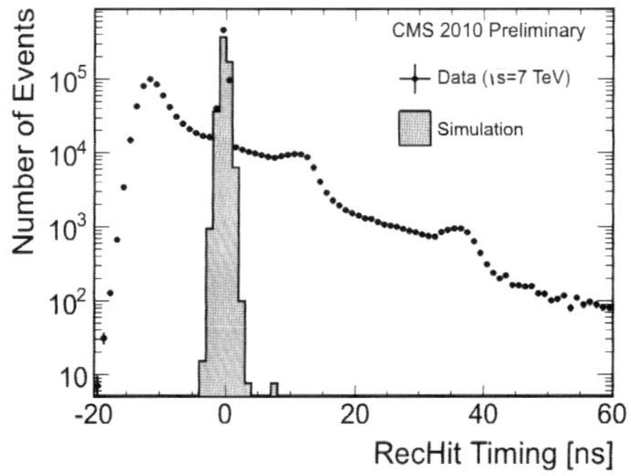

Figure 7. Distribution of the reconstructed time of ECAL signals for the highest energy deposit in each event for data and simulation ($\sqrt{s}$ = 7 TeV). Only events with an energy deposit with transverse energy > 3 GeV are plotted. The two distributions are normalized to the same total number of minimum-bias events with $(1 - E4/E1) < 0.9$.

5. Effect of spikes on CMS physics signatures

Events with large missing transverse energy (MET) are a sensitive tool to search for many sources of new physics at the LHC. ECAL spikes are one of the instrumental causes of anomalous MET measurements in CMS. If undetected, they contribute significantly to the tail of the MET distribution, degrading the performance of these physics searches.

An anomalous hit in an ECAL crystal is excluded from the reconstruction of jets and MET in CMS, using both the timing and topological cuts described in Section 4. This provides a reconstruction of jets and MET that are consistently cleaned of anomalous detector effects.

The calorimetric MET distribution (sum of the missing transverse energy in both the ECAL and the hadronic calorimeter, HCAL) for minimum-bias data and simulation at $\sqrt{s} = 7$ TeV is shown in Figure 8. Data are shown before (filled circles) and after (open circles) removal of the anomalous signals in the ECAL and HCAL[5] has been performed. Comparison of the "cleaned" data with simulation, which does not include anomalous energies, shows good agreement. Studies using simulations indicate a negligible probability of "over-cleaning" (energy due to bona fide particles produced in pp scattering removed by the cleaning).

will be reconstructed with a mean time of 0 ns.

[3] The long tail in this distribution is believed to be due to neutron-induced spikes, which propagate through the detector volume before interacting close to the APDs.

[4] The timing resolution for EM signals with energy > 1 GeV is less than 1 ns.

[5] A description of the removal of HCAL anomalous signals can be found in [11].

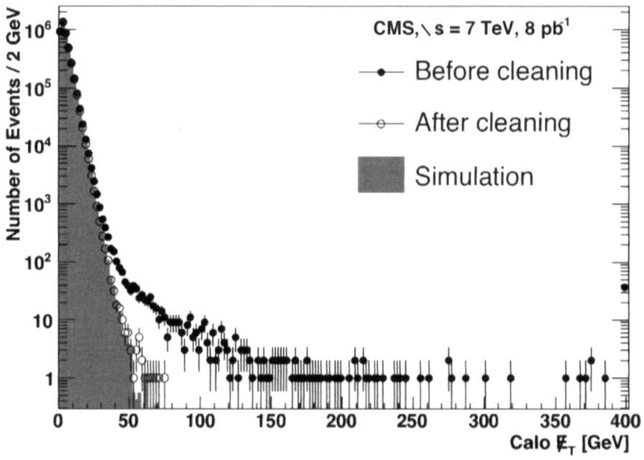

Figure 8. Missing transverse energy measured in the CMS calorimeters for minimum-bias data and MC pp collision events at $\sqrt{s} = 7$ TeV. Taken from [12].

6. Summary

The existence of large isolated signals in the CMS ECAL Barrel has been observed in pp collisions at the LHC. The origin of these signals due to direct ionization of the active silicon layers of the CMS APDs is supported by laboratory tests and Monte Carlo simulations.

The rate of anomalous signals in CMS is proportional to the minimum-bias collision rate, and to the number of charged tracks per event. Algorithms to efficiently remove these deposits in the online trigger and in the offline reconstruction of physics data have been developed and deployed in CMS. These algorithms have proved effective in rejecting spikes for LHC instantaneous luminosity above 10^{33} $cm^{-2}s^{-1}$, and remain efficient for the high event pile-up conditions experienced in 2012.

References

[1] CMS Collaboration 2008 *JINST* **3** S08004
[2] Evans, L and Bryant, P (eds) 2008 *JINST* **3** S08001
[3] CMS Collaboration 1997 *CMS ECAL Technical Design Report CERN/LHCC 97-33*
[4] CMS Collaboration 2010 *J. Instrum.* **5** T03010
[5] Renker D 2002 *Nucl. Instrum. Meth.* **A486** 164–169
[6] Cossutti F 2007 *CMS Conference Report CMS-CR-2007-043* URL `http://cdsweb.cern.ch/record/1061281/files/CR2007_043.pdf`
[7] Agostinelli S *et al.* 2003 *Nucl. Instrum. Meth. A* **506** 250
[8] Brown R, Deiters K, Ingram Q and Renker D 2011 *Nucl. Instrum. Meth. Section A: Accelerators, Spectrometers, Detectors and Associated Equipment* – ISSN 0168-9002 URL `http://www.sciencedirect.com/science/article/pii/S0168900211021644`
[9] Paganini P 2009 *J. Phys. Conf. Ser.* **160** 012062
[10] CMS Collaboration 2010 *CMS Note CMS-NOTE-2010-012* URL `http://cdsweb.cern.ch/record/1278160/files/CMS-Note-2010-012.pdf`
[11] Apresyan A 2012 Contribution to this conference
[12] CMS Collaboration 2011 *J. Instrum.* **6** P09001

Identification and Mitigation of Anomalous Signals in CMS Hadronic Calorimeter

Artur Apresyan

California Institute of Technology, Pasadena, California 91106
ON BEHALF OF THE CMS COLLABORATION

E-mail: apresyan@cern.ch

Abstract. The CMS HCAL detector occasionally records anomalous large energy signals that correspond to particles hitting the transducers. Anomalous signals in HCAL can also be produced by rare random discharges of the readout detectors. We present a summary of various sources of anomalous signals, the algorithms developed to identify them, and the performance of these algorithms in the CMS collision data.

1. The CMS hadronic calorimeter

The barrel and endcap hadronic calorimeter (HCAL) subdetectors of CMS detector [1, 2] completely surround the electromagnetic calorimeter (ECAL), and are immersed within the 3.8 T magnetic field of the solenoid. The barrel (HB) extends out to $|\eta| = 1.4$, and HE covers the overlapping range $1.3 < |\eta| < 3.0$. The forward calorimeters extend the pseudorapidity coverage from $|\eta| = 2.9$ to $|\eta| = 5$. The thickness of the HCAL sub-detectors varies from 7-11 λ_{I}, depending on $|\eta|$.

In HB and HE the active plastic scintillator tiles are interspersed between the stainless steel and brass absorber plates [3, 4, 5]. The front-end electronics of each wedge (72 channels) are housed in an enclosure referred to as a read-out box (RBX). Each RBX contains four hybrid photodiodes (HPD) [6] which register signals from eighteen towers at a fixed ϕ angle. The HF is made of steel absorbers and embedded quartz fibers, which provide a fast collection of Čherenkov light. Two types of quartz fibers are used: "long" fibers span the length of HF, and "short" fibers begin 22 cm into the detector. The differing fiber lengths allow for separation of electromagnetic and hadronic showers. Photomultiplier tubes (PMTs) [7] connected to the fibers convert detected light to electrical signals. Analog-to-digital converters (ADCs) digitize the signals, and the ADC output of each calorimeter channel is computed every 25 ns[1].

2. Characteristics of anomalous noise in HCAL sub-detectors

The CMS ECAL and HCAL occasionally record anomalous signals which can be misinterpreted as energy deposits[2], and it is essential to remove such anomalous signals so that they do not contribute to the measured energy attributed to collision events. Several sources of anomalous signals in HCAL have been observed (Figure 1). Electronics and detector noise observed in the

[1] These 25 ns intervals are later referred to as "time slices", or TS
[2] A description of the removal of ECAL anomalous noise can be found in Ref. [8]

HB and HE occurs randomly and at a constant rate, independent of the beam conditions. Beam induced anomalous signals are observed in signals recorded by the PMTs of the HF calorimeter.

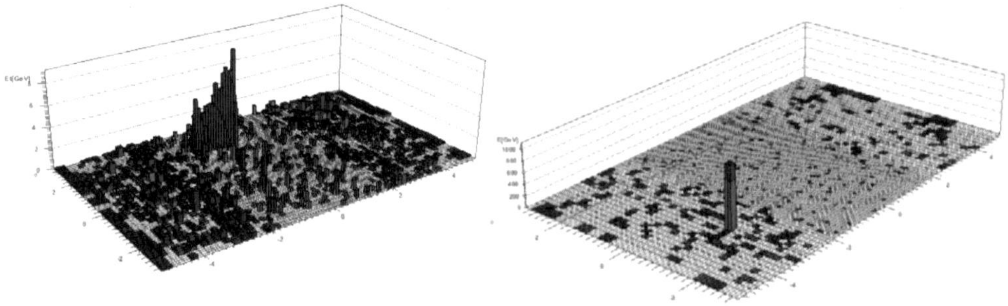

Figure 1. Event displays in η and ϕ showing noise in HB (left), and in HF (right), in events overlapping with a *pp* collision.

3. Identification of noise

Some of these anomalous noise signals were observed during past test beam and cosmic data taking [9], and detailed studies continued with the *pp* collision data [10]. Two approaches are used in the treatment of noise overlapping with a collision event. In the first approach, the energy reconstructed in channels having anomalous signals is removed during event processing, so that the affected channels do not contribute to the reconstructed physics objects ("noise cleaning"). In the second approach, if many channels are identified as having noise, the entire event is discarded from the analysis ("noise filtering").

3.1. Noise in barrel in endcap

Three categories of noise in HB and HE have been observed during both cosmic and collision data taking. Large signals are occasionally observed in the HPDs even when no light is incident on the photo-cathodes. These signals are originated by thermally emitted electrons which ionize the gas or surface molecules in the acceleration gap of the HPD, which in turn are accelerated back to the cathode, liberating additional electrons. This type of noise is referred to as "ion feedback", and manifests itself as a significant energy deposit in a single channel. Misalignments between the electric field within an HPD and the external solenoid field can lower the flashover voltage of the HPD, leading to an avalanche of secondary electrons, producing large signals in up to 18 channels of the HPD ("HPD noise"). Electronics noise in RBX can produce signals in channels of multiple HPDs within the RBX, which can affect up to all 4HPDs ("RBX noise").

3.1.1. Noise filtering in HB and HE The event filter to identify and reject noise events is based on channel multiplicity, total number of zero ADC counts, and the RBX "pulse shape ratio" ("R45"). An event is flagged as noise if any of the following conditions are met:

- at least 17 channels within an HPD with $E(\text{channel}) \geq 1.5$ GeV, or
- at least 10 channels within an HPD with $E(\text{channel}) \geq 1.5$ GeV, and there is no energy reported in the other HPDs contained in the same RBX, or

- at least 10 TSs within an RBX with zero ADC counts, and $E(\mathrm{RBX}) \geq 10$ GeV.

The R45 variable is defined as R45=(E4-E5)/(E4+E5), where E4 and E5 are the energies measured in channels of that RBX, in the triggered and subsequent TSs, respectively. For real signals $\sim 90\%$ of energy is contained within these two TSs, while noise pulses are generally broader. As can be seen from Figure 2, for events collected using a Minimum Bias trigger, majority of RBXs have the R45 distribution contained within the red *envelope*. The envelope defines the position of an energy dependent cut on the R45 variable: events with an RBX that has R45 above the upper curve, or below the lower curve are flagged as noise events.

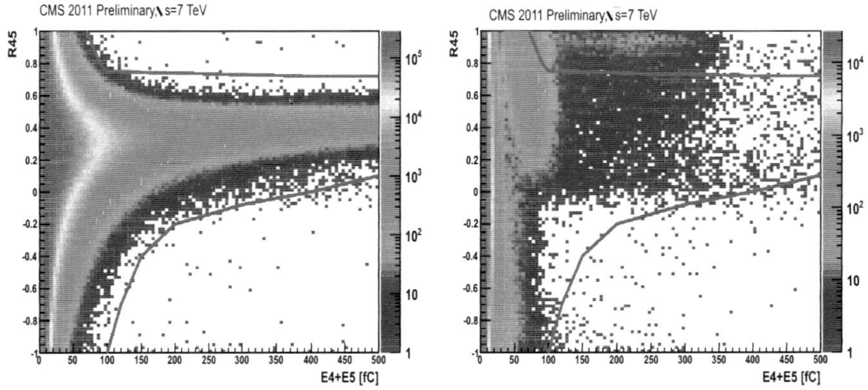

Figure 2. The distribution of R45 versus energy in E4+E5 for all RBXs in collision events (left), and events collected during periods with no circulating beams in LHC (right).

Events that fail any of the requirements listed above are marked as noise candidates, and discarded from physics analysis. The efficiency of the noise filter was measured in collision events, and was found to be >70-75%, with a misidentification rate of less than 0.3%. A modification of this filter that includes information from ECAL and tracker, and adds HCAL topological isolation requirements increases the efficiency to > 95%, with misidentification rate < 0.4%.

3.1.2. Noise cleaning in HB and HE A set of algorithms has been developed to further exploit the differences between noise and signal pulse shapes. The information used by these algorithms is largely uncorrelated to those described in Sec. 3.1.1, hence they provide additional identification power. In the fit-based algorithms a comparison of the measured and the ideal signal pulse shape is made, and a compatibility test to signal and noise hypothesis is performed. Two noise types are tested against the signal pulse, and the corresponding pulse-shapes are shown in Figure 3.

The linear discriminant is formed by comparing the χ^2 from signal and noise shapes hypotheses fits. The noise hypothesis here is that the noise pulse shape is a linear function of the TS index (middle in Figure 3). A log ratio of the χ^2 of the two hypotheses is then used as a discriminant variable to identify a channel as noisy. Another type of anomalous signals in HB and HE has the characteristics of all the charge being concentrated in one or two time slices, with very little activity in the remaining 8 TSs (right in Figure 3). "RMS8" is defined as the RMS value (in units of charge) of the lowest 8 TSs, divided by the maximum charge in all 10 TSs. The discriminant for this type of noise is then defined as the log ratio between the RMS8^2 and the χ^2 to the ideal pulse shape. An energy dependent cut is then chosen for both of these algorithms using events collected during runs with no circulating beams in LHC.

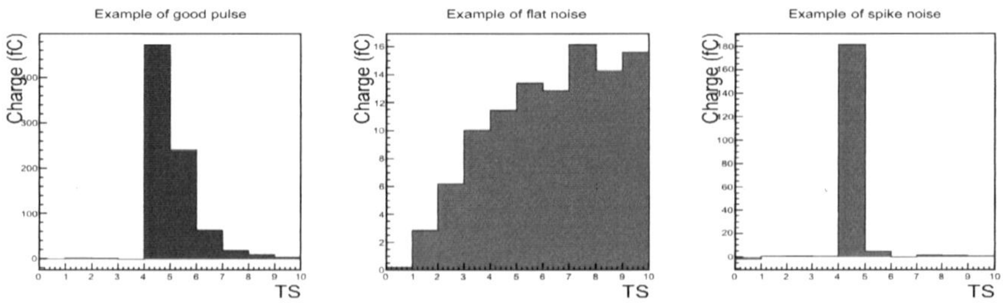

Figure 3. Pulse shapes of a good signal from hadronic shower (left) compared to common noise signals (middle and right). The events with noise pulses are collected during periods when there are no collisions in CMS.

The left plot Figure 4 shows a comparison of missing transverse (MET) energy reconstructed from channels only in HB and HE, before and after the noise cleaning is applied. As can be seen, the mis-identification rate is very low (below 0.01%). The middle plot in Figure 4 shows the distribution of energies of the noise signals identified by the fit-based algorithms, before applying the noise filter (Section 3.1.1). The right plot in Figure 4 shows the same distribution after the events identified by filter were rejected, showing the complementarity of the two algorithms.

3.2. Noise in HF

Anomalous signals in HF occur when particles interact in the HF PMT window, in the PMT device, or in the fibers. This type of noise is beam related and increases with increasing luminosity and is also $|\eta|$ dependent. PMT window events were first observed when HF modules were exposed to test beams, and the dominant source was found to be due to charged particles producing Čerenkov light in the thick glass window of the PMTs. A contribution is also observed from charged particles traversing the fiber bundle behind HF. Some of the high energy anomalous signals was found to originate from the material used to electrically isolate the PMT from the light guide ("sleeve"). This material produced scintillation light which was detected by the

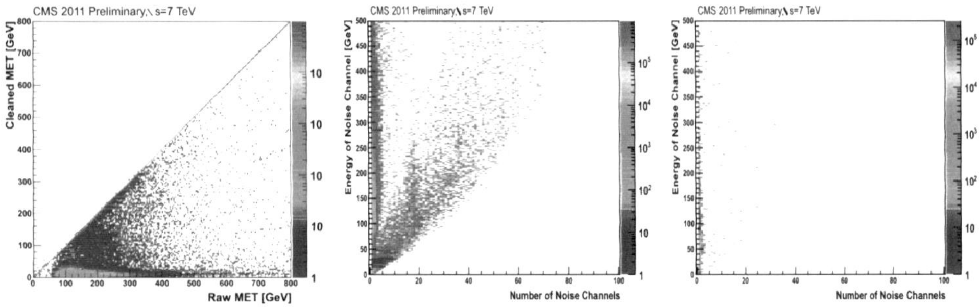

Figure 4. Comparison of MET computed using only HB and HE channels after cleaning (Y-axis) vs. after cleaning (X-axis) (left). Distribution of energies of the noise signals identified by the fit-based algorithms before applying the noise filter (middle), and the same plot after applying the noise filter (right). Events in all distributions are collected with a MET-based trigger.

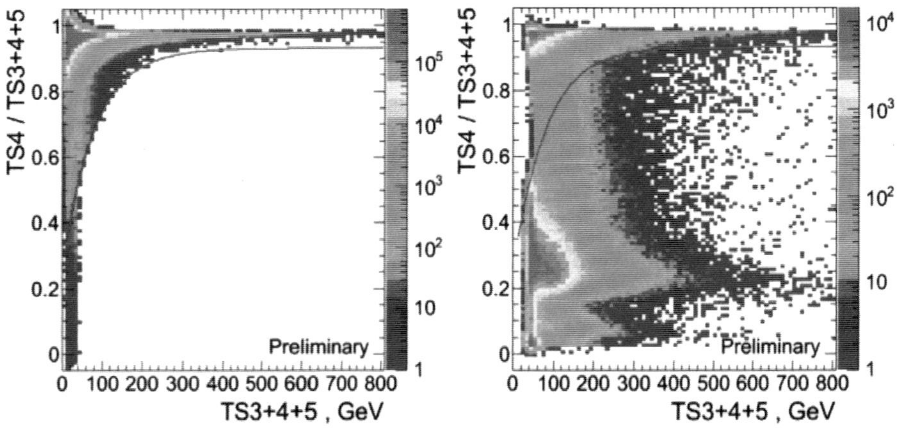

Figure 5. The distribution of PS1 ratio versus energy in TS3, 4 and 5, for all channels in HF in collision events (left), and events collected with a trigger based on jets in HF.

PMTs, and was replaced with a non-scintillating ones before 2011 data taking period.

The hits from relativistic charged particles impacting upon the PMT windows are subsequently detected and cause an early signal to be recorded for that individual channel. For relativistic particles the PMT window hits arrive significantly ahead of the Čerenkov light emitted in the HF absorber. The Čerenkov light captured by quartz fibers in the HF produces pulses having widths of only 10 ns. Therefore, the phase of HF electronics is setup so that the majority of the PMT window hits arrive in the TS previous to the triggering one, thus reducing the effect of noise on energy reconstruction.

3.2.1. Pulse-shape cleaning in HF The pulse shape based noise algorithm makes use of the ratio of energy E recorded in the triggered TS (E4) to the energy in the TSs around it ($\Sigma E = E3 + E4 + E5$). A discriminator, referred to as PS1, is defined as PS1=E4/ΣE. True signal hits are expected to have a PS1 value near one. As can be seen from Figure 5, for events collected using a Minimum Bias trigger, majority of hits have the PS1 distribution contained within the red *envelope*. The envelope defines the position of an energy dependent cut on the PS1, and channels that have PS1 below the red curve are flagged as noise events.

3.2.2. Topological cleaning in HF Topological cleaning of noise in HF is performed on all channels which are not flagged as noise by the pulse shape algorithm. Three different algorithms are used, referred to as the "long-short" ratio, S8S1, and S9S1 algorithms.

An asymmetric ratio (R) of the energy deposited in the long (E^L) and short (E^S) fibers for the same channel, is defined as $R = |E^L - E^S|/(E^L + E^S)$. The R ratio is computed for all channels, and checked whether it is greater than 0.8. If a channel's R value is greater than 0.8 and its energy is greater than an $|\eta|$-dependent threshold, then the channel is marked as noise.

Two isolation-based variables are formed to account for the lateral profile of real showers:

$$S8S1 = \frac{\sum_{i=1}^{4} E_i^L + \sum_{i=1}^{4} E_i^S}{E}, \quad S9S1 = \frac{E^S + \sum_{i=1}^{4} E_i^L + \sum_{i=1}^{4} E_i^S}{E^L} \tag{1}$$

where the index i runs over each channels adjacent cells. Any channel previously flagged as noise by the pulse shape filter or the PET algorithm is excluded from the construction of the

S8S1 or S9S1 variable. The S9S1 and S8S1 algorithms are applied to all channels with energy above 100 GeV, and energy dependent cut position is optimized on a sample of noise data. If the channel fails the cut, it's marked then the channel is marked as noisy, and discarded from the event reconstruction. The *S8S1* algorithm is applied to all short fiber hits , while the S9S1 algorithm is applied to all long fiber hits.

3.2.3. Pre-triggering from noise in HF Some of the noise in HF generate pulses that are reconstructed as early arriving signals. While this feature is exploited in rejecting the energies of such signals in event reconstruction, it can cause the trigger system to prematurely issue an "accept" signal, reading out the information from sub-detectors 25 ns earlier than the collision. An algorithm is implemented at the Level 1 trigger level, which vetos triggering in "BX-1", where BX is the colliding bunch crossing number. The BX number where proton bunches collide is identified using the Button Beam Pickup (BPTX) system [1]. The BPTX provides an accurate information on the timing and phase of each proton bunch, allowing one to calculate the interaction-point position from the relative phases of BPTX measurements on opposite sides of CMS. The "BX-1" veto was implemented in 2011 running, and was found to be fully effective in removing pre-triggers. Since *pp* collisions occur every 50 ns during the 2011-2012 running of the LHC, the BX-1 veto does not introduce any inefficiency at the trigger. The PMTs will be replaced during 2013-2014 LHC shutdown with thinner window PMTs that will significantly reduce the noise rates, and provide additional noise rejection due to multi-anode readout. This change will allow running with 25 ns bunch spacing without the BX-1 veto.

4. Summary

Several identification algorithms have been developed to suppress anomalous signals in CMS HCAL. The flexible software infrastructure allows to employ these algorithms both at the trigger level, and during the event reconstruction. The algorithms identify noise signals by exploiting the differences in pulse shape and energy isolation. It was shown that the efficiency in rejecting noise signals for high instantaneous luminosity of LHC data-taking in 2012 remains high.

This project was carried out with financial support from U.S. Department of Energy, U.S. National Science Foundation, RMKI-KFKI (Hungary), Russian Ministry of Education and Science, Russian State Committee for Atomic Energy, Scientific and Technical Research Council of Turkey (TUBITAK), Turkish Atomic Energy Agency (TAEK) and Bogazici University Research Fund.

References

[1] S. Chatrchyan *et al.*, "The CMS experiment at the CERN LHC," *JINST* **3**, S08004 (2008)
[2] S. Chatrchyan *et al.*, "The CMS hadron calorimeter project: Technical Design Report", CERN-LHCC-97-031
[3] S. Abdullin *et al.*, "Design, Performance, and Calibration of CMS Hadron-Barrel Calorimeter Wedges", *Eur. Phys. J.* C **55** (2008).
[4] S. Abdullin *et al.*, "Design, Performance and Calibration of the CMS Forward Calorimeter Wedges", *Eur. Phys. J.* C **53** (2008).
[5] G. Baiatian *et al.*, "Energy Response and Longitudinal Shower Profiles Measured in CMS HCAL and Comparison With Geant4," CERN-CMS-NOTE-2006-143.
[6] P. Cushman, A. Heering, A. Ronzhin, "Custom HPD readout for the CMS HCAL", *Nucl. Instrum. Methods* A **442** (2000).
[7] U. Akgun *et al.*, "Complete tests of 2000 Hamamatsu R7525HA phototubes for the CMS-HF Forward Calorimeter", *Nucl. Instrum. Methods* A **550** (2005).
[8] D.A. Petyt, "Mitigation of Anomalous APD signals in the CMS Electromagnetic Calorimeter", Contribution to this conference.
[9] S. Chatrchyan *et al.*, "Identification and filtering of uncharacteristic noise in the CMS hadron calorimeter", *JINST* **5** T03014 (2010).
[10] S. Chatrchyan *et al.*, "Missing transverse energy performance of the CMS detector", *JINST* **6** P09001 (2011).

XVth International Conference on Calorimetry in High Energy Physics (CALOR2012) IOP Publishing
Journal of Physics: Conference Series **404** (2012) 012045 doi:10.1088/1742-6596/404/1/012045

Pileup measurement and mitigation techniques in CMS

A Perloff, on behalf of the CMS collaboration

Graduate Student, Department of Physics and Astronomy, Texas A&M University, College Station, TX 77843-4242, USA

E-mail: aperloff@physics.tamu.edu

Abstract. When trying to reconstruct an event from a hard-scatter *pp* collision in CMS, it is of the utmost importance to correctly measure the energy from jets. The jet energy corrections (JEC) correct, on average, the energy of the reconstructed jets back to the energy of the final-state particles that initiated the jets. This effort is hindered by additional energy in the jets coming from other soft *pp* collisions. The additional energy is termed *pileup* or *offset* and comes from everything except the primary vertex (PV) and its underlying event (UE). In this paper, we describe how this pileup energy is measured and parametrized as well as the techniques used to remove this extra energy from the reconstructed jets.

1. Introduction

During the reconstruction of a *proton-proton* (*pp*) collision (event), jets are often reconstructed with a p_T that differs from that of the final-state particles that make up the jet. The jet energy corrections (JEC) correct the reconstructed jet energy back to the true energy of the final-state particles. All analyses at CMS which make use of jets must use the JEC. Therefore, the CMS collaboration has developed a factorized approach to these JEC which is composed of multiple levels representing corrections for various physics or detector effects and allows enough flexibility in the corrections to be applicable to many types of analyses [1–3].

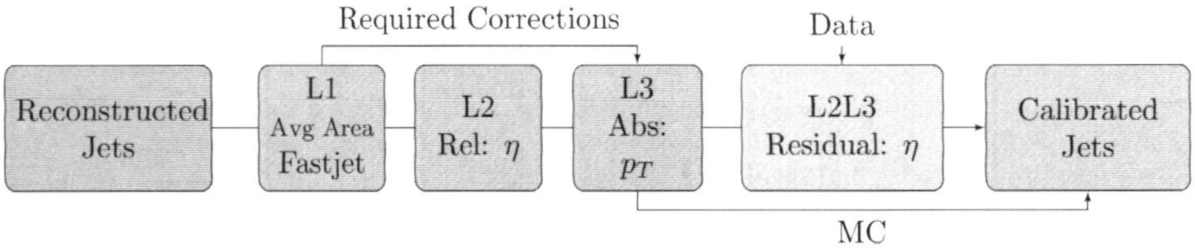

Figure 1. A graphical depiction of the jet energy correction levels required for all CMS analyses.

One of the physics effects that the JEC seek to correct for is the additional energy attributed to the jets which comes from *pp* interactions other than the hard-scatter event at the primary vertex (PV). The additional energy from these events is often called *pileup* or *offset* and will be the focus of this paper.

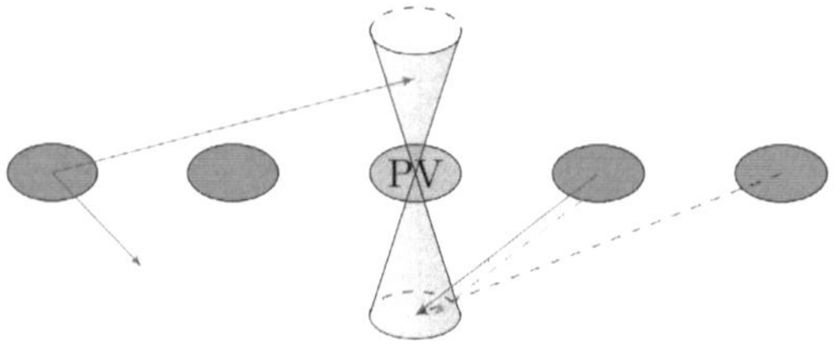

Figure 2. A basic graphic showing two jets coming from a PV with additional pp interactions in red. The arrows coming from the red pp interactions are particles which add additional energy to the jets during reconstruction. The different colors represent different types of particles (ex: photons, charged hadrons, etc.)

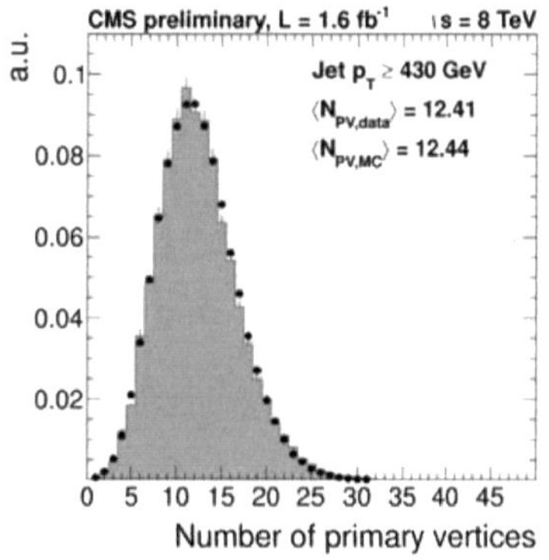

Figure 3. Distribution of the number of primary vertices (per event) in data and MC in 2012.

There are three major classifications of pileup based upon the time at which the additional energy enters the calorimeter system. In-time (IT) pileup refers to energy from pp collisions in the current bunch-crossing (BX) other than that at the hard scatter PV. This is the largest source of pileup energy. In addition, there is early out-of-time (EOOT) pileup, which refers to energy left in the calorimeters from previous BXs, and late out-of-time (LOOT) pileup, which refers to energy from later BXs that is integrated with the current event's energy.

2. Pileup Measurement

While pileup itself cannot be directly measured, it can be correlated to various other directly measurable quantities. As pileup comes from additional pp interactions, the number of PV (N_{PV}) is directly correlated to the amount of pileup; the greater the N_{PV}, the more pileup energy is added to the jets. The total offset for the event is equal to the average energy from a non-hard-scatter PV times the N_{PV}. The ability to correlate pileup with N_{PV} is predicated on the CMS detector having a high reconstruction efficiency for PVs. In creating the corrections we use the cuts $N_{dof} > 4$ and $IsFake = false$ to make sure we are measuring only good PV. Figure 3 shows the good agreement between the N_{PV} in data and in MC [4].

Another quantity which is directly correlated with the amount of pileup energy is the median energy per area (ρ), which is calculated using the Fastjet algorithm [5]. In order to make this correlation, pileup must be approximated as a homogeneous noise which can be subtracted off.

3. Pileup Corrections

The JEC corrections at CMS can be separated into two categories, those done at the hardware level and those done in software. Detector level corrections can be timing changes, threshold level modifications, or any other change to the hardware that might reduce the actual amount of pileup energy within the jets during reconstruction. Software level corrections are performed after reconstruction and scale jet p_T up or down based on the *measured* amount of pileup.

3.1. Detector Level

Detector level mitigation techniques have been extremely effective in reducing the amount of out-of-time (OOT) pileup. The barrel and endcap hadronic calorimeter system (HBHE) used four time-slices (TS=25 ns) as its integration time during 2011. The measured LOOT pileup contributed about 20% of the pileup energy with EOOT pileup contributing another 2-4%. In

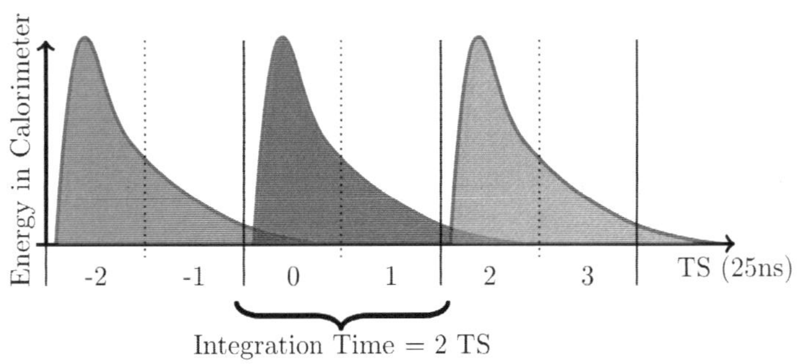

Figure 4. A graphic showing, on the most basic level, the energy deposited in the barrel and endcap hadronic calorimeters (HBHE). The central curve is the current event with the left and right curves being the events before and after, respectively.

2012, the integration time in the HBHE system was changed to two TS, which reduced the LOOT pileup to almost zero and also reduced the EOOT pileup. This change is depicted in figure 4. With four TS starting from TS 0, the energy that is integrated will contain all of the energy from the current event as well as a significant portion of the energy from the next event. The two TS scheme includes little to none of the LOOT pileup. This is a significant improvement as it not only reduces one of the contributors to the pileup to nearly zero, but it also means that the rest of the pileup energy is more easily correlated with the IT N_{PV}.

3.2. Software Level

Software level pileup mitigation techniques are essential to most analyses. The purpose of these offset correction techniques is to measure and remove the extra energy inside of the jets that is not associated with the high-p_T vertex and its UE [1–4; 6].

3.2.1. Average Offset The average offset (AO) method seeks to determine the average amount of energy added to an event due to low-p_T scatters (pileup). The correction is calculated using three different samples; a Zero-Bias data dataset, a Zero-Bias MC sample that is made by overlaying a neutrino gun sample with Minimum Bias events, and a QCD MC sample. In each of the three samples, the offset ($p_{T,offset}$) is measured on an event-by-event basis by finding the p_T deposited in a jet with a certain cone size (usually 0.5 or 0.7 at CMS), centered at a specific $\eta - \phi$ location. The ϕ direction of the *jet* is chosen at random and a scan of η, in steps of 0.1 within $|\eta| < 5.0$, is performed for each event. For calorimeter (Calo) jets, the energy in the jet is the sum of the energy inside the calorimeter towers whose $\eta - \phi$ location is inside the radius of the jet cone. For particle flow (PF) jets, the energy is that of the sum of the PF candidates inside the jet cone [6]. This procedure is called the *Random Cone Algorithm*.

These results are then binned in 1-D histograms, one for each value in $|\eta|$, with N_{PV} on the x-axis and the $< p_{T,offset} >$ on the y-axis. Figure 5 shows a sample plot of the AO versus N_{PV} for PF jets in $2.0 < |\eta| < 2.1$. The offset, shown in figure 5, is modeled by:

$$Offset(N_{PV}) = [0] + [1](N_{PV}) + [2](N_{PV})^2 \tag{1}$$

As expected, the value at the y-intercept of the line ($N_{PV} = 0$) is the p_T contribution of the OOT pileup. Whereas the first and second order terms in N_{PV} are dominated by IT pileup, with a small contribution from OOT pileup.

The correction to the raw, uncorrected jet p_T (p_T^{RAW}) is then derived to be a multiplicative scale factor which, when applied to the p_T^{RAW}, removes the added pileup energy. The scale factor for the AO method is:

$$C_{AverageOffset}(N_{PV}, p_T^{RAW}, \eta) = 1 - \frac{< Offset(N_{PV}, \eta) >}{p_T^{RAW}} \tag{2}$$

Once applied, the scale factor should correct the p_T^{RAW} of the jet back to the true (generator) p_T:

$$p_T^{COR} = p_T^{GEN} = C_{AverageOffset}(N_{PV}, p_T^{RAW}, \eta) \cdot p_T^{RAW} \tag{3}$$

3.2.2. Hybrid Jet Area When using the AO method, one assumes that every jet contains the same amount of additional energy from pileup. Therefore, the best that the AO corrections can do is to correction on an event-by-event basis. The jet area (JA) method takes into account variations in the energy from pileup on an event-by-event basis and on a jet-by-jet basis by using a new parametrization [5; 7]. The hope is to improve the jet energy resolution and gain adaptability for future pileup conditions.

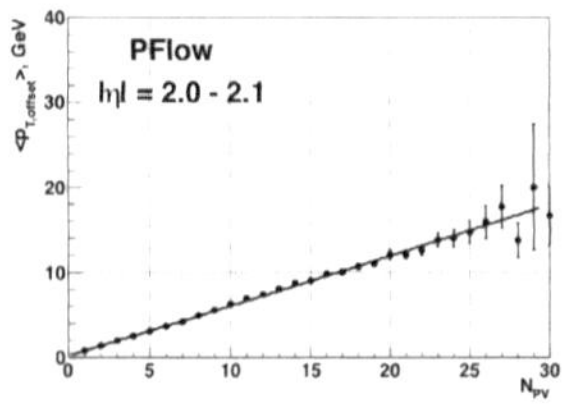

Figure 5. The average PF jet offset as a function of N_{PV} within $2.0 < |\eta| < 2.1$.

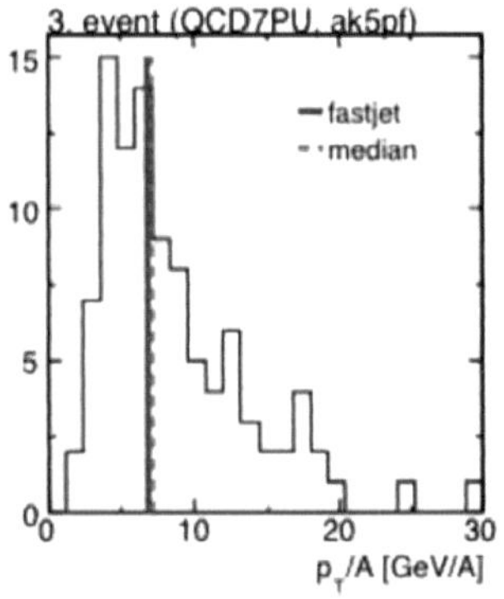

Figure 6. Distribution of p_T/A for all of the jets in a single event.

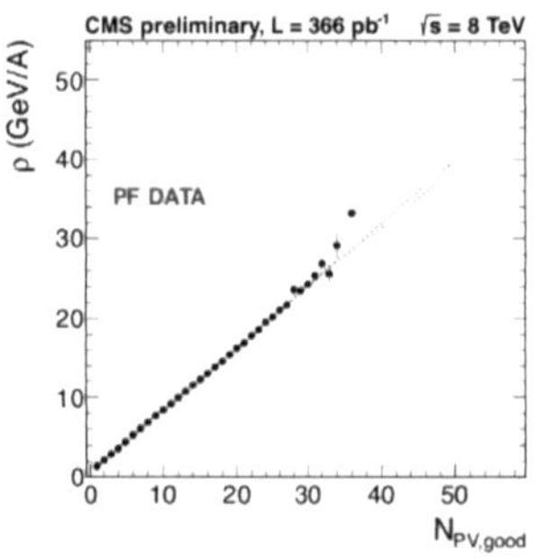

Figure 7. Distribution of ρ versus N_{PV} in data.

The new parametrization was developed based on the Fastjet algorithm. First the area (A_j) is calculated for each jet in the event. The next quantity needed is the median energy density of an event, ρ. This quantity is defined as the median of the distribution of p_{Tj}/A_j, where j is the jet index. The ρ computation is uses the k_T jet clustering algorithm (R=0.6) because of its well-defined energy clustering and is insensitive to jets from a hard scatter as seen in figure 6. The sparsely populated values around $p_T/A = 30$ are the hard jets in the event (higher p_T per unit area), whereas the large clump of jets lower in the spectrum represent the jets from pileup.

While the JA method is more granular and can correct on a jet-by-jet basis, the Fastjet algorithm used to parametrize such corrections does not know that the response of the detector is η dependent. In the AO corrections, this was taken into account by binning the corrections in η. Therefore, a method was developed that combines the η parametrization benefit of the AO method with the jet-by-jet granularity and parametrization of the JA method. This new method was called the hybrid jet area (HJA) method.

We start by parametrizing ρ in terms of N_{PV}, as is done in figure 7. Once this relationship is obtained, the formula is inverted to obtain $N_{PV}(\rho)$. This relationship is then inserted into the function for the offset derived above.

$$Offset(N_{PV}(\rho)) = [0] + [1](N_{PV}(\rho)) + [2](N_{PV}(\rho))^2 \tag{4}$$

This function has all of the same characteristics, irrespective of the parametrization, as the equation derived by the AO method. However, the HJA scale factor

$$C_{Fastjet-based}(\rho, A, p_T^{RAW}, \eta) = 1 - A\frac{< \frac{Offset(\rho),\eta}{A} >}{p_T^{RAW}} \tag{5}$$

uses the JA to correct on a jet-by-jet basis. As seen in figure 8, both of these methods measure the pileup energy almost equivalently, on average [4].

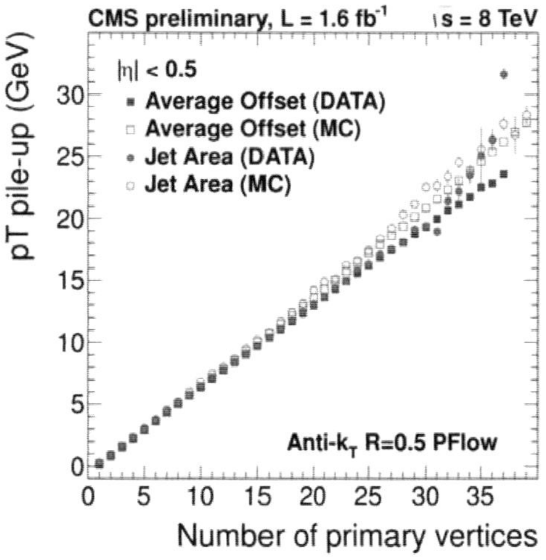

Figure 8. Comparison of the pileup measured in PF jets using AO method and the HJA method.

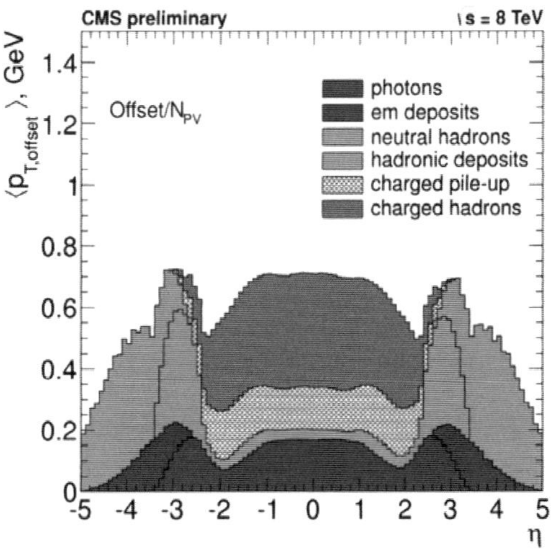

Figure 9. The particle flow composition of an $N_{PV} = 1$ event in MC.

3.2.3. Charged Hadron Subtraction Charged hadron subtraction (CHS) is an additional pileup subtraction tool that can be used in conjunction with either of the methods above, but which requires the use of particle plow jets. CHS uses the CMS detector's excellent tracking capabilities to identify and remove jet constituents (charged hadrons) which are known to have originated from pileup vertices. The algorithm does not remove unassociated tracks, as they may have originated from the high-p_T vertex.

The benefit to applying this additional algorithm to the jet corrections is that it lowers the amount of pileup energy that the AO or HJA methods must compensate for, thus lowering the multiplicative scale factors. Furthermore, CHS has the additional advanatage of being a particle-by-particle pileup subtractions technique, while at best the other algorithms work on an event-by-event or jet-by-jet basis.

3.3. Closure

In order to determine the efficacy of the corrections described above, We study the response of the detector to the jets, defined as $R =< p_T^{COR}/p_T^{GEN} >$. If the pileup corrections were ideal, we would expect the response for the jets to be identical at all N_{PV}, meaning that there was no response dependence upon the pileup conditions inside the detector. In figure 10 the left plot shows the response as a function of the GenJet p_T for PF jets. The plot clearly shows that the response for low p_T jets is not equal to 1 (R=1 being a perfect response). In addition, the

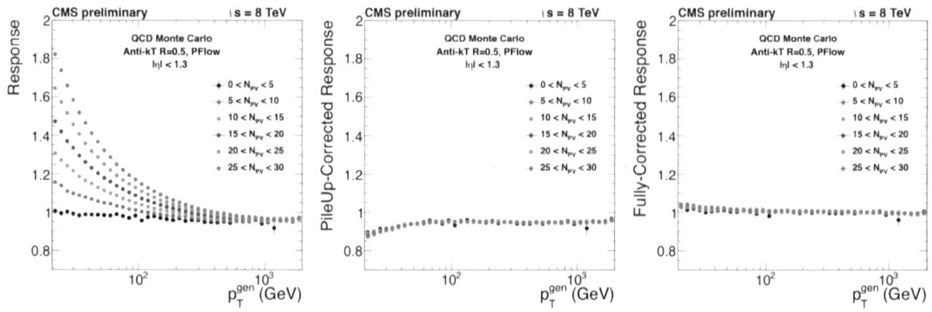

Figure 10. Response versus GenJet p_T for PF jets (left) **before** correcting for pileup, (middle) **after** correcting for pileup, and (right) after the required MC correction chain. The same plots can be made for the Calo jets with similar results.

response varies significantly for differing values of N_{PV}. However, the right plot show the same thing after the pileup corrections have been applied. One can see not only a better response at low p_T, but that there is no N_{PV} (pileup) dependence. Because all of the pileup energy has been removed from the jets, they also have a significantly lower response at low p_T. Any non-closure still present is due to various other effects, such as the detector's η and p_T dependencies, which are corrected for using additional steps in the factorized jet correction chain [3; 4].

4. Conclusion

All of the various pileup removal techniques discussed in this paper have one goal, remove the dependence on pileup energy from the jet response. The first level of mitigation against pileup comes during the detector level readout and reconstruction efforts (ex: time-slices in the HBHE system). These have been shown to provide significant improvements in reducing OOT pileup.

The second level of pileup mitigation was at the software level and came in several of varieties. The AO method is the recommended correction for Calo jets as there is empirical evidence for slightly higher resolution in Calo jets when using this method. The Fastjet-based, HJA method is used for most analyses in CMS and is the recommended correction for PF jets. The key abilities of this method were its adaptability and per event/jet granularity. CHS is an additional tool which is very promising when used in conjunction with the other pileup mitigation techniques.

Once applied, there is good closure over most of the jet p_T spectrum (in MC). Any remaining non-closure in the jet response is well understood and is removed at later stages of the factorized JEC chain [6].

References

[1] The CMS Collaboration 2010 CMS PAS JME-10-003

[2] The CMS Collaboration 2010 CMS PAS JME-10-010

[3] The CMS collaboration 2011 *J. of Inst.* **6** P11002 URL `http://stacks.iop.org/1748-0221/6/i=11/a=P11002`

[4] The CMS Collaboration 2012 CMS DP-2012/012

[5] Cacciari M and Salam G P 2008 *Phys. Lett.* B **659** 119 – 126 ISSN 0370-2693 URL `http://www.sciencedirect.com/science/article/pii/S0370269307011094`

[6] Kirschenmann H, on behalf of the CMS collaboration 2012 *JPCS*

[7] Cacciari M, Salam G P and Soyez G 2008 *JHEP* **2008** 005 URL `http://stacks.iop.org/1126-6708/2008/i=04/a=005`

Signal reconstruction performance with the ATLAS Hadronic Tile Calorimeter

Pawel Klimek, on behalf of the ATLAS Tile Calorimeter group
Stockholm University, Sweden
E-mail: pawel.klimek@cern.ch

Abstract. The Tile Calorimeter (TileCal) is the central section of the hadronic calorimeter of ATLAS. It is a key detector for the reconstruction of hadrons, jets, taus and missing transverse energy. TileCal is a sampling calorimeter with steel as absorber and scintillators as active medium. The scintillators are read-out by wavelength shifting fibers coupled to photomultiplier tubes (PMTs). The analogue signals from the PMTs are amplified, shaped and digitized by sampling the signal every 25 ns. The read out system is designed to reconstruct the data in real time fulfilling the tight constraints imposed by the ATLAS first level trigger rate (100 kHz). The signal amplitude for each channel and their phase are measured using Optimal Filtering techniques both at online and offline level. We present the performance of these techniques on the data collected in the proton-proton collisions at center-of-mass energy equal to 7 TeV. We will address the performance for the measurement on high pile-up environment and on various physics and calibration signals.

1. Introduction

ATLAS [1] is a general-purpose experiment, whose goal is to cover a broad range of experimental particle physics phenomena, with the long sought Higgs boson and supersymmetry on top of the wanted list. The Tile Calorimeter [2] is the hadronic calorimeter of ATLAS in the region $|\eta| < 1.7$. The purpose of TileCal is to identify hadronic jets and measure their energy and direction. It also provides vital information for the first level trigger, the measurement of the missing energy due to non-interacting particles and the identification of electrons and photons. It uses steel as absorber and scintillating plastic tiles as active material.

2. Read out architecture

Photomultiplier tubes (PMTs) read out the light from the scintillating tiles using wavelenght shifting fibers. The fibers are grouped to segment the calorimeter in cells with a granularity $\Delta\phi \times \Delta\eta \sim 0.1 \times 0.1$ and 3 radial layers. Each cell is read out by two PMTs, corresponding to two independent electronic channels. The PMT output is a current pulse with its amplitude proportional to the energy deposited in the associated cell. Electronics mounted on the PMT amplifies and shapes the output current pulse. Pulse shaping forms the signal width to 50 ns at half-maximum. In order to effectively extend the dynamic range a bigain read out is used with a ratio between the two of 64. The two analogue pulses (high gain and low gain) are digitized by two 10-bit ADCs at 40 MHz. The high gain chain is used until it saturate at about 12 GeV, for higher energies the low gain is used [3]. The analogue pulse is digitized by 10-bit analogue to digital converters with 7 samples at 25 ns intervals which are read out upon a trigger accept from the first level trigger and sent to the back-end electronics in a counting room 100 m off detector, the Read Out Drivers (ROD). The RODs use Digital Signal Processors (DSP) to compute the pulse amplitude, phase and quality factor.

3. Optimal Filtering algorithm

The goal of the energy reconstruction algorithms is to compute the energy deposited in a TileCal cell from the number of ADC counts measured in each of the two corresponding read out channels. The samples are referred to as S_i with $i = 1 - 7$ and are in units of ADC counts. The energy reconstruction combines the S_i to first obtain the amplitude in ADC counts and thereafter applies a calibration constant in MeV per ADC count. The S_i are linearly combined to provide the pulse amplitude A, the phase t with respect to the 40 MHz clock and the electronic pedestal P (figure 1 left panel). The quality factor QF is also computed to verify that the computed amplitude, phase, pedestal and the known pulse shape $g(t)$ are compatible with the observed data points S_i. In case of deviation between the actual shape and the expected shape, the quality factor takes large values which can be used to detect problems in the reconstruction procedure. These quantities are calculated as follows [4]:

$$A = \sum_{i=1}^{7} a_i \cdot S_i, \quad t = \frac{1}{A}\sum_{i=1}^{7} b_i \cdot S_i, \quad P = \sum_{i=1}^{7} c_i \cdot S_i, \quad QF = \sqrt{\sum_{i=1}^{7}\left(S_i - (A \cdot g_i - A \cdot t \cdot g_i' - P)\right)^2}$$

This method is called Optimal Filtering. Prior knowledge of the normalized pulse shape function is required to determine the constants a_i, b_i, c_i. The pulse shape was precisely measured in test beam and it was verified with data from the full installed TileCal that this function could be used for all channels. The linear coefficients are optimized, using the autocorrelation matrix, to minimize the effect of the noise on the reconstructed quantities. The coefficients are functions of the phase of the pulse with respect to the 40 MHz electronic clock. The pulse shape function and the linear constants are stored in a dedicated database for calibration constants.

Offline reconstruction can be performed with iterations (Iterative Optimal Filtering) and without iterations (Non-Iterative Optimal Filtering). In the iterative method the quantities are calculated with an initial assumed value of the phase, equal to the time of the maximum of the S_i. In later iterations, the phase is taken to be equal to t from the previous iteration. In absence of out-of-time pile-up the iterative algorithm always converges to the true value of the phase with an accuracy better than 0.5 ns. The iterative method is slower and more sensitive to out-of-time pile-up. In the non-iterative method the phase used is the best signal phase known, retrieved from the database for each channel. In this case no further iterations are performed.

The Optimal Filtering reconstruction is also performed online by the DSP which perform the above linear combinations in real time. Above a trigger rate of 30 kHz the Optimal Filtering must be performed without iterations due to the processing time in the DSP exceeding the limit allowed by the data acquisition system. The DSP reconstruction is limited by use of fixed point arithmetic and the internal precision available to describe the linear coefficients and calibration factors. Since the division is a time consuming operation in the DSP the phase is computed using a look-up table with the energy reciprocal pre-defined and stored in the DSP memory.

4. Validation

The ROD can be configured to transmit both the reconstructed quantities and the raw data samples S_i. Due to the limited bandwidth digital samples are transmitted only for pulses above a programmable threshold (currently $S_{max} - S_{min} > 5$ ADC counts). The raw data obtained in this way can be reconstructed using well tested offline algorithms and can be used to validate the reconstruction in DSP.

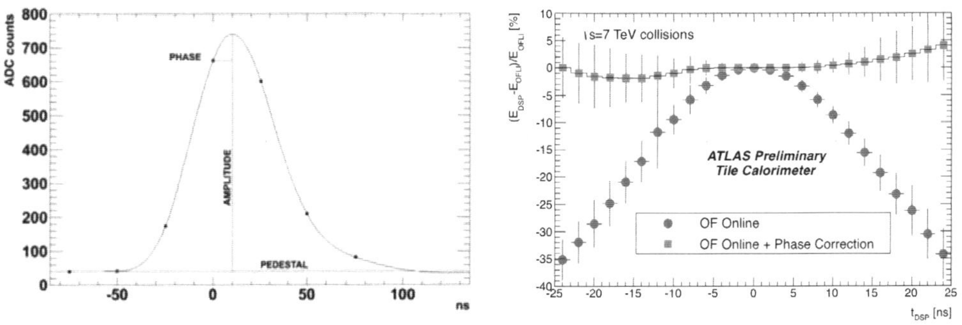

Figure 1. Left: Sketch of a typical Tile Calorimeter pulse shape with the ADC samples (dots) and the illustration of Optimal Filtering reconstructed quantities. Right: Relative difference between the energy reconstructed with DSP (E_{DSP}) and offline Iterative Optimal Filtering method (E_{OFLI}) as a function of the phase reconstructed by DSP (t_{DSP}) in collision data at 7 TeV. There is a bias in the energy reconstructed online with Optimal Filtering method (OF Online) due to phase variation (circles). The bias can be reduced applying a second order correction using the phase of the pulse (squares). The vertical error bars correspond to the RMS of the distributions.

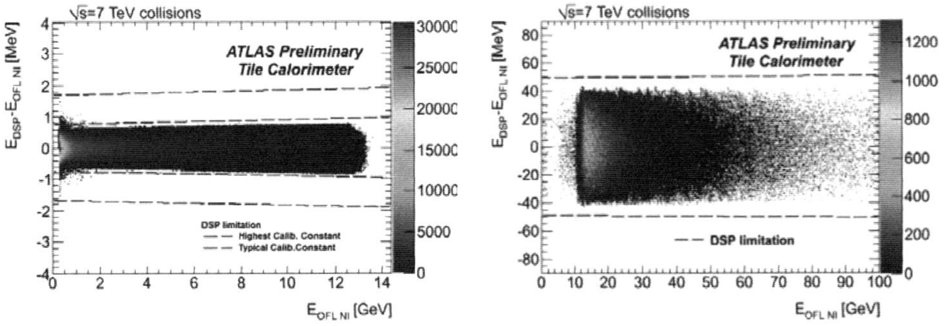

Figure 2. Absolute difference between the energy reconstructed with DSP (E_{DSP}) and offline Non-Iterative Optimal Filtering method (E_{OFLNI}) as a function of the energy reconstructed offline in collision data at 7 TeV in high gain (left) and low gain (right).

In order to validate the DSP results the consistency of online and offline implementations is checked. Figure 1 right panel shows performance of energy reconstruction with DSP in collision data at 7 TeV. In-time and out-of-time collision events populate the plot in order to evaluate the DSP reconstruction performance on a wide time window. Most of the pulses are in the time range $[-5, 5]$ ns. The circle points show the relative difference between the energy reconstructed with DSP and offline with Iterative Optimal Filtering method. The variation in the phase of pulses leads to an underestimation of the amplitude reconstructed online. This bias is parametrized by second order polynomial and reduced thanks to an amplitude correction function of the phase. The square points show the relative difference after the correction. In the time range $[-10, 10]$

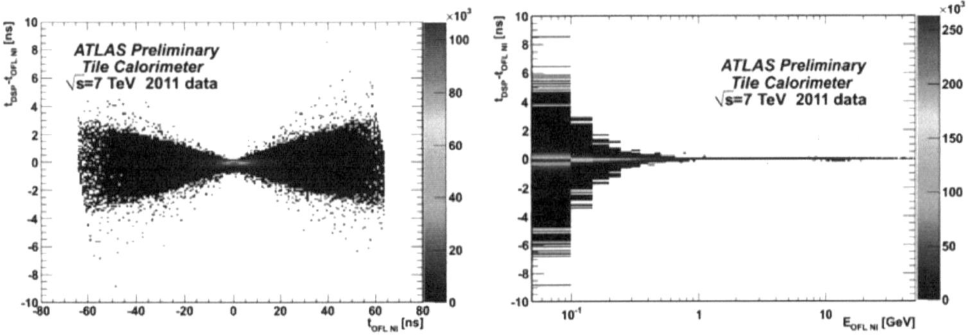

Figure 3. Left: Absolute difference between phase reconstructed with DSP (t_{DSP}) and offline with Non-Iterative Optimal Filtering method (t_{OFLNI}) as a function of phase reconstructed offline in collision data at 7 TeV. Right: Absolute difference between phase reconstructed with DSP and offline with Non-Iterative Optimal Filtering method as a function of energy reconstructed offline (E_{OFLNI}) in collision data at 7 TeV.

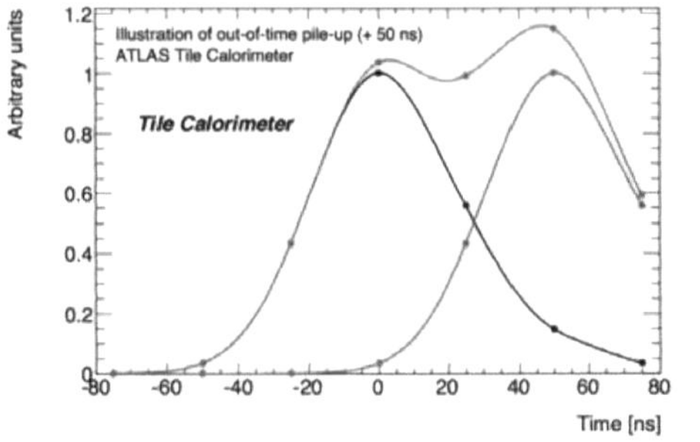

Figure 4. Illustration of out-of-time pile-up (+50 ns) in ATLAS Tile Calorimeter with pulse shapes obtained with the TileCal simulation [5].

ns the average difference between the offline and online reconstruction, after the correction, is smaller than 1%.

Figure 2 shows the residuals after applying the parabolic correction between the energy reconstructed with DSP and offline with Non-Iterative Optimal Filtering method in high gain (left panel) and low gain (right panel) in collision data at 7 TeV. The performance is slightly energy dependent. The maximum expected difference due to fixed point arithmetic is proportional to the calibration constant ADC to MeV and therefore changes from channel to channel. The worst cases arise in a few PMTs with abnormal high voltage settings. The response needs to be amplified by a large factor to compensate for the gain deficit. In left panel of figure 2 the red dashed lines indicate the maximum expected precision for standard functioning channels of about 1 MeV and contain 99% of the channels while the blue lines indicate the expected precision for the highest calibration constant of about 2 MeV. For comparison the electronic noise RMS level is about 30 MeV. In right panel of figure 2 the blue lines indicate the precision of about 50 MeV. This is fully adequate in the range where signals are larger than approximately 8GeV.

Figure 3 shows the absolute difference between phase reconstructed in DSP and offline with Non-Iterative Optimal Filtering as a function of offline reconstructed phase (left panel) and

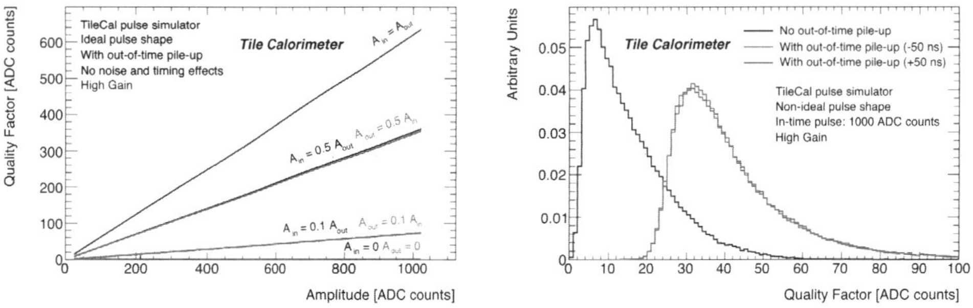

Figure 5. Left: Quality factor as a function of the amplitude in different pile-up scenarios obtained with the TileCal pulse simulation. A_{in} (A_{out}) is the amplitude of the in-time (out-of-time) pulse. The x-axis shows the amplitude in ADC counts before cell-dependent calibration constants are applied. The calibration factor is approximately 12 MeV per ADC count. Right: Normalized distributions of quality factor in TileCal simulator with non-ideal pulse shape. Generated amplitude of in-time pulse is 12 GeV. Amplitude of the out-of-time pulse follows the distribution in ZeroBias trigger with a cut on 34 ADC counts [5].

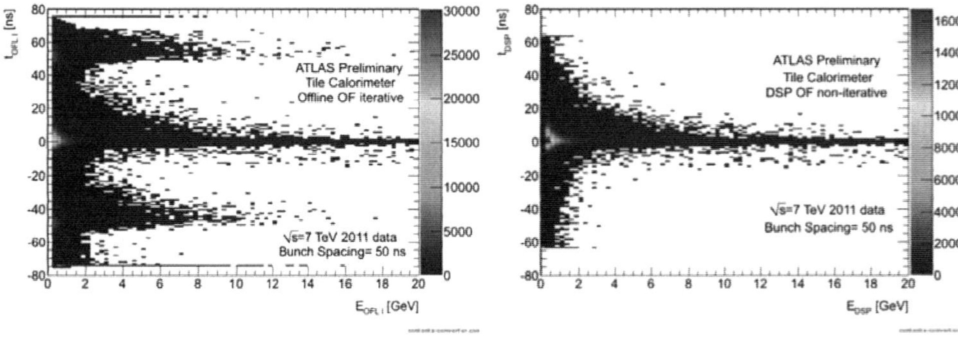

Figure 6. Left: The phase reconstructed offline with Iterative Optimal Filtering method as a function of the energy reconstructed offline in collision data at 7 TeV. Right: The phase reconstructed with DSP (Non-Iterative Optimal Filtering method) as a function of energy reconstructed with DSP in collision data at 7 TeV. A data sample with presence of out-of-time pile-up (bunch spacing 50 ns) is used.

energy (right panel) in collision data at 7 TeV. Since the same non-iterative algorithm is used online and offline the expected difference is due to fixed point arithmetic and the precision of the look up table (LUT) used in the DSP. The LUT effect is enhanced for larger phases but differences between phases reconstructed in DSP and offline are within 1.5 ns in the $[-25, +25]$ ns range.

5. Out-of-time pile-up

Signal pulses coming from different LHC bunch crossing (bunch spacing of 50 ns) can be measured in the readout windows used of ± 75 ns. The out-of-time pile-up results in the superposition of pulses shifted in time resulting in anomalous pulse shapes which can be detected thanks to large values of QF [5]. Figure 4 shows an illustration of an out-of-time pile-up pulse at +50 ns, in the case where the in-time and out-of-time pulses have the same amplitude.

Figure 5 left panel shows the dependence of the quality factor QF as a function of the in-time pulse amplitude A_{in} and the out-of-time pulse amplitude A_{out} for different values of the ratio between the in-time and out-of-time pulse amplitude. A Monte Carlo simulation is used for this study. Data samples are generated according to the expected pulse shape. The lines correspond to a linear fit to the mean value of the quality factor. It shows two important features, first that for a given ratio of A_{out}/A_{in}, the quality factor increases linearly with the amplitude of the in-time pulse, and second that the dependence on the amplitude A_{in} gets steeper when the ratio A_{out}/A_{in} gets closer to one. The worst case scenario occurs for in-time and out-of-time pulses of equal amplitude, in that case the quality factor becomes maximal. The introduction of an out-of-time pile-up pulse is equivalent to introducing a deviation between the ideal pulse shape and the real pulse shape.

Figure 5 right panel illustrates that for significant out-of-time pile-up pulses the distribution of quality factor is quite different in case of out-of-time compared to no out-of-time pile-up. A Monte Carlo simulation is used for this study. Data samples are generated according to the expected pulse shape with emulation of electronic noise, pulse shape and phase variation. There is a clear separation between the two cases. The effect of an out-of-time pulse at −50 ns is nearly equivalent to the effect of an out-of-time pulse at +50 ns. Based on these studies possible identify to select pile-up channels can be proposed.

Figure 6 shows the reconstructed phase as a function of energy reconstructed offline with Iterative Optimal Filtering method (left panel) and in DSP with Non-Iterative Optimal Filtering method (right panel) in collision data at 7 TeV. A run with train of bunches crossing every 50 ns was used. The iterative method is able to reconstruct signals generated by a different bunch crossings (peaks at 0 ns and ±50 ns on left panel of figure 6). The non-iterative algorithm as implemented in the DSP uses a well defined phase of triggered event. The reconstructed phases correspond to in-time pulses (peak at 0 ns only on right panel of figure 6). Therefore, this algorithm is robust against out-of-time pile-up.

6. Conclusions

The TileCal online signal reconstruction performed in the DSP has been validated with the LHC collision data at 7 TeV reconstructed offline. The results are compatible with the expected differences between the fixed point arithmetic used in the DSP and the floating point arithmetic used for the offline reconstruction. Currently LHC operates with 50 ns bunch spacing causing out-of-time pile-up signals. The Non-Iterative Optimal Filtering method is to a certain extent robust against out-of-time signals. It is shown that the quality factor is very sensitive to differences between expected and real pulses in the detector and that the quality factor can be used to flag cells with presence of pile-up.

References

[1] ATLAS Collaboration 2008 The ATLAS Experiment at the Large Hadron Collider *JINST* **3** S08003

[2] ATLAS Collaboration 2010 Readiness of the ATLAS Tile Calorimeter for LHC Collisions *Eur. Phys.* J **C70** 1193-1236

[3] Carvalho J et al. (on behalf of the ATLAS Collaboration) 2011 The TileCal/ATLAS calorimeter calibration systems *Proc. ICATPP* (Como)

[4] Fullana E et al. 2005 Optimal Filtering in the ATLAS Hadronic Tile Calorimeter *ATLAS Note* ATL-TILECAL-2005-001

[5] Clement C and Klimek P 2011 Identification of Pile-up Using the Quality Factor of Pulse Shapes in the ATLAS Tile Calorimeter *Proc. IEEE NSS MIC* (Valencia)

Data Quality system of the ATLAS hadronic Tile calorimeter

Stanislav Nemecek, on behalf of the ATLAS Tile Calorimeter group

Prague AS, Czech Republic

E-mail: nemecek@fzu.cz

Abstract. The Tile Calorimeter (TileCal) is the central section of the hadronic calorimeter of the ATLAS experiment. It is subdivided into a large central barrel and two smaller lateral extended barrels. Each barrel consists of 64 wedges, made of iron plates and scintillating tiles. Two edges of each scintillating tile are air-coupled to wave-length shifting (WLS) fibres which collect the scintillating light and transmit it to photo-multipliers. The total number of channels is about 10000. An essential part of the TileCal detector is the Data Quality (DQ) system. The DQ system is designed to check the status of the electronic channels. It is designed to provide information at two levels - online and offline. The online TileCal DQ system monitors continuously the data while they are recorded and provides a fast feedback. The offline DQ system allows a detailed study, if needed it provides corrections to be applied to the recorded data and it allows to validate the data for physics analysis. In addition to the check of physics data the TileCal DQ systems also operate with calibration data. The TileCal calibration system provides well defined signals and the response to the calibration signals allows checking the behaviour of the electronic channels in detail. The Monitoring and Calibration Web System supports data quality analyses at the level of channels. All online, offline and calibration versions of the TileCal DQ system also provide automatic tests, the results of which allow fast and robust feedback.

1. Introduction

ATLAS[1] is one of two general purpose experiments installed at Large Hadron Collider at CERN. Proton-proton collisions at a centre-of-mass energy of 7 TeV started in 2010; in 2011 ATLAS recorded 5.25/fb. The Tile calorimeter (TileCal)[2] is a sampling hadronic calorimeter of ATLAS with iron absorber and scintillating plastic tiles as active material which are read by WLS fibers. TileCal is located in the region $| \eta | < 1.7$ extending radially from an inner radius of 2.28m to an outer radius of 4.25m. Its main goal is to contribute to the determination of jet energy and missing E_T. Analog signals from TTileCal are used in level one trigger. It is divided into three cylindrical sections along the beam line, each cylinder is composed of 64 equal wedges, creating granularity in phi = 0.1. By appropriate grouping of WLS fibers the granularity in pseudorapidity η is also 0.1 except last third radial layer with a granularity of 0.2. Long central cylindrical section contains two Tile partitions – LBA and LBC, while partitions EBA and EBC is each contained in shorter extended (side) cylindrical section. Digitized signals from ~10 000 channels are transmitted by optical cables to back-end electronics in shielded against radiation technical cavern about 100m away. The TileCal DQ group is responsible for the Data Quality Assessment (DQA) of the detector and for developing the Data Quality tools. The assessment is done online during the data taking and offline from physics data (Tier0) and calibration data (CAF). Offline data quality assessment is performed on TileCal calibration runs and on the express stream of physics runs and is crucial for the go/no-go decision for bulk reconstruction. The decision must be taken within 48 hours from the end of the run and it involves both data and detector status checks. The DQA on calibration runs is called CAF DQ and the DQA on express stream data is called Tier0 DQ. The two assessments are very different in the way the data is processed and in the infrastructure involved

2. Tile DQ on calibration runs

Commissioning of TileCal detector at ATLAS cavern started in 2005. In parallel development of DQ tools started yielding the Tile Commissioning Web System - automated tasks needed to analyze the TileCal Commissioning runs through the Web. Today's version is called The Monitoring and Calibration Web System (MCWS) [3] and is used for TileCal calibration runs analysis and MCWS functionality manages also the calibration constants by updating their values in COOL database [4]. Calibration data is acquired and stored with ATLAS Trigger and Data AcQuisition (TDAQ) systems and reconstructed with ATLAS offline software - ATHENA[5] within CERN Analysis Facility (CAF) for ATLAS Users. The reconstruction generates a large number of detailed plots for analysis by the TileCal DQ group. An automatic system evaluates each of these plots as green (no problem), yellow or red (require inspection) reducing the number of plots requiring inspection by the DQ group. Figure 1 demonstrates the usage of Web Interface for Shifters (WIS) for evaluating calibration runs.

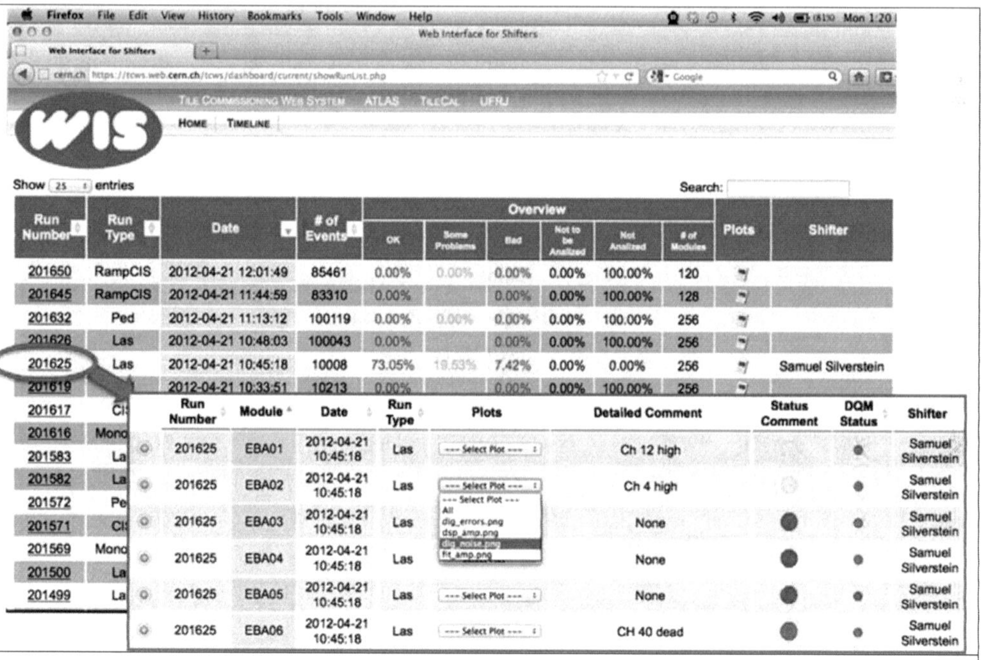

Figure 1. TileCal DQ WEB interface to analyze Calibration RUNs

Automated tests on each electronic channel mark those outside normal limits. If some channel problem is identified during DQ analysis process, COOL DB should be updated as fast as possible.

Outliers are inspected by the Tile DQ validator – he/she inspects them using detailed plots as on figure 2.

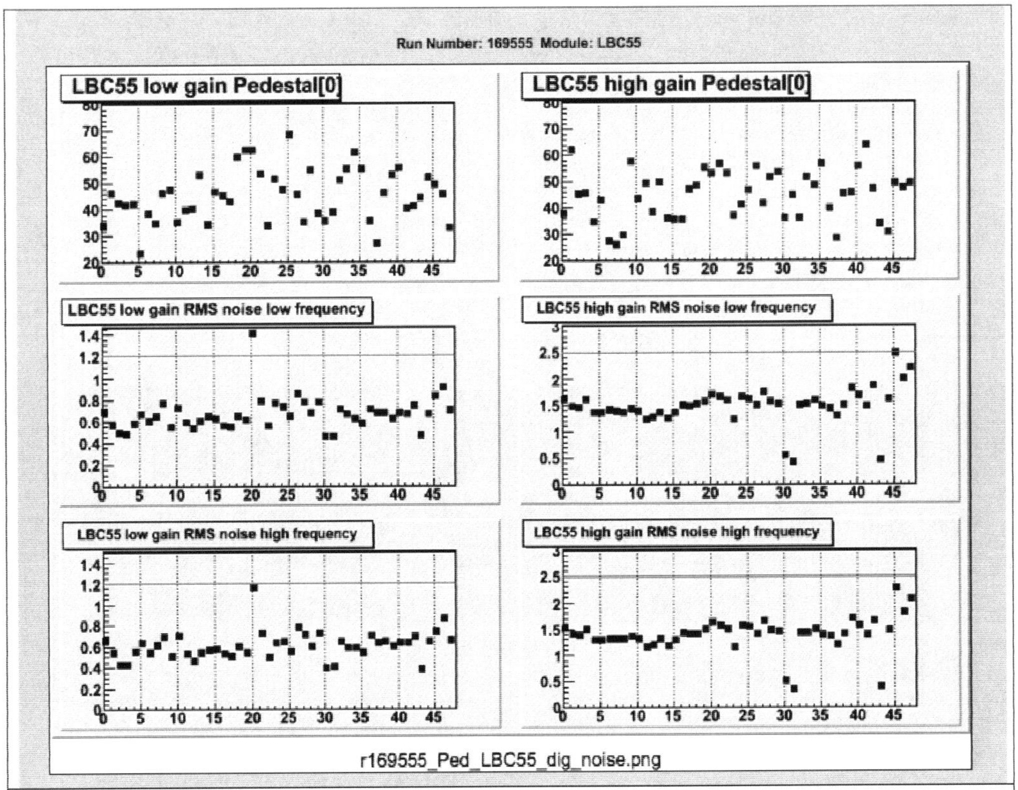

Figure 2. TileCal DQ WEB interface – automated tests on each electronic channel

3. Tile DQ on collision data

The ATLAS general DQ tool, Data Quality Monitoring Framework(DQMF), was successfully commissioned in the beginning of 2010 with wide scope allowing to analyze data and operational conditions, automatic checks and data archiving, visualization locally and remotely. Members of TileCal community are active in the continued development of this tool. This is a highly scalable distributed framework that is used to assess the quality of the data and the operational conditions of the detector, trigger and data acquisition system. DQMF provides quick feedback to the user about the correct functioning and performance of different parts of the detector. It quickly spots problems related with data quality and allows one to determine the origin of these problems. Many monitoring applications run online, this same DQMF framework is reused also offline.

3.1. Tile DQ ONLINE

Important part of DQMF is Data Quality Display - the visualization tool that allows great flexibility for displaying the histograms (with an overlay of reference histograms when applicable), configurations used for automatic checking of those histograms, the data quality flags, and much more. Single data quality tests are handled by DQParameters. Each DQParameter specifies what input histogram(s) to use, what algorithm to apply (DQAlgorithm), what are the thresholds to define good or bad result(DQResult). DQParameters are grouped in DQRegions, DQRegions also have DQResults associated - the mechanism to combine the results of the subparameters is specified in the configuration. DQRegions can be grouped in parent DQRegions thus creating a DQ tree. The whole TileCal detector is represented in "Data quality layout" – next part of DQMF allowing easier understanding of the actual status of the TileCal and faster navigation to problematic regions see figure 3.

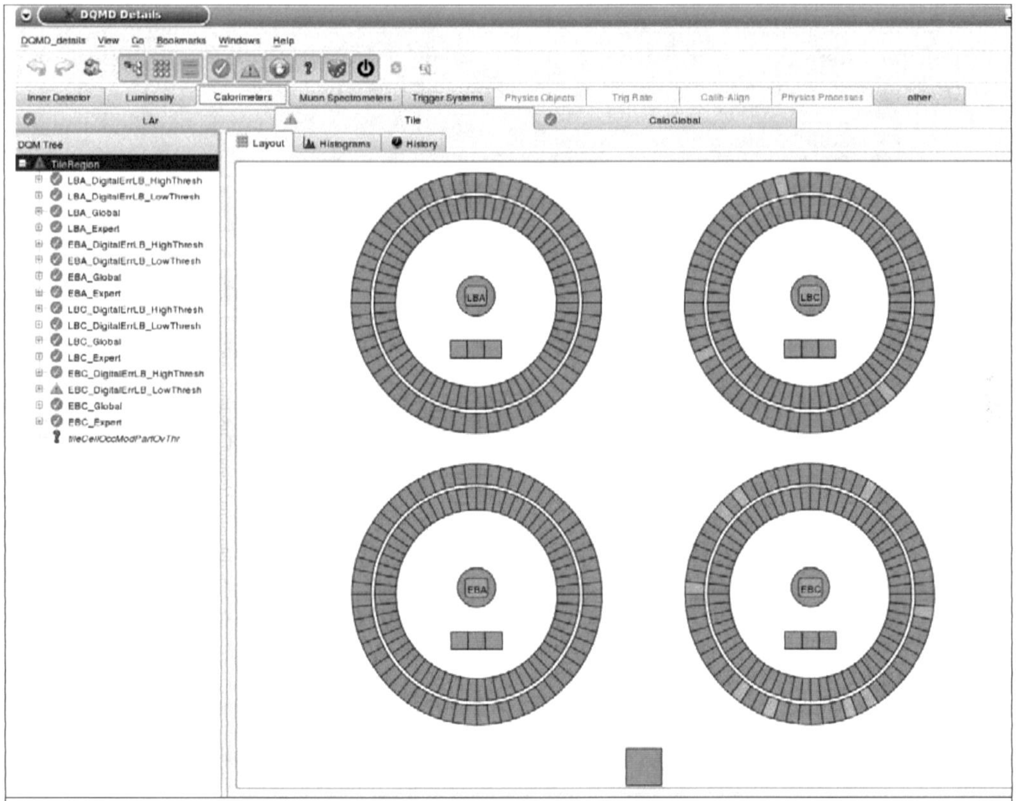

Figure 3 Tile Layout in DataQuality Monitoring Display

In 2010 frequent trips of low voltage power supplies (LVPS) located close to FE electronics of TileCal started appear in coincidence with collisions. Automatic procedures were worked out to restart LVPS and reconfigure the module as soon as possible. In rare cases this recovery procedure fails and reconfiguration has to be repeated. DQMD for TileCal was adapted to spot this problem and no module goes long time without sending good data. Special histograms showing fraction of digital errors during the data taking in each Lumi Block (usually a time period of one minute) were added. Automated evaluation of this histogram takes into account 7 Lumi Blocks i.e. the maximal time to recover the module after trip – if digital errors went up and down the result is YELLOW, only warning. The RED result is issued when fraction of digital errors did not decrease and shifter has to take the action, see figure 4.

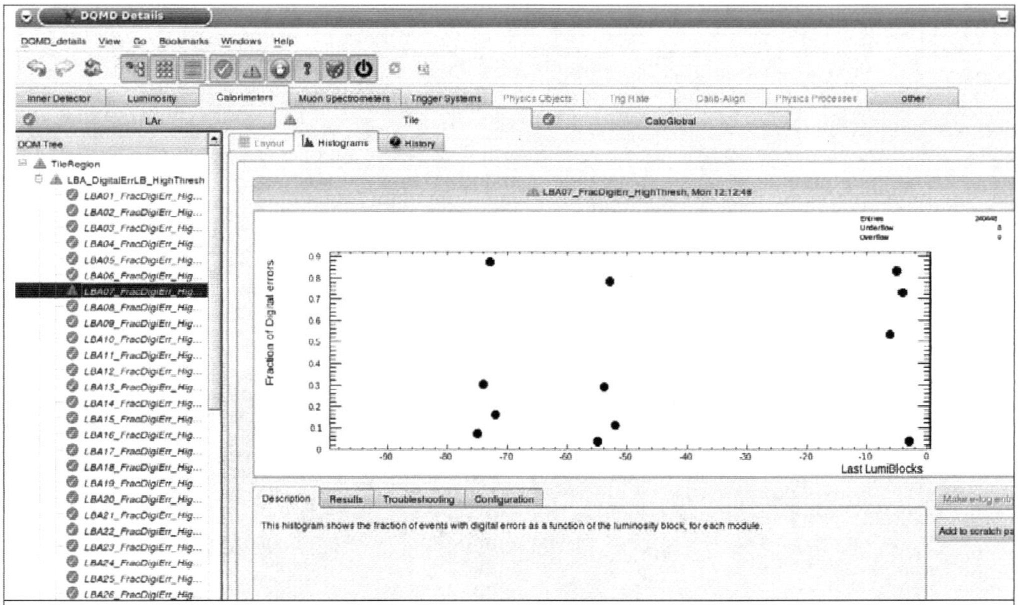

Figure 4 TileCal DQM histogram showing increase and decrease errors when module tripped

3.2. *Tile DQ OFFLINE*

The DQ group in TileCal is headed by two coordinators, DQ leader is assigned for one month period, while DQ validator for 1-2 weeks. Their goal is to validate raw data before reconstruction of entire data starts using a sampled data set called express-stream preferentially reconstructed at the computer farm Tier0. In 2011 Data Quality assessment system of ATLAS was upgraded in manner which allows users to extract as much information (of the data) for their particular analyses as possible. In the new system, the detector flags were replaced with defects[6]. A defect represents a specific problem with the detector rather than a final DQ decision. The essential aspects of the upgraded system are The Detector Control System (DCS) Calculator using conditions database information to automatically determine the status of sub detector components and calculate defects and The Defect Database allowing for categorization and storage of detector problems for a dynamic and flexible quality assessment. Defects come in two types: Primary defects are made directly from detector conditions; Virtual defects are logical combinations of primary and other virtual defects. Primary defects may be tolerable. The role of the virtual defect is to encapsulate the higher level logic which combines primary defects and leads to the production of good run lists for analyzers. A defect is a deviation from a nominal detector condition, a defect is either present or absent for a given Luminosity block. For the TileCal the defect "significant loss of detector covered" is INTOLERABLE, while trip of one module belongs to TOLERABLE defect. The DQ software within Tile group is all time evolving, i.e. recently was add method, when TileCal laser triggered (f=1 Hz) during the physics RUNs in empty bunch crossings is used to assess deviation from correct timing. Figure 5 illustrates the INTOLERABLE Tile defect, where several non running groups of modules (blue) create significant loss of detector coverage. Tile tolerable defect is a module trip, never more than 2 modules tripped at the same time, see detailed status of tripped module on figure 6.

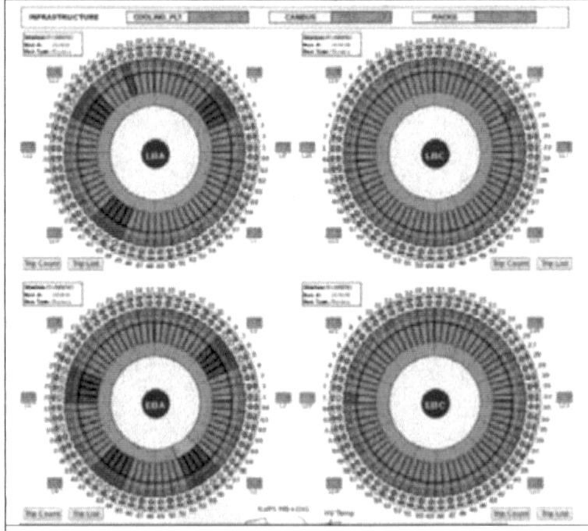

Figure 5 "significant loss of detector covered" TileCal INTOLERABLE defect

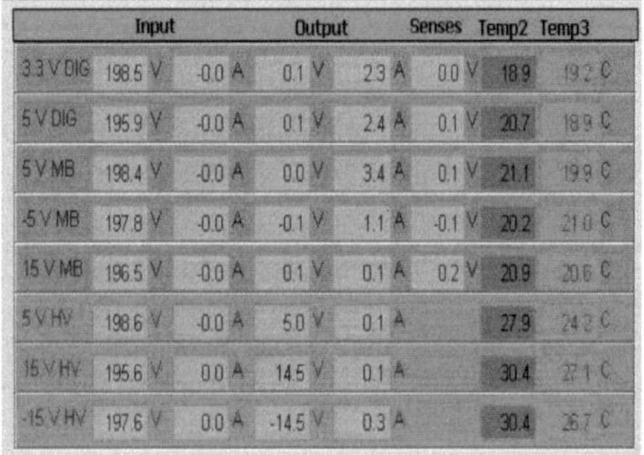

	Input		Output		Senses	Temp2	Temp3
3.3 V DIG	198.5 V	-0.0 A	0.1 V	2.3 A	0.0 V	18.9	19.2 C
5 V DIG	195.9 V	-0.0 A	0.1 V	2.4 A	0.1 V	20.7	18.9 C
5 V MB	198.4 V	-0.0 A	0.0 V	3.4 A	0.1 V	21.1	19.9 C
-5 V MB	197.8 V	-0.0 A	-0.1 V	1.1 A	-0.1 V	20.2	21.0 C
15 V MB	196.5 V	-0.0 A	0.1 V	0.1 A	0.2 V	20.9	20.6 C
5 V HV	198.6 V	-0.0 A	5.0 V	0.1 A		27.9	24.2 C
15 V HV	195.6 V	0.0 A	14.5 V	0.1 A		30.4	27.1 C
-15 V HV	197.6 V	0.0 A	-14.5 V	0.3 A		30.4	26.7 C

Figure 6 Status of fLVPS (Voltages, currents) in tripped module (TOLERABLE defect, see text)

4. Summary

TileCal has appropriate tools to assess the quality of the data, in 2011 the luminosity weighted relative detector uptime and good quality data delivery during stable beam in pp collisions reached 99.2% - see table below[7]. Automatic tools for LVPS trip recovery and to check correct function after trip are in usage – trip rate will substantially decrease after LVPS replacement during long shutdown in 2013-2014. The DQ tool of TileCal are under continuous development and improvement.

ATLAS 2011 p–p run												
Inner Tracking			Calorimeters				Muon Detectors				Magnets	
Pixel	SCT	TRT	LAr EM	LAr HAD	LAr FWD	Tile	MDT	RPC	CSC	TGC	Solenoid	Toroid
99.8	99.6	99.2	97.5	99.2	99.5	99.2	99.4	98.8	99.4	99.1	99.8	99.3

Luminosity weighted relative detector uptime and good quality data delivery during 2011 stable beam in pp collisions at $\sqrt{s}$=7 TeV between March 13th and October 30th (in %) .

References

[1] The ATLAS Collaboration. The atlas experiment at the cern large hadron collider. Journal of Instrumentation (JINST), 3, 2008.

[2] ATLAS/Tile Calorimeter Collaboration. Tile Calorimeter Calorimeter Technical Design Report, 1996. CERN/LHCC 96–42.

[3] A. Corso-Radu; H. Hadavand; M. Hautschild; R. Kehooe; S. Kolos. Data quality monitoring framework for the ATLAS experiment at he LHC. IEEE Xplore Digital Library, 55, 2008

[4] M. Verducci. Atlas conditions database experience with the lcg cool conditions database project. 2008 *J. Phys.: Conf. Ser.* **119** 042031

[5] Calafiuria et al., Proceedings of the CHEP'04 Conference, Interlaken, Switzerland 2004, http://indico.cern.ch/getFile.py/access?contribId=108&sessionId=6&resId=1&materialId=paper&confId=0

[6] Golling T, Hayward H S, Onyisi P S U, Stelzer H J and Waller P 2012 Eur. Phys. J. C. 72 1-6

[7] https://twiki.cern.ch/twiki/bin/view/AtlasPublic/RunStatsPublicResults2010#2011

Use of Particle Flow Algorithms in a Dual Readout Crystal Calorimeter

Stephen Magill

Argonne National Laboratory

Abstract. The ability to grow clear, dense scintillating crystals presents an opportunity for development of a total absorption calorimeter that could contain multi-GeV hadrons in a detector volume similar to that of present-day calorimeters. With appropriate crystals and optimized readout elements, both scintillation and Cerenkov photons can be produced and detected separately. This dual readout approach allows one to selectively correct particle energies, resulting in significant gains in energy resolution $\rightarrow 20\%/\sqrt{E}$ or even better for hadrons. An R&D program is underway to 1) develop appropriate clear, dense crystals, 2) test innovative readout methods for both scintillation and Cerenkov light, and 3) provide test beam capability for crystal and readout sensor testing and simulation verification. As part of this effort, simulation studies have been done assuming a dual readout crystal calorimeter implementation for a future e^+e^- linear collider detector. By using the dual readout correction, corrections for magnetic field effects on low momentum charged hadrons, and particle flow techniques, substantial improvements in dijet mass resolution are obtained.

1. Motivation

It may be possible to enhance the performance of a dual readout (DR) crystal calorimeter by using particle flow algorithm (PFA) reconstruction techniques as developed for pixel calorimeters for the ILC [1]. The high density of modern heavy metal crystals means that full containment of hadronic showers can be achieved by a total absorption calorimeter in a volume comparable to present collider detectors. In addition, development of on-detector readout devices, e.g., SiPMs, MCPs, etc., means that individual crystal calorimeter cells can be organized in a highly granular configuration. This high granularity enhances the purity of particle contribution to individual calorimeter cells - perfect clustering of cells then corresponds to individual particles with very high probability.

Applying PFA techniques to the DR calorimeter can result in enhanced performance in the following ways:

1) using PFA cluster merging algorithms can result in closer correspondence of the DR correction in real physics events to that used to generate the DR correction in a test beam (with single particles).

2) track momentum resolution is still much better than the DR-corrected calorimeter shower resolution so any track/shower association where the track 4-vector is used instead of the calorimeter cluster coordinates will improve the mass resolution in jets, optimizing the use of tracking and calorimetry.

3) even if no unique track/shower association can be found, determination of the shower starting points using PFAs for charged particles results in an improved mass correction to a jet.

An analysis program has been developed which combines corrections using the DR approach with PFA techniques in a highly granular homogeneous DR crystal calorimeter. In this report, it will be shown how the performance of the detector is improved by combining these 2 reconstruction and analysis approaches.

2. Simulated Dual Readout Crystal Calorimeter Parameters

A homogeneous crystal calorimeter was constructed in simulation to replace the W/Scin ECAL and the Fe/RPC HCAL of the SiD [1] concept. The barrel and endcap calorimeters were divided into "ECAL" and "HCAL" parts which were only differentiated by the crystal "pixel" size – 3 cm x 3 cm x 3 cm in the ECAL and 5 cm x 5 cm x 5 cm in the HCAL. The crystal selected was BGO. There were 8 layers of crystals in the ECAL representing ~21 X_0 (1.1 λ_I). The HCAL was made up of 17 layers comprising 4.6 λ_I. The total thickness of the crystal calorimeter was 5.7 λ_I both in the barrel and endcaps.

A threshold cut of 1 MeV was made on the scintillator hit energy which represented 1/30 of a mip signal in the ECAL and 1/60 in the HCAL. A similar energy threshold was also applied to the Cerenkov hits. No correlation in the resulting hit threshold cuts was required, so each hit collection (Cerenkov and scintillator) was independent of the other. A goal of the current photosensor R&D is to develop independent readouts for the Cerenkov and scintillator signals.

Separate DR calibrations were done in simulation as if each cell had been evaluated in an electron test beam. A calibration factor of 1.004 was obtained for the scintillator hits while ~7700 was obtained for the Cerenkov collection. In practice, it may be possible to calibrate a subset of the full crystal calorimeter in an electron test beam, obtaining calibration factors for untested cells through the use of CR muons, relating muon and electron calibrations in tested modules.

3. Cerenkov/Scintillator Ratio Correction for Hadrons

The scintillator response as a fraction of total energy expressed as S/E can be plotted versus the ratio of Cerenkov to scintillator reponse, C/S. The equations of a DR calorimeter can be written in the following form:

$$S/E = f_{em} + (1-f_{em})h_S/e_S$$

$$C/E = f_{em} + (1-f_{em})h_C/e_C$$

where f_{em} is the electromagnetic fraction of the shower, and h_S, e_S is the scintillator response to hadrons and electrons, and h_C, e_C is the Cerenkov response. After calibration with electrons, $e_S = e_C = 1$. Some algebraic manipulation of the above equations leads to an expression for the energy E equal to the scintillator measurement S divided by a term containing the hadronic response difference (h_S-h_C) times a polynomial function in powers of C/S. The electromagnetic fraction of the shower, f_{em}, can also be expressed as a polynomial in C/S(h_S-h_C). Figure 1 shows S/E in C/S bins and S/E vs C/S for simulated data along with several polynomial fits in C/S for the BGO crystal calorimeter.

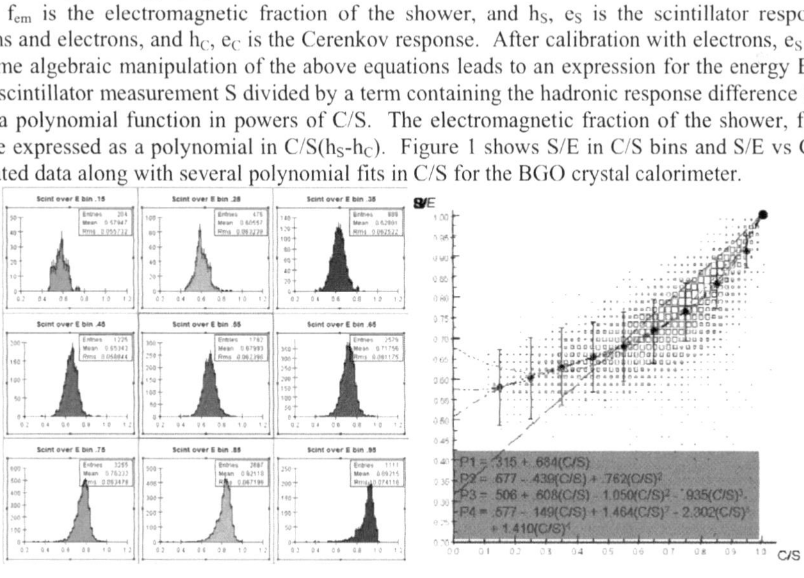

Figure 1. S/E in bins of C/S, and S/E vs C/S with E=S/Pn polynomials for the DR correction.

The calibrated point (response to electrons) is at S/E = C/S = 1. The data contour points are a collection of 5, 10, 20, 50, and 100 GeV charged pions. The energy dependence of the simulated data was checked and found to be much smaller than the fluctuations in the data points at any energy. The points represent the means in S/E for each bin in C/S and the error bars attached to the points are the σ/mean in each bin. The resolution improves with increasing C/S ratio. Also, at any point in S/E, the variation in C/S is larger than the variation in S/E, so by measuring C/S, the largest source of fluctuations can be reduced, thus improving the resolution of the corrected scintillator measurement.

There has been some discussion of whether the parameter h_S can be large enough in a crystal for the DR correction to be effective. Crystal EM calorimeters have been tested along with calibrated hadron calorimeters to measure the crystal response to hadrons [2]. After correcting for shower leakage, it appears that h_S is large (~0.7). This large value of h_S is also obtained in the simulation used in this report (see H. Wenzel in these proceedings).

Figure 2 shows the improvement in energy resolution for a 20 GeV charged pion using the C/S correction - σ/mean improves from 0.09 to 0.05.

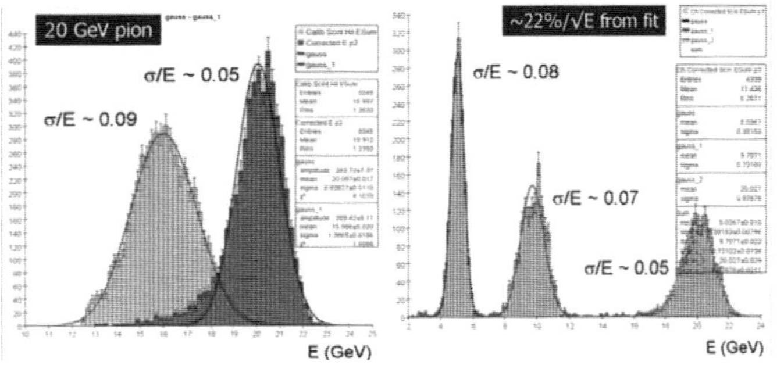

Figure 2. (right) Improvement in resolution of a 20 GeV pion using the DR correction, and (left) results of the DR correction on several pion energies.

For several charged pion energies, the improvement corresponds to an improvement in stochastic term resolution of 40%/√E → 22%/√E.

The C/S correction works as well for multiple clusters per particle – individual clusters can be corrected with their respective C/S ratios and then merged into particles with the same improvement in energy resolution as for single cluster particles. Using the DR correction, jets in e^+e^- events can be corrected using only the calorimeter objects. For a 100 GeV jet, with 80 GeV (charged and neutral hadrons) measured at ~25%/√E and 20 GeV (photons) measured at ~10%/√E, the resulting energy resolution should be ~3.5 GeV/100 GeV or ~3.5%. The left plot in Figure 3 shows the energy resolution obtained for these jets in simulated events corrected with the C/S ratio - σ/mean obtained is 3.5% for the total energy sum as expected.

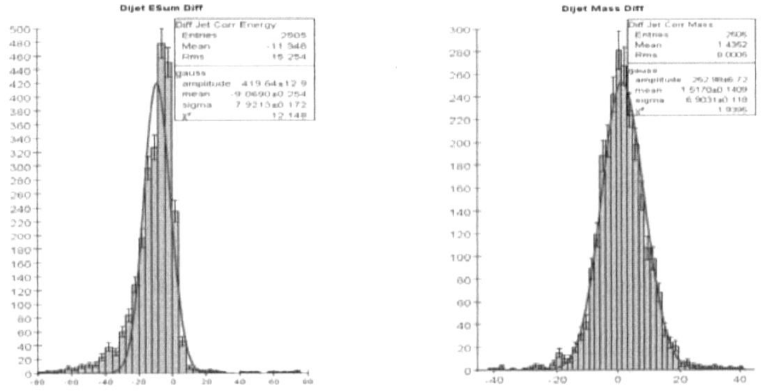

Figure 3. (left) $\Delta E/E$ for ~104 GeV jets, and (right) $\Delta M/M$.

The right plot shows the mass resolution, σ/mean = 7.6%, which in addition to the energy resolution contribution, includes effects of particles emitted at many angles in the presence of a strong magnetic field. Since the mass resolution is roughly twice the contribution of the energy resolution, it may be possible to improve the overall dijet mass resolution by reducing jet mass effects using PFAs.

4. PFAs in a DR Crystal Calorimeter

Reconstruction using PFAs may be appropriate if the particle contribution per cell is close to unity. Perfect reconstruction then yields objects representing final state particles with high purity. Figure 4 shows the particle contribution per hit for the simulated DR crystal calorimeter in a typical $e^+e^- \rightarrow ZZ \rightarrow qq\nu\nu$ event at E_{CMS} = 500 GeV.

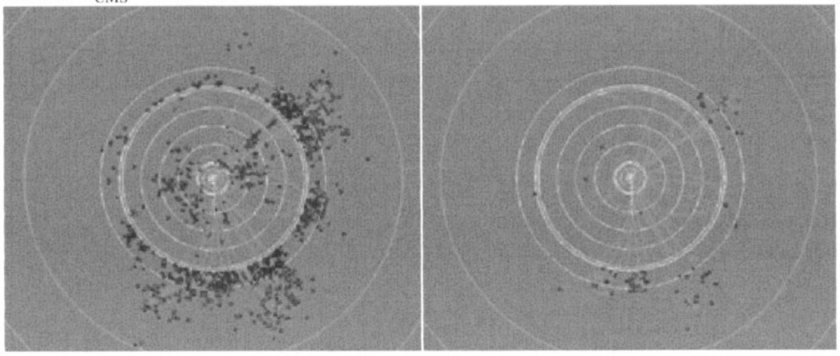

Figure 4. All hits (left) and the subset of those with particle contributions greater than 1 per hit (right) for the scintillator response (Cerenkov response is similar).

Since there is high particle purity per hit especially in the HCAL, PFAs will be used where possible to reconstruct events in addition to the DR corrections. In particular, a PFA which identifies the start of a shower in the calorimeter by linking a track with a calorimeter cluster is used in this analysis.

4.1. *Track/CAL Cluster Match PFA*

The first step in reconstruction is to try to match charged particle tracks to CAL clusters. The PFA algorithm that does this actually consists of several steps starting from track extrapolation into the calorimeter, linking hits from dE/dx losses (MIP-like hits) and adding core clusters, and finally adding other clusters to the matched shower while evaluating the E/p of the shower energy to track momentum. In this particular event, calorimeter clusters have been uniquely matched to 4 separate tracks while still maintaining high purity (>90%) in the track/cluster match as shown in Figure 6.

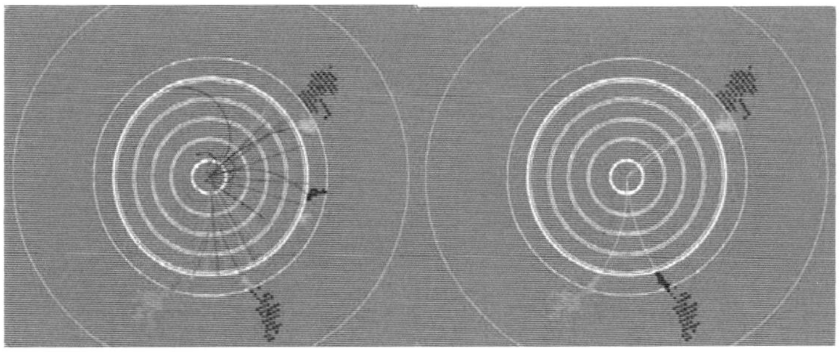

Figure 6. (left) Particles (lines) and CAL clusters used in track matching, and (right) CAL cluster/track matches obtained in this event.

By substituting the track momentum for the matched CAL cluster energy in events of this type, a 13% improvement in dijet mass resolution is obtained over that from the DR correction alone.

4.2. Correction to Jet Mass Due to Magnetic Field Effects

Even if some tracks can not be matched uniquely to CAL clusters, by extrapolating the tracks into the calorimeter, the starting point of the shower can be determined (using the same dE/dx hit linking algorithm as above). Using this information, the difference in the true track trajectory and the apparent trajectory that would be used if only CAL clusters were used can correct apparent individual jet masses. For the jets in the ZZ process at 500 GeV, an additional improvement of ~5% is obtained by correcting jet masses. Figure 7 shows dijet mass distributions (dijet mass-qq mass) for DR-corrected jets and for the PFA-enhanced jets.

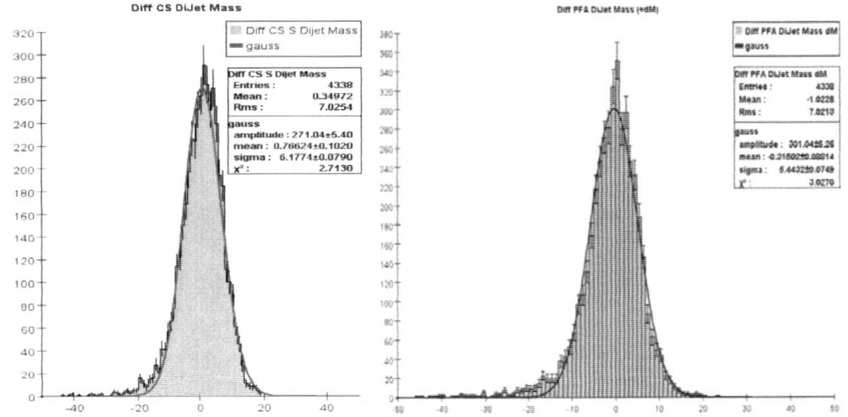

Figure 7. (left) Dijet mass difference for DR-only corrected jets, and (right) the same for PFA-enhanced jets.

The resulting enhancement obtained using PFA methods is ~14% overall – from σ/mean of 0.069 for the DR-only corrections to 0.059 for the PFA-enhanced reconstruction.

5. Summary

A total absorption dense crystal (BGO) calorimeter was developed in simulation capable of dual readout of both scintillation and Cerenkov light and used to study jets in e^+e^- -> ZZ -> ννqq at E_{CMS} of 500 GeV. Because of the high granularity of the calorimeter configuration, high particle purity per calorimeter cell was maintained, allowing for the use of PFA reconstruction methods in addition to the

Journal of Physics: Conference Series **404** (2012) 012048 doi:10.1088/1742-6596/404/1/012048

DR corrections. The DR corrections reduced the energy resolution of pions from 40%/√E to 22%/√E and was applied to calorimeter clusters. The PFA approach, used to match tracks with clusters and to find the starting point of calorimeter clusters resulting in corrections to the jet mass, enhanced the dijet mass resolution compared to that with only DR corrections by ~20% - from a σ/mean of 0.075 to 0.059.

References

[1] International Linear Collider, www.linearcollider.org/physics-detectors.
International Linear Collider, www.linearcollider.org/physics-detectors/Detectors, SiD Concept.
[2] Abdullin, S., et.al., Eur. Phys. J. C (2009) 60: 359-373.
Chen, C., et. al., Nucl. Instr. and Meth. A272 (1998) 713-721.

Journal of Physics: Conference Series **404** (2012) 012049 doi:10.1088/1742-6596/404/1/012049

Simulation Studies of a Total Absorption Dual Readout Calorimeter

Hans Wenzel
Fermilab, P.O. Box 500, Batavia, IL 60510-5011, USA
E-mail: wenzel@fnal.gov

Abstract. We have created a flexible, easy to use simulation framework based on GEANT 4 to perform detailed simulations of a crystal based total absorption calorimeter with dual readout. The correction to the observed scintillation signal can be determined by comparison of the scintillation signal with the beam energy as a function of the Cerenkov-to-scintillation C/S ratio. In this article we describe the features of the software we developed and show results for various crystal materials and physics lists. We show that applying an energy-independent correction results in an energy resolution of $\approx 10\%/\sqrt{E}$ for single π^-.

1. Introduction
In this article we describe the full GEANT 4 [1] simulation used to study the properties and performance of a homogeneous total absorption dual readout calorimeter. Such a calorimeter addresses the following principal contributions to hadron energy resolution and non-linearity:

- fluctuations in nuclear binding energy loss dominate the energy resolution resulting in a non-linear, non-Gaussian hadron response. We will demonstrate that this can be mitigated using dual readout where both Cerenkov C and scintillation S signal are read out from the same crystal and and the C/S ratio is used to correct the scintillation signal. We show that we can achieve linear and Gaussian response and improved hadronic resolution.
- Sampling fluctuations in the sharing of the shower energy between the active and passive materials in sampling calorimeters are eliminated by making the calorimeter homogeneous and totally active.
- Difference in the sampling fractions (i.e. ratio in the effective energy loss) between the different materials in the sampling calorimeters can be eliminated by making the calorimeters homogeneous.
- While leakage fluctuations due to escaping neutrinos and muons can not be avoided, tails of the hadronic shower escaping the detector can be minimized by using dense high density modern heavy metal crystals. The availability of such crystals means that full containment of hadronic showers can be achieved with a total absorption calorimeter with a volume fitting into a present collider detector. All materials are dense with a nuclear interaction length λ_I in a range from 21 cm (PWO and PbF_2) to 23 cm (BGO) for a complete list of crystal properties (BGO, PbF_2, $PbWO$) see [2].

Besides heavy crystals the other enabling technology is the development of on-detector readout devices, e.g., SiPMs, MCPs, etc., which allow that individual crystal calorimeter cells can be organized in a highly granular configuration that would allow the use of Particle Flow Algorithms [3] to further improve the performance.

2. The Software Environment:

We have developed a flexible simulation framework *CaTS* (Calorimeter and Tracker Simulation) [4] to study various aspects of a dual read out calorimeter. *CaTS* is a stand alone GEANT 4 application which allows for a large range of studies. For example one can do detailed studies of single crystals or small test beam setups where all effects like photon propagation, photon absorption, Rayleigh scattering, optical surface properties detection efficiency etc. are taken into account and optical photons are produced and traced until they are either absorbed or reach a photo-detector. The major components of *CaTS* include:

(i) Detector Description: *CaTS* uses the XML based Geometry Description Markup Language (GDML) [5], as the primary geometry implementation language. To change the detector setup no recompilation of *CaTS* is necessary one only needs to select a different GDML input file. GDML allows to define shapes, to place and rotate the volumes, to describe material properties like density, material composition as well as optical properties like e.g. absorption length and refraction index and surface properties [6]. In addition GDML allows to declare volumes as sensitive and to declare what hit collections should be produced and written out in the event stream. A library of various GDML detector descriptions ranging from single crystals to test beam setups is provided in the *CaTS* CVS repository.

(ii) Persistency: Events are stored as Root files [7]. Reflex is used to automatically generate dictionaries for all classes that we want to write out. This makes it easy to add new classes or to extend existing classes written out in the event stream. The resulting Root files can then be analyzed via Root macros or C++ routines. Several example macros and C++ routines are provided in the *CaTS* CVS repository.

(iii) Input modules:
- GPS: General Particle Source [8] allows the specifications of the spectral, spatial and angular distribution of the primary source particles.
- Particle Gun is a very simple particle gun that shoots a particle of given type into a given direction with a given kinetic energy or momentum.
- HEPMC [9] provides a C++ Event Record for Monte Carlo Generators like Pythia [10].

(iv) Physics Lists: *CaTS* allows to choose from all GEANT 4 Reference Physics Lists. This lists can be extended to include optical physics processes like the Cerenkov effect, Rayleigh scattering, the production of Scintillation Photons etc.. We found that FTFP_BERT is currently the best available physics list for the simulation of hadronic showers. This is also supported by studies of the CALICE collaboration [11] comparing results from various GEANT 4 models with test beam data.

(v) Sensitive Detectors and corresponding Hit collections:
- TrackerSD: is a sensitive detector class to simulate the response of a tracking detector. Each energy deposition in the detector material results in a different Tracker Hit.
- CalorimeterSD: is a sensitive detector class that registers the energy deposit of each particle traversing the sensitive volume. A hit is the sum of all energy deposits in a calorimeter cell.
- DRCalorimeterSD: is a sensitive detector class representing a dual read out calorimeter. Besides registering the energy deposit of each particle traversing the sensitive volume the number of produced Cerenkov photons is calculated for each charged particle above the Cerenkov threshold($\beta > 1/n$). The same algorithm as in the G4Cerenkov Class is used but here only the number of produced Cerenkov photons is of interest.

- StoppingCalorimeterSD: is a sensitive detector class that registers the properties of particles entering the sensitive Volume but then the particle is killed to save CPU time.
- PhotonSD: sensitive detector that registers optical photons.

(vi) <u>User Actions:</u> examples of user actions (EventAction, RunAction, StackingAction, SteppingAction...) are provided

3. Calorimeter response before dual readout correction

For this studies we use a very simple calorimeter with the total dimension of 3×3×3 m^3, longitudinally segmented in 1 mm thick sheets of Crystal material comprising 13/14 nuclear interaction length λ_I for BGO/(PWO and PbF2).

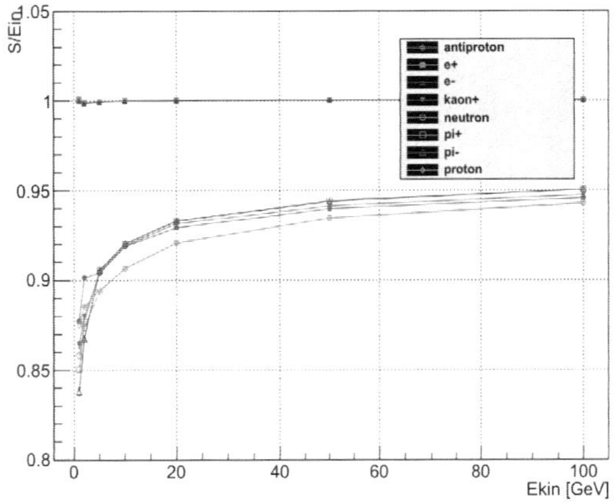

Figure 1. Average ratio of scintillation response to available energy S/E_{in} for mono-energetic particles as a function of E_{kin}. The calorimeter material is BGO. The Physics List is FTFP_BERT.

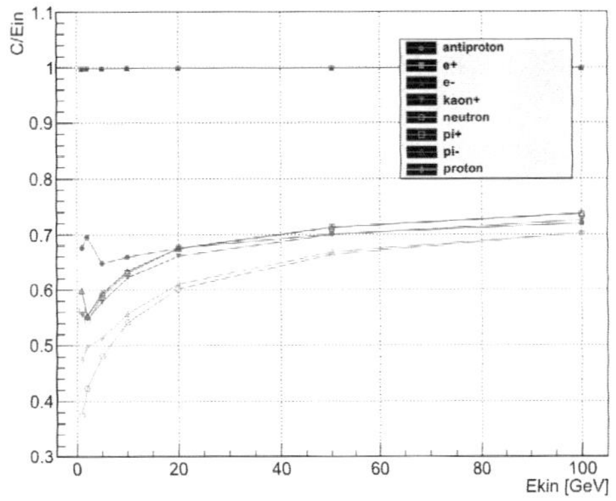

Figure 2. Average ratio of Cerenkov response to available energy C/E_{in} for mono-energetic particles as a function of E_{kin}.The calorimeter material is BGO. The Physics List is FTFP_BERT.

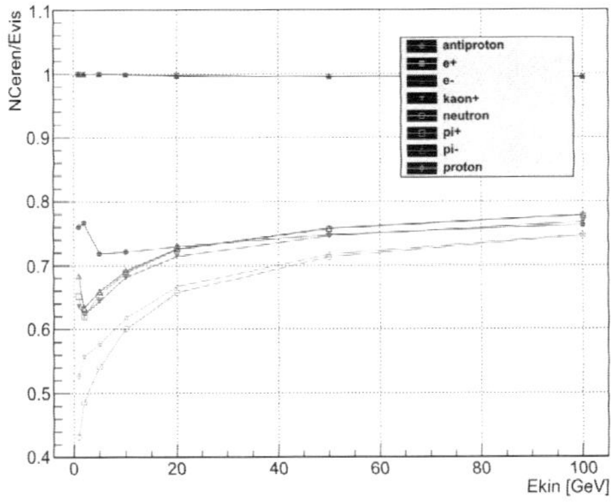

Figure 3. Average ratio of Cerenkov and ionization response(C/S) for mono-energetic particles as a function of E_{kin}.The calorimeter material is BGO. The Physics List is FTFP_BERT.

Figure 1 and 2 show the scintillation S and Cerenkov response C with respect to the deposited energy E_{in} as a function of the kinetic energy for different mono-energetic particles. The Cerenkov and ionization response was calibrated with electrons, the response is multiplied with a constant factor so that $S(e) = Ein(e) = C(e)$. Both S and C for $e^{\pm}$ are linear function of E_{in}. For the Cerenkov response this is due to the fact that the β ($\frac{v}{c}$) spectrum of charged tracks above the Cerenkov threshold in the electromagnetic shower is universal meaning composition and shape don't change with E_{in}.

For protons and neutrons E_{in} is equal to the kinetic energy E_{kin} of the incident particle ($E_{in} = E_{kin}$). For charged $K's$ and $\pi's$ in addition to E_{kin} the invariant mass is transformed into available energy when the particle decays ($E_{in} = E_{kin} + E_m(\pi^+, K^+)$). For anti protons ($\bar{p}$) twice the invariant proton mass is released when the $\bar{p}$ annihilates with a proton in the target material ($E_{in} = E_{kin} + 2 \times E_m(p)$). We observe the response S to hadrons (p, $\bar{p}$, n, $K^{\pm}$, $\pi^{\pm}$) is non-linear and lower than the response to electrons ranging in average from 0.85 to 0.95 for E_{kin} between 1 and 100 GeV. The response to different hadrons is slightly different ($\approx 1\%$). The hadronic Cerenkov response C is very nonlinear and compared to S highly suppressed ranging from 0.4 to 0.72 for E_{kin} between 1 and 100 GeV. Figure 3 shows the ratio of the average Cerenkov and ionization response C/S showing the different behavior of the two signals as a function E_{kin}. Using the differnce in the response of the two signals to correct the scintillation response is the basic idea of a dual read out calorimeter.

4. Hadronic Response

The hadronic Scintillation and Cerenkov response can be expressed in the following way:

$$ScintillationResponse : S/E_{in} = f_{em} + (1 - f_{em})\, h_S \tag{1}$$

$$CerenkovResponse : C/E_{in} = f_{em} + (1 - f_{em})\, h_C, \tag{2}$$

where f_{em} is the electromagnetic fraction in the cascade and h_s and h_C are the hadronic responses for Scintillation and Cerenkov readout respectively. We are using two different definitions to calculate f_{em} one is the traditional method of using the total energy of π^0 in the hadronic shower to estimate $f_{em}(\pi^0)$. We prefer to define $f_{em}(em)$ as the energy deposited by e^-, e^+or γ in the cascade. Combining equations 2 and 1 results in:

$$E_{in} = S\left[\frac{(1 - h_c) - C/S(1 - h_S)}{h_S - h_C}\right] \tag{3}$$

Where: $h_s > h_c$. If h_s and h_C are about the same, then S and C are about the same and the equations are degenerate and a dual read out correction would not help. But it appears that h_S is large compared to h_C as is shown in Figure 4.

5. Dual readout correction and hadronic energy resolution

Figure 5 shows the average S/E_{in} vs. C/S for simulated data of mono-energetic single $\pi^{\pm}$ and e^+ of (1,2,5,10,20,50,100) GeV kinetic energy. The response to e^+ is at $S/E_{in} = C/S = 1$. A 3rd degree polynomial $F_{corr}(C/S)$ is superimposed and the fitted function is used to apply the dual readout correction to the Scintillation response $S_{corr} = S/F_{corr}(C/S)$. We found that the correction function varies only weakly with energy. Here we used one correction function for all particle energies. Further improvement of the energy resolution and elimination of remaining non-linearity might be possible using energy dependent corrections. Also different particles probe different regimes in C/S (see Figure 3) and hence are sensitive to the imperfections of the correction function in different ways. This might also be mitigated by using energy dependent corrections.

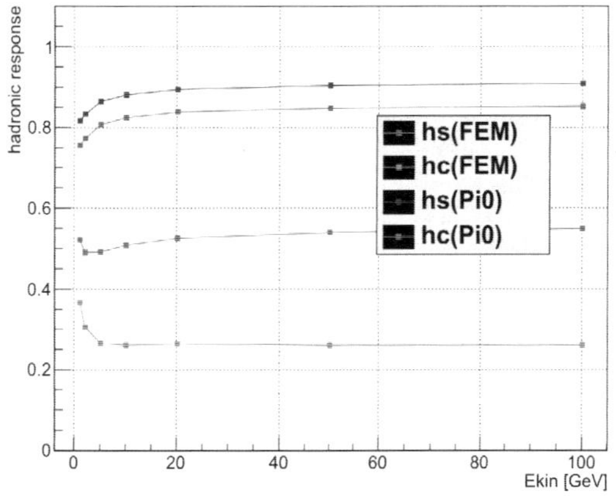

Figure 4. Hadronic scintillation ($h_s(em)$, $h_s(\pi^0)$) and hadronic Cerenkov response ($h_c(em)$, $h_C(\pi^0)$) as a function of E_{kin} of incident charged π^-.

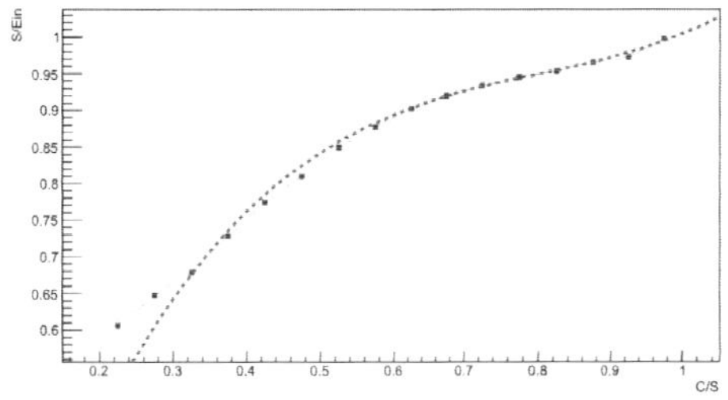

Figure 5. Dual readout correction function $f_{corr}(C/S)$ as estimated with mono-energetic single $\pi^{\pm}$ and $e^{\pm}$ of (1.2.5.10,20,50,100) GeV kinetic energy. The result of fitting a 3rd degree polynomial is superimposed. The calorimeter material is BGO. The Physics List is FTFP_BERT.

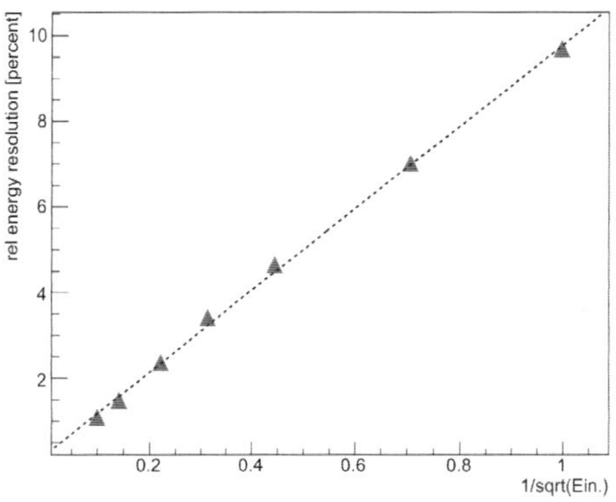

Figure 6. Relative hadronic resolution function $\frac{\sigma_{\pi^-}(E)}{E_{in}} = C_1 + \frac{C_2}{\sqrt{E}}[\%]$ as estimated with mono-energetic single $\pi^{\pm}$. A simple linear fit gives the following parameter values: $C_1 = 0.23 \pm 0.08\%$ and $C_2 = 9.55 \pm 0.15\%$. The calorimeter material is BGO. The Physics List is FTFP_BERT. This is before detector effects like noise or threshold cuts are taken into account.

Comparing S_{corr} with E_{in} for π^- of various kinetic energies and fitting Gaussian functions to the resulting distributions one can obtain the energy resolution. Figure 6 shows the relative resolution $\frac{\sigma_{\pi^-}(E)}{E_{in}}$ vs. $\sqrt{E_{in}}$. A simple linear fit gives the following parameter values: $C_1 = 0.23 \pm 0.08\%$ and $C_2 = 9.55 \pm 0.15\%$. Table 1 lists the fit results for various crystal materials and physics lists. In all cases using the energy-independent correction results in an energy resolution of $\approx 10\%/\sqrt{E}$ for single π^-. Figure 1 shows S_{corr}/E_{in} vs. E_{kin}. The non-linearity at lower energies are much reduced and at higher energies the response is linear. At higher energies the response to all particles is nearly identical and reproduces the deposited energy E_{in}. Again further improvement especially at lower energies might be possible using energy dependent corrections.

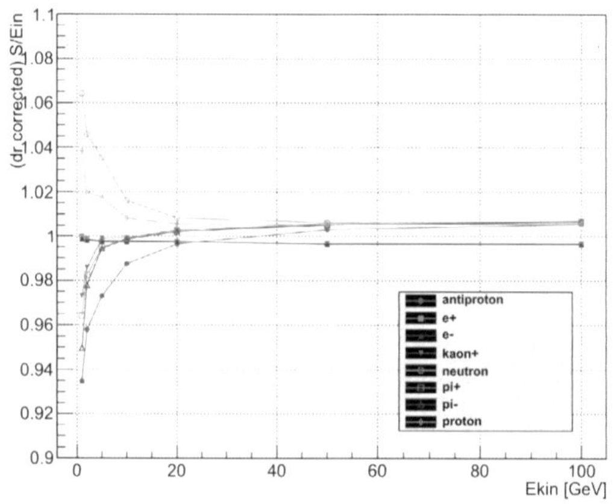

Figure 7. Average ratio of dual readout corrected scintillation response to available energy S_{corr}/E_{in} for mono-energetic particles as a function of E_{kin}.

Material	Physics List	$C_1[\%]$	$C_2[\%]$
BGO	FTFP_BERT	0.23 ± 0.08	9.55 ± 0.15
BGO	QGSP_BERT	1.16 ± 0.5	8.6 ± 1.0
PbWO	FTFP_BERT	0.89 ± 0.4	8.1 ± 0.9
PbF2	FTFP_BERT	0.84 ± 0.3	10.6 ± 0.5

Table 1. Relative hadronic resolution function $\frac{\sigma(E)}{E} = C_1 + \frac{C_2}{\sqrt{E}}\%$ as estimated with mono-energetic single π^- for various calorimeter materials and GEANT 4 physics lists. C_1 and C_2 were estimated with a simple linear fit.

6. Conclusion

We created a flexible, easy to use simulation framework. It allows detailed studies of single Crystals and full detector setups. Full GEANT 4 simulation supports the homogeneous, total absorption, dual readout concept and predicts excellent energy resolution of $\approx 10\%/\sqrt{E}$ for single π^- before detectors effects like noise and threshold cuts are taking into account. While the concept has not yet been demonstrated in a full size (containing hadronic showers) test beam setup there has been a lot of development recently by the GEANT 4 collaboration to improve the modeling of hadronic showers. Good agreement of CALICE test beam data with GEANT 4 simulation provides confidence in the GEANT 4 simulations. Also the availability of detailed test beam data will allow to further improve the models in the future.

[1] The two main reference papers for GEANT 4 are published in Nuclear Instruments and Methods in Physics Research A 506 (2003) 250-303, and IEEE Transactions on Nuclear Science 53 No. 1 (2006) 270-278. http://www.GEANT4.org/GEANT4/.
When not specified otherwise, results were obtained using GEANT4 9.5 (patch-01) released 20 March 2012 and the FTFP_BERT reference physics lists.
[2] Rihua Mao, Liyuan Zhang and Ren-Yuan Zhu, *Crystals for the HHCAL Detector Concept*, CALOR2012, Santa Fe, June 2012.

[3] Magill, Steve, *Use of Particle Flow Algorithms in a Dual Readout Crystal Calorimeter*, CALOR2012, Santa Fe, June 2012.
[4] http://home.fnal.gov/~wenzel/CaTS.html.
[5] http://lcgapp.cern.ch/project/simu/framework/GDML/gdml.html.
[6] Since GEANT 4.9.3 look up tables exist which describe various surface types and crystal surface conditions as measured by: Martin Janecek and William W. Moses, *Measuring Light Reflectance of BGO Crystal Surfaces* (LBNL-1790E (2009)) http://escholarship.org/uc/item/83v3b0gf.
[7] Rene Brun and Fons Rademakers, *ROOT - An Object Oriented Data Analysis Framework*, Proceedings AIHENP'96 Workshop, Lausanne, Sep. 1996, Nucl. Inst. & Meth. in Phys. Res. A 389 (1997) 81-86. See also http://root.cern.ch/.
[8] http://reat.space.qinetiq.com/gps/.
[9] http://lcgapp.cern.ch/project/simu/HepMC/.
[10] http://home.thep.lu.se/~torbjorn/Pythia.html.
[11] Marco Ramilli, *Hadronic models validation in GEANT4 with CALICE highly granular calorimeters*, CALOR2012, Santa Fe, June 2012.

Journal of Physics: Conference Series **404** (2012) 012050 doi:10.1088/1742-6596/404/1/012050

Hadronic models validation in GEANT4 with CALICE highly granular calorimeters

Marco Ramilli on behalf of the CALICE collaboration

Universität Hamburg, Institut für Experimentalphysik, Luruper Chaussee 149, 22761 Hamburg, Germany

E-mail: marco.ramilli@desy.de

Abstract. The CALICE collaboration has constructed highly granular hadronic and electromagnetic calorimeter prototypes to evaluate technologies for the use in detector systems at a future Linear Collider, and to validate hadronic shower models with unprecedented spatial segmentation. The electromagnetic calorimeter is a sampling structure of tungsten and silicon with 9720 readout channels. The hadron calorimeter uses 7608 small plastic scintillator cells individually read out with silicon photomultipliers. This high granularity opens up the possibility for precise three-dimensional shower reconstructions and for software compensation techniques to improve the energy resolution of the detector. We discuss the latest results on the studies of shower shapes and shower properties and the comparison to the latest developed GEANT4 models for hadronic showers. A satisfactory agreement at better than 5% is found between data and simulations for most of the investigated variables. We show that applying software compensation methods based on reconstructed clusters the energy resolution for hadrons improves by a factor of 15%. The next challenge for CALICE calorimeters will be to validate the 4th dimension of hadronic showers, namely their time evolution.

1. Introduction

The CALICE collaboration has been performing research and development on calorimeters intended for precision measurements at a future lepton collider. Prototypes of electromagnetic and hadronic calorimeters optimised for the Particle Flow approach have been built aiming for a jet energy resolution of 3-4% at the International Linear Collider [1, 2, 3]. As well as testing the hardware concepts, the CALICE data are able to test simulation models of particle showers in unprecedented detail owing to the highly granular readout of the calorimeters.

Extensive comparisons of hadron data obtained in the 8 GeV - 80 GeV range, with the most recent GEANT4 physics lists are presented, with particular interest in the 4 GeV - 30 GeV transition region between the *cascade models* (QGS, FTF) and the *string models* (Bertini, Binary) [4, 5, 6]. New observables are reported, including energy profiles after the identified interaction point, providing feedback for further improvement of the models.

Unless otherwise specified, version 9.3 of GEANT4 was used for all physics lists except for CHIPS, for which the patched version 4.9.3p01 was used.

2. The calorimeters

In this paper data taken in 2007 in the CERN H6 test beam are reported. The layout of the CALICE calorimeters and the beam instrumentation are discussed in [7, 8]. Two calorimeters

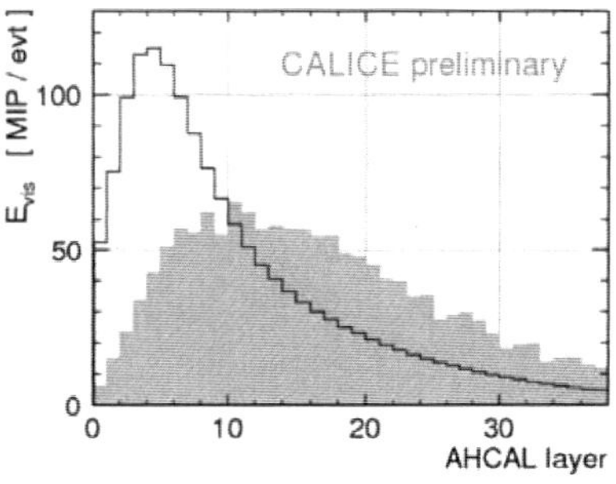

Figure 1: Longitudinal profile of a 45 GeV pion measured in the AHCAL relative to the first layer (filled) and to the shower starting point (black line).

had been used:

- the Si-W ECAL [9], an electromagnetic calorimeter composed of 30 layers of tungsten wrapped in carbon fibers as absorber and silicon wafers each segmented in a 6 × 6 array of 1 cm × 1 cm diode pads as sensitive detectors for a total depth of 23 radiation lengths (X_0) and about one interaction length (λ_{int});
- the Fe-AHCAL [8], an analog hadronic calorimeter, composed of 38 layers 18 mm thick iron absorber and plastic scintillator tiles (3 cm × 3 cm in the shower core)with analog readout (using Silicon Photomultipliers) as a sensitive detector, for a total depth of 47 X_0 and 5.3 λ_{int} .

The extremely high granularity of the CALICE prototype allows to acquire three-dimensional pictures of hadronic showers. An impression of the granularity is provided by the following numbers. One ECAL cell is about 1×1 Moliere radii (R_M) in size and the average longitudinal segmentation is 1 X_0 or 0.03 λ_{int}. In the AHCAL one cell has a size of about 0.85 x 0.85 R_M and a longitudinal segmentation of 1 X_0 or 0.15 λ_{int}. This granularity can be exploited to determine precise shower properties (e.g. the position of the first hadronic interaction, the energy density, the shower shape) and hence to validate different physics aspects implemented in Monte Carlo models.

3. Validation of models using testbeam data

The data presented in this publication had been acquired at CERN SPS H6 testbeam area, where positive and negative pion showers in the 8 - 80 GeV energy range had been recorded. Various observables are used to compare simulations to data, but the following analysis will focus on the shower spatial development, thus restricting to the study of radial and longitudinal profile and the identification of the position of the first hadronic interaction.
More details on the calorimeters calibration and comparison between data and Monte Carlo simulation can be found in [10, 11, 12].

3.1. First hadronic interaction

The high granularity of the CALICE calorimeters provides the capability for topological reconstruction within a shower, allowing the accurate localization of the first hard interaction layer using simple algorithms based on the energy deposition [11, 12]. Using the information of the starting point of the shower, we select only events which have their first interaction in the first λ_{int} of the calorimeter to minimize the effect of the leakage. As shown in figure 1, the shower

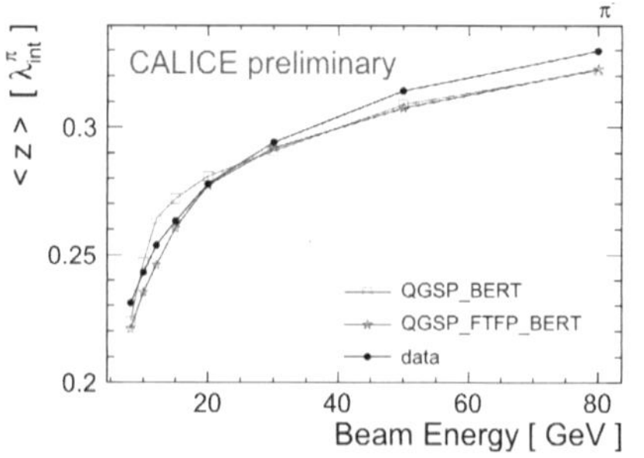

(a)

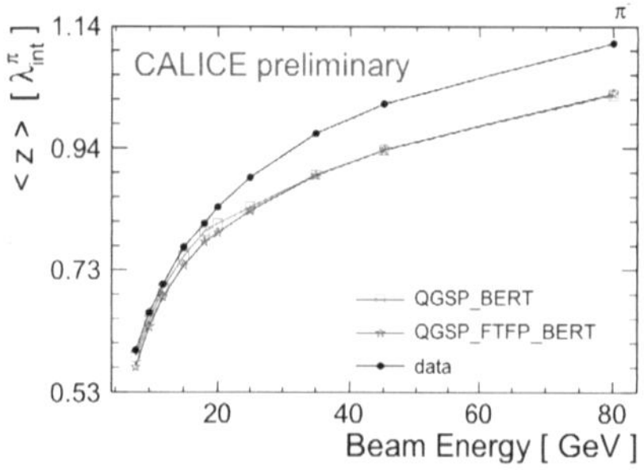

(b)

Figure 2: Mean shower depth in unit of interaction lenghts for the ECAL (a) and for the AHCAL (b) from data (black) for different pion energies, compared with mean Monte Carlo predictions obtained from QGSP-based physics lists (red and brown).

longitudinal profile can be studied either from the calorimeter front face (filled histogram) or from the layer of the first hadronic interaction (black line histogram): this second distribution is equivalent to the first one deconvoluted from the probability distribution of the first hadronic interaction, governed by the pion-nucleon cross-section. Therefore, the information of the first interaction point can easily be used to validate the cross-section implementation in the different physics lists [10, 11, 12].
Moreover, the calorimeters fine granularity allows a further analysis based on the energy deposition profile for incoming hadrons in the 2 GeV - 10 GeV range, providing a classification of the first hadronic interaction and a cross-check on the hadron-nucleon cross-section [11, 12].

3.2. Longitudinal development

In order to conveniently compare different physics lists at different energies, two energy-weighted parameters are introduced:

$$\langle z \rangle = \frac{\sum_i z_i E_i}{\sum_i E_i}, \quad (1)$$

$$\sigma_z = \sqrt{\langle z^2 \rangle - \langle z \rangle^2}, \quad (2)$$

representing respectively the mean shower depth, and the variance of the longitudinal distribution, where $\langle z^2 \rangle = \sum_i z_i^2 E_i / \sum_i E_i$.
In figure 2 (a) and figure 2 (b) mean shower depths for ECAL and AHCAL respectively are plotted: values increase non-linearly as expected up to about 1 λ_{int} in case of the AHCAL, while it reaches lower values of about 0.3 λ_{int} for the ECAL. This is due to the fact that, as mentioned, the total depth of the ECAL is ~ 1 λ_{int} and thus is not capable of full containment of a hadronic shower.

In comparing data with Monte Carlo results (see figure 3), it can be first of all noted how simulations in general predict shorter hadronic showers. FTF-based model show a less pronounced energy dependence and a global better agreement with data with respect to the QGSP-based models, whose agreement with data is nonetheless within the 10% and the energy dependence is at the 5% level. CHIPS physics lists, instead, exhibit a completely different behaviour, overestimating the shower length with a dependency from the energy of the 15% in case of the ECAL and sligthly larger than 10% in the AHCAL. A similar beahviour can be noted studying the variances of the longitudinal distribution as shown in figure 4.

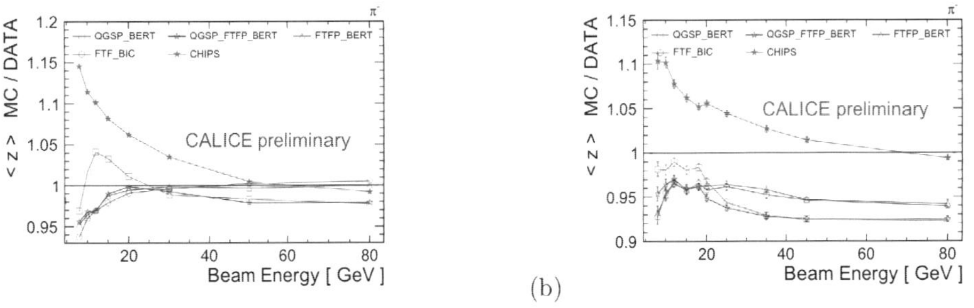

Figure 3: Ratio between mean shower depth from data and from Monte Carlo from different physics lists for the ECAL (a) and for the AHCAL (b) for different pion energies.

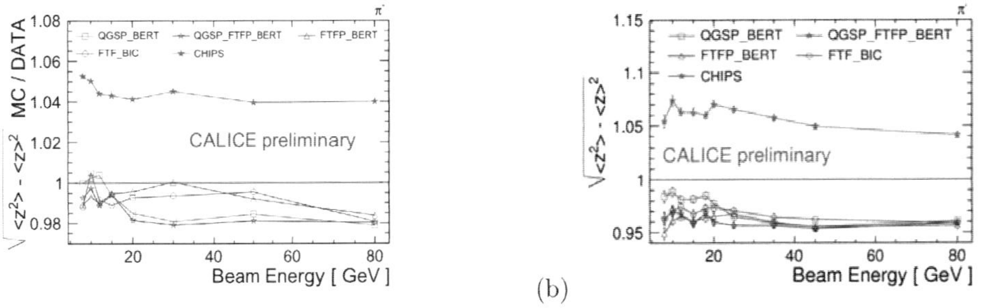

Figure 4: Ratio between variance of the longitudinal profile from data and from Monte Carlo from different physics lists for the ECAL (a) and for the AHCAL (b) for different pion energies.

3.3. Radial development

Good modeling of the transverse shower width is of importance for the development of particle flow algorithms, since it effects the degree of overlap between showers, and therefore the efficiency for separating them. For each calorimeter hit, the radial coordinate is computed as $r_i^2 = (x_i - x_C)^2 + (y_i - y_C)^2$, where (x_C, y_C) correspond to the center of gravity coordinates in case of the ECAL or the incoming pion track coordinates as measured by the tracking system in case of the AHCAL and (x_i, y_i) are the coordinates of the hit cell. As in case of the longitudinal profile, energy-weighted first and second momenta of the radial distribution have been introduced, to better compare different physics lists at different energies with data:

$$\langle R \rangle = \frac{\sum_i r_i E_i}{\sum_i E_i}, \tag{3}$$

$$\sigma_R = \sqrt{\langle R^2 \rangle - \langle R \rangle^2}, \tag{4}$$

where $\langle R^2 \rangle = \sum_i r_i^2 E_i / \sum_i E_i$.

In figure 5 (a) and (b) the mean radial shower extension for ECAL and AHCAL respectively are plotted: once again ECAL shows sensibly lower values due to the fact that the shower is not fully contained neither in the radial direction.

Looking at the comparison between data and Monte Carlo for the mean shower radii in figure 6, it can be seen how FTF-based models are, in general, closer to data and show a less pronounced energy dependence with respect to the QGS-based models, though both underestimate the shower radial extension. CHIPS physics list, instead, shows the best agreement with data (within

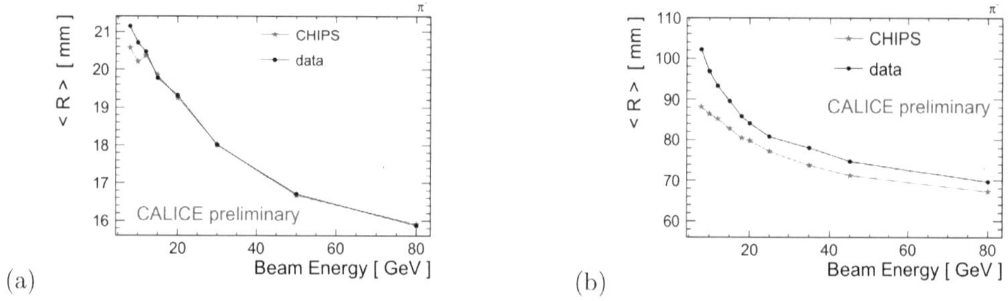

Figure 5: Mean shower radius in millimeters for the ECAL (a) and for the AHCAL (b) from data (black) for different pion energies, compared with mean Monte Carlo predictions obtained from CHIPS physics lists (green).

3% for the ECAL and 5% for the AHCAL) and a less pronunced variation with incoming pion energy. Similar consideration can be drawn studying the variances of the radial profile as shown in figure 7.

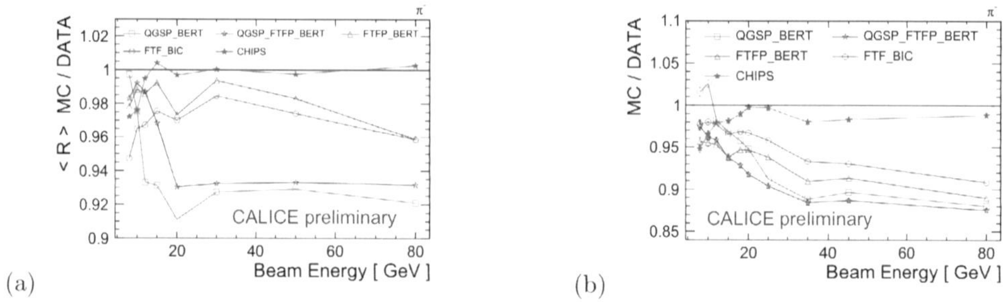

Figure 6: Ratio between mean shower radius from data and from Monte Carlo from different physics lists for the ECAL (a) and for the AHCAL (b) for different pion energies.

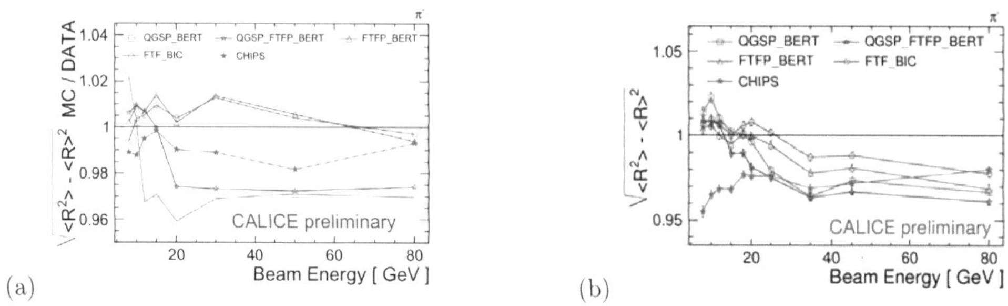

Figure 7: Ratio between variance of the radial profile from data and from Monte Carlo from different physics lists for the ECAL (a) and for the AHCAL (b) for different pion energies.

4. Conclusions

The CALICE collaboration has built finely granular electromagnetic and hadronic calorimeters which allow in-depth studies of hadronic shower properties and validation of Monte Carlo models on an unprecedented level. With these imaging calorimeters it is possible to investigate shower profiles and to determine the position of the first hard interaction in a shower. Summarizing the comparison between data and GEANT4 9.3p01 Monte Carlo simulations for different physics list it can be concluded that:

- the best agreement at the level of 5% is found for most of the selected observables in both ECAL and AHCAL data with the FTF-based physics lists, with the prediction for the longitudinal shower extension narrower than data only slightly less than 5%;
- QGSP-based lists are performing worse than FTF-based ones on the whole energy range covered: in particular the QGSP-FTFP-BERT physics list shows no clear improvement over the combination which still includes LEP as a stop-gap between QGSP and BERT;
- in the description of the radial extension of the hadronic shower, the CHIPS model is closest to data in terms of mean shower radius, showing in general less artifacts related to the transition between different models and therefore proving to be an interesting model, although still experimental and in a testing phase.

The following step will be the commissioning of new technological prototype of the CALICE calorimeters with timestamping capabilities, which will allow the exploration of the time structure of hadronic showers.

References

[1] Brient J C and Videau H 2001 The calorimetry at the future e+ e- collider *Proceedings of APS/DPF/DPB Summer Study on the Future of Particle Physics (Snowmass 2001).*

[2] Morgunov V L 2002 Calorimetry design with energy-flow concept (imaging detector for high-energy physics). *Proceedings of the Tenth International Conference Pasadena, California, USA*

[3] Thomson M A 2009 *Nucl. Instrum. Meth. A* **611** 25

[4] Amelin N S, Staubo E F and Csernai L P 1992 *Phys. Rev. D* **46** 4873

[5] Heikkinen A, Stepanov N, Weillish H P 2003 Bertini intra-nuclear cascade implementation in geant4 *Proceedings of 2003 Conference for Computing in High-Energy and Nuclear Physics (CHEP 03), La Jolla California*

[6] Andersson B, Gustafson G and Nilsson-Almqvist B 1987 *Nuclear Physics B* **281** 289

[7] Adloff C et al. 2010 Study of the interactions of pions in the CALICE silicon-tungsten calorimeter prototype. *JINST*,5:P05007

[8] Adloff C et al. 2010 Construction and commissioning of the CALICE Analog Hadron Calorimeter prototype *JINST*,5:P05004

[9] Adloff et al. 2009 Response of the Si-W Electromagnetic Calorimeter physics prototype to electrons *J. Phys. Conf. Ser.*160:012065

[10] Apostolakis J, Dotti A, Garutti E, Kaplan A, Kosov M, Uzhinskiy V and Ward D on behalf of the CALICE Collaboration 2011 Validation of GEANT4 hadronic models using CALICE data *EUDET Memo*

[11] Kaplan A for the CALICE Collaboration 2011 Pion showers in the CALICE AHCAL prototype *CALICE Analysis Note CAN-026*

[12] Doublet P, Faucci-Giannelli M, Pöschl R and Richard F for the CALICE Collaboration 2011 Interactions of hadrons in the CALICE Si-W ECAL prototype *CALICE Analysis Note CAN-025*

On the Energy Measurement of Jets in High-Energy Physics Experiments

Sehwook LEE and Richard WIGMANS

Department of Physics, Texas Tech University, Lubbock, TX, USA

E-mail: sehwook.lee@ttu.edu

Abstract. As the energies at which the elementary structure of matter is studied increased, the emphasis in scattering experiments has shifted from detecting individual hadrons to fragmenting quarks and gluons, which manifest themselves as particle jets. We investigate and quantify some systematic effects that affect the precision with which the properties of the fragmenting constituents can be determined. We concentrate on the calorimeters that are used to measure the energies, and in particular how the non-compensating nature of a calorimeter affects the energy measurement of different types of quarks and gluons.

1. Introduction

In modern high-energy physics experiments, calorimeters detect the jets which carry primary information about the collisions, and measure their properties. But the calorimeter has a non-trivial problem, its response is different for the hadronic and non-hadronic jet constituents, and energy dependent. At this conference, we present results on how compensating and non-compensating calorimeters respond to different types of quark-jets, and how the charged particle multiplicity and jet energy affect the response of non-compensating calorimeters.

2. Calorimeter Properties

High-energy particles experience the strong interaction with the nuclei in the absorbing calorimeter medium. Typically, π^0s, ηs, $\pi^{\pm}$s, kaons, protons and neutrons are produced in such nuclear reactions. Of these products, π^0s and ηs decay into γs and develop em showers. The fraction of the total shower energy carried by this electromagnetic component is defined as the em shower fraction, f_{em}. This em shower fraction fluctuates strongly from event-to-event. Figure 1a shows that $< f_{em} >$ increases with the pion energy and depends on the type of absorber material. This energy dependence implies that a non-compensating calorimeter is non-linear for hadron detection. Figure 1b shows the distribution of f_{em} for showers produced by 150 GeV π^- in a lead-based hadron calorimeter. The event-to-event fluctuations are large and non-Gaussian.

In contrast, charged pions, kaons, protons, neutrons, and the supplied energy to break up nuclei are classified as the non-electromagnetic component. This binding energy is called "invisible" since it doesn't contribute to the calorimeter signals. Quantum number conservation leads to an important leading-particle effect in showers induced by baryons and kaons. Figure 2 shows the signal distributions produced by 300 GeV pions and protons. In the pion signal distribution, the high-side tail is populated by events in which a large fraction of the energy

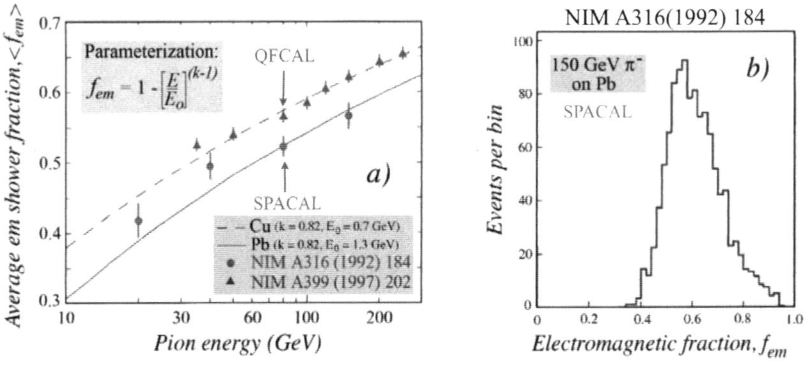

Figure 1. The average em shower fraction as a function of the pion energy. Showers are produced by pions in lead and copper-based calorimeters (a). Event-to-event fluctuations of the em shower fraction measured with showers produced by 150 GeV π^- in a lead-based calorimeter (b) [1, 2, 3].

carried by the incoming pion is transferred to a π^0 in the first nuclear interaction. In proton-induced showers, the leading particle in the first nuclear interaction is always a baryon, π^0s can only be produced in later stages of the fragmentation process. A similar effect is expected for kaons due to conservation of strangeness. As a result, pions generate larger calorimeter signals than protons (or kaons) of the same energy, and the signal distribution of pions is broader and asymmetric, on average (Figure 2a).

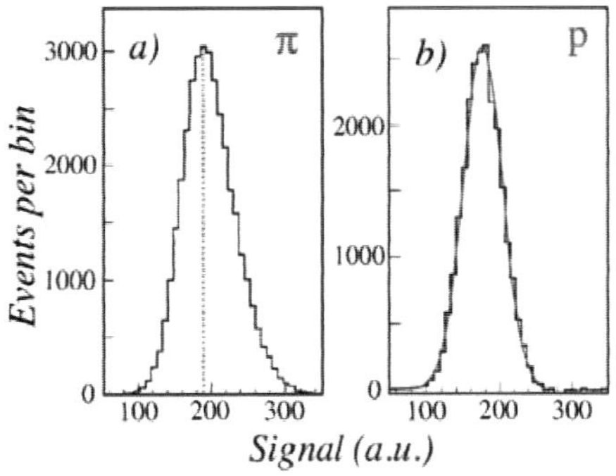

Figure 2. Signal distributions produced by 300 GeV pions and protons in the CMS HF quartz-fiber calorimeter [4].

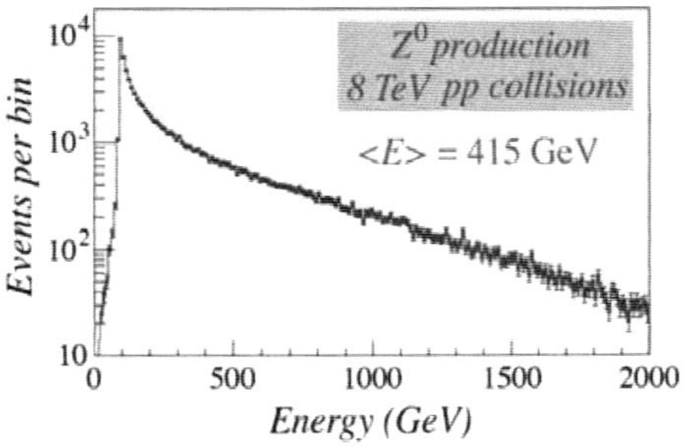

Figure 3. Energy spectra of the Z^0 bosons produced in pp collisions at 8 TeV.

3. The Calorimeter Response

3.1. Jet Simulations

Pythia 8.162 [9] was used for the generation of matrix elements, fragmentation, hadronization and decay. Hadronically decaying Z^0 bosons were produced with initial-, final-state parton showering and multi-parton interactions at LEP I ($\sqrt{s} = 91$ GeV) [5], at the Tevatron ($\sqrt{s} = 1.96$

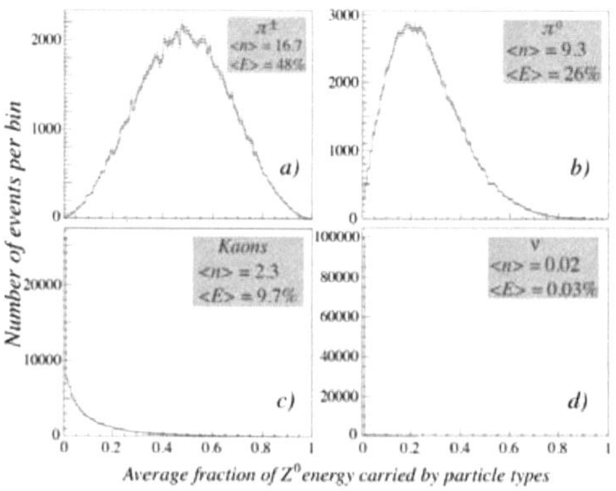

Figure 4. Distributions of the fractions of the Z^0 energy carried by $\pi^\pm$, π^0, kaons and $\nu(\bar{\nu})$ in $Z \rightarrow u\bar{u}$ processes.

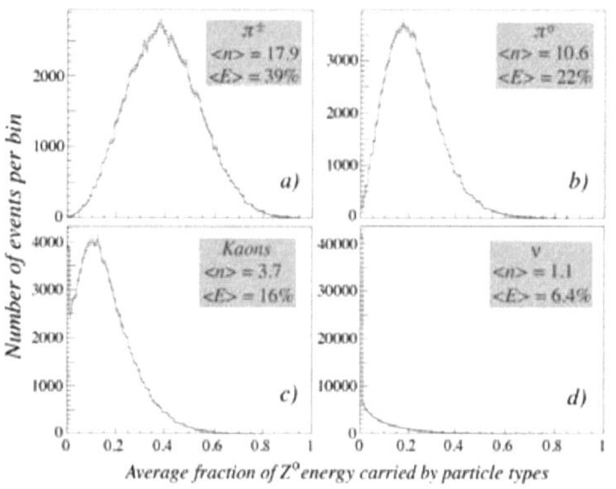

Figure 5. Distributions of the fractions of the Z^0 energy carried by $\pi^\pm$, π^0, kaons and $\nu(\bar{\nu})$ in $Z \rightarrow b\bar{b}$ processes.

TeV) [6] and at the LHC ($\sqrt{s} = 8$ TeV) [7]. The purpose of our study is to investigate the parton dependence of the calorimeter response, so we didn't reconstruct quark and anti-quark jets separately, and considered all final-state particles coming from a Z^0 boson. The final-state particles were grouped into 8 categories: (1) $\pi^\pm$s, (2) π^0s, (3) kaons, (4) baryons, (5) γs from sources except π^0, (6) $e^\pm$, (7) $\mu^\pm$, (8) ν, $\bar{\nu}$. Figure 3 shows that the average energy of Z^0 bosons produced at the LHC at $\sqrt{s} = 8$ TeV was 415 GeV. Figures 4 and 5 are simulation results for $Z \rightarrow u\bar{u}$ and $Z \rightarrow b\bar{b}$ produced in pp collisions at $\sqrt{s} = 8$ TeV, respectively. These are the distributions of the fractions of the Z^0 boson energy carried by $\pi^\pm$s, π^0s, kaons and $\nu(\bar{\nu})$. In the grey boxes, $< n >$ is the average multiplicity and $< E >$ the average fraction of the Z^0 boson energy carried by the different jet fragments. The main difference between $Z^0 \rightarrow u\bar{u}$ and $Z^0 \rightarrow b\bar{b}$ concerns the average energy carried by kaons and neutrinos. The semileptonic decay of b-quarks and the strange particles produced in the decay chain of the heavy quarks lead to a substantial increase of kaon and neutrino production.

3.2. The Hadronic Calorimeter Response

The calorimeter response to hadrons, defined as the average calorimeter signal per unit deposited energy and normalized to 1.0 for electrons, can be written as the sum of the em and non-em shower components:

$$R = < f_{em} > + (1 - < f_{em} >)\frac{h}{e} \quad (1)$$

The average em shower fraction is described with Grooms's parameterization (Eq. 2), where E_0 is the average energy to produce a pion in the shower development and k is a measure of the average multiplicity in the nuclear reaction. We selected 0.7 GeV and 0.82 for E_0 and k, respectively [10].

$$< f_{em}(E) > = 1 - \left[\frac{E}{E_0}\right]^{(k-1)} \quad (2)$$

Figure 6 shows the response of the CMS calorimeter system to $\pi^\pm$, $K^\pm$, p and $\bar{p}$ as a function of the energy of the incoming particles [8]. It shows that the hadronic response depends on the

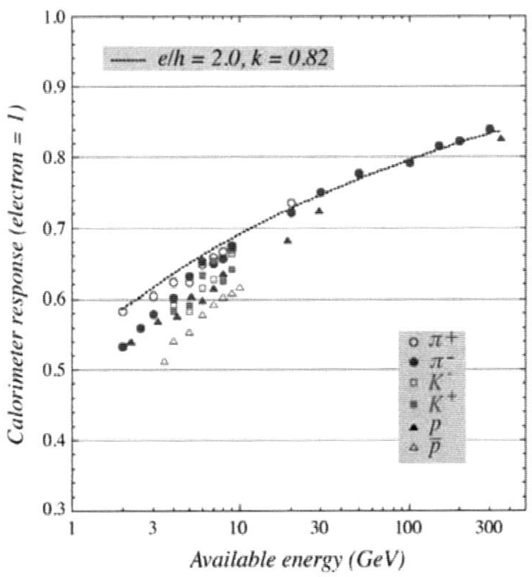

Figure 6. The response of the CMS calorimeter system to different types of particles as a function of energy [8].

energy of the incoming particles and that the response to kaons and protons is smaller than that to pions. The dashed line is the theoretical hadronic response for a $e/h = 2.0$ calorimeter and k is 0.82 (Eqs. 1,2), and agrees well with the π^+ data. Based on these experimental data, we assume that the response to kaons and baryons is reduced by 20 % compared to that for charged pions.

We determined the response of a calorimeter to a given type of jet as follows. For example, we generated a Z^0 boson at the Tevatron and made it decay into an $s\bar{s}$ pair. The average energies of the final state particles were found to be 4.4 GeV ($\pi^{\pm}$), 4.2 GeV (π^0), 12.9 GeV (K), 8.7 GeV (baryons), 3.3 GeV (γs from sources other than π^0 decay), 2.4 GeV ($e^{\pm}$), 3.4 GeV ($\mu^{\pm}$) and 3.6 GeV ($\nu, \bar{\nu}$). Next, the calorimeter responses to these fragments were calculated. The final-state particles, π^0s, $e^{\pm}$ and γs from sources other than π^0 decay, have a calorimeter response 1. The response to charged pions, kaons and baryons was calculated with Eq. 1 and reduced by 20% in the case of baryons and kaons. In this way, the individual responses of the $e/h = 2.0$ calorimeter to charged pions, kaons and baryons were found to be 0.64, 0.66 and 0.65, respectively. (Anti)neutrinos escape the calorimeter (response 0). We assumed that muons deposit on average 2 GeV in a typical calorimeter. The calorimeter response to muons was in this case thus $2/3.4 = 0.59$. The overall response of this $e/h = 2.0$ calorimeter to Z^0s produced at the Tevatron and decaying into $s\bar{s}$ pairs was found by considering the relative contribution of all final state particles. We found 0.64×0.40 ($\pi^{\pm}$) + 1×0.21 (π^0) + 0.66×0.26 (kaons) + 0.65×0.11 (baryons) + 1×0.02 (γs from sources other than π^0 decay) + 1×0.00033 ($e^{\pm}$) + 0.59×0.00018 ($\mu^{\pm}$) + 0 ($\nu, \bar{\nu}$) = 0.73. This method was applied to all hadronic Z^0 decay modes at LEP I, the Tevatron and the LHC. In addition, a subset of Z^0 bosons with an energy between 0.8 and 1.2 TeV produced in pp collisions at 8 TeV were selected for the highest energy point in this study. We investigated the responses for $e/h = 2$, $e/h = 1.6$ and $e/h = 1$ calorimeter systems.

3.3. Parton dependence

The parton dependence of the response was studied for all hadronic Z^0 decay modes for calorimeters with $e/h = 1$, $e/h = 1.6$ and $e/h = 2.0$, for various Z^0 boson energies. Figure 7

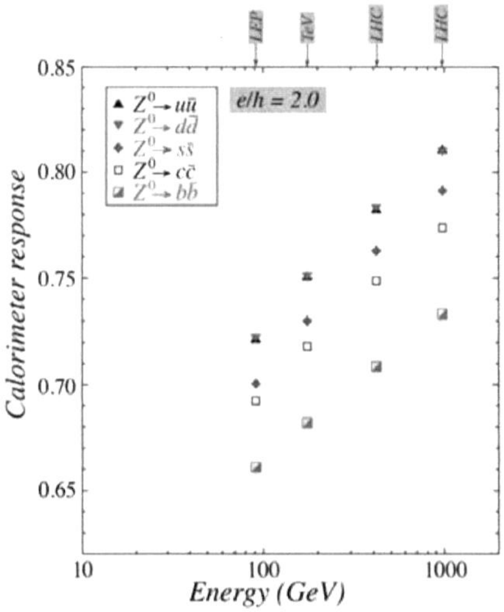

Figure 7. The response of the non-compensation calorimeter with $e/h = 2$ to all hadronic Z decay modes as a function of the Z boson energy.

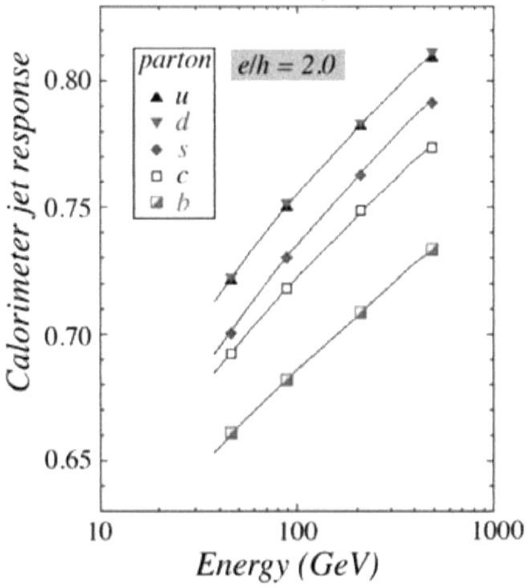

Figure 8. The response of the non-compensating calorimeter with $e/h = 2$ to jets as a function of jet energy.

shows the response of a calorimeter with $e/h = 2.0$ to all hadronic Z^0 decay modes as a function of the Z^0 energy. Figure 8 shows the response of the same calorimeter as a function of the average jet energy. The response increases as a function of energy because of the increased em component. The response to s jets is smaller than that to u, d jets by about 3% because kaons in s jets carry a larger fraction of the shower energy than those in u and d jets, and kaon showers have a reduced $< f_{em} >$ value. Also, the response to c and b jets is smaller than that to light quark jets due to the combination of strangeness conservation and μ, ν leakage. All response curves in Figure 8 were parameterized as follows:

$$R_p = 1 - \left[\frac{E}{p_0}\right]^{(p_1 - 1)} \tag{3}$$

The coefficients p_0 and p_1 for different types of quark jets and different calorimeters are listed in Table 1. Figure 9 shows the e/h dependence of the calorimeter response to different types of quark jets. The response of the compensating calorimeter to s jets is same as that to u, d jets because the calorimeter response to kaons is 1, and the different kaon contents for u, d and s jets don't affect the calorimeter response. The response to c, b jets is smaller because of semi-leptonic decays, in which particles escape detections. For the non-compensating calorimeters, the response to s jets is smaller than that to u, d jets because of the reduced em energy fraction of kaon showers. In addition, for c and b jets, a combination of the kaon content and semileptonic decay leads to a smaller calorimeter response than for the light quark jets.

Table 1. The values of the coefficients p_0 and p_1 that parameterize the calorimeter response to different types of quark jets as a function of jet energy for the calorimeters with $e/h = 1.6$ and 2.0 (Eq. 3).

$e/h = 1.6$

Parameter	u	d	s	c	b
p_0 (GeV)	0.0033	0.0031	0.0031	0.0004	0.00004
p_1	0.836	0.837	0.845	0.877	0.910

$e/h = 2.0$

Parameter	u	d	s	c	b
p_0 (GeV)	0.0168	0.0185	0.0178	0.0056	0.00115
p_1	0.838	0.836	0.846	0.869	0.898

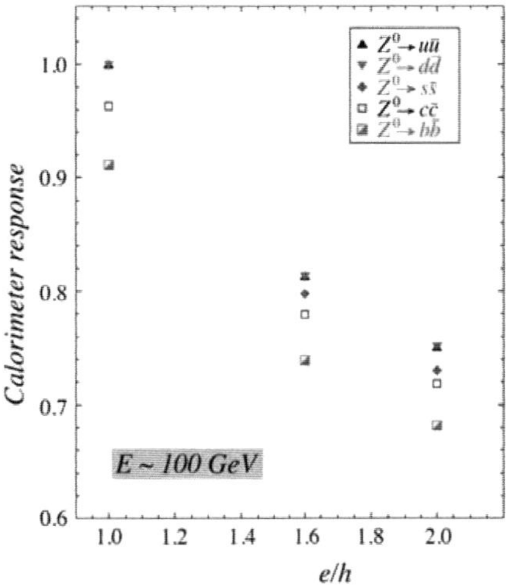

Figure 9. The responses of the three e/h values of the calorimeters to the different types of quark jets.

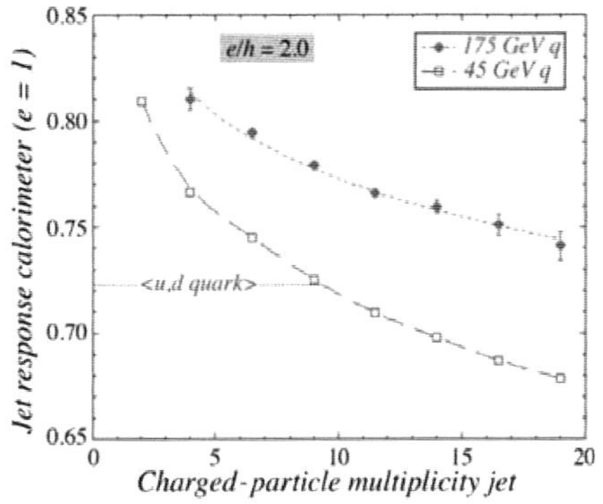

Figure 10. The response of a calorimeter with $e/h = 2.0$ to jets as a function of the charged-particle multiplicity of jets.

3.4. Multiplicity dependence

To investigate the charged particle multiplicity dependence of the calorimeter response, u, d quark jets in a calorimeter with a $e/h = 2.0$ calorimeter were used. The effects were studied for Z^0s produced either at LEP I or at the LHC ($\sqrt{s} = 8$ TeV). In the latter case, a sample of Z^0s with energies between 300 and 400 GeV was selected. Figure 10 shows the response to these jets as a function of the charged-particle multiplicity. The response to LHC jets is larger than to 45 GeV jets. Also, high multiplicity jets have smaller response than low multiplicity jets because of the lower average energy of the charged fragments, and the corresponding smaller em shower fraction. The effects of non-compensation on the calorimeter response thus depend on both the charged particle multiplicity and the jet energy.

4. Conclusions

The calorimeter response to jets depends on the type of fragmenting parton, the final-state multiplicity, the jet energy and the e/h value of the calorimeter. The effects of jet energy and final-state multiplicity are eliminated if $e/h = 1$. If $e/h \neq 1$, then the response increases with the jet energy and decreases with increasing final state multiplicity. The response to gluon jets is also reduced compared to u and d jets of the same energy since gluon jets have a larger final state multiplicity than quark jets. If $e/h \neq 1$, then the response depends on the relative contribution of baryons and strange mesons to the final state, because the response to these jet fragments is reduced w.r.t. the response to charged pions. The response to c and b quark jets is reduced because of the semi-leptonic decay processes.

References

[1] D. Acosta *et al.*, Nucl. Instr. and Meth. A316, 184 (1992).
[2] N. Akchurin *et al.*, Nucl. Instr. and Meth. A399, 202 (1997).
[3] T.A. Gabriel *et al.*, Nucl. Instr. and Meth. A338, 336 (1994).
[4] N. Akchurin *et al.*, Nucl. Instr. and Meth. A408, 380 (1998).
[5] D. Brandt *et al.*, *Accelerator physics at LEP*, Rep. Prog. Phys. 63, 939-1000 (2000).
[6] Fermi National Accelerator Laboratory, *Design Report Tevatron 1 Project*, FERMILAB-DESIGN-1984-01 (1984); L. M. Lederman, *The Tevatron*, Scientific American 264, 26-33 (1991).
[7] L. Evans and P. Bryant, *The LHC Machine*, JINST3:S08001 (2008).
[8] N. Akchurin *et al.*, *The Response of the CMS Combined Calorimeters to Single Hadrons, Electrons and Muons*, CMS Note 2007/012, CERN, Geneve, Switzerland (2007).
[9] Tobjörn Sjöstrand *et al.*, *A Brief Introduction to PYTHIA 8.1*, Comput. Phys. Comm. 178, 852-867 (2008).
[10] R. Wigmans, *Calorimetry - Eneergy Measurement in Particle Physics*, International Series of Monographs on Physics, vol. 107, Oxford Unversity Press (2000).

The design and performance of the electromagnetic calorimeters in Hall C at Jefferson Lab

V. Tadevosyan[1], H Mkrtchyan[1], A Asaturyan[1], A Mkrtchyan[1] and S Zhamkochyan[1] (for the Hall C Collaboration)

[1] A. I. Alikhanyan National Science Laboratory, Yerevan 0036, Armenia

Abstract.

The design and performance of the electromagnetic calorimeters in the magnetic spectrometers in Hall C at Jefferson Lab are presented. For the existing HMS and SOS spectrometers, construction information and comparisons of simulated and experimental results are presented. The design and simulated performance for a new calorimeter to be used in the new SHMS spectrometer is also presented.

We have developed and constructed electromagnetic calorimeters from TF-1 type lead-glass blocks for the HMS and SOS magnetic spectrometers at JLab Hall C. The HMS/SOS calorimeters are of identical design and construction except for their total size. Blocks of dimension 10 cm × 10 cm × 70 cm are arranged in four planes and stacked 13 and 11 blocks high in the HMS and SOS respectively. The energy resolution of these calorimeters is better than $6\%/\sqrt{E}$, and pion/electron (π/e) separation of about 100:1 has been achieved in energy range 1 – 5 GeV. Good agreement has been observed between the experimental and GEANT4 simulated energy resolutions. The HMS/SOS calorimeters have been used nearly in all Hall C experiments, providing good energy resolution and a high pion suppression factor. No significant deterioration in their performance has been observed in the course of use since 1994.

For the SHMS spectrometer, presently under construction, details on the calorimeter design and accompanying GEANT4 simulation efforts are given. A Preshower+Shower design was selected as the most cost-effective among several design choices. The preshower will consist of a layer of 28 modules with TF-1 type lead glass radiators, stacked in two columns. The shower part will consist of 224 modules with F-101 type lead glass radiators, stacked in a "fly's eye" configuration of 14 columns and 16 rows. The active area of 120 × 130 cm^2 will encompass the beam envelope at the calorimeter. The anticipated performance of the new calorimeter is simulated over the full momentum range of the SHMS, predicting resolution and yields similar to the HMS calorimeter. Good electron/hadron separation can be achieved by using energy deposition in the Preshower along with total energy deposition in the calorimeter. In this case the PID capability is similar to or better than that attainable with HMS calorimeter, with a pion suppression factor of a few hundreds predicted for 99% electron detection efficiency.

1. Introduction

The experimental program at Jefferson Lab focuses on the studies of the structure of nucleons and nuclei. The initial base equipment of Hall C is well suited to the JLab scientific program that required high luminosity, intermediate detector acceptances and resolution [1]. A magnetic spectrometer pair points to a common pivot with scattering chamber. The Short Orbit Spectrometer (SOS) accesses a momentum range of 0.3 - 1.7 GeV/c, and an angular range of 13.3° - 168.4°. The High Momentum Spectrometer (HMS) covers a momentum range 0.5 -

7.3 GeV/c, and angular range between 10.5° - 80°. The JLab 12-GeV Upgrade [2] envisages a Hall C scientific program focused again on high luminosity measurements, with detection of high energy reaction products at small forward angles. The SOS will be superseded by the newly built Super High Momentum Spectrometer (SHMS), capable of achieving a minimum scattering angle of 5.5° and a maximum momentum of 11 GeV/c.

The detector packages in the HMS and SOS are comprised of a pair of drift chambers for particle tracking, two pairs of x-y scintillator hodoscopes for triggering and time-of-flight measurements, and a gas Čerenkov detector and an electromagnetic calorimeter for electron/hadron separation. Hall C experiments typically demand detection efficiencies of better than 99%, and background particle suppression of 1,000:1 in e/π separation. This can be achieved by combining 100:1 suppression in the calorimeter, with the remaining suppression in the gas Čerenkov counter.

The detector package of the SHMS will be a near-clone of the HMS, and includes a heavy gas Čerenkov for hadron selection, and a noble-gas Čerenkov and lead-glass electromagnetic calorimeter for electron/hadron separation. The approved experiments demand a suppression of pion background for electron/hadron separation of 1,000:1, with suppression in the electromagnetic calorimeter alone on the level of 100:1.

2. HMS and SOS Calorimeters

The calorimeters for the HMS and SOS spectrometers were installed in 1994, becoming the first operational detectors at JLab. The detectors are of identical design and construction (see Fig. 1) except for their total size. The thickness, ~14.6 radiation lengths, is enough to absorb the major part of energy of electrons within the HMS/SOS momentum ranges. To avoid shower leakage, the lateral dimensions of the calorimeters were extended by 5 cm beyond the sizes required by spectrometer acceptance. This gave calorimeter physical areas of 70×130 cm^2 for the HMS and 70×110 cm^2 for the SOS.

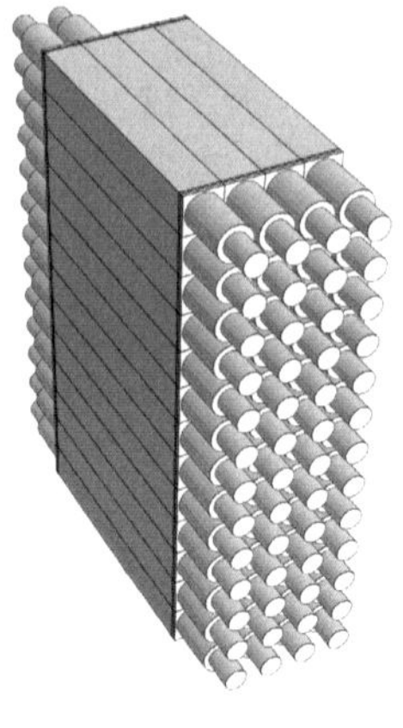

Figure 1. A sketch of the HMS calorimeter. The front of detector is at left. The far side PMTs in the first two layers were added in 1998.

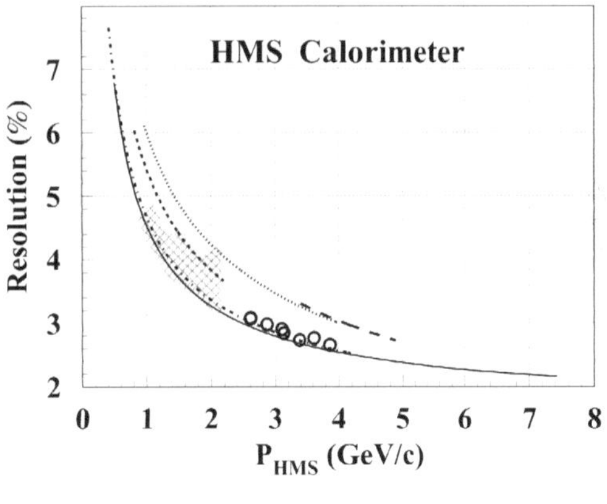

Figure 2. Resolution of the HMS calorimeter from the GEANT4 calculations (solid line) and from Hall C experiments E89-008 (· · · · · ·), E00-116 (- - - -), E04-001 (— · —), E01-006 (— —), E01-004 (○) and E03-003 (cross-hatched area).

The Čerenkov radiator is of TF-1 type lead glass blocks (index of refraction ~1.65, radiation length 2.74 cm, density of 3.86 g/cm^3). The light attenuation length of TF-1 varies significantly in the range of sensitivity of the PMT photocathode, and is ~100 cm at the peak of sensitivity

~400 nm. The blocks are wrapped in 25 μm thick aluminized Mylar and 40 μm thick Tedlar type film for light-tightness. High viscosity silicone grease ND-703 from Bicron is used for the PMT – block optical contact (index of refraction ~1.46).

The calorimeter signals are read out by 3" Philips XP3462B PMTs, with semitransparent bi-alkaline photocathode with peak quantum efficiency of ~29% at 400 nm, and 8-stage linear focused cube dynode structure. The operating voltages were set in the range ~1.4-1.8 kV to match the gain ~10^6. The PMT base is of a purely resistive, high current (2.3 mA at 1.5 kV), surface mounted (~0.640 $M\Omega$) design, operating at negative HV and linear to within ~2% up to the peak anode current of 120 μA (~5×10^4 pe). The dark current is typically less than 3 nA.

The readout electronics were identical for both calorimeters. The raw anode signals from the phototubes were taken to the electronics room through ~480 feet RG coaxial cables, then split 50/50, with one output sent through 400 ns delay cable to a LeCroy 1881M Fastbus ADC module, and the other to a Philips 740 linear fan-in modules to be summed. Data from the Fastbus modules were acquired in the "sparsified" mode, in which only data above a threshold were read from each ADC channel. Signals from the whole calorimeter and from the front layer alone are summed for use as an option in the first level electronic trigger for e/π discrimination [3].

The calorimeter calibration is performed with electrons selected in the gas Čerenkov detector. The standard calibration algorithm [4] is based on minimization of the variance of the estimated energy with respect to the calibration constants, subject to the constraint that the estimate is unbiased. The calorimeter energy corresponding to a track is divided by the track momentum and used for particle identification.

The simulation code for the calorimeters is based on the GEANT4 package, version 9.1 [11], and the QGSP_BERT physics list [5]. The code closely emulates the geometry and the

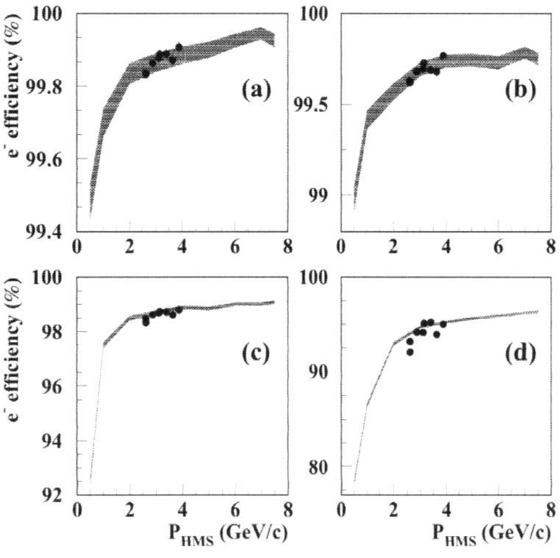

Figure 3. Efficiency of electron detection in HMS calorimeter at different momenta, for cuts on the normalized energy deposition 0.7 (a), 0.8 (b), 0.9 (c), 0.95 (d). The shaded areas represent results from GEANT4 simulations. The symbols are data from Hall C experiment E00-104.

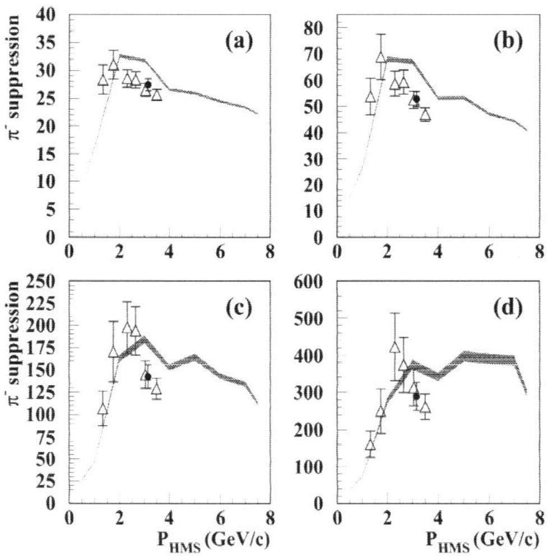

Figure 4. Pion suppression factor of HMS calorimeter versus momentum of the spectrometer, for cuts on the normalized energy deposition 0.7 (a), 0.8 (b), 0.9 (c), 0.95 (d). The GEANT4 simulation (shaded area) is compared to data from Hall C experiments E00-108 (△) and E00-104 (●).

composition of the detector, and electronic effects in readout electronics. The projectiles are generated at the focal plane of the spectrometer, then pass through the material between the focal plane and calorimeter.

The resolution of the HMS calorimeter (shown in Fig. 2) is close to the resolution of the lead-glass calorimeters of similar thicknesses. The 3-parameter fit [8] to the simulated resolution (in %) gives dependence on energy in the form $3.75/\sqrt(E) \oplus 1.64 \oplus 1.96/E$. Experiments before the modification of the counter in 1998, such as E89-008 shown in the figure, report a resolution of $\sim 6\%/\sqrt{E}$ (E in GeV) ([3], [6]). Experiments carried out afterward found improved energy resolution. As for the SOS calorimeter, on-line data analyses of a number of Hall C experiments gave resolution $6\%/\sqrt{E} + 1\%$ within the range of SOS momentum 0.5 – 1.74 GeV/c. Another experiment reported a resolution of $\sim 5\%/\sqrt{E}$ for momenta 1.2 – 1.7 GeV/c [7].

The experimental efficiency of electron detection, at momenta within the range 2.8–4.1 GeV/c, for different cuts on energy deposition is in reasonable agreement with the simulation (see Fig. 3). The simulation predicts a steady rise of e^- detection efficiency with energy due to the improvement in resolution. Both experiment and simulation show a momentum dependence of the π^- suppression factor peaking at several GeV/c (see Fig. 4).

The HMS/SOS calorimeters' resolution shows only slight changes during the years of usage (see Fig.2). These changes include variations in electronics, calibration technique and possible degradation of the calorimeter components. Stability of both calorimeters has also been evaluated by tracking changes in the ADC pedestals and PMT gains. No significant degradation of HMS/SOS calorimeters' performances after 15 years of operation have been noticed.

3. SHMS Calorimeter

The newly designed SHMS calorimeter consists of two parts (see Fig. 5): the main part at the rear (Shower), and Preshower before the Shower to augment particle identification capability of the detector. The Shower is 18.2 radiation length deep, which almost entirely absorbs showers from ~10 GeV electromagnetic projectiles, and the Preshower is 3.6 radiation length thick.

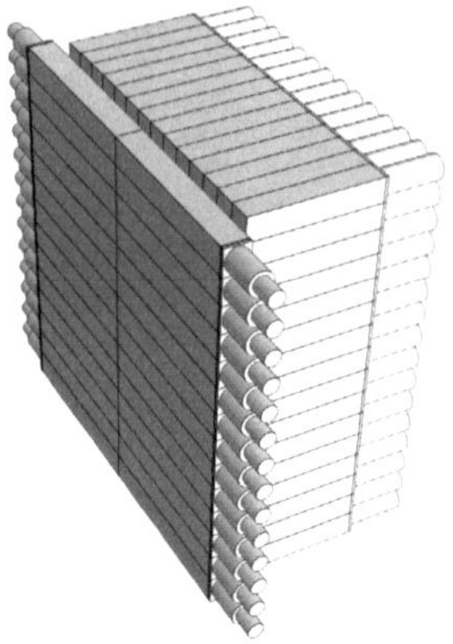

Figure 5. A sketch of SHMS calorimeter, consisting of a lateral Preshower detector and a projective Shower detector.

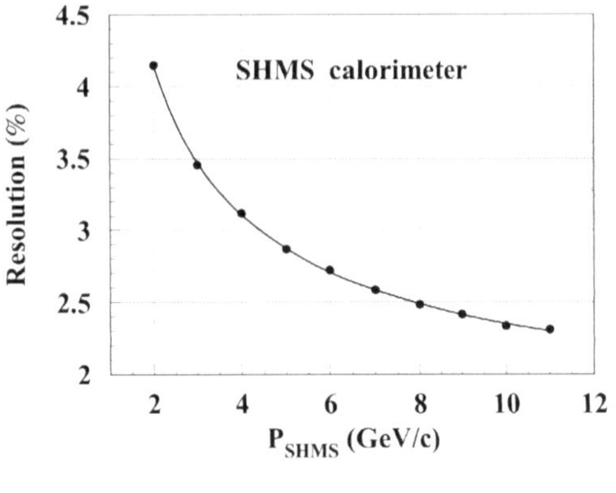

Figure 6. Resolution of the modeled SHMS calorimeter. The bullet symbols are data from the GEANT4 simulation, the line is a fit to them in the form $5.04/\sqrt{E} \oplus 1.71 \oplus 2.46/E$.

The Preshower radiator consists of a layer of TF-1 type lead glass blocks from the calorimeter of the retired SOS spectrometer, stacked in an aluminum enclosure (not shown in Fig. 5). 28 PMT assemblies, one per block, are attached to the left and right sides of the enclosure.

The Shower part consists of 224 modules from the decommissioned HERMES detector [9]. $\sim 120 \times 130$ cm^2 of effective area of detector covers the beam envelope at the calorimeter.

The optical insulation of the TF-1 blocks in the Preshower is optimized to minimize the dead material between them. The PMT assembly tubings are screwed in 90 mm circular openings on both sides of the enclosure. The 3" XP3462B PMTs are optically coupled to the blocks using ND-703 type Bicron grease of refractive index 1.46.

The HERMES modules to be used in the Shower part are similar in construction to the HMS/SOS modules but differ in details. The radiator is an optically isolated $8.9 \times 8.9 \times 50$ cm^3 block of F-101 lead-glass, which is similar to TF-1 in physical parameters and fractional composition. The small amount of Cerium, added in F-101 for the sake of radiation hardness ([10]), absorbs light at small wavelengths, and thus restricts the band of optical transparency to a higher wavelengths. Each block is coupled to a 3" XP3461 PMT from Photonis, with a green extended bialkali photocathode, of the same sizes and internal structure as the XP3462B the Preshower. Silgard-184 silicone glue of refractive index 1.41 is used for optical coupling of the PMTs to lead-glass blocks. See [10] for details on the construction of the HERMES module.

As both SOS and HERMES modules have been in use for more than 14 years under conditions of high luminosity, the lead glass blocks and the PMTs were checked for possible degradation. No sign of noticeable damage had been observed.

The simulation code is again based on GEANT4 package (version 9.2) [11] and the QGSP_BERT physics list [5]. The calibration algorithm is similar to the HMS calorimeter.

Resolution of the modeled SHMS calorimeter (Fig. 6) is analogous to the other lead-glass shower counters, though it is somewhat lower compared to the HMS calorimeter due to the excess of the light absorption in the F-101 lead glass in the Shower.

Electron detection efficiency improves with momentum (see Fig. 7, top), which is consistent with better resolutions at higher energies. Meanwhile, pion rejection tends to worsen because

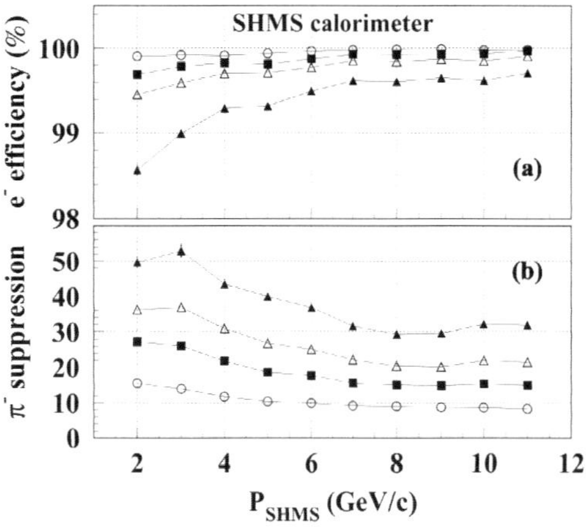

Figure 7. Electron detection efficiency (a) and pion suppression factor (b) of the SHMS calorimeter versus spectrometer's momentum setting for cuts on the normalized total energy deposition at 0.7 (○), 0.8 (■), 0.85 (△) and 0.9 (▲).

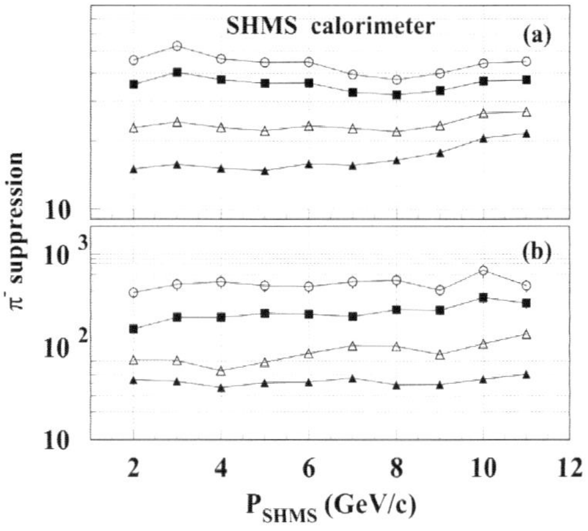

Figure 8. Pion suppression factor versus SHMS momentum setting for electron detection efficiencies 99% (○), 99.5% (■), 99.8% (△) and 99.9% (▲), by cutting on the normalized total energy deposition only (a), and by use of energy deposition in the Preshower in addition (b).

of the increase in electromagnetic component of hadron induced cascades (Fig. 7, bottom).

Combining the total energy deposition with deposition in the Preshower significantly improves the pion rejection (compare top and bottom panels in Fig. 8). A gain in suppression by a factor of 2 - 10 is achievable, dependent on momentum and the chosen electron detection efficiency.

4. SUMMARY AND CONCLUSIONS

We have developed and constructed electromagnetic calorimeters from TF-1 type lead-glass blocks for the HMS and SOS magnetic spectrometers at JLab Hall C. An energy resolution better than $\sigma/E \sim 6\%/\sqrt{E}$ and a pion suppression ~100:1 (for ~99% e^- detection efficiency) have been achieved in the 1 – 5 GeV energy range. Performance of the HMS calorimeter within the full momentum range of the spectrometer is modeled. Within the limited momentum range, the calculated resolution and π^- suppression factor are in good agreement with experimental data. The HMS/SOS calorimeters have been used in nearly all the Hall C experiments, providing good energy resolution and high pion suppression factor. No significant deterioration in the performance is observed in the course of operation since 1994.

Design and construction of the electromagnetic calorimeter for the newly built SHMS spectrometer in Hall C have been finalized. Preshower+Shower configuration was selected. A Monte Carlo program for the counter was developed, and simulations have been conducted with realistic parameters of the detector. The predicted resolution is slightly lower than that for the HMS calorimeter. Good electron/hadron separation can be achieved by using energy deposition in the Preshower along with total energy deposition in the calorimeter. A pion suppression factor of a few hundreds is predicted for 99% electron detection efficiency.

Acknowledgments

The authors wish to thank Ts. Amatuni for the work on hardware and software of HMS/SOS calorimeters in the development and construction stages. We thank A. Gasparyan for the work in the early stages of the hardware development. We thank Hall C technical staff for helping in assembling and installation of the detectors. We acknowledge the help from the collaborations of the many Hall C experiments in obtaining data on the HMS/SOS calorimeters.

The Southeastern Universities Research Association operates the Thomas Jefferson National Accelerator Facility under the U.S. Department of Energy contract DEAC05-84ER40150.

References

[1] 1990 *Conceptual Design Report (CDR). CEBAF Basic Experimental Equipment* (Newport News, VA: CEBAF)
[2] 2005 *Conceptual Design Report (CDR) for The Science and Experimental Equipment for the 12 GeV upgrade of CEBAF (Prepared for the DOE Science Review, April 6-8, 2005)* (Newport News, VA: Jefferson Lab)
[3] Arrington J 1998 *PhD thesis: Inclusive Electron Scattering From Nuclei at $x > 1$ and High Q^2* (California Institute of Technology, CA, USA) (Preprint arXiv:nucl-ex/0608013)
[4] Amatuni Ts 1995 *On the calibration of segmented full absorption calorimeters* (Unpublished)
[5] Apostolakis J *et al* 2009 Progress in hadronic physics modeling in GEANT4. XIII Int. Conf. on Calorimetry in High Energy Physics (CALOR2008) *J. Phys.: Conf. Series* **160** 012073
Apostolakis J *et al* 2007 Hadronic Shower Shape Studies in GEANT4 (CERN-LCGAPP-2007-02)
[6] Niculescu M I 1999 *PhD thesis: Inclusive resonance electroproduction data from Hydrogen an Deuterium and studies of quark–hadron duality* (Hampton University, VA, USA)
[7] Navasardyan T 2007 *PhD thesis: Quark-Hadron Duality in Mesons Electroproduction* (Yerevan Physics Institute, Armenia)
[8] Amsler C *et al* (Particle Data Group) 2008 *Phys. Lett.* B **667** 1
[9] Avakian H *et al* 1998 *Nucl. Instrum. Meth.* A **417** 69–78
[10] Avakian H *et al* 1996 *Nucl. Instrum. Meth.* A **378** 155–61
[11] Agostinelli S *et al* 2003 *Nucl. Instrum. Meth.* A **506** 250–303;
Allison J *et al* 2006 *IEEE Transactions on Nuclear Science* **53** 270–278

Simulation work on calorimetric energy resolution for the TAC-PF Detector

Ilhan Tapan, Fatma Kocak

Uludag Universitesi, Fen Edebiyat Fakultesi, Fizik Bolumu, 16059, Bursa, Turkey

E-mail: ilhan@uludag.edu.tr

Abstract. According to the Turkish Accelerator Center (TAC) project, a tau-charm factory is proposed based on colliding a 1 GeV electron beam against a 3.5 GeV positron beam. The Particle Factory (TAC-PF) detector will be constructed for the detection of the producing particles from this collision. $PbWO_4$ and CsI(Tl) crystals are considered for the construction of the electromagnetic calorimeter. The photons generated from incident particles in the crystal material are detected by Avalanche photodiodes (APD) or PIN photodiodes placed at the end of the crystal. In this work, the contribution to the calorimetric energy resolution from both the shower fluctuations in the crystal and photoelectron statistics in the detectors have been simulated for $PbWO_4$-APD and $PbWO_4$-PIN combinations.

1. Introduction

The Turkish Accelerator Center project was started as a regional facility for accelerator based fundamental and applied research [1]. The project has been ongoing by eleven Turkish Universities collaboration with the support of State Planning Organization (DPT). The TAC project will include: light sources (synchrotron radiation and free electron laser), an electron-positron collider, a GeV scale proton accelerator. A proposed tau-charm factory will be based on colliding a 1 GeV electron beam against a 3.5 GeV positron beam. This factory will have an important role in the TAC project for a deeper understanding of QCD, as well as for detailed investigations of new physics. On the other hand, detector design studies are ongoing for the factory.

Electromagnetic calorimeters made of the crystals have been used in many years for their good energy resolutions. For this reason, $PbWO_4$ (PWO) and CsI(Tl) crystals are considered to use for the electromagnetic calorimeter part of the TAC-PF detector. PWO is a high density inorganic scintillator with sufficient energy and time resolution. The drawbacks of this crystal are a high sensitivity to a temperature variation and poor light yield. CsI(Tl) is widely used in the calorimeters due to the its high light output, short radiation length, small Molière radius and good mechanical properties.

The crystals act as both the shower development medium and light producer. The produced photons from incident particles in these crystals could be detected from a Hamamatsu S2744-08 PIN photodiode or a Hamamatsu S8664-55 (S8148) avalanche photodiode (APD) which is attached at the rear end of the crystals. While Hamamatsu S2744-08 PIN photodiodes are used as photodetectors for the BaBar [2], BELLE [3] and BES III [4] CsI(Tl) calorimeters, Hamamatsu S8664-55 APDs are used for the CMS [5] PWO calorimeter. Active area of the PIN photodiode is 1 x 2 cm^2 and the thickness of depletion region is 300 μm. For the APD, active area is 5 x 5 mm^2 and the thickness of depletion region is 50 μm which have a 5 μm thick high field avalanche region and a 45 μm thick low field drift

Published under licence by IOP Publishing Ltd

region [6]. Both type of the photodiodes can also be used as photodetectors for the proposed the TAC-PF electromagnetic calorimeter.

2. Calorimetric Energy Resolution

The energy resolution of a calorimeter can be parameterized as $\sigma_E/E = a/\sqrt{E} \oplus b \oplus c/E$, where a is the stochastic term, b is the constant term, c is the noise term, E is the incident particle energy in GeV and $\oplus$ represents addition in quadrature. The stochastic term of the energy resolution for crystal–photodiode combination is composed of a contribution from event to event fluctuations in the lateral shower containment ($a_{lateral}$) and a contribution from photoelectron statistics (a_{pe}) given as $a = a_{lateral} \oplus a_{pe}$. The lateral shower shape determines the distribution of the energy deposition in a cluster of crystals around the impact point. The photoelectron statistics contribution is related with fluctuations in the photodetector signal [7]:

$$a_{pe} = \sqrt{\frac{F}{N_{pe}}} \qquad (1)$$

Here, F is the emission weighted excess noise factor of the APD coming from the fluctuations in the avalanche gain process. Since the PIN photodiode has no internal gain, there is no excess noise *(F = 1)*. N_{pe} is the number of primary photoelectrons resulted from photoabsorption in the photodiode and calculated from:

$$N_{pe} = N_{ph} \cdot QE \qquad (2)$$

N_{pe} is the number of incident photons collected by the photodiode which is related to the number of photons leaving at the end of the crystal and QE is the emission weighted quantum efficiency.

3. Simulation and Results

Geant4 simulation code [8] has been used to simulate photons at the different energies passing through the PWO crystals. The PWO crystal has a 200 mm length with cross section 20x20 mm^2 (22X_0). Photons in the energy region from 250 MeV to 2 GeV were injected into the central crystal of the matrice as shown in figure 1.

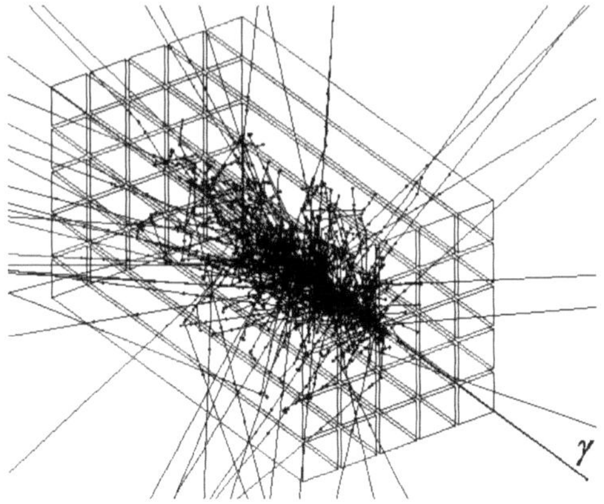

Figure 1. Electromagnetic shower for a single incident photon at 2 GeV

The energy of the incident particle is deposited in active medium of the crystals. Figure 2 shows the energy deposition spectra in PWO crystal matrices, at 2 GeV for example, obtained in the nine and twenty-five crystals.

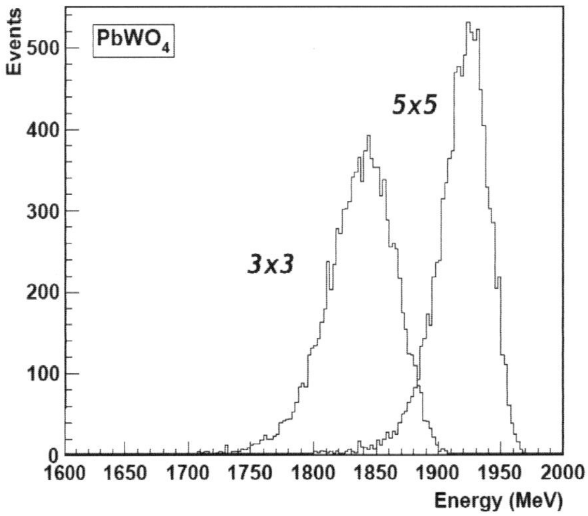

Figure 2. Energy spectra of PWO crystal matrices for 2 GeV photons injected into the center of the central crystal.

The energy resolution (σ_E / E) can be directly obtained from the simulated response functions as shown in figure 2. Figure 3 shows simulated energy resolutions for PWO crystal matrices as a function of incident photon energy. The resolution of the simulated response function includes the effect of the shower fluctuations and the leakage out of the crystal volumes. Mostly the shower leakages in the transverse directions outside the crystal matrix contribute to stochastic term while the shower leakages from back of the crystals contribute to constant term. Monte Carlo simulations show that the shower leakages for 5x5 PWO crystal matrix contribute to constant term 0.5%. The results are consistent with Ref. [9].

Monte Carlo simulations show that shower fluctuations give a contribution of 1.11% to stochastic term for 5x5 PWO crystal matrix. To estimate the other major contribution in the stochastic term we need to know the excess noise factor for the crystal-APD combination and the photoelectron yields. Combining these contributions we would estimate the stochastic term. The emission weighted APD excess noise factor has been calculated as 2.2 for the PWO emission spectrum. Here, the wavelength-dependent APD excess noise value obtained from Ref. [10]. In order to get photoelectron yields contribution to the stochastic term, photons at 1 GeV energies were also injected from the front side of the crystal. The photons produce charged particles in the crystal which interact with the crystal and emit scintillation photons. The number of collected photons at the rear end of the PWO crystal are 71 photons/MeV. The number of primary photoelectrons (N_{pe}) in the photodetector can be calculated by considering back face cross section of the crystals, active area of a pair photodetectors and quantum efficiency of the photodetector. The emission weighted APD and PIN quantum efficiencies have been calculated as 77% and 61% for the PWO emission spectrum at the rear end of the crystal. The calculated emission weighted quantum efficiency value for the crystal-APD system is consistent with the experimental measurements given in Ref. [5]. So the photoelectron statistics contributions on the stochastic term have been calculated as 1.8% for PWO-APD, 0.48% for PWO-PIN combinations by using Eq.1.

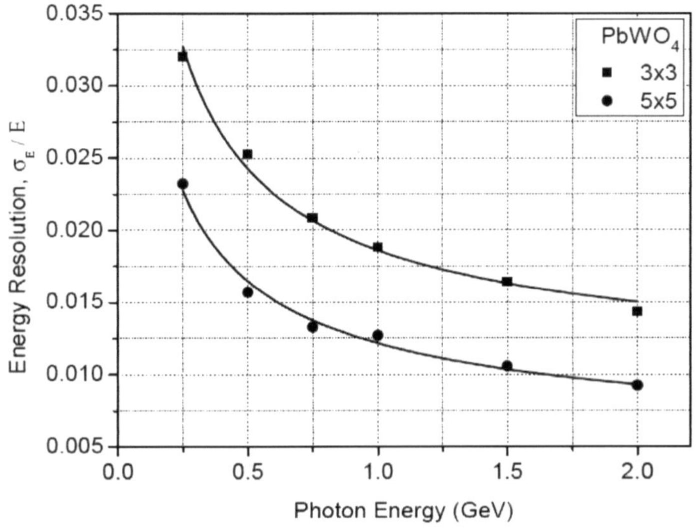

Figure 3. Energy resolutions as a function of incident photon energies for PWO crystal matrices. Results of Geant4 simulations incorporate only the pure energy deposit. Solid lines are the results of fits.

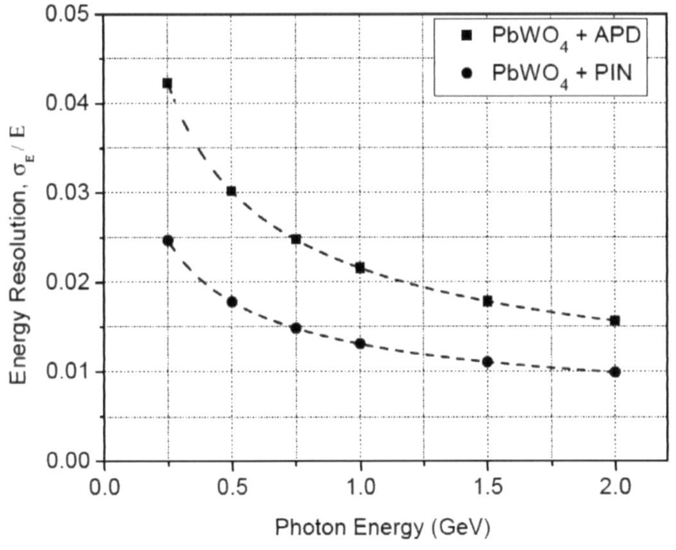

Figure 4. Energy resolutions as a function of incident photon energies for the different photodiode combinations.

Thus, the stochastic term have been calculated by adding photoelectron statistics contributions to the shower fluctuations and plotted in figure 4. Then results of the energy resolution have been obtained as $\sigma_E/E = 2.11\%/\sqrt{E} \oplus 0.5\%$ for the PWO-APD combination and $\sigma_E/E = 1.21\%/\sqrt{E} \oplus 0.5\%$ for the PWO-PIN combination.

As the APD has lower active area and bigger excess noise factor, the photoelectron statistics contribution to the stochastic term is bigger for PWO-APD combination. So that PWO-PIN combination gives better energy resolution values. In order to compare these results, similar studies should be made for CSI-APD and CSI-PIN combinations for the TAC-PF electromagnetic calorimeter.

Acknowledgment
This work was supported by Turkish State Planning Organization (DPT) under the Grant DPT2006K-120470.

References

[1] Sultansoy S et al. The Status of Turkic Accelerator Complex Proposal *Proceedings of PAC 05* 449
[2] Lewandowski B 2002 *Nucl. Instr. and Meth. in Phys. Res. A* **494** 303-307
[3] Miyabayashi K 2002 *Nucl. Instr. and Meth. in Phys. Res. A* **494** 298-302
[4] Ablikim M et al. 2010 *Nucl. Instr. and Meth. in Phys. Res. A* **614** 345
[5] Renker D 2002 *Nucl. Instr. and Meth. in Phys. Res. A* **486** 164
[6] Brown R M et al. *Nucl. Instr. and Meth. in Phys. Res. A* In Press http://dx.doi.org/10.1016/j.nima.2011.12.005
[7] CMS Collaboration 2008 *JINST* **3** S08004
[8] Agostinelli S et al. 2003 *Nucl. Instr. and Meth. in Phys. Res. A.* **506** 250-303
[9] Novotny R W et al. June 2008 *IEEE Transactions on Nuclear Science*, Vol.**55** No.3 1295
[10] Pilicer E et al. 2005 *Nucl. Instr. and Meth. in Phys. Res. A* **552** 146-151

Detection of Clustering Instabilities for Sequential Recombination Algorithms

C. S. Cowden and I. Volobouev

Texas Tech University, Physics Department, Box 41051, Lubbock TX 79409-1051

E-mail: christopher.cowden@ttu.edu

Abstract. We explore clustering stability of sequential recombination jet reconstruction algorithms. Events are reconstructed many times, using random variations of kinematic properties of the jet fragmentation process. Sensitivity of different algorithms to initial conditions are quantified by introducing probabilistic assignment of initial particles to jets (fuzzy clustering). A criterion detecting unstable configurations (bifurcation points) is proposed, based on the overall fuzziness of the event.

1. Introduction

Studies of jet substructure can provide insights into a number of interesting phenomena at high energy colliders by providing such tools as identification of boosted resonances or differentiation of quark and gluon jets. In a typical current jet substructure analysis (see [1, 2, 3] for a few examples and [4] for a recent comparison of different methods), one finds large jets by some sequential recombination algorithm then un-clusters the target jet by another sequential recombination algorithm (Cambridge/Aachen is the usual choice for the second stage). In this work we consider a different approach in which the sequential recombination is performed in a manner appropriate for substructure studies without the necessity of unclustering the results of the initial cluster finding. We call a sequential recombination algorithm "multiscale" (or multiresolution) if it has the following features:

- Stopping condition is implemented as a cutoff on the recombination distance function.
- The distance function is dimensionless.
- The minimum pairwise distance is almost always increasing in the recombination process[1].

An algorithm with such properties will result in identical clusterings no matter whether we cluster to a distance cutoff D_1 or whether we interrupt the clustering process at at some smaller cutoff D_2 and then resume and continue to D_1. Also, identical clusterings will be obtained by a class of distances which differ from each other only by a monotonous transformation (with the same transformation applied to the stopping criterion). Thus we associate the distance cutoff with the concept of resolution scale (or jet size). The use of dimensionless distances allows us to compare events with different partonic center of mass energy on equal footing which is essential in hadronic collisions.

Multiscale jet clustering algorithms provide a wealth of new information regarding the hadronic portion of collider events by adding the scale dimension. In other words, instead of returning just a set of four-momenta for the reconstructed jets, multiscale algorithms return a clustering tree (a.k.a. hierarchical

[1] Ideally the minimum distance should monotonically increase, but no known algorithms employing four-vector recombination have such a property. Monotonous clustering methods without pairwise recombinations do exist: see for example [5].

clustering dendrogram). A good choice of distance function ensures that the recombination scale is almost always increasing in the process of tree construction. In a true multiscale jet clustering algorithm, one could stop at any distance by merely navigating the history of a more inclusive clustering. Adopting a multiscale jet clustering algorithm, one obviates the need of the two phase analysis in jet substructure studies. In the context of global jet finding algorithms, the ideas of multiscale jet reconstruction were pioneered by FFTJet [6].

Not all commonly used sequential recombination algorithms have the necessary properties to adapt to a multiscale application nicely. Most notably, algorithms akin to k_T [7] and anti-k_T [8] do not adapt well because these algorithms employ dimensionful distance functions which do not compare well event to event in hadron collider events. For the following analysis, the jet clustering algorithms employed do not perform any beam removal. The algorithms simply work by comparing inter-particle distances defined by a given distance function and selecting the particles with the smallest distance to recombine. This repeats until the algorithm encounters a predefined stopping criterion. The Cambridge/Aachen [9] algorithm used in this study defines a distance function based on separation of a pair of particles i and j in the η-ϕ space given by

$$d_{ij} = \Delta R_{ij}^2 \tag{1}$$

where $\Delta R_{ij}^2 = \Delta\eta_{ij}^2 + \Delta\phi_{ij}^2$.

In addition to the algorithm which employs the distance defined by Equation 1, we consider more general distances represented by

$$d_{ij} = \Delta R_{ij} f\left(\frac{\min(p_{Ti}, p_{Tj})}{\max(p_{Ti}, p_{Tj})}\right), \tag{2}$$

where f is some continuous function defined on the $[0,1]$ interval. In particular, we can choose the function f in Equation 2 so that d_{ij} becomes the minimum width of a Gaussian filter in the η-ϕ space for which only a single peak remains when two initial energy deposits are smeared by the filter[2]. One can also think of this distance as being proportional to the square root of the time for the two energy deposits to "diffuse" into a single blob, so we call the resulting distance function the "diffusion distance". This distance function is inspired by the scale-space theory [10], and we expect it to perform well in multiscale applications. Figure 1 shows the function f versus the input argument $x = \frac{\min(p_{Ti}, p_{Tj})}{\max(p_{Ti}, p_{Tj})}$.

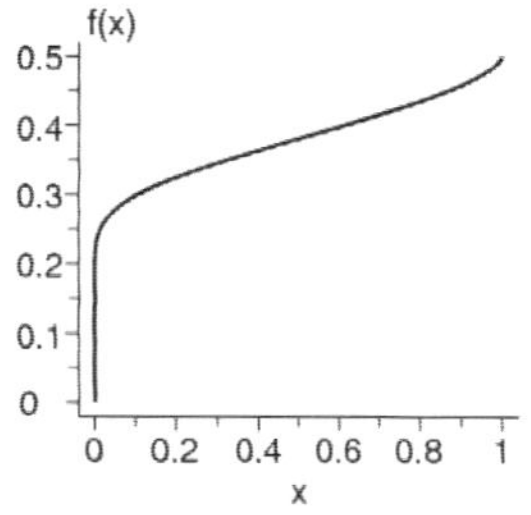

Figure 1. The function f defined by the diffusion distance versus the input argument $x = \frac{\min(p_{Ti}, p_{Tj})}{\max(p_{Ti}, p_{Tj})}$.

A number of studies have suggested using clustering stability as a tool for model selection (see [11] for a review). In what follows we study the clustering stability as a function of scale, *i.e.*, jet size or distance cutoff. We intend to associate stable clustering configurations with the physical structures

[2] `C++` implementation of this distance function is available upon request.

present in high energy events (such as inter-jet distances in the η-ϕ space). Intuitively, one expects the stability, defined in some sensible way, to decrease when the distance cutoff exists in a band where two or more jets merge together. Conversely, one expects a relatively high degree of stability when jets are well separated.

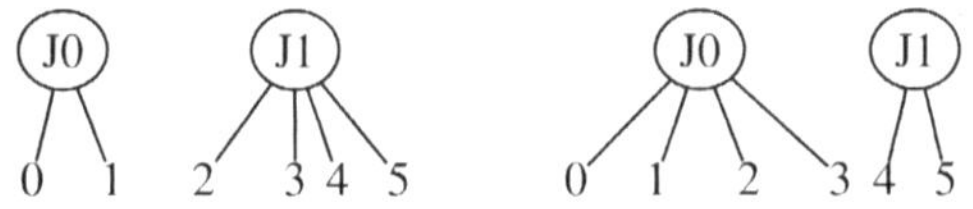

Figure 2. An illustration showing two possible clusterings in which objects may be assigned to two clusters, J0 and J1. Clustering C_1 is on the left while clustering C_2 is on the right.

One can effectively measure the stability of a particular jet configuration by perturbing the system a number of times and comparing the clustering outcomes at the same distance cutoff after the perturbations. To quantify the clustering stability, we use modifications of the Rand [12] and Jaccard [13] indices. These indices are defined by considering what happens to each pair of input particles in two different clusterings C_1 and C_2 (Figure 2 shows an example of such a comparison). Each pair of particles falls into one of the following four categories:

I Both particles belong to same cluster in both clusterings (for example, particles 0 and 1 in Figure 2).

II Particles belong to different clusters in both clusterings (particles 0 and 4).

III Particles belong in the same cluster in clustering C_1, but belong to different clusters in clustering C_2 (particles 3 and 4).

IV Particles belong to different clusters in clustering C_1, but belong to the same cluster in clustering C_2 (particles 1 and 2).

In the statistical literature, the Rand and Jaccard indices are defined via the number of pairs in these categories; we instead weight each pair by $w_{ij} = p_T^i p_T^j$. The p_T-weighted Rand index is defined by:

$$R = \frac{\Sigma^I + \Sigma^{II}}{\Sigma} \tag{3}$$

where Σ is the sum of all weights, Σ^I is the sum of weights w_{ij} in category I, and so on. The p_T weighted-Jaccard index is defined by:

$$J = \frac{\Sigma^I}{\Sigma^I + \Sigma^{III} + \Sigma^{IV}} \tag{4}$$

The subsequent analysis also makes use of another stability measure referred to as the "p_T-weighted minimal matching distance", or MMD. This measure is defined by forming an $M \times N$ matrix of the M clusters in C_1 and N clusters in C_2. The matrix element in row i and column j is set to the scalar sum of the p_T of the particles which belong simultaneously to the cluster i in clustering C_1 and to the cluster j in clustering C_2. Then, in case $M < N$ we find a set of M matrix elements, all from different rows and columns, with maximum sum of the elements (N such elements are summed in case $N \leq M$). The MMD is defined as one minus this maximized sum divided by the scalar sum of p_T values of all initial particles[3]. Since all stability measures so far considered (R, J, and MMD) range between zero and one, we compute the corresponding instability as one minus the stability measure.

[3] An approximate greedy implementation of this algorithm is used in practice as the computational complexity of the exact algorithm scales factorially with the number of clusters.

Fuzziness, on the other hand, quantifies how the clustering changes as a result of reasonable jet energy flow fluctuations. We employ the following fuzziness definition [6]:

$$F_j = \frac{\sqrt{\sum_i w_{ji}(1-w_{ji})(p_T^i)^2}}{\sum_i w_{ji} p_T^i} \tag{5}$$

where the weight w_{ij} represents the probability of particle i to belong to jet j in the ensemble of perturbed clustering outcomes. Subsequently, we define the event fuzziness as

$$F = \frac{\sum_j F_j \sum_i w_{ji} p_T^i}{\sum_i p_T^i} \tag{6}$$

which can be understood as the average p_T-weighted fuzziness of the individual jets. One also has a choice regarding the outer summation in the numerator of Equation 6 to be a regular sum or in quadrature. The results shown below were found using the regular summation.

2. Event Generation and Simulation

We study clustering stability and fuzziness in the context of a controlled dijet toy MC environment as well as hadronic $t\bar{t}$ events in high energy pp collisions. We generate events using the Pythia event generator. For the toy dijet MC sample, we generate the events by randomly selecting a point in the η-ϕ space and shooting a parton for which Pythia performs the showering and hadronization. Then we select a point at a given ΔR of that first parton to shoot the second parton for which Pythia also simulates the showering and hadronization. We also separately generate a sample of $t\bar{t}$ events using standard Pythia cards.

The stability and fuzziness measures employed in this analysis are defined with respect to the statistical ensemble formed by multiple perturbations of the initial event. We perturb the event according to one of two schemes. In the first scheme (referred to as "Scheme 1") we perturb the jet showering process and detector response in a semi-realistic manner. We treat each particle as if it were a gluon and calculate the probability for each particle pair that the pair originated from the $g \to gg$ splitting:

$$p \sim C_A 4\pi \alpha_S ({P_T^{soft}}^2) \frac{p^{mother} \,.\, p^{hard}}{p^{mother} \,.\, p^{soft} p^{hard} \,.\, p^{soft}}, \tag{7}$$

where $P^{mother} = P^{hard} + P^{soft}$ and p^{soft} is the 4-momentum of the particle with the smaller p_T in the pair[4]. Then we re-decay the mothers of the most probable pairs. The scheme then excludes the particle in the pair with lower p_T from further consideration in the fluctuation process. More specifically, this scheme of QCD like fluctuation works in the following manner.

- For each pair of particles, estimate the probability that these particles have the same mother according to Equation 7.
- Sort all pairs of particles in the decreasing order of probability.
- For the highest probability pair, re-decay the mother particle according to a flat phase space.
- Mark the softer particle in the pair as "used" to exclude it from further consideration.
- Repeat the previous two steps until the 4-momentum of half of the event particles has been fluctuated.

We also apply simple efficiency and resolution simulation with effective parameters similar to those of the CMS detector. Typically we execute the perturbation sequence and re-cluster 100 times per event.

The second scheme "Scheme 2" is simpler: in this scheme we perturb the event by adding noise to each particle's η, ϕ, and energy according to a set of Gaussian distributions. Some results below are shown for the case where $\sigma_\eta = \sigma_\phi = 0.087$ and $\sigma_E = 0.15$.

[4] The authors thank Steve Mrenna for guidance regarding Equation 7.

3. Results

3.1. Toy Dijet MC Stability

We first tested this method on a sample of toy MC dijet events in which we controlled the initial partons' ΔR and p_T ratio (this ratio is denoted r in the figures). Figure 3 illustrates these results by plotting instability averaged over many events vs. the distance cutoff. One can see from Figure 3(a) that the three stability measures employed (Rand, Jaccard, and p_T-weighted minimal matching distance) all share a peak of instability at the same location but have varying degrees of sensitivity to the instability. Also as one may expect, applying Scheme 2 to perturb the system results in a magnified instability peak found at the unstable points as shown by Figure 3(c). One can also compare the behavior of these stability measures in the C/A algorithm from Figures 3(b) and Figure 3(d). One can observe from these Figures that instability peak magnitude remains roughly the same for both clustering algorithms studied if events are perturbed by the same perturbation Scheme. One may also note that the location of instability peaks for events clustered using the diffusion distance depends upon the p_T ratio of the jets, unlike the C/A algorithm.

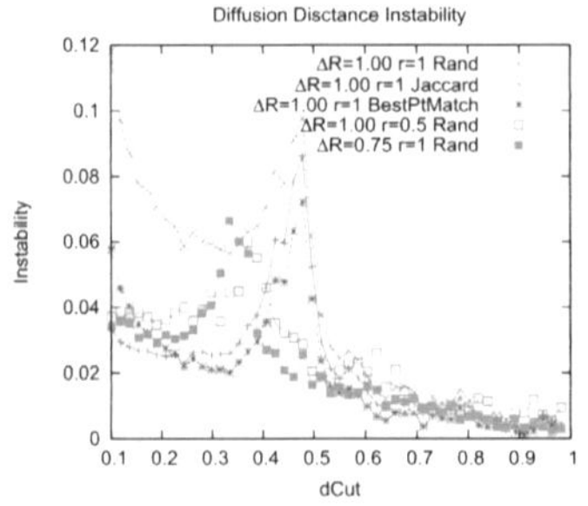

(a) Instability of dijet configurations clustered with the diffusion distance averaged over many events perturbed by Scheme 1.

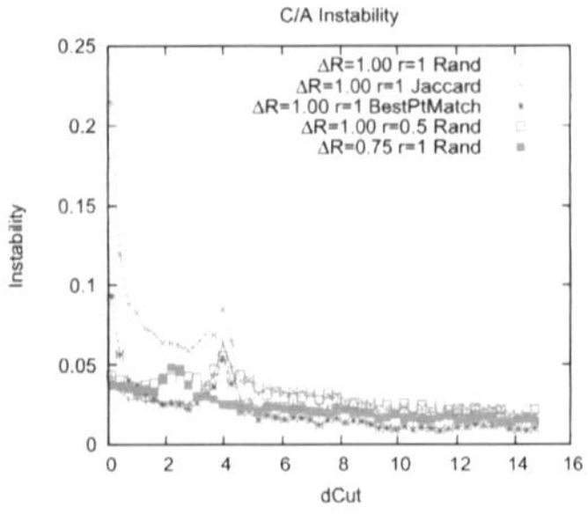

(b) Instability of dijet configurations clustered with the C/A distance averaged over many events perturbed by Scheme 1.

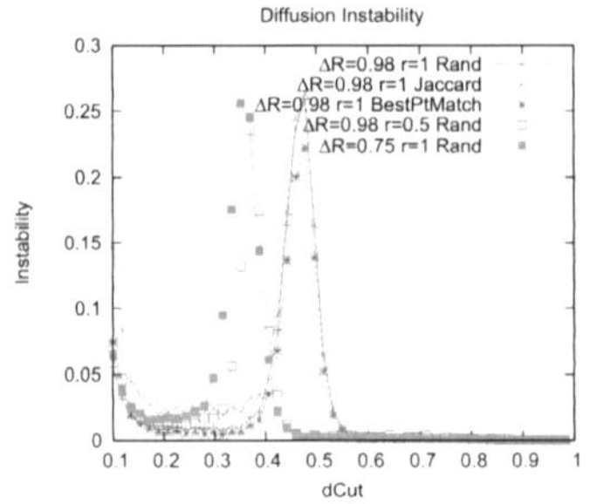

(c) Instability of dijet configurations clustered with the diffusion distance averaged over many events perturbed by Scheme 2.

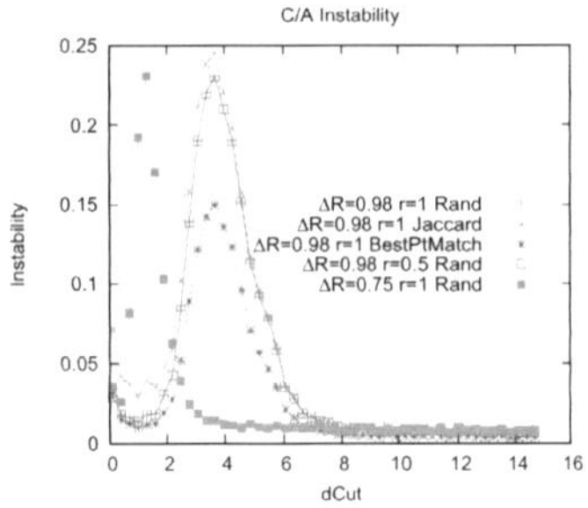

(d) Instability of dijet configurations clustered with the C/A distance averaged over many events perturbed by Scheme 2.

Figure 3. These plots illustrate the instability in a few toy MC dijet configurations and the two clustering algorithms previous described. One can see excesses in the instability near cutoff values where the two jets merge. One may also notice differences in stability sensitivity by choice of smearing scheme but not quite such a dependence upon choice of algorithm.

Figure 4 illustrates the fuzziness in these dijet events for the smearing and clustering algorithms discussed. Upon comparing Figure 4(a) to Figure 4(c), one observes a similar structure of peaks but events perturbed by Scheme 1 show instability peaks with a suppressed significance compared to events perturbed by Scheme 2. The C/A algorithm exhibits a similar, but more drastic, suppression of average fuzziness structure for events perturbed by Scheme 1 as opposed to Scheme 2.

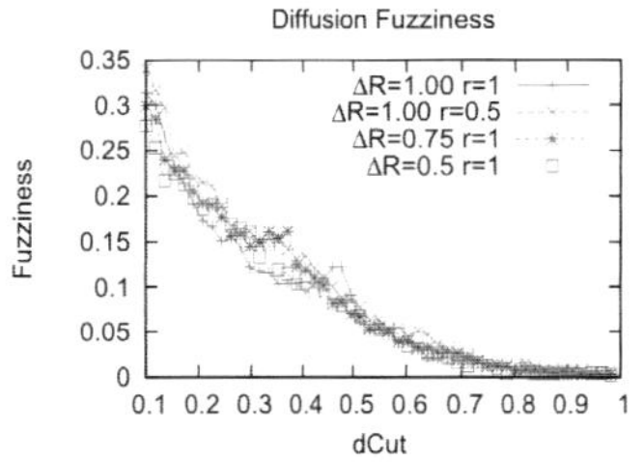

(a) Fuzziness of dijet configurations clustered with the diffusion distance averaged over many events perturbed by Scheme 1.

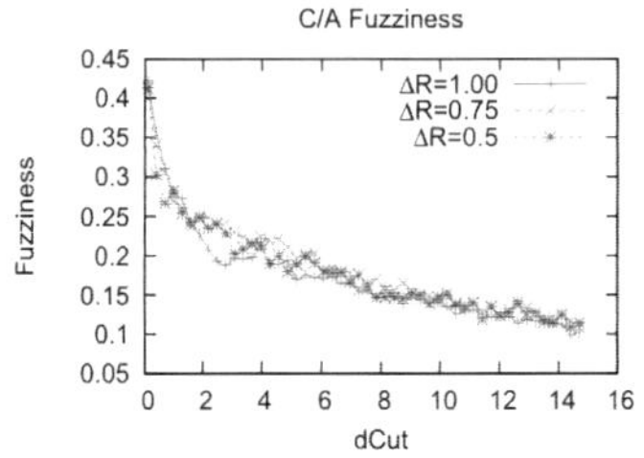

(b) Fuzziness of dijet configurations clustered with the C/A distance averaged over many events perturbed by Scheme 1.

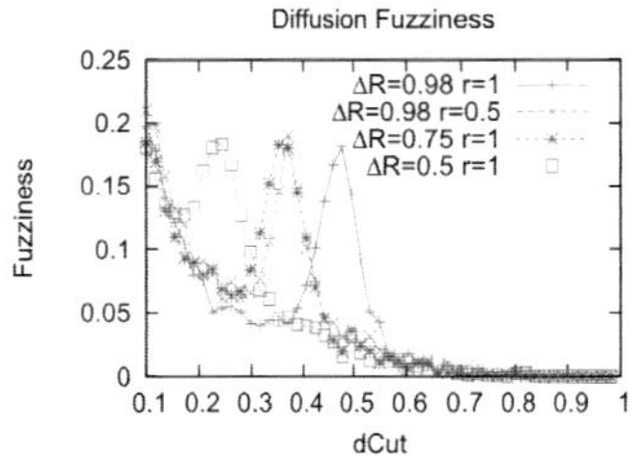

(c) Fuzziness of dijet configurations clustered with the diffusion distance averaged over many events perturbed by Scheme 2.

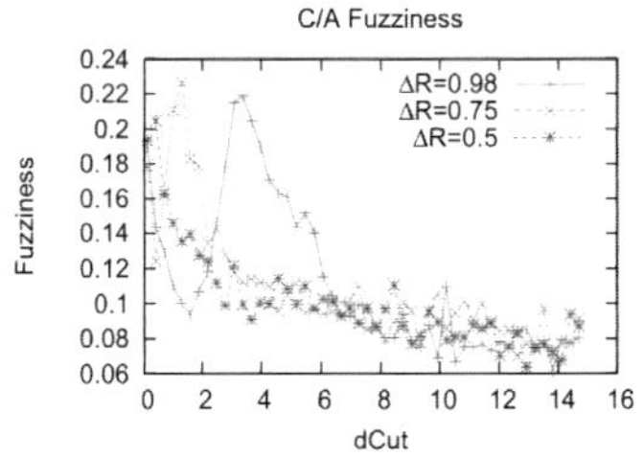

(d) Fuzziness of dijet configurations clustered with the C/A distance averaged over many events perturbed by Scheme 2.

Figure 4. These plots illustrate the fuzziness in a few toy MC dijet configurations and the two clustering algorithms previously described. The fuzziness shows peaks around distance cutoffs similar to the instability.

3.2. $t\bar{t}$ MC Stability

Since jets produced in $t\bar{t}$ events have a wide range of ΔR separations and p_T ratios, it makes sense to consider the stability and fuzziness of single events rather than averages over many events. Figure 5(a) shows the instability of a $t\bar{t}$ event using the diffusion distance for jet clustering and perturbed by Scheme 1. In this figure, peaks emerge indicating points where jets merge together; the peak occurring at dCut ~ 3.75 shows the point at which the entire event has been reconstructed into a single jet. Figure 5(b) depicts the fuzziness in the same $t\bar{t}$ event; this plot shows a similar pattern of peaks as Figure 5(a). The peak in instability at dCut ~ 1 corresponds to a transition from a 6 jet configuration to a 3 jet configuration. Likewise, over the range $0 \leq$ dCut $\lesssim 0.5$ there is a transition from some initial unstable high multiplicity clustering to a more stable 6 jet configuration one very well expects in a fully hadronic $t\bar{t}$ event.

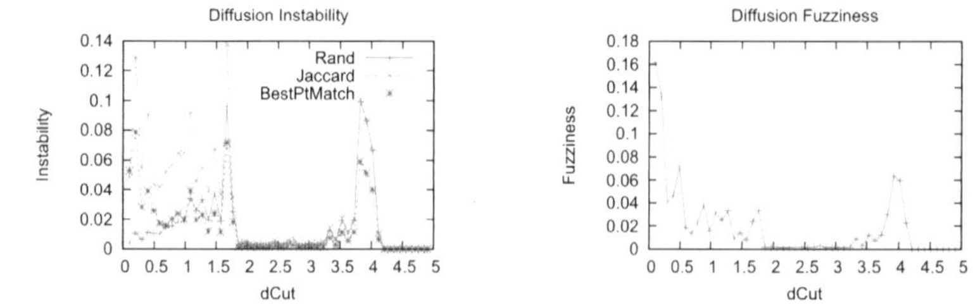

(a) Instability of a single $t\bar{t}$ event clustered with the diffusion distance perturbed by Scheme 1.

(b) Fuzziness of a single $t\bar{t}$ event clustered with the diffusion distance perturbed by Scheme 1.

Figure 5. Plots showing jet stability and funzziness results for a single $t\bar{t}$ event.

4. Conclusions

In this note we have developed the concept of multiscale sequential recombination algorithms appropriate for jet reconstruction at hadron colliders. The jet substructure information is produced by these algorithms in the most natural manner. We have also introduced the diffusion distance clustering algorithm based on ideas from the scale-space theory. The use of clustering stability as a function of scale was explored as a tool for identifying physically meaningful jet configurations. Such configurations are characterized by a substantial range of scales free of bifurcation points (clustering instabilities). We construct the clustering stability measures in such a way as not to be affected by infrared or collinear divergences. The concepts presented here can potentially become powerful tools in the analysis of event energy flow in hadronic collisions and deserve further study.

References

[1] Jonathan M. Butterworth, Adam R. Davison, Mathieu Rubin, and Gavin P. Salam. Jet substructure as a new higgs-search channel at the large hadron collider. *Phys. Rev. Lett.*, 100:242001, Jun 2008.

[2] David Krohn, Jesse Thaler, and Lian-Tao Wang. Jet Trimming. *JHEP*, 02:084, 2010.

[3] Stephen D. Ellis, Christopher K. Vermilion, and Jonathan R. Walsh. Recombination Algorithms and Jet Substructure: Pruning as a Tool for Heavy Particle Searches . *Phys.Rev.*, D81:094023, 2010.

[4] Paloma Quiroga-Arias and Sebastian Sapeta. A comparative study of jet substructure taggers. 2012, 1209.2858.

[5] Anil K. Jain and Richard C. Dubes. *Algorithms for Clustering Data*. Prentice Hall, 1988.

[6] I. Volobouev. FFTJet: A Package for Multiresolution Particle Jet Reconstruction in the Fourier Domain. 2009, 0907.0270.

[7] S. Catani, Yuri L. Dokshitzer, M.H. Seymour, and B.R. Webber. Longitudinally invariant K_t clustering algorithms for hadron hadron collisions. *Nucl.Phys.*, B406:187–224, 1993.

[8] Matteo Cacciari, Gavin P. Salam, and Gregory Soyez. The Anti-k(t) jet clustering algorithm. *JHEP*, 0804:063, 2008, 0802.1189.

[9] Thomas H. Cormen, Clifford Stein, Ronald L. Rivest, and Charles E. Leiserson. *Introduction to Algorithms*. McGraw-Hill Higher Education, 2nd edition, 2001.

[10] Tony Lindeberg. Scale-space theory: A basic tool for analysing structures at different scales. *Journal of Applied Statistics*, pages 224–270, 1994.

[11] U. von Luxburg. Clustering Stability: An Overview. *ArXiv e-prints*, July 2010, 1007.1075.

[12] William M. Rand. Objective criteria for the evaulation of clustering methods. *JASA*, 66:846, 1971.

[13] P. Jaccard. Nouvelles recherches sur la distribution florale. *Bull. Soc. Vaud. Sci. Nat*, 44:223–270, 1908.

Precision Timing of the ATLAS Level-1 Calorimeter Trigger: From Beam Splashes to High Luminosity Proton-Proton Collisions

Y. Davygora[1], G. Unal[2] on behalf of the ATLAS Collaboration

[1] Heidelberg University, Kirchhoff Institute for Physics, Im Neuenheimer Feld 227, D-69120 Heidelberg, Germany

[2] European Organization for Nuclear Research CERN CH-1211, Genève 23, Switzerland

E-mail: [1] Yuriy.Davygora@cern.ch, [2] Guillaume.Unal@cern.ch

Abstract. The ATLAS Level-1 Calorimeter Trigger uses trigger tower signals from the ATLAS calorimeter as input. In real-time, it identifes high-pT objects, determines total and missing transverse energy sums and assigns bunch-crossing identification. Reliable operation requires collision signals to be synchronised at the nanosecond level. This timing was first established through the analysis of beam splash events and subsequently refined with data from LHC proton-proton collisions. In this contribution, details of the timing synchronization method as well as selected results from the timing adjustments are presented.

1. Introduction

The Large Hadron Collider (LHC) at CERN is designed to collide protons at a centre-of-mass energy of 14 TeV with an instantaneous luminosity of $10^{34}\,\mathrm{cm^{-2}s^{-1}}$. The ATLAS detector [1] measures particles which are created in those collisions and looks for new physics both within and beyond the Standard Model. Online selection of the relevant collision events is performed by a 3-stage trigger system. The hardware-based Level-1 Trigger (L1) operates at 40 MHz, synchronous with the LHC bunch crossings. Within a fixed latency of 2.5 μs, a decision based on fast processing of data with coarsened detector granularity is provided, so that the event is either rejected completely or transferred to the following trigger levels. The decision is generated by the Central Trigger Processor (CTP) which receives information mainly from the Level-1 Calorimeter Trigger (L1Calo) and the Level-1 Muon Trigger (L1Muon). Altogether, the input rate of 40 MHz is reduced to $\lesssim$ 75 kHz.

The L1Calo system [4] processes 7168 analogue signals from the Liquid Argon (LAr) and Tile Calorimeters of the ATLAS detector [2] [3]. These signals are created by summing up to 60 calorimeter cells into Trigger Towers (TT) most of which are of size 0.1×0.1 in $\Delta\eta \times \Delta\phi$[1]. The L1Calo receivers apply a gain, independently adjustable for each TT, to the analogue signals. In the end, each TT is finally calibrated to transverse energy (E_T). The signals are then conditioned and digitized in the PreProcessor Module (PPM) with a digitization step of 25 ns. In addition, their bunch-crossing is identified (BCID) and the E_T is determined. This

[1] Here, $\eta = -\ln[\tan(\theta/2)]$ is the pseudorapidity, θ being the polar angle, and ϕ is the azimuthal angle.

information is then processed in parallel by the Cluster Processor (CP) and the Jet/Energy-Processor (JEP) which search for and identify high-p_T electrons/photons, taus/hadrons and jets. The JEP also provides a total and missing E_T measurement. The object multiplicities per threshold are then transmitted to the CTP.

In L1Calo, the timing of the input signal digitization directly affects the identification of the correct bunch-crossing and measured energy. For a stable operation of BCID for non-saturated signals, a sampling of analogue pulses with a precision of ± 10 ns is required. Even higher precision is required for an accurate E_T measurement (2%), namely ± 5 ns. Within the PPM, the signal digitization can be adjusted for every TT independently with respect to the LHC clock. A TT-specific and individually-adjustable delay register (FIFO) is used for a coarse timing adjustment in steps of 25 ns. The so-called PHOS4 chip allows a finer adjustment in 25 steps of 1 ns. Together, both procedures allow to set the sampling of the analogue TT signal directly to the maximum of the analogue signal peak for the central digitized time-slice. In 2009, LHC splash events were used to cross-check and re-calibrate the initial timing of the L1Calo system. Since the start of proton- proton collision data taking at 7 TeV in 2010, timing calibration has been done with those data.

2. Fit method for L1Calo timing synchronization

Since the analogue pulse shape of the signals is not available in recorded data, a fit of the digitized pulse shape is needed to extract the timing shift. For the fit method, TTs with similar pulse shape features are grouped together according to their calorimeter division. One of such divisions is, for example, the so-called electromagnetic barrel calorimeter (EMB), which is the central part of the electromagnetic calorimeter layer. All divisions and their respective location are listed in Table 1.

Depending on the calorimeter division, TT signals, sampled every 25 ns, are fitted with a combination of a Gaussian and a Landau function (GLu) or of two Landau functions (LLu). Due to the asymmetry of the pulses, combined functions which are joined at the maximum (such as GLu and LLu) have been found to describe the pulses best. The mathematical expression for the GLu fit function is given in equation (1):

$$
\begin{aligned}
f(x \leqslant x_{max}) &= A \cdot \exp\left(-\frac{(x-x_{max})^2}{2\cdot\sigma^2_{Gauss}} - \frac{1}{2}\right) + C, \\
f(x > x_{max}) &= \left(A + e^{+\frac{1}{2}} \cdot D\right) \cdot \exp\left(-\frac{1}{2}\left(\frac{x-x_{max}}{\sigma_{Landau}} + \exp\left(-\frac{x-x_{max}}{\sigma_{Landau}}\right)\right)\right) - D + C.
\end{aligned}
\tag{1}
$$

Here, x_{max} denotes the maximum position of the signal in units of time and A is the amplitude of the signal fit in the hardware units of digitization, i.e. in ADC counts. σ_{Gauss} (σ_{Landau}) denotes the fit width, in ns, for the rising (trailing) edge of the pulse. C stands for a constant offset or pedestal of the pulse (usually 32 ADC counts), and D allows the fit to fall below its pedestal value. In the LAr calorimeter, the electronic shaping generates a long negative signal part, the so-called undershoot, which reaches, for example in the EMB, a depth of approximately 20% of the positive signal height. For such pulses, the parameter D becomes relevant. The LLu function is obtained from (1) by replacing the Gaussian by another Landau function like the one for the trailing edge, but with parameter $D = 0$.

Table 1 summarizes the fit procedures used for different calorimeter divisions. The choice between GLu and LLu functions is based on their performance and stability. An example of how the fit method works is given in Figure 1. Here is shown a calibration pulse which was sampled with two different digitization settings in the L1Calo PPr system. In Figure 1(a), the pulse is well-timed. Its maximum is unambigiously identifiable and it lies in the central time-slice of

Calorimeter division	Fraction of TTs [%]	η-range	Fit function	Fit range
EMB	25.0	$\|\eta\| < 1.4$	GLu	4
emOverlap	1.8	$1.4 < \|\eta\| < 1.5$	GLu / LLu	4
EMEC	21.4	$1.5 < \|\eta\| < 3.2$	LLu	4
FCal	3.6	$\|\eta\| > 3.2$	LLu	3
Tile	26.8	$\|\eta\| < 1.4$	GLu	4
HEC	21.4	see EMEC	LLu	4

Table 1. List of the calorimeter divisions and the fit procedures for the individual calorimeter divisions, including the range of the fit.

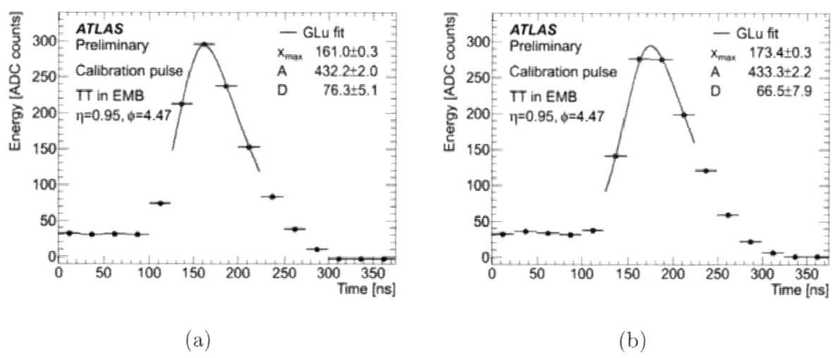

Figure 1. Calibration pulse for one EMB TT as recorded by L1Calo for each of two digitization settings, and fitted with the GLu function [6]. The deposited energy (in units of ADC counts) is plotted versus the time, using an arbitrary time origin. (a) A well timed pulse. (b) A pulse with the timing shifted by 12 ns.

the digitized pulse. The pulse in Figure 1(b) is obviously mistimed, because it shows no clear maximum but rather two values of approximately the same height. This leads to ambiguity in BCID and an E_T mismeasurement. Here, the timing was shifted by 12 ns.

To both pulses in Figure 1, a GLu fit has been applied. The fit reconstructs the shape and, especially, the maximum position and the amplitude of the original analogue pulse. The challenge of the fit method is to obtain the same pulse shape independent of current digitization settings. In particular, in Figure 1, the signal peak position and the amplitude of both pulses are expected to be identical, apart from a 12 ns shift. Indeed, within small uncertainties, this is the case. Thus, in the timing synchronization procedure, no correction would be required for a pulse like the one in Figure 1(a) and a correction of 12 ns would be needed for a pulse like the one in Figure 1(b).

While for calibration pulses 15 time-slices of a TT pulse are read out from the L1Calo trigger system as displayed in Figure 1, during normal data-taking only 5 time-slices can be handled by the Data Acquisition System (DAQ). Thus, the fit range for GLu and LLu functions is restricted significantly. This is the reason why only 4 or 3 (in FCal) time-slices around the signal peak are used by the fit method (see Table 1). With 6 free parameters, the GLu and LLu are under-constrained, so to reduce the number of degrees of freedom the values of σ_{Gauss}, σ_{Landau} and the undershoot-to-amplitude ratio (D/A) are fixed to predefined values derived from calibration data. In addition, the pedestal parameter (C) is fixed at the TT signal baseline as set in the L1Calo hardware before digitization. Thus, only the maximum position x_{max} and the amplitude A are allowed to vary freely within certain constraints.

The main source of systematic uncertainty of the method comes from the shape differences between the calibration and real physics pulses. They are estimated to be of the order of magnitude of 5-15% in the EMEC and in FCal they even reach to some 20-30%. In the application of the fit method to collision data, these shape differences have been corrected for. Overall, however, the timing synchronization determined by the fit method has a ± 1 ns systematic uncertainty and a ± 2 ns statistical uncertainty [6]. With respect to the precision of 1 ns in the digitization point setting in the L1Calo hardware, the method's accuracy lies well within the targeted precision of 5 ns.

3. Timing calibration with 2009 LHC beam splashes

During autumn 2009, the LHC provided splash events during commissioning, prior to physics (collision) operation. A splash event is produced by beam packets interacting with the collimators located in the beam pipes at ± 145 m from the interaction point at the centre of the ATLAS detector. The timing of such events does not correspond to that of beam collision timing. However, given the distances from collimator to each particular TT, and from that TT to the nominal interaction point, a correction for time-of-flight can be performed to estimate the collision timing. Initially, this has been done with data from Run 140370 (20th November 2009) where 55 of the splash events provided by the LHC were used [5].

Figures 2(a), 2(b) show the fitted signal peak positions which have not been corrected for time-of-flight, while Figures 2(c), 2(d) show the corrected values which were used as the timing corrections with an expected timing accuracy of ± 5 ns (for more details, see [5]).

4. Timing calibration with proton-proton collision data

At the beginning of 2010, the LHC started delivering collision data with a proton-proton centre-of-mass energy of $\sqrt{s} = 7$ TeV. This allowed the possibility of a timing update derived from actual physics data.

4.1. Event and pulse selection

The precise event and pulse selection requirements have evolved, largely to adapt to LHC increases in luminosity. In all cases, events are required to pass certain basic data quality criteria. They have to be taken during stable beam conditions when the detector, especially the calorimetry, is fully operational. A good primary vertex within a maximum displacement along the beam axis, as well as a minimum number of tracks from the vertex are required. These requirements enhance the probability that the signals originate from head-on proton-proton collisions. The pulse selection criteria are mainly there to reduce the number of noise pulses contaminating the analyzed sample. For calorimeter divisions which are especially susceptible to calorimeter noise, e.g. the HEC, a special noise suppression is applied. Also, a minimum transverse energy of approximately 7 GeV in a given TT as measured by L1Calo is required. For details on event and pulse selection, see [6].

4.2. First timing calibration with collision data

The first timing calibration with collision data was done with data from 4-5 July 2010, corresponding to an integrated luminosity of $\approx 30\,\mathrm{nb}^{-1}$. Following the described pulse selection, in each TT, the pulses were fitted with a GLu/LLu function. The differences between the fitted maximum positions and the central time-slice are expected to be gauss-distributed around the best timing correction. Figure 3 shows such a distribution for one of TTs in the EMB and a corresponding gaussian fit. However, the mean value of the distribution itself and not the gaussian fit mean is used as the timing correction, because this has been found to be a more stable procedure.

Figure 4 shows timing corrections for all TTs in both electromagnetic and hadronic layers of the calorimeter determined as described above. Most of the corrections are within ± 2 ns which

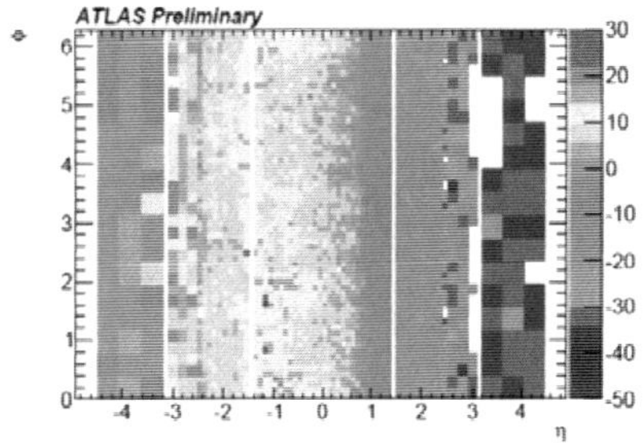

(a) Electromagnetic layer

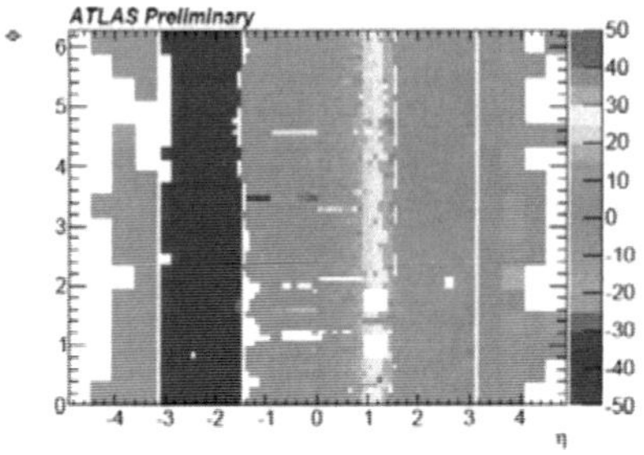

(b) Hadronic layer

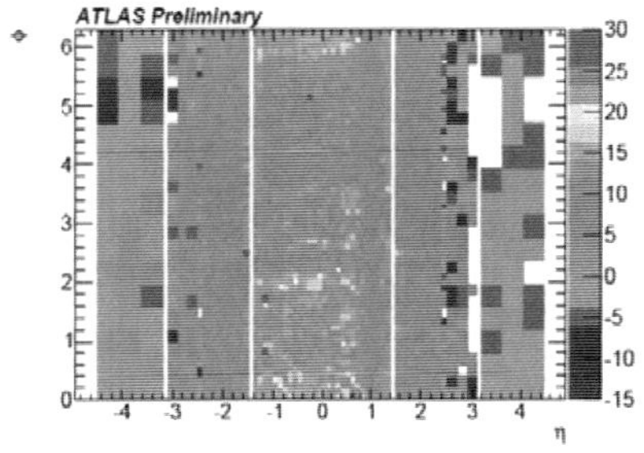

(c) Electromagnetic layer

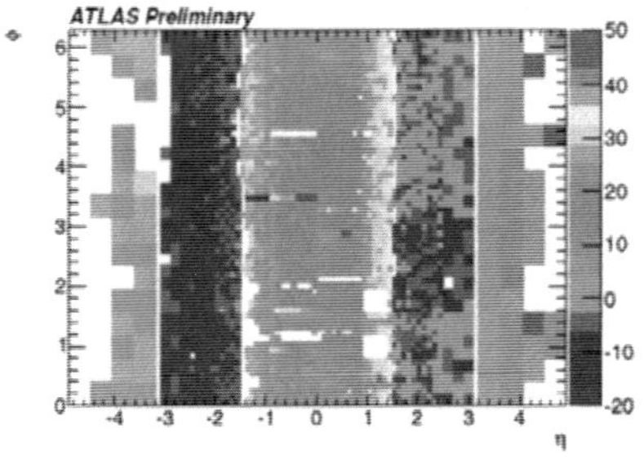

(d) Hadronic layer

Figure 2. (a), (b) $\eta - \phi$ maps of the peak time (ns) plotted on the z-axis, for typical splash events [5]. The peak times (t_0) are measured by fitting each trigger tower signal with a GLu/LLu function. The timing reference was taken as 175 ns. In white regions, no reliable fit could be made, hence no correction computed. In (c), (d) the corrected (for time-of-flight) peak time (ns) for each calorimeter layer is plotted. The electromagnetic layer (a), (c) has beam-1 approaching from $+\eta$ direction. The hadronic (b), (d) has beam-2 approaching from $-\eta$ direction.

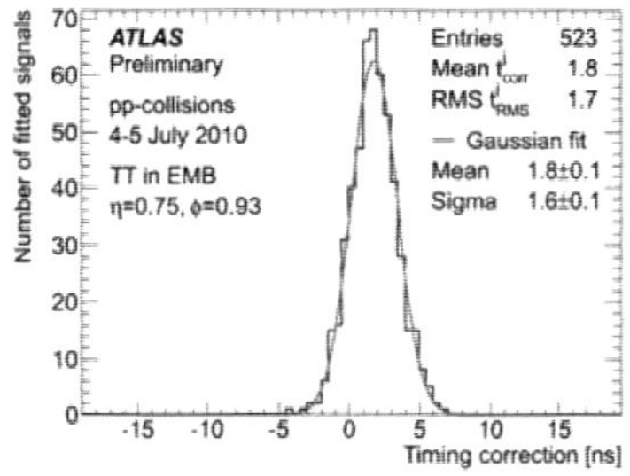

Figure 3. Distribution of the difference between the fitted maximum position and the middle of the central read-out time-slice for one TT in the EMB from the first physics data based timing synchronization in July 2010 [6]. The mean value of the distribution determined from 523 fits to collision signals indicates a timing correction of 1.8 ns for the given TT. For comparison only, a simple gaussian fit is shown.

illustrates the consistency of the method. In 2010 and 2011, the L1Calo timing synchronization was checked nine times using the same method, resulting in six timing adjustments from proton-proton collision data. For a full list of updates, see [6]. The magnitudes of the corrections were usually less than 2 ns.

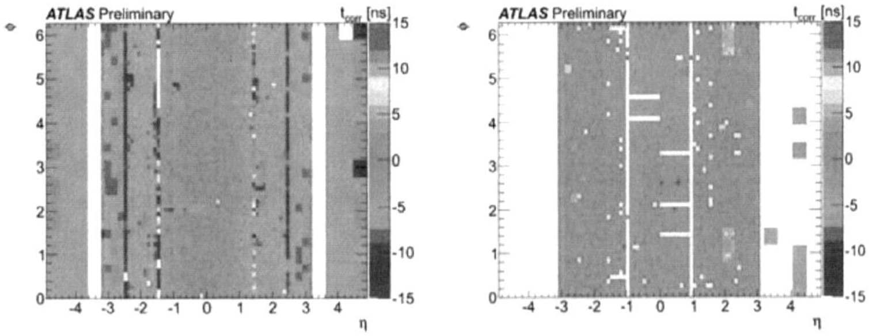

(a) Electromagnetic layer (b) Hadronic layer

Figure 4. Timing corrections in ns for the electromagnetic calorimeter layer (a) and the hadronic calorimeter layer (a) [6]. The timing corrections per TT were derived from proton-proton collision data of July 2010.

5. Conclusion

Synchronizing the trigger tower (TT) signals of the Level-1 Calorimeter Trigger (L1Calo) with respect to the LHC clock and with respect to each other, as well as maintaining and refining the timing synchronization, are important to ensure a stable and reliable operation of the ATLAS trigger system. Bunch-crossing identification of signals (BCID) and energy measurement in L1Calo necessitate an accuracy of ± 5 ns in the timing synchronization. To accomplish the goal, a special fit method has been developed. It has been applied to beam splash data and to collision data. Over the course of 2010 and 2011, it has been applied nine times to collision data with mean timing corrections of $\lesssim \pm 1.5$ ns. The method has demonstrated the stability of the L1Calo timing synchronization to within ± 1 ns systematic uncertainty and a ± 2 ns statistical uncertainty, as required for reliable trigger operation.

References

[1] The ATLAS Collaboration, *The ATLAS Experiment at the CERN Large Hadron Collider*, JINST **3** (2008) 8003-8440.
[2] The ATLAS Collaboration, *ATLAS Liquid Argon Calorimeter Technical Design Report*, CERN-LHCC 96-041, https://cdsweb.cern.ch/record/331061/.
[3] The ATLAS Collaboration, *ATLAS Tile Calorimeter Technical Design Report*, CERN-LHCC 96-042, https://cdsweb.cern.ch/record/331062/.
[4] R. Achenbach et al., *The ATLAS Level-1 Calorimeter Trigger*, JINST **3** (2008) P03001.
[5] The ATLAS Collaboration, *Timing Calibration with 2009 LHC Beam Splashes*, ATLAS note, ATL-DAQ-PUB-2010-001, https://cdsweb.cern.ch/record/1262858/.
[6] The ATLAS Collaboration, *Precision Synchronization of the ATLAS Level-1 Calorimeter Trigger with Collision Data in 2010 and 2011*, ATLAS note, ATL-DAQ-PUB-2012-001, https://cdsweb.cern.ch/record/1454684/.

XVth International Conference on Calorimetry in High Energy Physics (CALOR2012) IOP Publishing
Journal of Physics: Conference Series **404** (2012) 012056 doi:10.1088/1742-6596/404/1/012056

Triggering on electrons and photons with the CMS experiment at the LHC

Alexandre Zabi

Laboratoire Leprince Ringuet CNRS-IN2P3, Ecole polytechnique, 91128 Palaiseau, France

E-mail: Alexandre.Zabi@cern.ch

Abstract. Throughout the year 2011, the Large Hadron Collider (LHC) has operated with an instantaneous luminosity that has risen continually to around 4×10^{33}cm^{-2}s^{-1}. With this prodigious high-energy proton collisions rate, efficient triggering on electrons and photons has become a major challenge for the LHC experiments. The Compact Muon Solenoid (CMS) experiment implements a sophisticated two-level online selection system that achieves a rejection factor of nearly 10^6. The first level (L1) is based on coarse information coming from the calorimeters and the muon detectors while the High-Level Trigger (HLT) combines fine-grain information from all sub-detectors. In this intense hadronic environment, the L1 electron/photon trigger provides a powerful tool to select interesting events. It is based upon information from the Electromagnetic Calorimeter (ECAL), a high-resolution detector comprising 75848 lead tungstate ($PbWO_4$) crystals in a "barrel" and two "endcaps". The performance as well as the optimization of the electron/photon trigger are presented.

1. Introduction

The CMS detector has been designed to study proton-proton and heavy ion collisions produced by the LHC, to search for new particles and physics processes [1]. From the millions of collisions produced per second only 300 events per second can be stored offline. The search for new physics crucially relies on the trigger system performance used to select them [2]. The CMS trigger system is organised in two consecutives steps [3] : the L1 trigger utilizes coarse energy deposits in the calorimeters and signals in the muon systems to reduce the rate from about 20 MHz to 100 kHz; this followed by the HLT, implementing selection algorithms (in commercial computers) based on finer granularity and higher resolution information from all sub-detectors in regions of interest identified at L1 (output rate about 300 Hz). The CMS ECAL provides a precise measurement of the energies and positions of incident electrons and photons for both triggering and offline analysis purposes. The energy measured by the hadronic calorimeter (HCAL) is used to help identify and isolate electromagnetic signals. A set of configuration parameters enables the performance of the electron/photon trigger to be optimized for the wide range of luminosities experienced at the LHC.

2. From ECAL to the Level-1 trigger

2.1. ECAL and the trigger primitive generation

The CMS ECAL, composed of a Barrel (EB) and two Endcaps (EE), comprises 75848 lead tungstate ($PbWO_4$) scintillating crystals equipped with avalanche photodiode (APD) or vacuum

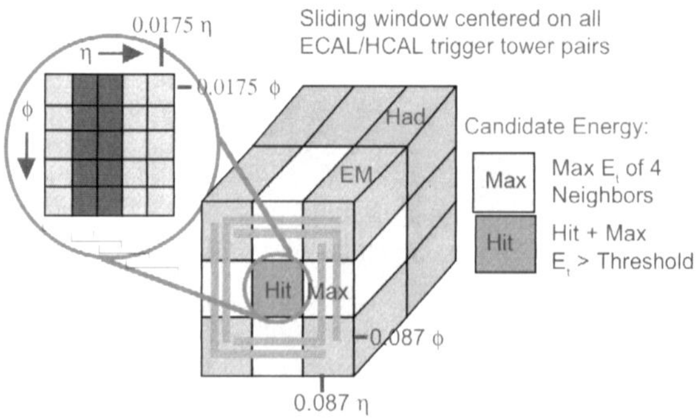

Figure 1. The L1 electron/photon trigger algorithm. The candidate E_T is the sum of the central TP (orange) and highest E_T TT from the 4 broadside neighbours (yellow). The Fine Grain profile (left box) and the ratio with HCAL TT energy H/E (green) are used as vetoes while the quiet corners (orange L-shapes) are used to separate isolated from non-isolated candidates. The HCAL TT are aligned with the ECAL TT in pseudorapidity.

phototriode (VPT) light detectors in the EB and EE respectively. A Preshower detector (ES), based on silicon sensors, is placed in front of the endcap crystals to aid particle identification. The ECAL is highly segmented, radiation tolerant and has a compact and hermetic structure, covering the pseudorapidity range to $|\eta| < 3.0$. Its target energy resolution is 0.5% for high-energy electrons/photons. In the EB, 5 strips of 5 crystals (along the azimuthal direction) are combined into trigger towers (TT) corresponding to a 5×5 array of crystals in η/ϕ. The arrangement in the EE is similar but more complicated due to the X-Y layout of the crystals. The transverse energy (E_T) measured by the crystals in a single TT is summed into a trigger primitive (TP) by the front-end electronics and sent to off-detector Trigger Concentrator Cards (TCC) via optical fibres.

2.2. Electron/Photon trigger path and algorithm

The TCCs [4] transmit groups of TPs to the Regional Calorimeter Trigger (RCT), which in turn combines pairs of TPs into L1 trigger candidates in each region of interest (4×4 TT). The Global Calorimeter Trigger (GCT) then sends the four most energetic candidates to the Global Trigger (GT), which generates the final L1 decision by applying E_T threshold cuts (named EG thresholds in the case of ECAL-based candidates). The electron/photon algorithm is based on a 3×3 trigger tower sliding window as shown in Figure 1. The E_T of an electron/photon candidate corresponds to the central TP of the sliding window summed with the largest deposit in one of its 4 adjacent towers. Electromagnetic showers are characterized by a compact lateral extension: candidates must have their central tower containing 2 adjacent strips with a significant fraction of the tower E_T (typically 90%). This criterion is characterized by the Fine Grain (FG) veto bit that is enabled only for candidates with $E_T > 4$ GeV. Moreover, the associated HCAL energy contribution is required to be below a threshold (typically H/E < 5% only for the central tower and for candidates with $E_T > 2$ GeV). Non-isolated electron/photon candidates must pass these criteria. In addition, isolated candidates must have a quiet neighbourhood characterized by at least five adjacent TT among the 8 nearest ones with their E_T below 3.5 GeV.

3. Online anomalous signals and their suppression

Anomalous signals were observed in the EB shortly after collisions began in the LHC: these were identified as being due to direct ionization within the APDs, thus producing fake isolated signals, with high apparent energy. These "spikes" can induce large trigger rates at both L1 and HLT if not removed from the trigger decision. On average, one spike with $E_T > 3$ GeV is observed per 370 minimum bias triggers in CMS at $\sqrt{s} = 7$ TeV. If untreated, 60% of the EG trigger candidates, above a threshold of 12 GeV, would be caused by spikes. At high luminosity these would be the dominant component of the 100 kHz CMS L1 trigger rate bandwidth [5].

3.1. Spike identification and removal

In the CMS ECAL the energy of an electromagnetic (EM) shower is distributed over several crystals, with up to 80% of the energy in a central crystal (where the electron/photon is incident) and most of the remaining energy in the four adjacent crystals. This lateral distribution can be used to discriminate spikes from EM signals. A "Swiss-cross" topological variable $s = 1 - E_4/E_1$ (E_1 : E_T of the central crystal; E_4 : summed E_T of the 4 adjacent crystals) has been implemented offline to serve this purpose. A similar topological variable has also been developed for the on-detector electronics: the "strip Fine Grain Veto Bit" (sFGVB). Every TP has an associated sFGVB that is set to 1 (signifying a true EM energy deposit) if any of its 5 constituent strips has at least two crystals with E_T above a programmable "sFGVB threshold", of the order of a few hundred MeV. If the sFGVB is set to zero, and the trigger tower E_T is greater than a "killing threshold", the energy deposition is considered spike-like. The trigger tower energy is set to zero and the tower will not contribute to the triggering of CMS for the corresponding event.

3.2. Optimisation of the online spike removal algorithm

As the sFGVB threshold is a single value, the electron/photon efficiency depends upon the particle energy: the higher the threshold, the more low-energy real EM deposits would be initially flagged as spikes. However, these fake spikes may not pass the killing threshold so they would still be accepted. With a very low sFGVB threshold, spikes could be accepted due to neighbouring crystals having noise. A detailed emulation of the full L1 chain has been developed in order to optimize the two thresholds to remove as large a fraction of the anomalous signals as possible whilst maintaining excellent efficiency for real electron/photon signals.

In order to determine the removal efficiency, data were taken in 2010 without the killing thresholds active. Spike signals identified offline (with the "Swiss cross") were then matched to L1 candidates in the corresponding RCT region and the emulator used to evaluate the fraction of L1 candidates that would have been eliminated. In a similar fashion the efficiency for triggering on real electrons/photons could be estimated.

Three killing thresholds have been emulated ($E_T = 8$, 12, and 18 GeV), combined with six sFGVB thresholds (152, 258, 289, 350, 456, 608 MeV). Figure 2 shows the electron efficiency (fraction of electrons triggered after spike removal) versus the L1 spike rejection fraction, for all sFGVB thresholds mentioned above (one point for each threshold value) and a killing threshold of 8 GeV. The optimum configuration was chosen to be an sFGVB threshold of 258 MeV and a killing threshold of 8 GeV. This corresponds to a rejection of 96% of the spikes, whilst maintaining a trigger efficiency for electrons above 98%. With these thresholds the efficiency for higher energy electrons is even larger: 99.6% for electrons with $E_T > 20$ GeV.

Table 1 summarises the rate reduction factors obtained for L1 EG algorithms considering the working point discussed above. This optimized configuration was tested online at the beginning of 2011. It gave a rate reduction factor of about 3 (for an EG threshold of 12 GeV), and up to a factor of 10 for E_T sum triggers (which calculate the total EM energy in the whole calorimeter system).

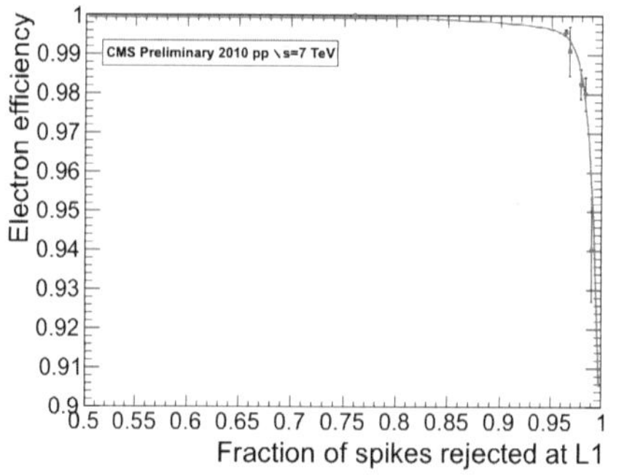

Figure 2. Electron efficiency as a function of the spike rejection at L1. Each point corresponds to a different spike removal "sFGVB" threshold. The "killing threshold" is set to 8 GeV).

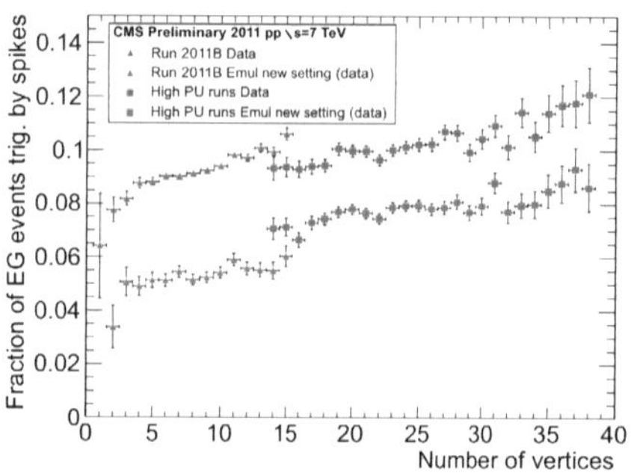

Figure 3. Fraction of spike-induced EG triggers as a function of the number of reconstructed vertices for the 2011B data (triangles) and High pileup runs (squares). The red points represent the spike removal working point used in 2011 and the green points the optimized working point for 2012.

Table 1. Rate reduction factors obtained for L1 EG algorithms considering a 258 MeV sFGVB threshold and an 8 GeV killing threshold on the Ecal Trigger Primitives.

EG Threshold (GeV)	12	15	20	30
Rate reduction factors	3.4	4.3	6.0	9.6

4. Level-1 electron/photon trigger efficiency

The trigger efficiency has been measured with electrons from Z→ee events, using a tag and probe method. The dataset represents the full 4.98 fb^{-1} collected in 2011. Both the tag and the probe are required to pass tight identification cuts in order to reduce significantly the background contamination. The tag electron must also trigger the event at L1 while the probe electron is used for the efficiency studies. The invariant mass of the tag and probe system should be consistent with the Z boson mass ($60\ GeV/c^2 < M_{ee} < 120\ GeV/c^2$). This approach results in a pure unbiased electron sample. The trigger efficiency is given by the fraction of probes above a given EG threshold, as a function of the probe E_T. In order to trigger, the location of the highest energy TT within the electron supercluster must match a corresponding region of an L1 candidate in the RCT.

The trigger efficiency curves are shown in Figure 4 for an EG threshold of 15 GeV. The E_T on the x-axis is obtained from the fully reconstructed offline energy. In the EE this includes the preshower energy that is not available at L1. As a consequence the trigger efficiency turn-on point for the EE is shifted to the right with respect to the EB. For both EB and EE, corrections for crystal transparency changes were not included at L1 in 2011, which further affects the turn-on curve (section 5.2). The width of the turn-on curves is partly determined by the coarse trigger granularity, since only pairs of TTs are available for the formation of L1 candidates, which leads to lower energy resolution at L1. An unbinned likelihood fit has been used to derive the efficiency curves. Parameters of the turn-on curves are given in Table 2.

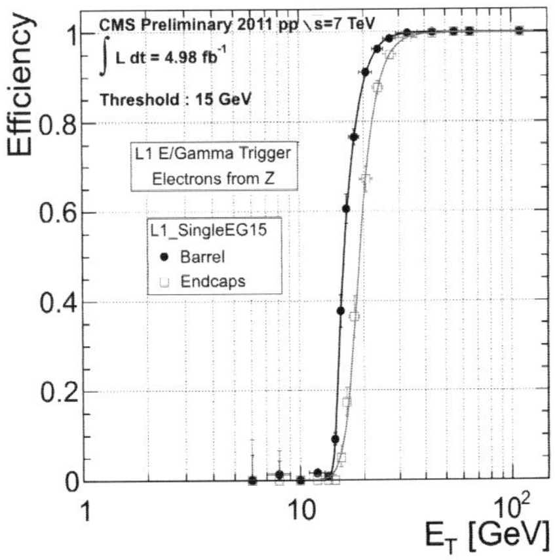

Figure 4. Electron trigger efficiency at L1 ("EG" threshold : 15 GeV E_T), as a function of E_T for electrons in the EB (black dots) and EE (red dots).

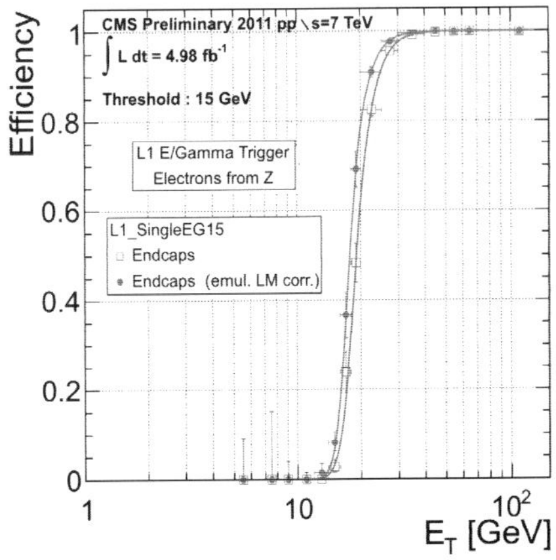

Figure 5. EE L1 electron triggering efficiency before (red) and after (green) laser transparency corrections are applied at the ECAL TP level.

Table 2. L1 electron trigger turn-on curve parameters. This table gives the electron E_T thresholds for which an efficiency of 50 %, 95 % and 99 % are reached for EB and EE separately. The last entry corresponds to the efficiency obtained at the plateau of each curve shown on Figure 4.

EG15	EB	EE
50%	$16.06^{+0.01}_{-0.01}$ GeV	$19.05^{+0.05}_{-0.06}$ GeV
95%	$22.46^{+0.04}_{-0.05}$ GeV	$27.06^{+0.58}_{-0.43}$ GeV
99%	$28.04^{+0.07}_{-0.10}$ GeV	$34.57^{+1.48}_{-1.10}$ GeV
100 GeV	$99.95^{+0.01}_{-0.88}$ %	$99.84^{+0.10}_{-0.28}$ %

Table 3. EE L1 electron trigger turn-on curve parameters. This table gives the electron E_T thresholds for which an efficiency of 50 %, 95 % and 99 % are reached before and after laser corrections applied. The last entry corresponds to the efficiency obtained at the plateau of each curve shown on Figure 5.

EG15	EE	EE (corr)
50%	$19.11^{+0.03}_{-0.06}$ GeV	$17.79^{+0.03}_{-0.06}$ GeV
95%	$27.05^{+0.01}_{-0.01}$ GeV	$24.46^{+0.10}_{-0.23}$ GeV
99%	$34.36^{+0.01}_{-0.01}$ GeV	$30.78^{+0.21}_{-0.48}$ GeV
100 GeV	$99.84^{+0.06}_{-0.60}$ %	$99.89^{+0.01}_{-0.67}$ %

In the EE the material budget in front of the detector causes more bremsstrahlung which, together with the more complex TT geometry, causes the turn-on curve to be wider than that for the EB. Some masked or faulty regions (0.2% in EB and 1.3% in EE) result in the plateaus being slightly lower than 100% (99.95 % in EB and 99.84 % in EE) – Table 2. The effect on efficiency of the L1 spike removal has been verified to be negligible, but this will require further optimization as the number of collisions per bunch crossing increases in the future (section 5.1).

5. Further optimization of the electron and photon trigger

The full dataset recorded in 2011 has been used to study the effect of pileup on the spike removal procedure, the impact of crystal transparency changes on the trigger efficiency as well as the implementation of a new L1 EG calibration.

5.1. Spike removal optimisation for higher pileup

At the end of 2011 the pileup had reached a maximum of about 40, with an average around 25, exceeding the original design criteria of CMS. This is expected to increase throughout 2012. Efficient identification of EM showers at trigger level becomes more and more challenging. As pileup events act as noise in the calorimeter, they can degrade trigger object resolution and reduce the probability to observe isolated spikes. The fraction of spike-induced EG triggers has been measured as a function of the number of vertices (roughly equivalent to the number of pileup events) in Figure 3. The fraction of spike-induced EG triggers reaches 10% for collisions including more than 20 pileup events (red points). Using the L1 trigger emulator, a more efficient working point (sFGVB threshold=350 MeV, killing threshold = 12 GeV) for the spike removal algorithm reduces this fraction to 6% (green points), but still preserves the same high efficiency for real electrons and photons.

5.2. Mitigation of crystal transparency changes at the trigger level

Under irradiation the ECAL crystals lose some of their transparency, part of which is recovered when the radiation stops (between LHC fills, for example). The effect of this is that the response of the ECAL varies with time. This is mitigated by the use of a sophisticated laser system that frequently monitors the transparency of each crystal [6] and allows offline corrections to the measured energies to be made [7]. In 2011 the levels of radiation in ECAL were quite small, especially in the EB, but improvements in L1 efficiency could be made if corrections to the TT energies had been included, to reflect the changes in response. Figure 5 shows the improved turn-on curve for the EE if corrections had been included at L1 in 2011, while Table 3 summarizes the parameters of the turn-on curves and compares them with the actual EE turn-on curve in 2011 (Figure 4). From 2012 the corrections to the TT energies are calculated on a weekly basis and applied at L1 in order to keep the high trigger efficiency and low triggering threshold energies.

5.3. Implementing a Level-1 E_T calibration

At the RCT level, input TPs are corrected by a calibration factor before being summed to form L1 candidates. The calibration factors are η dependent they are intended to correct for material effects upstream of ECAL (Section 4). Figure 6 shows the calibration factors used during the 2011 data taking period. They were obtained by compiling the ratio of the offline reconstructed electron E_T with its L1 E_T counterpart as a function of η using simulated Z→ee events. The same matching procedure used in Section 4 was implemented here. The introduction of these corrections had the effect of sharpening the efficiency curves and moving the turn-on points closer to the actual EG algorithm thresholds. Using the full 2011 dataset, new RCT calibration factors were derived using the same procedure. Figure 6 shows the new RCT calibration factors obtained after correcting for the transparency change over the duration of the 2011 physics data taking period.

6. Conclusion

The results presented here display excellent overall performance of the electron/photon trigger and demonstrate the flexibility of this system. A reprogramming of the front-end electronics and ECAL TCC has allowed the implementation and optimization of a spike removal algorithm at L1, which rejects a majority of spikes (>96%) whilst having a negligible impact on electron/photon triggering efficiency. Further optimizations were achieved on the spike removal procedure in high pileup conditions, the calibration of the EG candidates as well as the mitigation of the ECAL crystal response change with time. These improvements have been implemented in 2012 in order to guarantee the best possible physics performance.

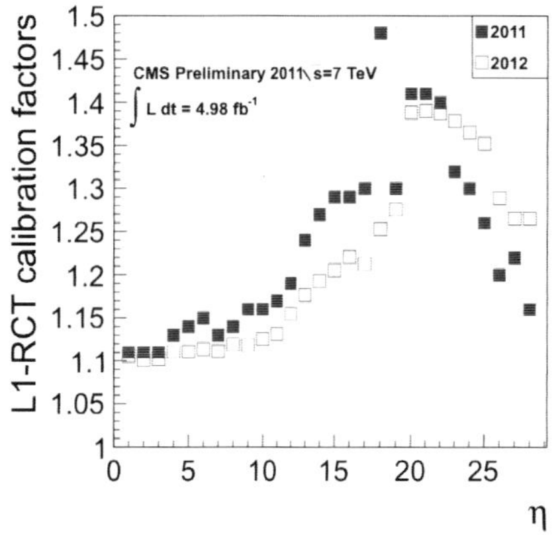

Figure 6. 2011 L1 EG calibration factors as a function of η for Z→ee simulated events (blue points). 2012 EG calibration factors (red points) derived from Z→ee selected events using the full 2011 dataset. ECAL crystal transparency changes have been corrected at the TP level prior to compute this ratio (Section 5.2).

[1] The CMS collaboration (S. Chatrchyan et al.) 2008 , "The CMS experiment at the CERN LHC", JINST 3 S08004, doi:10.1088/1748-0221/3/08/S08004

[2] The CMS collaboration (S. Chatrchyan et al.) 2000, "CMS TRIDAS project technical design report, volume 1, the trigger systems", CMS TDR CERN/LHCC 2000-38, CMS-TDR-006-1, http://cdsweb.cern.ch/record/706847

[3] CMS Collaboration (S. Chatrchyan et al.) 2010,"Performance of the CMS Level-1 Trigger during Commissioning with Cosmic Ray Muons", CMS-CFT-09-013, FERMILAB-PUB-10-145-CMS, Nov 2009. (Published Mar 19, 2010). 49pp. Published in JINST 5:T03002,2010.

[4] P. Paganini for the CMS Collaboration 2008, "CMS Electromagnetic Trigger commissioning and first operation experiences", XIII International Conference on Calorimetry in High Energy Physics, Journal of Physics: Conference Series 160 (2009) 012062.

[5] D. A. Petyt, for the CMS Collaboration 2011,"Anomalous APD signals in the CMS Electromagnetic Calorimeter", Nuclear Instruments and Methods in Physics Research Section A: Accelerators, Spectrometers, Detectors and Associated Equipment, http://dx.doi.org/10.1016/j.nima.2011.10.025.

[6] Anfreville M et al., "Laser monitoring system for the CMS lead tungstate crystal calorimeter", Nucl. Instr. and Meth. A 594 (2008) 292-320.

[7] Tabarelli de Fatis T for the CMS collaboration, "Role of the CMS electromagnetic calorimeter in the hunt for the Higgs boson in the two-gamma channel", CMS CR-2012/251 (2012) and these Proceedings

XVth International Conference on Calorimetry in High Energy Physics (CALOR2012) IOP Publishing
Journal of Physics: Conference Series **404** (2012) 012057 doi:10.1088/1742-6596/404/1/012057

Scintillating Glasses for Total Absorption Dual Readout Calorimetry

V. Bonvicini[1], A. Driutti[2], D. Cauz[2], G. Pauletta[2], P. Rubinov[3], L. Santi[2], H. Wenzel[3]

[1]I.N.F.N. Trieste, TS, Italy.
[2]Universita di Udine and I.N.F.N. Udine, UD, Italy,
[3]Fermilab, Batavia, IL, U.S.A.
E-mail: giovanni. pauletta@uniud.it

Abstract. Scintillating glasses are a potentially cheaper alternative to crystal - based calorimetry with common problems related to light collection, detection and processing. As such, their use and development are part of more extensive R&D aimed at investigating the potential of total absorption, combined with the readout (DR) technique, for hadron calorimetry. A recent series of measurements, using cosmic and particle beams from the Fermilab test beam facility and scintillating glass with the characteristics required for application of the DR technique, serve to illustrate the problems addressed and the progress achieved by this R&D. Alternative solutions for light collection (conventional and silicon photomultipliers) and signal processing are compared, the separate contributions of scintillation and Cherenkov processes to the signal are evaluated and results are compared to simulation.

1. Introduction

Energy resolution in Hadron Calorimetry is limited by:
(a) event-to-event fluctuations in the relative fractions of e.m. (f_{em}) and hadronic (1-f_{em}) of the energy deposited by the hadronic shower ($< E_{had} >= e < f_{em} > +h(1-< f_{em} >)$, coupled with differences in the calorimeter responses (*e* and *h*, respectively) to these components;
(b) fluctuations in the component of *invisible energy* E_i (which corresponds to nuclear binding energy and invisible reaction products). Corrections for $<E_i>$ are generally incorporated in *h* with its consequent fluctuation;
(c) fluctuations in leakage corrections which, because the interaction length $l_{int} >> X_0$, are difficult to eliminate completely.

Setting e/h = 1 (*compensation*) by alternating the active material with appropriate passive material will help, while improving containment, but at the expense of introducing *sampling fluctuations* . Measurement of deposited energy by independent means, e.g. "Dual readout" (DR) of Cerenkov (C) and scintillation (S) signals, is an alternative means of obtaining compensation but it has, as yet, only been tested in sampling calorimeters. Assuming, however, that DR can be applied to total absorption calorimeters one could obtain compensation while avoiding the sampling fluctuations. A high degree of segmentation would facilitate the correction of leakage fluctuations [1] and might also allow for the application of PFA methods (Magill, this conference [2])

This is all, in principle, possible using scintillating crystals and those suitable for Total Absorption Hadron Calorimetry (TAHCAL) have been reviewed by Ren-Yuan Zhu at this conference [3].

Scintillating glasses might afford a cheaper alternative but heavy glasses with the required characteristics have yet to be developed. A lighter scintillating glass [4] which appears to have the required properties (except for l_{int}) is presently available at FNAL as a legacy of E705 (characterics are summarized in fig.1) . Although the glass is too light for implementation of TAHCAL in compact collider detectors it can be used to investigate problems related to light collection and separation of the C and S signals in homogenous calorimeters and, assuming the glass characteristics are confirmed to be suitable , one might use the available glass to configure a module for a proof of principle test of TAHCAL. The work reported in this paper is aimed at verifying the properties of the glass (shown in

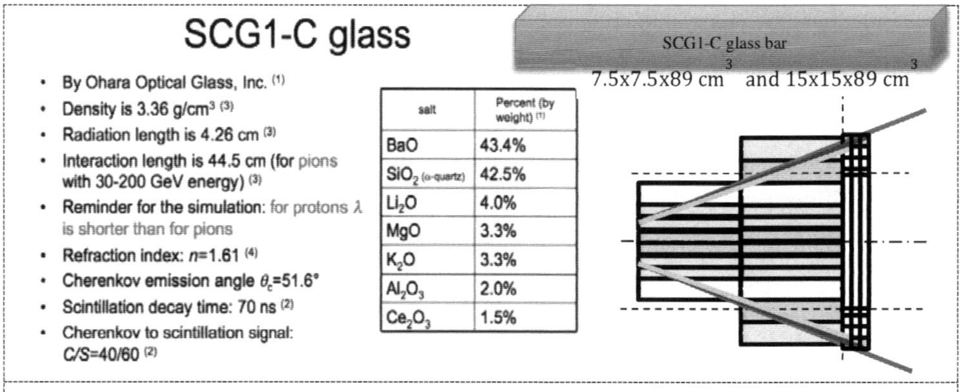

Figure 1. SGC1-C glass characteristics , with **an** illustration of glass bars (top right) and of how such blocks might be used to construct a test calorimeter module (lower right).

fig. 1), determining unknown properties such as the relative intensities of the G and S signals for both minimally ionizing particles (mips) and hadron showers, determining under which conditions they may be separated and identifining the best combination of photodetector and readout electronics for the purpose.

2. Tests using Cosmics

Preliminary tests were performed at FNAL using cosmics. A schematic representation of the setup is shown in fig. 2. The average of a typical digititized signal (amplitude vs time) triggered by cosmics is shown in the insert on the right. The distinction between the fast Cerenkov (C) contribution and the slower (decay time ~ 70 ns) Scintillation (S) is evident and evaluation of their respective contributions is clearly possible. Absolute light yields were measured by calibrating the PMT gains and their line shapes were measured using a picosecond laser and by observing signals due to single photoelectrons.

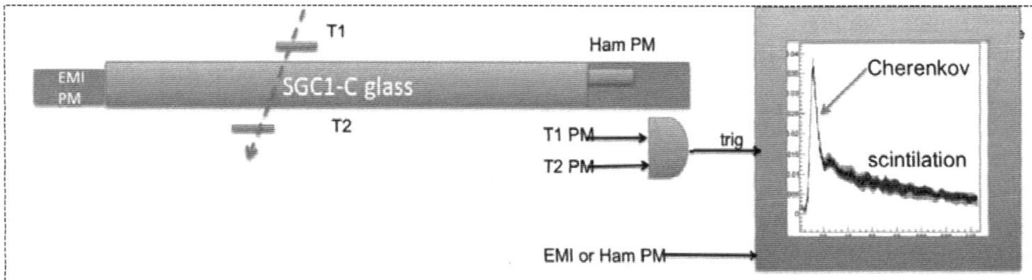

Figure 2. Schematic illustruation of lay-out for acquisition of cosmic data. The path of cosmics traversing the SGC1-C glass is defined by counters T1 and T2. Their coincidence triggers a GHz scope which digitizes signals from EMI and Hamamatsu PMs viewing opposite ends of the glass.

Both Cherenkov and scintillation signals were simulated using GEANT 4 and their combinations were fitted to the data to extract the ratio C/S. Given the number of Cherenkov photons generated by the simulation, the number of scintillation photons per MeV was found to be about 600.

3. Tests using using particle beams

Particle beams from the Fermilab Meson Test Beam Facility (MTBF) were used to test the response of these glasses to hadrons and to investigate readout with silicon-based photodetectors (because of their much smaller sensitive areas, cosmic rates in these detectors were much too small to be useful) The experimental setup at MTBF is shown in fig 3. Several silicon-based photodetectors [6] viewed

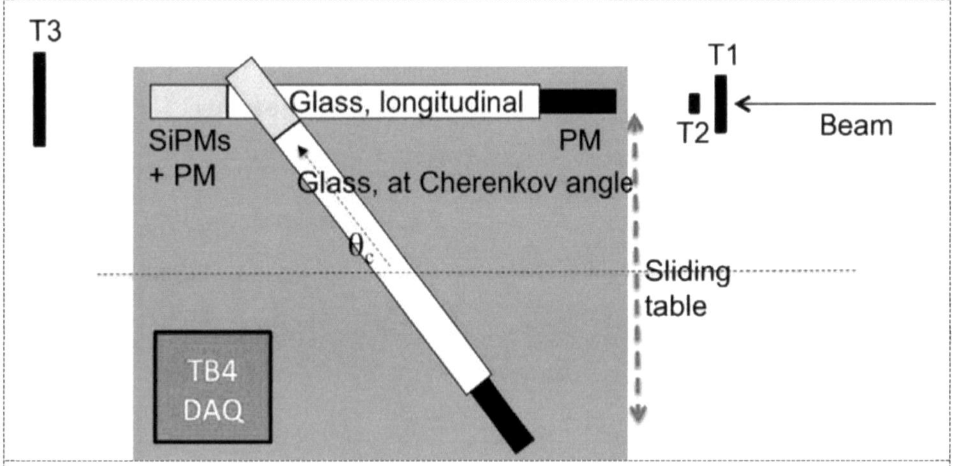

Figure 3 is a schematic illustration of the experimental layout at MTBF. Two glass orientations are illustrated: parallel to the beam direction ("longitudinal") for maximal harmonic shower production and at the Cherenkov angle θ_c w.r.t. the beam direction.

the signals from the same end of the glass as the Hamamatsu PM (see fig. 3). Their signals were read out and digitized by TB4 [7] DAQ system developed in-house at FNAL. PM signals were read out and digitized commercial system, [8] with higher (~500 MHz) bandwith and much higher (up to 5 GHz) sampling rate.

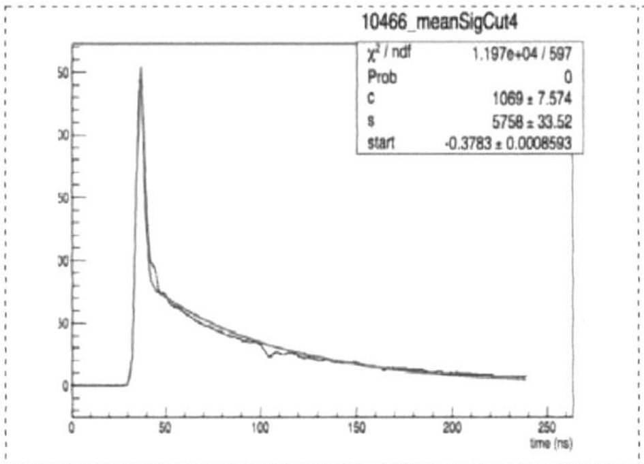

Figure4. Mean signal and simulation with PM"downstream" and glass at Θc w.r.t beam

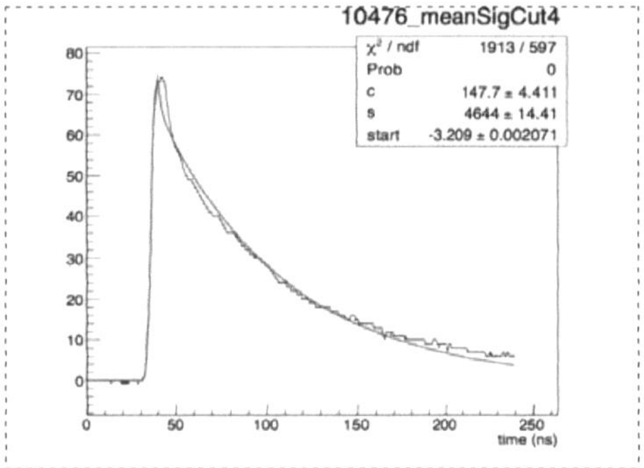

Figure 5. Mean signal and simulation with PM "upstream" and glass at Θc w.r.t beam

Measurements were made with both proton and muon beams. With the glass at the Cerenkov angle θ_c (see fig. 3) relative to the beam direction, the responses of all detectors were first measured for different points of incidence of a 120 Gev/c beam. The glass was then rotated by 180 degrees and the process repeated. Mean signals from the EMI PM before and after glass rotation (i.e., with EMI downstream and upstream of beam impact point) are shown in figs. 4 and 5 together with their simulations. Although the simulations still need perfecting, it is clear that, despite a 20% contribution to the signal from hadronic showers (when the glass is oriented at the Cherenkov angle w.r.t. the beam), one will be able to separate the Cherenkov (C) and scintillation (S) signals reliably on the basis of their different time dependence. Scintillatore light output is much smaller (~7%) than that of BGO , which facilitates the separation of C and S components while remaining sufficient for good energy resolution.

With the glass oriented in the beam direction , i.e. "longitudinally" (L), the showering probability increases to ~ 92.7% and, although much of the shower energy escapes the glass, the residual energy still dominates as was confrimed by the distributions of signal integrals (fig.9). Average signals together with fits (in red) obtained using the simulated C and S contributions are shown in figs. 6 & 7. It is evident that a time-based analysis allows for determination of the C/S ratio even in the presence of a predominant hadronic shower contribution to the total energy deposited.

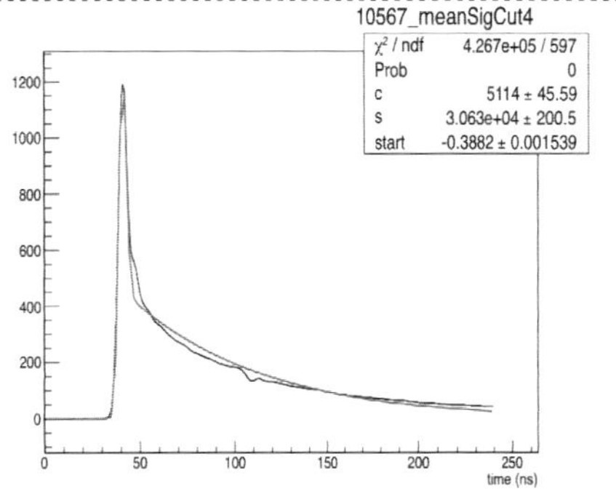

Figure 6. Mean signal and simulation with PM"downstream" and glass // beam.

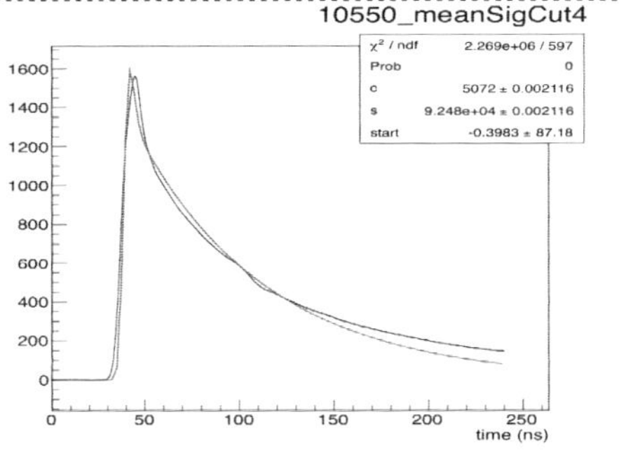

Figure 7. Mean signal and simulation with PM "upstream", and glass // beam.

The signals from the silicon – based photodetectors show similar characteristics as shown below. However, the bandwidth (<100 MHz) and sampling rate (212 MS/sec) of theTB4 DAQ used to acquire this data were not sufficient for as clean a time-based analysis as was possible for the PM signals with the PM signals to extract C/S. However, there is no reason to expect it will not be possible using the same DAQ as for PMs and this will be verified in future measurements.

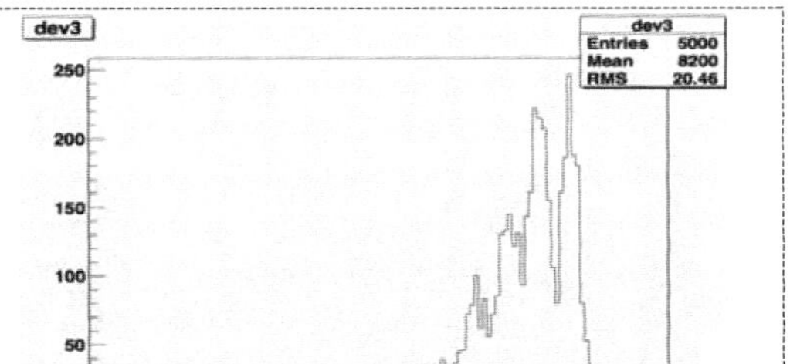

Figure 8 Distribution of signal integrals for mips when the glass is oriented at θ_c w.r.t. the beam direction and the mppc viws the downstream end.

On the other hand, the MPPCs were able to resolve the response for different numbers of photoelectrons in the mip peaks, as shown by fig. 8, and this allowed for direct calibration of the light collected by them. After subtraction of the dark count contribution, the light collected per mip by each 3x3 mm^2 of mppc pad, when the glass was oriented at θ, w.r.t. the beam direction, was found to vary between 0.5 and 1.0 photoelectrons (pe) depending on whether the beam was incident closer to the downstream or upstream end of the glass. With the glass parallel to the beam direction and the mppc viewing the downstream end of the glass, the predominance of shower over mip signals is evident in fig.9. A blow up of the region around the mip peak is shown in fig. 10. After subtraction of the dark count component, the the ammount of light collected by one 3x3 mm^2 mppc pad corresponds to 2.3 pe/mip (1.8 pe/mip with 32 GeV muons).

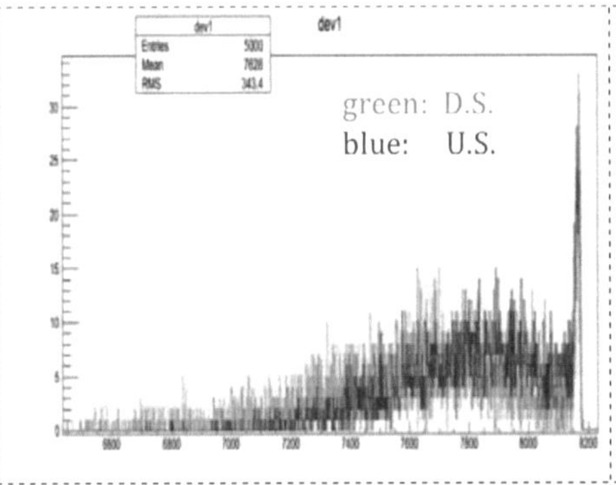

Figure 9. Distribution of signal integrals with glass // beam and mppc upstream (blue) and downstream (green) of the beam direction.

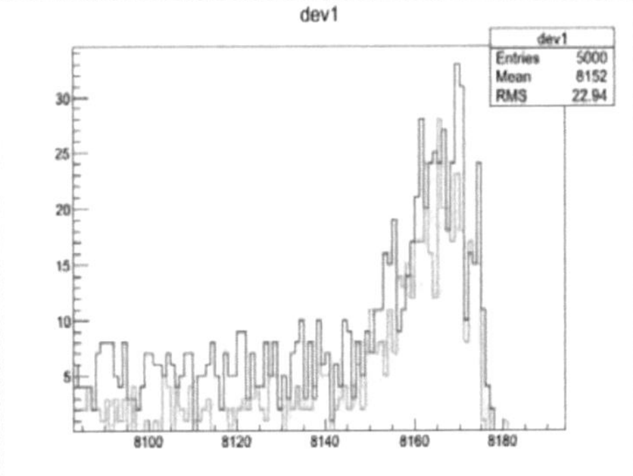

Figure 10. Blow-up of the region around the mip peak in fig. 9.

5. Conclusions

1) From a preliminary analysis of data taken with cosmics and particle beams , it appears that the timing characteristics, the C/S ratio (~ 10 – 30%) and the light output (~ 7 % BGO) of SCG1-C glass are appropriate for Dual readout Total Absorption Calorimetry.

2) The sampling rate of 2.5 GHz used to digitize the PM signals is more than adequate for a time – based off – line separation of C and S components, even when the hadronic shower predominates. Given the circumstances, simple cuts on the signal time distribution would seem sufficient for an on-line DR analysis.

3) A 2 x 2 array of 3 x 3 mm^2 silicon, faced up against the glass surface, collects enough light to measure the energy deposition of single mips (for energy calibration) and, with sufficient bandwidth and sampling rate, one will be able to perform the same time – based analysis as for PMs to detemine C/S.

4) When the thermal noise (dark count) is sufficiently low (below ~ 1 MHz as was the case for the MPPCs), one is able to separate single photoelectrons at the mip level so that the device is auto-calibrating (in photoelectrons). This is an important property for cross-calibration of calorimeter elements and monitoring stability and saturation.

5) Analysis of data taken with other conditions and SiPMs is still in progress but it already appears clear that, though they are not heavy enough for use in a compact collider calorimeter, glasses with the SCG1-C characteristics may be used for a test of the principle of Total Absorption Dual Readout Hadron calorimetry, assuming leakage corrections can be reduced to a tolerable level [1]. It also appears that ~ 1 cm^2 of silicon – based photodetector, faced up against the glass, collects enough light and that it may be substituted for conventional PM.

References

[1] Driutti A, Pauletta G., et al. (2011). Towards jet reconstruction in a realistic dual readout total absorption calorimeter. In: J.Phys.Conf.Ser. 293 (2011) 012034

[2] Magill, Steve, " Simulation Studies of a Dual Readout Crystal Calorimeter" , CALOR2012, Santa Fe, June 2012.

[3] Zhu, Ren-Yuan: " Crystals for the HHCAL Detector Concept" , CALOR2012, Santa Fe, June 2012

[4] (a) B. Cox, et al., IEEE Transactions on Nuclear Science, Vol. NS-30, No 1, 1983. (b) G.E. Theodosiou, et al., IEEE Transactions on Nuclear Science, Vol. NS-31, No 1, 1984. (c) L. Antoniazzi, et al., The Experiment 705 Electromagnetic Shower Calorimeter, FERMILAB-Pub-93/001 E705, 1993.

[5] See MOU for experiment T1015 ("Dual Readout Calorimetry with Glasses") at the Meson Test Beam Facility (MTBF): http://www-ppd.fnal.gov/FTBF/MOUpage.html.

[6] SiPMs from AdVanSid (model ASD-SiPM3S-P and other models developed in collaboration with them) and and a 2x2 MPPC array from Hamamatsu (model S1095-025C-050C-100C) wer used.

[7] Paul Rubinov, Henry Fisk, "Custom electronics for SiPM devices: making life in the test beam easier", CALOR2010, IHEP, Beijing, 2010.

[8] CAEN N6742 , 16 ch+1 digitizer, based on DRS4 chip: switched capacitor ADC up to 5 GS/S.

XVth International Conference on Calorimetry in High Energy Physics (CALOR2012) IOP Publishing
Journal of Physics: Conference Series **404** (2012) 012058 doi:10.1088/1742-6596/404/1/012058

Performance and Improvements of the ATLAS Jet Trigger System

P Conde Muíño* **on behalf of the ATLAS Collaboration**

*LIP, Lisbon, Portugal

E-mail: Patricia.Conde.Muino@cern.ch

Abstract. In 2011-2012 the Large Hadron Collider (LHC) provided proton bunch crossings every 50 ns with up to 40 interactions every crossing. These unprecedented conditions posed stringent demands on the trigger system of the ATLAS experiment, which must provide a fast rejection of background events while maintaining a high efficiency for physics signals of interest.

This note focuses on the jet trigger system in ATLAS. Jets are the most prevalent high-p_T objects produced at the LHC, and are an important component of a wide range of physics analyses. The challenges faced by the jet trigger system, and how its initial limitations were overcome, will be described. Performance results, including jet efficiency and resolution, will be presented.

1. Introduction

ATLAS [1] is a multipurpose detector designed for the LHC proton-proton collider. At the nominal LHC luminosity of 10^{34} cm^{-2}s^{-1}, the ATLAS bunch crossing rate will be 40 MHz, with about 23 interactions per bunch crossing (pile-up) and a 25 ns separation between successive proton bunches. These conditions impose stringent requirements on the trigger system, which must retain a high efficiency for physics signatures of interest while reducing the rate to about 400 Hz, limited by bandwidth and storage capacity. The actual running conditions of the LHC in 2011-12 were even more demanding than originally anticipated, with the peak number of interactions per bunch crossing reaching up to 60.

In order to have an unbiased and efficient event selection, the trigger identifies generic objects, like leptons, photons or jets with large transverse momentum (p_T) or large missing transverse energy.

The ATLAS jet trigger [2] has been designed to select final states corresponding to a wide spectrum of physics processes, ranging from Standard Model signatures, such as high-p_T jets, top quark production, and vector boson production, to searches for the Higgs boson and physics beyond the Standard Model, such as Supersymmetry (SUSY) or Extra Dimensions. The cross sections for these processes span many orders of magnitude, from $\mathcal{O}(100$ nb) for jets to $\mathcal{O}$(pb) for some SUSY models.

In order to fulfill the trigger requirements at the LHC, the ATLAS trigger system is divided into three levels. The first level trigger (L1) is based on custom hardware, and must form a decision with a 2.5 μs latency, reducing the input rate from 40 MHz to about 75 kHz. In order to achieve a fast processing time while identifying high p_T objects, the L1 makes use of a coarse granularity in the ATLAS calorimeter and muon systems.

Table 1. LHC nominal and running conditions in 2011/12.

	Nominal	2011	2012
Energy ($\sqrt{s}$)	14 TeV	7 TeV	8 TeV
Peak luminosity ($cm^{-2}s^{-1}$)	10^{34}	3×10^{33}	6×10^{33}
Interactions per bunch crossing	23	11	up to 60
Bunch crossing rate	40 MHz	20 MHz	20 MHz
Offline storage rate	200 Hz	400 Hz	400 Hz
Bunch spacing	25 ns	50 ns	50 ns

The second level trigger (L2) is software based. It can use the full granularity of all detectors and it is based on the concept of Regions of Interest (RoIs). This means that it uses the geometrical position of the L1 triggered objects as seeds, performing reconstruction in a limited region around the L1 seed and thereby avoiding the need to read out the full detector. This scheme reduces the amount of detector data accessed to about 2%. The L2 uses simplified algorithms to take a decision in about 40 ms processing time, reducing the rate to 3 kHz.

The third level trigger, called the Event Filter (EF), uses offline-like algorithms that may operate in RoI-based reconstruction or with the full event readout (referred to as full scan), using the offline calibrations available at the time of data taking.

2. ATLAS Jet Trigger description

The L1 jet reconstruction algorithm consists of a sliding window of programmable size applied to calorimeter towers of 0.2×0.2 in pseudorapidity (η) and azimuthal angle (ϕ). Jets are identified when a local energy maximum above a given adjustable threshold is found. Although the same algorithm is used in the forward region ($3.2 < |\eta| < 4.9$), the L1 trigger tower granularity is larger, 0.4×0.4 in $\eta \times \phi$, and for jets, trigger towers are summed in η, thereby reducing the resolution.

In the original design, the L2 jet trigger algorithm requested the calorimeter data corresponding to a configurable RoI, centered around the L1 seed, be transferred from the detector readout drivers (RODs). It then converted this byte stream raw data into more refined objects such as calorimeter cells, for processing by jet algorithms. The early design executed a simplified cone-like jet algorithm with radius R=0.4 that determined the energy-weighted center of the RoI within 3 iterations. The outcome was a jet defined by its (η,ϕ) position and total energy. At this point, the calorimeter energy scale was set to be the electromagnetic (EM) energy scale determined at testbeams. Due to the non-compensating nature of the ATLAS calorimeters, the EM scale underestimates the hadronic component of jets, but for the results presented in this paper no additional hadronic energy corrections are applied.

The Event Filter can run any of the offline jet algorithm configurations, either RoI-based or with the full event reconstruction, as implemented for the first time in the trigger in 2011. By default the EF runs in full scan mode. In the first step it reconstructs three dimensional calorimeter clusters that are used as input to the jet reconstruction, performed using an anti-k_t algorithm [3], with radius parameter R=0.4 or R=1.0.

The RoI-based approach described above suffered from reduced performance in the case of multi-jet events, due to the lower efficiency of the L1 sliding window algorithm to identify close-by jets. To overcome this a full calorimeter reconstruction was introduced in 2011 at the EF level, and in 2012 at L2, as indicated schematically in figure 1. The readout time of all calorimeter cells is considerably high [5] and the time spent running any jet algorithm using cells as constituents

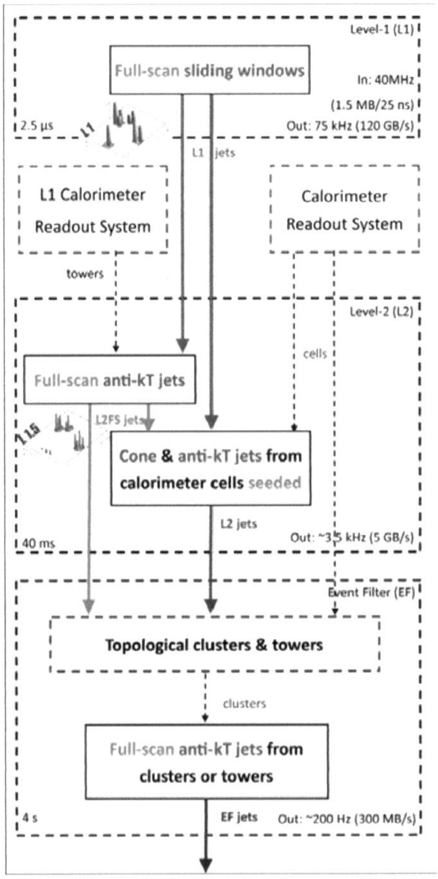

Figure 1. Architecture of the ATLAS Jet Trigger System.

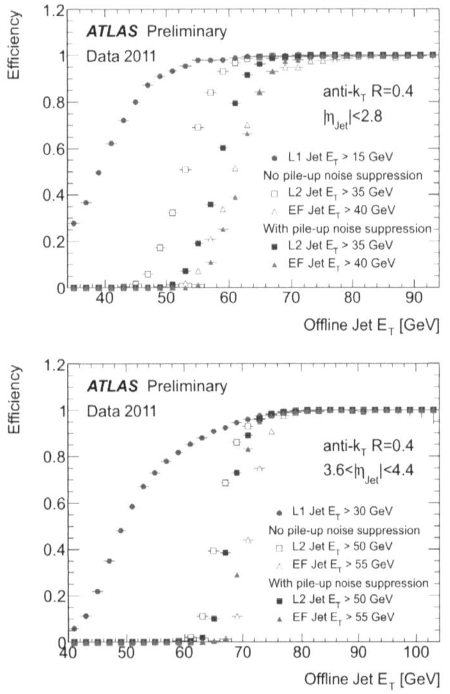

Figure 2. The efficiency of the jet trigger for a single L1→L2→EF trigger chain as a function of the offline calibrated jet transverse energy (E_T) for central (upper plot) and forward (lower plot) jets in two different data-taking scenarios: before (empty markers) and after (full markers) pile-up noise suppression was applied to both L2 and EF jets [4].

is too large for the available L2 processing time. Therefore, in the new implementation the L1 trigger towers are read out directly from the L1 calorimeter readout system, enabling access to the entire event at the L2 and enhancing the L2 input rate. This scheme has the advantage that it is possible to access either the trigger tower granularity available to the electron/photon triggers (0.1×0.1 in the central part of the detector) or to the L1 jets trigger (0.2×0.2). The L2 full event reconstruction using trigger towers is also called *L1.5*.

In addition, the FastJet [6] algorithm was introduced at L2 in 2012. FastJet is a package that can be configured easily to run different jet algorithms on a variety of inputs, allowing for the deployment of more sophisticated algorithms at L2. By default it uses the same configuration as the EF, anti-k$_t$ with R=0.4 or R=1.0, since this is the default in the ATLAS offline jet reconstruction.

The introduction of the full scan at L2 opens up a range of possibilities for configuring the ATLAS jet trigger system, since it can be used to seed L2 RoI-based algorithms or the EF directly. The result is a highly-configurable system that can be adapted to different LHC beam conditions and requirements for physics analysis.

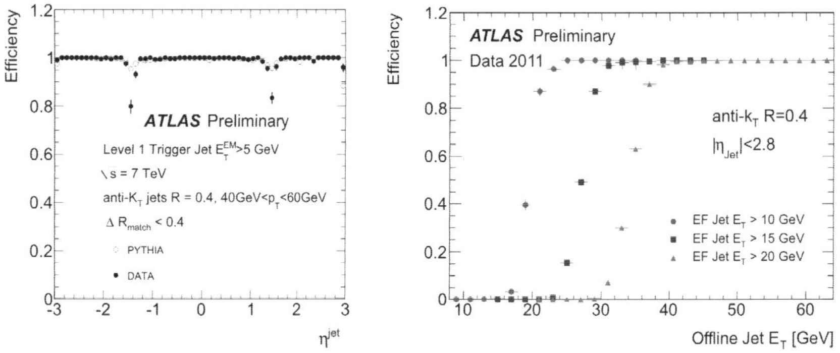

Figure 3. Left: comparison of the L1 trigger efficiency as a function of the jet pseudorapidity in data (full circles) and in PYTHIA[7] simulated events (open circles) [5]. Right: efficiency of three EF jet triggers as a function of the offline calibrated jet E_T for the case when the EF is run on L1 random-triggered events [4].

3. Performance

Detailed jet trigger efficiency studies have been performed for different data taking conditions, including increasing levels of pile-up. The jet identification in the trigger is affected by pile-up, due to the overlap of energy depositions in the calorimeter, causing an increase in the reconstructed transverse energy (E_T) of the jet. Since May 2011 onwards, a noise suppression tool has been implemented at both L2 and the EF. The noise suppression tool applies a symmetric, configurable energy threshold to cells, taking into account the measured noise from pile-up and calorimeter front-end electronics.

The efficiency of the jet trigger has been evaluated for offline jets reconstructed with the anti-k_t algorithm with radius parameter R=0.4. The efficiency is defined as the fraction of offline reconstructed jets that match a trigger jet with $\Delta R = \sqrt{\Delta\phi^2 + \Delta\eta^2} < 0.4$ and which pass the corresponding trigger threshold. An example of the efficiencies obtained for central and forward jets is shown in figure 2, before and after the introduction of the pile-up noise suppression tool. Different thresholds are applied at each level of the trigger to increase rejection of low p_T jets while maintaining a high efficiency for events satisfying the overall jet trigger. The overall 99% efficiency point improved by ~ 5 GeV by suppressing the pile-up noise. The shift due to noise suppression is larger at L2 than at the EF, since the EF jets use three dimensional topological clusters as constituents, which already include some noise suppression. The efficiency curve for the L1 in the forward region has a slower turn-on due to the reduced granularity. The rejection is considerably improved at L2 and EF.

A comparison of the L1 trigger efficiency as a function of the offline jet η in data and Monte Carlo (MC) simulation is shown in figure 3 (left), as obtained with 2010 data. The agreement between data/MC is very good, within 2%, for the full range shown, except in the transition regions of the calorimeter near $|\eta|$=1.7, where the efficiency is lower in the data by about 10%.

3.1. Performance for full scan triggers

The introduction of the full event reconstruction mode in the EF in 2011 permitted the lowering of jet thresholds. As shown in figure 2, a 15 GeV threshold at L1 becomes 100% efficient for an offline jet of about 50 GeV. This is due to the higher noise thresholds at L1 and to the non-compensating nature of the ATLAS calorimeter, which is not corrected for at L1. To recover low E_T jets, the EF full event reconstruction is run on random L1 triggers (called un-seeded EF

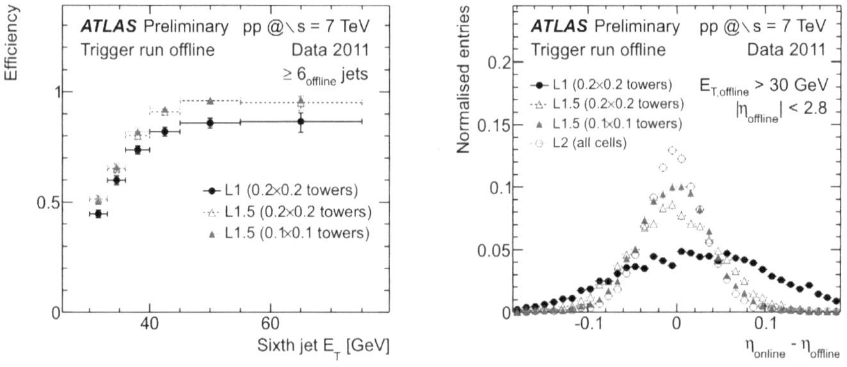

Figure 4. Left: efficiency for L1 and L2 (anti-k_t R=0.4, full scan) jets to satisfy a six jet trigger (trigger run offline) as a function of the sixth offline jet E_T. Right: the jet position resolution (in pseudorapidity) of the L1, L1.5 and L2 jet triggers in 2011 (trigger run offline) evaluated with respect to offline anti-k_t R=0.4 jets [4].

reconstruction). An example of the EF efficiency curves for three different un-seeded triggers is shown in figure 3 (right). The plateau of the turn-on curve for the lowest threshold is reached at around 20 GeV, improving the acceptance for low E_T jets.

The L1 sliding window algorithm has a low efficiency for close-by jets, as shown in figure 4 (left), where the efficiency for L1 jets to satisfy a six jet trigger is plotted as a function of the sixth offline jet E_T. Here events are selected in which at least six anti-k_t R=0.4 offline jets have been reconstructed offline within $|\eta| < 2.8$ and with $E_T > 30$ GeV. The events were pre-selected using a four jet trigger and the trigger was re-run offline for this study. The efficiency is improved by about 10% by the L2 full event reconstruction, thanks to the use of the anti-k_t algorithm.

The spatial resolution for the L2 full scan jets, calculated with respect to the offline anti-k_t R=0.4 jets, is shown in figure 4 (right) for the two possible granularities available at L2, in comparison to the L1 jets and the standard L2 cone jets. The L2 full scan improves considerably the L1 jet pseudorapidity resolution, for both cases in which either the lower or higher trigger tower granularity is used. The L2 cone jets based on cells have the best resolution, as expected, due to the better granularity.

3.2. Performance in Heavy Ion Collisions

The flexibility of the jet trigger configuration in the EF, in which any of the offline configurations can be used, allowed for the use of dedicated underlying event subtraction algorithms during the LHC Heavy Ion run in November 2011, with Pb-Pb collisions at a nucleon-nucleon centre-of mass energy of $\sqrt{s_{NN}} = 2.76$ TeV.

Heavy ion collisions are characterized by a parameter called centrality, which describes the central, versus peripheral, nature of the impact between colliding nuclei. More central collisions involve a larger number of colliding nucleons, resulting in a larger number of particles created and a larger activity in the underlying event. In figure 5 the position resolution (left) and efficiency (right) for the main jet trigger in Heavy Ion collisions are shown for different values of centrality. The comparison is made to offline anti-k_t jets with R=0.2. One observes the same performance over a range of centrality values.

Furthermore, the average readout time of the L1 trigger towers was measured in Pb-Pb collisions to be 7.5 ms for the smallest calorimeter granularity (0.1×0.1), with tails extending

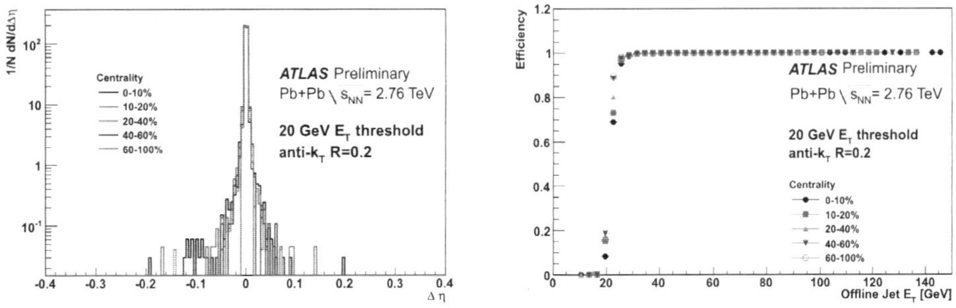

Figure 5. Jet position resolution in pseudorapidity (left) and efficiency (right) of the primary jet trigger used for the 2011 heavy ion run calculated with respect to offline anti-k_t R=0.2 jets [4].

up to 20 ms. For the larger granularity of (0.2×0.2), the average readout time was 2.7 ms, with tails extending to 14 ms. Both of these times are well within the limits of the L2 trigger design.

4. Summary and conclusions

The ATLAS Trigger system was designed to cope with the challenging LHC conditions by dividing the trigger into three levels and using a Region of Interest based approach, in which the first level identifies high p_T objects that are verified by the subsequent levels. For the jet triggers, the seeded mechanism introduced performance limitations for very low p_T jets and multi-jet events. These were overcome by the introduction of a full event reconstruction at the EF in 2011 and at L2 in 2012. The L2 full event reconstruction is only possible using the L1 trigger towers, which are accessed directly from the L1 calorimeter readout. The result is a highly flexible system that can run many different algorithms and configurations, adapting to different requirements of physics analyses and changing data-taking conditions. The jet trigger has demonstrated an excellent performance since the first collisions of the LHC.

Acknowledgments

The author acknowledges the support by FCT co-financed by QREN/COMPETE of European Union ERDF fund, Portugal.

References

[1] ATLAS Collaboration 2008 *JINST* **3** 08003
[2] ATLAS Collaboration 2008 CERN-OPEN-2008-020
[3] M Cacciari G S and Soyez G 2008 *JHEP* **04** 063
[4] https://twiki.cern.ch/twiki/bin/view/AtlasPublic/JetTriggerPublicResults
[5] ATLAS Collaboration 2011 ATLAS-CONF-2010-094 https://cdsweb.cern.ch/record/1299109
[6] M Cacciari G S and Soyez G 2011 CERN-PH-TH/2011-297
[7] T Sjostrand S Mrenna and P Z Skands 2006 JHEP **05** 026

The ATLAS hadronic tau trigger

Mansoora Shamim on behalf of the ATLAS collaboration
University of Oregon, Eugene, OR, 97403, U.S.A
E-mail: mshamim@cern.ch

Abstract. The extensive tau physics programs of the ATLAS experiment relies heavily on trigger to select hadronic decays of tau lepton. Such a trigger is implemented in ATLAS to efficiently collect signal events, while keeping the rate of multi-jet background within the allowed bandwidth. This contribution summarizes the performance of the ATLAS hadronic tau trigger system during 2011 data taking period and improvements implemented for the 2012 data collection.

1. Introduction
Hadronic tau decays play a crucial role in precisions measurements of Standard Model (SM) processes as well as in the searches for physics beyond the SM e.g. through searches for the Higgs boson and supersymmetric and exotic (W' and Z') particles.

Due to its short lifetime ($c\tau = 87.11$ μm), the tau lepton decays inside the beam pipe. The identification of taus is, therefore, done through their decay products inside the detector. The tau decays into two neutrinos and an electron or muon 35% of the time, while 65% of its decays involve a neutrino and prompt charged hadrons, mostly pions. A dedicated tau trigger has been designed and implemented at the ATLAS experiment [1] to select events where a tau decays into one or more hadrons. A decay like this results in a hadronic jet which is characterized by the presence of a well collimated calorimeter cluster with only a small number of associated tracks.

It is challenging to keep the rates for these triggers low due to the high production rate of multi-jet events. Nevertheless it is advantageous to implement tau triggers to increase the sensitivity of searches for new physics.

2. The ATLAS trigger system
The ATLAS trigger system [2] is divided into the hardware-based level 1 (L1) and the software-based level 2 (L2) and event filter (EF). L2 and EF are referred to together as the high level trigger (HLT). The L1 trigger identifies regions-of-interest (RoI) using the information from calorimeter and muon systems. The decision time at L1 is ~ 2 μs. L2 takes these RoIs as input and refines the object identification using the information from all subsystems. The average latency at L2 is ~ 40 ms. In the EF, algorithms similar to the offline reconstruction are used to select interesting events with an average latency of ~ 4s.

3. The ATLAS tau trigger
The tau trigger is designed to select hadronic decays of the tau lepton, which are characterized by the presence of one or three charged pions accompanied by a neutrino and possibly neutral pions.

Journal of Physics: Conference Series **404** (2012) 012059 doi:10.1088/1742-6596/404/1/012059

At L1, the tau trigger uses the electromagnetic (EM) and hadronic (HAD) calorimeter trigger towers to calculate the energy in a *core* and an *isolation* region outside the core. The trigger towers have a granularity of $\Delta\eta \times \Delta\phi = 0.1 \times 0.1$. The core region is defined as the two-by-two trigger tower region of $\Delta\eta \times \Delta\phi = 0.2 \times 0.2$. The isolation region is defined as a four-by-four trigger tower region around the core subtracting the two-by-two region in the center. At HLT, selection criteria are applied using tracking and calorimeter-based information. This takes advantage of the narrow energy deposition and low track multiplicity to discriminate taus from the multi-jet background. The selections at HLT are separately optimized for tau decays with one (1-prong) or several (multi-prong) charged particles. EF uses more refined algorithms, which are similar to the offline tau reconstruction algorithms.

4. Performance of the ATLAS tau trigger in 2011

This section describes the performance of the tau trigger using a tag-and-probe method in $Z \rightarrow \tau_\mu \tau_h$events, where one tau decaying into muon is used as the tag, while the hadronically decaying tau is used as the probe. The trigger efficiency is measured w.r.t to the taus identified by the offline algorithms [3]. Figure 1 shows the efficiency of the tau20_medium1 trigger, with

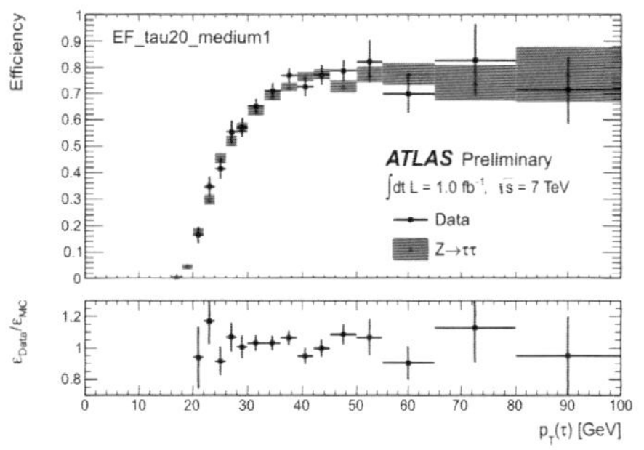

Figure 1. Efficiency of the tau20_medium1 trigger, with respect to the offline identified tau candidates, measured using a tag and probe analysis with $Z \rightarrow \tau_\mu \tau_h$events.

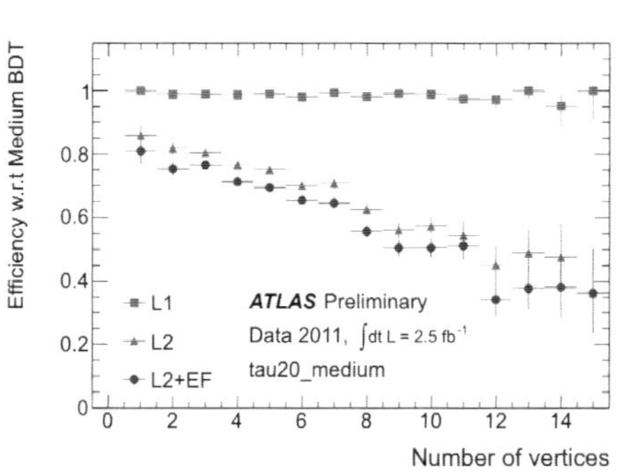

Figure 2. Efficiency of tau20_medium with respect to the offline identified tau candidates as a function of number of vertices measured in 2011.

respect to the offline identified tau candidates, as a function of the offline tau p_T. The ratio of the efficiency in data and Monte Carlo (MC) simulation is also shown, demonstrating that the performance of the trigger in data and MC agree. The term tau20_medium1 implies a 20 GeV requirement on the transverse energy, medium selections on the shower shape variables and strict requirement on the number of tracks for an EF tau candidate. Figure 2 shows the performance of a trigger, with 20 GeV threshold at EF but less stringent requirement on the number of tracks, as a function of number of vertices measured in data. The number of primary vertices is a measure of the overlapping interactions in the same bunch crossing (BC) which is also known as pile-up. At L1, the trigger efficiency shows no dependence on pile-up while at L2 and EF efficiencies decrease with increasing number of pile-up interactions. This is mainly due to two reasons. Firstly, the shower shape variables used at the trigger level in 2011 were highly pile-up dependent with no procedure implemented to correct for pile-up dependence. Secondly, the optimization of selection requirements at the HLT for the shower shape variables was performed using only the MC samples with fewer number of pile-up interactions than observed in data.

5. Improvements in tau trigger algorithms for 2012

In 2011, the average number of collisions per BC was 10-12 corresponding to a peak instantaneous luminosity of 3.65×10^{33} cm^{-2} s^{-1} at $\sqrt{s} = 7$ TeV and 50 ns bunch spacing. While in 2012, the peak instantaneous luminosity of 7×10^{33} cm^{-2} s^{-1} at $\sqrt{s} = 8$ TeV and 50 ns bunch spacing would be expected. This would result in an average of ~28 pile-up events while the peak value could be upto 35. A large number of improvements were added to the tau trigger algorithms and selection criteria before the start of data taking in 2012. This is to avoid degradation in the performance as a function of pile-up.

- At HLT, the energies of the cells in a cone of radius $\Delta R = \sqrt{(\Delta\eta)^2 + (\Delta\phi)^2}$ around the cluster direction are summed up to calculate the total energy of the cluster, after applying noise suppression. The value of ΔR will be changed from 0.4 to 0.2.
- Shower shape variables, such as the energy weighted radius in the EM calorimeter will be removed from the set of selection variables.
- New variables including the ratio of the energy in a cone of radius 0.1 to that in 0.2, the p_T weighted distance of tracks from the cluster direction, and the ratio of tau cluster energy to the p_T of the leading track will be added to the list of selection variables.

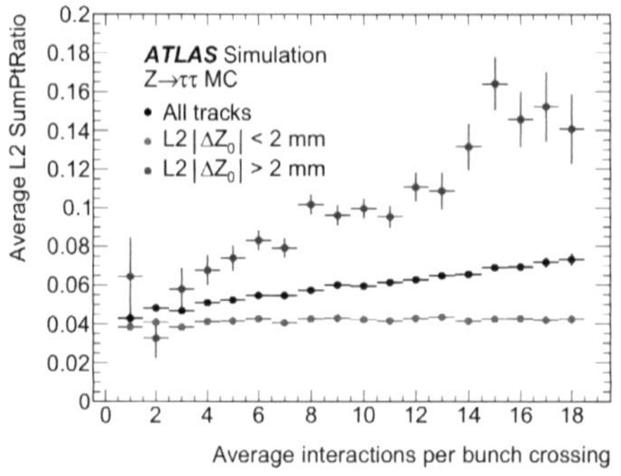

Figure 3. Average value of SumPtRatio at L2 as a function of the average number of interactions per BC in $Z \rightarrow \tau\tau$ simulated events.

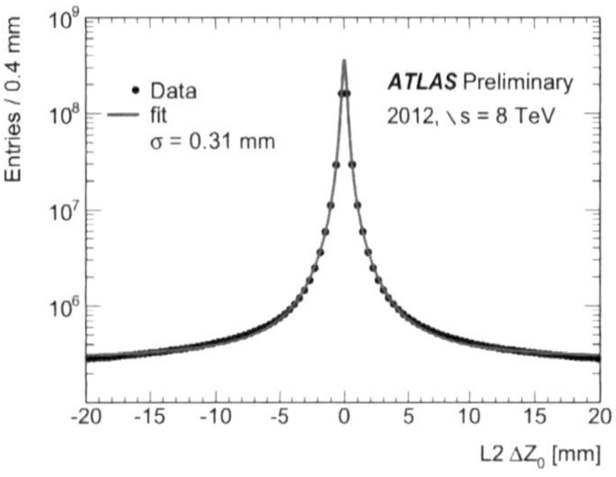

Figure 4. ΔZ_0 distribution as observed in 2012 (8 TeV) data, with an average value of eighteen interactions per BC. A Lorentzian function is fitted to the distribution and the quoted value of σ integrates 68% of the area of the central peak.

- To select tracks associated with the tau candidate a requirement will be applied on the $|\Delta Z_0|$ of each track. The $|\Delta Z_0|$ of a track associated with trigger object is defined as the longitudinal coordinate difference with respect to the leading track. Large values of $|\Delta Z_0|$ correspond to tracks coming from additional vertices created in overlapping interactions. Figure 3 shows the average SumPtRatio at L2 as a function of the average number of interactions per BC in simulated $Z \rightarrow \tau\tau$ events. The SumPtRatio is the ratio of the scalar sum of the p_T of all tracks in the isolation region, defined as annulus between $0.1 < \Delta R < 0.3$ with respect to the leading track in the RoI, to the scalar sum of the p_T of all tracks in the core region, defined as the $\Delta R < 0.1$ cone around the leading track. A requirement of 2mm on the $|\Delta Z_0|$ provides robustness against pile-up.
Figure 4 shows the ΔZ_0 distribution, as observed in 2012 data at $\sqrt{s} = 8$ TeV with an

average value of eighteen interactions per BC. This distribution justifies the requirement of 2 mm with no additional loss of signal events in data due to resolution effects.

- At EF, the simple cut based selections will be replaced by the Multivariate (MV) algorithms known as Boosted Decision Tree (BDT) and Log Likelihood (LLH). A decision tree is a structure which defines a recursive sequence of cuts on multiple variables which separate signal from background. The averaging technique known as "boosting" is used to deal with statistical fluctuations. The likelihood function, for signal and background is defined as the product of the one-dimensional probability density functions, of each identification variable. A set of pile-up robust variables is used as input to the MV algorithms.
  Figures 5 and 6 demonstrate the expected performance of the MV based tau triggers (BDT and LLH) at EF in simulated events for one-prong and multi-prong taus, respectively. Signal MC samples for $Z \rightarrow \tau\tau, W \rightarrow \tau\nu$ and $Z' \rightarrow \tau\tau$ were combined to evaluate the performance of MV algorithms. The signal efficiency is defined with respect to the offline identified tau candidates and the background rejection is evaluated using simulated multi-jet events. The operational point, along the curves, for a set of medium selections is chosen to have 85% and 80% signal efficiency for one-prong and multi-prong taus, respectively.

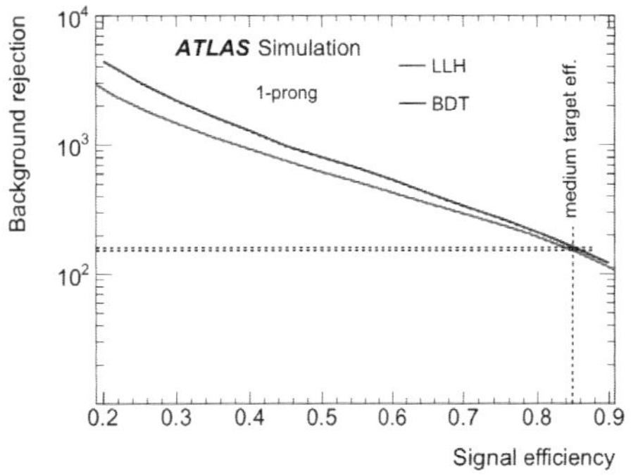

Figure 5. Performance of the multivariate based tau triggers for one prong taus at EF.

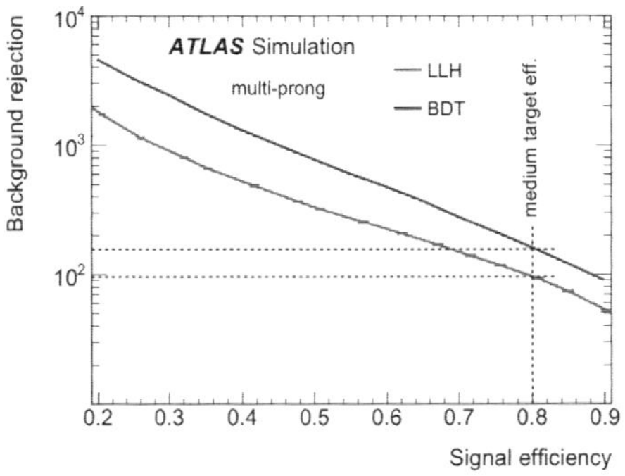

Figure 6. Performance of the multivariate based tau triggers for multi-prong taus at EF.

6. Performance of the ATLAS tau trigger in 2012 Data

ATLAS has been collecting data with the latest and improved version of the tau trigger since May 2012. The performance of the tau trigger has been verified in data. Figure 7 shows the efficiency of tau20_medium1 trigger, with respect to the offline tau candidates identified by the BDT algorithm, as a function of number of vertices measured in 2012 data while Figure 8 shows the same as a function of the p_T of the offline tau candidate. The trigger efficiency is measured using a tag and probe analysis with $Z \rightarrow \tau_\mu\tau_h$events following the offline tau identification efficiency measurement [3]. Comparing Fig. 2 and Fig. 7 one can see that with the new selections, the loss of efficiency as a function of pile-up has been recovered to a large extent.

7. The tau trigger thresholds in 2012

In order to make maximum use of the data collected by the tau trigger, it is important to apply loose selections and keep the p_T thresholds at L2 and EF lower than those used in the offline analyses. Although the LHC is expected to deliver peak luminosity which is a factor of two

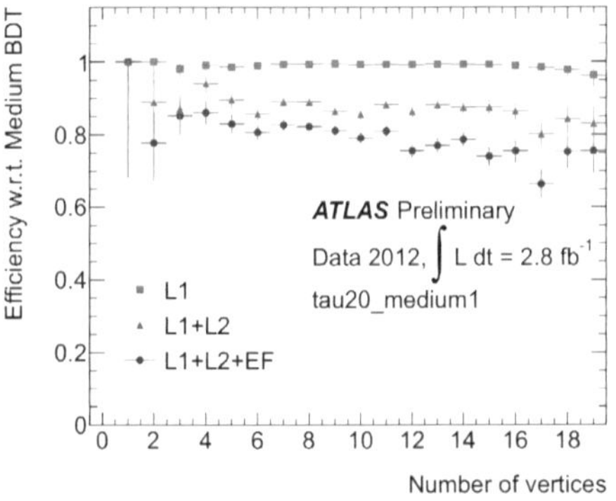

Figure 7. Efficiency of tau20_medium1 trigger, with respect to the offline tau candidates identified by the BDT algorithm, as a function of number of vertices measured in 2012 data.

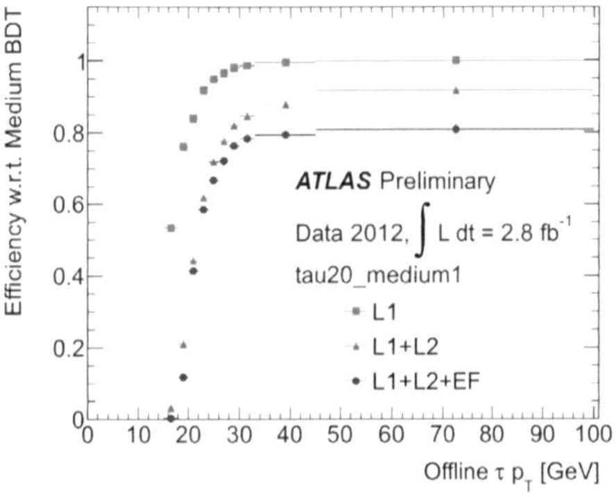

Figure 8. Efficiency of tau20_medium1 trigger, with respect to the offline tau candidates identified by the BDT algorithm, as a function of offline tau p_T measured in 2012 data.

higher than the one delivered in 2011, yet due to the improvements added to the tau trigger algorithms for 2012, it is possible to keep the p_T thresholds at HLT similar to those used in the 2011 run. Furthermore, combining the tau trigger with other triggers such as electrons, muons and missing transverse energy (MET) helps with controlling the rates while keeping the threshold low. Table 1 shows a list of single and combined tau triggers along with the thresholds at EF. The combinations and thresholds are optimized for a specific analysis as shown in the first column of the table.

Table 1. Single and combined tau triggers being used in 2012.

Process	Trigger	$p_T(\tau)$ at EF [GeV]	[GeV]
$H^+ \to \tau_h \nu$	τ+MET	29	MET > 40
$H^0(Z^0) \to \tau_h \tau_{e,\mu}$	$\tau + e$	20	$p_T(e) > 18$
	$\tau + \mu$	20	$p_T(\mu) > 15$
$H^0_{\text{MSSM}} \to \tau_h \tau_h$	$\tau + \tau$	38, 38	
$H^0_{\text{SM}} \to \tau_h \tau_h$	$\tau + \tau$	29, 20	
$Z' \to \tau_h \tau_h$	$\tau + \tau$	100, 70	
$W' \to \tau_h \nu$	τ	115	

8. Conclusions

The ATLAS tau trigger has performed remarkably well over the 2011 LHC run, collecting data for physics analysis [4, 5, 6, 7], monitoring and calibration purposes. The efficiency of the trigger has been analyzed with tag-and-probe methods showing good agreement between data and MC simulations. Pile-up robust selections have been implemented for the 2012 run in order to recover the efficiency loss observed in 2011 data. The performance of newly implemented algorithms has been verified in data collected in 2012 at $\sqrt{s} = 8$ TeV.

References

[1] ATLAS Collaboration, The ATLAS Experiment at the CERN Large Hadron Collider, JINST 3 (2008) S08003.
[2] ATLAS Collaboration, Performance of the ATLAS Experiment-Detector, Trigger and Physics, CERN-OPEN-2008-020; arXiv: 0901.0512[hep-ex]; 3 Volumes; ISBN:978-92-9083-321-5 (2009).
[3] ATLAS Collaboration, Performance of the Reconstruction and Identification of Hadronic Tau Decays with ATLAS, ATLAS-CONF-2011-152, https://cdsweb.cern.ch/record/1398195.
[4] ATLAS Collaboration, A search for high mass resonances decaying to $\tau^+\tau^-$ in the ATLAS detector, ATLAS-CONF-2012-067, https://cdsweb.cern.ch/record/1460263.
[5] ATLAS Collaboration, Search for the Standard Model Higgs boson in the $H \rightarrow \tau\tau$ decay mode with 4.7 fb^{-1} of ATLAS data at $\sqrt{s} = 7$ TeV, ATLAS-CONF-2012-014, https://cdsweb.cern.ch/record/1429662.
[6] ATLAS Collaboration, Measurement of Tau Polarization in $W \rightarrow \tau\nu$ Decays with the ATLAS Detector in pp Collisions at $\sqrt{s} = 7$ TeV, ATLAS-CONF-2012-009, https://cdsweb.cern.ch/record/1428549.
[7] ATLAS Collaboration, $Z \rightarrow \tau\tau$ cross section measurement in proton-proton collisions at 7 TeV with the ATLAS experiment, ATLAS-CONF-2012-006, https://cdsweb.cern.ch/record/1426991.

Upgrade for the ATLAS Tile Calorimeter Readout Electronics at the High Luminosity LHC

A. S. Cerqueira on behalf of the Tile Calorimeter group

Electronics Engineering Department, Federal University of Juiz de Fora (UFJF), Juiz de Fora, Brazil

E-mail: augusto.santiago@ufjf.edu.br

Abstract. The Tile Calorimeter (TileCal) is the hadronic calorimeter covering the most central region of the ATLAS experiment at LHC. The TileCal readout consists of about 10000 channels. The main TileCal upgrade will occur when preparing for the high luminosity operation. This (Phase 2) upgrade is scheduled around 2022. The upgrade aims at replacing the majority of the on- and off-detector electronics so that all calorimeter signals are directly digitized and sent to the off-detector electronics in the counting room. An ambitious upgrade development program is pursued to study different electronics options. Three options are presently being investigated for the front-end electronic upgrade. The first option is an improved version of the present system built using discrete components, the second alternative is based on the development of a dedicated ASIC, and the third is the development of a new version of the "QIE" based on the one developed for Fermilab. For the off-detector electronics a new back-end architecture is being developed, the so-called "super" Read-Out Driver (sROD). A demonstrator prototype read-out for a slice of the calorimeter with most of the new electronics, but also compatible with the present system, is planned to be inserted in ATLAS already in mid 2014 (at the end of the phase 0 upgrade).

1. Introduction

TileCal is a sampling calorimeter composed of steel plates (tile shape) as absorber material interleaved with plastic scintillating plates as sampling material. It is divided in a central barrel (covering $|\eta| < 1.0$) and two extended barrels (covering $0.8 < |\eta| < 1.7$), where each part is formed by 64 modules in order to complete the entire cylinder (see Figure 1).

When high energy particles interact with the steel, showers of lower energy particles are created, which in turn produce light when passing through the scintillating tiles. The light is transmitted through wavelength shifting fibers to photomultipliers (PMTs), which convert the light into electrical signals. Adjacent tiles and their WLS fibers are grouped together to form TileCal cells. For each cell the fibers are read-out from two sides by two PMTs, one for each side. Thus, each cell is readout via two different electrical signal paths. The central barrel modules are divided in 46 cells each, while the extended barrels modules are divided in 14 cells. Therefore, TileCal is comprised of more than 10000 readout channels.

The TileCal front-end electronics, located inside the outermost part of the modules, is responsible for processing the PMT signals and transmitting them to the back-end electronics, which is responsible for calorimeter signal acquisition.

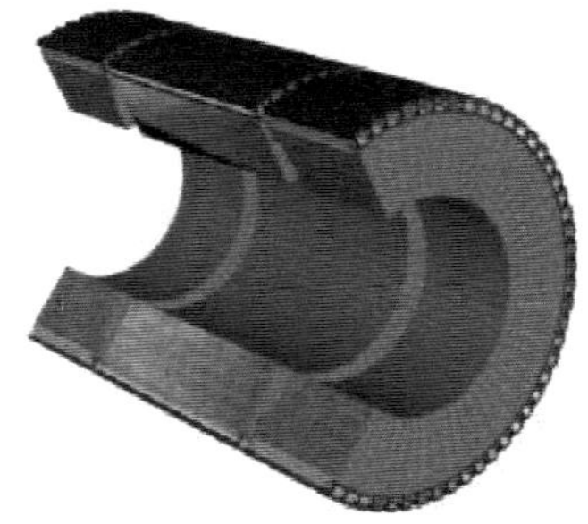

Figure 1. . Illustration of the Tile Calorimeter.

Currently, ATLAS is investigating necessary changes in order to prepare for the proposed high luminosity upgrade of the LHC during the next decade (HL-LHC). Several detector components should be replaced (e.g. front-end electronics of the calorimeters) and major changes on the trigger system are required to cope with the new luminosity requirements.

In 2013-2014, the first long shutdown (LS1) at LHC is planned and will allow LHC to continue its operation reaching nominal peak luminosity of 10^{34} cm^2s^{-1} (Phase 0). A second long shutdown (LS2) is planned for 2018-2019 to prepare for peak luminosities around $2-3\times10^{34}$ cm^2s^{-1} corresponding to 55 to 80 interactions per bunch-crossing with 25 ns bunch interval, well beyond the initial design goals (Phase 1). Finally, the Phase 2 upgrade is foreseen for 2022-2023 after which the peak luminosity will reach $5-7\times10^{34}$ cm^2s^{-1} (HL-LHC).

This paper aims to present the TileCal front-end and back-end electronics upgrade program for the HL-LHC. Next section briefly describes the TileCal readout electronics. The TileCal electronics upgrade program is summarized in section 3. Section 4 presents details about the front-end and back-end upgrade projects. Finally, the conclusions are stated in section 5.

2. The Current TileCal Readout Electronics

The TileCal signal chain can be seen in Figure 2. The light produced by energy depositions in the detector is collected by optical fibers and sent to the light mixers, where several fibers are grouped together in order to form the detector cells. The light is converted to an electrical signal in the photomultiplier tube (PMT) and is processed by the 3-in-1 card (Front-End Board), which is responsible for signal conditioning and amplification providing three analog signals as outputs, two for the detector readout (high and low gain) and another for triggering purpose. The low and high gain signals are then digitized at 40 MHz by 10 bit Analog to Digital Converters (ADCs) in the digitizer boards. Digital signals of all calorimeter cells in a module are merged and formatted into packages and sent via high speed optical links (Interface Board) that connect the on- and off-detector electronics. The on-detector electronics is located in the outermost part of the TileCal module, in the electronics "drawers".

In the back-end electronics, the main component is the Read-Out Driver (ROD) which performs preprocessing and gathers data coming from the front-end electronics at a maximum level 1 trigger rate of 100 kHz. The ROD then sends these data to the Read-Out Buffers (ROB) in the second level trigger [1].

Figure 2 also shows the signal path to the first level trigger. The TileCal first level trigger signal is produced by analog summation of up to six signals on the trigger board [2] and its analog output is sent to the level one receiver by means of long cables (around 70 m). The first level trigger reduces the event rate from 40 MHz to a maximum of 100 kHz.

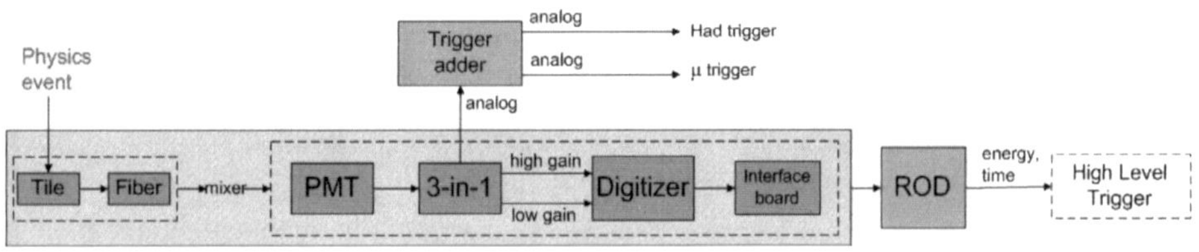

Figure 2. TileCal Signal Chain.

3. Upgrade Plans for High Luminosity

The main goals of the TileCal electronics upgrade program for high luminosity are the replacement of the aging electronics (2022-2023), the increase of radiation tolerance, the improvement of system reliability (less connectors - split Main Board design mitigates against single point failure causing loss of cell), to increase data precision and to improve the level one trigger system by the availability of the full detector resolution.

A new on-detector electronics architecture is under design and is being incrementally tested, where three different design options for the new TileCal front-end board are under evaluation. The increased event rate also requires larger currents in the PMT voltage divider chains. New active dividers and new high voltage power supply are under development. Concerning the off-detector electronics, a "super" Read-Out Driver (sROD) is being designed.

Along with the development of new electronics a modification of the TileCal mechanics is being considered. The initial development work has been carried out by Laboratoire de Physique Corpusculaire in Clermont-Ferrand (LPC). The aim is to split the present drawers into two "mini-drawers". This is compatible with the new electronics architecture. The mini-drawers will simplify handling of the drawers and improve the access to the TileCal electronics since it will be easier to open the detector to replace mini-drawers. Practical solutions to insertion and cooling and electrical connections are being tested with different prototypes.

The TileCal electronics upgrade program is making progress where an early slice prototype of the new system had already been tested with one version of the new front-end board. The other two front-end options are being designed and are scheduled for radiation test this year. In the end of 2013, one upgraded drawer should be ready for tests together with the sROD. This new "Demonstrator" drawer will be backward compatible with the present system so that it can be seamlessly installed in the present system. In the end of the first long shutdown, one Demonstrator drawer should be installed in the detector in order to be tested under production conditions. If it proves to operate well three additional Demonstrator drawers will be installed in later Christmas shut downs. Between 2014-2018 extensive tests should also be performed with the three versions of the new system in test beam after which the decision about the front-end board to choose should be taken. The production of new electronics will take place between 2019-2020. Finally, the installation is foreseen for 2022-2023.

4. Readout Electronics Upgrade Projects

4.1. On-Detector Electronics

The foreseen archtecture for the new on-detector electronics can be seen in Figure 3. The new readout electronics is composed of: new Front-End boards (FE) that provide conditioning for the PMT signals as well amplification, digitization (depending on the Front-End option) and calibration functionalities; new Main Boards (MB) providing digitization (depending on the front-end board) and control; and new Daughter Boards (DB) which provide data processing and interface with the back-end electronics via optical links at 5 or 10 GHz rate with redundance.

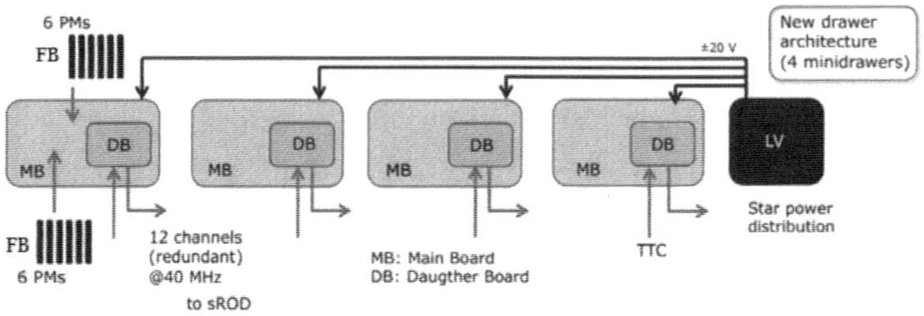

Figure 3. New drawer archtecture.

The star power distribution along with local point-of-load voltage regulators in the new drawer reduces the voltage deviations and the noise coupling along the drawer. Another important change on the drawer archtecture is the replacement of a single interface board (actual design) by four Daughter Boards, improving the system robustness and decouples the drawer electronics into independent units.

4.1.1. Front-End - Modified 3-in-1 Card: The Enrico Fermi Institute (University of Chicago) is developing a modified version of the present 3-in-1 card [3]. This Front-End Board is composed of discrete components and can be divided in three stages: the fast signal processing chain, the slow signal processing chain and the calibration electronics and the control bus interface.

The fast signal processing chain includes a 7-pole passive LC shaper, bi-gain clamping amplifiers with a gain ratio of 16 and a pair of differential drivers feeding the analog signals from the low-gain channel and the high-gain channel to the ADCs which are placed in the Main Board. The slow signal processing chain includes a programmable 3-gain integrator which monitors the PMT current induced by a Cesium source and the minimum bias current induced during the collisions. Finally, the last stage includes a precise charge injection circuit, integrator gain control and the control bus interface. This modified version has better linearity and a lower noise level than the previous version. The prototype of the modified 3-in-1 card has been built using COTS components and has passed initial radiation tests.

4.1.2. Front-end - QIE Chip: The Argonne National Laboratory is working on the design of a front-end board which includes a new version of the Charge (Q) Integrator and Encoder (QIE) chip developed in collaboration with Fermilab and CMS HCAL. The QIE includes a current splitter composed of 23 splitter transistors, providing 4 different ranges (16/23, 4/23, 2/23, 1/23), followed by a gated integrator and an on-board 6 bit flash ADC to cover a dynamic range of 17 bits. In this way, only a simple digital interface is needed to communicate with the Main Board. The QIE also includes a charge injection circuit for calibration and an integrator for source calibration. The QIE does not perform pulse shaping, minimizing pile up problems and allowing raw PMT pulses to be measured.

At the present time two QIE prototypes have been produced and tested. The QIE version 10.4 is expected by November 2012.

4.1.3. Front-end - FATALIC ASIC: FATALIC means Front-end for Atlas TileCal Integrated Circuit [4] which is being designed at Laboratoire de Physique Corpusculaire in Clermont-Ferrand (LPC). FATALIC includes a multi-gain current conveyor (CC) with three different gains (1, 8, 64) which cover the full dynamic range of the PMT signal, followed by a shaper in order to improve the Signal-to-Noise Ratio (SNR). The readout chain is completed using an external 12 bit pipelined ADC with a sampling rate of 40 MHz also developed at LPC and called Twelve bits ADC for s-ATLAS TileCal Integrated Circuit (TACTIC). Moreover FATALIC includes an integrator and a 10 bit ADC with a low sampling rate for calibration purposes. Both chips are designed using the CMOS IBM 130 nm technology.

The first prototypes of the FATALIC, version 1 and 2, have been produced and tested. The FATALIC version 3 has been delivered in March 2012 and it includes an integrator amplifier. LPC is currently designing the amplifier stage of the TACTIC which is planned to be submitted to foundry by the beginning of August 2012.

4.1.4. Main Board: The Main Board [5] is responsible for the digital control of the FEBs, data organization and for the transmission of the data to the Daughter Board. The current prototype design digitizes the signals coming from four Modified 3-in-1 Cards by using four 12 bit ADCs working at a sampling rate of 40 MHz. The final version of the Main Board should process 12 PMT signals.

4.1.5. Daughter Board: The Daughter Board [5] is intended to serve as a processing board in the next TileCal electronics drawer and is designed for redundancy and high data throughput readout. The two separately programmable FPGAs are responsible for reading out signals originating from the same tile cells but from different sides of the scintillating tiles. This means that they process equivalent data. If one chain fails it can be replaced by the other (there is a loss of statistics though). The Daughter Boards sends the digitized data to the super Read Out Driver (sROD) via high-speed links using the GBT protocol [6]. In order to perform these functions, the first version of the Daughter Board includes two Xilinx Virtex 6 FPGAs, one 12×5 Gb/s transmitter using a SNAP12 format connector and two 5 Gb/s SFP+ format connectors. Future versions of the Daughter Board will include two Xilinx Kintex 7 FPGAs, one 4×10 Gb/s QSFP+ and one transmitter using SNAP12 format connector, to help selecting the optimum link strategy.

The Daughter and the Main boards have been designed by the Stockholm University and the Enrico Fermi Institute (University of Chicago). First prototypes of the Main Board and the Daughter Board have been produced and tested. A more final version of these boards is currently being designed based on the test results. Figure 4 shows the Main Board prototype (1), which is connected to the Daughter Board prototype (2) and the modified 3-in-1 cards (3).

4.2. Off-Detector Electronics: sROD

The sROD demonstrator board [7] consists of one Xilinx Virtex 7 and one Xilinx Kintex 7 FPGAs as processing core, 4 receiver Avago MiniPOD connectors, 2 transmitter Avago Mini-POD connectors, one QSFP+ format connector and one SFP format connector. This board is a double mid-size Advanced Mezzanine Card (AMC) and it is conceived to be plugged in a Advanced Telecommunications Computing Architecture (ATCA) carrier or in a Micro Telecommunications Computing Architecture (μTCA) system.

The sROD performs several functions. These include: data processing and data reception from Daughter Boards; Timing, Trigger and Control (TTC); Detector Control System (DCS) management and transmission to Main Boards; data reconstruction and transmission to the ReadOut Subsystem (ROS); as well as data preprocessing and transmission to the Level-1 Calorimeter trigger system (L1Calo).

The Instituto de Física Corpuscular - Universidad de Valencia (IFIC-UV), the Laboratório de Instrumentação e Física Experimental de Partículas (LIP) and the Stockholm University are involved in the development of this project. The first prototypes of the sROD demonstrator board are expected by November 2012.

Figure 4. Test setup with the new on-detector eletronics prototypes. (1) Main Board, (2) Daughter Board and (3) Front-end Boards (Modified 3-in-1).

5. Conclusions

In order to cope with the higher data rates and radiation levels, the on- and off-detector electronics must be redesigned. For the on-detector eletronics, three different Front-End boards approaches are being considered. Additionally, in order to provide sufficient data processing, control and interface with the new back-end electronics, a Main board and Daughter board combination are being designed. For the back-end processing, the sROD is under design and should be ready for tests by the end of this year.

A slice of the new upgraded drawer is currently being tested providing promising results. Extensive laboratory tests will be performed in 2013, when one full Demonstrator drawer will be available.

At the end of Phase 0, one Demonstrator drawer should be installed into the detector. The final decision about the Front-End Board design should be taken by the end of 2018. The production of the new Tilecal readout electronics will take place during 2019-2020 and should be prepared for installation during Phase 2.

Acknowledgments

I would like to thank my colleagues from the Tile Calorimeter group for the fruitful discussions and UFJF, CNPq, CAPES and FAPEMIG from Brazil for the financial support to this work.

References

[1] ATLAS Collaboration 2011 *Eur. Phys. J. C* **72** 1849. 63 p
[2] Cerqueira A S, Seixas J M and Calôba L P 2007 *Nuclear Instruments and Methods in Physcs Research A* **570** 117–125
[3] Tang F, Anderson K, Drake G, Genat J, Oreglia M, Pilcher J and Price L 2010 *Real Time Conference (RT), 2010 17th IEEE-NPSS* pp 1–5
[4] Pillet, N et al 2011 *Topical Workshop on Electronics for Particle Physics Conference 2011*
[5] Eriksson D, Anderson K, Bohm C, Kavianipour H, Muschter S, Oreglia M and Tang F 2011 *Topical Workshop on Electronics for Particle Physics Conference 2011*
[6] Baron S, Cachemiche J, Marin F, Moreira P and Soos C 2009 *Topical Workshop on Electronics for Particle Physics Conference 2009*
[7] Castillo V, Ferrer A, Fiorini L, Gonzalez V, Hernandez Y, Higon E, Moreno P, Sanchis E, Solans C, Valero A and Valls J 2011 *Nuclear Science Symposium and Medical Imaging Conference NSS/IEEE*

Upgraded Readout Electronics for the ATLAS Liquid Argon Calorimeters at the High Luminosity LHC

Timothy R. Andeen on behalf of the ATLAS Liquid Argon Calorimeter Group

Nevis Laboratory, Columbia University, New York, USA

E-mail: timothy.robert.andeen@cern.ch

Abstract. The ATLAS liquid-argon calorimeters produce a total of 182,486 signals which are digitized and processed by the front-end and back-end electronics at every triggered event. In addition, the front-end electronics sum analog signals to provide coarsely grained energy sums, called trigger towers, to the first-level trigger system, which is optimized for nominal LHC luminosities. However, the pile-up background expected during the high luminosity phases of the LHC will be increased by factors of 3 to 7. An improved spatial granularity of the trigger primitives is therefore proposed in order to improve the identification performance for trigger signatures, like electrons or photons, at high background rejection rates. For the first upgrade phase in 2018, new Liquid Argon Trigger Digitizer Boards are being designed to receive higher granularity signals, digitize them on detector and send them via fast optical links to a new, off-detector digital processing system. The digital processing system applies digital filtering and identifies significant energy depositions. The refined trigger primitives are then transmitted to the first level trigger system to extract improved trigger signatures. The general concept of the upgraded liquid-argon calorimeter readout together with the various electronics components to be developed for such a complex system is presented. The research activities and architectural studies undertaken by the ATLAS Liquid Argon Calorimeter Group are described, particularly details of the on-going design of mixed-signal front-end electronics, of radiation tolerant optical-links, and of the high-speed off-detector digital processing system.

1. Introduction

The liquid-argon (LAr) calorimeters of the ATLAS experiment [1] have functioned with excellent reliability since installation in 2006 [2]. In the last year they have played a pivotal role in the search for the Higgs boson, particularly in the diphoton and four electron channels [3]. Looking forward there are several constraints: The existing front-end electronics limit the granularity, bandwidth and latency at the level-1 trigger. Additionally, the on detector electronics contain many technologies leading to many opportunities for failure with a limited number of spares. Furthermore, the front-end electronics were qualified for radiation levels corresponding to 10 years of Large Hadron Collider (LHC) operations (with a safety factor of 10). In the high luminosity running of the LHC (HL-LHC) [4], with instantaneous luminosities of 5×10^{34} cm^{-2}s^{-1} and 3000 fb^{-1}, this safety factor will be substantially exceeded. Already by 2018 the peak instantaneous luminosity of 3×10^{34} cm^{-2}s^{-1} will put pressure on the existing trigger thresholds. By 2022 the readout electronics will need to be replaced entirely in order to enhance the physics reach of the experiment.

Maintaining low thresholds at the level-1 trigger in order to maximize our sensitivity for measurements at the electroweak scale and to probe for new physics at higher energies is the most important motivation for the upgrade. As can be seen in Figure 1, studies of the simulated detector response show that in order to stay within the level-1 bandwidth allowance of 20 kHz for calorimeter triggers, the threshold of non-isolated electromagnetic (EM) objects rises nearly linearly with luminosity. At 2.5×10^{34} $\mathrm{cm}^{-2}\mathrm{s}^{-1}$, approximately half of the expected HL-LHC luminosity, the threshold is $E_\mathrm{T} = 45$ GeV. Alternatively, one could prescale triggers with a lower threshold. In either case the trigger will fail to select some low transverse energy (E_T) electrons and photons, such as those produced in the decay of a low mass Higgs boson. In order to record these events the rejection in the hardware trigger at level-1 must be improved. This can be achieved by increasing the granularity available to the level-1 trigger system, which will allow for the implementation of more complex algorithms.

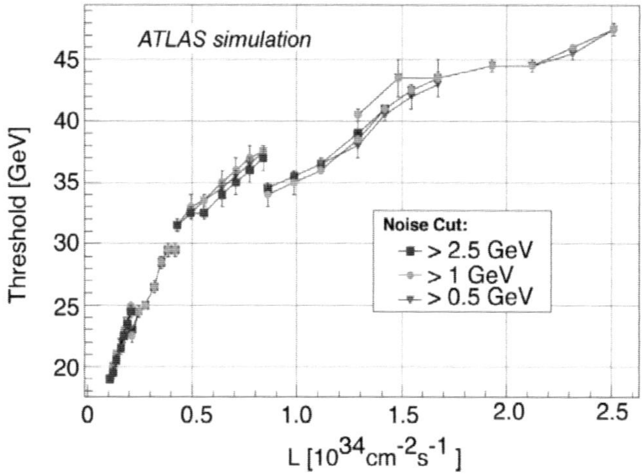

Figure 1. Thresholds for non-isolated EM objects vs. instantaneous luminosity at 20kHz level-1 trigger rate.

2. Upgrade Strategy

Currently, to provide the fast level-1 calorimeter trigger decision only a fraction of the available readout granularity is utilized. The four calorimeter layers are reduced to one "trigger tower" about 0.1×0.1 in$\Delta\ \eta \times \Delta\phi$. This tower includes 4×1 $(\Delta\eta \times \Delta\phi)$ cells from the presamplers, 32×1 cells from the first layer, 4×4 cells from the second layer and 2×4 cells in the back layer, as shown in Figure 2. The analog sum of the signals from these cells is sent offdetector, digitized and clustered for EM objects and (combined with Tile Calorimeter signals) jets. The full precision readout of all the cells is only done after the level-1 trigger accepts the event, with a latency of ~ 2.1 μs.

To reduce the level-1 trigger rate an increase in the granularity, by adding both lateral and longitudinal readout segmentation, is planned. In this scheme the individual four layers will be read out separately. Both the presampler and the back layer will be summed in the existing 0.1×0.1 manner, but the first and second layers will be summed into finer 0.025×0.1 cells, as seen in Figure 3. This allows the application of a trigger algorithm that is seeded with the highest energy cell, combined with the next highest energy cell to define a 0.025×0.2 cluster core, with the neighbors added in η,ϕ to form the cluster. Further neighbors provide an environment for studying isolation and rejecting un-isolated objects. The effect of this improvement is studied in simulation and the result is shown in Figure 4 at luminosities expected in 2018.

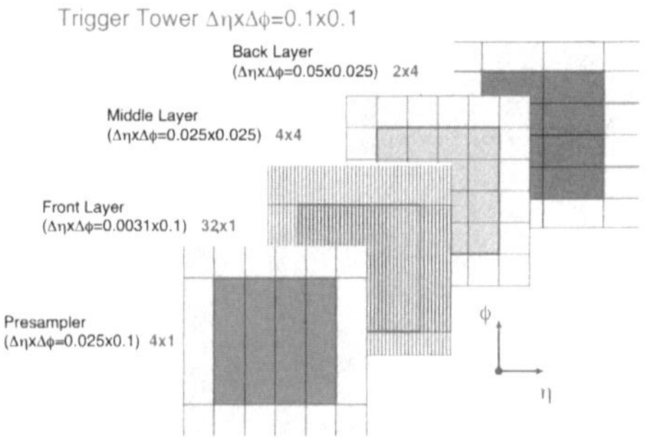

Figure 2. Diagram of existing trigger towers.

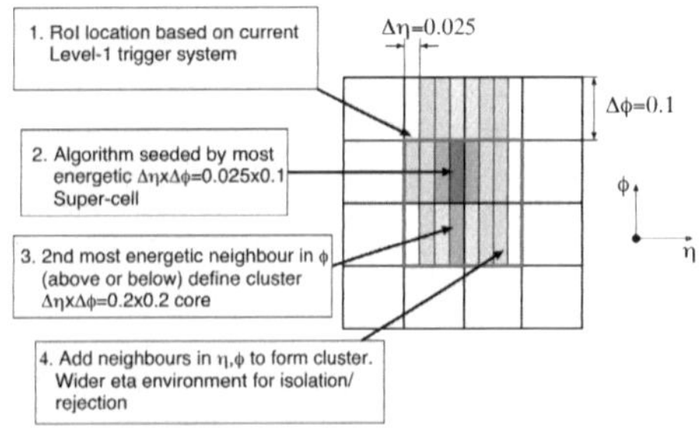

Figure 3. Diagram of proposed super-cell with upgraded trigger electronics.

The black points show a level-1 trigger based only on the $E_{\rm T}$ of the cluster, in pink is the existing triggering method, well above the 20 kHz rate limit by 2018. The blue and red points show an isolation requirement applied using the increased granularity. With this increase in granularity an unprescaled level-1 trigger near an $E_{\rm T}$ of 25 GeV is maintained.

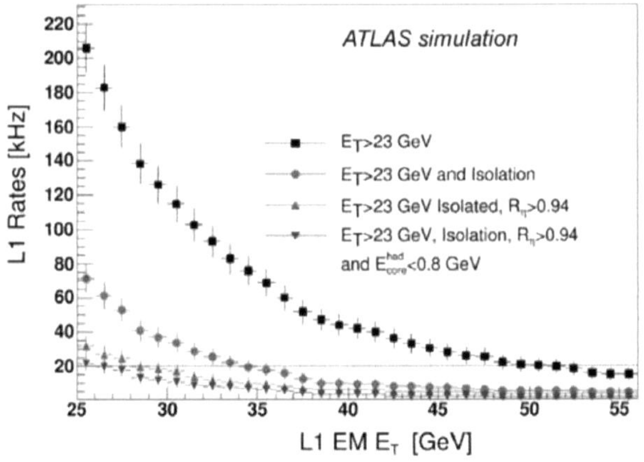

Figure 4. Simulation of the reduction in the rate for a given threshold at the level-1 trigger with high granularity input. Finer lateral and longitudinal segmentation of the EM shower shape for electrons and photons is used to reduce rates.

For the HL-LHC, planned for 2022, it will be necessary to upgrade all of the more than 1500 front end boards (FEBs), preparing the detector for the expected 3000 $\rm fb^{-1}$. This complete replacement allows the design of a more flexible system that utilizes the full precision and granularity of the LAr calorimeters at trigger level. This simpler, "free running" architecture results in an effectively infinite pipeline and level-1 bandwidth with little or no impact on the latency. Technically, this requires 40 MSPS digitization for all channels over a 16 bit dynamic range within the existing power, cooling and space constraints in a radiation environment. Once digitized the data will be moved off each board at 100 Gbps, or 150 Tbps for the LAr calorimeters. With this the maximum granularity of the calorimeter is provided to the trigger.

3. Existing Electronics

The current readout electronics of the calorimeter are shown schematically in Figure 5. The analog signals from the detector are captured by the FEBs and are prepared for digitization by the preamplifier and shaper chips. Additionally the Layer Sum Boards (LSBs) combine the analog, cell-level signals into a single signal for each layer and send the analog sums to the Tower Builder Board (TBB). The TBB combines the four layers into one tower which is digitized, converted to transverse energy and sent to the level-1 calorimeter trigger. If the event passes the trigger selection an accept signal is sent to the FEBs to digitize the full readout of the calorimeter. The digital counts are then sent off detector, converted into energy and sent to the DAQ system. Though adequate for the present data taking the system is limited by the granularity of the signals sent to the level-1 trigger.

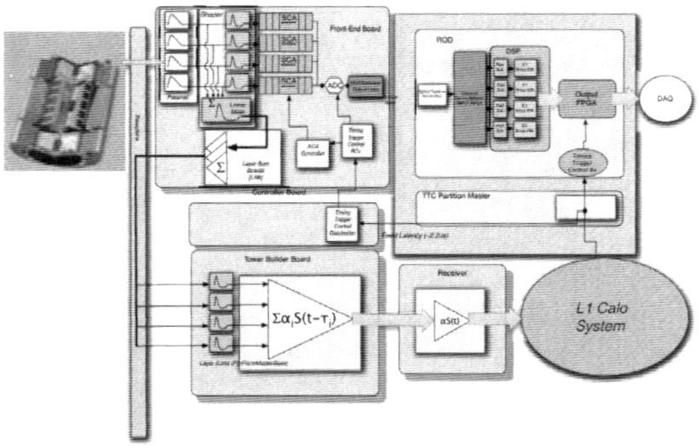

Figure 5. Existing LAr calorimeter readout electronics.

4. Upgrade Implementation Phase 1

The upgrade foreseen for Phase 1 (2018) sends additional information to the level-1 trigger by making a minimal change to the overall system and maintaining compatibility with the existing trigger (Figure 6). Four layer sums (the middle two with higher granularity than the existing readout) are sent to the new LAr Trigger Digitizer Board (LTDB) which digitizes these signals. At the same time it combines the layer sums to provide the correct granularity input to the existing TBB. The digital signals from the LTDB are sent to the off detector Digital Processing System (DPS), which converts the digital signals to energy and timing information. The L1 calorimeter trigger system then receives information from LTDB and the TBB. These can be used simultaneously during commissioning, eventually phasing out the TBB.

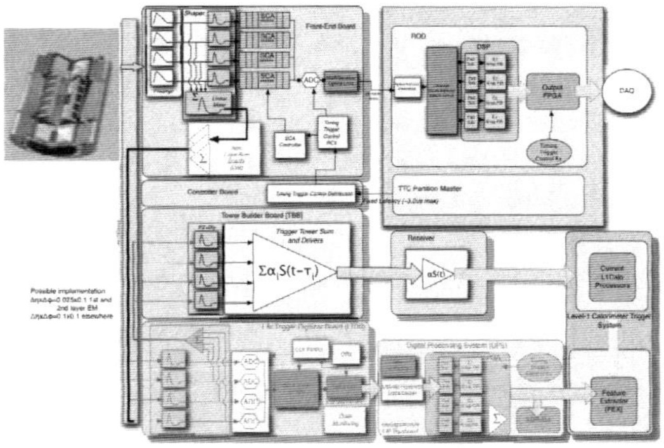

Figure 6. Proposed block diagram for Phase 1 electronics upgrade.

The construction of the LTDB and the DPS requires the development of several new components. For the LTDB a radiation hard ADC and a serializer and optical link are necessary, as well as the design of the high-speed off-detector FPGA-based DPS. For the LTDB the ADC must be capable of digitizing at 40 MSPS with 12 bit precision. This chip should consume less than 100 mW of power per channel and fit in a small footprint, send serialized output and be radiation tolerant to withstand $\sim 10^{14}$ hadrons/cm^2. The specification for the new FEBs in Phase 2 are similar and the same chip may be used for both boards. For Phase 2 there is the additional requirement of 16 bit precision which is achieved with the same 12 bit ADC and adding a gain selection. A test chip has been prepared using the IBM 8RF 130 nm technology. The test chip itself contains two pipeline channels each containing a four stage ADC with 1.5 bits/stage, sample-and-hold and gain selection structures, and additional support structures. These four stages digitize the most significant bits with the remaining eight bits digitized by a commercial ADC. The additional half bit at each stage is used to determine and apply a digital error correction. Testing has been completed and the analog performance is found to be equivalent to a commercial 12 bit ADC. The cross talk between the two channels is found to be below the noise level and the device has been tested to be radiation tolerant to greater than 10 MRad and 2×10^{14} protons/cm^2. Two test chips are being designed for submission in 2012/2013. The first is a four channel 12 bit ADC with the first four bits digitized as described above, and the last eight bits by a SAR ADC. The other is a new design for a two channel 12 bit SAR ADC which has the advantages of being low power and low latency.

The data are sent off detector using a radiation hard optical connection. A test chip has been built using CMOS 0.25 μm Silicon-on-Saphire technology. Each optical link transfers data at $\sim$5 Gbps. The data is multiplexed and serialized, reducing 16 parallel data streams to one serial output. The power dissipated is 460 mW and the radiation hardness has been tested. An 8 Gbps test chip has been submitted for fabrication.

The new LTDB signals require off-detector electronics to integrate with the level-1 calorimeter trigger system. The digitized high granularity layer sums will be calibrated and filtered to suppress pileup by the Digital Processing System (DPS). Each LDTB will send $\sim$200 Gbps/board over optical links to the DPS which must process this large volume of data and send the resulting information to the trigger system. The DPS is an ACTA board that receives the high bandwidth optical signals on $4 \times 40 \times \sim 5$ Gbps links from LTDBs. Internally, high-end FPGAs each receive $40 \times \sim 5$ Gbps signals and apply digital filtering to convert raw ADC counts to calibrated energy and timing information. These data are sent to the new Feature EXtractor (FEX) in the level-1 calorimeter trigger system. Several FPGAs are being considered, including the Xilinx Virtex-7. This FPGA is being used to optimize the firmware for the energy calibration.

5. Upgrade Implementation Phase 2

As the existing FEBs come to the end of their expected operational lifetime in 2022 there will be an opportunity to switch to a fully digital architecture. The new FEBs will digitize continuously at the full precision. This will be read out to a new digital off detector system that will provide input for a fully digital level-1 trigger system. The LTDB and DPS from the Phase 1 upgrades may then be used for a new level-0 trigger. Schematically these changes can be seen in figure 7. Here all the channels are immediately digitized by the FEB, the TBB has been removed completely, and a new off detector Read Out Driver (ROD) provides the link to the level-1 trigger. Many of the new technologies that are being developed for Phase 1 will be applied in Phase 2.

One outstanding ASIC needed for the new FEB but not the LTDB is the preamplifier/shaper chip. This ASIC receives, shapes and amplitude adjusts the analog signals for digitization. This must be radiation tolerant with a dynamic range of 16 bits achieved in three stages (gains). A test chip has been prepared and the power consumption was measured with the preamplifer consuming less than 50 mW and the shaper less than 300 mW, with a maximum signal of 5 mA with an INL of 0.1%. A new test chip has been prepared using the IHP SiGe 25 H3P technology

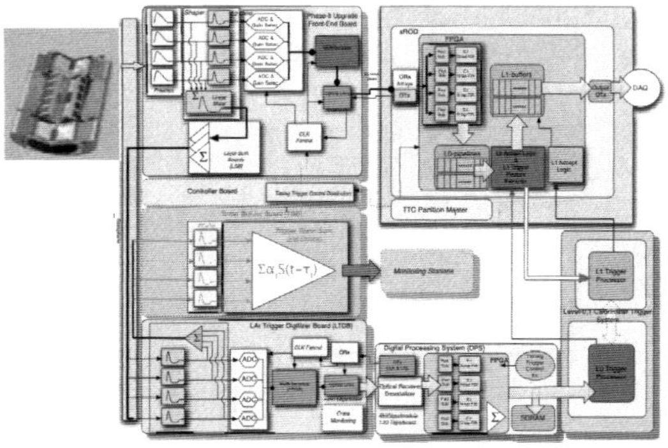

Figure 7. Proposed block diagram for Phase 2 electronics upgrade.

with 2 preamplifier blocks and 1× and 10× gain/shaper blocks.

Off detector the RODs will be replaced to handle 150 Tbps of data from the new detector readout (100 times more than the current system). This requires high speed data processing and transmission. The use of ATCA blades/boards is being planned, which will use high-end FPGAs to apply calibration for energy, time alignment, and provide data monitoring. This work overlaps extensively with the DPS design of Phase 1, which will form the core of the new RODs. The data output from the new RODs will send the full granularity of the calorimeter to trigger system, allowing for the most flexible trigger algorithms possible, and the same calibration is applied for both trigger and reconstruction. Complex algorithms such as the extraction of the features of electromagnetic shower shapes or fast tagging of π^0s may be possible.

6. Outlook

Over the last two years the current detector and readout have performed exceptionally well, resulting in exceptional physics. Over 98% of 182k channels are operational [2]. However, for 2018 an upgrade is needed in order to maintain the sensitivity to physics measurements at luminosities beyond 10^{34} cm^{-2}s^{-1}. The upgrades will happen in two stages: Phase 1 and Phase 2. In Phase 1 the trigger path will be upgraded. Test devices have been produced and have been shown to meet the required specifications. Phase 2 will be a full upgrade of on- and off-detector electronics and the level-1 trigger to a completely digital system. All signals from the LAr will be digitized at the beam crossing rate. This requires the design of a system for digitizing and moving the data, while fitting into existing power, cooling, and space requirements in a radiation environment. In the near term a Phase "0" quarter size demonstration board is planned and organization of the Phase 1 construction, testing, installation and commissioning is underway. The last year has witnessed the excellent performance of the LHC and ATLAS. With this success now in the past, the LAr readout and trigger systems are being preparing for highest LHC luminosities and interesting physics at the HL-LHC.

References

[1] ATLAS Collaboration 2008 *JINST* **3** S08003.
[2] Abouzeid, H. 2012 Status of the ATLAS Liquid Argon Calorimeter and its Performance after Two Years of LHC Operation, *XVth International Conference on Calorimetry in High Energy Physics Conference Proceedings.*
[3] ATLAS Collaboration 2012 *Phys. Lett.* B **716** 1.
[4] ATLAS Collaboration 2011 Letter of Intent for the Phase-I Upgrade of the ATLAS Experiment, CERN-LHCC-2011-012.

Neutron and proton tests of different technologies for the upgrade of cold readout electronics of the ATLAS Hadronic End-cap Calorimeter

M Nagel

Max-Planck-Institut für Physik (Werner-Heisenberg-Institut), Föhringer Ring 6, 80805 Munich, Germany, www.mpp.mpg.de

E-mail: nagel@mpp.mpg.de

on behalf of the HECPAS Collaboration
(IEP Košice, Univ. of Montréal, MPI Munich, IEAP Prague, NPI Řež)

Abstract. The expected increase of total integrated luminosity by a factor ten at the HL-LHC compared to the design goals for LHC essentially eliminates the safety factor for radiation hardness realized at the current cold amplifiers of the ATLAS Hadronic End-cap Calorimeter (HEC). New more radiation hard technologies have been studied: SiGe bipolar, Si CMOS FET and GaAs FET transistors have been irradiated with neutrons up to an integrated fluence of $2.2 \cdot 10^{16}\,\mathrm{n/cm^2}$ and with 200 MeV protons up to an integrated fluence of $2.6 \cdot 10^{14}\,\mathrm{p/cm^2}$. Comparisons of transistor parameters such as the gain for both types of irradiations are presented.

1. The ATLAS Hadronic End-cap Calorimeter

The hadronic end-cap calorimeter (HEC) of the ATLAS experiment [1, 2] at the CERN Large Hadron Collider (LHC) is a copper-liquid argon sampling calorimeter in a flat plate design [3, 4]. The calorimeter provides coverage for hadronic showers in the pseudorapidity range $1.5 < |\eta| < 3.2$. The HEC shares each of the two liquid argon end-cap cryostats with the electromagnetic end-cap (EMEC) and forward (FCAL) calorimeters, and consists of two wheels per end-cap, as illustrated in Figure 1.

A HEC wheel is made of 32 modules, each with 40 liquid argon gaps, which are instrumented with active read-out pads. The signals from the read-out pads are sent through short coaxial cables to preamplifier and summing boards (PSB) mounted on the perimeter of the wheels inside the cryostat. The PSB boards carry highly-integrated preamplifier and summing amplifier chips in Gallium-Arsenide (GaAs) technology. The signals from a set of preamplifiers are summed to one output signal, which is transmitted to the cryostat feed-through [5]. Figure 2 shows a PSB board mounted and connected on the perimeter of a HEC wheel.

Journal of Physics: Conference Series **404** (2012) 012062 doi:10.1088/1742-6596/404/1/012062

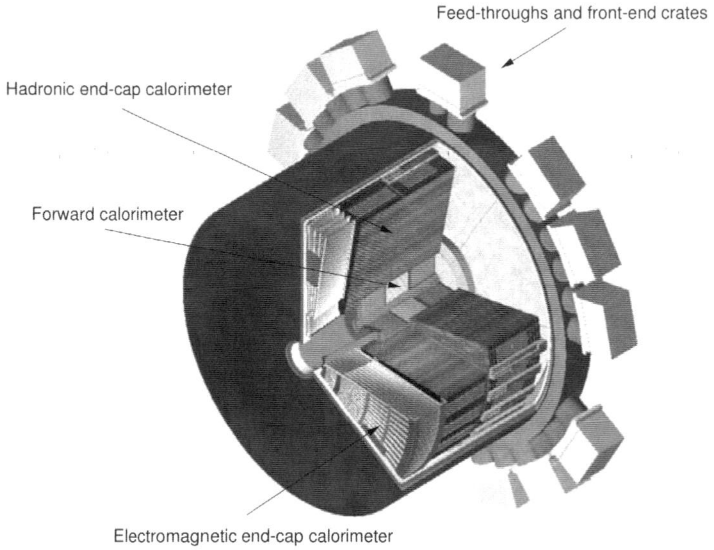

Figure 1. A liquid Argon end-cap cryostat, containing the hadronic and electromagnetic end-cap calorimeters, and the forward calorimeter [2].

Figure 2. A detailed view of a PSB board mounted on the perimeter of a HEC wheel [5].

2. Requirements of the HEC cold electronics for the HL-LHC upgrade

The GaAs technology currently employed in the HEC cold electronics has been selected for its excellent high frequency performance, stable operation at cryogenic temperatures, and radiation hardness. The radiation hardness specifications were defined for ten years of operation at the LHC design luminosity of $10^{34}\,\mathrm{cm}^{-2}\,\mathrm{s}^{-1}$, including a safety factor of ten. For the high-luminosity upgrade of the LHC (HL-LHC), the luminosity is foreseen to increase by a factor of 5–10, effectively eliminating the safety factor. The ATLAS collaboration therefore decided to re-examine the radiation hardness of the current HEC cold electronics and of potential alternative technologies [6, 7]. Detailed studies of the expected radiation levels after ten years of running under HL-LHC conditions yielded the following requirements (including a safety factor of 10) for the HEC cold electronics [8]:

- 1 MeV equivalent neutron fluence (NIEL in Si) of $1.7 \cdot 10^{15}\,\mathrm{n/cm^2}$
- Total hadron fluence (SEE) of $1.8 \cdot 10^{14}\,\mathrm{h/cm^2}$ with $E_h > 20$ MeV
- Total ionization dose (TID) of 22 kGy

3. Tests

The neutron irradiation tests were performed at the Fast Neutron Facility at the Nuclear Physics Institute of the ASCR in Řež near Prague, Czech Republic, up to an integrated fluence of $2.2 \cdot 10^{16}\,\mathrm{n/cm^2}$ (in terms of the 1 MeV equivalent NIEL in Si described in the next section). A 37 MeV proton beam incident on a D_2O target created a beam of neutrons with a continuous energy spectrum up to a maximum neutron energy of 34 MeV and a flux density up to $10^{11}\,\mathrm{n/cm^2/s}$ with a $1/r^2$ decrease in flux density. The proton irradiation test were performed at the Proton Irradiation Facility at the Paul Scherrer Institute in Villigen, Switzerland, with a monoenergetic beam of 200 MeV protons up to an integrated fluence of $2.6 \cdot 10^{14}\,\mathrm{p/cm^2}$. The transistors were bonded in ceramic casings and mounted on boards, which were then placed in a standard aluminum frame with 17 mm distance between slots and carefully aligned in the particle beams. The three different transistor technologies being tested were Si CMOS FET in SGB25V 250nm technology from IHP, SiGe Bipolar HBT (IHP SGB25V 250nm and IBM 8WLBiCMOS 130nm), and the GaAs FET currently used in ATLAS, either the Triquint CFH800 250nm transistors themselves or integrated into the HEC BB96 Preamplifiers and Systems. The performance of the transistors was monitored *in situ* during irradiation with a vector network analyzer recording a full set of S-parameters [9], which were converted to standard transistor parameters using suitable small signal circuit models.

4. Flux determination

A combination of radiation films and radmon diodes, placed at various slot positions along the beam, were used to determine the beam profile and, together with the beam current measurements, the particle flux. The films were exposed to a given beam current for various amounts of time and subsequently scanned. An iterative calibration procedure was used to obtain the dose as a function of net optical density [10], by combining the known relative dose measurements with the absolute normalization obtained from MC simulations and from the radmon diodes.

For the neutron test with its continuous neutron energy distribution, the fluence was scaled to the 1-MeV equivalent neutron fluence in Silicon, employing the non-ionizing energy loss (NIEL) scaling hypothesis [11], independent of the actual technology under irradiation. The divergent nature of the neutron beam emerging from the heavy water target yielded a total fluence per slot given by $\Phi = 3.22 \cdot 10^{16} \cdot (\mathrm{slot\#} + \frac{3}{17})^{-2.11}\ n/cm^2$. At PSI in contrast, the almost collimated proton beam together with an offset in the alignment of the boards yielded a basically constant

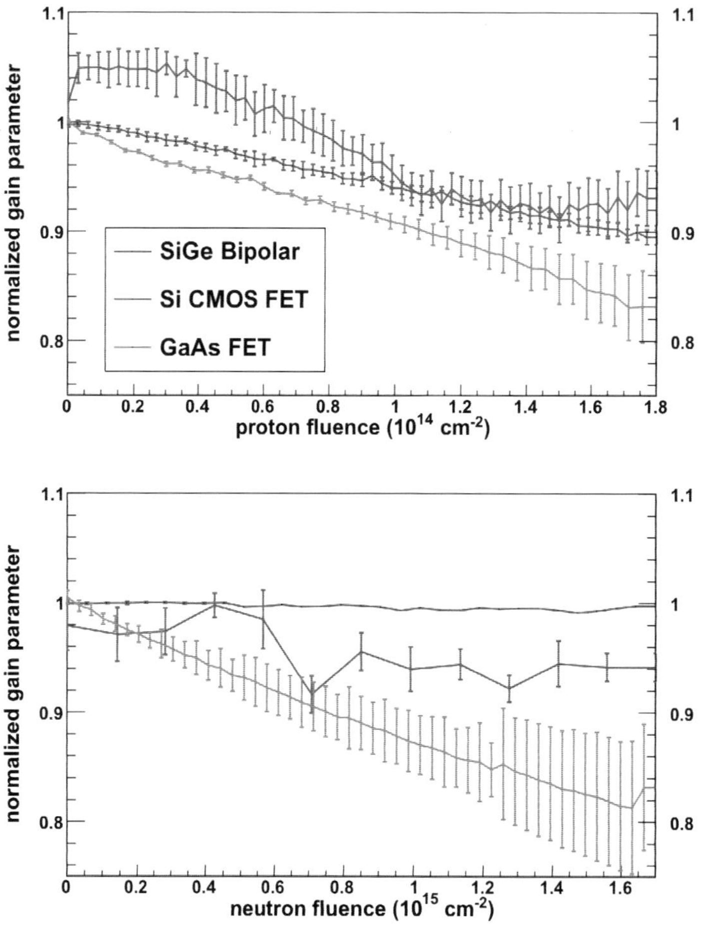

Figure 3. Relative gain loss as a function of proton (top) and neutron (bottom) fluence for the various technologies described in the legend. The respective gain parameters are described in the text. The data points represent the average of all the devices of a particular technology that were irradiated up to a given particle fluence; the error bars are the sample RMS [12].

fluence for all slots of $\Phi = (2.6 \pm 0.4) \cdot 10^{14}$ p/cm^2, where the main uncertainty stems from the unknown exact positions of the test structures inside the chip casing.

5. Results

The various transistor parameters were calculated from the measured S-parameters using suitable small signal circuit models. The transistor parameters were averaged over a certain frequency range to obtain their mean and rms-values for every set of S-parameters. This frequency range extended from 300 kHz to 100 MHz, unless a certain parameter was unstable at low or high frequencies, in which case appropriate cuts were applied. These averaged transistor values were then used to characterize their behaviour as a function of particle fluence.

Table 1. Loss of gain of various transistor technologies under neutron and proton irradiation [12].

Technology	Neutron fluence $1.7 \cdot 10^{15}\,\mathrm{n/cm^2}$	Proton fluence $1.8 \cdot 10^{14}\,\mathrm{p/cm^2}$
Si CMOS FET	0 %	-11 %
SiGe Bipolar	-5 %	-7 %
GaAs FET	-20 %	-17 %

Figure 3 summarizes the results for the proton (top) and the neutron (bottom) tests in terms of the appropriate gain parameters as a function of the corresponding particle fluence up to the required limit, normalized to the corresponding value before irradiation. The various gain parameters are the real part of the transconductance g_m for the FET transistors, the differential current gain β for the Bipolar transistors, and in case of the HEC BB96 preamplifiers and systems, where our simple transistor models are not applicable, the transresistance $r_m = \frac{\Delta V_{\mathrm{out}}}{\Delta I_{\mathrm{in}}} = |S_{21} \cdot Z_{\mathrm{in}}|$, which relates the output voltage to the input current, where Z_{in} is the input impedance. Displayed are the sample averages of all the devices of a particular technology under test at each of the two test facilities, that were irradiated up to a given particle fluence; the error bars represent the sample RMS. The relative change of the gain parameter is quoted in Table 1, where we also used a linear extrapolation of the gain parameter up to the required fluence for those devices that did not reach that fluence.

6. Summary

Our first observation is that proton irradiation seems to cause more damage to the electronics than neutron irradiation (at least for the Si CMOS and SiGe Bipolar transistors), despite the fact that an order of magnitude more neutrons than protons are expected to impact the HEC cold electronics. In the case of the GaAs FETs, the gain degradation is roughly the same for both types of irradiation, although a quantitative assessment suffers from the fact that two different gain parameters (transconductance vs. transresistance) are being compared.

It can be concluded that both the Si CMOS FETs and the SiGe Bipolar transistors considered as alternative technologies are more radiation hard than the currently used GaAs FETs. Among the alternative technologies, a preference would be given to the Si CMOS FETs, since the SiGe Bipolar transistors require a stabilization of their operation point, which would lead to a more complex circuitry to meet the same radiation hardness requirements.

The GaAs technology currently used in the HEC cold electronics degrades significantly at the expected radiation levels, but it does not break down yet. The observed level of degradation ($\sim$20%) could conceiveably be compensated for by some sort of calibration procedure. But before one can come to a decision whether it is safe or not to keep the current HEC cold electronics for HL-LHC conditions, the impact of this level of degradation on the overall HEC performance (e.g. in terms of jet energy resolution) needs to be better understood. In addition to that, a measurement of the performance degradation in the cold is needed. A decision has thus been reached to perform a more detailed study of the current HEC BB96 ASICs, including characterization of the linearity of their response with realistic pulses as they appear in the ATLAS detector, and a detailed simulation of the propagation of the 'degraded' signals through the summation, digitization, and reconstruction stages. This, and the measurement of the performance degradation in the cold are both ongoing efforts.

References

[1] 1994 *ATLAS: technical proposal for a general-purpose pp experiment at the Large Hadron Collider at CERN* LHC Tech. Proposal (Geneva: CERN) cern-lhcc-94-43, lhcc-p-2
[2] Aad G *et al.* (ATLAS Collaboration) 2008 *J.Inst.* **3** S08003 doi:10.1088/1748-0221/3/08/S08003
[3] ATLAS Collaboration (ATLAS) 1996 Atlas liquid argon calorimeter: Technical design report Tech. rep. CERN cern-lhcc-96-41, atlas tdr 2
[4] Gingrich D M *et al.* (The ATLAS Hadronic End-Cap Calorimeter Group) 2007 *J.Inst.* **2** P05005 atl-larg-pub-2007-009, atl-com-larg-2007-006, mpp-2007-237
[5] Ban J, Brettel H, Cwienk W D, Fent J, Kurchaninov L, Ladygin E, Oberlack H, Schacht P, Stenzel H and Strizenec P 2006 *Nucl.Instrum.Meth.A* **556** 158–168 mpp-2005-193
[6] Vankov P 2012 *arXiv:1201.5469* ATL-UPGRADE-PROC-2012-003
[7] Schacht P 2010 *Proceedings of the 11th ICATPP Conference on Astroparticle, Particle, Space Physics, Detectors and Medical Physics Applications* (World Scientific) pp 392–403 ISBN 9789814307529 doi:10.1142/9789814307529_0065
[8] Baranov S, Bosman M, Dawson I, Hedberg V, Nisati A and Shupe M 2005 Estimation of radiation background, impact on detectors, activation and shielding optimization in atlas Tech. Rep. ATL-GEN-2005-001, ATL-COM-GEN-2005-001,CERN-ATL-GEN-2005-001 CERN Geneva (and unpublished updates)
[9] Dicke R H 1947 *J.Appl.Phys.* **18** 873 doi:10.1063/1.1697561
[10] Martišíková M and Jäkel O 2010 *Phys.Med.Biol.* **55** N281–290 doi:10.1088/0031-9155/55/10/N03
[11] Vasilescu A 1997 The niel scaling hypothesis applied to neutron spectra of irradiation facilities and in the atlas and cms sct Tech. Rep. ROSE/TN/97-2 CERN
[12] Oberlack H, Dannheim D, Fischer A, Hambarzumjan A, Pospelov G, Reimann O, Rudert A and Schacht P 2009 *Nuclear Science Symposium Conference Record (NSS/MIC), 2009 IEEE* pp 758–762 ISSN 1095-7863

The Electromagnetic Calorimetry of the PANDA Detector at FAIR

R. W. Novotny and for the PANDA collaboration

2nd Physics Institute, University Giessen, Heinrich-Buff-Ring 16, D-35392 Giessen, Germany

r.novotny@exp2.physik.uni-giessen.de

Abstract. The PANDA collaboration at FAIR, Germany, will focus on undiscovered charm-meson states and glueballs in antiproton annihilations to study QCD phenomena in the non-perturbative regime. For fixed target experiments at the storage ring HESR a 4π-detector for tracking, particle ID and calorimetry is under development and construction to operate at high annihilation rates up to 20 MHz. The electromagnetic calorimeters are composed of a target spectrometer (EMC) based on $PbWO_4$ crystals and a shashlyk-type sampling calorimeter at the most forward region. The EMC, comprising more than 15,000 crystals, is operated at a temperature of -25 °C and read-out via large-area avalanche photo-diodes or vacuum photo-triodes/tetrodes. The photo sensor signals are continuously digitized by sampling ADCs. More than 50% of the high quality PWO-II crystals are delivered and tested. The excellent performance with respect to energy, time and position information was determined over a shower energy range from 10 MeV up to 15 GeV by operating several prototype detectors. In addition, the concept of stimulated recovery has been investigated to recover radiation damage on- and off-line during the calorimeter operation. Besides the overall concept of the target spectrometer the response function of the shashlyk spectrometer down to photon energies even below 100 MeV is presented.

1. The overall concept

Lead tungstate ($PbWO_4$ or PWO) has become a widely used scintillator material in medium and high energy physics [1,2,3] due to its compactness, fast response, short decay time and radiation hardness. For applications at medium energies the initially low luminescence yield has been significantly improved and optimized, leading to a second generation PWO-II [4], which was developed in collaboration with the *Bogoroditsk Technical-Chemical Plant* (BTCP), Bogoroditsk in Russia.

The PANDA detector, one of the approved major experiments at the future FAIR facility at GSI (Darmstadt, Germany), is dedicated to the hadron-physics program exploiting the stored and cooled beam of anti-protons. The research program is focusing on charmonium spectroscopy, gluonic excitations, open and hidden charm in nuclei and gamma-ray spectroscopy of hypernuclei [5]. Figure 1 provides an overview of the target and forward section of the PANDA detector and marks the two electromagnetic calorimeters EMC and FSC, respectively.

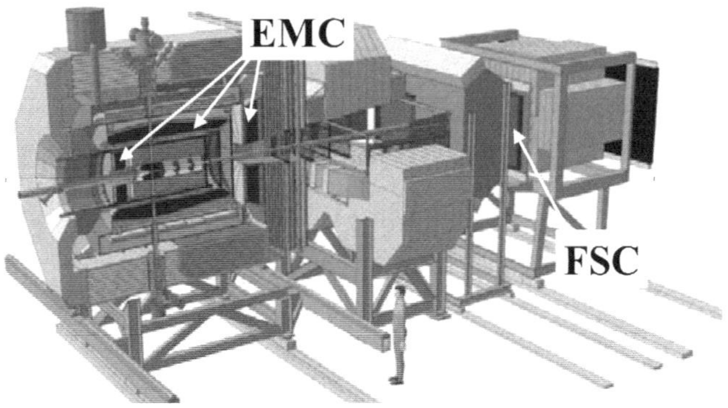

Figure 1. Schematic view of the PANDA detector consisting of a Target and Forward Spectrometer. The electromagnetic calorimeters consist of the EMC in the target and the FSC in the forward region, respectively.

2. The detector concept of the EMC in the target region

The electromagnetic calorimeter (EMC) of the target spectrometer has to provide high-resolution photon and meson spectroscopy [6,7]. It will have to cope with photons in the energy range starting at a few tens of MeV up to several GeVs. A pair of large area avalanche photo diodes (LAAPD) or a vacuum photo triode/tetrode (VPT, VPTT) are foreseen as photo-sensor of the individual crystal to allow operating in the magnetic field of the superconducting solenoid. Figure 2 illustrates two of the three major components of the calorimeter, the barrel and forward endcap. Irrespectively of the collected percentage of the light yield, additional cooling reduces the thermal quenching of the scintillation processes and allows an improvement of the energy resolution, when photon statistics dominates. A temperature of $T = -25^{o}C$ as the operating temperature provides a light increase of a factor of ~ 4 compared to room temperature.

The quality of PWO was improved primarily due to a reduced concentration of defects achieved by raw material selection and growing technology. The co-doping with La- and Y-ions could be reduced to a level <40 ppm. Therefore, full size crystals of 200mm length and nearly rectangular shape deliver at least 16 photoelectrons per MeV measured at room temperature using a phototube with bialkali photocathode. The selection of the raw material has reduced those impurities leading to slow decay components. The response is dominated by the decay component of $\tau = 6.5$ns.

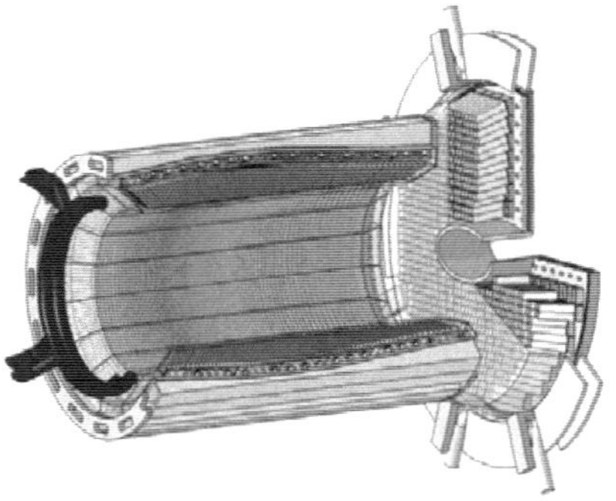

Figure 2. Schematic figure of the barrel and forward end cap as major components of the electromagnetic calorimeter of the PANDA target spectrometer (EMC).

Journal of Physics: Conference Series **404** (2012) 012063 doi:10.1088/1742-6596/404/1/012063

The layout of the barrel comprises 16 identical slices. Each one consists of 710 tapered PWO crystals of 200mm length shaped in eleven geometries of two different symmetries. All crystals have a transversal size slightly above the Moliére radius of 2.1cm. The 11.360 crystals are pointing slightly off the asymmetrically located target position. Each crystal is contained in a carbon fiber alveole including the two rectangular LAAPDs (active area $7x14mm^2$), the pre-amplification and shaping circuits. Low noise and power consumption and a dynamic range of 10,000 are achieved by a custom designed ASIC (*APFEL*). The concept of the front-end electronics is based on count rates of 100kHz per module and all components will be integrated into the cold section temperature stabilized on the level <0.1°C. The mechanical support structure will house the electronics for control and digitization to minimize the space for signal transfer out of the detector.

Both endcaps are designed in a similar manner as illustrated in figure 2. However, crystals are of identical and slightly tapered (forward endcap) or straight (backward endcap) shape pointing off the target. In order to cope with count rates well above 500kHz vacuum photo triodes/tetrodes are chosen in the most forward region due to the momentum boost in a fixed target experiment.

3. The quality of PWO crystals

The so far available PWO crystals were produced by the Bogoroditsk Technical Chemical Plant, Bogoroditsk, in Russia. The mass production was performed with very selective quality limits for optical transmittance, homogeneity, light yield, scintillation kinetics and radiation hardness.

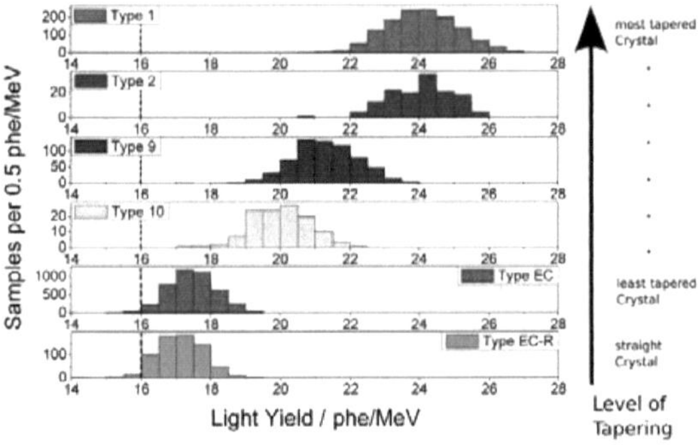

Figure 3. Distribution of the light yield for different crystal shapes ordered by the level of tapering.

The general acceptance tests were performed in close collaboration with the CMS/ECAL collaboration at CERN exploiting the ACCOS machine, which was adapted to the PANDA geometry and specifications. The measurements of radiation hardness and absolute light yield including cross checks at low temperature as well as the final analysis are done at Giessen. So far, more than 9.000 crystals have been delivered and are tested.

The optical transmissions at the relevant wavelengths of 360, 420 and 620nm, respectively, are measured longitudinally and perpendicularly to the crystal axis in order to control the homogeneity. The obtained values are well above the specification limits (T≥35% @ 360nm; T≥60% @ 420nm and ≥70% @ 620nm). The significantly improved light yield is documented in figure 3. The detected number of photoelectrons per MeV deposited energy (> 16 p.e./MeV) is determined at room temperature (T=18°C) with a standard photomultiplier with fused silica window and bialkali photocathode. Light collection in a more tapered crystal shape further enhances the effective light yield compared to the nearly straight geometry of the endcap modules. More than 90% of the light is collected within a time gate of 100ns, which even holds when crystals are cooled down to T=-25°C.

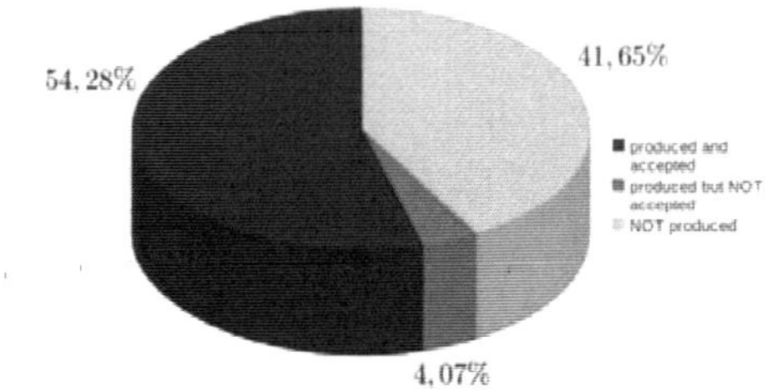

Figure 4. Present status on the crystal delivery for the PANDA EMC.

The most critical parameter is the change of the absorption coefficient Δk due to radiation damage. The crystals are irradiated at room temperature at the facility at Giessen with an integral dose of 30Gy using a set of ^{60}Co sources (dose rate 4Gy/minute). The longitudinal transmission is measured up to a wavelength of 900nm delayed by 30 minutes after irradiation in order to exclude fast recovery processes. The required value $\Delta k \leq 1.1 m^{-1}$ at 420nm guarantees that light losses due to radiation damage within an experimental period of 6 months remain tolerable even if the detectors are operated at low temperature. The rejection rate is on the level of ~5%. Figure 4 provides an overview of the delivered crystals. Unfortunately, BTCP went out of business recently. As an alternative manufacturer for the yet missing PWO crystals prototypes produced by SICCAS at Shanghai, China, are under investigations and fulfill presently most of the required specifications.

4. Prototype tests

In the past years several experiments with full size prototype arrays in PANDA geometry (tapered, 200mm length) have been performed to determine the response functions to photons and hadrons over a wide dynamic energy range [5, 7, 8]. The achievable energy resolution for electro-magnetic probes up to 15GeV energy is the most relevant parameter to enable the experimental program.

The tagged photon facilities at MAXLab (Lund, Sweden) and at MAMI (Mainz, Germany), respectively, have provided energy-marked photons between 20MeV and 1.5GeV with high intrinsic resolution. The measurements at a temperature of T=-25°C were performed either using a photomultiplier tube at the lowest photon energies or a single LAAPD. The further amplification, digitization and data acquisition was done with commercial electronics.

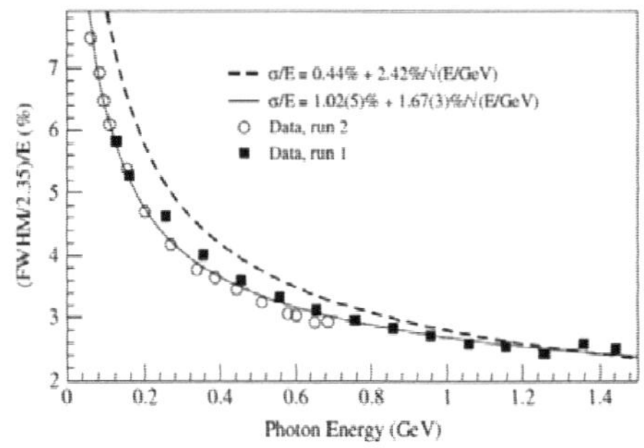

Figure 5. Cluster energy resolution as a function of the tagged photon energy obtained in two different measurements for a 3x3 matrix. The solid line is a fit to the data obtained after digitization in a sampling ADC. The dashed line corresponds to a fit to data converted with conventional readout electronics.

At photon energies as low as 20 MeV an excellent resolution σ/E<13% was achieved. The corresponding results up to 1.5GeV, as shown in figure 5, were obtained with a single LAAPD as photo-sensor. Exploring the upper energy limit figure 6 shows the line shape of a reconstructed shower created by 15 GeV positrons measured at CERN. Slightly depending on the point of impact, an energy resolution of $\sigma/E<1.5\%$ was derived. Even further improvements are expected by adding a second LAAPD, the preamplifier ASIC and performing digitization with a customized sampling ADC. Timing resolutions well below σ = 1ns for photon energies above 100MeV are confirmed already.

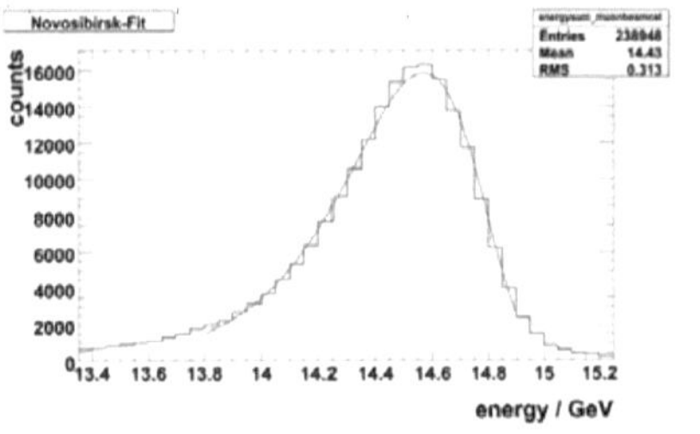

Figure 6. Line-shape of the response of a 5x5 PWO matrix (PROTO60) to the EM shower induced by 15GeV positrons of a CERN test beam. A 150GeV/c muon beam has been used for relative and absolute calibration.

5. Radiation damage and stimulated recovery

As investigated within the collaboration [9], the time constants of the recovery mechanisms become extremely slow at T=-25°C in the order of τ >400h. During irradiation the effective light yield deteriorates asymptotically towards a value depending on the radiation hardness as expressed by Δk. The inner part of the forward endcap will cope with a dose rate of 20mGy/h for the maximum event rate of $2\cdot 10^7$ $p\bar{p}$- annihilations. If crystals with $\Delta k<0.6m^{-1}$ are selected, the detector module will lose <50% of the signal response. Color centers created during irradiation due to point structure defects or traps for electrons and holes cause absorption bands in a wide spectral region but can be spontaneously relaxed afterwards via thermo-activation. Therefore, heating of the crystal is the standard annealing procedure, which requires the disassembly of the detector system. Our R&D program for PWO-II has invented the process of *stimulated recovery* [10], which enhances significantly the recovery by illumination with external light even in the infrared region. Figure 7 illustrates the effect observed at room temperature after an irradiation with an integral dose of 30Gy. The recovery with illumination time is expressed by the change of the induced absorption coefficient Δk at 420nm using light of different wavelength.

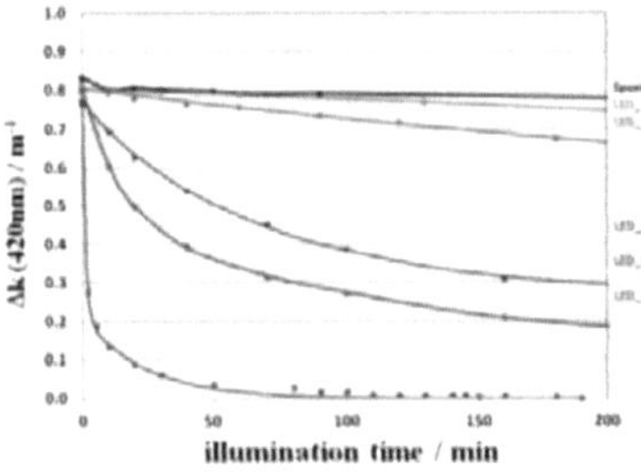

Figure 7. Recovery of a PWO-II crystal after irradiation with 30Gy at room temperature.

More relevant for the PANDA-EMC application is the behavior at low temperature. For blue light, nearly 90% of the original signal amplitude can be restored within the first 200 minutes imposing a photon flux of ~10^{16} photons/s. Even at longer wavelength (~950nm) a significant recovery is observed with longer time constants and indicates sensitivity to different color centers. Depending on the wavelength sensitivity of the quantum efficiency of the different photo sensors, *stimulated recovery* can be applied on- or off-line without bringing the calorimeter up to room temperature.

6. The forward calorimeter FSC

In order to reach a nearly 4π detection efficiency for electromagnetic probes the forward solid angle determined by the forward spectrometer is covered by a sampling calorimeter of 4.6 m^2 surface comprising 54x28 cells of a shashlyk type design. Each sub-module has a cross section of 5.5x5.5cm^2 (~ $1R_M$) composed of 380 layers of lead absorbers (0.275mm thick) and plastic scintillator tiles (1.5mm thickness), respectively, with a total length of 68.4cm corresponding to 19.6 X_o. The scintillation light is collected and accumulated via 18 wavelength shifting fibers (Bicron BC91, 1mm^{Ø}), which are inserted through holes and form loops on the front side of the module. Photomultipliers (19mm^{Ø}) are used as photo sensors since the calorimeter is placed at sufficient distance to the tracking magnets. Figure 8 illustrates the mechanical design of the two parts surrounding the beam pipe.

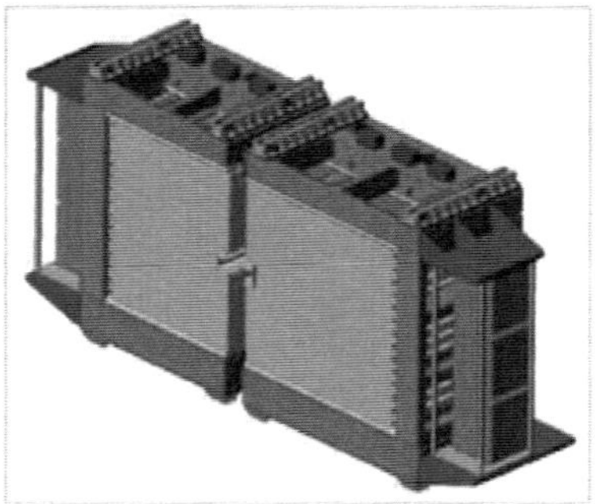

Figure 8. Schematic view of the forward shashlyk calorimeter comprising in total 54x28 detector cells arranged in two sections on both sides of the beam pipe.

The design of the shashlyk calorimeter is optimized for electromagnetic probes and a very low detection threshold to cover the full photon energy range down to a few tenths of MeV. Initial tests with electrons between 1 to 20 GeV, performed at the test beam facility at IHEP Protvino, provided first experience to optimize the detector structure, the overall performance with respect to energy, time and position response and the envisaged read-out chain including photo-sensors and signal digitization.

However, the most sensitive criteria to explore the detector limitations were performed recently using energy marked photons between 100MeV and 750MeV, respectively. The high-resolution photon beam was provided by the Bremsstrahlung tagging facility at the electron accelerator MAMI at Mainz in Germany.

Photomultiplier tubes of two different types (Philips XP1912, Hamamatsu R7899) were used as photo sensors and a sampling ADC (Wiener AVM16) was performed the digitization of the complete signal shape. All detector units composing a 6x6 matrix were relatively calibrated by impinging the direct photon beam of approximately 10mm diameter onto the center of each module. Acceptable energy, position ($\sigma < 2$cm) and time resolutions ($\sigma \sim 100\text{ps}/\sqrt{E/\text{GeV}}$) have been obtained so far from the reconstruction of the electromagnetic shower. However, the energy response of the individual detector modules show a not acceptable dependence on the point of impact of the photons. Therefore, a significant modification of the detector design has been initiated and to submit the required final technical design report. Figure 9 shows as an example the obtained promising energy resolution for different module clusters, however for a fixed point of impact.

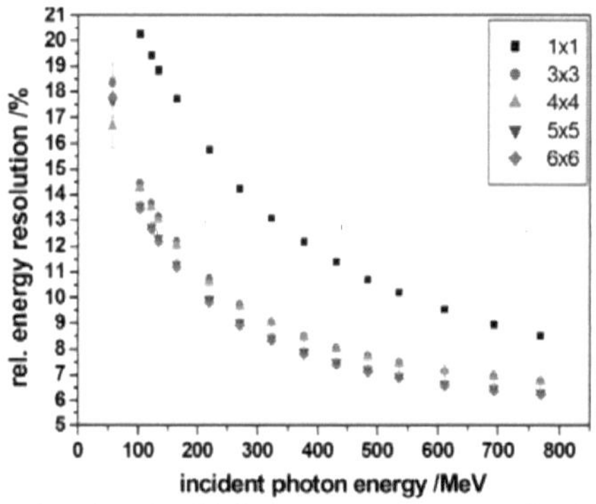

Figure 9. Energy resolution of a the shashlyk prototype measured with high-energy photons. The resolution of the reconstructed electromagnetic shower is shown for different clusters of modules. The beam was hitting the center of the central module.

7. Conclusions and outlook

The two high-resolution electromagnetic calorimeters, which represent one of the major detector components of the PANDA detector, will provide for the first time excellent performance over an extremely wide energy range as required by the physics program. The final construction of parts of the target EMC will be started soon in order to install the PANDA detector on time to be ready for commissioning in 2017/18.

References

[1] CERN/LHCC 97-33, CMS TDR 4/15 December 1997
[2] Dafinei I 2005, *Nucl. Instr. and Meth. in Phys. Res.* A 537 368
[3] Aleksandrov D V, et al. 2005 *Nucl. Instr. and Meth*, A 550 169
[4] Novotny R W, Döring W M, Hjelm F, Melnychuk D, Makonyi K, Reiter A, Salz C, Steinacher M, Thiel M and Zwieglinski B 2005 *Proc. IEEE Nuclear Science Symposium*, Puerto Rico, Conference Record, ISBN: 0-7803-9222-1, N12-4
[5] Physics Performance Report 2009 *arXiv:0903.3905v1*
[6] Novotny R W 2004, *IEEE Trans. on Nucl. Sci.* 51 no. 6 3076-3080
[7] Technical Design Report for: PANDA Electromagnetic Calorimeter (EMC) 2008 *arXiv:0810.1216v1*
[8] Kavatsyuk M 2011, *Nucl. Instr. and Meth. in Phys. Res.* A 648 77
[9] Novotny R W 2008, *IEEE Trans. on Nucl. Sci.* 55 1283-1288
[10] Dormenev V 2010, *Nucl. Instr. and Meth. in Phys. Res.* A 623 1082

XVth International Conference on Calorimetry in High Energy Physics (CALOR2012) IOP Publishing
Journal of Physics: Conference Series **404** (2012) 012064 doi:10.1088/1742-6596/404/1/012064

Crystals for dual-readout calorimetry

Gabriella Gaudio[1]
Istituto Nazionale di Fisica Nucleare, Sezione di Pavia, via Bassi 6, 27100 Pavia
E-mail: gabriella.gaudio@pv.infn.it

Abstract. Dual-Readout calorimetry was proposed as a technique to eliminate the factors that limit the performance of hadron calorimeters such as electromagnetic fraction fluctuations and invisible energy fluctuations in the shower development. By comparing the signals generated by the Čerenkov and scintillation light components, it is possible to determine the electromagnetic shower fraction on an event-by-event basis. Experimental tests with the Cu-fiber DREAM calorimeter have clearly demonstrated the validity of this principle. Some high-Z scintillating crystals offer the possibility of measuring independently the contributions of the scintillation and Čerenkov lights exploiting the different characteristics of the generated signals. Among these crystals there are *BGO* and $PbWO_4$. We have tested matrices of these crystals as electromagnetic calorimeters and studied the properties of the Čerenkov and scintillation components of the signals generated by high-energy electrons showering in these detectors. We evaluated the performance of such matrices as a part of the electromagnetic section of a detector for the dual readout technique.

1. Introduction

The Dual-Readout technique is based on the simultaneous measurement, on an event-by-event basis, of Čerenkov and scintillation light components. Čerenkov light is only produced by relativistic particles, dominated by the electromagnetic component of the hadron shower, while scintillation light is a measure of the visible deposited energy. This allow a direct measurement of the electromagnetic fraction (fem) of the hadron shower, thus correcting for one of the most important sources of fluctuations that limits the hadronic calorimetry.
High-Z scintillating crystals are widely used in particle physics experiments since they ensure excellent energy resolution for electromagnetic showers. On the other hand, calorimeters using a crystal electromagnetic compartment usually have a poor hadronic resolution due to fluctuation of the shower starting point in the EM section and to the different response to the em and non-em component of the shower in the two calorimeters. Application of the Dual-Readout Method to an hybrid system may improve the performance of such a calorimeter since measuring the fem on an event-by-event basis allows to correct for such fluctuations and thus to eliminate the main reasons for poor hadronic resolution.
Over the past few years many measurements have been done by the DREAM collaboration on individual crystals to assess the possibility to split the signals in the Č and S components using four different techniques. In this paper the tests performed with crystal calorimeters large enough to contain em showers are described. The two lights components have been separated and the performances of each of these has been studied.

[1] On behalf of the Dream - RD52 collaboration

Crystals for dual readout calorimetry must allow for a good Čerenkov vs Scintillation separation, a good response uniformity and a high Čerenkov light yield to reduce contributions of photo-electron (p.e.) statistics to the resolution. The aim of the current study is to determine if crystals optimized for dual readout still guarantee a good em resolution. We used two matrices, one made of BGO crystals and one of $PbWO_4$ [1].

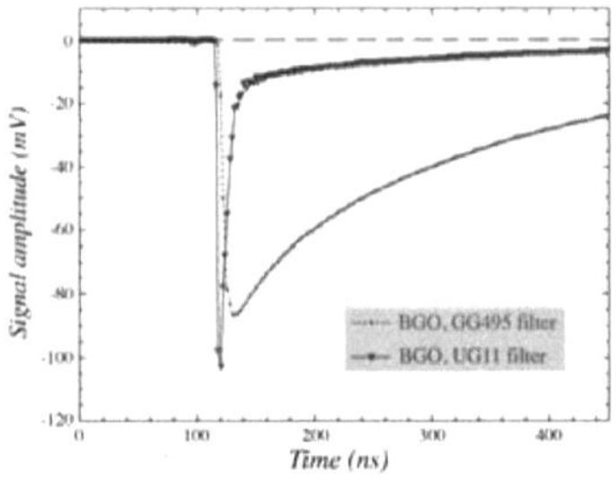

Figure 1. The time structure of typical signals measured in a single BGO crystal. The crystal was equipped with a yellow filter on one side and with a UV filter on the other.

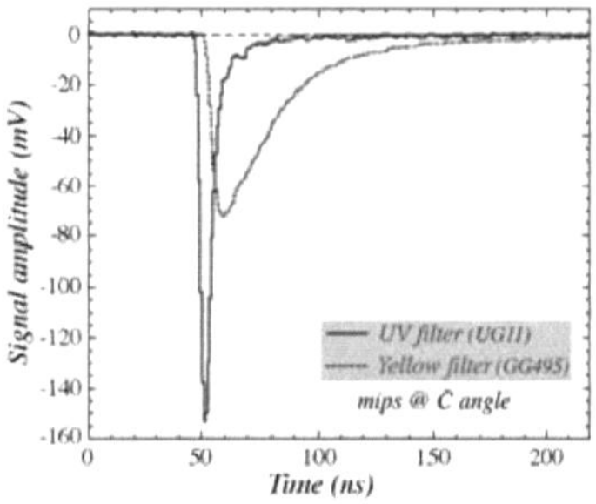

Figure 2. Average time structure of the signals from a single Mo-doped $PbWO_4$ crystal, equipped with a yellow filter on one side and with a UV filter on the other.

2. Experimental setup

In previous studies four techniques were found to allow for separation of the Čerenkov and Scintillation light component, based on the different characteristics of their emission [1]. In the present study we only exploit differences in the optical spectra and time structure.
Čerenkov light is mainly emitted in the UV region and shows a spectral characteristic of the type $\frac{dN_C}{d\lambda} = \frac{k}{\lambda^2}$ while scintillation emission is strongly dependent on the crystal type and it's usually concentrated in a (narrow) wavelength range. Since the crystal we used are quite bright scintillators, the scintillation component needed to be suppressed in order to extract the Čerenkov light by means of an UV filter. A yellow filter is also used to select a pure scintillation signal.
Moreover Čerenkov is emitted instantaneously and the signal duration is usually short, using fast enough PMT . On the contrary, scintillation light emission is characterized by one or several time constant, often showing the presence of long tails (slow components). By using a multi-sample digitizer to record the time structure it is possible to optimize the signal separation by integrating on only a part of the pulse shape. We used the CAEN V1742 digitizer, based on the DRS-IV chip, which allows for GHz range sampling [2]. Examples of the separation of the Čerenkov and Scintillation signals achievable with the two different techniques described are visible in figures 1 and 2, for BGO and $PbWO_4$ respectively.

3. BGO matrix measurements

The BGO matrix consisted of 100 crystals from a projective tower of the L3 experiment. They are 24 cm long and tapered in shape, with end faces of $2.4 \times 2.4\ cm^2$ and $3.2 \times 3.2\ cm^2$, respectively. The crystals are arranged as shown in figure 3 and readout by 16 PMTs (Hamamatsu R1355), each collecting light produced by clusters of at least 9 adjacent crystals [3]

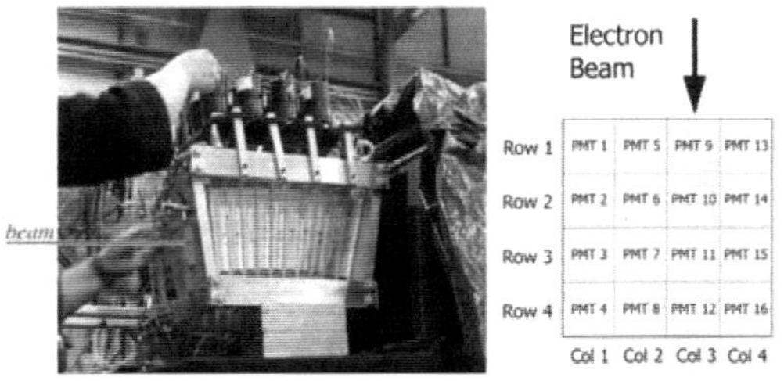

Figure 3. The BGO matrix and the PMT readout schema.

Figure 4. The $PbWO_4$ crystals positioned in the matrix

Table 1. Properties of the different optical transmission filters that have been used in the studies of the BGO and Mo-doped $PbWO_4$ crystals.

Filter type	Filter name	> 90% transmission for
UG11	"UV"	$\lambda < 400nm$
U330		$\lambda < 410nm$
UG5	"Blue"	$\lambda < 460nm$
GG495	"Yellow"	$\lambda > 495nm$

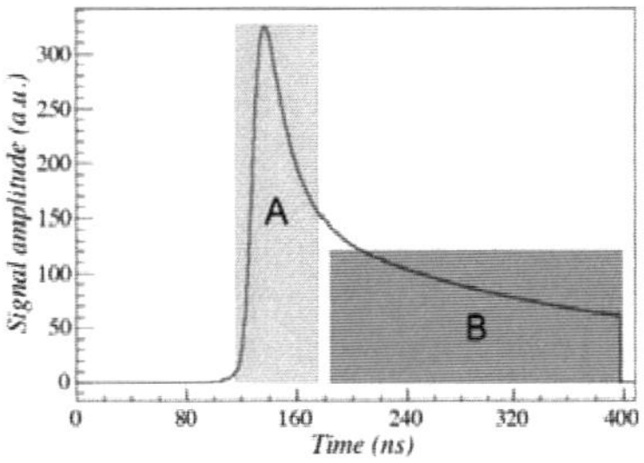

Figure 5. Average waveform for 100 GeV electrons in the BGO crystal matrix. The data measured in areas A and B were used to decompose the signals into Čerenkov and scintillation components

The BGO crystals are readout on a single side, equipped with a UG11 filter (see table 1) transmitting less than 1% of the scintillation light. This reduction is needed in order to extract the Čerenkov light. The inter-calibration of PMT is done both with LED and with the electron beam steered in each PMT column (see figure 3 right).
The signal obtained event-by-event from the digitizer, is integrated offline over different time windows in order to extract the fast component, the Čerenkov light, (gate A of figure 5) and the slow component, scintillation (gate B). To subtract the residual scintillation component in gate A, the scintillation pulse shape is modeled on pure scintillation pulses (yellow filter signals) and folded to region B of the UV filtered signal
The measurements have been performed with electron beams of 30, 60, 100, 150 GeV. The energy resolution is shown in figure 6 for the Čerenkov and scintillation light separately, and for the sum of the two contributions. For the Čerenkov light the resolution is about $36\%/\sqrt{E}$, mainly due to the low light yield of about 6 p.e./GeV. It also shows a constant term of about 1.5%. The scintillation resolution obtained is of order of $19\%/\sqrt{E}$, caused by the low photo-statistic left after the light filtering. Both signals, on the other hand, show good linearity, within $\pm 3\%$,

as shown in figure 7.

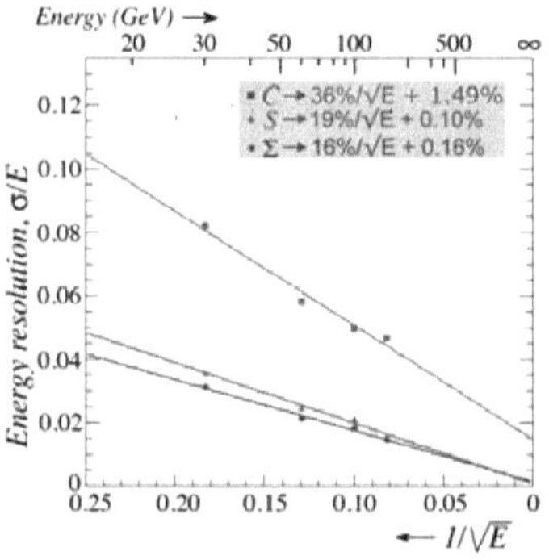

Figure 6. The energy resolution for electrons in the BGO matrix, as a function of the energy of the showering particles. Results are given for the total charge collected by the PMTs, and for the Čerenkov (C) and scintillation (S) components of the signal.

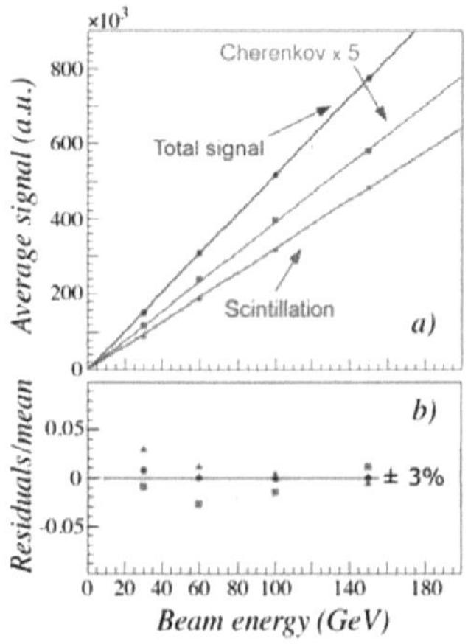

Figure 7. Linearity of the BGO matrix for electron detection. Shown are the average total integrated signal, as well as the Čerenkov and scintillation components of that signal, as a function of energy.

3.1. Comparison BGO and BSO

Other tests were performed to compare of BGO and BSO single crystals in terms of properties for applications in dual-readout calorimetry [4]. Both crystals are 18 cm long and $2.2 \times 2.2\ cm^2$ in cross-section and were tested with a 180 GeV pion beam. Due to the fact that in the spectral range the self-absorption edge is at lower wavelengths for the BSO, the Čerenkov light yield, in terms of p.e. detected per unit of deposited energy, is 2-3 times larger than in BGO. Also the purity of the Č signal obtained with filters achieved a separation power better by a factor of 6 for BSO, while the light attenuation length for Čerenkov light is almost the same in both crystals. Looking at these properties, BSO turns out to be a promising crystal for dual readout even though no further test has been performed at present.

4. $Mo : PbWO_4$ measurement

The $PbWO_4$ matrix is constituted by 7 custom made crystals doped with 0.3% Molybdenum. They were developed to optimize their properties for dual-readout calorimetry [5]. Crystals are $3 \times 3 \times 20\ cm^3$, corresponding to 22.5 X_0 in the longitudinal direction. Each crystal is read out at both ends, with Hamamatsu 8900 and 8900-100 (SBA), for the Scintillation and Čerenkov light, respectively. Calibration has been performed with 30 and 80 GeV electron beam steered in each crystal and by using Geant4 simulation to get the calibration constant: it was found that 77% of energy was deposited in the hit crystal, while 93% of the energy was contained in the entire matrix.

Different filter combinations (table 1) were used during the $PbWO_4$ matrix test, in order to optimize various aspects of the readout. In the first configuration we used a yellow filter upstream and the U330 downstream. This allows for good scintillation resolution of order of ~1% for 100

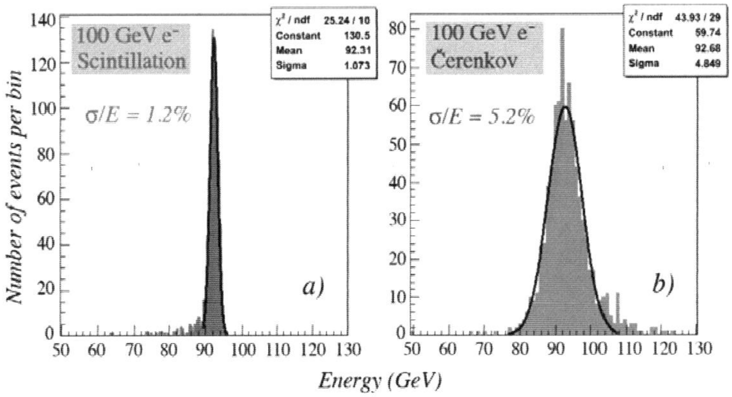

Figure 8. Signal distributions for 100 GeV electrons detected in the $PbWO_4$ crystal matrix, for the scintillation (a) and Čerenkov (b) components of the signals. The scintillation signal was obtained from a GG495 filter mounted at the upstream end while the Čerenkov signal was measured by using the light selected with a U330 filter.

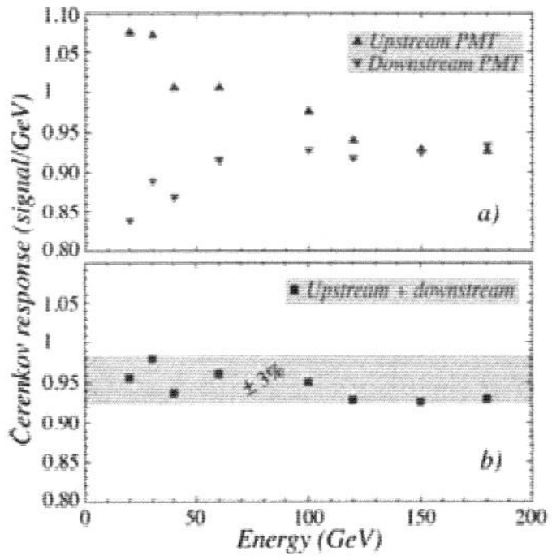

Figure 9. Signal linearity for electrons detected in the Mo-doped $PbWO_4$ crystal matrix for the Čerenkov signals. The matrix was equipped at both end faces with U330 filters and the signals are shown both separately (a) and added together (b).

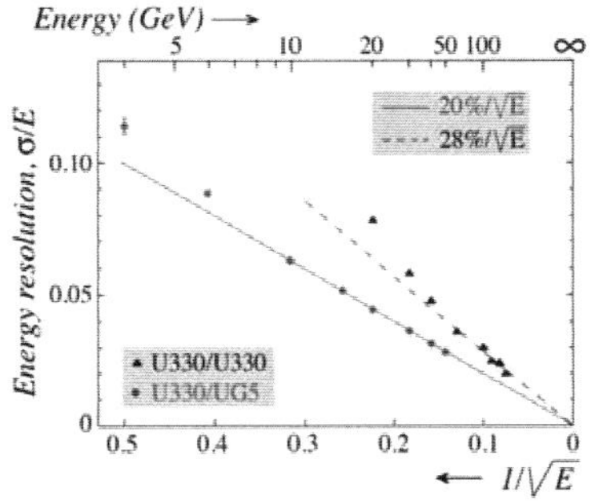

Figure 10. Energy resolution for electrons showering in the $PbWO_4$ crystal matrix, as a function of energy for two different filter configurations (see text for details).

GeV electrons (figure 8a). Unfortunately the small number of photoelectrons and the vicinity to the self absorption edge cause poor resolution (figure 8b) and strong non linearity for the Čerenkov light.

A second configuration uses U330 filters on both sides. This allows to add the signals of the two PMTs on event-by event basis, thus eliminating the effect of light absorption and therefore of the non linearity (figure 9). On the other hand, the scintillation light can only be extracted by the tail of the waveform, with the method shown in figure 5 for the BGO, and the number of photoelectrons turned out to be too low.

A compromise configuration implies the use of an U330 filter on the downstream side, for detecting Čerenkov light, and of the UG5 filter on the upstream side, to allow for more scintillation light transmission. The linearity for Čerenkov light is restored also in this case, and the energy resolution slightly improves due to the higher number of photoelectron detected (figure 10). The resolution for scintillation light is still quite poor, at least a factor of two worse with respect to the one obtained with the yellow filter (figure 11).

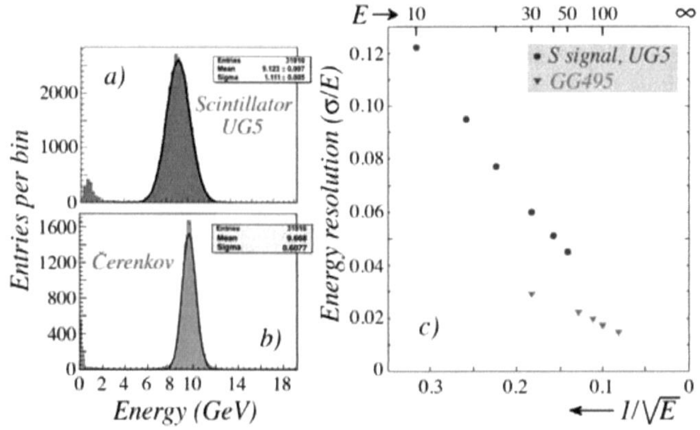

Figure 11. Signal distributions for the scintillation (a) and Čerenkov (b) components of the light generated in the $PbWO_4$ matrix by 10 GeV electron showers and filtered with the U330/UG5 combination. The energy resolution of the scintillation signal is compared with that obtained with the yellow (GG495) filter in diagram c.

5. Conclusions

In order to use crystals in the context of dual readout calorimetry, they have to be read out in a non conventional way, and this leads to results in terms of electromagnetic energy resolution of separated Č and S components that, at present, are far from being optimal and not as good as the ones obtained in standard em calorimetry. In fact, extracting sufficiently pure Č signals from these scintillating crystals implies that a large fraction of the potentially available Č photons needs to be sacrificed (by means of optical filters). Therefore the light that does contribute to the Č signals is strongly attenuated (by UV self absorption).

Our results show that the stochastic fluctuations in the Čerenkov channel are at best $20\%/\sqrt{E}$ in the case of our Mo-doped $PbWO_4$ crystal matrix. Assuming that these fluctuations are completely determined by photoelectron (p.e.) statistics, this would mean that the Č light yield for the electron showers was 25 p.e./ GeV of deposited energy.

Crystals in combination with filters does not seem to offer a benefit in terms of the Č light yield in dual-readout calorimeters. We recently measured a light yield in excess of 50 Č p.e./GeV in our new dual-readout fiber calorimeter. Nonetheless there is room for improvement by optimizing both the crystals and the readout.

References

[1] N. Akchurin, et al. 2012 *Nucl. Instr. and Meth.* A **686** 125
[2] F. Scuri, 2010 Proc. Int. Conf. IEEE 2010 Knoxville, USA.
[3] N. Akchurin, et al. 2009 *Nucl. Instr. and Meth.* A **610** 488
[4] N. Akchurin, et al. 2011 *Nucl. Instr. and Meth.* A **640** 91
[5] N. Akchurin, et al. 2010 *Nucl. Instr. and Meth.* A **621** 212

Journal of Physics: Conference Series **404** (2012) 012065 doi:10.1088/1742-6596/404/1/012065

The Super$B$ Factory Electromagnetic Calorimeter

S. Germani[1,2], M. Bizzarri[2,3], C. Cecchi[2,3], P. Lubrano[2], E. Manoni[2], A. Papi[2], G. Scolieri[2], A. Rossi[2], G. Eigen[4], Z. Zhou[4], D. Chao[5], C. H. Cheng[5], B. Echenard[5], K. T. Flood[5], D. G. Hitlin[5], F. C. Porter[5], R. Y. Zhu[5], G. De Nardo[6,7], C. Sciacca[6,7], V. Bocci[9], G. Chiodi[9], I. Dafinei[9], R. Faccini[8,9], S. Fiore[8], P. Gauzzi[8,9], G. Martellotti[9], F. Pellegrino[9], V. Pettinacci[9], D. Pinci[8,9], L. Recchia[9], A. Zullo[9], P. Branchini[10], A. Budano[10]

[1] CNRS LAL
[2] INFN Sezione di Perugia
[3] Universitá degli Studi di Perugia
[4] University of Bergen, Institute of Physics
[5] California Institute of Technology
[6] Universita' di Napoli Federico II
[7] INFN Sezione di Napoli
[8] Universita' di Roma La Sapienza
[9] INFN Sezione di Roma
[10] INFN Sezione di Roma Tre

E-mail: stefano.germani@pg.infn.it

Abstract. The Super$B$ project is an asymmetric e^+e^- accelerator of $10^{36}cm^{-2}s^{-1}$ luminosity, capable of collecting a 50 - 75 ab^{-1} data sample in five years of running. The Super$B$ electromagnetic calorimeter (EMC) provides energy and direction measurement of photons and electrons and identification versus other charged particles for electrons. A matrix of 25 LYSO crystals has been tested at the Beam Test Facility at Frascati in May 2011 at energies between 100 MeV and 500 MeV. Results from this test will be presented. Design and Monte Carlo studies for the general EMC system will also be presented.

1. Introduction

The Super$B$ project is an asymmetric e^+e^- accelerator of $10^{36}cm^{-2}s^{-1}$ luminosity [1], capable of collecting a data sample of 50-75 ab^{-1} in five years of running (two orders of magnitude higher than the previous B-factories). The Super$B$ detector [2] concept is based on the *BABAR* detector [3], with modifications required to operate at a luminosity of 10^{36} or more, and with a reduced center-of-mass boost. Further improvements are needed to cope with higher beam-beam and other beam-related backgrounds, as well as to improve detector hermeticity and performance. A number of *BABAR* components are reused: the flux-return steel, the superconducting coil, the barrel of the EMC and the Cherenkov detector fused silica bars while the other parts will be replaced with new detectors. The hermeticity of the Super$B$ detector, and, thus, its performance for certain physics channels will be improved by including a backward veto-quality EMC detector comprising a lead-scintillator stack and a forward particle identification detector consisting of a fast Cherenkov light based time-of-flight system.

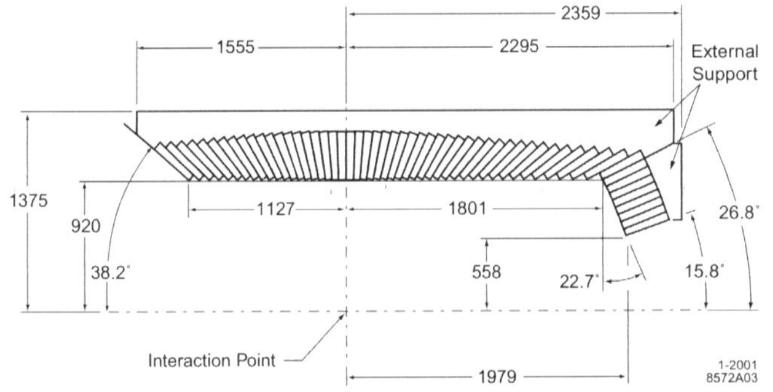

Figure 1. Schema of the *BABAR* calorimeter with dimensions and crystal arrangement. The forward endcap covers the angular range $15.8^\circ < \theta < 26.8^\circ$ with respect to the beam axis.

Figure 1 shows a sketch of the *BABAR* calorimeter, made of CsI(Tl) crystals arranged in a central part (barrel) and a forward endcap. While the barrel, with the necessary upgrades to match the new requirements of an accelerator operating at higher luminosity, can also be used in the Super$B$ environment, a new forward electromagnetic calorimeter has to be conceived due to the high rate of luminosity related background, especially radiative Bhabhas, and the radiation damage of the BaBar CsI(Tl) crystals.

2. The Forward Electromagnetic Calorimeter

The forward electromagnetic calorimeter (FwdEMC) designed for Super$B$ is a new device replacing the *BABAR* CsI(Tl) forward calorimeter. Because of the increased background levels, a faster and more radiation hard material is required in the forward region. One proposed solution assumes the full or partial replacement of CsI(Tl) crystals with LYSO (Lutetium Yttrium Orthosilicate, with Cerium doping) crystals [4]. The advantages of LYSO include a much shorter scintillation time constant (LYSO: 40 ns, CsI(Tl): 680 ns and 3.34 μs), a smaller Molière radius (LYSO: 2.1 cm, CsI: 3.6 cm), and a better tolerance to radiation.

For the option of partial replacement for the endcap crystals, the *BABAR* mechanical structure is reused and each cell in the inner rings is filled with four LYSO crystals. A full LYSO endcap the FwdEMC will be composed by 20 rings of crystals, arranged in four groups of 5 layers each. The crystals maintain the almost projective geometry of the barrel. Each group of five layers is arranged in modules of 5 crystals each. Each crystal is up to $2.5 \times 2.5\ cm^2$ at the back end, with a projective taper to the front. The maximum transverse dimensions are dictated by the Moliere radius. The length of each crystal is approximately 20 cm, or 17.5 X_0. The readout sensors for the LYSO crystals are Avalanche Photodiodes (APD). A new Very Front End (VFE) board has been designed and it incorporates a dual-gain amplifier (1×, 16×); the VFE output signals are shaped by a Gaussian shaper with a shaping time of 100 ns. Finally signals are digitalized by a twelve-bit Analog-to-Digital Converter (ADC).

LYSO crystals are widely used in medical applications like PET but they have never been used in large High Energy Physics experiments which require much larger crystal size. Several years of R&D have been performed [5] in order to obtain from manufacturers large size crystals with high light yield (LY); special care has been devoted to the study and improvement of the longitudinal LY uniformity, which is related to Ce doping concentration.

2.1. Beam Test

A test beam has been performed with a prototype LYSO matrix (see Figure 2) at the Beam Test Facility (BTF) [6] in Frascati in May 2011. The prototype matrix is composed by 25 LYSO crystals of pyramidal shape inserted in a support structure assembled by the RIBA company (Faenza, Italy). Each crystal is coupled to an APD and the same VFE boards and Gaussian shaper designed for the FwdEMC are used in the test. The processed signals are acquired by four 12bit sampling ADC (Caen V1720 250 MS/s 54 sampling ADC [6]). Beside the effect of the variation of Ce doping concentration, the crystals show an intrinsic longitudinal LY non-uniformity related to the focusing effect of the tapering. Due to this focusing effect and to the fact that all the crystals have been cut with the same orientation from the boule, all of them show an LY increasing from the direction of the large face with the photosensors to the small face. To obtain a crystal LY longitudinal uniformity within 5% a 15 mm band of black paint has been applied on the small side close to the face opposite to the photosensor as shown in Figure 3. This operation causes an average light yield loss of about 40%.

Figure 2. Prototype mechanical structure.

Figure 3. Crystal uniformisation technique.

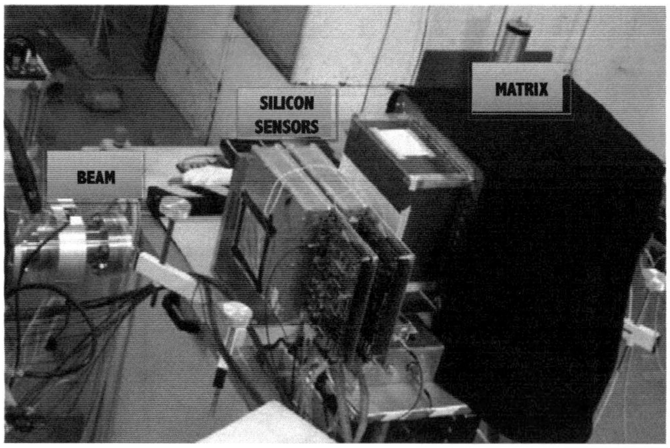

Figure 4. Test beam setup at BTF.

The setup for the beam test of the matrix at the BTF is shown in Figure 4, where the end of the electron beam line, two planes of silicon strip detector (one horizontal and one vertical position measurement) and the box containing the matrix with the crystals and the VFE boards are shown. The silicon strip detector is composed by two planes with orthogonal strip orientation with digital readout, each plane is composed by 384 silicon strip with a pitch of 228 μm. Data are taken at 5 different electron energies: 486, 398, 297, 198 and 99 MeV, with the beam pointing to the center of the matrix. The dimension and the energy spread of the beam depend on the electron momentum becoming larger at lower energies.

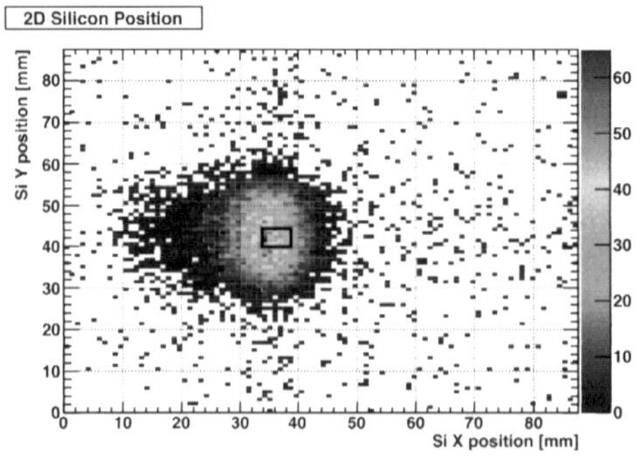

Figure 5. Beam intensity profile at 486 MeV as measured by the silicon detectors and the corresponding X-Y(horizontal-vertical) selection highlighted by the black box.

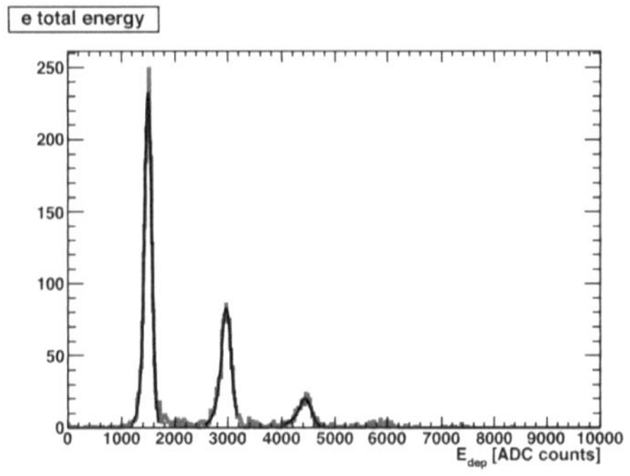

Figure 6. Measured energy distribution, in ADC counts, for the 198 MeV beam where the peaks for 1, 2 and 3 e^- have been fitted with Crystal Ball functions.

In order to select particles hitting the LYSO matrix face approximately at the same point, a selection on the electron position measured by the silicon detector is applied before the total energy deposit is evaluated. The selected area is about 6×4 mm^2. The plot in Figure 5 shows the beam profile at 486 MeV, the black box represents the selected area at this energy. Figure 6 shows the total energy distribution for the selected events at 198 MeV. Here the deposited energy of 1, 2 and 3 electrons are evident and each one is fitted with a Crystal Ball function. Since the beam energy spread is not known with precision, it is estimated directly from data. Events with more than one electron of the same energy are used in order to have more than one independent measurement at the same energy (i.e. 200 MeV could be measured with one e^- of 200 MeV or two e^- of 100 MeV). Starting from the assumption that the calorimeter prototype resolution for two particles of energy E is the same of one particle with twice the energy (2E), we can estimate the beam energy spread at different beam line energy configurations, and then subtract these values from the total resolution in order to extract the intrinsic LYSO EMC prototype resolution. All the results are extracted with a minimization procedure where the matrix resolution is assumed to be: $\frac{\sigma(E)}{E} = \frac{a}{\sqrt{E[GeV]}} \oplus \frac{b}{E[GeV]} \oplus c$ %. The beam energy spread at 486 MeV is assumed to be 1%, similar to the nominal LINAC spread at 510 MeV. The results are shown in Figure 7, the resolution as a function of energy obtained with this technique is:

$$\frac{\sigma(E)}{E} = \frac{1.1}{\sqrt{E[GeV]}} \oplus \frac{0.4}{E[GeV]} \oplus 1.2\ \%.$$

A Geant4 Monte Carlo simulation has been developed encompassing all the dead materials on the beam hall and with a detailed representation of the crystals and the support structure

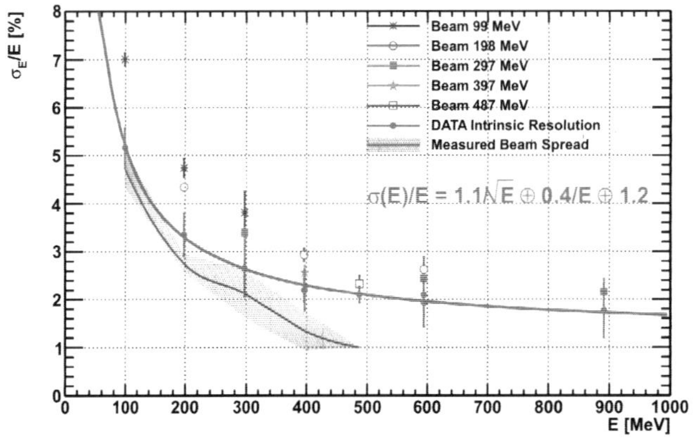

Figure 7. Resolution as a function of energy. The symbols display the measurements (∗ 99 MeV, ○ 198 MeV, ■ 297 MeV, ⋆ 397 MeV, □ 487 MeV), the solid line the fitted resolution and the shaded area the estimated beam energy spread.

geometries. The effects of electronic noise as measured on the beam, electronic cross-talk, LY non uniformity and crystals intercalibration uncertainties have been added to the simulation. The output of the simulation includes the sampling of the signal shape from the ADC emulating the real data output format so that after the first step, beam and Monte Carlo data are processed with the same analysis chain. Once the beam energy spread, as evaluated from the data, has been included in the simulation, the Monte Carlo results are well in agreement with the measured matrix resolution as shown in Figure 8.

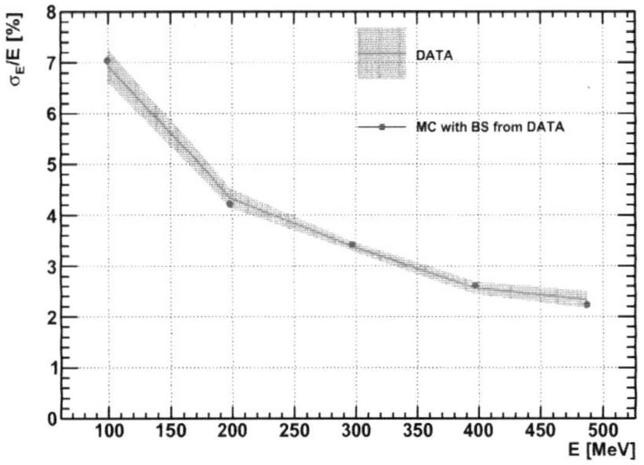

Figure 8. Comparison between data and Monte Carlo energy resolution with the beam energy spread in the simulation as estimated from the data using the minimization technique. The shaded area displays the data and the dots the Monte Carlo.

3. Barrel optimization

In order to be confident that the barrel can cope with the higher radiation and background level expected at Super*B*, several studies about the CsI(Tl) crystals are being carried out both on the background simulation and readout electronics sides. The aim of the combined effort is to find a compromise where the VFE electronics can extract short enough signals from the slow CsI(Tl) crystals to avoid signal pile-up and mitigate the background effect but is still retaining a good signal to noise ratio.

4. Conclusions

The Super*B* electromagnetic calorimeter will reuse the *BABAR* barrel while the forward endcap need to be replaced, in part or in toto, with a detector more suitable to cope with the level of radiation expected at higher luminosity. We presented the results of a beam test, performed after years of studies and R&D on the crystals, showing that the LYSO is a very good solution for the Super*B* forward endcap. The studies on the readout for the CsI(Tl) crystals from *BABAR* are still ongoing to find a good compromise between the mitigation of machine backgrounds effects and a safe electronic noise level.

References

[1] M. E. Biagini et al. [Super*B* Collaboration], Super*B* Progress Reports: The Collider, [arXiv:physics.acc-ph/1009.6178].
[2] E. Grauges et al. [Super*B* Collaboration], Super*B* Progress Reports: Detector, [arXiv:physics.ins-det/1007.4241].
[3] B. Aubert et al. [BABAR Collaboration], The BaBar detector, Nucl. Instrum. Meth. A **479** (2002) [arXiv:hep-ex/0105044].
[4] J. M. Chen, et al., Large Size LSO and LYSO Crystals for Future High Energy Physics Experiments, IEEE Trans. Nucl. Sci. **54** (2007) 718724. 101
[5] R. Mao, L. Zhang and R. -Y. Zhu, "LSO/LYSO crystals for future HEP experiments," J. Phys. Conf. Ser. **293** (2011) 012004.
[6] G. Mazzitelli and P. Valente, Commissioning of the DAFNE Beam TestFacility, LNF-03/003(P)
[7] Caen Documentation on V1720 ADC, `http://www.caen.it/csite/CaenProd.jsp?parent=11&idmod=570`

Status of the CALICE analog calorimeter technological prototypes

Mark Terwort on behalf of the CALICE collaboration
DESY, Notkestrasse 85, 22607 Hamburg, Germany
E-mail: mark.terwort@desy.de

Abstract. The CALICE collaboration is currently developing engineering prototypes of electromagnetic and hadronic calorimeters for a future linear collider detector. This detector is designed to be used in particle-flow based event reconstruction. In particular, the calorimeters are optimized for the individual reconstruction and separation of electromagnetic and hadronic showers. They are conceived as sampling calorimeters with tungsten and steel absorbers, respectively. Two electromagnetic calorimeters are being developed, one with silicon-based active layers and one based on scintillator strips that are read out by MPPCs, allowing highly granular readout. The analog hadron calorimeter is based on scintillating tiles that are also read out individually by silicon photomultipliers. The multi-channel, auto-triggered front-end chips are integrated into the active layers of the calorimeters and are designed for minimal power consumption (power pulsing). The goal of the construction of these prototypes is to demonstrate the feasibility of building and operating detectors with fully integrated front-end electronics. The concept and engineering status of these prototypes are reported here.

1. Introduction

One of the physics goals at a future linear collider (LC) is the separation of hadronic W- and Z-boson decays. To achieve this goal a relative jet energy resolution of ~3% is necessary, while typical jet energies are of the order of 100 GeV and typical single particle energies are about 10 GeV. One of the strategies to achieve such a jet energy resolution is to measure the details of the shower development in order to separate the showers of individual jet particles. This information is then combined with the information obtained from the tracking detectors, such that only the energies of neutral particles are determined with the calorimeter system. The approach is known as *particle flow*. This implies a number of challenges for the design and the construction of the calorimeters:

- High granularity: ECAL (HCAL) cell size $<$ ($\sim$) Molière radius, longitudinally $<$ ($\sim$) $1\,X_0$,
- Explosion of channel count: ECAL $\sim 10^8$, HCAL $\sim 10^7$,
- Compactness: the calorimeters are placed inside the magnet coil,
- Heat development: power pulsing the integrated front-end electronics becomes necessary,
- 4th dimension: information about the time of each hit with respect to the bunch clock is needed.

The CALICE collaboration [1] is developing and testing new technologies for calorimeters for a future LC experiment. Technological prototypes are meant to provide answers and solutions

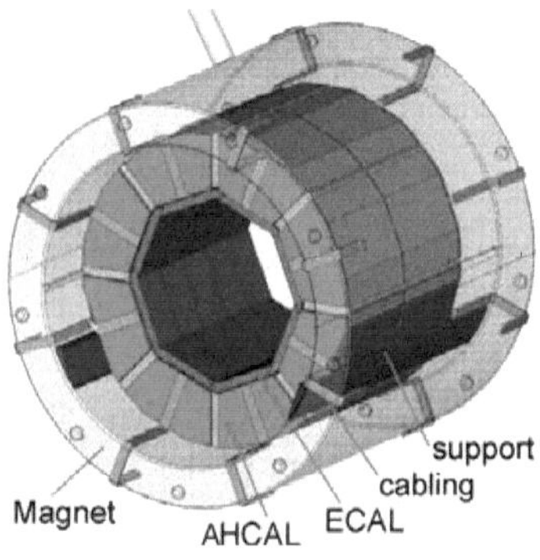

Figure 1. Layout of the barrel calorimeter system of a LC detector. The ECAL is shown in blue and the AHCAL in green, while the structure is surrounded by the magnet.

to the challenges listed above. These technological prototypes are developed in a phase where physics prototypes already provided valuable information about the physics of hadron showers from extensive beam test programs [2]. Therefore the main goal is to demonstrate that the construction of a realistic detector with fully integrated front-end electronics is feasible. Figure 1 shows a layout of the calorimeter system of a LC detector. The ECAL is shown in blue, the analog HCAL option is shown in green, while the structure is surrounded by the magnet.

Common developments in the collaboration, like a data acquisition system for all CALICE detectors or readout chips and electronics boards that are similar for different detectors, allow for a parallel and cost effective development of different technology options. In the following, three detectors will be described in more detail. These are the analog scintillator tile hadron calorimeter (AHCAL) and the SiW ECAL, as well as the scintillator strip ECAL. For a review of the physics prototypes of these detectors see [3, 4, 5]. For a review of the latest status of (semi-) digital hadron calorimeters see [6].

2. The AHCAL technological prototype

The CALICE collaboration is currently developing a technological prototype for an AHCAL [7]. It is based on scintillating plastic tiles read out by silicon photomultipliers (SiPMs), while steel or tungsten can be used as absorber material with a thickness of 16 mm and 10 mm, respectively. Figure 1 shows how the AHCAL is placed inside the magnet and outside the ECAL system. It is divided into four sections along the beam direction, the two endcaps and two half-barrels. Each half-barrel is further divided into 16 sectors in ϕ-direction. Each sector consists of 48 layers. The total thickness is 110 cm, while the total length of the barrel is 220 cm. A single active layer is made of three parallel slabs of HBUs (HCAL base units). Each HBU has a size of $36 \times 36\,\mathrm{cm}^2$ and has 144 detector channels. The latest HBU version is shown in Fig. 2(a). It is equipped with four SPIROC2b ASICs [8] each to read out the photo detectors. On the backside of the module the scintillating tiles are attached to the PCB via alignment pins with a nominal distance of $100\,\mu$m, as shown in Fig. 2(b). Synergies between different CALICE detector technologies are achieved by deriving the PCBs for an alternative AHCAL option (by mounting the SiPMs on top of the PCB), as well as for the scintillator strip ECAL, from the HBU design.

The tiles have a size of $30 \times 30 \times 3\,\mathrm{mm}^3$ and are equipped with a wavelength shifting fiber to guide the light to a SiPM with a size of $1.26\,\mathrm{mm}^2$ and 796 pixels. The latest batch of SiPMs from CPTA that are used in the current HBUs have a gain in the range of 1 to 3 million with a noise frequency around 50 Hz at a 0.5 MIP threshold. An alternative option for the tile design is currently under investigation, where the tiles are read out directly by SiPMs, e.g. without a wavelength shifting fiber. This reduces the mechanical complexity, since otherwise the SiPM has to be aligned precisely to the fiber. First tests already showed that the tiles are very uniform.

(a)

(b)

Figure 2. (a) Photo of the latest basic AHCAL module, equipped with four SPIROC2b ASICs. The DAQ interface modules are also shown. (b) Photo of tiles that are assembled below an HBU. The orientation of the wavelength shifting fibers depends on the position of the holes for the pins. This is determined by the details of the PCB design.

The 36-channel SPIROC2b ASIC is specifically designed for LC operation. It comprises an input DAC for channel-wise bias voltage adjustment for the SiPMs, it can be power pulsed in order to reach a power consumption of not more than 25 μW per channel (for details see e.g. [9]) and it can be operated in a self-triggering mode. Besides setting a global threshold per chip, a channel-wise tuning of the threshold is possible with a dynamic range of about 0.25 MIP. A dual-ramp TDC allows for precise time measurements of individual hits with a resolution of less than 1 ns [7]. Strong synergies between the different CALICE calorimeter technologies are also achieved here, since the SPIROC2b is also used in the scintillator strip ECAL and a similar chip, called SKIROC, is used in the SiW ECAL. Detailed laboratory and test beam measurements have lead to a detailed understanding that allows further developments. For example, the next generation chip SPIROC2c is currently investigated in the HBU environment.

Four fully equipped new HBUs are currently available and are extensively tested in the DESY laboratory and test beam facility. Figure 3(a) shows the setup of a light-tight aluminum HBU cassette as it is mounted on a movable stage in order to scan all channels with a 2 GeV electron beam. The MIP signals are measured in self-triggering mode and a typical spectrum is shown in Fig. 3(b).

There are several important steps to be done in the near future. Since there are enough PCBs available to built a full slab, it is possible now to test the power pulsing mode in a more realistic environment. The mechanical prototype structure is also in place and therefore realistic temperature measurements can be performed as well. On the other hand it is possible now to built a 2×2 HBU layer that can be used in a hadron beam test. Besides important technical aspects that can be tested with such a device, it offers the possibility to measure the radial time distribution of hadron showers when used as an additional layer behind another CALICE calorimeter.

(a)

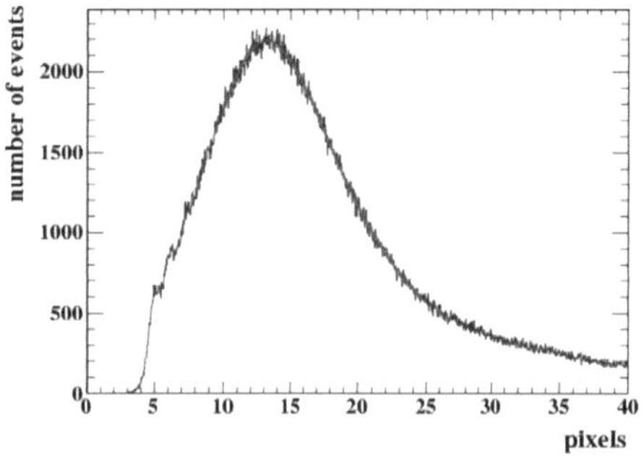

(b)

Figure 3. (a) Setup of the light-tight HBU cassette as it is mounted on a movable stage at the DESY test beam facility. (b) Typical MIP spectrum obtained from a 2 GeV electron beam.

3. The ECAL technological prototypes

The basic requirements for an electromagnetic calorimeter at a particle flow experiment are similar to the requirements for the hadron calorimeter. It has to have an extremely high granularity, while being compact and hermetic. In order to achieve this, tungsten has been chosen as the absorber material for narrow showers ($X_0 = 3.5\,\mathrm{mm}$, $R_M = 9\,\mathrm{mm}$). Two different options for the active layers are currently under development.

3.1. SiW ECAL

One of the options for the ECAL active layers is to use silicon as active material. The key parameters of the technological prototype that is currently developed by the CALICE collaboration are [10]:

- Individual cell size of $5.5 \times 5.5\,\mathrm{mm}^2$,
- Depth of $24\,X_0$,
- Thickness of an individual layer of 3.4 mm and 4.4 mm according to the position within the calorimeter.

Figure 4(a) shows the geometry of the mechanical construction that is realized as a tungsten carbon composite to serve simultaneously as the absorber structure, where individual detector slabs are inserted. Figure 4(b) shows a detailed view of the structure.

Six of the latest modules (FEV8, front-end version) have been tested recently in the DESY test beam facility. Each FEV8 has been equipped with four 64-channel SKIROC ASICs, that are specifically designed for the readout of the ECAL silicon cells. The wafers have a size of $9 \times 9\,\mathrm{cm}^2$ and comprise 324 pixels. Figure 5(a) shows a photo of the current setup, including the DAQ interface cards. The main goals for the beam test are the determination of the signal-to-noise ratio, the establishment of a calibration procedure for a large number of cells and the measurement of the homogeneity of the response. During the beam test it was possible to insert tungsten absorber plates between the active layers and measure electron showers between 1 and 6 GeV. Additionally, this beam test has also successfully demonstated the functionality of the common CALICE DAQ concept and hardware that ran smoothly over the complete data taking period. Further beam tests will follow in the near future, where power pulsing tests and tests with magnetic field will be performed.

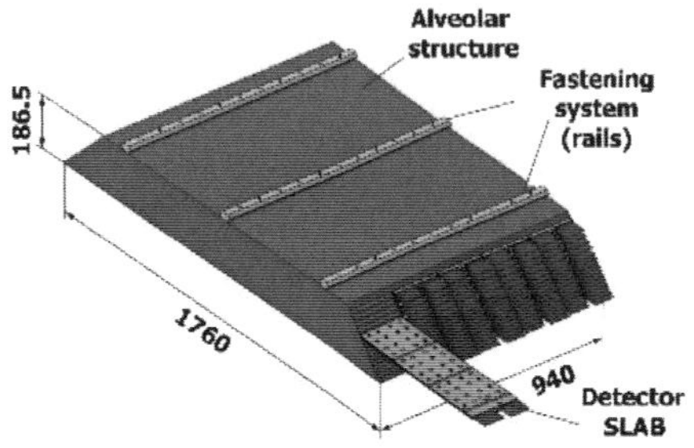

(a)

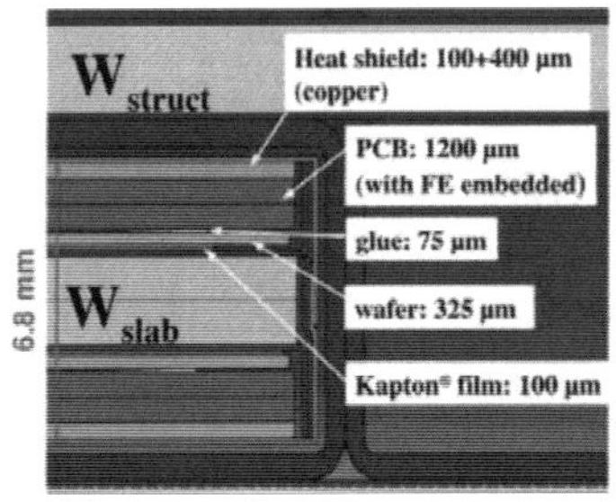

(b)

Figure 4. (a) Schematic of the ECAL alveolar structure, which shows how the slabs are inserted into the mechanical housing. (b) Cross section of two ECAL layers. The sensitive material and the front-end electronics are mounted on both sides of a tungsten carbon composite plate.

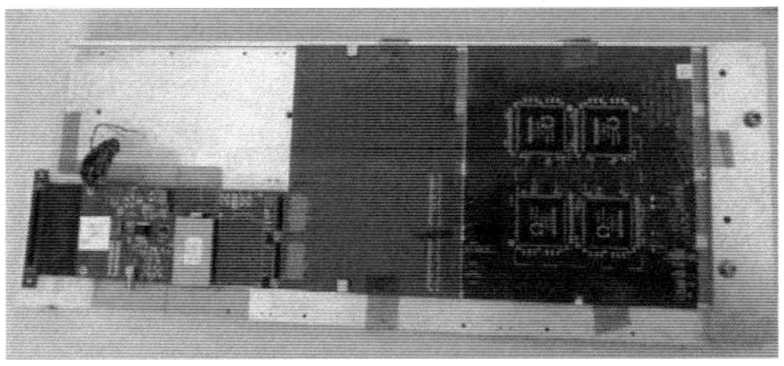

(a)

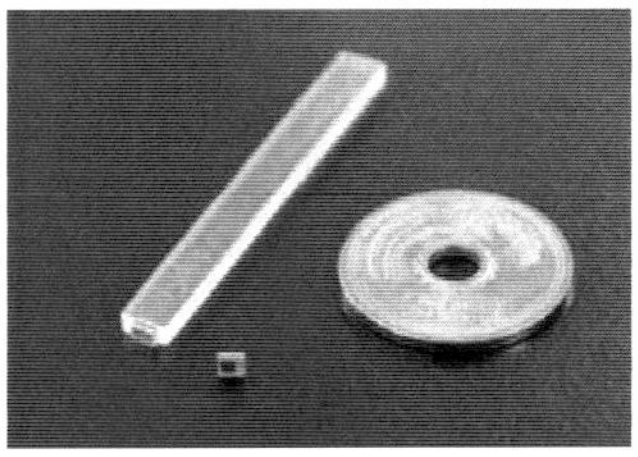

(b)

Figure 5. (a) Photo of the FEV8 together with the DAQ interface. (b) Scintillator strip coupled to an MPPC as it is used in the active layers of the scintillator strip ECAL.

Besides the testing of the existing modules, there are further current and future challenges. Among these are the optimization of the guard ring (to suppress cross talk between the modules), the low cost production of silicon wafers (3000 m^2 for LC detector), the glueing of the wafers to the PCB, the optimization of the PCB thickness and the interconnection of individual units (in order to reduce mechanical stress to the wafers).

3.2. Scintillator stip ECAL

A second option for the active ECAL layers is to use scintillator strips with MPPC readout [5]. The idea is to use alternating layers with fine segmentation in x- or z-direction, which leads to an effective channel size of $5 \times 5\,mm^2$. After the successful validation of the concept with a physics prototype, the CALICE collaboration is currently constructing a technological prototype with fully integrated front-end electronics. The EBU (ECAL base unit) is therefore adapted from the AHCAL design and also the SPIROC2b ASIC is used to read out the MPPCs that are coupled to the strips. Strips are developed that do not incorporate a wavelength shifting fiber, but are read out directly with the MPPC, see Fig. 5(b). The next important step is to integrate all components and start testing the complete system in the laboratory and in the DESY test beam environment.

4. Summary

The CALICE collaboration has extensively and successfully tested physics prototypes for particle flow calorimeters in the past. Therefore the phase of the validation of particle flow concepts has finished and lead to the construction and development of technological prototypes. These prototypes are used to prove the feasibility of the technological concepts to built realistic linear collider detectors with fully integrated front-end electronics. Prototypes for electromagnetic as well as hadronic calorimeters have already taken data and although the development is still ongoing, basic concepts could already be addressed and verified:

- Operation of fully integrated electronics,
- Power pulsing,
- Mechanical engineering and
- Costs and mass production issues.

In this report the current status of two electromagnetic and an analog hadron calorimeter were presented. The tests that have been performed so far, have lead to an increased understanding of the details of the systems that allow further developments and larger beam test campaigns in the near future.

Acknowledgments

The author gratefully thanks Karsten Gadow, Erika Garutti, Peter Göttlicher, Oskar Hartbrich, Benjamin Hermberg, Katsushige Kotera, Roman Pöschl, Mathias Reinecke, Felix Sefkow and the whole CALICE collaboration for very useful discussions and valuable contributions to the results presented here.

References

[1] CALICE home page: https://twiki.cern.ch/twiki/bin/view/CALICE/CaliceCollaboration
[2] M. Ramilli, *Hadron showers from CALICE test beam data*, these proceedings.
[3] C. Adloff et al., *Construction and commissioning of the CALICE analog hadron calorimeter prototype*, JINST 5 (2010) P05004, arXiv:1003.2662.
[4] J. Repond et al., *Design and Electronics Commissioning of the Physics Prototype of a Si-W Electromagnetic Calorimeter for the International Linear Collider*, JINST 3 (2008) P08001, arXiv:0805.4833.
[5] K. Kotera, *Study of the granular electromagnetic calorimeter with PPDs and scintillator strips for ILC*, Proc. 2010 12th Vienna Conference on Instrumentation, NIM, A628, 347 (2011).
[6] B. Bilki, *Tests of a Digital Hadron Calorimeter*, J.Phys.Conf.Ser. 293 (2011) 012075, arXiv:1005.0409; M.C. Fouz, *The 1m^3 Semidigital Hadronic Prototype*, Proc. 2011 LCWS, arXiv:1202.5567; J. Yu, *Application of Large Scale GEM for Digital Hadron Calorimetry*, arXiv:1110.4326.
[7] O. Hartbrich et al., *Recent Advances of the Engineering Prototype of the CALICE Analog Hadron Calorimeter*, Proc. 2011 LCWS, arXiv:1201.5264; M. Reinecke, *Towards a full scale prototype of the CALICE tile hadron calorimeter*, Proc. 2011 IEEE Nuclear Science Symposium (NSS11); M. Terwort, *Concept and status of the CALICE analog hadron calorimeter engineering prototype*, Proc. 2011 Technology and Instrumentation in Particle Physics (TIPP), arXiv:1109.0927.
[8] L. Raux et al., *SPIROC Measurement: Silicon Photomultiplier Integrated Readout Chips for ILC*, Proc. 2008 IEEE Nuclear Science Symposium (NSS08), NSSMIC.2009.5401891; M. Bouchel et al., *SPIROC (SiPM Integrated Read-Out Chip): dedicated very front-end electronics for an ILC prototype hadronic calorimeter with SiPM read-out*, Proc. 2010 TWEPP, JINST 6 (2011) C01098.
[9] P. Göttlicher, *A concept for power cycling the electronics of CALICE-AHCAL with the train structure of ILC*, Proc. 2011 Technology and Instrumentation in Particle Physics (TIPP).
[10] R. Pöschl, *A large scale prototype for a SiW electromagnetic calorimeter for a future linear collider*, Proc. 2011 LCWS, arXiv:1203.0249.

Journal of Physics: Conference Series **404** (2012) 012067 doi:10.1088/1742-6596/404/1/012067

A Silicon-Tungsten Electromagnetic Calorimeter with Integrated Electronics for the International Linear Collider

J. Brau[b], M. Breidenbach[a], R. Frey[b], D. Freytag[a], C. Gallagher[c], N. Graf[a], G. Haller[a], R. Herbst[a], B. Holbrook[c], J. Jaros[a], R. Lander[c], P. Radloff[c], D. Strom[b], M. Tripathi[c], M. Woods[c]

[a] SLAC National Accelerator Laboratory
[b] Department of Physics, University of Oregon
[c] Department of Physics, UC Davis

E-mail: Norman.Graf@slac.stanford.edu

Abstract. We present an update of the development of an electromagnetic calorimeter for the Silicon Detector concept for a future linear electron-positron collider. After reviewing the design criteria and related simulation studies, we discuss progress in the research and development of the detector. This concept has from the outset made the case for highly integrated electronic readout with small (1 mm) readout gaps in order to maintain a small Molière radius for electromagnetic showers and to avoid active heat removal. We now have fully functioning 1024-channel readout chips which have been successfully bonded to 15 cm silicon sensors. We present initial results from these assemblies.

1. Introduction

The Silicon Detector (SiD) is a general purpose detector designed to conduct precision physics measurements of high energy e^+e^- collisions at the International Linear Collider (ILC). The baseline design uses a Particle Flow Algorithm-based (PFA) approach to calorimetry. PFAs have been successfully applied to existing detectors, such as CDF, ZEUS, and CMS and have resulted in significant improvements of the jet energy resolution compared to methods based on calorimetric measurement alone. However, these detectors were not originally designed with the application of PFAs in mind. The SiD baseline design on the other hand considers a PFA approach necessary to reach the goal of obtaining a measurement uncertainty on the jet energy resolution of the order of 3% or better. SiD is therefore optimized assuming the PFA approach and the major challenge imposed on the calorimeter by the application of PFAs is not the intrinsic calorimetric energy measurement of either the electromagnetic calorimeter (ECAL) or the hadronic calorimeter (HCAL), but the association of energy deposits with either charged or neutral particles impinging on the calorimeter. This results in several requirements on the calorimeter design:

- To minimize the lateral size of electromagnetic showers the Molière radius of the ECAL must be minimized. This enables efficient separation between photons and electrons and charged hadron tracks.

- Both ECAL and HCAL must have imaging capabilities which allow assignment of energy cluster deposits to charged or neutral particles. This implies that the readout of both calorimeters needs to be finely segmented transversely and longitudinally.
- The calorimeter needs to be extendable to small angles to ensure hermeticity.

2. Requirements

The major challenge imposed on the calorimeter by the application of PFAs is the unambiguous association of energy depositions in the calorimeters with the showers originating from individual particles. For the ECAL, this implies that electromagnetic showers be confined to small volumes in order to avoid overlaps. Effective shower pattern recognition is possible if the segmentation of readout elements is small compared to the showers. This level of transverse segmentation then also facilitates the separability of the electromagnetic showers from charged particle tracks due to un-interacted charged hadrons (and muons). The longitudinal segmentation is chosen not only to achieve the required electromagnetic energy resolution, but also to provide discrimination between electromagnetic showers and those hadrons which interact (typically deeper) in the ≈ 1 interaction length of the ECAL. Finally, there should be a sufficient number of longitudinal readout layers to provide charged particle tracking in the ECAL. This is important not only for the PFA algorithms, but also to aid the tracking detectors, especially for tracks which do not originate from the primary beam interaction point.

3. Global ECAL Design

A sampling ECAL provides adequate energy resolution for the ILC physics, as discussed above. Because of its small radiation length and Molière radius, as well as its mechanical suitability, we have chosen tungsten as the absorber/radiator. Due to practical considerations for ease of production of large plates and machining, the tungsten will be a non-magnetic alloy. The currently chosen alloy includes 93% W with radiation length 3.9 mm and Molière radius 9.7 mm. An additional benefit of tungsten is that it has a relatively large interaction length, which helps to minimize confusion between electromagnetic and hadron showers in the ECAL.

The longitudinal structure we have chosen has 30 total layers. The first 20 layers each have 2.50 mm tungsten thickness and 1.25 mm readout gap. The last 10 layers each have 5.00 mm tungsten plus the same 1.25 mm readout gap. This configuration attempts to compromise between cost, shower radius, sampling frequency, and shower containment. The cost is roughly proportional to the silicon area, hence the total number of layers. We chose finer sampling for the first half of the total depth, where it has the most influence on electromagnetic resolution for showers of typical energy. The total depth is $26X_0$, providing reasonable containment for high energy showers. Simulations have shown the energy resolution for electrons or photons to be well described by $0.17/\sqrt{E}$.

Silicon detectors are readily segmented, and in the baseline design we have chosen there is little penalty for segmenting the silicon sensors much more finely than typical shower radii. As discussed above, the scale for this is set by the shower size, which we wish to be as small as feasible. A useful figure of merit for this is the Molière radius, which is 9 mm for pure tungsten. Since showers will spread in the material between tungsten layers, it is crucial to keep the readout gaps as small as possible. Our design results in an *effective* Molière radius of 14 mm.

4. Sensor and Readout Technology

In the baseline design, the ECAL readout layers are tiled by large, commercially produced silicon sensors (presently from 15 cm wafers). The sensors are segmented into pixels which are individually read out over the full range of charge depositions. The complete electronics for the pixels is contained in a single chip, the kPiX ASIC, which is bump bonded to the wafer. We take advantage of the low beam-crossing duty cycle (10^{-3}) of the ILC to reduce the heat load using power pulsing, thus allowing passive thermal management within the ECAL modules. The realization of this technology has been the subject of an intensive, ongoing R&D program. The main parameters associated with the baseline technology choice are given in Table 1.

Table 1. Parameters of baseline silicon-tungsten ECAL.

readout gap (incl. Si sensor)	1.25 mm
effective Molière radius	14 mm
pixel area	13 mm^2
pixels per silicon sensor	1024
dynamic range requirement	$\sim$ 0.1 to 2500 MIPs
heat load requirement	20 mW per sensor

Figure 1 shows a sensor with 1024 pixels. Not shown in the drawing are the signal traces, part of the second layer metallization of the sensors, which connect the pixels to a bump-bonding pad at the center of the sensor for input to the kPiX readout chip. The pixels are DC-coupled to the kPiX, thus only two metallization layers are required for the sensors. The pixels near the bump-bonding array at the center are split to reduce capacitance from the large number of signal traces near the sensor center. The electronic noise due to the resistance and capacitance of the traces has been minimized within the allowed trace parameters. The cutouts at the corners of the sensor are to accommodate mechanical stand-offs which support the gaps between the tungsten layers.

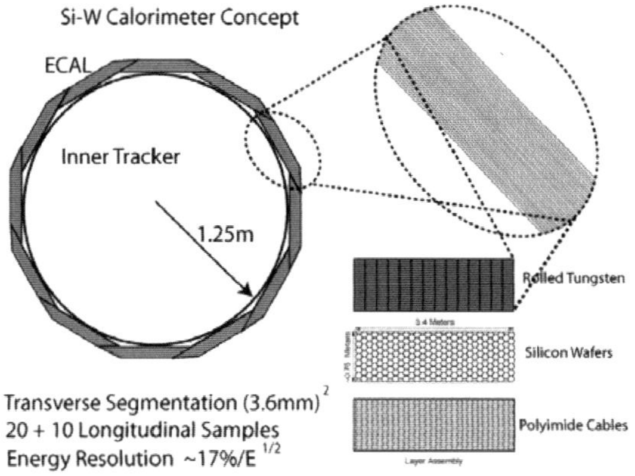

Figure 1. Overview of the SiD ECAL design, showing the overlapping wedge geometry and basic sampling calorimeter layout.

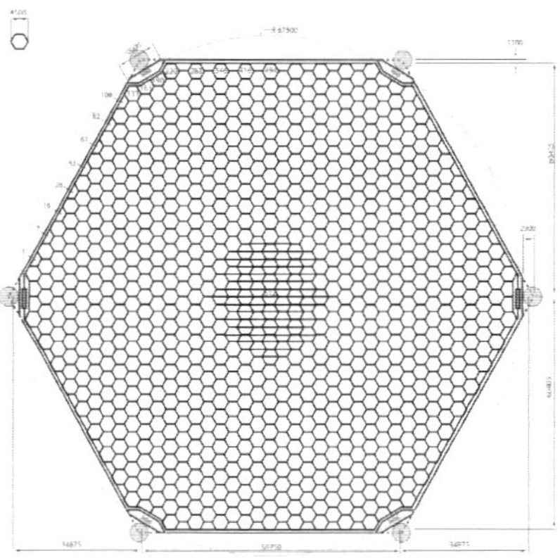

Figure 2. Drawing of an individual silicon sensor for the ECAL. The sensors are segmented into 1024 13 mm^2 pixels.

After several iterations with the R&D, in early 2012 a full 1024-channel kPiX was successfully bump-bonded to a sensor by IZM Company. Following this, a kapton cable was successfully bump-bonded to the sensor assembly at UC Davis. The cable bonding uses a lower temperature solder than that used for the kPiX bonding. Figure 3 shows the fully bonded assembly, along with a closeup of the kPiX chip in Figure 4.

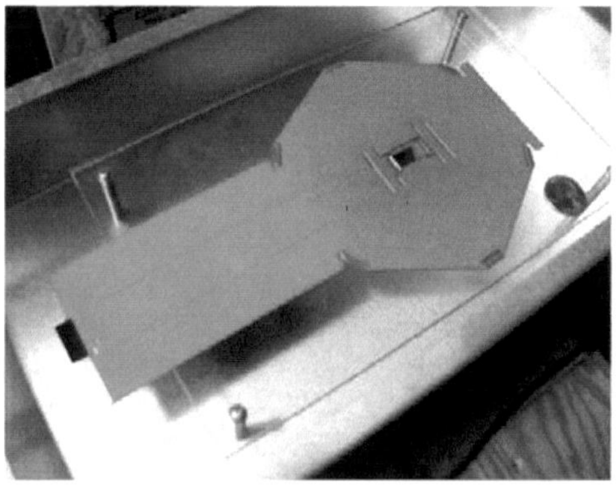

Figure 3. Photograph of a fully bonded sensor. The kPiX chip is bump-bonded to the sensor. Power, control and readout signals are provided via the kapton cable.

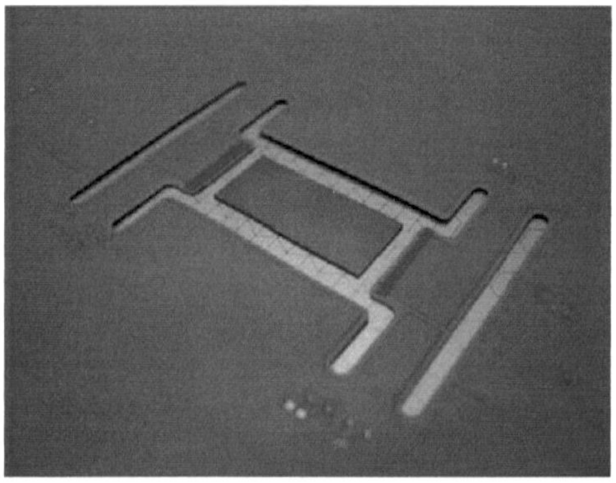

Figure 4. The central region of a sensor showing the kPiX chip. The slots in the kapton allow for differential thermal expansion.

5. Bonded sensor results

Initial bench tests of the bonded sensor of Figure 3 have been carried out and the results are quite promising. A cosmic ray telescope was used to trigger kPiX and the charge of the pixel having the maximum charge was entered in the distributions shown in Figure 5. The red distribution resulted when the ECAL sensor was placed within the telescope acceptance, while the blue distribution resulted when the sensor was outside the telescope acceptance. A Landau distribution (black) is fit to the red signal. The peak of the signal at about 4 fC is consistent with our expectation for minimum-ionizing particles (MIP) passing through the fully-depleted 320 μm thick sensors.

With the highly integrated design we have chosen, a potential worry is crosstalk between channels. Figure 6 indicates no evidence for crosstalk in any other channel when a large 500 fC signal is injected. The noise distribution is nicely fit by a gaussian with RMS 0.2 fC. This is to be compared with the 4 fC MIP signal. This noise level exceeds our requirements for the ECAL.

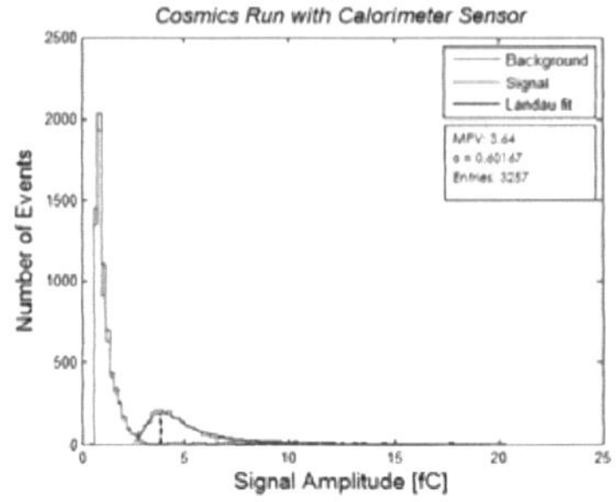

Figure 5. Distribution of charge depositions in a bonded sensor for cosmic ray triggered events. The MIP peak is clearly visible above the noise peak.

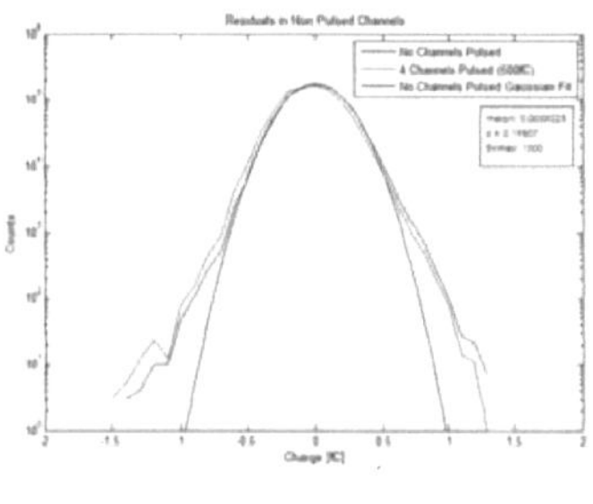

Figure 6. Crosstalk test of a bonded sensor. The charge distributions for all non-pulsed pixels are compared for a large pulse injection (red) and no pulse injection (blue). Also shown is a gaussian fit with RMS 0.2 fC.

6. Overall Mechanical Design

Figure 1 displayed the overall mechanical structure of the ECAL barrel, showing the overlapping stave geometry which is designed to minimize any uninstrumented regions, especially any projective gaps. Figure 7 depicts a cross-sectional view of the readout gap in the vicinity of the center of a wedge module. Kapton flex cables connect near the center of the sensors, bringing power and control signals into the kPiX chip and returning the single digital output line for the 1024 channels. These signals are then multiplexed at the ends of the modules.

Thermal management is a crucial feature of this design. The most power hungry elements of the kPiX chip, particularly the analog front end, are switched off for most of the interval between bunch trains. Our requirement is to hold the average power dissipation per wafer to less than 40 mW. This will allow the heat to be extracted purely passively, providing a much simpler design, less subject to destructive failure modes. The design of the kPiX chip in fact gives average power less than 20 mW, resulting in a total heat load per wedge module of 115 Watt. A cold plate with water pipes routed laterally at the outside of the wedge as shown in Figure 8 results in a maximum temperature differential of 1.35 °C.

Figure 7. Cutaway of the SiD ECAL Barrel wedge, showing the Si sensor and readout cable layout.

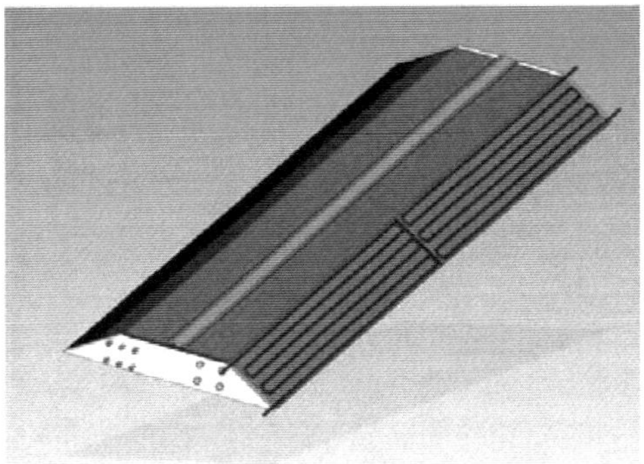

Figure 8. SiD ECAL Barrel wedge, showing the cooling elements, which are located at the outer radius of the barrel.

Referring to Figure 7, the construction of a barrel "wedge" module is carried out as follows. Because tungsten plates are only available with a maximum size of 1×1 m^2, the wedge assembly is done by interconnecting the plates with a screw-and-insert network, which transfers the load from the bottom of the stack to the rail. The design is self-supporting and it does not require additional material to provide the required stiffness. The assembly procedure for a single wedge is sequential with the sensors permanently captured in the gap between tungsten plates, which are specified to have high planarity, achieved at the vendor site by grinding. This specification has been verified on a batch of 15×15 cm^2 plates procured for the beam test module, which have planarity tolerances of $\pm 10 \mu m$, and have been confirmed by interviewing several tungsten vendors/producers. Because of the trapezoidal cross-section of the wedge, the assembly sequence is bottom up, with the wider plate at the base. The first layer of tungsten will be laid down on a jig tool to set the basic tolerances of the stack. Spacing inserts are placed at the locations of the cutouts at the sensor edges (see Figure 2), followed by the sensors with flex cables.

The control of the gap tolerances relies on the flatness of the tungsten plate and on the spacers, whose quality is individually controlled by metrology. The positioning tolerances of the sensor modules in the plane rely on the quality control of not only the spacers, but also on the flex-cable, which will have mounting pads which mate with the inserts. The assembly of the sensors on the flex-cable will be done on a precision jig, which will guarantee the repeatability of the assigned tolerances. The second layer of tungsten will be overlaid on the sensors, once mechanical and electrical connection are tested. This process is repeated 30 times along the stack, which is the number of the active layers of a single wedge module. The last plate on the top has rails, which will allow the insertion and the support from the HCAL. Prior to insertion, each individual wedge will be equipped with a cold plate for thermal management, running along z on one side of the wedge. The boxes on the two opposite sides at $\pm z$ contain the data concentrator electronics, which completes the assembly.

7. Prototype Module and Test Beam

Given the positive initial results of the first bonded sensors, we are moving forward with our plans to build a full-depth test module. This is shown in Figure 9. The test stack is to have a width of one sensor, easily sufficient to contain electromagnetic showers. The longitudinal structure closely matches that of the SiD ECAL. The main difference is that we will have 1.5 mm readout gaps for the test stack, rather than the nominal 1 mm gaps of the SiD design, in order to allow clearance for sensor assemblies to be slid in or out of the stack.

Since the operation of kPiX has been optimized for the bunch timing structure of the ILC, the optimal test facility would be an electron beam having a similar timing structure. Fortunately, SLAC is presently restoring a test beam capability at End Station A. We expect to have the sensors for the test module prepared and first data from this facility in 2013.

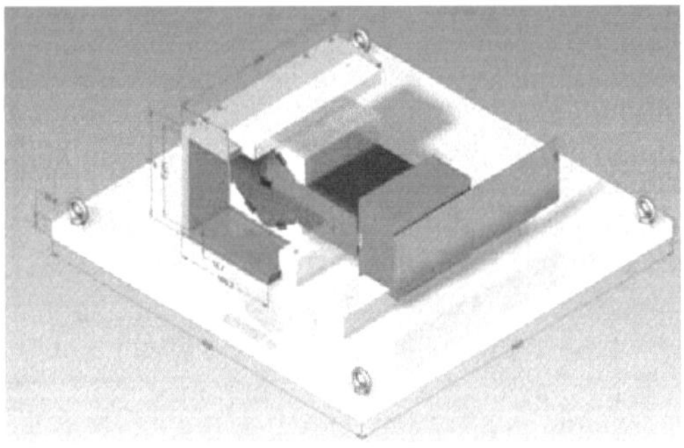

Figure 9. Schematic of a test module to be tested in a beam. The module has a width of one sensor and a depth of 30 layers. The kapton cables attached to each sensor feed concentrator boards, which in turn are connected to a mother board.

8. Calibration and alignment

Silicon detectors are inherently insensitive to gain variations with time and should not have significant inter-pixel gain differences. Pixel to pixel gain differences in the electronic readout are calibrated by dedicated calibration circuitry within the kPiX chip. Perhaps the main calibration issue will be sensor to sensor gain differences. These are not expected to be large, but we are investigating different options for this calibration.

Alignment within ECAL modules and between modules should not be difficult to control with careful fabrication. Alignment to the inner detectors can be sufficiently established using charged particle tracks.

The new RD52 (DREAM) fiber calorimeter

Richard Wigmans[1]
Department of Physics
Texas Tech University
Lubbock, TX 79409-1051, U.S.A.

E-mail: wigmans@ttu.edu

Abstract. Simultaneous detection of the Čerenkov light and scintillation light produced in hadron showers makes it possible to measure the electromagnetic shower fraction event by event and thus eliminate the detrimental effects of fluctuations in this fraction on the performance of calorimeters. In the RD52 (DREAM) project, the possibilities of this dual-readout calorimetry are investigated and optimized. In this talk, the first test results of prototype modules for the new full-scale fiber calorimeter are presented.

1. Introduction - Dual-Readout Calorimetry

High-precision measurements of the four-vectors of fragmenting quarks and gluons will be increasingly important in experiments at the high-energy frontier, and in particular at a future Linear electron-positron Collider. RD52 is a *generic* detector R&D project, carried out at CERN by the DREAM Collaboration, in which we investigate, and to the extent possible eliminate, the factors that prevent calorimetric measurements of the four-vectors of jets and single hadrons with the same level of precision as for electrons and photons.

The key aspect of DREAM detectors is the simultaneous measurement of scintillation light and Čerenkov light generated in the shower development process. By comparing these two signals, the electromagnetic shower fraction can be measured event by event, and the effects of fluctuations in this fraction on the hadronic calorimeter performance can be eliminated.

The merits of this technique were first illustrated with a calorimeter in which the two signals were provided by two different types of optical fibers. More recently, we have also concentrated on crystals ($PbWO_4$, BGO and BSO), which have the potential of eliminating (or at least reducing) the contributions of the next two important sources of fluctuations: photoelectron statistics and sampling fluctuations. We have identified and investigated four different methods to split the signals from such crystals into scintillation and Čerenkov components. These results have been reported at previous editions of the CALOR conference series [1, 2] and in the scientific literature [3].

At this conference, Gabriella Gaudio has reported our recent results on electromagnetic dual-readout calorimetry with two different matrices of crystals that are large enough to contain em showers at a sufficient level [4]. In this talk, I will describe the progress on the new dual-readout fiber calorimeter that is being built and tested by the RD52 Collaboration. In Section 2, I describe the rationale for the design choices that were made, and in Section 3, I present the first results of beam tests of prototype modules of this calorimeter.

[1] Talk presented on behalf of the RD52 Collaboration (Gagliari, Cosenza, Pavia, Pisa, Roma, Iowa State, TTU)

2. Why a Copper-Fiber Calorimeter?

The energy resolution of a calorimeter is determined by *fluctuations*. This means, for example, that the fact that two thirds of the energy of a jet is typically carried by charged particles is completely irrelevant in that context. Even though this jet component can be measured with very high precision, a detector that *only* would be sensitive to this component would not be capable of measuring jet energies with a precision better than 20%, at *any energy*, because of event-to-event fluctuations in this fraction [5]. Fluctuations that determine the hadronic energy resolution of a calorimeter system include:

(i) Fluctuations in the energy fraction carried by the *electromagnetic shower component*, $f_{\rm em}$. These fluctuations are non-Gaussian and typically dominate the hadronic energy resolution of the calorimeters used in the current generation of experiments, especially at high energies. They can be eliminated either by making the calorimeter intrinsically compensating ($e/h = 1$), or by measuring the value of $f_{\rm em}$ event by event. DREAM calorimeters use the latter approach, which offers essentially the same advantages as compensation, without the inherent limitations in terms of the choice of absorber medium, sampling fraction, integration time and volume [6].

(ii) Fluctuations in the *invisible* energy. These are dominated by the nuclear binding energy losses in the numerous nuclear reactions through which the shower absorption proceeds. Efficient detection of the nuclear reaction products, and in particular of the abundantly produced neutrons, makes it possible to limit the effects of these fluctuations on the hadronic energy resolution. These fluctuations determine the ultimate resolution limit one may hope to achieve ($\sim 15\%/\sqrt{E}$ [6]), and also explain the different single-hadron resolutions obtained with the ZEUS [7] and D0 [8] calorimeters. These devices were both (approximately) compensating and stochastic fluctuations contributed similarly to the resolution. Yet, the plastic scintillator active medium of ZEUS was a much more efficient detector of the nuclear reaction products than the liquid argon used in D0, which led to a much better resolution for ZEUS.

(iii) *Stochastic* fluctuations. In the DREAM calorimeters, these include sampling fluctuations and Poisson fluctuations in the number of Čerenkov photoelectrons.

Based on our results, we believe that a copper-fiber dual-readout calorimeter offers the best opportunities to achieve hadronic energy resolutions at the 1% level. The first two types of fluctuations are minimized by design in such a calorimeter, and the stochastic fluctuations can be reduced considerably compared to the original DREAM calorimeter, in which sampling fluctuations and Poisson fluctuations in the numbers of Čerenkov photoelectrons each contributed $\sim 35\%/\sqrt{E}$ to the measured

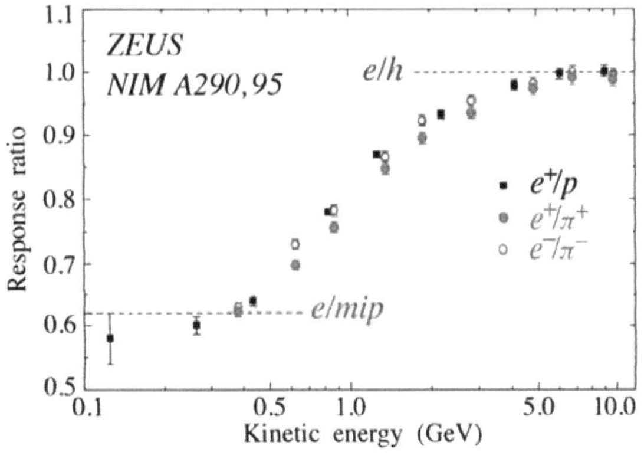

Figure 1. The response of the (compensating) ZEUS uranium/plastic-scintillator calorimeter to low-energy hadrons [10].

resolution, and we expect to bring this down to $\sim 10\%/\sqrt{E}$. One could argue that these stochastic fluctuations could be reduced even more in homogeneous dual-readout calorimeters. However, in such calorimeters fluctuations in invisible energy would be much larger because of the insensitivity of such devices to the nuclear shower component, and thus dominate the resolution. Using complementary arguments, Don Groom has shown that the performance of such devices, even if the funds could be found to build them, would most likely be disappointing [9].

Also the choice of copper as absorber material is an essential ingredient of our plans. As mentioned above, one advantage of the dual-readout technique over intrinsically compensating calorimeters is the fact that it allows the use of low-Z absorber material such as copper. This is especially important for obtaining excellent energy resolutions for *jets*. This is illustrated in Figure 1, which shows the response of the ZEUS calorimeter to low-energy single hadrons, relative to the response to electrons. For energies above 5 GeV, this response ratio is 1.0, since the calorimeter is compensating. The strong non-linearity observed for lower-energy hadrons is a consequence of the fact that, as the energy decreases, an increasing fraction of the hadrons do not develop showers, but rather range out, *i.e.*, behave as muons. In comparison with mips, em showers are inefficiently sampled in calorimeters with high-Z absorber material [6]. In the ZEUS calorimeter, which used depleted uranium (^{238}U) as absorber material, the e/mip ratio was only ~ 0.6. And since soft hadrons are typically a major component of fragmenting quarks and gluons, the jet resolution in ZEUS was significantly worse than for single hadrons. This effect could be avoided in dual-readout calorimeters that use copper as absorber material, since the e/mip value is in that case ~ 0.9.

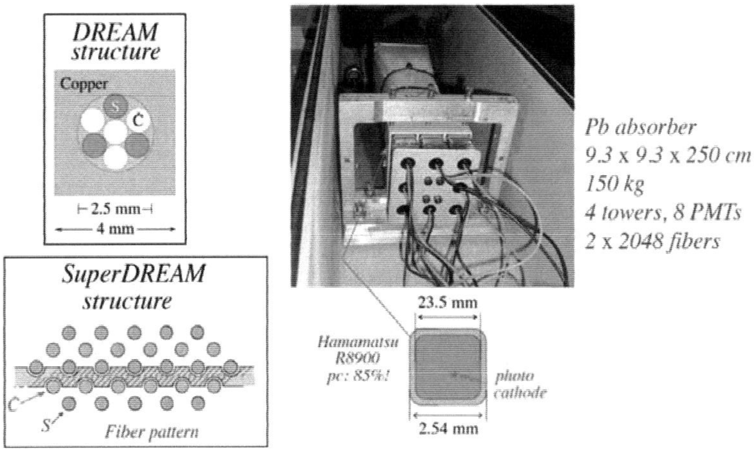

Figure 2. The fiber patterns used in the original DREAM calorimeter and in the new SuperDREAM modules (*left diagrams*). Also shown is how the fibers sticking out from the rear end of the module are connected to 8 PMTs. These PMTs (Hamamatsu R8900) have a very large effective photocathode area, 85% of the outer envelope of the tube.

Our design to limit the contributions of stochastic fluctuations as much as possible is based on the following elements:

- Embed the fibers *individually* in the absorber structure. In the original DREAM modules, seven fibers were grouped together and split at the rear end of the module into C and S bunches. This difference is illustrated in Figure 2.
- Maximize the *fiber density* in the detector modules. The limit in this respect is determined by the fact that the readout structure has to fit within the detector envelop. To that end, we selected PMTs with a large ratio of the effective photocathode area and the outside perimeter. In the R8900

Hamamatsu tubes this ratio is 85%, which allowed us to increase the fiber packing fraction by more than a factor of two compared to the original DREAM module (Figure 2).

- Use Čerenkov fibers with a considerably *larger numerical aperture* (N.A.) than the quartz ones used in the DREAM module. Initially, we had selected multi-clad polystyrene (PS) Kuraray fibers for this purpose, which have a N.A. of 0.72. However, beam tests of the first SuperDREAM module equipped with these fibers revealed an unacceptably strong light attenuation ($\lambda_{att} \approx$ 3m), as a result of Rayleigh scattering in the PS. We then decided to use PMMA based fibers instead. These fibers, made by Mitsubishi, have a N.A. of 0.50, but thanks to the much longer light attenuation length the signals were found to be about twice as large as in the PS case (Figure 3). A direct measurement

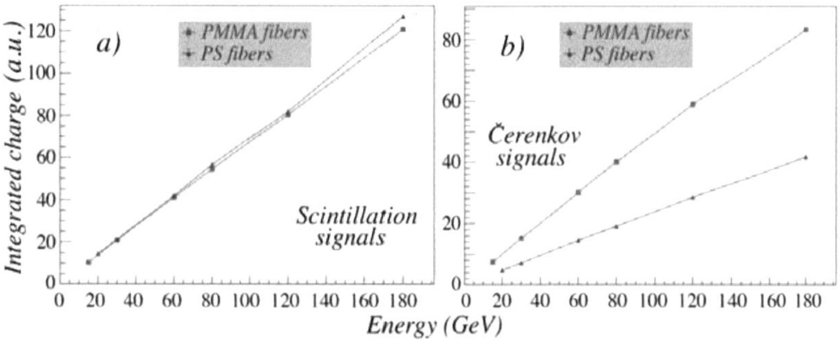

Figure 3. The average electron signal as a function of energy for a SuperDREAM module equipped with polystyrene and another module equipped with PMMA fibers for the detection of the Čerenkov light. The scintillating fibers were of the same type in both modules. The 8 readout PMTs were exactly the same in both modules. Shown are the results for the scintillation (a) and the Čerenkov (b) signals.

of the Čerenkov light yield gave 32 photoelectrons per GeV deposited energy in the PS module. In the PMMA module, this light yield was a factor two larger. For comparison, we mention that the Čerenkov light yield in the original DREAM module was only 8 photoelectrons per GeV. We expect to be able to increase this number to well over 100 p.e./GeV in the final SuperDREAM modules.

- The Čerenkov light yield is further increased by aluminizing the upstream ends of the fibers and by using PMTs with a super-bialkali photocathode.

We expect to test a calorimeter module in which all these improvements have been implemented in July.

3. Results of beam tests of prototype modules

The first two prototype modules, results of which are shown in Figure 3, were build with lead as absorber material. The absorber structure needed for the high-sampling-frequency SuperDREAM calorimeter is relatively easy to extrude using this material. In the meantime, we have also found a way to make this absorber structure out of copper, and the first two copper modules are being prepared for our beam tests in July. The results shown in this section were obtained with the lead module equipped with PMMA Čerenkov fibers. Figure 4 shows the em energy resolution in the Čerenkov channel as a function of energy. For comparison, the published resolution obtained with the original DREAM module [11] is shown as well. The improvement is especially important at the highest energies, due the absence of a significant deviation from $E^{-1/2}$ scaling (the so called "constant term"). Further improvements may be expected as a result of the use of copper absorber, fiber aluminization and better shower containment.

Measurements of the em energy resolution are a good way to assess the reduction obtained in the stochastic fluctuations, since such fluctuations are the only ones that matter in a well designed sampling calorimeter. However, since SuperDREAM is also intended to be a very good hadron calorimeter, we also

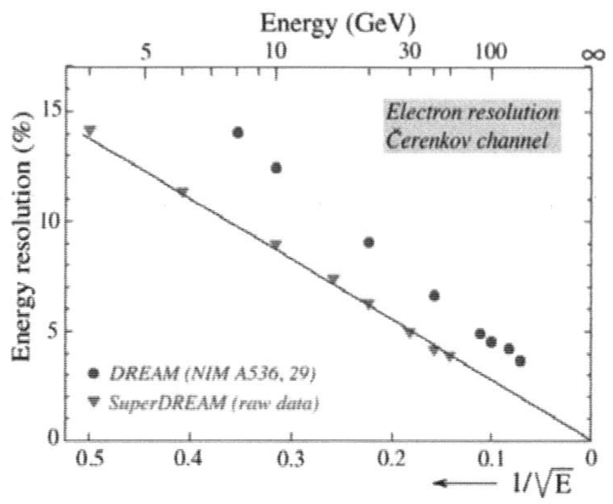

Figure 4. Comparison of the em energy resolution in the Čerenkov channel, measured with the first SuperDREAM modules and with the original DREAM calorimeter [11].

made some measurements with hadrons. Of course, side leakage is a major factor for the performance in that case. In order to limit the effects of such leakage out of the 150 kg calorimeter module, we surrounded it by 16 plastic scintillator counters (see Figure 5), which provided a crude measurement of the energy of the particles escaping from the calorimeter. Results of these measurements are shown

Figure 5. The setup in which individual SuperDREAM modules are being tested in the H8 beam. The module is contained in a box surrounded by 16 leakage counters.

in Figure 6. Figures 6a and 6b show the scintillation signal distribution for 180 GeV pions sent into this module and the (anti-)correlation between these signals and those observed in the leakage shield, respectively. The virtue of the dual-readout principle is illustrated in Figure 6c, which shows a scatter plot of the (corrected) scintillation signal versus the Čerenkov/scintillation signal ratio. As expected, the smaller this ratio, the smaller $f_{\rm em}$ and, therefore, the smaller the scintillation signal. After an event-by-event correction based on the measured $f_{\rm em}$ value, this scatter plot is converted into the one shown in Figure 6d, and the projection of this scatter plot on the horizontal axis (Figure 6e) yields a Gaussian distribution with approximately the right energy and a fractional width of less than 5%. This is an encouraging result, which we expect to improve considerably at the end of this year, when all available SuperDREAM modules will be assembled to form a calorimeter with a mass of $\sim$ 2000 kg, and tested in hadron beams at CERN. Figure 7 shows some pictures of the production of SuperDREAM modules at the University of Pavia.

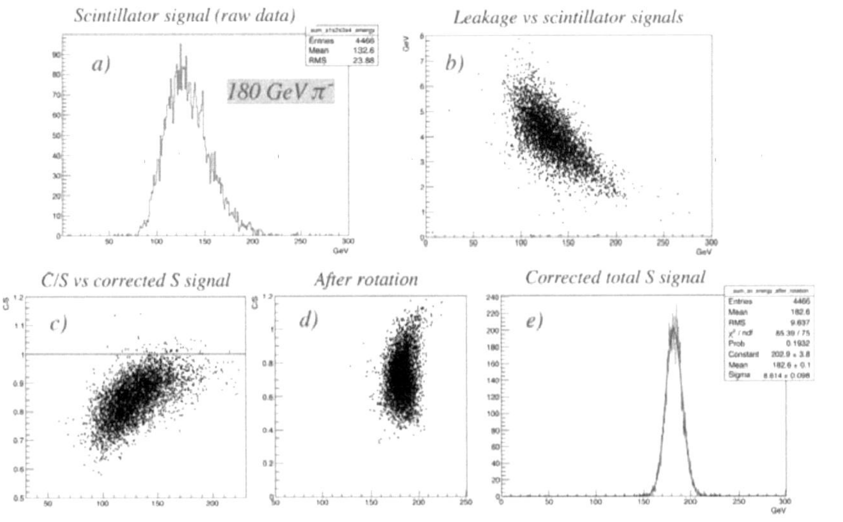

Figure 6. First measurements of pions in a single module of the new SuperDREAM fiber calorimeter, surrounded by leakage detectors. See text for details.

A crucial aspect of this type of calorimeter is its *longitudinally unsegmented* structure. This allows a compact construction and avoids the complications usually encountered with the intercalibration of different longitudinal sections in a longitudinally segmented detector system. Potential disadvantages of this structure concern the identification of electrons and photons, and the measurement of the direction of neutral incoming particles.

However, a fine lateral segmentation combined with information on the detailed time structure of

Figure 7. New SuperDREAM modules under construction at the University of Pavia (top), and at INFN Pisa (bottom). The Pavia modules use lead as absorber material, the Pisa ones copper.

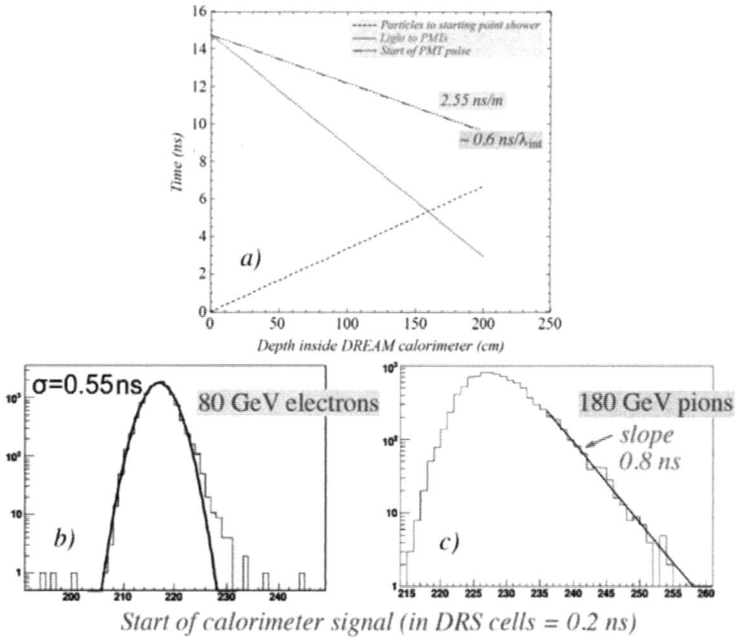

Figure 8. Principle of the measurement of the depth of the light production through the starting time of the signal (a). Distribution of the starting times of signals from 80 GeV electrons (b) and 180 GeV pions (c) in the fiber calorimeter.

the events greatly alleviates these problems. Measurement of the time structure of the signals makes it possible to obtain crucial information about the depth at which the light is produced. By using the fact that light travels at a speed of c/n in the fibers, while the particles producing the light travel at c, the starting time of the signals makes it possible to measure the depth at which the light is produced, as illustrated in Figure 8a. The deeper the light is produced, the earlier the PMT signal starts. The distribution of the starting points of the signals generated by 80 GeV electron showers (Figure 8b) indicates a time resolution of 0.55 ns, which corresponds to a depth resolution of $\sim$ 20 cm (Figure 8a). The varying depth at which pion showers start is clearly illustrated in Figure 8c.

This feature makes it possible to correct for small effects of light attenuation in the fibers. It is also an important tool for electron/photon identification in this type of calorimeter. As demonstrated by SPACAL, a simple measurement of the duration of the signals (the full width at 20% of the amplitude) made it possible to separate electrons from pions at the 99.99% level [12]. Finally, the time structure is an important tool for measuring the neutron contribution to the scintillation signals [13].

We measure the time structure of each calorimeter signal with a Domino Ring Sampler chip [14]. An array of 1024 switching capacitors samples the input signal at a frequency determined by the Domino Wave. Once a trigger is received, the capacitors are sequentially read out by a pipeline 12-bit ADC, starting from the leading edge of the trigger signal, moving backward. In this way, the time history of the signal is stored in the DRS output, while the trigger itself is stored in a DRS channel. In our measurements, we operate this circuit at 5 GHz, for a time resolution of 0.2 ns. Figure 8 clearly illustrates the benefits of this approach.

References

[1] Wigmans R 2009, Proceedings of the XIIIth Conference on Calorimetry in High Energy Physics, J. Phys. Conf. Series **160**, 012018; *ibid.* Gaudio G, 012068; Carosi R, 012069; Voena C, 012070: Pinci D, 012071; Hauptman J, 012072.
[2] Gaudio G 2011, Proceedings of the XIVth Conference on Calorimetry in High Energy Physics, J. Phys. Conf. Series **293**, 012006; *ibid.* Hauptman J, 012080.
[3] Akchurin N *et al.* 2005 Nucl. Instr. and Meth. **A537**, 537; *ibid.* **A550**, 185; **A581**, 643; **A595**, 359; **A621**, 212; **A638**, 47.
[4] Gaudio G 2012, Contribution to these Proceedings.
[5] Lobban O, Shriharan A and Wigmans R 2002, Nucl. Instr. and Meth. **A495**, 107.
[6] Wigmans R 2000, *Calorimetry - Energy Measurement in Particle Physics*, International Series of Monographs on Physics, vol. 107, Oxford University Press.
[7] Behrens U *et al.* 1990, Nucl. Instr. and Meth. **A289**, 115.
[8] Aronson S *et al.* 1988, Nucl. Instr. and Meth. **A269**, 492.
[9] Groom D 2012, Contribution to these Proceedings.
[10] Andresen A *et al.* 1990, Nucl. Instr. and Meth. **A290**, 95.
[11] Akchurin N *et al.* 2005, Nucl. Instr. and Meth. **A536**, 29.
[12] Acosta D *et al.* 1991, Nucl. Instr. and Meth. **A302**, 36.
[13] Akchurin N *et al.* 2009, Nucl. Instr. and Meth. **A598**, 422.
[14] Ritt S *et al.*2010 Nucl. Instr. and Meth. **A623**, 486.

Micromegas for imaging calorimetry

Maximilien Chefdeville on behalf of the LAPP/LC group
LAPP, Université de Savoie, CNRS/IN2P3, Annecy-le-Vieux, France
E-mail: chefdevi@lapp.in2p3.fr

Abstract. Gaseous detectors have been used in sampling calorimeters for several years and still remain an attractive option for future collider experiments. The use of Micromegas chambers as active elements of a hadronic calorimeter was proposed a few years ago and stirred up exciting developments of this technology. Basic building blocks of large size chambers acting as signal generating and processing units have been designed and fabricated. They were used to construct two chambers of 1 m^2 size which after test in muon and pion beams, showed excellent performance. The most important results are presented followed by a discussion of the prospects towards realistic calorimeter modules at a future collider experiment.

1. Introduction

1.1. *Calorimetry at a future linear collider*

The detailed study of electroweak symmetry breaking and of the properties of a hypothetical standard model Higgs boson are some of the physics goals motivating the construction of a linear electron collider. The physics case is now enhanced with the discovery at LHC of a Higgs-like new particle. At ILC or CLIC [1, 2], several interesting physics channels will appear in multi-jet final states, often accompanied by charged leptons and missing transverse energy. The di-jet energy resolution should be such that Z and W bosons can be identified with an accuracy comparable to their natural decay width. This translates into a jet energy resolution of 3–4 % over the whole energy range. The approaches followed to meet this requirement are the *Dual Readout* and the *Particle Flow*, respectively studied by the DREAM and CALICE collaborations [3, 4]. The first one is a powerful compensation technique that combines measurements of cherenkov and scintillation light produced by showering particles [5]. On the other hand, the CALICE collaboration develops calorimeters with high 3D imaging capability to identify the jet's charged and neutrals components. Thanks to the granularity, the dominant charged component can be measured very precisely with the tracking system resulting in improved jet energy resolution [6].

1.2. *Micromegas for calorimetry*

Several calorimeter technologies are studied by the CALICE groups such as silicon, scintillators and gas detectors for the active parts with tungsten or steel absorbers. Three HCAL projects presented at this conference are documented in the proceedings: the scintillator analog HCAL and the Glass Resistive Plate Chamber (GRPC) and Micromegas semi-digital HCAL (or SDHCAL). Micromegas is a fast, position sensitive Micro Pattern Gas Detector (MPGD [7]) operating in the proportional mode [8]. Essentially free of space charge effects and thus high rate capable, it suffers very little from aging because it functions in simple gas mixtures (*e.g.* Ar/CO_2) and at relatively low electric fields and voltages (< 500 V).

2. The 1 m^2 Micromegas chamber

2.1. Micromegas principle

Micromegas is a parallel plate detector that uses a thin micro-mesh to separate the gas volume into two regions of low and high field. Electrons liberated in the low field region drift towards the mesh and are multiplied in the high field region, inducing a signal with a fast electron component ($\sim$ 1 ns) and an ion tail ($\sim$ 150 ns). The signal is proportional to the gas gain, itself controlled by the mesh voltage. The fast ion collection at the mesh makes Micromegas free of space charge effects which is useful to obtain some information on cell occupancy (semi-digital calorimetry).

2.2. Active Sensor Unit

The basic building block of a 1 m^2 chamber is a 32×48 cm^2 8 layer Printed Circuit Board (PCB) dubbed Active Sensor Unit (or ASU). One PCB side consists of the mesh laminated on a anode matrix of 1536 pads of 1 cm^2 at a 128 μm distance. The pad's readout is performed by 24 ASICs located on the backside. Passive components are also present on this side to provide protection against gas discharges. The ASICs are read out with 2 Detector Interface boards (DIF, inter-DIF) which distribute voltage to ASICs and mesh. The ASUs can be chained through flexible connectors which allows to read out several of them with one pair of DIF/inter-DIF board.

The 64 channel ASIC, called MICROROC, was developed by LAPP/LC and LAL/Omega groups [9]. Its input stage consists of a diode network for discharge protection, a low noise charge preamplifier (1500 ENC) and 2 shapers of variable peaking time (75–200 ns) and different gains. Shaper signals are compared to 3 thresholds and the result is stored in a 127-event depth memory. Thresholds are common to the 64 channels but individual pedestals can be adjusted with respect to the low threshold. This feature allows to minimize the detector non-uniformity. The ASIC is synchronised to the 5 MHz DIF clock for time-stamping with a 200 ns precision.

2.3. Chamber design

The 1 m^2 chamber consists of 6 ASUs assembled in a one gas volume (Fig. 1). Small spacers (1 mm wide, 3 mm high) are inserted between ASUs and support the cathode cover, defining precisely the drift gap. Plastic frames are closing the chamber sides, leaving openings for 2 gas pipes and flexible cables. The chamber is eventually equipped with readout boards and a patch panel for voltage distribution. The total chamber thickness amounts to 9 mm which includes 2 mm for the cathode cover, 3 mm of drift gap and 4 mm for PCB and ASICs. With this mechanical design, we achieve less than 2 % of inactive area.

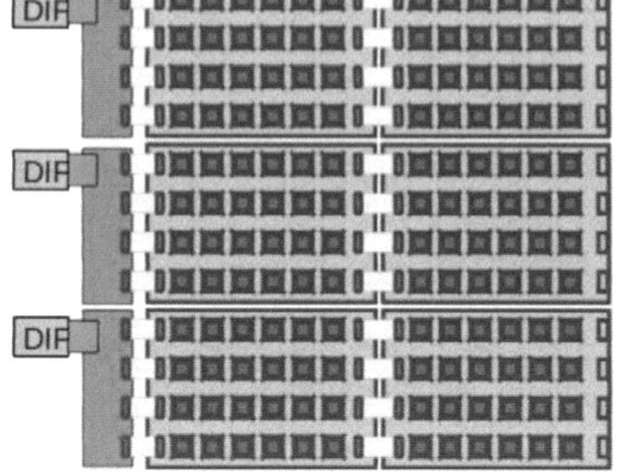

Figure 1. Drawing and photograph of the ASIC and mesh sides of the 1 m^2 chamber.

3. Chamber response to minimum ionising particles

The response to minimum ionising particles (MIPs) was studied in a 150 GeV/c muon beam at CERN/SPS. The 1 m^2 chamber was flushed with a non-flammable mixture of $Ar/CF_4/iC_4H_{10}$ 95/3/2. The mesh voltages were varied between 300–420 V (gas gain of 100–8000) and the drift field set to 300 V/cm at the local maximum of the drift velocity. A telescope of small chambers [12] with a point resolution of ~ 3 mm was used to measure efficiency and hit multiplicity at various positions. A trigger signal was provided by a set of 3 scintillators and photomultiplier tubes to discriminate between cosmic background and beam particles.

3.1. Noise and channel occupancy

The noise rate over all channels is equalised to a target value of 0.1 Hz by tuning pedestals and thresholds. This is done with HV set to zero in order to avoid bias from cosmic or beam particles, and is achieved by masking 16 / 9216 channels only. When HV is applied, particles crossing the chamber gas dominate the channel occupancy. Cosmic muons are responsible for a uniform background above which the beam profile clearly separates (Fig. 2 (left)). Using time information (time of trigger and time of hits), the background is removed, leaving a very clean pattern of the beam (see presentation slides).

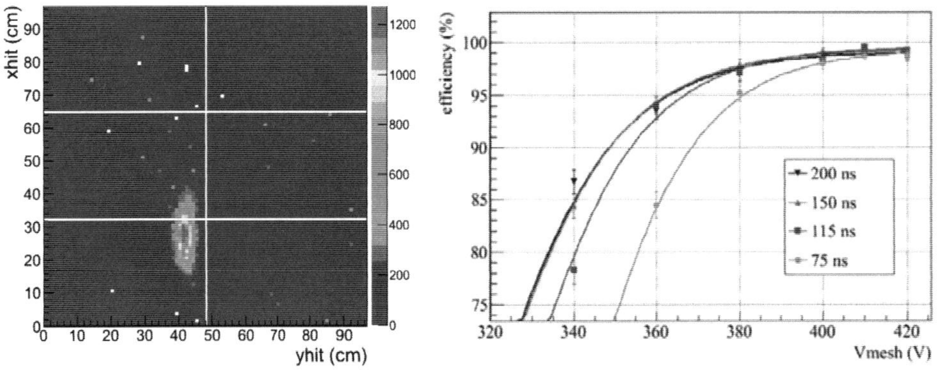

Figure 2. Profile of a muon beam crossing the 1 m^2 chamber before applying a cut on the trigger time (left). Efficiency versus mesh voltage curves for various peaking times (right).

3.2. Efficiency study

Beam muons (150 GeV/c) are a good approximation of MIPs which liberate in the drift region a most probable number of 14 electrons. These are multiplied in the amplification region by factors up to 10^4 depending on the mesh voltage. The strong dependence of the detection efficiency on the applied voltage is shown in Fig. 2 (right) for voltages of 340–420 V and peaking times of 75-200 ns. The curves indicate that Micromegas signals last between 115–150 ns, as expected for a 128 μm gap. For peaking times larger than 150 ns, a gas gain as low as 10^3 ($V_{\text{mesh}} = 365$ V) is sufficient to detect MIPs with an efficiency larger than 95 %. This is due to very low readout thresholds of 1–2 fC achieved with the MICROROC chip. Upon full exposure of two chambers, detailed efficiency maps over 8×8 cm^2 regions were produced revealing an efficiency of (96 ± 2) %. Such a little variation indicates a good control of the chamber dimensions (gas gaps) as well as of the electronics parameters (gains, thresholds).

3.3. Multiplicity study

A benefit of Micromegas w.r.t. other gas detector technologies is the limited spatial extension of the avalanche signals. This is a result of the little diffusion experienced by the electrons in the gas. The electrons liberated in the drift gap arrive at the mesh with a transverse spread of 110 μm RMS [10]. Moreover, electrons liberated by one avalanche are distributed at the anode with a 20–30 μm spread [10]. It is thus likely that all electrons from a crossing particle are collected on one pad. As showed in the presentation slides, the hit multiplicity is below 1.15 up to 390 V (*i.e.* gain $G = 3000$). At higher gains, neighbouring pads become sensitive to single electrons, increasing the multiplicity to 1.35 at 420 V (*i.e.* $G = 8000$). There is however no reason to work in that regime as high MIP efficiency is reached at lower gains (95 % at $G = 1000$).

4. Chamber response to hadron showers

Measurements in a muon beam allow to define the benchmark performance of the chamber. The various particles generated in a hadron shower, however, are mostly far more ionising than MIPs. The particle density is of course larger, especially in the shower core and where π^0 induced electromagnetic showers develop. Larger dE/dx and multiplicity significantly increase the discharge probability. It is thus mandatory to investigate the response and stability of the chamber in such an environment. Two experimental setups were used.

(i) The first is the one described in the previous section with an additional 20 cm long ($\sim 1\,\lambda_{\text{int}}$) and 10×10 cm^2 area Fe block placed 50 cm upstream of the Micromegas chamber.

(ii) The second setup is a 50 layer sampling GRPC/Fe calorimeter in which 2 Micromegas chambers were inserted at layer 48 and 49 (*i.e.* behind $\sim 5\,\lambda_{\text{int}}$ of Fe).

In both cases, a pion beam was used with an energy of 150 GeV and 60–180 GeV respectively.

4.1. Gas gain study

The average pad signal in a hadron shower is larger than the one induced by a MIP. It can hence be expected that the gas gain necessary to image 95 % of the shower is lower than that necessary to detect a MIP with the same efficiency. Using experimental setup (i), the distribution of the number of hits in the chamber was measured at various mesh voltages. A statistics of 50×10^3 events was recorded at 325 V, 350 V and 375 V which translates into gas gains of about 350, 800 and 1700. The number of hit distributions, shown in Fig. 3, exhibit a peak at $N_{\text{hit}} = 1$ and a long tail from penetrating and showering pions respectively. The distributions at 350 V and 375 V yield different efficiency to penetrating pions but remarkably, have both a similar tail. It can thus be concluded that a gas gain as low as 800 is sufficient to image most of the shower.

4.2. Beam energy study

Compensation is difficult to achieve in sampling hadron calorimeters which are thus non linear devices [11]. With a digital readout, saturation effects add to the complexity of the response. Additional thresholds can probably be used to mitigate saturation effects but the feasibility of this technique still needs to be demonstrated. While this will be the goal of future beam tests, a first measurement of the response of a Micromegas chamber to pions has been carried out with the calorimetric setup (ii). At pion energies from 60 to 180 GeV, the distribution of hits passing the three thresholds after $5\,\lambda$ of Fe was measured at 390 V. The mean value for the low threshold is plotted versus beam energy in Fig. 3. Because most of the shower particles have been stopped at this depth of the calorimeter, the mean values are small: *e.g.* 5.5 hits et 100 GeV. This presumably explains the linear trend, which is also observed for medium and high thresholds. Fortunately, more measurements at various depths should become available by the end 2012 so the linearity of a Micromegas SDHCAL can be assessed.

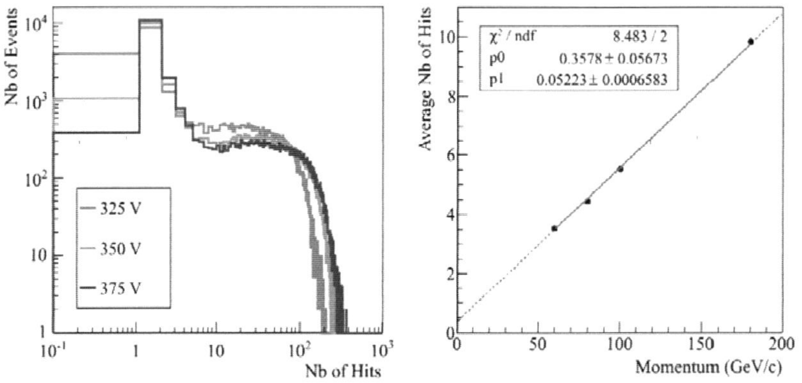

Figure 3. Hit distribution from 150 GeV/c pions at various mesh voltages measured with setup (i) (left). Mean value versus pion momentum measured with setup (ii) at 390 V (right).

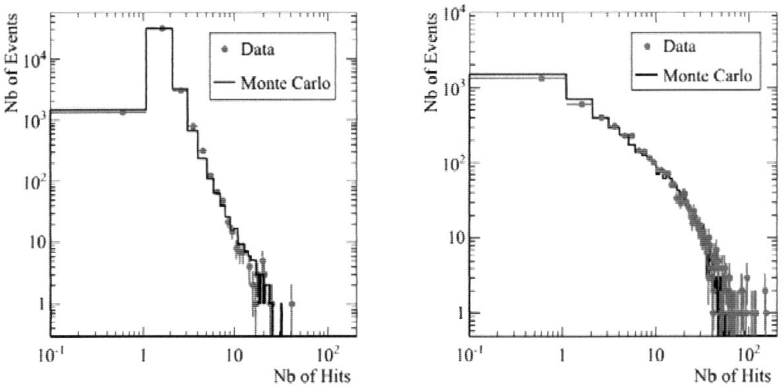

Figure 4. Hit distribution from 100 GeV muons (left) and pions (right) as measured at layer 48 of a hadron calorimeter (setup (ii)). Data points are compared to Monte Carlo predictions from GEANT4 LHEP physics list. The mesh voltage was 390 V during both runs.

4.3. Monte Carlo simulation

Computer programs can be used to simulate the development of showers and therefore predict the response of a full Micromegas calorimeter. As a first step in this direction, the number of hit distribution of a single chamber at a given depth of the calorimeter was simulated and compared to test beam data collected with setup (ii). Doing so, a straightforward digitisation procedure was implemented where simulated energy deposits are turned into a number of electrons which diffuse, multiply and are compared to a threshold. A preliminary result of our analysis is presented in Fig. 4 (right) which shows the distribution after $5\,\lambda_{\text{int}}$ of Fe for 100 GeV pions. The readout threshold was tuned so as to reproduce the efficiency to muons (Fig. 4 (left)). A satisfactory agreement is obtained, both for muons and pions. It should be stressed that no noise was introduced in the simulation, meaning that data are essentially free of noise.

5. Prospects towards a technological module

5.1. Scalability to large area

The design of the 1 m^2 chamber is scalable in both transverse directions. It is now such that 2 ASUs are chained (forming a slab) but that can in principle be extended to 6 along a total length of 3 m. The current design consists of 3 slabs, each one read out by a pair of DIF/inter-DIF boards. The number of slab can be extended as well.

5.2. Chamber thickness and spark protection

At a future linear collider, the calorimetric system will fit inside the magnet. To minimize the solenoid radius and cost, the chamber should be as thin as possible. The Micromegas chamber is 9 mm thick, including the 2 mm thick steel cover. At an ILC, the HCAL absorbers will be made out of steel so the effective chamber thickness is actually 7 mm, already matching the baseline value.

Ways to reduce the thickness further are nevertheless considered. One is to shrink the 3 mm gas gap to 2 mm. This would yield a signal loss that can be compensated by a 50 % increase in gas gain. Another way is to slim the PCB components by adopting thinner ASIC packages and by removing/embedding the protection networks. These supplement MICROROC diodes and may not be mandatory for operation at low gas gains. Also, they could be replaced by a thin resistive layer deposited onto the anode plane. Both options are currently under investigation.

5.3. Pressure and temperature effects

Previous studies [12] showed that the gas gain sensitivity to P/T variations is small (-0.5 %/mbar and 1 %/K with a slight gas mixture dependence), so the detector response is slightly affected by P/T variations. The MIP efficiency variation under usual atmospheric changes is negligible, provided that the readout threshold is low enough (*e.g.* 1–2 fC). With additional thresholds at higher values, P/T changes should probably be corrected for. An on-line solution is to monitor P/T and adjust the voltage according to the known sensitivity of $\sim$3 %/V.

6. Conclusion

Micromegas is a relatively new technology which is studied for calorimetry for the first time. In view of this application, an innovative design of Micromegas detector has been proposed and used to construct two chambers of 1 m^2. Experimental results on the chamber performance to MIPs are excellent: an efficiency larger than 95 % is achieved while keeping the pad multiplicity below 1.15. The chamber shows a stable behaviour in pion showers and a quantitative understanding of the data is well underway. Although two chambers are insufficient to measure the energy resolution of a 5 λ_{int} Micromegas calorimeter, a test program with four chambers has been established to measure linearity and leakage inside the CALICE Fe structure at the end of 2012.

References

[1] ILC Reference Design Report, 2007, (http://arxiv.org/PS_cache/arxiv/pdf/0709/0709.1893v1.pdf).
[2] CLIC Conceptual Design Report, 2011, (https://edms.cern.ch/document/1177771).
[3] The RD52 experiment web-page: http://highenergy.phys.ttu.edu/dream.
[4] The CALICE collaboration web-page: https://twiki.cern.ch/twiki/bin/view/CALICE.
[5] R. Wigmans, *Nucl. Instr. and Meth.* **A617** (2010) 129133.
[6] M.A. Thomson et al., *Nucl. Instr. and Meth.* **A611** (2009) 25-40.
[7] The RD51 collaboration web-page: http://rd51-public.web.cern.ch/rd51-public/.
[8] Y. Giomataris et al., *Nucl. Instr. and Meth.* **A376** (1996) 29-35.
[9] C. Adloff et al., *JINST* **7** (2012) C01029.
[10] S.F. Biagi, *Nucl. Instr. and Meth.* **A421** (1999) 234240.
[11] R. Wigmans, Calorimetry, *Scientifica Acta* **2**, No. 1, (2008) 18-55.
[12] C. Adloff et al., *JINST* **4** (2009) P11023.

Journal of Physics: Conference Series **404** (2012) 012070 doi:10.1088/1742-6596/404/1/012070

Concluding Remarks

Nural Akchurin

Texas Tech University, Department of Physics, Lubbock, Texas, USA

E-mail: Nural.Akchurin@ttu.edu

Abstract. The International Conferences on Calorimetry in High Energy Physics, or the CALOR series, have always been where the calorimeter experts come together to review the state of calorimetry and bring forth new ideas every two years. The fifteenth conference, CALOR2012, in Santa Fe was no exception. Although they were built roughly a decade ago, we are now witnessing the exceptional power of the LHC calorimeters and the crucial role they have been playing in the discovery of the 125 GeV Higgs-like boson. As we ruminate on the coming generation of experiments at the next (linear) collider and on the upgrades at the LHC, we are heartened by the substantial advances we made in calorimetry in the last decade. These advances will certainly help uncover new physics in the years to come, not only at colliders but also in astroparticle experiments that take advantage of natural elements such as air, water, and ice.

I do not dare to review all the excellent contributions to this conference in a few pages, but instead offer a heuristic perspective of where we stand and the challenges we face in calorimetry today as depicted in Figure 1.

Many calorimeters were built based on scintillation light from solids, liquids, and gasses. Similarly, when Cherenkov radiation formed the basis of signal in em calorimeters, heavy crystals and glasses were often used. These two processes are in fact complementary, and if we put them together appropriately in a single calorimeter, one of the larger sources of fluctuations that degrade performance is naturally eliminated. The ratio of the Cherenkov (Q) to scintillation (S) signal is proportional to the electromagnetic fraction (f_{em}) in a shower, given that e/h ratios for the scintillation and the Cherenkov part of the calorimeter are known separately. Because the fluctuations in f_{em} can be measured on an event-by-event basis in this way, we are now able to eliminate its contribution to the energy resolution by simply measuring Q/S. This so-called dual-readout technique was pioneered by the DREAM collaboration in the last decade.

The separation of scintillation and Cherenkov photons can be accomplished in different ways. One can instrument an absorber with two different types of fibers (scintillating and clear) where the separation is accomplished by design, or one can exploit the different spectral characteristics of Cherenkov and scintillation light from a single transparent medium by means of optical filters. The pulse shape may also be exploited if the temporal features of these signals are sufficiently different, *e.g.* since scintillation pulses are relatively slow. The directional aspect of Cherenkov light is another possibility, but it is difficult to implement in most practical calorimeters. The Cherenkov polarization, as opposed to unpolarized scintillation, may also be used but needs more research.

A large fraction of the energy is lost through nuclear interactions that do not materialize in a measurable signal in calorimeters. If an appropriate scheme is developed to turn these MeV neutrons into a signal, we will further reduce fluctuations in energy carried away by these poorly sampled neutrons. It may require special materials in the detector and/or algorithms that exploit the (late) pulse shape. When achieved, the triple- or multiple-readout will certainly advance us towards the ultimate limit in hadronic energy resolution.

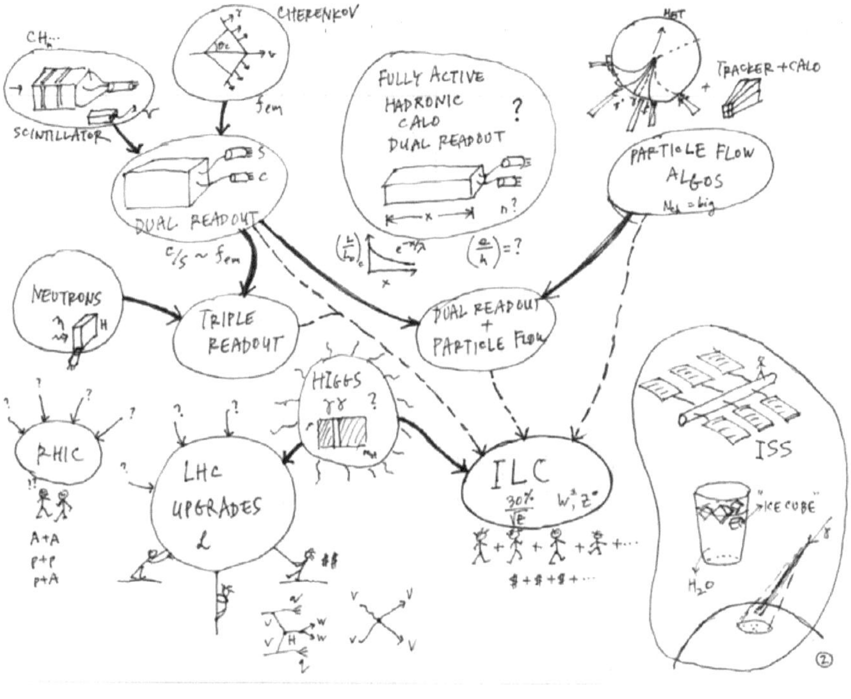

Figure 1. A snapshot of where we are today with the imminent projects in particle physics.

Particle flow algorithms (PFA) use the information from a highly segmented calorimeter and the information from a tracker system to achieve resolutions that would separate Ws from Zs in future lepton colliders. Much progress has been reported in this conference and elsewhere by the CALICE collaboration. Although it is not clear, but it may be possible to employ PFA in combination with a multiple-readout calorimeter in the future. We have to contrast the advantages against technical challenges but it is worth investigating this possibility.

The LHC experiments are developing upgrade plans to better cope with higher luminosities and higher center of mass energies. Some of these plans were reported in this conference. Nearly $\sim$20 fb^{-1} of LHC data per experiment taught us many lessons (see Figure 2). The radiation damage to detectors continues to plague us. Although we have developed techniques to "correct" for it, we have yet to develop fundamental understanding of radiation damage mechanisms. Similarly, we have developed clever techniques to mitigate "noise" in the calorimeters where "noise" means anything that fakes a true signal, such as low energy neutrons or charged particles hitting photosensors, photosensors misbehaving, *etc.* Large mismatches in the e/h ratios of ECAL and HCAL combinations pose a difficulty in correct energy reconstruction because the calorimetry response depends on where the shower starts. In addition, the ECAL readout tends to be in the neighborhood of the shower maximum and thus prone to produce fake signals. The challenge is perhaps to invent a single calorimeter (CALO) that would match the performance of excellent ECAL systems but at the same time measure jets and single hadrons with precision.

The precision of calorimeter simulations has been improving over the last few years. Increasingly sophisticated "physics lists" have been introduced to get the shower physics right. A good validation test for these simulation packages is to verify if they can reproduce the experimental data from highly non-compensating calorimeters ($e/h \sim 4 - 5$), especially at low energies. As discussed in this conference, jet reconstruction algorithms have also become a crucial part of calorimetry, and they constitute an essential link between the detector and the physics.

The pulse shape from a calorimeter of any kind reveals the interaction history of an event in that detector. As higher digitization rates and logic processing become more common-place, the front-end electronics may be able to process information such that only critical information (f_{em}, f_{had}, ΣE, *etc*) is transmitted to the counting room, changing our usual paradigm.

The field of high-energy calorimetry is as rich and exciting as ever. We certainly know more about calorimetry now than we did a decade ago. As we embark on developing the next generation of calorimeters, I am certain we will discover more and have the opportunity to use our creativity to uncover more physics.

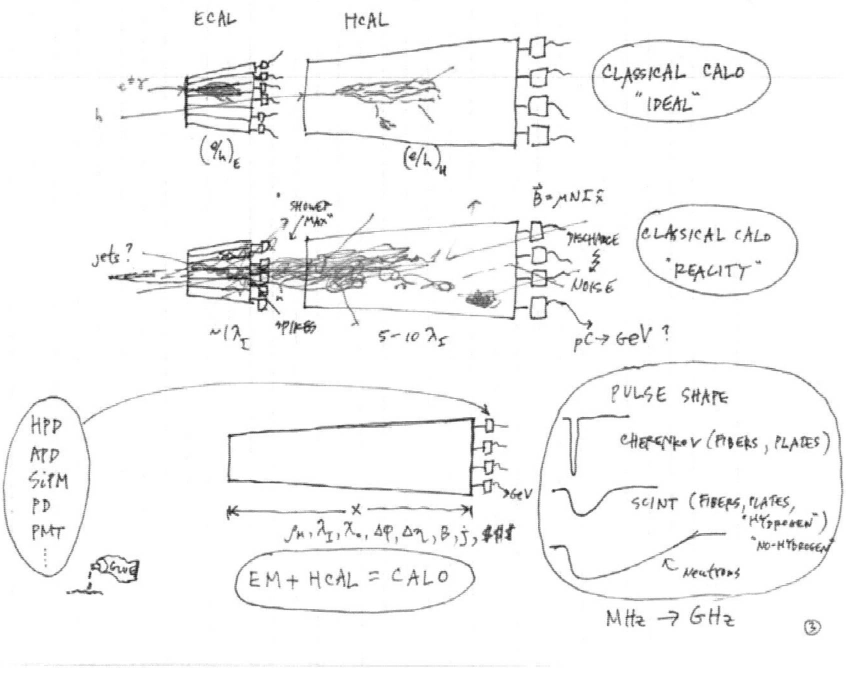

Figure 2. We have learned many things in the last decade...

Acknowledgments

We all thank the participants for their hard work and excellent presentations at CALOR 2012, and for their written contributions to these proceedings. These proceedings will serve as a good source of calorimeter information, as have precious CALOR proceedings, as current calorimeters are upgraded and future calorimeters are being designed.

Institute of Physics
Dirac House, Temple Back
Bristol BS1 6BE UK